Shortwave Receivers Past and Present
Communications Receivers 1942-1997

Third Edition

Edited by
Fred Osterman N8EKU

Universal Radio Research
6830 Americana Parkway
Reynoldsburg, Ohio 43068
United States of America

Contents

Contents

38 Hammarlund

39 Harvey-Wells

40 Heathkit

103 Watkins-Johnson

104 Yaesu

105 Briefly Mentioned

106 Receivers That Never Were

107 Additional Information

108 Model Number Index

1 Introduction

This book is designed to provide the radio hobbyist or receiver collector with concise information on the value, features, specifications and performance of current and former shortwave communications receivers. Strictly defined, a shortwave communications receiver is one that covers the entire shortwave spectrum (1.6 to 30 MHz) in amplitude modulation and at least Morse code. In reality most communications receiver additionally cover the medium wave band (.540 to 1.6 MHz) and often long wave as well. Such receivers manufactured after the early 1960's typically receive the single sideband transmission mode as well. This book also features shortwave broadcast receivers which receive only select international shortwave broadcasting frequencies, and also amateur band receivers that cover only the ham bands. Some marine receivers covering LF and MF frequencies are covered. This book does not attempt to cover the plethora of "parlor" radios that incidently receive one or two bands of shortwave. These receivers, usually found in wood cases, where produced in the 1950's and 1950's. For examples of "parlor" radios please see the Concord radios in *Chapter 105 Briefly Mentioned.*

The First Edition of this book covered 150 table top receivers and 75 portables made for the preceding twenty years. A questionnaire was included with the first two printings asking readers what they wanted to see in the next edition. Nearly all respondents felt that the next edition should go back further in time to include receivers they grew up with. Portable receivers were not judged as important as table top receivers. A stronger emphasis on commercial receivers was noted. To a lesser extent, readers also wanted to learn more about foreign, especially European, receivers. Some respondents also felt receiver kits should receive some attention. The current edition attempts to respond to these requests. While the Second Edition was more comprehensive than the First, it feel short in covering receivers manufactured in Europe. This current Third Edition, now begins to properly represent the vast assortment of interesting radios produced in this region. While the First and Second Editions were produced largely from the research archives of the author, the new entries for this Third Edition are almost exclusively the result of information contributed by readers. The author is grateful to the many readers in Europe and Australia who sent material on missing models.

Some truly excellent values can be realized by purchasing a used shortwave radio. Receivers built in the last twenty years are solid state. Solid state means the circuit is composed of transistors and integrated circuits rather than vacuum tubes. The problems of tube aging, such as heat and wear, are not a factor with solid state sets. In fact, most solid state receivers work as well five or ten years after they were made as the day they came out of the factory. Therefore, purchasing a quality used solid state shortwave radio can afford substantial savings with no loss in performance.

Unlike computers, stereos and other consumer electronic items, shortwave receivers have a long product life. Shortwave receivers do not become obsolete after five or even ten years. In fact, a twenty year old receiver is probably more useful and usable than a two year old personal computer.

A key objective of this book is to help the reader make an informed choice when purchasing a used shortwave receiver. The following questions are addressed:

◆ When was the radio introduced?
◆ What did it sell for when new?
◆ What is the current market value for the radio in used condition?
◆ Where was the radio made?
◆ What frequencies does it cover?
◆ What modes does it receive?
◆ What is the frequency display ... analog or digital?
◆ If a digital display, what is its resolution?
◆ What voltages (or batteries) are required?
◆ Where can I find a review of the radio?
◆ What accessories were made for it?
◆ How was the receiver performance rated when new?
◆ What is the used value rating of the receiver?

This Third Edition illustrates many commercial receivers that are not available on the used market today. They were included for two reasons. The first reason is that they may be on the used market tomorrow! Used Racal RA6790's can now be easily found, but this was clearly not the case ten years ago. Who knows what will be available ten years from now? The second reason such receivers were included is simply because they are interesting.

Every effort to insure accuracy in this book has been made, but the publisher cannot be responsible for errors, omissions or misprints. Conflicting information was often encountered in researching material especially on older models. Specifications would vary depending on whether you were reading a specification sheet, owners guide or service manual. Specifications and features can also change during the production cycle of a receiver. In some cases the author had to make an educated judgement (a.k.a. "a guess") when encountering such contradictory data.

European receivers are also better represented (although still not fully) in this edition. Again there are two reasons for this. The first is that this book will be sold in Europe. The second reason is simply because they are interesting. Few Americans are aware of the scope and selection of the amazing Eddystone line. Rohde & Schwarz, called the Collins of Germany, produces fabulous receivers.

Some receiver entries will have less information than others. Manufacturers in the 1940's, 1950's and 1960's did not provide the profusion of technical specifications we have become accustomed to in more recent times. The further one goes back, the less technical information is supplied. Specifications on lower-end models are particularly hard to come by. Perhaps the manufacturers felt such information would only confuse the entry-level buyer, or maybe the specifications were so unbecoming as to better be left unsaid.

Other entries are incomplete simply because the information was not available to the author at press time. A certain "critical mass" of information was required before a receiver was included in this book. Unfortunately information on several receivers was so incomplete that they were not included at all. If you have information on an H.F. tabletop receiver manufactured since 1942 that is not shown in this book, the author would like to hear from you. Any help with missing information and corrections will be appreciated. Information and corrections from readers of the Second Edition were valuable in the preparation of this Third Edition. Readers who have information on communications receivers not shown in this book are encouraged to send this material to the author for use in future editions. Readers contributing significant new material will be acknowledged.

I would like to especially thank **Matt Stutterheim** of Long Island, New York for sharing his receiver library, photographs, insights and extensive knowledge for the preparation of this book. Much of the commentary beyond the facts and figures was supplied by Matt. I am also indebted to **John F. Tuppen** and **John Wright** of Australia for supplying information on receivers from their part of the world. **Rolf Folkesson** and **Karl-Arne Markström** provided extensive information on Swedish receivers. **Andrea Antonini** supplied material for the Italian manufacturer, Geloso. **Guido Roels** provided information on Belgian receivers. **David W. Whiting** and **Herman Cone III** also assisted with many unusual models.

I would also like to thank the contributors on the following page for their generosity in supplying materials, corrections and ideas for this book.

Jay Agan
Andrea Antonini *IK2JEQ*
Geoffrey C. Arnold
Ed Barton
Richard T. Bennett *K8MZ*
Paul A. Bernhardt
Jean-Louis Beroud *F6EFM*
Lars Bergström
Marvin Born *KF8XU*
Dr. Martin Boesch
Alan Bosch
Nigel Boyd *K4UGD*
David Thomas Briely
Dave Clark
David Colburn *AA1FA*
Herman Cone III *N4CH*
James A. Conrad
J. Michael Cox *K3GEG*
Loren Cox Jr.
Michael Crestohl *W1RC*
David Crystal
John Davis *KA8ZNH*
Robert S. Davey
Hubert Eisner
Rolf Folkesson *SM0HP*
Michael Freedman *VE3BGE*
Dr. Hans H. Friedrich
William A. Frost
Dennis Gibbs
John H. Gibson
Rudolf Grabau
Gary B. Hanke
John S. Hassan
Paul F. Herman *W8SF*
Hiroshi Hirasaki *JA1AFT*
Howard H. Hood *WA7QQI*
Sandro O. Jaeger *HB9DBJ*
Alan Johnson
Hideo Kanemich
Michael G. Kamp
John Kelly *N3GVF*
Tom Kennedy *WA6SBN*
Dick King

Ramon Klein *WB8VDS*
Jorge Klingenfuss
Richard A. Kumada
Leslie A. Locklear
Uwe Lorenzen
R.A. Lytle *W8THU*
Larry Magne
Raymond E. Makul
Lee Martin
Thomas Marcotte *N5OFF*
Pete A. Markavage *WA2CWA*
Karl-Arne Markström *SM0AOM*
William A. Matthews *KB8PIZ*
Thomas Nilsson
Walt Novinger
Wayne Owsley *W4AX*
Gerald L. Park *W8QS*
John Poland *AE4EN*
Sam Powrie
Ralph R. Radermacher *DL9KCG*
George Rancourt
Dan Robinson
Guido Roels *ON6RL*
Michael W. Roslowski
Hank Scharfe *W6SKC*
Thomas M. Salvetti *KC3NF*
Robert Scheel
Harold Sellers
Russell F. Sievert *W8OZA*
Bill Sorsby *N5BU*
Matt Stutterheim
L. Bruce Sumner
Mike Tara
John F. Tuppen *VK6XJ*
Joe Veras *N4QB*
John Wagner
Charles W. Werblan
David W. Whiting
Dr. David Wilson
Larry Wolken *N3OJD*
John Wright

The Ontario DX Association

The author welcomes your comments, corrections or additions for the preparation of a future editions. The author is pleased to announce that a volume similar to this covering **portable** shortwave receivers from 1948 to 1998 is in production for printing in 1999.

Fred Osterman *N8EKU*
Universal Radio Research
6830 Americana Pkwy.
Reynoldsburg, OH 43068 U.S.A.
dx@universal-radio.com

2 Buying A Used Receiver

Why purchase a used receiver? The answer is simple: savings! You can typically expect to pay 30 to 40 percent off the price of a current model radio. On a discontinued model, the savings can be 50% or more.

In the era of vacuum tube receivers, purchasing a used receiver could be quite risky. As the tubes would age and become weaker, the performance of the receiver would gradually decline. The older tube-type sets would only work as well as the tubes in them. Tube receivers were also more mechanically intensive, and would use either gears or strings for tuning the VFO. Band switching was accomplished by complex rotary switches. Older receivers required far more maintenance than today's solid-state digital radios.

The performance of most solid-state (i.e. no vacuum tubes) radios does not diminish with time. A Kenwood R-5000 that was built in 1987 will work almost as well as one purchased last week. This fact makes purchasing a used R-5000 at a 40% savings quite tempting. With a little knowledge and preparation you can confidently purchase a used shortwave receiver.

What about tube-era receivers? Safety is one concern. Many lower priced tube radios were built without a power transformer. All the filament voltages added up to 117 VAC, thus precluding the need for the manufacturing expense of a transformer. Such receivers can be dangerous if capacitors fail. Be very careful when examining or servicing any elderly sets. A serious shock, or electrocution can result from "live" chassis, shorted transformers and even bad tube sockets. Unless you are technically qualified to evaluate such equipment, you should avoid it.

With some notable exceptions, most tube receivers do not perform on par with today's solid-state rigs in terms of stability, selectivity and dial accuracy. However, tubes radios usually exceed their contemporaries in terms of sound quality (and character). Audio fidelity is seriously lacking in the majority of today's communications receivers. It you enjoy music on shortwave, or are primarily a "content listener" rather than a "DXer" a tube radio may be a reasonable choice. Many say that these metal monsters also have an allure and ambiance simply not found in today's smaller more efficient sets.

Older tube sets may also pose potential parts and service difficulties. Most of the major manufacturers of tube radios (Hammarlund, Hallicrafters, National, etc.) are no longer in business.

Most tube radios today are more sought after by collectors than listeners. A few models such as the Collins R-390A, 51S-1 and Hammarlund HQ-180 and SP-600 still perform admirably and can compete on some performance parameters with contemporary solid state models.

SOURCES:
Private Sales

You have done your technical research. You have limited your target radio down to two or three different models, and have also developed a good feel for what you should spend for each. Now it is time to find the radio.

One equipment source is other hobbyists. They have bought a new receiver and wish to sell their old one. Or perhaps they have decided to explore amateur radio, and are looking to raise capital for a transceiver. Still others may be leaving the hobby. Used receivers also occasionally show up at auctions and yard sales. Purchasing a radio (or most anything else) from a individual has advantages and disadvantages:

Advantages:

✦ When purchasing a radio from an individual you can normally expect to pay less than if you bought it from a radio store.

✦ An individual will not charge you sales tax.

Disadvantages:

✦ You must take the seller's word as to the history, condition, and performance level of the radio. If you are making the purchase in person, you can check the condition and performance level.

✦ If purchasing by mail, you cannot determine the integrity of the seller.

✦ You cannot determine if the radio is stolen property.

✦ An individual will not permit payment by credit card.

✦ An individual will not generally be in a position to offer a warranty.

SOURCES:
Hamfests

Attending a hamfest can be a fun and educational experience. The flea-market at a major hamfest can be like a trip through a radio museum. It is an excellent opportunity to see many interesting and sometimes rare radios. Also be prepared to see an immense amount of unadulterated junk. But one man's garbage is another man's treasure!

A salesman at a leading East coast dealer suggests that you can never go wrong at a radio flea market ... as long as you take only $5.00 with you ... and that includes parking fees! That cautionary note may be an exaggeration, but one should be prepared.

When you enter the hamfest flea market, ask if a *test bench* has been set up. Most of the larger events provide a *test bench* as a service for those wishing to checkout a used equipment purchase. AC voltage (and 12 VDC) plus an antenna is provided. You can *fire-up* the radio and confirm that it is operational.

Come prepared to test the radio, even if a test bench is not available.

■   A Very Long Extension Cord
You may not be able to find an outside AC outlet. With an extension cord, you can connect to an outlet near the door and bring the extension out.

■   A *Cheater* Antenna
Obtain a 15 to 20 foot length of plastic coated wire (18 or 20 gauge, preferably stranded). Connect a banana plug at one end and an alligator clip at the other.

With this configuration you will be able to add an antenna to virtually any radio. If the radio has a standard SO-239 jack (for a PL259-type plug) you can fit the banana plug end perfectly into the center of this jack. Use the alligator end to conveniently attach the antenna to any nonconductive item. If the receiver only has screw terminals, use the alligator end to attach to the radio. Such an antenna is not designed for DXing, but is adequate for receiver testing.

Special Notes ...

● For safety reasons, very old, low-end, tube receivers without power transformers should not be plugged in until it can be established that they do not present a shock hazard.

● Purchasing tubes and electrolytic capacitors at hamfests can be occasionally risky. Sometimes "dud" tubes and "leaky" capacitors are sold, resold and sold again at hamfests. This is one case where "recycling" is not a good thing.

If you are seeking an older tube-radio, bring mono headphones (with a ¼ inch plug), as some older radios did not have internal speakers.

———————

Testing a digital communications receiver will not take long, but should include the following considerations:

1. Look at the frequency display. Rotate the knob to ensure all the segments of all the digits work. The numeral **"8"** will test all segments.

2. Check to be sure all indicators are working. For example, if an LED is used to indicate mode, check all modes. Also be sure the lamp in the S-meter is working. The failure of an LED indicator or S-Meter lamp may not discourage you from buying the radio, but this information may help when you negotiate the best price.

3. Rotate the gain knob(s). If there is a *scratchy* sound, the potentiometer is dirty. If it is not serious, a "shot" of TV-tuner spray (without carbon tetrachloride) may clear it up. If it is very scratchy, the potentiometer may have to be replaced.

4. If there are rotary-switch knobs, try turning them. Do they snap in place and make a positive, clean contact or do they have a *mushy* feel and make a scratchy sound? Do you have to fiddle with the knob to help it make solid contact? Again, if the problem is minor, tuner spray can help. If the problem is advanced, the rotary switch would have to be replaced. This is a costly repair. If the *band switching* rotary knob has to be replaced, this is a very costly repair ... assuming the rotary switch can even be found.

5. Listen to the audio for signs of *hum.* A noticeable hum will indicate that the electrolytic capacitors in the receiver's power supply are going bad. This condition is especially prevalent in tube sets which are now reaching the age of twenty-plus years. Equipment that has been stored and not used for two or more years, will often suffer from capacitor failure. This condition can occur in radios bought from an estate sale.

Please note that some older tube radios with 600 ohm headphone output will exhibit some *hum* if used with low impedance headphones.

6. Now it is time to actually listen to the receiver for signals. You are probably listening to the radio during the day and probably with a modest antenna. As a shortwave listener, you already know that you will not be able to hear a lot of stations. However, you should hear some. Start with the 19 meter band 15000 to 15450 kHz. Look for WWV at 15000 kHz and broadcast stations above. Monitoring just one frequency is not an adequate test. It is important to remember that even general coverage digital receivers are *electronically* divided into bands. This is not normally apparent, but perhaps you have noticed the sensitivity of the receiver changes when going from 1600 to 1601 kHz; or perhaps on your receiver it might also change when going from 4000 to 4001 kHz. Within the range of shortwave (1600 to 30000 kHz), there may be five or six bands. Therefore, try to sample the radio over every 5 MHz. You should be able to at least hear background static at roughly the same level throughout the range of the receiver. If the last 5 or 6 MHz of the receiver are totally dead (not even hiss), the receiver may have been used by a CB'er to monitor his own signal. If the shortwave and CB antenna were too close, this may have damaged the highest *band* of the receiver.

If you are interested in purchasing the radio do not be shy about making an offer at substantially lower than the marked price (maybe even one half). If the seller is tired of carrying the radio around, he or she may let it go cheaply.

Advantages:

+ When purchasing a radio at a hamfest or flea market, you normally will pay less than if you bought it from a radio store.

+ At most hamfest flea markets you will not be asked to pay sales tax.

+ At a flea market, you may be able to negotiate the price ... especially at the end of the day!

Disadvantages:

+ You must often take the seller's word as to the history and performance level of the radio.

+ You cannot determine if the radio is stolen property.

+ An individual will not permit payment by credit card.

+ An individual will not generally be in a position to offer a warranty.

SOURCES:
Radio Stores

Most stores that sell new amateur and shortwave equipment also sell used equipment. Most have come in to the store when customers trade-up. Unfortunately, not every city has an amateur radio store. Buying used gear from an established and reputable radio store should not be a problem. The hobbyists radio publications *QST, CQ, Monitoring Times, Radio Bygones* and *Popular Communications*, feature ads for radio stores. Many have a toll-free 800 number that you can call to check for a particular model, or request a used list. Some radio stores also post available used equipment on their Internet web sites.

Advantages:

+ Most radio stores will thoroughly test a used receiver to be sure it meets its specifications before putting it out for sale. (Or they will mark it "as-is".)

+ Most radio stores will offer a warranty including parts and labor. This is usually a thirty or sixty day limited warranty.

+ Some radio stores will permit a return without penalty within seven to ten days.

+ Most radio stores are available to answer any operational questions you have on the unit.

+ You may use a credit card.

Disadvantages:

+ When purchasing a used radio at a radio store you will often pay more than at a flea market or private sale.

+ If you actually purchase the radio in the store you will have to pay sales tax. If you are purchasing mail order, out of state, you will not have to pay sales tax in most cases.

A Reminder ...

When trading **in** a radio, expect to receive 20 to 25% less than the selling amount. Also understand that you will seldom recover the cost of special modifications and options when selling or trading a radio.

SOURCES:
The Internet

The Internet is an increasingly popular place to find used and collector radios. All the above advantages and disadvantages of "Private Sales" listed above are applicable to the Internet ... plus the normal caveats of any business transaction over the Internet. Some related usenet news groups include:

```
rec.radio.amateur.Boatanchors
rec.radio.amateur.shortwave
rec.radio.amateur.swap
```

Some interesting mailing lists include:

```
WUN World Utility Network
VSS Vintage Solid State
Boatanchors
```

A few Web sites to try for information and/or actual equipment include:

```
http://www.antiqueradio.com
http://www.radiofinder.com
http://www.users.fast.net/~wa3key/collins.html
http://www.universal-radio.com
http://chide.bournemouth.ac.uk
http://www.rnw.nl/realradio/antique_index.html
http://www.torontosurplus.com/
http://www.mindspring.com/~johnmb/bawebpg.htm
http://www.users.fast.net/~wa3key/collins.html
http://cayman.ebay.com/aw/index.html
http://www.snafu.de/~wumpus/index.html
http://www.cyberventure.com/heathkit/ham
http://ouvaxa.cats.ohiou.edu/~post/Pix/BA.html
http://alpha.wcoil.com/~fairadio/
```

You can also use search engines like AltaVista to search for receiver manufacturers and sometimes even models.

3 Using This Book

Each manufacturer is represented by a Chapter in this book. This simple scheme turns out to be not so simple. In today's environment of corporate acquisitions, consolidations and joint ventures, making such classifications becomes difficult. Should one have a separate chapter for Collins and Rockwell? Sometimes the same model can be found under the Rohde & Schwarz label and the Siemens label. And there is GEC, Marconi and Eddystone. Should the DX-1000 be listed under Bearcat or Electra or Uniden?

The author, somewhat arbitrarily, elected to list the receiver under its most commonly accepted moniker. Therefore Collins models are listed under Collins, not Rockwell. Where receivers continue to be marketed under their traditional name plate, despite corporate consolidations, there will be separate chapters (e.g. Cubic and Swan, Marconi and Eddystone). In cases where a receiver is known primarily by its namesake, this will be used as the Chapter heading. The DX-1000 is seldom referred to as the Electra DX-1000. Therefore this model will be listed under Bearcat. Multi-Elmac will be a Chapter heading rather than Multi-Products. A review of the following list may avoid some of the potentially confusing Chapter names:

Name	See Chapter	Name	See Chapter
AEG	Telefunken	Mechanical Lab	Videoton-Mechlabor
American Electrola	Quality U.S. Tech.	Micon	Vigilant
Arvin	see Collins	MIMCO	Eddystone
Automation Elect.	see Pierson-Holt	Moradco	Morrow
Bendix	see General Dynamics	Multi-Products	Multi-Elmac
Canadian Marconi	Marconi	National Panasonic	Panasonic
Cubic	also see Swan	Northrop	Hallicrafters
Daimler-Benz	Telefunken	Realistic	Radio Shack
Datron	Transworld	Realistic	also see Allied
Electra	Bearcat	Rediffusion	Redifon
Electromekano	Dansk Radio	Polarad Elec.	see Rohde & Schwarz
Harris	R.F. Communications	Sommerkamp	see Yaesu
Hygain	Galaxy	SoftWave	ComFocus
Inoue	Icom	Stoner	McKay Dymek
Intl. Marine Radio	Standard Radio & Tel.	Terma Electronik	Dansk
ITT	see Mackay	Trio	Kenwood
Knight Kit	also see Allied	WinRadio	Rosetta Labs
Liniplex	Phase Track	World Access	Quality U.S.A. Tech.
Marconi	also see Eddystone	Zenith	Heathkit

A very brief company description or profile will introduce each Chapter. It is not within the scope of this book to provide detailed corporate histories. Many fine books covering the corporate histories of Heath, Zenith, Radio Shack, Hallicrafters, etc. are now available.

To the right of the profile will be shown the manufacturer's address for those wishing to obtain additional information on current models. Addresses will also be shown for companies that may be inactive. The dates associated with these address histories are derived largely from advertisement and are approximate at best. No attempt to indicate address histories prior to 1942 was made as this is outside the scope of this book.

We believe that we have covered most of the known Western manufacturers. However, there may be dozens of manufacturers in former Communist bloc countries and other parts of the world. The author welcomes further information on these models for the next edition.

①

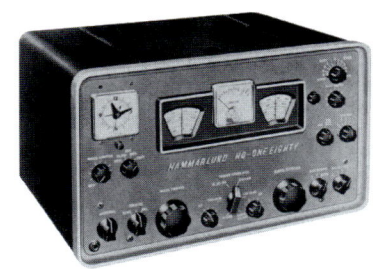

❶

② **HQ-180**

③ **General Coverage Communications Receiver**
④ Triple Conversion Superheterodyne. 18 Tubes.

⑤ **Features:**
- ¼" Head. Jack
- RF Gain
- Spkr. Terminals
- Mute Line

- S Meter
- ANL
- BFO ±2 kHz.
- Flywheel Tuning

- Fiduciary Adjust
- Antenna Trimmer
- Bandspread
- Slot Filter ±5 kHz

- Calibrator 100 kHz
- Vernier Tune ±3 kHz
- AVC OFF/SLO/MED/FST

⑥ **Specifications:**

Coverage .... 540 - 30000 kHz
Selectivity .... 6/4/3/2/1/.5 kHz -6dB.
Sensitivity .... <.7μV -10 dB S/N CW
Audio Out .... 2.5 W 4 ohm 5% dist.

Modes .... AM/LSB/USB/CW
I.F. .... 3035, 455, 60 kHz
Image Rej .... >25 dB at 22 MHz
Antenna Input .... SO239 and Terminals

⑦ **Circuit Complement:**
6BZ6 RF Amp, 6BE6 1st Mixer, 6BE6 Converter, 6BA6 455 IF Amp, 6BE6 Converter, 6BA6 60 kHz IF Amp, 6BA6 60 kHz IF Amp, 6BV8 60 kHz IF Amp/AVC/AM Detector, 12AU7 SSB Product Detector, 6AL5 NL, 6BZ6 Crystal Cal, 6C4 HF Osc, 12AU7 60 kHz BFO/S Meter, 0A2 Voltage Regulator, 5U4GB Rectifier, 6AV6 1st AF Amp/AVC, 6AQ5 AF Output and 6BA6 455 kHz Gate. Double conversion below 7.85 MHz.

⑧ **Accessories:**   S-200 Speaker   24 Hour Clock/Timer (shown)   Noise Silencer

⑨ **Comments:** Ranges: .54-1.05, 1.05-2.05, 2.05-4.04, 4-7.85, 7.85-15.35 and 15.35-30 MHz. Bandspread: 3.44-4.04, 6.81-7.3, 13.98-14.425, 20.925-21.6 and 27.89-29.7 MHz. Requires a speaker. An outstanding tube radio. [¥70,000]

⑩ **Variants:** Model **HQ-180C** (shown) has clock. Model **HQ-180RC** is the rack version with clock. Model **HQ-180XE** has 11 fixed crystal positions and adds 230 VAC.

❷ **Made In:** United States 1959-1962

❸ **Voltages:** 105-125 VAC 50/60 Hz 120 W

❹ **Readout:** |⎯|⎯|⎯|▮⎯|⎯| Analog

❺ **Physical:** 19x10.5x13" 38 Lbs. 482x266x330mm 17.2 kg

❻ **Status:** Inactive Manufacturer Discontinued Model

❼ **Rarity:** Common

❽ **Reviews:** *QST* June 1960
CQ April 1960
QST July 1960
NASWA July 1995
Electric Radio Jan. 1996

	New	Used
❾ **Price:**	$429	$210-310
❿ **Rating:**	★★★★★	★★★★

① The Company Logo

The company logo or trademark is indicated in this upper-left position. An attempt has been made to show the logo or logos that were contemporaneous with the indicated model. This is not always a simple proposition. Often a different logo would be used in advertising than would be embossed on the receiver itself. Long running models (e.g. Collins 51S-1) may use two or more logos. All logos and trademarks illustrated are the property of the respective companies.

② Model Number

Receivers are sometimes produced in several versions. Usually the most common model will be indicated and illustrated. However, as in the example above, a variant may be shown. By illustrating the HQ-180C, rather than the HQ-180, the clock option can be shown to the reader. If a variant model is pictured, rather than the primary model, this will indicated by the words "(shown)" in the **Variants** section, later in the entry.

If the receiver also had a name, it will be shown under the model number. For example the Hallicrafters S-41G was referred to as the S-41G Skyrider Junior.

Some models will also indicate a military nomenclature below. Many will end in /URR. U stands for general utility. R stands for radio. R stands for receiving.

③ **Receiver Type**
The following classification system is used in this publication:

◆ Shortwave Broadcast Receiver
This type of receiver will not cover the entire shortwave spectrum and will not be capable of Morse code (or S.S.B.) reception. Such receivers only receive the international broadcast bands and are only capable of amplitude modulation detection. Most also include medium wave (AM band).

◆ General Coverage Broadcast Receiver
A general coverage broadcast receiver will cover the entire shortwave spectrum, but will not be capable of Morse code or single sideband (S.S.B.) reception. Most also include medium wave.

◆ General Coverage Communications Receiver
A general coverage communications receiver will tune the entire shortwave spectrum and will be capable of amplitude modulation detection and at least Morse code reception. Most better general coverage communications receivers manufactured after 1960 are also able to copy S.S.B.

◆ Amateur Band Communications Receiver
This type of receiver will cover the amateur bands and will be capable of at least Morse code reception. Once again, most sets made after 1960 were also able to copy single sideband. By default, some shortwave broadcast stations can also be heard on such receivers where the amateur and broadcast stations share common frequencies such as the 40 meter band. Some amateur band communications receivers also include one or more additional shortwave broadcast bands. From the mid 1960's, the trend was to offer crystal positions so those hams wishing to add additional coverage could do so. For example, the famous Drake R-4 family of receivers offered ten additional 500 kHz crystal positions.

◆ Fixed Channel Communications Receiver
A fixed channel receiver is designed to operate on a set frequency. For example, a merchant vessel may a have dedicated fixed channel receiver for monitoring the international distress frequency of 2182 kHz. Some fixed channel receivers may be capable of multiple channels. The advantage to such receivers is the speed and simplicity of tuning. The obvious disadvantage is that one is limited to the frequencies the receiver is fixed to. Many such receivers are crystal controlled. They are most popular with maritime users.

◆ Marine Band Communications Receiver
This type of receiver may not cover the entire H.F. spectrum, but will usually only tune in limited bands and often only up to 5 or 13 MHz. Some marine receivers only have a position for upper sideband (USB). To tune lower sideband (LSB) you had to use the BFO. This was for safety.

◆ Kit
The word "Kit" will be appended to the receiver type to indicate that the receiver was sold unassembled.

④ **Circuit Type**
This general circuit type will identify the receiver as Single, Double, Triple or Quadruple Conversion; Superheterodyne or Regenerative. The number of conversions is the same as the number of intermediate frequencies (IFs) a receiver has. Very generally speaking, the more IF's a receiver has, the better the selectivity and immunity from image problems. Images are the reception of a signal other than the transmitted signal. Following this indication will be the total number of tubes used in the receiver. Although perhaps not technically correct, rectifiers and regulators will also be counted as tubes. If the receiver does not utilize tubes, "Solid State" will be indicated. Some receivers will have further circuit descriptions in the **Circuit Complement** section later in the entry.

⑤ Features

This section will list the key features of the receiver. A brief description of the more common features is shown below:

• ¼" Head. Jack	Nearly all listed receivers include a ¼" (6.3 mm) headphone jack. Until the last few years this was strictly a monaural jack. Recently some receivers will additionally support stereo headphones. Most hobbyist receivers will disengage the internal speaker when the headphones are plugged in. Many commercial and marine receivers feature a switch to accomplish this. The switch does permit the speaker and headphones to operate concurrently.
• 4 Tuning Rates	Many receivers offer multiple tuning rates selected by a rotary knob or buttons.
• AFC	Automatic Frequency Control is a feature of the radio which compensates for variations in received signal frequency. Usually applies to FM broadcast band, but certain models follow AM carriers to correct for drift.
• AF Gain	The AF gain control adjusts the gain of audio frequency circuits. This is typically just referred to as "Volume". Also see • RF Gain.
• AGC OFF/FST/SLO	Automatic gain control is a feature of the radio which compensates for variations in received signal strength. Less expensive sets have a single built-in AGC setting. This entry indicates that the receiver's automatic gain control can be set three ways: Off (manual gain control), Fast or Slow.
• Antenna Trimmer	Older general coverage receivers typically employed a variable capacitor to roughly match the antenna to the receiver.
• APTR	Automatic Progressive Tuning Rate. With this feature, the faster the tuning control is spun, the higher is the rate of frequency change per revolution.
• Attenuator -20/40 dB	Under special circumstances, it may be desirable to reduce the sensitivity of the receiver. This can be useful in combating overloading, intermod or even noise. In this example, a switch may insert 20 or 40 dB of attenuation.
• Aux. Tuning Scale	An arbitrary index scale, usually marked in 100 or 1000 increments that is connected to the main tuning scale. These logging scales make returning to a given frequency very easy. Popular with Eddystone and others.
• AVC	See • AGC.
• Bandspread	A true bandspread system involves two calibrated dials. The main or band set dial is used to set the major range of frequencies. The bandspread dial spreads a small portion of the main dial over a wide range. For example, the 20 meter ham band may occupy a half inch on the main dial and 4 or 5 inches on the bandspread dial. This system provides for increased frequency accuracy but is predicated on the main dial having been properly set in order for the bandspread band to align itself. A crystal calibrator is often useful in precisely setting the main dial. See • Calibrator below.
• Bandspread 0-100	Less elaborate receivers may have a bandspread dial arbitrarily marked from 0 to 100. This system does not provide improved frequency accuracy. This arrangement is nothing more than a repeatable, electrical fine tuning.
• BFO ±2 kHz	The Beat Frequency Oscillator is circuit designed to clarify CW and/or single sideband signals. If the range of the B.F.O. is known, it will be indicated.
• B.I.T.E.	BITE stands for Built-In Testing Equipment. Receivers with this feature have the ability to run a self-diagnostic test and determine the source of the problem, usually to the board level. This information is then shown on the receiver's display.
• Break-In Switch	This switch activates the mute line.

• Calibrator 25/100 kHz	Crystal calibrators were frequently featured in receivers before the advent of digital displays. When turned on, the calibrator would generate a precise carrier every 100 kHz. Some calibrators also produced a signal every 25 kHz. These tones were useful in setting bandspread scales and determining frequency.
• Cal Reset	This fine adjustment permits the receiver dial or fiduciary to be tuned to a standard frequency or crystal calibrator tone.
• CAT Jack	A Yaesu proprietary interface port to assist in the control of the receiver from a computer. Other hardware may be required.
• Carry Handle	A handle for the convenient moving of the radio.
• Clock	An analog or digital clock.
• Dial Drag Adjust	This mechanical adjustment permits the user to adjust how easily the tuning knob turns. This may be referred to as Dial Brake Adj.
• Dial Lamp	One or more lamps that illuminate the radio dial from the back or from the side.
• Dial Lamp Switch	A few receivers feature a switch to turn off the dial lamp. This extends battery life when the radio is not operating from the mains.
• Dial Lock	This feature disables the tuning knob by physical or electrical means. This prevents the receiver from accidently being tuned off frequency.
• Dial Set	See • Fiduciary Adj.
• Dimmer	The dimmer feature allows the user to adjust the intensity of the frequency display. This may be set in two or more steps or may be continuously adjustable by a knob.
• DSP	Digital signal processing.
• Duplex Filter	A filter that automatically protects the shipboard receiver from overloading by blocking the ship's transmit frequency.
• Ferrite MW Antenna	Some table top receivers and most portables feature a ferrite bar antenna to improve Medium Wave (AM 550-1600 kHz) performance.
• Fiduciary	This is a fixed, hairline pointer on the front of a receiver. The tuning dial rotates behind this pointer.
• Fiduciary Adj.	On some receivers, the fiduciary could be slightly adjusted or angled off the perpendicular to compensate for receiver drift and/or circuit aging.
• Fine Tuning	A mechanical or electrical adjustment to make small tuning adjustments.
• Fixed Channel	A crystal position providing one unchangeable frequency.
• Flywheel Tuning	A heavily weighted, well balanced wheel behind the tuning knob that allows for smooth and rapid analog tuning.
• Hinged Top Cover	Many older receivers featured a hinged top cover to facilitate tube replacement and adjustments to the radio.
• IF Gain	A few receivers permit the IF gain to be adjusted, as well as the RF and AF gain.
• IF Out Jack	A jack, typically on the back of the receiver, to allow access to the receiver's I.F. This may facilitate the connection of a spectrum display or other ancillary device.
• IF Shift	An IF Shift control allows the listener to shift the IF passband of the receiver without changing the actual center frequency of the receiver. This control is useful when there is interference on one side of the signal.
• ISB	Independent sideband is the transmission of separate intelligence on each sideband. A military station may send voice on lower side band and radioteletype on the upper side band. A feeder may use each sideband to send different audio.

• Jack for S-Meter	A jack, usually on the back panel of the set, that permits the easy connection of an optional, external S-Meter.
• Line Out Jack	This output jack provides a low level audio signal that is not affected by the volume control of the receiver. This fixed-level jack is typically used to drive a tape recorder or radioteletype decoder.
• Keypad	A numeric keypad is used to quickly tune the receiver to the desired frequency.
• Line Out	A low level audio output jack of fixed level. A line out jack is used to drive a tape recorder or external demodulator. In some cases a line output can also be used to connect the receiver to a phone line at 600 ohms.
• Memories	Memories or presets are a relatively new convenience feature in communications receivers. Early memory systems often include only 4, 6 or 10 presets and stored frequency information only. Other operational parameters such as mode and bandwidth were added later. Recently some receivers, such as the Drake R-8A store the station name or callsign along the other parameters. This labeling feature is particularly useful when memory capacity exceeds 100 channels.
• Memory Pass	A feature that lets you lock out a memory location from scanning.
• Memory Scan	The receiver has the ability to scan the frequencies stored in memories. (Also see • Sweep)
• Modular Construction	The receiver has some circuits built onto plug-in printed circuit cards for easy service and/or replacement.
• Monitor Level	Monitor is used in conjunction with a transmitter to listen to the transmitted signal on the receiver.
• Mute Line	A jack or terminal that connects the receiver to a transmitter or transceiver. This connection will quiet or mute the receiver during times that the transmitter is actually transmitting, thus protecting the receiver from damage from high R.F.
• Notch Filter	A notch filter suppresses a very narrow band on frequencies within the passband. An adjustable notch can be particularly effective at the reduction or elimination of heterodynes. A good notch filter will permit the very fine adjustment required to match the notch with the offending tone. If the notch is narrow enough, it may virtually eliminate the het with minimal disruption to the signal. May also be referred to as a Rejection Filter.
• NB	Noise blanker. A device to reduce noise. Usually most effective on man-made pulse type noise.
• P.B.T.	Pass band tuning permits the tuning of the passband without changing the receiver's frequency. It is very similar to IF Shift and is useful in reducing adjacent channel interference.
• Osc. Trimmer	An oscillator trimmer is a vernier control to permit the exact electrical adjustment of the oscillator frequency, to bring it into coincidence with the frequency indicated by the pointer and dial after checking against the crystal calibrator.
• Phono Input Jack	In the 1930's and 40's, the shortwave radio was often the center of the family's home entertainment. These receivers, which often had substantial audio amplifiers in them, could serve as the amplification source and speaker system for a phonographic turntable. The monaural turntable would plug directly into this input jack on the rear. A switch on the radio would disengage the radio's receiver circuitry and allow it to operate strictly as an audio amplifier.

• Preamp	A preamplifier applies an extra measure of amplification to the RF stage of the receiver.
• Preselector	A preselector peaks the receiver circuits for maximum sensitivity on the frequency being received.
• Presets 500/2182 kHz	Marine receivers often have the H.F. international distress frequency of 2182 kHz built into the receiver for immediate access. In this example the HF frequency and the longwave distress frequency of 500 kHz are both available.
• Rack Handles	Two handles on either side of the receiver that serve two purposes. Rack handles facilitate the mounting or removing of the receiver from an equipment rack. Rack handles also can be helpful during service. They allow the receiver to be laid on repair bench "facedown" without damage to the faceplate, knobs, etc.
• Record Out Jack	(See • Line Out Jack)
• Recorder Activation	Relay contacts that are usually normally open, that close when the timer turns on the radio. These contacts, when connected to the "remote" jack of a tape recorder, will start the tape recorder when the radio comes on.
• RF Gain	The RF gain control adjusts the gain of radio frequency circuits, thus controlling the sensitivity of the set. (A separate volume control adjusts the gain in the audio section of the radio). Sometimes the RF gain must be decreased to avoid receiver overload or to adjust for some types of noise.
• R.I.T.	Receiver incremental tuning allows tuning plus or minus from the main VFO frequency. This is typically used in amateur radio transceive situations.
• RS-232	A jack or port on a receiver that permits communications and control of the receiver at RS232 voltage levels with a terminal or computer.
• Scan	As used in this book, Scan indicates the receiver can scan the memory channels. See also • Sweep.
• S Indicator	Typically an LED bargraph that indicates signal strength.
• S Meter	A meter that measures the relative strength of the incoming signal.
• S/AF Meter	A meter that can indicate signal strength or audio output. A switch on the receiver selects which function of the meter is active.
• Sensitivity	Same as RF Gain.
• Speaker	A built-in speaker. Most early communications receivers did not feature a speaker built into the radio. It was thought that the vibration associated with the speaker would contribute to receiver instability or misalignment.
• Speaker Switch	A switch to turn the speaker on or off.
• Spinner Knob	An indent or protrusion of the main tuning knob to facilitate the rapid rotation of the knob.
• Spkr. Terminals	Terminals for connection of an external speaker
• Squelch	This control is used to eliminate unwanted background noise when monitoring an inactive frequency.
• Sweep	As used in this book, Sweep indicates that the receiver can automatically tune the spectrum between two defined frequency limits. For example, a receiver with Sweep capability could be programmed to sweep from 9.6 to 10 MHz. In some cases the speed of the sweep may be adjusted. In a carrier-operated Sweep the receiver will simply stop when a signal is detected. The sweep will not continue as long as the signal's carrier is detected. In a time-operated Sweep the receiver will pause, or "dwell", for a period of time and then resume, whether the signal is persists or not. Some sophisticated receivers will permit the user to adjust the dwell time. This function is sometimes called "Search".

• Sync. Detection	An amplitude modulated signal (such as used by medium wave and shortwave broadcast stations) consists of a carrier plus a lower and an upper sideband. Propagational fading may destabilize the carrier. Frequently one sideband may be distorted by a nearby interfering signal. Synchronous detection replaces the fading carrier with a pure carrier frequency with no level variation. The circuit then reconstitutes the signal from the stronger of the two sidebands. A stable signal with less distortion is the result.
• TCXO	Temperature compensated crystal oven for improved stability.
• Telescopic Antenna	A multi-segmented, monopole antenna.
• Tilt Bar	A tilt bar or tilt feet that allow the receiver to be angled up from the desk for more convenient viewing.
• Tilt Handle	A tilt handle has the same function as a tilt bar and additionally can be used as a carry handle.
• Tip Head Jack	Early economy receivers did not feature a ¼" headphone jack. Instead a jack with two small holes on it was usually mounted in the back of the receiver. Inexpensive headphones, terminating in two tips, would be used.
• Tone	Changes the balance between highs and lows on the audio output. In some cases this was a simple switch that cut the high tones.
• Two VFOs	A VFO or variable frequency oscillator is simply a tuner. Some receivers feature two tuners or VFOs within the set. This has a variety of uses.
• Voice Scan Control	This function pauses only when a voice signal is received. Scanning will not stop when a beat signal (het) or noise is encountered.
• VRIT Tuning	Variable Rate Incremental Tuning. This is another name for Automatic Progressive Tuning Rate. See • APTR

⑥ Specifications

This portion of the entry will list key specifications for the receiver. These specifications are gleaned from manufacturer's manuals, ads and sales literature. Unfortunately some manufacturers are conservative in their specs., others are, shall we say, "optimistic". Generally speaking commercial manufacturers like Japan Radio, Rohde & Schwarz, etc. will provide specifications that will withstand the greatest scrutiny. Companies that are consumer oriented may be more generous with their performance specs.. This book shows .3 μV sensitivity for the Bearcat DX-1000 and .5 uV for the J.R.C. NRD-525. One should not assume the NRD-525 is less sensitive. Ideally one should acquire each of the 770 receivers in this book and test them all in the same lab, on the same frequencies and with the same technician. This however is not possible. Specifications are nonetheless a good guide. But remember, it would be difficult to select a car based solely on horsepower, miles-per-gallon and acceleration figures.

◆ Coverage

The coverage for the receiver will be shown here and will be usually expressed in kiloHertz (kHz). Where space is short in this field, the frequency may be indicated in MegaHertz. To convert MHz to kHz, simply multiply by 1000. (To convert MHz to meters divide 300 by the MHz. Using 7.3 MHz as an example: $300/7.3 = 41$ meters). For non-general coverage receivers this field might indicate "Ham. See Comments." Please then refer to the Comments field further down for a detailed description of the coverage.

◆ Modes

This field indicates the modes of reception that the receiver is designed and promoted as being able to copy. Please note that an "FM" indication in this field does **not** mean the receiver will cover the FM Broadcast Band 88 - 108 MHz. FM in the mode field indicates that the receiver is capable of demodulating a frequency modulated signal within part or all of the receiver's range. FM mode (or more correctly NBFM or narrow band FM) is used on shortwave in the upper portion of the 10 meter amateur band (29 MHz). A radio with FM mode could properly demodulate these signals.

The "AM" indication in this field does not refer to what Americans call the "AM band". AM in this context refers to any amplitude modulated signal. This book will refer to the "AM" broadcast band 540 to 1600 kHz as the Medium Wave (MW) band. Just recently the medium wave band has been extended to 1700 (or 1710) in some parts of the world.

This section generally only identifies the modes of reception based on the radio's switch positions. Other modes of reception may be possible. For example, many quality receivers do not have a specific mode button or position for FSK or RTTY (radioteletype). However, such receivers may be perfectly suited to handle a RTTY signal while the receiver is set in the LSB or USB mode. Independent sideband is a similar issue. A quality communications receiver can individually tune one or the other sidebands of an ISB signal. However, a receiver with true ISB will usually be able to simultaneously demodulate both sidebands. Most receivers of the 1940's and 1950's were capable of only AM and CW detection. These receivers typically do not make good single sideband receivers.

AM	Amplitude modulation (A3 or A3E)
CW	Continuous Wave, or more commonly; Morse code (A1 or A1A)
MCW	Modulated Continuous Wave, or modulated Morse code (A2 or A2A)
SSB	Single Sideband (A3H, J3E, H3E)
LSB	Lower Sideband
USB	Upper Sideband
RTTY	Radioteletype
FAX	Facsimile
ISB	Independent Sideband
FM	Frequency modulation

◆ Selectivity

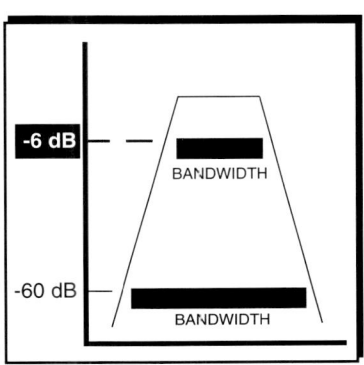

Selectivity is the ability of a receiver to discriminate between the desired signal and undesired adjacent signals. Or put more simply, selectivity is the receiver's ability to pass one wanted signal while rejecting others. In today's crowded band conditions, this is a very important specification. In our Hammarlund HQ-180 example, we see the entry as: 6/4/3/2.1/1/.5 kHz -6dB. This indicates that the receiver generously features six different bandwidths. The 6 and 4 kHz positions would be suitable for amplitude modulated voice signals. The 3 and 2.1 kHz positions would be good for single sideband signals and the 1 and .5 kHz positions would typically be used for radioteletype and Morse code. The entry for the J.R.C. NRD-515 found later in the book, shows 6/2.4/_/_ kHz -6dB. This indicates that the receiver comes standard, or stock, with a 6 and 2.4 kHz filter plus has two empty slots for one or two optional filters.

Most receiver bandwidths are indicated at -6 dB, some at -3dB. It is difficult to interpret the quality of a filter strictly from its width at -6 dB down. Ideally we would want to indicate the filter's bandwidth at -6dB *and* -60 dB. The ratio of these two numbers is referred to as the shape factor. The closer this number is to the ideal "1", the steeper are the filter's "skirts" and the better the filter is at simultaneously passing the desired signal and rejecting others. Unfortunately this important second number is seldom provided on anything but the latest models.

◆ I.F.

The receiver's intermediate frequency or frequencies, will be shown if known. The first number shown will be the first I.F., the second number the second I.F. and so forth. Single conversion receivers will show just one number (usually 455 kHz), a double conversion receiver, two numbers, and so forth.

◆ Sensitivity

Sensitivity is the receiver's ability to hear weak signals. It is the characteristic of the radio which determines minimum usable input. For the Hammarlund HQ-180 we show: .7µV S/N 10 dB CW. Sometimes this will be indicated in its more technically complete form of: .7µV S+N/N 10 dB CW. This simply means that .7µV signal plus noise is 10 dB above noise alone. The lower the number of microvolts (µV), the more sensitive the receiver is. Some manufacturer's use 6, 15 or 20 dB rather than 10 dB. Making accurate comparisons becomes difficult.

The original way for specifying sensitivity involved showing the number of microvolts required to produce "normal test output". This method was inconsistent.

◆ Stability

Stability is a factor that we have come to almost take for granted in solid state radios. Even today's modestly priced receivers display virtually no drift when listening to a broadcast station. In the days of tube receivers this was not the case. It was common to turn your receiver on a half hour *before* you planned to listen to let it stabilize. Stability *is* an important specification, even today, for listeners that wish to copy Morse code, radioteletype or facsimile. These modes demand far more stability than an amplitude modulated voice signal. Stability is now usually indicated as ± a number (for example ± 10 Hz). The smaller the number the better.

◆ Image Reject

The receivers ability to suppress spurious images. The higher the number, the better the receiver's image rejection.

◆ IF Reject

IF rejection is not as great a concern, in most cases, as image rejection. IF rejection is the ability of the receiver to reject false signals from nearby transmitters that operate near the same frequency as the receiver's first IF. Again the more negative the number, the better. One reason that the IF of 455 was selected by early radio designers was that it was below the medium wave broadcast band.

◆ Audio Output

This entry of 2.5 W 4 ohm 5% dist. indicates the receiver has an audio amplifier that delivers 2½ watts into a 4 ohm output with 5 percent distortion.

◆ Antenna Input

In our example, "Terminals 72 ohms" indicates that this receiver uses screw terminals for connecting the antenna. Most current amateur and shortwave communications receivers feature an SO-239 jack. This is the female receptacle for a PL-259 connector. Other common antenna inputs are RCA (sometimes also referred to as PHONO or CINCH). Modern military receivers use the more expensive and lower loss BNC jack. The female chassis mounted BNC jack may be referred to as a UG-1094/U. The male BNC may be referred to as a UG-88B/U. Some military receivers use N connectors. The female chassis mounted N jack may be referred to as a UG-556. The male N may be referred to as a UG-21. Your local radio dealer or Radio Shack can provide the necessary adapters to change the antenna jack as needed.

⑦ **Circuit Complement**

This section will provide additional details on the receiver's circuit. This will show the tube line up, if known, on non-solid state receivers.

⑧ **Accessories**

This section will list optional accessories associated with the receiver. Such items as speakers, filters, converters and other options will be listed.

⑨ **Comments**

This section will be used to provide additional information, specifications or commentary on the receiver.

The Japanese are very keen collectors of quality American-made communications receivers. Models that are commonly collected in Japan will show an entry in this format: [¥120,000]. This indicates the market price for this receiver in the Japanese used market. For a very rough conversion to dollars, simply move the decimal point two places to the left. In this example: ¥120,000 = $1,200.

⑩ **Variants**

This portion of the entry will list model variants. Variants may be very similar or quite different from the primary receiver listed. If the differences are significant, this book will likely have a separate entry for the variant. In this case the variant listed will be printed in normal type. If the variant model number is listed in bold type then there will not be a separate entry. In most cases, the distinguishing features of the variant will be explained. Variants are common with commercial manufacturers.

① HAMMARLUND

❶

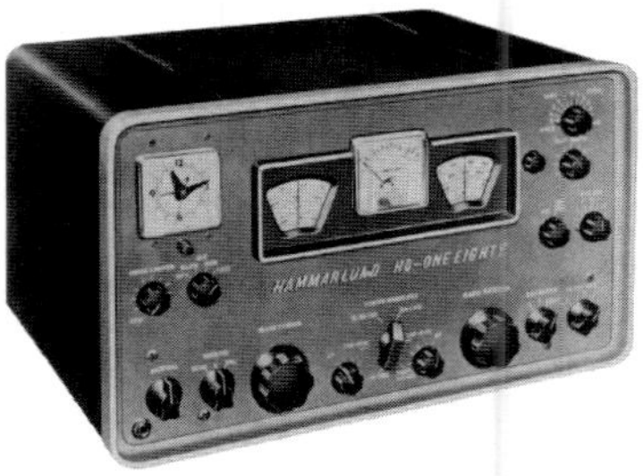

② HQ-180

③ **General Coverage Communications Receiver**

④ Triple Conversion Superheterodyne. 18 Tubes.

⑤ **Features:**

• ¼" Head. Jack	• S Meter	• Fiduciary Adjust	• Calibrator 100 kHz
• RF Gain	• ANL	• Antenna Trimmer	• Vernier Tune ±3 kHz
• Spkr. Terminals	• BFO ±2 kHz.	• Bandspread	• AVC OFF/SLO/MED/FST
• Mute Line	• Flywheel Tuning	• Slot Filter ±5 kHz	

⑥ **Specifications:**

Coverage	540 - 30000 kHz	Modes	AM/LSB/USB/CW
Selectivity	6/4/3/2/1/.5 kHz -6dB.	I.F.	3035, 455, 60 kHz
Sensitivity	<.7μV -10 dB S/N CW	Image Rej	>25 dB at 22 MHz
Audio Out	2.5 W 4 ohm 5% dist.	Antenna Input	SO239 and Terminals

⑦ **Circuit Complement:**
6BZ6 RF Amp, 6BE6 1st Mixer, 6BE6 Converter, 6BA6 455 IF Amp, 6BE6 Converter, 6BA6 60 kHz IF Amp, 6BA6 60 kHz IF Amp, 6BV8 60 kHz IF Amp/AVC/AM Detector, 12AU7 SSB Product Detector, 6AL5 NL, 6BZ6 Crystal Cal, 6C4 HF Osc, 12AU7 60 kHz BFO/S Meter, 0A2 Voltage Regulator, 5U4GB Rectifier, 6AV6 1st AF Amp/AVC, 6AQ5 AF Output and 6BA6 455 kHz Gate. Double conversion below 7.85 MHz.

⑧ **Accessories:** S-200 Speaker 24 Hour Clock/Timer (shown) Noise Silencer

⑨ **Comments:** Ranges: .54-1.05, 1.05-2.05, 2.05-4.04, 4-7.85, 7.85-15.35 and 15.35-30 MHz. Bandspread: 3.44-4.04, 6.81-7.3, 13.98-14.425, 20.925-21.6 and 27.89-29.7 MHz. Requires a speaker. An outstanding tube radio. [¥70,000]

⑩ **Variants:** Model **HQ-180C** (shown) has clock. Model **HQ-180RC** is the rack version with clock. Model **HQ-180XE** has 11 fixed crystal positions and adds 230 VAC.

❷ **Made In:** United States 1959-1962

❸ **Voltages:** 105-125 VAC 50/60 Hz
120 W

❹ **Readout:** Analog

❺ **Physical:** 19x10.5x13" 38 Lbs.
482x266x330mm 17.2 kg

❻ **Status:** Inactive Manufacturer
Discontinued Model

❼ **Rarity:** Common

❽ **Reviews:** *QST* June 1960
CQ April 1960
QST July 1960
NASWA July 1995
Electric Radio Jan. 1996

	New	Used
❾ Price:	$429	$210-310
❿ Rating:	★★★★★	★★★★

❶ Photograph

A photograph of the receiver will be shown in this top-right position. The quality of the photographs will vary significantly. Manufacturers have provided high quality original photographs for most current models. Such photographs are not typically available for receivers that are 40 or 50 years old. In such cases, the image had to be scanned from old advertising materials and may lack clarity. It is the belief of the publisher that a poor photograph is still more instructive than none at all.

❷ Made In

The country in which the receiver was manufactured will be indicated in this position. A year or range of years will follow the country. This indicates the approximate years that the receiver was actively marketed. This will not necessarily be concurrent with the years that the receiver was manufactured.

❸ Voltages

This field indicates the voltage and usually the line frequency that the receiver operates from. The wattage will also be shown if known. Battery operation, if any, will be indicated here.

D cells = UM-1, AA cells = UM-3

❹ Readout

This portion of the entry shows an icon and label indicating the type of readout or frequency display used on the receiver.

Icon	Label	Description
[analog dial icon]	Analog	This receiver uses an analog pointer that is moved along a nonlinear horizontal dial or has a fixed fiduciary (hairline) behind which is rotated a dial. Analog dials generally are the least accurate.
[analog linear dial icon]	Analog Lin.	An analog linear dial is similar to a regular analog dial. However, the dial behind it is different in that all portions of the radio dial take up a uniform amount of space. On an analog linear dial if 3.5 to 4 MHz occupies one inch, so will 14 to 14.5 MHz. Such linear dials usually are configured to yield much more accuracy than a traditional analog dial. In some cases dial readout better than ±1 kHz can be achieved.
00005.	Digital LED	This icon indicates the receiver has a digital display with 5 kHz accuracy. The display utilizes light emitting diodes (LED).
00001.	Digital LED	This icon indicates the receiver has a digital display with 1 kHz accuracy. The display utilizes light emitting diodes (LED).
00000.1	Digital Phs.	This icon indicates the receiver has a digital display with 100 Hz (.1 kHz) accuracy. The display utilizes a phosphorescent display.
00000.01	Digital LCD	This icon indicates the receiver has a digital display with 10 Hz accuracy. The display utilizes liquid crystal display.
00000.001	Digital LCD	This icon indicates the receiver has a digital display with 1 Hz accuracy. The display utilizes liquid crystal display.
00001.	Digital Mech.	This icon indicates the receiver has a mechanical digital display with 1 kHz accuracy. Often referred to as an "odometer" type display.
[knobs icon] .0	Dig. Knobs	This receiver has an individual rotary knob to select (and display) each frequency digit, in this case, out to 100 Hz accuracy.
[knobs icon] .	Dig. Knobs	This receiver has an individual rotary knob to select (and display) each frequency digit, in this case, out to 5 kHz accuracy.
00000.	Digi. Switch	This receiver uses ten position (decadic) thumbwheel switches to select and display each frequency digit to 1 kHz accuracy.
00000.0	Digi. Switch	This receiver uses ten position thumbwheel switches to select and display each frequency digit out to 100 Hz accuracy.
00000.00	Digi. Switch	This receiver uses ten position thumbwheel switches to select and display each frequency digit out to 10 Hz accuracy.
00000.000	Digi. Switch	This receiver uses ten position thumbwheel switches to select and display each frequency digit out to 1 Hz accuracy.

00001.	Digital Mech. & Ana. Linear	This icon represents a receiver with a mechanical digital display (odometer-type) for all positions to 1 kHz. Then there is a vertical analog vernier dial for readout less than 1 kHz (usually to 100 or 200 Hz). See Collins R-390.
001	Digital Mech. to 100 kHz then Analog Lin.	This receiver has a mechanical digital display (odometer-type) for the 10 MHz, 1 MHz and 100 kHz positions. The remaining digits are displayed via a linear analog dial. See Collins 51S-1.
00 000.01		This receiver has a mechanical knob to display the 10 MHz and 1 MHz positions. The remaining digits are displayed via a digital readout. See Racal RA1772.
●●● 00.1		This receiver utilizes rotary switches to select the 10 MHz, 1 MHz and 100 kHz positions. The 10 kHz, 1 kHz and 100 Hz positions are mechanically digitally displayed.
●●● 00.1		This receiver utilizes rotary switches to select the 10 MHz, 1 MHz and 100 kHz positions. The 10 kHz, 1 kHz and 100 Hz positions are mechanically digitally displayed. Then there is a vertical analog vernier dial for readout less than 100 Hz
	Coil Chart & Micrometer	This receiver uses a micrometer-type calibrated multi-turn rotary knob. A coil chart is then used to interpolate the frequency. See early National receivers.
		These icons indicate that the receiver uses an analog dial in conjunction with a micrometer dial. See National.
500.	Digital Mech. & Micrometer	A combination of mechanical digital readout (10 500 kHz) and micrometer dial is used to determine frequency. See National HRO-500.
Channel	Channel	This receiver does not have a frequency display. It is a fixed, multichannel receiver. The channel number is visible when the channel is selected.
	No Display	This receiver has no display. It is a fixed channel, single frequency receiver.
1 Hz.	Digital	This receiver is controlled exclusively by computer. The receiver itself has no frequency display. The frequency (and other operating parameters) are displayed on the screen of the computer controlling the receiver. The frequency display resolution is to the nearest 1 Hz.
10 Hz.	Digital	This receiver is controlled exclusively by computer. In this example frequency display resolution is to the nearest 10 Hz.
1 kHz	Digital	This receiver is controlled exclusively by computer. In this example frequency display resolution is to the nearest 1 kHz (1000 Hz.).

❺ Physical

This portion of the entry shows the size and weight of the receiver. The dimensions are shown in the format of: **width** x **height** x **depth** (WxHxD) in inches ("). These dimensions should be considered approximate. The height and especially the depth may not be exact. There is no standardized method of measuring a receiver's size. Some manufacturers include the receiver's "feet" in the height, others do not. Some manufacturers include the protrusion of the knobs, and rear panel connectors, others do not. The weight will then be shown in pounds (Lbs.). This data is also provided in metric format for the benefit of those readers living in the civilized world.

❻ Status

The first line of this entry indicates whether the manufacturer is still actively producing H.F. communications receivers. Note that an inactive indication does not necessarily mean the company is out of business. In some cases the company may have been acquired by another concern or involved in a different product line.

The second line of this entry indicates whether the model shown is still actively manufactured and marketed.

❼ Rarity

This entry indicates the rarity of the radio. Five categories are used:

Abundant	This model is continuously available on the used market.
Common	This model is frequently available and can be found with a minimum of patience and persistence.
Scarce	This model is not common. One may have to look for several weeks or months to find this model.
Very Scarce	This model only appears on the used market occasionally.
Extremely Scarce	This model is not normally found in the used market. Extreme patience and a large element of luck would be required to find this model.
Typically Not Available	This model is virtually unobtainable on the normal consumer (amateur, shortwave) market. This classification usually applies to exotic, commercial receivers.

These rarity classifications are for the American market. The Eddystone EB-35 is "Very Scarce" in the United States, but is "Common" in the United Kingdom. Many high-end commercial receivers, rarely seen on the American market, appear regularly on the Japanese used market (...at equally rarefied prices).

❽ Reviews

This entry will be of interest to those seeking additional information on the indicated model. The articles sited here vary from simple product announcements to comprehensive technical reviews. In some cases the book or magazine indicated may be out of print. Consult your library for such titles. Please refer to *Chapter 4 More Information* later in this book for further details on most of the referenced publications.

One highly respected, and often mentioned review source is *Passport to World Band Radio*. The first annual edition was actually titled *Radio Database International 1985-86*. The second annual edition was titled *Radio Database International 1987*. All editions since are titled *Passport to World Band Radio* followed by the year. For consistency the first annual edition will show as *Passport 1985/86*. The second annual edition will be shown as *Passport 1987* and so forth. This publication usually provides a detailed technical analysis the first year the model appears in the book. The same model in subsequent years is treated in a more concise manner. However, these subsequent reviews, although short, often contain valuable additional information regarding updates, improvements and changes and should therefore not be ignored. The *World Radio TV Handbook* is another annual publication that now also includes receiver reviews nearly every year.

❾ Price New and Price Used

The radio's new price will be shown first. In the 1940's and 1950's, models typically sold in a very tight price range. In most cases, a list price was established by the manufacturers and that was the going price for everyone. In the 1960's and 1970's receivers were sold both at a list price and discounted. For such receivers a price range will be shown. This range will be from a number lower than the list price up to the list price. In the 1980's and 1990's, most noncommercial receivers tend to consistently sell below the manufacturer's suggested list price. Therefore the price or price range will likely be under the list price. A small **A** next to the value denotes the amount is approximate.

One should keep inflation in mind when noting new prices. A 1950 price of $500 would translate to an equivalent buying power of perhaps $3000 or $4000 today. Many commercial receivers will not show a new price. In many cases receivers are manufactured in production runs for a specific client, often under a private contract. In some cases the author simply could not obtain a new price.

The used price or price range is also shown. This price is for a unit in good working order, with owner's manual and having no cosmetic problems. The price indicated is what you might expect to pay for this model, from a radio store, providing a 30 to 60 day warranty. You should expect to pay 10% to 25% less when purchasing from a private individual or at a hamfest where no warranty is supplied. At a flea market or garage sale, virtually anything goes.

Prices shown are approximate, subject to change, and vary from region to region. Used prices are also subject to normal market forces of supply and demand. A desirable, scarce model may decrease in if a large number of that model are suddenly dumped on the market. If a radio is listed in this book with a used value of $350-$380, and someone is selling it at $400, it does not necessarily mean that $400 is an unfair price. The prices indicated are just a general guide. However, if you see a Davco DR-30 at a hamfest for $70 or a Squires-Sanders SS-1R (with SS-1V!) at the Dayton Hamvention fleamarket for $200, the information in this book should help you act quickly (both true examples!).

The following symbols are utilized in the New and Used Price entries:

① New price unknown.
② This model is typically unavailable on the used market. Used price undetermined.
③ This price will vary dramatically depending on condition.
④ Too new for a used price to be determined.
⑤ Used price unknown.

⑩ New Value Rating and Used Value Rating.

It is difficult to rate receivers. Nearly every receiver has a combination of good and bad qualities. When evaluating radios, cars, cameras or computers, many attributes have to be weighted and then combined.

When rating an entire price range of products for "value" the issue becomes even more complicated. It is easy to say that a JRC NRD-93 at $7000 is a better receiver than a Drake SW-8 at $700. But which is the better *value* for the dollar?

The first rating is the **New** rating. A scale of one to five stars is used to indicate the value score of the receiver when sold **new**. This score is weighted for the price class. A $300 radio rated five stars will typically not perform as well as a $3000 radio rated three stars.

The second rating is the **Used** rating. This rating is designed to assist those looking to purchase a **used** radio for **listening** and **DX'ing**. This is <u>not</u> a rating for collectors. A radio may have a low rating, but be highly collectable because of its historical significance or rarity. The composite used rating is designed strictly for those looking for performance, value, reliability and serviceability. A receiver made by a defunct manufacturer will not merit five stars (unless it is extremely good such as the National HRO-500). Parts and service may be a problem. In fact virtually all receivers prior to 1970 will be unrated. These receivers may make a poor choice for some listener simply from a parts and maintenance perspective. Radios of this vintage are better left to radio collectors or those with the skills to maintain them.

The New and Used ratings in this book do not necessarily reflect the personal opinion of the author! The author, having been in the radio and electronics business for 25 years, has had the opportunity to personally use many, but certainly not all the receivers covered in this book. The ratings are derived from *a combination* of personal experience and from a consensus of other opinions. In cases where the author has not personally tested a radio, the comments of other reviewers is used exclusively. Unfortunately there are some receivers (usually foreign or commercial models) that the author has not used and have not been reviewed by anyone. Such models are unrated. Receiver kits are also not rated due to their unknown construction quality.

The rating system is as follows:

    ★★★★★   **Excellent Value**
      ★★★★   **Very Good Value**
        ★★★   **Good Value**
          ★★   **Fair Value**
            ★   **Poor Value**

In the absence of a rating, the following symbols will be shown:

    ⑥   Insufficient information available for a value rating.
    ⑦   Too old for a value rating.
    ⑧   Too new for a value rating.
    ⑨   Kits are not value rated.
    ⑩   Too specialized for a value rating.

There is certainly room to debate any rating. I apologize in advance for upsetting or offending any reader with a poor rating for their receiver. One can become emotionally attached to one's receivers and love is in fact, sometimes blind. Although I have used receivers costing five figures, none can match the "excitement" value of my first receiver, a Hallicrafters S-120. The reality, however, is that the S-120 was only a fair value new and would be a poor value as a used receiver investment today.

4 More Information

■ Books

Buyer's Guide to Amateur Radio
    by Angus McKenzie
    Radio Society of Great Britain ©1985
This book covers the whole range of amateur radio equipment including receivers, transceivers, amps and accessories.

Collins S-Line Compendium of Factory Modifications
    by Marvin Born *KF8XU*
    Worthington, Ohio
This compilation includes all authorized modifications for the Collins S-Line including the 75S-1 to 75S-3 and the 51S-1 series (also KWM1/2. 32S-1/2/3).

Communications Receivers Principles and Design
    by Ulrich L. Rohde and T.T.N. Bucher
    McGraw-Hill Book Company, New York, New York. ©1985
This highly technical book is written for system and design engineers.

Communications Receivers. The Vacuum Tube Era
    by R.S. Moore
    R.S.M., Key Largo, Florida. ©1997
The golden age of vacuum tube receivers is revisited in this comprehensive book covering the period 1932 to 1981. Key facts, features and photographs of Breting, Collins, Echophone, Gonset, Hallicrafters, Howard, National, Hammarlund, Drake, Sargent, RME and many more. Many photos. Fourth Edition.

Fine Tuning's Proceedings
    by John H. Bryant
    Fine Tuning
This annual compendium of scholarly articles delves deeply into all aspects of shortwave listening and DXing including propagation, equipment, antennas and techniques. Extensive evaluations of current high-end receivers are usually featured.

Ham Equipment Buyer's Guide
    by Bernarr Wixon W9MFL
    Barbara Brand Wixon, Glen Ellyn, Illinois
This publication is produced every few years and features amateur equipment for the covered period. It also includes shortwave receivers.

Heathkit A Guide to the Amateur Radio Products
    by Chuck Penson WA7ZZE
    Electric Radio Press, Inc., Durango, Colorado. ©1995
The definitive examination of all amateur, shortwave and related accessories offered by Heathkit. An outstanding compilation with large sharp photographs.

Passport to World Band Radio
    by Larry Magne
    International Broadcasting Services, Penn's Park, Pennsylvania
This annual publication is the leading guide to shortwave broadcast stations, schedules and frequencies. It is also a highly respected source for hard-hitting, objective commentary on shortwave receivers currently on the market.

The Pocket Guide to Collins Amateur Radio Equipment 1946-1980
>by Jay H. Miller KK5IM
>Jay H. Miller. ©1995

A well organized guide to the famous Collins line of amateur equipment.

The Racal Handbook - Racal Communications Equipment 1956-1975
>by Rinus Jansen
>G C Arnold Partners. ©1993

A survey of Racal receivers, transmitters and accessories made 1956-1975.

Radio Receiver - Chance or Choice
>by Rainer Lichte
>Gilfer Associates, Inc., Park Ridge, New Jersey. ©1985

Reviews many popular shortwave portables and table radios from the period 1975 to 1985.

More Radio Receiver - Chance or Choice
>by Rainer Lichte
>Gilfer Associates, Inc., Park Ridge, New Jersey. ©1987

Reviews 14 more radios produced in the mid 1980's.

Radios By Hallicrafters
>by C. Dachis
>Schiffer. Atglen, Pennsylvania. ©1996

A marvellous book with over 1000 sharp photos of radio receivers, transmitters, speakers, early TV sets and accessories from the famous Hallicrafters label. Technical descriptions of every known model including dates with new prices and current values.

Radio Communications Receivers
>by Cornell Drentea
>Tab Books Inc. Blue Summit, Pennsylvania. ©1982

An historical and technical guide to HF and VHF receiver design. Out of print.

Radio Nederlands Receiver Shopping List
>by Jonathan Marks
>Radio Nederland

An informative pamphlet providing information and evaluation of current portable and tabletop shortwave receivers.

World Radio TV Handbook
>by Andrew G. Sennitt
>Billboard Publications.

This venerable annual publication primarily covers shortwave, medium wave, FM and television broadcasting. Most years also include equipment reviews.

WRTH Equipment Buyers Guide 1993
>by Willem Bos and Jonathan Marks
>Billboard Publications.

This publication reviewed shortwave portables, communications receivers, car radios, antennas and accessories that were available in 1993. No edition beyond the 1993 has been announced as of press time.

■ Periodicals, Clubs & Journals

The following publications are valuable sources for one or more of the following:

- Current equipment reviews.
- Advertising of new equipment.
- Classified advertising of used equipment.
- Historical or restoration articles for collectors.

All are published monthly unless otherwise noted. As one would expect, the publications that focus on radio listening will have a greater emphasis on receivers than general amateur radio publications.

Antique Radio Classified
P.O. Box 2
Carlisle, MA 01741

Emphasis: Collecting and Restoration
- Historical articles on all types of receivers
- Classified section often includes H.F. rx.

Amateur Radio Trader
P.O. Box 3729
Crossville, TN 38557

Emphasis: Amateur Radio Sales
- Ads for new equipment.
- Classified section for used radios.
- Published 24 times per year.

Collins Collectors Assoc.
P.O. Box 840924
Pembroke Pines, FL 33084

Emphasis: Collecting and Restoration
- Focus is on Collins equipment.

Collins Journal
David A. Knepper W3BJZ
Box 34
Sidman, PA 15955

Emphasis: Collecting and Restoration
- Focus is on Collins equipment.

CQ
76 North Broadway
Hicksville, NY 11801

Emphasis: Amateur Radio
- Ads for new equipment.
- New equipment reviews (mostly ham).
- Small classified section for used radios.

DX News
National Radio Club
P.O. Box 118
Poquonock, CT 06064-0118

Emphasis: Medium Wave DXing.
- Occasional receiver reviews.

DX Ontario
Ontario DX Association
P.O. Box 161, Station A
Willowdale, ON M2N 5S8
Canada

Emphasis: Shortwave DXing.
- Occasional receiver reviews.

Eddystone Newsletter
Eddystone User Group
c/o Eddystone Radio
Alvechurch Road
Birmingham B31 3PP, England

Emphasis: Collecting and Restoration
- Focus is on Eddystone equipment.

Electric Radio
14643 County Road G
Cortez, CO 81321-9575

Emphasis: Collecting and Restoration
• Focus is on vintage ham & military gear.

Ham Trader Yellow Sheets
P.O. Box 2057
Glen Ellyn, IL 60138

Emphasis: Amateur Radio Sales
• No articles or reviews
• Big classified section for used gear.
• Published 24 times per year.

Hollow State Newsletter
Ralph Sanserino
P.O. Box 1831
Perris, CA 92572-1831

Emphasis: Collecting and Restoration
• Focus is on tube equipment, especially
  the Collins R-390 and variants.

I.R.C.A.
Intl. Radio Club of America
P.O. Box 1831
Perris, CA 92572-1831

Emphasis: Medium Wave DXing.
• Occasional receiver reviews.

Monitoring Times
P.O. Box 98
Brasstown, NC 28902

Emphasis: Radio Listening.
• New equipment reviews.
• Ads for new equipment.
• Some historical articles.
• Small classified section for used radios.

N. American SW Assoc.
45 Wildflower Rd.
Levittown, PA 19057

Emphasis: Shortwave Broadcast DXing.
• Frequent receiver reviews.

Popular Communications
76 North Broadway
Hicksville, NY 11801

Emphasis: Radio Listening.
• New equipment reviews.
• Ads for new equipment.
• Some historical articles.
• Small classified section for used radios.

Practical Wireless
PW Publishing Ltd.
Arrowsmith Court
Station Approach
Broadstone, Dorset BH18 8PW
England

Emphasis: Amateur Radio.
• New equipment reviews (mostly ham)
• Ads for new equipment.
• Some historical articles.

QST
American Radio Relay League
225 Main St.
Newington, CT 06111-9965

Emphasis: Amateur Radio
• Ads for new equipment.
• New equipment reviews (mostly ham).
• Classified section for used radios.

Radio & Electronics World
Sovereign House
Brentwood
Essex
England

Emphasis: Radio
• New equipment reviews.
• Ads for new equipment.

Radio Bygones
9 Wetherby Close
Broadstone
Dorset BH18 8JB England

Emphasis: Collecting and Restoration

Radio Communications
Lambda House
Cranborne Road
Potters Bar, Herts, EN6 3JE
England

Emphasis: Amateur Radio
• New equipment reviews (mostly ham).
• Ads for new equipment.

Shortwave Magazine
PW Publishing Ltd.
Arrowsmith Court
Station Approach
Broadstone, Dorset BH18 8PW
England

Emphasis: Radio Listening
• New equipment reviews.
• Ads for new equipment.
• Some historical articles.

73 Amateur Radio
70 Route 202 N.
Peterborough, NH 03458

Emphasis: Amateur Radio
• Ads for new equipment.
• New equipment reviews (mostly ham).

I.B.S. Whitepaper Reviews
by Larry Magne
International Broadcast Services,
P.O. Box 300
Penn's Park, PA 18943

These authoritative reports, 12 to 20 pages in length, offer complete and unbiased evaluations and ratings on a single radio model. These exhaustive reviews are available for most current communications receivers (and some worldband portables). I.B.S. also publishes a special White Paper titled *How To Interpret Receiver Specifications and Lab Tests.* The I.B.S. White Paper reports can be obtained directly from the publisher and selected radio dealers.

The following publications are cited in this book, but are no longer printed:

Electronics Illustrated
Ham Radio Magazine

5 Restoration & Repair

There are many rewards to owning and collecting older radios. But there is a downside too. Obtaining parts and service for vintage radios can be difficult. If the manufacturer has been out of business for thirty years, it can be a real challenge. It is unlikely your neighborhood TV-VCR repair shop will be able to help you with your ailing HQ-150. Fortunately, there are companies that specialize in servicing and restoring older shortwave and amateur receivers. It is important to realize and appreciate that these special skills do command a price. A professional restoration can preserve history and create an heirloom that can be appreciated for generations.

The following companies have recently placed ads in leading radio magazines offering radio service or restoration. It is important to contact them by mail or phone **before** sending in your receiver. When shipping anything electronic, remember to "double box" the device with plenty of extra packing ("foamies"). An extra dollar of packing material may save costly repairs and hours of aggravation. Please be sure to fully insure the package.

Please note that the author is **not** personally familiar with all the listed companies and cannot be responsible for their workmanship, ethics or business practices.

Antique Radio Restoration and Repair
Bob Eslinger
20 Gary School Rd.
Pomfret Center, CT 06259

Specialty: Tube Radios

Email: radiodoc@neca.com
http://www.neca.com/~radiodoc
Telephone: 860 928-2628

Tom Miller Electronics
22516 S. Normandie SP41B
Torrance, CA 90502

Specialty: General

Telephone: 310 320-8980

RTO Electronics
5585 Hochberger
Eau Claire, MI 49111

Specialty: Heathkit

Telephone: 616 461-3057

Great Northern
Alan Jesperson
P.O. Box 17338
Minneapolis, MN 55417

Specialty: General
Email: mte612@aol.com

Telephone: 612 727-2489

W3HM Radio Labs
Rt. #3, Box 712
Harpers Ferry, WV 25425

Specialty: Collins 75A-4

Telephone: 304 876-6483

Bavarian Radio Works
Ross A. Hochstrasser
162 Beulah St.
Whitman, MA 02382

Specialty: Grundig, Philips, Siemens, etc.

Telephone: 617 447-4299

Sound In Mind   *Specialty:* *Tube sets.*
Dee Almquist W4PNT
534 W. Main St.
Waynesboro, VA 22980   Telephone: 800 755-2365

Miltronix   *Specialty:* *Collins R-390/A*
Rick Mish
P.O. Box 80041
Toledo, OH 43608   Telephone: 419 255-6220

David Knepper   *Specialty:* *Collins*
P.O. Box 34
Sidman, PA 15955   Telephone: 814 487-7855

Chuck Rippel *WA4HHG*   *Specialty:* *Collins R-390A*
2341 Herring Ditch Rd.   Email: crippel@exis.net
Chesapeake, VA 23323   Telephone: 757 485-9660

Jim Clark   *Specialty:* *General*
1292 Starboard
Okemos, MI 48864   Telephone: 517 349-2249

Dennis Brothers *WAØCBK*   *Specialty:* *Collins 51S-1*
H.C. Box 1
Potter, NE 69156-9501   Telephone: 308 879-4552

Vintage Restoration Services   *Specialty:* *General*
Joe Reminder
228 Buckman Rd.
Rochester, NY 14615   Telephone: 716 663-0438

    If you elect to perform your own service an excellent source for tubes, parts and supplies is:

                Antique Electronic Supply
                6221 S. Maple Av.
                Tempe, AZ 85283
                Telephone: 602 820-5411

Even the technically competent collector will find service difficult without a manual and/or schematic. Fortunately there are companies that specialize in providing out-of-print owner's and service manuals. A typical owner's manual includes instructions on the operation of the receiver. Many, but not all, owner's manuals also include a basic schematic. A service or technical manual will always include a schematic plus additional diagnostic and technical information, instructions for restringing dials and alignments, parts list, circuit photographs, etc. Expect to pay more for a service manual.

The Manual Man
P.A. Markavage
27 Walling St.
Sayreville, NJ 08872-1818

Catalog: 3 First class stamps.
Tel./FAX: 732 238-8964

HI-Manuals
P.O. Box 802
Council Bluffs, IA 51502

Catalog: $2 (Outside U.S.$3)
E-Mail: himan@radiks.net
Web: www.hi-manuals.com

A.G. Tannenbaum
P.O. Box 386
Ambler, PA 19002

Catalog: Free
Telephone: 215 540-8055
Web: www.agtannenbaum.com

W7FG Vintage Manuals
3300 Wayside Dr.
Bartlesville, OK 74006

Catalog: $1
Telephone: 918 333-3754
E-Mail: w7fg@w7fg.com
Web: www.w7fg.com

Richard Prester
131 Ridge Rd.
West Milford, NJ 07480

Catalog: $1
Telephone: 973 728-2454

Puett Electronics
P.O. Box 28572
Dallas, TX 75228

Catalog: $5
Telephone: 214 321-0927

The **Allied Radio Company** was established over 75 years ago and has been a leading mail-order supplier of radios, electronics and parts. Until fairly recently, Allied carried most major lines of shortwave and amateur equipment, plus their own private label, Allied line and their famous line of Knight-Kit radios. Although Knight-Kit, was an exclusive property of Allied, they have their own Chapter in this book.

By 1970 Allied had 41 company stores. It was purchased from then owners, LTV Ling Altec, by Tandy Corporation and combined with the Radio Shack chain to become Allied Radio Shack. Department of Justice antitrust concerns made this a short marriage. Tandy quickly sold the stores to Schaark Electronics.

The Allied house-brand receivers were fairly unexciting as a group. The Allied AX-190 and SX-190's however, were fairly state of the art for their time. These models were sturdily built and performed well. They are considered the best models ever to carry the Allied label. For someone on a budget, desiring strictly shortwave broadcast or strictly amateur radio analog coverage, they still represent a good value and can be bought for a song.

Today Allied produces one of the most comprehensive and informative electronic parts catalogs in the world (1 800 433-5700).

Allied Radio
100 N. Western Ave.
Chicago, IL

Allied Radio Shack
100 N. Western Ave.
Chicago, IL

Allied Electronics
A Subsidiary of Hall-Mark Elec.
7410 Pebble Dr.
Fort Worth, TX 76118

A2508

Shortwave Broadcast Band Receiver
Single Conversion Superheterodyne. Solid State.

Features:
- ¼" Head. Jack
- S-Meter
- Tone Control
- MW Ferrite Antenna
- Dial Lamp
- Speaker 4"
- AFC [FM]

Specifications:
Coverage ...... SWBC. See Comments.   Modes .... AM/FM-W/FM-N
Antenna Input Terminals

Circuit Complement:
1 Integrated circuit, 1 MOSFETs, 12 transistors, 8 diodes, 1 zener diode and 2 thermistors.

Comments:
Coverage: .15-.4, .55-1.6, 5.9-6.25, 9.45-9.85, 11.45-12 and 15.05-15.55 MHz. The A2508 also covers the FM broadcast band 88-108 MHz and VHF "high" band 152-176 MHz. Allied stock number 20-5508.

Made In:	Japan	1970-1971
Voltages:	117 AC 15W or 12 VDC	
Readout:	Analog	
Physical:	12x4.75x8.75" 11 Lbs. 325x121x223mm 5 kg	
Status:	Inactive Manufacturer Discontinued Model	
Rarity:	Scarce	

	New	Used
Price:	$119	$40-60
Rating:	★★★	★★

A2509

Shortwave Broadcast Band Receiver
Single Conversion Superheterodyne. Solid State.

Features:
- ¼" Head. Jack
- S-Meter
- Tone Control
- MW Ferrite Antenna
- Dial Lamp
- Speaker 4"

Specifications:
Coverage ...... SWBC. See Comments.   Modes .... AM/FM-N
Antenna Input Terminals

Comments:
Coverage: .15-.4, .55-1.6, 5.9-6.25, 9.45-9.85, 11.45-12 and 15.05-15.55 MHz. The A2509 also covers 152-176 MHz which is called the VHF High Band. This is a public service band.
Allied stock number 20-5509

Made In:	Japan 1971
Voltages:	117 AC 50W or 12 VDC
Readout:	Analog
Physical:	12x4.75x8.75" 9 Lbs. 325x121x222mm 4 kg
Status:	Inactive Manufacturer Discontinued Model
Rarity:	Scarce

	New	Used
Price:	$79	$40-45
Rating:	★★★	★★

A2515

General Coverage Communications Receiver
Single Conversion Superheterodyne. Solid State.

Features:
- ¼" Head. Jack
- S-Meter
- Tone Control
- Bandspread
- RF Gain
- ANL
- AVC
- Antenna Trimmer
- Dial Lamp
- Mute Terminal
- BFO
- Flywheel Tuning

Specifications:
Coverage ...... 150 - 30000 kHz.
Selectivity ...... 1.8 kHz. -6dB.
Sensitivity ...... 2µV 10 dB S/N
Audio Out ...... 1.3 W
IF .... 455 kHz

Modes .... AM/CW/SSB
IF Rejection .... 40 dB
Noise Ratio ..... 30 dB down
Antenna Input . Terminals

Circuit Complement:
10 Transistors, 12 diodes and 1 zener diode. Two FET RF stages, product detector and push-pull audio amp.

Accessories:
Speaker

Comments:
Ranges: .15-.4, 550-1.6, 1.6-4.8, 4.8-14.5 and 10.5-30 MHz. Reception gap from 400 kHz to 550 kHz. Requires a speaker.

Variants:
Model **A2515A** $120 utilizes two MOSFETS and has a 1.8 kHz mechanical filter.

Made In:	Japan 1967-1971
Voltages:	117 AC 50/60 Hz 10W or 12 VDC
Readout:	Analog
Physical:	15x7.7x10" 20 Lbs. 381x195x254mm 9 kg
Status:	Inactive Manufacturer Discontinued Model
Rarity:	Scarce
Reviews:	*QST* February 1969 *CQ* July 1969 *Ham Radio* January 1969

	New	Used
Price:	$100	$40-60
Rating:	★★★	★★★

A2516

Amateur Band Communications Receiver
Double Conversion Superheterodyne. 7 Tubes plus Semiconductors.

Features:
- S-Meter
- Preselector
- ¼" Head. Jack
- Line Out 500 ohm
- RF Gain
- ANL
- AVC
- Mute Terminal
- Standby
- 28:1 Tuning Ratio

Specifications:
Coverage ...... Ham. See Comments.
Selectivity ...... 1.5 kHz. -6dB.
Sensitivity ...... 2 µV 10 dB S/N
Audio Out ...... 1 W
IF .... 8900-9500, 455 kHz

Modes .... AM/CW/SSB
IF Rejection .... >50 dB
Image Ratio .... >50 dB
Dial Accuracy . ± 1 kHz.
Antenna Input . Terminals

Circuit Complement:
6BZ6 RF Amp, 6BL8 1st Mixer/Osc., 6BE6 2nd Mixer, 6BA6 IF, 6BA6 IF, 6AQ8 Product Detector/BFO, 6BM8 AF Output plus semiconductors. Crystal controlled first oscillator, mechanical filter.

Accessories:
A-2514 Speaker   Calibrator

Comments:
Coverage: 3.5-4, 7-7.3, 14-14.35, 21-21.45, 28-28.5, 28.5-29.1 and 29.1-29.7 MHz plus 9.6-10 MHz for WWV. Provides 1 kHz dial accuracy.
Allied stock #20-5516

Variants:
See Kenwood JR-500SE.

Made In:	Japan	1969-1971
Voltages:	117/230 AC 50/60 Hz 75W or 12 VDC	
Readout:	⨳ Analog Lin.	
Physical:	13x7x10" 18 Lbs. 330x178x254mm 8.16 kg	
Status:	Inactive Manufacturer Discontinued Model	
Rarity:	Scarce	
Reviews:	QST January 1970	

	New	Used
Price:	$170	$70-160
Rating:	★★★	★★

AX-190

Amateur Band Communications Receiver
Double Conversion Superheterodyne. Solid State.

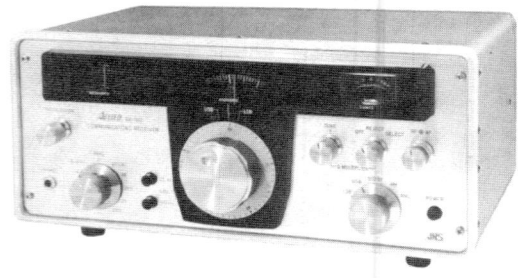

Features:
- ¼" Head. Jack
- S-Meter
- Preselector
- Record Jack
- RF Gain
- Noise Limiter
- AGC
- Calibrator 25/100 kHz
- Mute Terminal
- VFO Out Jack
- HFO Out Jack
- Speaker Jack
- Q-Multiplier

Specifications:
Coverage ...... Ham. See Comments.
Selectivity ...... 4 kHz. -6dB.
Sensitivity .... 5µV SSB/CW 10 dB S/N
Inter. Freqs. ... 2420-2920 & 455 kHz
Audio Out ...... 1 W 8 ohm
Stability .... 500 Hz after warm-up

Modes .... AM/LSB/USB(CW)
Spur.Reject. ... >50 dB
Image Ratio .... >60 dB
Image Rej. ...... >60 dB
Antenna Input . SO-239 50-75 ohm

Circuit Complement:
4 FETs, 22 transistors, 13 diodes, 2 zener diodes and 2 thermistors. Crystal controlled first oscillator.

Accessories:
SP-190 Speaker

Comments: Coverage: 3.5-4, 7-7.5, 14-14.5, 15-15.5, 21-21.5, 27-27.5, 28-28.5, 28.5-29, 29-29.5 and 29.5-30 MHz. One crystal position available for optional 500 Hz band (3.5-10 MHz.). Excellent dial accuracy of ±500 Hz. Requires a speaker. European version is 230 VAC 50/60 Hz. May be labelled as Realistic or Radio Shack AX-190. Allied-Radio Shack stock #20-5155

Made In:	Japan	1971-1973
Voltages:	110-120 AC 60 Hz 10 W or 12 VDC	
Readout:	⨳ Analog Lin.	
Physical:	15x7x10" 20 Lbs. 381x178x254mm 9 kg	
Status:	Inactive Manufacturer Discontinued Model	
Rarity:	Abundant	
Reviews:	CQ May 1972 Ham Radio June 1972 73 January 1972	

	New	Used
Price:	$250	$90-110
Rating:	★★★	★★★

SX-190

Shortwave Broadcast Communications Receiver
Double Conversion Superheterodyne. Solid State.

Features:
- ¼" Head. Jack
- RF Gain
- Q-Multiplier
- S-Meter
- Noise Limiter
- Mute Terminal
- Preselector
- AGC
- Speaker Jack
- Record Jack
- Calibrator 25/100 kHz
- VFO Output Jack

Specifications:
Coverage ...... SWBC. See Comments.
Selectivity ...... 4 kHz. -6dB.
Sensitivity .... 5μV SSB/CW 10 dB S/N
Stability .... 500 Hz after warm-up
Inter. Freqs. ... 2420-2920 & 455 kHz
Audio Out ...... 1 W 8 ohm
Modes .... AM/LSB/USB(CW)
Spur.Reject. ... >50 dB
Image Ratio .... >60 dB
Image Reject .. >50 dB
Antenna Input . SO-239 50-75 ohm

Circuit Complement:
4 FETs, 22 transistors, 13 diodes, 2 zener diodes and 2 thermistors. Crystal controlled first oscillator.

Accessories:
SP-190 Speaker

Comments: Coverage: 3.5-4, 5.7-6.2, 7-7.5, 9.5-10, 11.5-12, 14-14.5, 15-15.5, 17.5-18 and 27-27.5 MHz. Two crystal positions available for two optional additional 500 Hz bands (one in the range of 3.5 to 10 MHz. and one in the range of 10 to 30 MHz.). Dial accuracy is ±500 Hz. Requires a speaker. European version is 230 VAC 50/60 Hz. The most serious receiver ever to be sold under the Allied label. May be labelled as Realistic or Radio Shack SX-190. Allied-Radio Shack stock #20-5190.

Made In: Japan   1971-1973

Voltages: 110-120 AC 60 Hz or 12 VDC

Readout: Analog Lin.

Physical: 7x15x10" 20 Lbs.
381x178x254mm 9 kg

Status: Inactive Manufacturer Discontinued Model

Rarity: Common

Reviews: *CQ* May 1972
Popular Elec November 1972
NASWA December 1991
Elem. Elec. Nov/Dec 1972

	New	Used
Price:	$250	$100-140
Rating:	★★★★	★★★

7 Amalgamated Wireless Australia

Amalgamated Wireless Australia Ltd.
Sydney, Australia

Amalgamated Wireless Australia Ltd. was established in Sydney, New South Wales, Australia on July 11th, 1913 by the merger of the interests of Marconi and Telefunken. The company's technical manager was Ernest Thomas Fisk, who soon rose to the position of managing director. With the outbreak of the First World War in 1914, the links with Telefunken were broken and with partial government ownership, A.W.A. became an entity in it's own right.

For decades, A.W.A. featured prominently in almost all aspects of the Australian radio scene. Controlling the R.C.A. tube patents for Australia, its tube manufacturing subsidiary, Amalgamated Wireless Valve Co., provided a large proportion of Australia's tube requirements from the 1930's to the end of the tube era. Practically all of Australis's radio needs were catered for by A.W.A. from domestic receivers, military communications, through to broadcast transmitters and sophisticated test equipment. The company made quite a number of communications receivers in addition to the C52820 and CR-6 shown. Other models include the 3BZ, AMR-101 (yet another HRO copy) and the AR8. They even built a sophisticated receiver for the Australian military called the CRH-11. (The author welcomes information on these models). The company also made avionics equipment, VHF transceivers and NDB radio navigation ground stations. A.W.A. no longer makes communications receivers, but their remaining areas of interest include broadcasting, telephone and computer equipment.

C52820

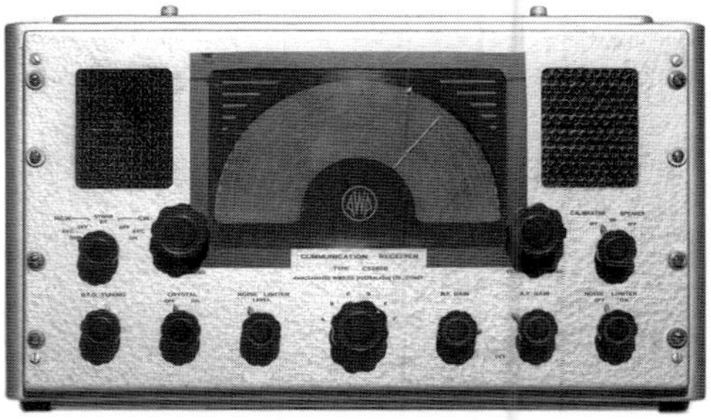

General Coverage Communications Receiver
Single Conversion Superheterodyne. 13 Tubes.

Features:
- Speaker Switch
- Speaker
- NL
- RF Gain
- BFO
- Bandspread 0-180
- Standby
- Dial Lamp
- AVC ON/OFF
- Calibrator 500 kHz

Specifications:
Coverage ...... 100 - 32000 kHz
Selectivity ...... 7/2.5 -6 dB
Sensitivity ...... 2 µV 1.6-32 MHz
Modes .... AM/CW/MCW
IF .... 565 kHz (See Comments)
Audio Out 2W

Circuit Complement:
6BA6 RF Amp, 6BE6 Mixer, 6C4 HF Osc., 6BA6 1st IF, 6BA6 2nd IF, 6AL5 Detector/AVC, 6AL5 Noise Limiter, 6AV6 1st Audio, 6AQ5 Audio Output, 12AU7 BFO, 5Y3 Rectifier, 0D3/VR150 Voltage Regulator and 6BA6 Calibrator.

Comments: Ranges: .1-.22, .2-.545, .535-1.62, 1.6-4.9, 4.8-16 and 16-32 MHz. The IF is 565 kHz on all bands except .535-1.6 MHz when it is changed to 515 kHz. As a result of this separate IF strip, the BFO and crystal filter do not operate on this band. The mechanical bandspread uses a separate small pointer encompassing the outer most logging scale of 0 to 180. It is visible in the photo at the one o'clock position, with the large main pointer at the two o'clock position. This model may be mounted in a desktop cabinet or 19" rack mount. This model is not well known, even in Australia. It is believed only a few were made for government applications.

Variants: The model **C55163** (1950) has a similar dial and speaker placement, but with a few less dials.

Made In: Australia 1952-1953

Voltages: 240 VAC 50 Hz

Readout: Analog

Physical:

Status: Active Manufacturer
Discontinued Model

Rarity: Typically Unavailable

	New	Used
Price:	①	⑤
Rating:	⑥	⑥

CR-6B

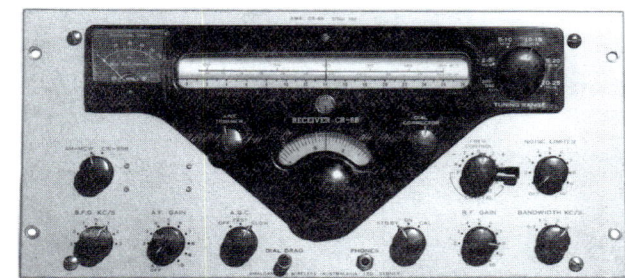

General Coverage Communications Receiver
Double Conversion Superheterodyne. 12 Tubes plus Semicond.

Features:
- ¼" Head. Jack
- S-Meter
- Standby
- AVC OFF/FST/SLO
- Speaker
- BFO ±3 kHz
- Dial Lamp
- Calibrator 500 kHz
- NL Variable
- RF Gain
- Logging Scale
- Line Out 600 Ohm
- Dial Adjust
- Dial Drag Adjust
- Speaker Switch
- Antenna Trimmer
- Flywheel Tuning

Specifications:
Coverage ...... 100 - 32000 kHz
Selectivity ...... 6/3/1.5/.7 -6 dB
Sensitivity ...... 3 µV AM 10 dB S/N

Modes .... AM/CW/MCW/SSB
IF .... 1800, 100 kHz
Audio Out 1W

Circuit Complement: 6BY7 RF Amp, 6AJ8 Mixer/Osc., 6BA6 1st IF, 6BA6 2nd IF, 6AU6 3rd IF, 6AU6 BFO, 12AU7 Product Detector, 6AU6 Audio Driver, 6AQ5 Audio Output, 6AJ8 1st Mixer/Osc., 6AU6 Calibrator and 0B2 Voltage Regulator.

Accessories:
1D60608 Speaker   1C60604 Crystal Osc. Unit for 6 Fixed Crystal Positions

Comments: Ranges: .2-.54, 2-5, 5-10, 10-15, 15-20 and 20-25 MHz. Designed for 19" rack mounting. The CR-6A and CR-6B are fairly common in Australia and sell for approximately $300 USD. Silver gray Hammertone enamel finish. Requires a speaker.

Variants: Model **CR-6A** covers 2-30 MHz in six ranges: 2-5, 5-10, 10-15, 15-20, 20-25 and 30 MHz. The CR-6B was introduced with low frequency coverage, to enable Flight Service Units to monitor longwave radio navigations aids (NDBs).

Made In: Australia   1958-1962

Voltages: 220-250 VAC 50-60 Hz
50W

Readout: ⊔⌁⌁⌁▉⌁⊔ Analog

Physical: 19x8.75x12.5"
483x222x317mm

Status: Active Manufacturer
Discontinued Model

Rarity: Typically Unavailable

	New	Used
Price:	①	⑤
Rating:	⑥	⑥

8 Ameco

For most of **Ameco's** long history, they have produced accessory items for the amateur and shortwave markets. In the 1960's however, they produced a couple of receivers and even manufactured some transmitters.

The entry level R5 was introduced in 1967, followed by the improved R5A in 1969. These early solid-state receivers covered medium wave, shortwave and even extended to 54 MHz to include the 6 meter amateur band.

For a period of time Ameco and Gonset were owned by Aerotron. Aerotron was briefly owned by Siemens.

Ameco no longer manufactures receivers, but does produce preamps, amateur accessories, code keys, filters and offers training materials and books for amateur radio and electronics enthusiasts.

Ameco Div. of Aerotron
U.S. Highway 1N
Raleigh, NC 27608   1968

Ameco Corporation
224 East Second St.
Mineola, NY 11501

R5

General Coverage Communications Receiver
Double Conversion Superheterodyne. Solid State.

Features:
- ¼" Head. Jack
- ANL
- RF Gain
- Q-Mult Input Jack
- Fine Tuning
- 6M Coverage
- BFO
- Bandspread 0-10
- Dial Lamp

Specifications:
Coverage ......540 - 54000 kHz   Modes .... AM/CW
Antenna Input Terminals

Accessories:
S-5 Speaker (shown below)   BK-5 Battery Kit

Comments:
Ranges: .54-1.35, 1.35-3.5, 3.5-9, 9-23 and 23-54 MHz. The BK-5 Battery Kit holds six D cells. Not a sensitive receiver. Gray metal case.

Variants:
Model **R5-K** was kit form version $50-70.

Made In:	United States 1967-1969
Voltages:	105-125 AC 50/60 Hz

Readout:	Analog
Physical:	12x6x9" 12 Lbs. 305x152x229mm 5.4 kg
Status:	Active Manufacturer Discontinued Model
Rarity:	Scarce
Reviews:	*Electronics Illustr.* Sept. 1968

	New	**Used**
Price:	$70-90	$50-100
Rating:	★★★	★

R5A

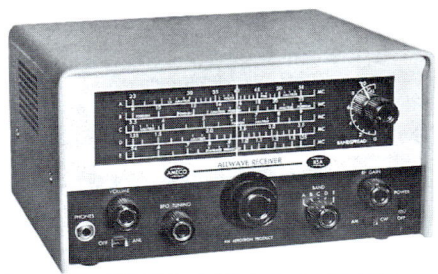

General Coverage Communications Receiver
Double Conversion Superheterodyne. Solid State.

Features:
- ¼" Head. Jack
- ANL
- RF Gain
- Bandspread 0-10
- Fine Tuning
- 6 M coverage
- Mute Line
- Q-Mult Input Jack
- Speaker
- Dial Lamp
- BFO

Specifications:
Coverage ......540 - 54000 kHz   Modes .... AM/CW
Antenna Input Terminals

Accessories:
S-5 Speaker (shown below)   BK-5 Battery Kit

Comments:
Ranges: .54-1.35, 1.35-3.5, 3.5-9, 9-23 and 23-54 MHz.
The Owner's Manual proposes that this receiver is capable of
single sideband and narrowband FM reception. Gray metal
case.

Variants:
Model **R5A-K** was kit form version.

| **Made In:** | United States 1969-1970 |
| **Voltages:** | 105-125 AC 50/60 Hz |

Readout:	Analog
Physical:	12x6x9" 12 Lbs. 305x152x229mm 5.4 kg
Status:	Active Manufacturer Discontinued Model
Rarity:	Scarce

	New	**Used**
Price:	$100	$50-100
Rating:	★★★	★

SWL-4

General Coverage Broadcast Receiver
Single Conversion Superheterodyne. Solid State.

Features:
- ¼" Head. Jack
- Dial Lamp
- Bandspread 0-10
- Speaker

Specifications:
Coverage ......550 - 23000 kHz   Modes .... AM
Antenna Input Terminals

Comments:
The simple SWL-4 was sold as an introductory set. Please note that coverage stops
at 23 MHz. It has a metal, walnut colored case.

| **Made In:** | United States 1969 |
| **Voltages:** | 105-125 AC 50/60 Hz |

Readout:	Analog
Physical:	12x6x9" 10 Lbs. 305x152x229mm 4.5 kg
Status:	Active Manufacturer Discontinued Model
Rarity:	Very Scarce

	New	**Used**
Price:	$70	$50-60
Rating:	★★★	⑥

9 Anritsu

The **Anritsu Corporation** of Tokyo, Japan was established in 1931. They are a worldwide supplier of communications and electronic equipment employing over 3,000 people. Their main areas of production include: transmission and switching terminals, radio communications, measuring equipment, data processing, industrial automation and components and device manufacture. Anritsu radio products include: marine radars, INMARSAT terminals, SSB receivers and transmitters, VHF/UHF radios and automated maritime subscriber telephones.

Anritsu has subsidiaries in Brazil, England, Germany, Italy France and the United States. The company enjoyed sales of $709 million (USD) in 1990.

Over the past decades Anritsu has produced a prodigious number of commercial H.F. communications receivers. Due to a lack of original reference material available to the author, only a few are illustrated in this edition. The author welcomes assistance on Anritsu information for future editions of this book.

The Japanese language book titled *Japanese Professional Receivers* written by Mr. Hideo Kanemich does cover the entire line of Anritsu receivers and is highly recommended.

Anritsu Corporation
5-10-27, Miaamiazabo
Minato-ku
Tokyo 106, Japan

Anritsu

RG01A

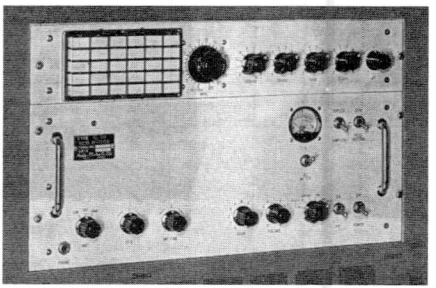

General Coverage Communications Receiver
Double Conversion Superheterodyne. Tubes.

Features:
- ¼" Head. Jack
- RF Gain
- 30 Presets
- S-Meter
- Mute
- IF Out Jack 1.5 MHz
- BFO
- Rack Handles
- Fine Tuning
- Line Out 600 ohm

Specifications:

Coverage ...... 270-540, 1000-30000 kHz	Modes .... SSB/CW
Sensitivity ...... 1 µV 20 dB S/N	Stability .... 1×10^{-6}
Selectivity ...... 2.4/.5 kHz -6dB	IF Rejection .... >70 dB
Image Rej ...... >80 dB	Antenna Input . 50 ohm
Audio Out ...... 5/600 ohms	

Circuit Complement:
6BZ6 RF Amp, 6CB6 1st Mixer, 6BE6 2nd Mixer, 6CB6 3rd Mixer, 6BA6 1st IF Amp, 6BA6 2nd IF Amp, 6BA6 3rd IF Amp, 12AU7 IF/AF Amp, 6AV6 Detector/AF, 6AQ5 PA, 12BH7A PA, 12AU7 PHI, 12AX7 AFA, 12BH7A PA, 12AU7 PHI, 12AX7 AFA, 35K23 AFA, plus 1S315 rectifier.

Comments:
Normal tuning is accomplished with the six rotary knobs on the top of the front panel. The large knob on the left selects MHz 1 through 29 (or the .27-.56 MHz band). The next four knobs select 100 kHz, 10 kHz, 1 kHz and 100 Hz. The last knob is Fine Tuning. The 30 channel preset arrangement can be switch selected by the six by five matrix of channel switches. This model is typically custom installed into a commercial marine communications panel as shown.

Made In:	Japan	1969-1974
Voltages:	80-120 VAC 50/60 Hz	
Readout:	●●●●●●.● Dig. Knobs	
Physical:	20x13.4x17.5" 110 Lbs. 510x340x445mm 50 kg	
Status:	Active Manufacturer Discontinued Model	
Rarity:	Typically Unavailable	

	New	Used
Price:	①	②
Rating:	⑥	⑥

Anritsu

RG11A

General Coverage Communications Receiver
Double Conversion Superheterodyne.

Features:
- ¼" Head. Jack
- S-Meter
- BFO
- Speaker 4"
- RF Gain
- Mute
- Rack Handles
- Line Out 600 ohm
- Zero Adjust
- AGC ON/OFF
- Speaker Switch

Specifications:
Coverage ...... 100 - 30000 kHz
Sensitivity ...... 1 µV 20 dB S/N
Audio Out ...... 8/600 ohms 1W
Modes .... AM/SSB/CW
Antenna Input . 50 ohm

Comments:
The frequency is displayed digitally to the 100 kHz position. Then the analog linear dial is tuned to the nearest 1 kHz.

Variants:
Model **RG22B** is very similar, but with 192 presets (called spot channels), and FAX mode (25 kg).

Made In:	Japan	1973-1979

Voltages: 80-120/200-240 VAC
50/60 Hz 70 VA

Readout: `001` ▮▮ Digital to 100 kHz, then Analog Lin.

Physical: 16.7x7.7x16.7" 53 Lbs.
426x195x425" 24 kg

Status: Active Manufacturer
Discontinued Model

Rarity: Typically Unavailable

	New	**Used**
Price:	①	②
Rating:	⑥	⑥

Anritsu

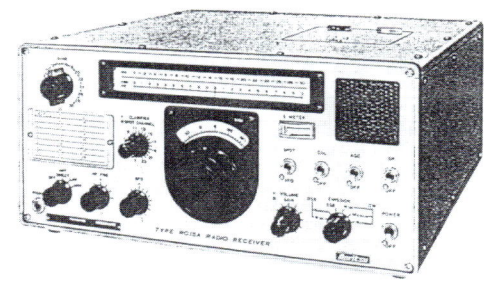

RG15A

General Coverage Communications Receiver
Quadruple Conversion Superheterodyne. Solid State.

Features:
- ¼" Head. Jack
- S-Meter
- BFO ±3.5 kHz
- Calibrator 100 kHz.
- RF Gain
- Mute
- Rack Handles
- Line Out 600 ohm
- Zero Adjust
- AGC ON/OFF
- Speaker Switch
- Fine Tuning
- 23 Presets
- Speaker 4"
- Clarifier ±400 Hz
- IF Output

Specifications:
Coverage ...... 100 - 30000 kHz
Sensitivity ...... 1 µV 20 dB S/N
Selectivity ...... 6/2.4/.6 kHz -6dB
Audio Out ...... 8/600 ohms 1W
Modes .... AM/SSB/CW
Stability .... <500 Hz 10-30 mins.
Image Rej .... >70 dB 2-14 MHz
Antenna Input . 50 ohm

Circuit Complement:
Quadruple conversion up to 2 MHz, triple conversion from 2 to 14 MHz and double conversion from 14 to 30 MHz.

Accessories:
F1 Option   F4 Option

Comments:
The linear analog yields 1 kHz accuracy. One rotation of the tuning knob corresponds to 100 kHz.

Variants:
The model **RG16A** is similar (18 kg).

Made In:	Japan	1974-1980

Voltages: 80-120/200-240 VAC
50/60 Hz 40 VA

Readout: ▯▯▯▯▮ Analog Lin.

Physical: 16.8x7.7x16.7" 44 Lbs.
426x195x425mm 20 kg

Status: Active Manufacturer
Discontinued Model

Rarity: Typically Unavailable

	New	**Used**
Price:	①	②
Rating:	⑥	⑥

Anritsu

RG52A

General Coverage Communications Receiver
Double Conversion Superheterodyne. Solid State.

Features:
- ¼" Head. Jack
- S-Meter
- BFO
- 500/2182 kHz Presets
- RF Gain
- Mute
- Rack Handles
- Line Out 600 ohm
- Attenuator
- Dial Lock
- Preselector
- AGC OFF/FST/SLO
- Dimmer
- * IF Output
- Clarifier

Specifications:
Coverage ...... 100 - 35000 kHz	Modes .... AM/CW/MCW/LSB/USB
Sensitivity ...... 0.5uV CW 12 dB S/N	Stability .... ±5 Hz
Selectivity ...... 6/3/.5 kHz -6dB	IF Rejection .... 70 dB
Image Rej ...... >80 dB	Environment ... -10 to +50°C
Audio Out ...... 8/600 ohms 1W	Antenna Input . 50 ohm

Accessories:
ISB Option   FAX Option
Memory Option

Comments: A commercial-maritime receiver. The RG52A has a separate BFO and Clarifier.

Variants: The **RG53A** (right) has a 10 Hz display and 6/3/.5/.2 kHz bandwidths [¥250,000-400,000].

Made In:	Japan	1979-1994

Voltages: 100/110 200/220 VAC 50/60 Hz 70 VA or 24 VDC

Readout: `00000.1` Digital LED

Physical: 19.2x5.9x15.9" 53 Lbs. 488x149x405" 24 kg

Status: Active Manufacturer Discontinued Model

Rarity: Typically Unavailable

	New	Used
Price:	$5700	$2400-2600
Rating:	⑥	⑥

Anritsu

RG81A

General Coverage Communications Receiver
Double Conversion Superheterodyne. Solid State.

Features:
- ¼" Head. Jack
- S/AF-Meter
- BFO ±2.5 kHz
- 500/2182 kHz Presets
- RF Gain
- Mute
- Rack Handles
- Line Out 600 ohm
- Memory Scan
- IF Output
- Dimmer (5 step)
- Attenuator -20 dB
- 128 Memories
- Keypad
- 3 Tuning Speeds
- AGC OFF/FST/SLO

Specifications:
Coverage ...... 100 - 35000 kHz	Modes .... AM/CW/USB/FAX/RTTY
Sensitivity ...... 0.5uV CW 12 dB S/N	Stability .... 1×10^{-9}
Selectivity ...... 6/3/1/.3 kHz -6dB	IF Rejection .... >80 dB
Image Rej ...... >70 dB	Environment ... -10 to +50°C
Audio Out ...... 8/600 ohms 1W	Antenna Input . 50-75 ohms
IFs .... 80.455 MHz, 455 kHz	

Accessories:
EIA RS232 Bus	GP-1B IEEE Bus	1200 Baud Modem
ZR101B ISB Demod.	SR1044R FSK Demod	ZA81A Auto Ant. Tuner
ZN62A 512 Ch. Controller	ZC177A T/R Interface	ZN81A Preset Timer (shown)

Comments:
The memories store: frequency, mode, bandwidth and AGC setting. The RG81A has a 5 Hz VFO. A sophisticated communications receiver for the marine commercial market. [¥450,000].

Variants:
Model **RG81B** has both LSB and USB, but no FAX mode.

Made In:	Japan	1983-1994

Voltages: 100/110 200/220 VAC 50/60 Hz 70 VA or 24 VDC

Readout: `00000.01` Digital LED

Physical: Cabinet Version: 19.2x6.1x15.9" 59.2 Lbs 488x157x405mm 27 kg Rack Version: 19x5.9x15.3" 44.1 Lbs 480x149x389m 20 kg

Status: Active Manufacturer Discontinued Model

Rarity: Typically Unavailable

	New	Used
Price:	$11900	$4500
Rating:	⑥	⑥

10 A.O.R. Ltd.

A.O.R. is relatively new to the North American receiver scene. They entered the market with VHF-UHF scanners that initially earned them a spotted reputation. In recent years the quality and design of A.O.R. scanners has improved substantially.

The AR2515 was AOR's first successful offering of a wideband receiver. Covering 500 kHz to 1500 MHz, this radio offered a big chunk of the spectrum to the listener at an affordable price and in a very compact package. As with most wideband receivers of this class, the medium wave and shortwave performance was marginal. Such radios are satisfactory for the scanner enthusiast who wants to dabble in shortwave. However the serious shortwave listener who wants to dabble in VHF-UHF is better off buying a dedicated shortwave receiver and a separate scanner. These wideband receivers did improve with subsequent models leading to the current AR3000A.

The AR3030 was the first traditional shortwave receiver offered by AOR. This compact and capable receiver performs well on shortwave but never found a large following outside the portable user market.

AOR's most sophisticated, interesting and acclaimed offering is the new AR7030. This somewhat radical receiver, designed by John Thorpe of Lowe HF-225E fame, is manufactured in England.

A.O.R. Ltd.
2-6-4 Misuji, Taito-Ku
Tokyo 111, Japan

AR2515

Wideband Broadcast Receiver (Scanner)
Quadruple Conversion Superheterodyne. Solid State.

Features:
- Mini Head. Jack
- Priority
- Keypad
- Ext. Spkr Jack
- S-Indicator
- Dial Lamp
- 1984 Memories
- RS232 Jack
- Squelch
- Key Lock
- Scan/Sweep
- Backlit LCD
- Atten. -10 dB
- Speaker

Specifications:
Coverage ......5 - 1500 MHz.
Sensitivity ......<3μV (1SW AM).
Audio Out ......1 W 10% distortion

Modes .... AM/FM-N/FM-W
IF .... 750, 45.03, 5.5, .455 MHz
Antenna Input . BNC

Accessories:
MS-2515 Speaker   BFO-1 BFO Option   EG-1 Control Software

Comments:
The memories store: frequency, mode, bandwidth, AGC and attenuator settings. Tuning steps: 5, 12.5 and 25 kHz. The AR2515 scans at 36 channels per second. Please note that shortwave coverage does not start until 5000 kHz. The filters are a bit wide for shortwave. A stronger performer as a scanner than as a shortwave receiver.

Made In:	Japan	1989-1991
Voltages:	12 VDC or supplied 110 VAC Adapter	
Readout:	`00001.`	Digital LCD
Physical:	5.25x3.12x7.875 2.8 Lbs. 133x79x200mm 1.3 kg	
Status:	Active Manufacturer Discontinued Model	
Rarity:	Common	
Reviews:	*Monitoring Times* May 1989	

	New	Used
Price:	$599-799	$420-440
Rating:	★★★	★★

AR3000

Wideband Broadcast Receiver (Scanner)
Quadruple Conversion Superheterodyne. Solid State.

Features:
- Mini Head. Jack
- Priority (4)
- Keypad
- Ext. Spkr Jack
- S-Indicator
- Dial Lamp
- 400 Memories
- RS232 Jack
- Squelch
- Key Lock
- Scan/Sweep
- Record Jack
- 2 Tuning Speeds
- Atten. -10 dB.
- Speaker
- Clock - Timer

Specifications:
Coverage ....1 - 2036 MHz.
Selectivity ...... 180/12/2.4 kHz -6dB.
Sensitivity ......<.25µV (2-1800 MHz).
Audio Out ......1.2 W 4 ohm

Modes .... AM/CW/LSB/USB/FM-N/W
IF Rejection .... 70 dB
Antenna Input . BNC 50 ohm

Accessories:
MM-1 Mobile Bracket

Comments:
The memories store: frequency, mode, bandwidth, AGC and attenuator settings.
Four scan banks of 100 channels each. Tuning increments as low as 50 Hz. Scan
rate is 20 channels per second. Sweep (search) rate is 20 steps per seconds.

Made In: Japan   1989-1992

Voltages: 12 VDC .5 A or supplied
                110 VAC Adapter

Readout: `00001.`   Digital LCD

Physical: 5.5x3x7.75' 2.6 Lbs.
                138x80x200mm 1.2 kg

Status: Active Manufacturer
              Discontinued Model

Rarity: Common

Reviews: *Monitoring Times* July 1990

	New	Used
Price:	$799-1095	$425-450
Rating:	★★★★	★★★

AR3000A

Wideband Broadcast Receiver (Scanner)
Quadruple Conversion Superheterodyne. Solid State.

Features:
- Mini Head. Jack
- Priority (4)
- Keypad
- Ext. Spkr Jack
- Priority Channel
- S-Indicator
- Dial Lamp
- 400 Memories
- RS232 Jack
- Sleep Timer
- Squelch
- Key Lock
- Scan/Sweep
- Record Jack
- Remote Switch
- Atten. -20 dB.
- Speaker
- Clock - Timer

Specifications:
Coverage ....1 - 2036 MHz.
Selectivity ...... 180/12/2.4 kHz -6dB.
Sensitivity ......<.25µV (2-1800 MHz).
Audio Out ......1.2 W 4 ohm

Modes .... AM/CW/LSB/USB/FM-N/W
IF Rejection .... 70 dB
Stability .... +5 ppm (-10 to 50°C).
Antenna Input . BNC 50 ohm

Accessories:
MM-1 Mobile Bracket

Comments:
The memories store: frequency, mode, bandwidth, AGC and attenuator settings.
Four scan banks of 100 channels each. Tuning increments as low as 50 Hz.
Provides faster scanning than the AR3000 Scan rate is 50 channels per second.
Sweep (search) rate is 50 steps per seconds. Quadruple conversion on WBFM only.
Later production units sold in America will have the cellular frequencies blocked.

Made In: Japan   1993-1998

Voltages: 12 VDC .5 A or supplied
                110 VAC Adapter

Readout: `00001.`   Digital LCD

Physical: 5.5x3x7.75" 2.6 Lbs.
                138x80x200mm 1.2 kg

Status: Active Manufacturer
              Active Model

Rarity: Common

Reviews: *Monitoring Times* June 1994

	New	Used
Price:	$899-1099	$425-475
Rating:	★★★★	★★★★

AR3030

General Coverage Communications Receiver
Double Conversion Superheterodyne. Solid State.

Features:

- ¼" Head. Jack
- S-Meter
- Squelch
- AGC FST/SLO
- RF Gain
- ANL
- Tone (2 Pos.)
- Atten. -10/-20 dB
- Keypad
- 100 Memories
- Scanning
- Dual VFOs
- Ext. Spkr Jack
- IF Out Jack
- Tilt Stand
- Sync. AM
- BFO
- RS232 Port
- Record Jack
- Record Activation
- Speaker 2.6"
- Memory Pass
- Backlit Display
- Backlight On/Off

Specifications:

Coverage ...... 30 - 30000 kHz.
Selectivity ...... 6/2.4/_ kHz -6dB [15 FM]
Sensitivity ...... <.5µV (1.8-30 MHz SSB)
Audio Out ...... 1.8 W 10% distortion

Modes .... AM/LSB/USB/FAX/FM
IF Rejection .... 70 dB
Dyn. Range .... 100 dB @ 25 kHz spacing
Antenna Input . BNC & Terminals

Accessories:

.5 kHz Filter   2.5 kHz Filter   VHF Converter 108-140 MHz.
Control software   4.0 kHz Filter   VHF Converter 140-170 MHz.

Comments:
The LCD is backlit. The memories store: frequency, mode, bandwidth, AGC, tone and attenuator settings. Tuning steps: 1 MHz, 1 kHz, 100 Hz, 10 Hz and 5 Hz. The 6 kHz filter is Collins. Battery life from the internal AA cells is less than one hour. A very well built and capable receiver for its compact size.

Made In:	Japan	1994-1997
Voltages:	11-16 VDC .7 A or 8 AA cells internal	
Readout:	00000.01	Digital LCD
Physical:	10x3.55x9.5" 4.8 Lbs. 254x90x241mm 2.2 kg	
Status:	Active Manufacturer Discontinued Model	
Rarity:	Abundant	
Reviews:	*Passport* 1995-1998 *WRTH* 1995 *Pop. Comm.* Sept. 1994 *Mon. Times* January 1995 *R.D.I. Whitepaper* *SW Magazine* Jan. 1994 *SW Magazine* Sept. 1994	

	New	**Used**
Price:	$599-849	$450-550
Rating:	★★★★	★★★★

AR5000

Wideband Communications Receiver
Triple Conversion Superheterodyne. Solid State.

Features:

- ¼" Head. Jack
- S-Meter
- Squelch
- Alphanumeric Mem.
- RF Gain
- ANL
- Keypad
- Atten. -10 dB
- Keypad
- 1000 Memories
- Scanning
- Speaker
- Ext. Spkr Jack
- 10.7 IF Out Jack
- Mute Line
- 10 MHz Stnd. Input
- BFO
- RS232 Jack
- Tilt stand
- Dial Torque Adj.
- HF Preamp
- Lock
- Priority
- Sleep 1-120 min
- Voice Scan
- Duplex Offset
- Memory Pass
- Detector Output
- 5 VFOs
- Sweep
- Audio Filter
- AGC [FM]   • Tone

Specifications:

Coverage ...... 10 kHz - 2600 MHz
Selectivity ...... 220/110/40/15/6/3/_ kHz -6dB.
Sensitivity ...... <.56µV (1.8-30 MHz).
Audio Out ...... 1 W 8 ohms 10% dist.

Modes .... AM/FM-N/FM-W/LSB/USB/CW
IF .... 622.2 MHz, 10.7 MHz, 455 kHz
Stability .... ±1 PPM
Antenna Input . SO-239 and N

Accessories: CTCSS Decode   DTMF
MF500 Collins 500 Hz MF2.5 Collins 2500 Hz   MF6.0 Collins 5500 Hz

Comments: Tuning steps: 1, 10, 50, 100, 500 Hz, 1, 5, 6.25, 9, 10, 12.5, 20, 25, 30, 50, 100 kHz and user defined. Scan rate: 25 or 50 channels/second. Non-governmental units sold in the United States have the cellular frequencies blocked. A highly configurable receiver.

Variants: The **AR5000+3** model introduced in early 1998 adds: Noise Blanker, AFC, AM Synchronous Detection and 1000 more memories for a total of 2000.

Made In:	Japan	1996-1998
Voltages:	11-16 VDC 1A or supplied 120 VAC Adapter	
Readout:	00000.01	Digital LCD
Physical:	8.5x3.5x10" 10A Lbs. 216x89x254mm 4.5 kg	
Status:	Active Manufacturer Active Model	
Rarity:	Too new for used market.	
Reviews:	*Pop. Comm.* February 1997	

	New	**Used**
Price:	$1900-2000	$750-950
Rating:	★★★★	★★★★

AR7030

General Coverage Communications Receiver
Double Conversion Superheterodyne. Solid State.

Features:
- ¼" Head. Jack
- IF Gain
- P.B.T. ±5 kHz
- Ext. Spkr Jack
- Clock-Timer

- S-Indicator
- Synchronous Det.
- 100 Memories
- LCD Contrast
- Timer Contacts

- Squelch
- Tone
- Infrared Remote
- Record Jack
- RS-232 DIN Port

- AGC
- Atten. (4 Pos.)
- Speaker
- Tilt Stand
- Scan

Specifications:
Coverage ...... 0 - 32000 kHz.
Selectivity ...... 10/7/4.5/2.2/_/_ kHz -6dB.
Sensitivity ...... <.3µV 1.5-32 MHz SSB
Audio Out ...... 1.8 W 10% distortion

Modes .... AM/LSB/USB/CW/FM/FSK
Stability .... ±2.5 PPM
Dyn. Range .... >100 dB AM 7 kHz
Antenna Input . SO-239 and Terminals

Accessories:
Various Filters   NB7030 CPU/Notch/NB
BP123 Battery   FL124 Daughter Board   TW7030 Whip Antenna

Comments: All receiver parameters may be displayed on the two-line, dot matrix alphanumeric, backlit LCD. This receiver is highly adjustable despite the dearth of buttons. The memories store: frequency, mode, bandwidth, PBT, squelch BFO and scan settings. Tuning steps: 1 MHz, 1 kHz, 100 Hz, 10 Hz and 5 Hz. The supplied infrared remote control allows tuning, volume, tone, keypad, memory functions, PBT and filter selection.

Variants: The **AR7030+** features a high tolerance components, CPU upgrade and a total of 400 memories. Optional UPNB7030 upgrade adds NB/Notch to AR7030+.

Made In:	England	1996-1998
Voltages:	15 VDC 500 ma. Supplied with AC adapter.	
Readout:	`00000.01` Digital LCD	
Physical:	9.45x3.54x10" 4.9 Lbs 240x90x255mm 2.2 kg	
Status:	Active Manufacturer Active Model	
Rarity:	Scarce	
Reviews:	*Passport* 1998-1998 *Monitoring Times* April 1996 *NASWA* December 1996 *NASWA* February 1997 *SW Mag.* March 1996 *WRTH* 1997	

	New	Used
Price:	$1149-1349	$750-900
Rating:	★★★★★	★★★★★

11 Atlas Radio

The **Atlas Radio** Company was a significant producer of American made amateur radio equipment in the late 1970's. They are best known for their Atlas 210x/215x and 350-XL H.F. ham transceivers. They produced only one receiver, that being the RX-110 shown below. Examples of the "RX-110 Special" variant do exist, but are extremely scarce.

Facing increasing offshore competition, the Atlas President, Herbert G. Johnson and Vice President Les. J. Johnson discontinued operations on October 31, 1979.

The Atlas name was revived in the early 1990's when another transceiver was advertised under this famous label. Unknown difficulties arouse and this project did not materialize.

Atlas Radio Inc.
417 Via Del Monte
Oceanside, CA 92054   1978

RX-110

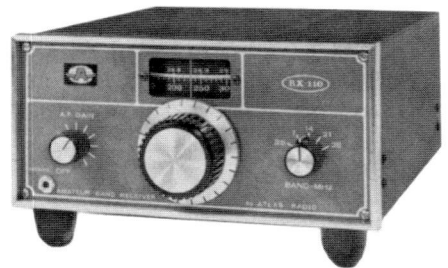

Amateur Band Communications Receiver
Single Conversion Superheterodyne. Solid State.

Features:
• Mini Head. Jack • Speaker 3"   • Mute Line

Specifications:

Coverage ......Ham. See Comments	Modes .... CW-SSB
Selectivity ......2.7 kHz -6 dB	IF .... 5595 kHz
Sensitivity ......0.25µV 10 dB <28 MHz	Image Rej. ...... >60 dB
Sensitivity ......0.40µV 10 dB >28 MHz	Stability .... <500 Hz/hour after 30 min
Audio ....2W 4 ohms	Antenna Input . RCA Phono Jack

Circuit Complement:
Four integrated circuits, 8 transistors and 25 diodes. Crystal filter.

Comments: This simple, low cost receiver was designed for the beginner. Ranges: 3.5-4, 7-7.5, 14-14.5, 21-21.5 and 28-29 MHz. The tuning rate is 22 kHz/rev., except on 10 meters where it is 44 kHz/rev. The TX-100 transmit modules may be added to the RX-110 to form a five band transceiver. The TX-100L module provides 10-15 watts input and the TX-100H module provides 150-200 watts input. This transceiver configuration is shown below.

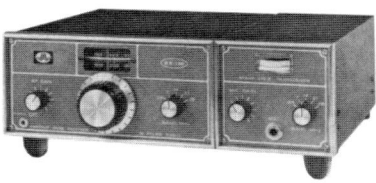

Variants:
The model **RX-110 Special** features additional controls for RF Gain, VBT and RIT. It is not certain whether this variant was ever produced in significant quantities.

Made In: United States 1977-1979

Voltages: 110-130 VAC 50-60 Hz
10W or 12-15 VDC .2W

Readout: [⌷⌷⌷⌷⌷⌷⌷⌷] Analog Lin.

Physical: 8.125x3.75x9.75" 7 Lbs
206x95x248mm 3.2 kg

Status: Inactive Manufacturer
Discontinued Model

Rarity: Very Scarce

	New	Used
Price:	$229	$150
Rating:	⑥	⑥

The **Electra Company**, best known for their famous Bearcat scanners, decided to test the waters in 1983 with a shortwave communications receiver, the Bearcat DX-1000. The DX-1000 was not able to gather the same respect that Bearcat scanners had earned.

The Electra Company was acquired by Uniden in 1985.

Electra Company
Division of Masco Corp.
300 E. County Line Rd.
Cumberland, IN 46229   1983-1984

Bearcat

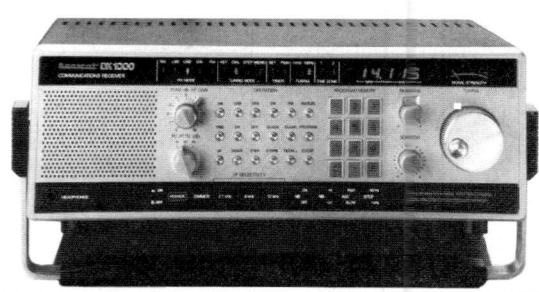

DX-1000

General Coverage Communications Receiver
Double Conversion Superheterodyne. Solid State.

Features:
- ¼" Head. Jack
- S-Meter
- Tone Control
- Attenuator -20/40 dB
- Noise Blankers
- Mute Terminal
- Record Jack
- Recorder Activation
- Keypad Entry
- 10 Memories
- Fine Tuning
- Tilt Carry Handle
- AGC FST/SLO
- Squelch
- Dual 24 Hr Clock
- Telescopic Whip
- Dimmer
- 2 Tuning Rates
- Speaker
- Ext. Spkr. Switch
- Ext. Speaker Switch

Specifications:

Coverage ...... 10 - 30000 kHz.
Selectivity ...... 12/6/2.7 kHz -6dB.
Sensitivity ...... <.5 μV 1-30 MHz CW
Image Ratio ... >70 dB .15-30 MHz
Audio Out ...... 2W 10% distortion

Modes .... AM/LSB/USB/CW/FM
Stability .... ±100 Hz
IF Rejection .... >60 dB (@7 MHz).
IF .... 40.455 MHz, 455 kHz
Antenna Input . SO-239 & Terminals.

Made In:	Japan	1983-1984
Voltages:	120/240 AC or 12 VDC or 8 D cells. 3 AA cells for memory	
Readout:	00001. Digital LED	
Physical:	14.5x5x9" 17.6 Lbs. 370x130x240mm 8 kg	
Status:	Active Manufacturer Discontinued Model	
Rarity:	Common	
Reviews:	Passport 1985/86-1987 WRTH 1985 Ham Radio Today Mar. 1987	

	New	Used
Price:	$500-550	$150-250
Rating:	★★	★

Comments:
Memories store frequency only. Two speed tuning: 1 kHz or 100 Hz on the main dial or ±2 kHz or ±150 Hz on the fine-tuning knob. Filter may be selected independent of mode. This receiver did not live up to the manufacturer's specifications. Electra produced an amusing advertising chart to illustrate why the DX-1000 was superior to the Japan Radio NRD-515 (... true if you rate a radio by the number of buttons on it). The manufacturer at the time was Electra (now Uniden).

13 Bharat Electronics

Bharat Electronics Limited was founded in 1954 and is a government of India enterprise. Bharat Electronics produces a wide range of products in the areas of defence communications, radars and sonars, broadcasting equipment, medical electronics, satellite equipment and electronic components. Nine major manufacturing plants are located in Bangalore, Ghaziabad, Pune, Machilipatnam, Panchkula, Kotdwara, Taloja, Madras and Hyderabad.

Bharat Electronics employes over 17,000 workers and enjoys annual sales of over 10000 million Rs.

Bharat Electronics Ltd.
Trade Centre
116/2 Race Course Road
Bangalore 560 001,
India

HS412

General Coverage Communications Receiver
Double Conversion Superheterodyne. Solid State.

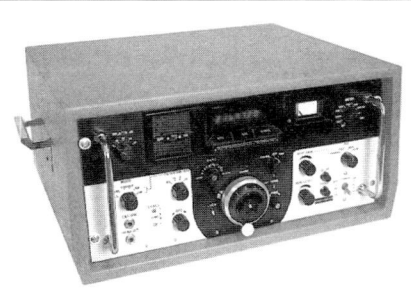

Features:
- ¼ Head. Jack
- S-Meter
- Speaker
- Speaker Switch
- RF Gain
- BFO ±1.5 kHz
- 3 Tuning Speeds
- Line Out 600 ohm
- IF Out Jack
- Dial Lock
- Rack Handles
- AGC MAN/FST/MED/SLO
- Speaker Out
- Spinner Knob
- Fine Tuning
- Modular Construction

Specifications:
Coverage ...... 1500 - 30000 kHz
Selectivity ...... 10/6/3/1.3/.3 kHz -6dB.
Sensitivity ...... <2µV SSB
I.F.s .... 35.75 MHz, 1.75 MHz
Audio Out .... 5 W 15 ohm 5% dist.

Modes .... AM/CW/SSB
Stability .... <±15 Hz
IF Rejection .... >80 dB
Image Rej. ...... >90 dB 1.5-30 MHz
Environment ... -20° to + 55° C 95% Humid

Accessories:
SRE 32 ISB Option

Comments:
This model is currently in production with a lead time of six months. The cabinet version is shown.

Variants:
The model **HS412A** has ISB. Model **HS412B** has an Low Frequency adapter to permit reception down to 15 kHz. Model **HS412C** has both ISB and the LF adapter. The **HS412D** is designed with a TCXO.

| **Made In:** | India | 1990-1998 |

Voltages: 115/230 VAC 50 Hz
90W or 24 VDC

Readout: `00000.01` Digital LED

Physical: Cabinet Version:
21.85x11x19.125" 121 Lbs
555x280x486mm 55 kg
Rack Version:
19x8.7x16.3" 88 Lbs
483x221x415mm 40 kg

Status: Active Manufacturer
Active Model
Rarity: Typically Unavailable

	New	**Used**
Price:	$13,500	②
Rating:	⑥	⑥

14 Cardwell

The **Allen D. Cardwell Manufacturing Corporation** was primarily a manufacturer of capacitors and electronic parts. The Cardwell Fifty-Four was their first and only attempt to manufacture a complete receiver. It was only on the market for two years.

Cardwell Capacitor is owned by the Kjeldsen family that also owns Pax Manufacturing which bought out the Hammarlund Company assets.

The Allen D. Cardwell Mfg. Corp.
97 Whiting St.
Plainville, Conn. 1945-1947

Cardwell

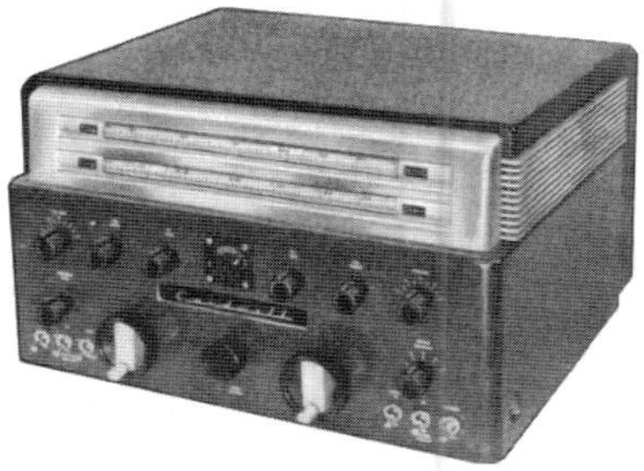

CR-54

General Coverage Communications Receiver
Double Conversion Superheterodyne. 18 Tubes.

Features:

• ¼" Head. Jack	• S-Meter	• Squelch	• Turret Type Display
• RF Gain	• ANL	• Mute Line	• Cal. 100/1000 kHz.
• Ext. Spkr Jack	• Panoramic Jack	• Bandspread	• Spinner Knobs

Specifications:

Coverage ...... 540 - 40000 kHz	Modes .... AM/CW
Selectivity ...... 5 Position	Stability .... <25 PPM per degree C.
Audio Out ...... 8 W	Antenna Input . Terminals

Circuit Complement:
The five position selectivity is three with crystal and two without.

Accessories:
Coil Strip for coverage to 54 MHz.

Comments:
Supplied with external speaker. An extremely rare and collectable receiver.

Made In: United States 1946-1947

Voltages: 115 VAC
50/60 Hz

Readout: Analog

Physical: 18.25x11x16" 70 Lbs.
463x279x406mm 31.7 kg

Status: Inactive Manufacturer
Discontinued Model

Rarity: Extremely Scarce

	New	Used
Price:	①	⑤
Rating:	⑥	⑥

The **Collins Radio Company** was founded in 1932 by Arthur A. Collins WØCXX. The company's high quality amateur transmitters gained a reputation that led to the successful expansion of the Company into broadcast, commercial and military markets. The Collins company grew rapidly in the 1940s and became a major communications equipment supplier during World War II. Many military designs would later find an application in amateur models.

The first post War amateur receiver was the 75A (75A-1). This model produced stability and frequency accuracy not previously available. This design would prevail for the ten year period of 1945 to 1955 covering models 75A-1, 75A-2, 75A-3 (introducing the famous Collins mechanical filter) and finally the venerable 75A-4. The general coverage version of this series began with the 51J-1 through 51J-4. The 51J-1 was a highly respected radio when introduced in 1949, but with a price of $875 (a very large sum in 1949!), it could only be a acquired by the most dedicated.

In 1958 the legendary amateur S line was introduced. It represented the finest amateur equipment of the time. The general coverage 51S-1 produced from 1959 to 1972 can legitimately compete with many of today's finest receivers. Rockwell International acquired Collins Radio in 1974. Since 1980 the Company has focused on the military-commercial, rather than the amateur-shortwave market.

Collins Radio Company
Cedar Rapids, IA 52498

51J

General Coverage Communications Receiver
Double Conversion Superheterodyne. 16 Tubes.

Features:
- ¼" Head. Jack
- S-Meter
- ANL
- Antenna Trimmer
- RF Gain
- BFO
- Mute
- AVC ON/OFF
- Dial Set Adjust
- IF Out Jack
- Standby
- Calibrator 100 kHz.
- Dial Lamp

Specifications:
Coverage ......500 - 30500 kHz	Modes .... AM/CW
Selectivity ......3 - .2 kHz 5 steps	Stability .... 2 kHz (After warm-up)
Sensitivity ......<3μV 6 dB S/N CW	Dial Accuracy . ±1 kHz
Audio Out ......1.5 W 4/600 ohms	Antenna Input . Terminals 300 ohms

Circuit Complement:
Uses 70E-7A VFO

Accessories:
Speaker Table Top Cabinet

Comments: This radio is organized with thirty 1 MHz bands. The bands are displayed on a drum or turret type mechanism. Designed for 19" rack mount. Requires speaker. An outstanding general coverage receiver for its time. This model is also referred to as the **51J-1.**

Variants: Model **51J-2** 1950 is very similar but has a toggle switch mounted to the left of the meter which selects the function of the meter as input or output (i.e., S or AF). The military version of the 51J-2 is the **R-381**. The faceplate for the 51J-1 and 51J-2 is simply "51J".

Mfg. In: United States 1949-1951

Voltages: 115 VAC 50/60 Hz 85W

Readout: ⌊⌊⌊⌊⌊⌊█⌊⌊⌋ Analog Lin.

Physical: Cabinet Version:
21.125x12.5x13.88" 55 Lbs.
536x318x352mm 25 kg
Rack Version:
19x10.5x13" 35 Lbs
482x266x330mm 15.9 kg

Status: Active Manufacturer
Discontinued Model

Rarity: Very Scarce

	New	Used
Price:	$875	$250-350
Rating:	★★★★	⑦

51J-3

General Coverage Communications Receiver
Triple Conversion Superheterodyne. 18 Tubes.

Features:
- ¼" Head. Jack
- RF Gain
- Dial Set Adjust
- Standby
- S/AF Meter
- AVC ON/OFF
- IF Out Jack
- ANL
- Antenna Trimmer
- Dial Lamp
- Rack Handles
- Mute
- Dial Lamp
- BFO
- Calibrator 100 kHz.

Specifications:
Coverage ...... 540 - 30500 kHz
Selectivity ...... 6/2-.2 kHz 5 steps
Sensitivity ...... <5µV 10 dB S/N CW
Audio Out ...... 1.5 W 4/600 ohms
Environment .. -40° to +65°C. 95%

Modes .... AM/CW/MCW
IF Rejection .... >50 dB
Image Rej .... >40 dB
Antenna Input . SO-239 50 ohms

Circuit Complement: 6AK5 RF Amp, 6BE6 1st Mixer, 6BE6 Band 1 Mixer, 6BA6 Cal Osc, 6AK5 Crystal HF Osc, 6BE6 2nd Mixer, 6BA6 1st 500 kHz IF Amp, 6BA6 2nd 500 kHz IF Amp, 6BA6 3rd 500 kHz IF Amp, 12AX7 Detector/AVC Rectifier, 12AU7 AVC Amp/IF Out Cathode Follow, 12AX7 NL/1st Audio Amp, 6AQ5 Audio Power Amp, 6BA6 BFO, 5V4 Power Rectifier, 0A2 Volt. Reg., 6BA6 VFO Osc, 6BA6 Osc Isolation Amp. Double conversion from 3.5-30.5 MHz. Uses 70E-15 VFO.

Accessories: 270G-3 Speaker

Comments: Designed for 19" rack mount. Requires speaker. Produced for the military under the R-388 number (shown later in this chapter). Versions built for the Navy were tropicalized and had nonmagnetic cabinets. Could be converted to 51J-4 via the 354A-1 Collins Mechanical Filter Conversion Kit.

Mfg. In:	United States 1952-1956
Voltages:	115/230 VAC 45-70 Hz 85W
Readout:	Analog Lin.
Physical:	Cabinet Version: 21.125x12.5x13.13" 43 Lbs. 536x317x333mm 19.5 kg
Status:	Active Manufacturer Discontinued Model
Rarity:	Common

	New	**Used**
Price:	$1000	$200-410
Rating:	★★★★★	⑦

51J-4

General Coverage Communications Receiver
Triple Conversion Superheterodyne. 19 Tubes.

Features:
- ¼" Head. Jack
- RF Gain
- Dial Set Adjust
- Dial Lamp
- S/AF Meter
- BFO
- IF Out Jack
- Standby
- ANL
- Mute
- Rack Handles
- Antenna Trimmer
- AVC ON/OFF
- Speaker Jack

Specifications:
Coverage ...... 540 - 30500 kHz
Selectivity ...... Dependent on filter.
Sensitivity ...... <3µV 6 dB S/N CW
IF Rejection ... >50 dB
Audio Out ...... 2.5 W 4/600 ohms

Modes .... AM/CW/MCW
Environment ... -20° to 50°C.
Image Rej .... >50 dB <7 MHz
Image Rej .... >70 dB 7-14, 40 dB>14 MHz
Antenna Input . SO-239 50 ohms

Circuit Complement:
Same tube complement as the 51J-3 except for an additional 6BA6 4th 500 kHz IF Amp. Conversion: triple .54-1.5, single 1.5-3.5, double 3.5-30.5 MHz. Uses 70E-15 VFO.

Accessories:
270G-3 Speaker   F500B-08 800Hz Filter   F500B-14 1400Hz Filter
F500B-31 3100Hz Filter F500B-60 6000Hz Filter   Tabletop cabinet

Comments:
Similar to model 51J-3, but now with mechanical filters. Note the concentric filter selection switch on BFO knob. Designed for 19" rack mount (most common). The inside top cover features a schematic. Requires a speaker.

Mfg. In:	United States 1957-1963
Voltages:	115/230 VAC 45-70 Hz 85W
Readout:	Analog Lin.
Physical:	Cabinet Version: 21.125x12.5x13.13" 55 Lbs. 536x317x333mm 24.9 kg Rack Version: 19x10.5x13" 47 Lbs 482x266x330mm 21.3 kg
Status:	Active Manufacturer Discontinued Model
Rarity:	Scarce
Reviews:	*Proceedings* 1989

	New	**Used**
Price:	$1200-1464	$490-850
Rating:	★★★★★	⑦

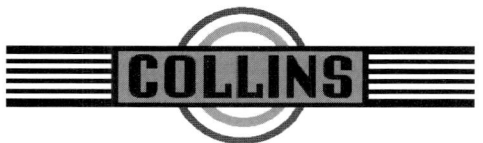

51N-7

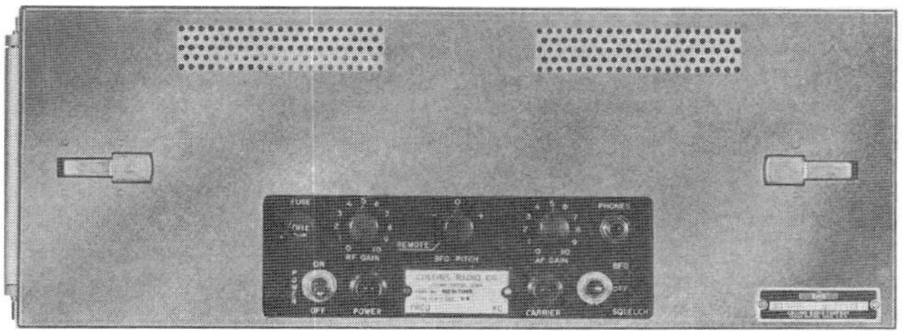

Fixed Frequency Communications Receiver
Single Conversion Superheterodyne. Tubes.

Features:
• ¼" Head. Jack • Squelch • BFO ±2 kHz • RF Gain

Specifications:
Coverage ......2 - 24 MHz Range
Selectivity ......4 kHz -6dB.
Stability 001%
IF ....455 kHz
Audio Out ....5 W 4/600 ohms
Modes .... AM/CW
Image Rej .... >60 dB <16 MHz
Image Rej .... >50 dB >16 MHz
Antenna Input . 100 ohm
Environment ... 0° to 55° C. 0-95% Hum.

Comments:
This fixed frequency HF receiver is designed to operate on one frequency in the range of 2 to 24 MHz. Adjustment made by internal slugs. Utilizes Collins mechanical filter F455K40. It is designed primarily for ground-to-air and point-to-point communications. Designed for 19" rack mount. Requires a speaker.

Variants:
Model **51N-5** 1951-1952. Model **51N-7F** is a flush mount version, **51N-7R** is recessed mount and **51N-7H** is hinged mount. (Model 51M-8 is a VHF version covering 108-152 MHz.).

Mfg. In: United States 1956

Voltages: 115/230 VAC 50/60 Hz 45W

Readout: None

Physical: 19x7x7" 17.8 Lbs. 482x178x178mm 8.1 kg

Status: Active Manufacturer Discontinued Model

Rarity: Very Scarce

	New	Used
Price:	[1]	[5]
Rating:	[6]	[8]

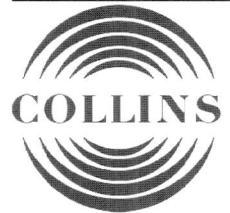

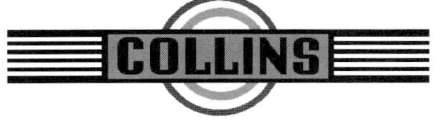

51S-1
(R-1122/GR)

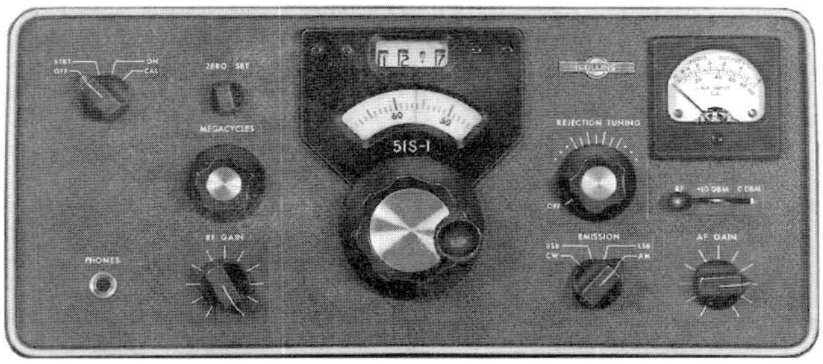

General Communications Receiver
Triple Conversion Superheterodyne. 17 Tubes plus Semi.

Features:
• ¼" Head. Jack • S-Meter • Hinged Top Cover • Calibrator 100 kHz
• RF Gain • Rejection Tuning • Mute • Ext. VFO Jack
• IF Out Jack • Dial Lock • Zero Set • Atten. 10 dB
• Sidetone Jack • Dial Brake (late prod.)

Specifications:
Coverage ......200 - 30000 kHz
Selectivity ......5/2.75/.8 kHz -6dB.
Sensitivity ......<.6µV 10 dB S/N SSB
Image Rej ......>50 dB 2-25 MHz
Audio Out ......3 W 4 ohms
Modes .... AM/LSB/USB/CW
IF .... 14.5-15.5, 3-2, .5 MHz
Stability .... 100 Hz/wk (After warm-up)
IF Rejection .... >70dB except 4.8-5.2 MHz
Antenna Input . RCA 50 ohms
Environment ... 0° to 50° C. 0-90% Hum.

Accessories:
55G-1 LF/MF Presel/Spkr. 351R-1 Rack Mount Kit
CC-2 Carry Case 350D-5 Shock Mount 312C-1 Rack Mount Speaker
312B-3 Speaker 200Hz Filter 3200Hz Filter 6000Hz Filter

Comments: Double conversion above 7 MHz. ±400 Hz dial accuracy. Tunes in thirty 1 MHz bands. Requires a speaker. An outstanding receiver and a very long lived model. Early units had white S-Meter, later had brown S-Meter, dial-lock and round logo. Very late units substituted mechanical AM filter for T4 & T5. [¥250,000]

Variants: Model **51S-1A** (R-1430/UR) is 28VDC. Model **51S-1B** is equipped with an IF Out Jack on the rear. Model **51S-1F** is rack mounted with 2.4/.8 kHz (R-1156/GR), also model **51S-1F** with 2.75 kHz (R-1156A/GR). Model **51S-1AF** is 28 VDC, rack mounted version with 2.4/.8 or 2.75 kHz. Also see LTV model G133.

Made In: United States 1959-1975

Voltages: 115/230 VAC 50-400 Hz 125W

Readout: 001 Digital Mech. to 100 kHz then Analog Lin.

Physical: 14.75x6.6x13.18" 26 Lbs. 375x167x332mm 11.8 kg

Status: Active Manufacturer Discontinued Model

Rarity: Scarce

Reviews: *WRTH* 1976
Proceedings 1992-93

	New	Used
Price:	$1500-2567	$1100-2100
Rating:	★★★★★	★★★★★

75A-1

Amateur Band Communications Receiver
Double Conversion Superheterodyne. 14 Tubes.

Features:
- ¼" Head. Jack
- S-Meter
- ANL
- Hinged Cover
- RF Gain
- BFO
- Mute
- AVC ON/OFF
- Dial Lamp
- Standby
- Fiduciary Adj.

Specifications:
Coverage ...... Ham. See Comments. Modes .... AM/CW/MCW
Selectivity ...... 4-.2 kHz 5 steps Image Rej .... >50 dB
Sensitivity ...... <1µV to 1W IF Rejection .... >70 dB
Audio Out ...... 2.5 W 4/500 ohms Antenna Input . Terminals 300 ohms

Circuit Complement: 6AK5 RF Amp, 6SA7 1st Mixer, 6SK7 IF Amp, 6L7 2nd Mixer, 6AK5 Crystal Osc, (2) 6SG7 500 kHz IF Amp, 6H6 Detector/NL, 6SJ7 AVC, 6SJ7 BFO, 6SJ7 1st Audio Amp, 6V6 Audio Power Amp, 6SJ7 VFO and 5Y3GT Rectifier. Uses 70E-7 sealed VFO.

Accessories:
353C-14 1.4 kHz Mechanical Filter Adapter 270G-1 Speaker
353C-31 3.1 kHz Mechanical Filter Adapter 307E-1 Gear Reduction Knob
353C-60 6 kHz Mechanical Filter Adapter

Comments: One view is that the 75A-1 was simply referred to as the Collins 75A until the 75A-2 came out in 1950. Another theory suggests that a small run of <200 75As were made in 1946. Ranges: 3.2-4.2, 6.8-7.8, 14-15, 20.8-21.8, 26-28 and 28-30 MHz. Only the band in use is illuminated. Dial accuracy is to 1 kHz (2 kHz above 26 MHz). Virtually no backlash. St. James gray wrinkle finish. Requires a speaker.

Mfg. In:	United States 1947-1950	
Voltages:	115 VAC 50-60 Hz 85W	
Readout:	[analog dial] Analog Lin.	
Physical:	21.125x12.25x13.9" 57 Lbs. 536x311x352mm 25.8 kg	
Status:	Active Manufacturer Discontinued Model	
Rarity:	Scarce	
Reviews:	QST September 1947 Electric Radio Sept. 1992	

	New	Used
Price:	$375	$250-490
Rating:	★★★★★	⑦

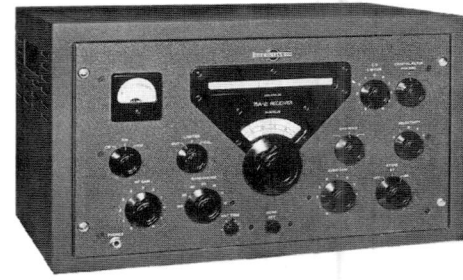

75A-2

Amateur Band Communications Receiver
Double Conversion Superheterodyne. 17 Tubes.

Features:
- ¼" Head. Jack
- S-Meter
- ANL
- Antenna Trimmer
- RF Gain
- BFO
- Mute
- AGC OFF/SLO/FST
- Dial Set Adjust
- CW Limiter
- Dial Lamp

Specifications:
Coverage ...... Ham. See Comments. Modes .... AM/CW/MCW
Selectivity ...... 2.4-.2 kHz 5 steps Image Rej .... >50 dB
Sensitivity ...... 3 V 6 dB S/N CW IF Rejection .... >50 dB
Audio Out ...... 2.5 W 4/500 ohms Antenna Input . Terminals & SO-239 hole

Circuit Complement:
Uses 70E-12 sealed VFO. Nine tuned circuits at the 455 kHz IF plus improved crystal filter.

Accessories:
270G-2 Speaker 148C-1 NBFM Adapter 8R-1 Crystal Cal. 100 kHz
307E-1 Gear Reduction Knob

Comments:
Frequency bands: 1.5-2.5, 3.2-4.2, 6.8-7.8, 14-14.5, 20.8-21.8, 26-28 and 28-30 MHz. Drum type MHz dial. Dial accuracy is to 1 kHz (2 kHz above 26 MHz). St. James gray wrinkle finish. Requires a speaker.

Variants:
Model **75A-2A** 1953 added mechanical filters.

Mfg. In:	United States 1950-1952	
Voltages:	115 VAC 50-60 Hz 85W	
Readout:	[analog dial] Analog Lin.	
Physical:	Cabinet Version: 21.125x12.5x13.13" 50 Lbs. 536x317x333mm 22.7 kg Rack Version: 19x10.5x13.325" 483x267x338mm	
Status:	Active Manufacturer Discontinued Model	
Rarity:	Scarce	
Reviews:	QST July 1950	

	New	Used
Price:	$420	$275-430
Rating:	★★★★★	⑦

75A-3

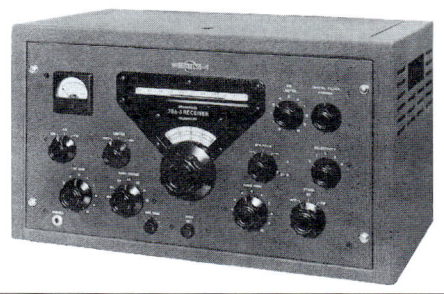

Amateur Band Communications Receiver
Double Conversion Superheterodyne. 18 Tubes.

Features:
• ¼" Head. Jack   • S-Meter   • Antenna Trimmer   • AGC OFF/SLO/FST
• RF Gain   • Hinged Cover   • Mute   • Line Out 500 ohms
• Dial Set Adjust   • CW Limiter   • Dial Lamp   • ANL   • BFO

Specifications:
Coverage ......Ham. See Comments.   Modes .... AM/CW/MCW/FM
Selectivity ......3/_ & 2.4-.2 kHz 5 steps   Image Rej .... >50 dB
Sensitivity ......<2μV 6 dB S/N CW   IF Rejection .... >50 dB
Audio Out ......2.5 W 4/500 ohms   Antenna Input . SO-239 & Terminals

Circuit Complement:
6CB6 RF Amp, 6BA7 1st Mixer, 6BA7 2nd Mixer, 12AT7 Crystal Osc, (3) 6BA6 455 kHz IF Amp, 6AL5 Det/AVC Rect. 12AX7 AVC Amp/AF Amp, 6AL5 ANL, 6AQ5 Audio Power Amp, 6BA6 BFO, 5Y3GT Power Rect, 6BA6 VFO, 6BA6 VFO Isolation, 6AL5 CW NL, 0A2 Voltage Reg., 6BA6 455 kHz IF Amp. Uses 70E-12 sealed VFO.

Accessories:
270G-2 Speaker   148C-1 NBFM Adapter   8R-1 Calibrator 100 kHz
F445B-08 800 Hz Filter   F455B-60 6 kHz Filter   307E-1 Gear Reduction Knob

Comments: Frequency bands, tuning and dial accuracy is the same as the 75A-2. St. James gray wrinkle finish. Requires speaker. The 75A-3 saw the introduction of the famous *Collins Mechanical Filter*. A two position concentric switch under the BFO knob selects the supplied 3 kHz or the optional 800 Hz mechanical filter.[¥100,000].

Mfg. In:	United States 1952-1954	
Voltages:	115 VAC 50-60 Hz 85W	
Readout:	Analog Lin.	
Physical:	21.125x12.5x13.13" 50 Lbs. 536x317x333mm 22.7 kg	
Status:	Active Manufacturer Discontinued Model	
Rarity:	Scarce	

	New	Used
Price:	$530	$320-500
Rating:	★★★★★	⑦

75A-4

Amateur Band Communications Receiver
Double Conversion Superheterodyne. 22 Tubes.

Features:
• ¼" Head. Jack   • S-Meter   • PBT   • Calibrator 100 kHz
• RF Gain   • BFO   • ANL   • AGC OFF/SLO/FST
• Dial Set Adjust   • Dial Drag Adjust   • Mute   • Antenna Trimmer
• Hinged Cover   • Notch Filter   • Standby

Specifications:
Coverage ......Ham. See Comments.   Modes .... AM/CW-SSB
Selectivity ......3.1/_/_ kHz   Stability .... <300 Hz after 15 min.
Sensitivity ......<1μV 6 dB S/N CW   IF Rejection .... >70 dB
Audio Out ......0.75 W 4/500 ohms   Image Rej .... >50 dB
Dial Accuracy ±300 Hz   Antenna Input . N & Terminals

Circuit Complement:
6BA6 Crystal Cal., 6DC6 RF Amp, 6BA7 HF Mixer, 12AT7 Crystal Osc, 6BA7 LF Mixer, 6BA6 IF Amp, 12AX7 Q-Mult, 6BA6 IF Amp, 6BA6 IF Amp, 6AL5 AM-MCW Det, 12AU7 SSB/CW Det, 6AL5 NL, 12AT7 AF Amp, 6BA6 VFO, 6BA6 VFO, 6AL5 AVC Rectifier, 5Y3 Power Rectifier, 0A2 Voltage Regulator, 6AL5 Gain Gate/Bias Rectifier, 6BA6 BFO, 6BA6 AVC Amp and 6AQ5 AF Output. Uses 70E-24 VFO.

Accessories:
312A-1 Speaker/Lamp   F455J-08 800Hz Filter   F455J-60 6000Hz Filter

Comments: Coverage: 1.5-2.5, 3.2-4.2, 6.8-7.8, 14-15, 20.8-21.8, 26.5-27.5, 28-29 and 29-30 MHz. Matches KWS-1 transmitter. Units with serial numbers > 4200 have a vernier tuning knob. This is a highly sought after model. [¥210,000].

Mfg. In:	United States 1955-1958	
Voltages:	115 VAC 50-60 Hz 85W	
Readout:	Analog Lin.	
Physical:	17.25x10.5x15.5" 35 Lbs. 438x266x393mm 15.9 kg	
Status:	Active Manufacturer Discontinued Model	
Rarity:	Scarce	
Reviews:	*QST* April 1955	

	New	Used
Price:	$495-695	$530-840
Rating:	★★★★★	⑦

75S-1

Amateur Band Communications Receiver
Double Conversion Superheterodyne. 10 Tubes plus Semiconductors.

Features:
- ¼" Head. Jack
- S-Meter
- Preselector
- Crystal Cal. 100 kHz.
- RF Gain
- Mute Line
- Anti-Vox Jack
- CW Sidetone Jack
- Fiduciary Adjust

Specifications:

Coverage ...... Ham. See Comments.
Selectivity ...... 4.5/2.1/_ kHz
Sensitivity ...... <1µV 15 dB S/N CW
Audio Out ...... 1.85 W 4 ohms
Dial Accuracy 200 Hz

Modes .... AM/LSB/USB/CW
Stability .... <100 Hz after warm-up.
IF Rejection .... >70 dB
Image Rej .... >50 dB
Antenna Input . RCA

Circuit Complement:
6DC6 RF Amp, 6U8A 1st Mixer/Crystal Osc, 6U8A 2nd Mixer/VFO Isolation Amp, 6BA6 1st IF Amp, 6BA6 2nd IF Amp, 6U8A Product Detector/BFO, 6AT6 AM Detector/AVC Rectifier/AF Amp, 6BF5 AF Output Amp, 6DC6 Crystal Cal., 6AU6 VFO, (2) 1N1084 Power Rect., 1N34A Freq. Shift switch. Uses 70K-2 sealed VFO.

Accessories: 312B-3 Speaker   312B-4 Station Control
136A-1 NB   F455Q-5 500Hz CW Filter

Comments: Ranges: 3.4-3.6, 3.6-3.8, 7-7.2, 7.2-7.4, 14-14.2, 14.2-14.4, 14.8-15, 21-21.2, 21.2-21.4, 21.4-21.6 and 28.5-28.7 MHz plus 2 additional 200 kHz positions for the 10 meter band. Matches the 32S-1 transmitter. The supplied 2.1 kHz filter is mechanical.

Variants: Model **75S-2** has 14 additional 200 kHz positions.

Mfg. In:	United States 1958-1962
Voltages:	105-125 VAC 50-60 Hz 90W
Readout:	Analog Lin.
Physical:	14.75x6.675x11.5" 20 Lbs. 374x169x292mm 9 kg
Status:	Active Manufacturer Discontinued Model
Rarity:	Common

	New	Used
Price:	$495-575	$200-400
Rating:	★★★★★	★★★★

75S-3

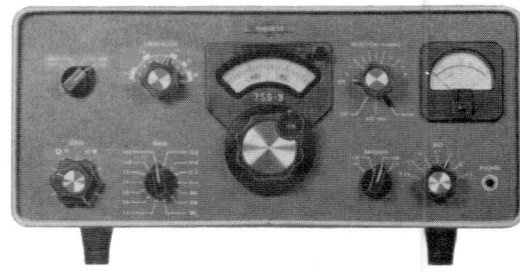

Amateur Band Communications Receiver
Double Conversion Superheterodyne. 11 Tubes plus Semiconductors.

Features:
- ¼" Head. Jack
- S-Meter
- Preselector
- Calibrator 100 kHz
- RF Gain
- Rejection Tuning
- Mute
- AGC OFF/SLO/FST
- BFO
- Anti-Vox Jack
- Dial Adjust
- CW Sidetone Jack

Specifications:

Coverage ...... Ham. See Comments.
Selectivity ...... 4.5/2.1/.25/_ kHz -6dB.
Sensitivity ...... <.5µV 10 dB S/N SSB
Audio Out ...... 3-4 ohms

Modes .... AM/LSB/USB/CW
Stability .... 100 Hz (After warm-up)
Image Rej .... >50 dB
Antenna Input . RCA

Circuit Complement:
6DC6 Crystal Cal., 6DC6 RF Amp, 6U8A 1st Mixer/Crystal Osc, 6U8A 2nd Mixer/Cathode Follow, 12AX7 Q-Mult., 6BA6 1st IF Amp, 6BA6 2nd IF Amp, 6U8A Prod. Det./BFO, 6AT6 AM Detector/AVC Rectifier/AF Amp, 6BF5 AF Output Amp, 6AU6 VFO, (2) 1N1084 Power Rect, 1N34A Freq. Shift switch. Uses 70K-2 sealed VFO.

Accessories:
312B-3 Speaker   312B-4 Station Console   F455Y-31 3100Hz Filter
F455Y-40 4000 Hz Filter F455Y-60 6000 Hz Filter

Comments:
Frequency bands: 3.4-3.6, 3.6-3.8, 3.8-4, 7-7.2, 7.2-7.4, 14-14.2, 14.2-14.4, 14.8-15, 21.2-21.4, 21.4-21.6, 28.5-28.7 MHz. Matches 32S-1 transmitter. [¥100,000]

Variants: Model **75S-3A** has 14 additional 200 kHz positions and an additional knob above the band select switch.

Mfg. In:	United States 1961-1964
Voltages:	115 VAC 50-60 Hz 90W
Readout:	Analog Lin.
Physical:	14.75x6.675x11.5" 20 Lbs. 374x169x292mm 9 kg
Status:	Active Manufacturer Discontinued Model
Rarity:	Scarce
Reviews:	QST February 1962

	New	Used
Price:	$680	$300-450
Rating:	★★★★★	★★★★★

75S-3B

Amateur Band Communications Receiver

Double Conversion Superheterodyne. 11 Tubes plus Semiconductors.

Features:
- ¼" Head. Jack
- S-Meter
- Preselector
- Calibrator 100 kHz
- RF Gain
- Rejection Tuning
- Dial Lamp
- AGC OFF/SLO/FST
- Standby
- Anti-Vox Jack
- Sidetone Jack
- Mute
- BFO
- Fiduciary Adjust

Specifications:
Coverage ......Ham. See Comments.
Selectivity ......2.1/_/_/_ kHz -6dB.
Sensitivity ......<.5µV 10 dB S/N SSB
Image Rej ......>50 dB
Audio Out ......3 W 4 ohms

Modes .... AM/LSB/USB/CW
Stability .... 100 Hz (After warm-up)
IF Rejection .... >50 dB
Antenna Input . RCA 50 ohms
Environment ... 0° to 50° C. 0-90% Hum.

Circuit Complement:
6DC6 Crystal Cal., 6DC6 RF Amp, 6EA8 1st Mixer/Crystal Osc, 6EA8 2nd Mixer/Cathode Follow, 12AX7 Q-Mult, 6BA6 1st IF Amp, 6BA6 2nd IF Amp, 6AT6 AGC Rectifier/AM Detector/AF Amp, 6BF5 AF Output, 6DC6 BFO, 6AU6 VFO plus semiconductors.

Accessories:
312B-3 Speaker   312B-4 Station Console
X455KQ200 200Hz Filter   F455FA-05 500Hz Filter   F455FA-08 800Hz Filter
F455FA-15 1500Hz Filter   F455FA-40 4000Hz Filter   F455FA-31 3100Hz Filter

Comments: Ranges: 3.4-3.6, 3.6-3.8, 3.8-4, 7-7.2, 7.2-7.4, 14-14.2, 14.2-14.4, 14.8-15, 21-21.2, 21.2-21.4, 21.4-21.6 and 28.5-28.7 MHz. Matches the 32S-3 transmitter. Requires a speaker. [¥100,000-140,000].

Mfg. In:	United States 1963-1975
Voltages:	115/230 VAC 50-400 Hz 85W
Readout:	Analog Lin.
Physical:	14.75x7.75x12.5" 20 Lbs. 375x197x317mm 9.1 kg
Status:	Active Manufacturer Discontinued Model
Rarity:	Scarce

	New	Used
Price:	$620-795	$450-750
Rating:	★★★★★	★★★★★

Rockwell International

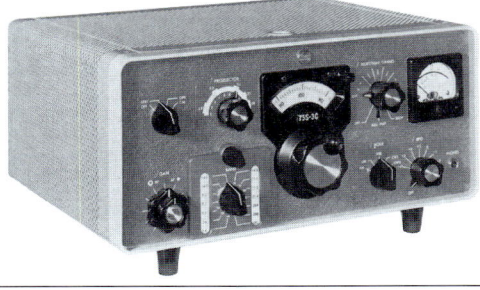

75S-3C

Amateur Band Communications Receiver

Double Conversion Superheterodyne. 11 Tubes plus Semiconductors.

Features:
- ¼" Head. Jack
- S-Meter
- Preselector
- Calibrator 100 kHz
- RF Gain
- Rejection Tuning
- Mute
- AGC OFF/SLO/FST
- Standby
- BFO
- Anti-Vox Jack
- CW Sidetone Jack
- Fiduciary Adjust

Specifications:
Coverage ......Ham. See Comments.
Selectivity ......2.1/_/_/_ kHz -6dB.
Sensitivity ......<.5µV 10 dB S+S/N SSB
Image Rej ......>50 dB
Audio Out ......3 W 4 ohms

Modes .... AM/LSB/USB/CW
Stability .... 100 Hz (After warm-up)
IF Rejection .... >50 dB
Antenna Input . RCA 50 ohms
Environment ... 0° to 50° C. 0-90% Hum.

Circuit Complement:
6DC6 Crystal Cal., 6DC6 RF Amp, 6EA8 1st Mixer/Crystal Osc, 6EA8 2nd Mixer/Cathode Follow, 12AX7 Q-Mult, 6BA6 1st IF Amp, 6BA6 2nd IF Amp, 6AT6 AGC Rectifier/AM Detector/AF Amp, 6BF5 AF Output, 6DC6 BFO, 6AU6 VFO plus semiconductors.

Accessories:
312B-3 Speaker   312B-4 Station Console
X455KQ200 200Hz Filter   F455FA-05 500Hz Filter   F455FA-08 800Hz Filter
F455FA-15 1500Hz Filter   F455FA-40 4000Hz Filter   F455FA-31 3100Hz Filter

Comments: Ranges: 3.4-3.6, 3.6-3.8, 3.8-4, 7-7.2, 7.2-7.4, 14-14.2, 14.2-14.4, 14.8-15, 21-21.2, 21.2-21.4, 21.4-21.6 and 28.5-28.7 MHz. Matches 32S-3A transmitter. Same as the 75S-3B but with an additional HF crystal board (14 sockets) and front panel selector switch. Requires a speaker. Most production was sold in the price range of $690 to $1200. The very last run sold for $3390. [¥200,000].

Mfg. In:	United States 1964-1975
Voltages:	115/230 VAC 50-400 Hz 85W
Readout:	Analog Lin.
Physical:	14.75x7.75x12.5" 20 Lbs. 375x197x317mm 9.1 kg
Status:	Active Manufacturer Discontinued Model
Rarity:	Common

	New	Used
Price:	$690-3390	$600-850
Rating:	★★★★★	★★★★★

95S

Wideband Communications Computer Receiver
Rockwell Direct Conversion. Solid State.

Features:
- Headphone Out
- RS-232C Control
- BFO ±6.4 kHz
- S-Level Reporting
- RF Gain
- Speaker Jack
- Line Out Jack
- Preselector >20 MHz
- Ext. Freq. Stand
- Two Ant. Inputs
- BITE
- AGC Programmable
- DSP

Specifications:

Coverage ......5 kHz - 2000 MHz	Modes .... AM/FM/LSB/USB/CW/ISB
Selectivity ......See Comments.	Stability .... ±1 PPM 0° to 50° C
Sensitivity ......<.4µV 10 dB S/N	IF Rejection .... >80 dB <30 MHz
Image Rej......>100 dB <30 MHz	Antenna Input . 50 ohms
Audio Out ......1 W 8 ohms 10% Dist.	Environment ... 0° to 50° C. 0-95% Hum.

Circuit Complement:
Per Rockwell literature: The 95S utilizes homodyne receiver architecture ... there is only one mix in the receiver: a single synthesizer local oscillator is tuned to the signal frequency, and directly converts the signal to baseband, or a zero-frequency "IF signal". The 16-bit DSP section utilizes twin microprocessors to perform all FIR filtering, demodulation, AGC and ancillary functions.
Comments: Bandwidths: .2, .4, .8, 1.6, 3.2, 6.4, 12.8, 25.6, 51.2, 91, 102.4 and 150 kHz -3dB. The built-in tracking preselector operates from 20 to 2000 MHz. The receiver may be purchased in a case or unpackaged (as shown).
Variants: Model **95S-1A** is a rack-mounted version (19x1.172x14.75"). Model **95V-1** is a 6U Versa Module Europa (VME) version.

Mfg. In:	United States 1995-1998
Voltages:	9-16 VDC
Readout:	 Digital PC
Physical:	13.4"x9.21"x.9" 340x234x23mm
Status:	Active Manufacturer Active Model
Rarity:	Typically Unavailable
Reviews:	*Mon. Times* November 1995

	New	Used
Price:	$5000A	④
Rating:	⑥	⑥

COLLINS **Rockwell International**

451S-1

General Coverage Communications Receiver
Double Conversion Superheterodyne. Solid State.

Features:
- ¼" Head. Jack
- S/AF-Meter
- PBT
- Modular Construction
- RF Gain
- Speaker
- Mute Line
- AGC SLO/FST
- IF Out 455 kHz
- Dial Lock
- Atten. -20 dB
- Speaker Jack
- Line Out Jack
- Keypad Input Jack (later production)

Specifications:

Coverage ......200 - 30000 kHz	Modes .... AM/SSB/CW
Selectivity ......6/2.1/.36/_/_ kHz -3dB.	Stability .... ±20 Hz
Sensitivity ......<.5µV 10 dB S/N SSB	IF Rejection .... >60 dB
Image Rej......>60 dB	Antenna Input . BNC 50 ohms
Audio Out ......3 W 4 ohms	Environment ... 0° to +50° C. 95% Humid

Accessories:

AC-3801 NB	AC-3811 140 Hz Filter	AC-3812 1700 Hz Filter
AC-3814 3.2 kHz Filter	AC-3807 Hi-Stability Oven	AC-3805 Wired Keypad
Rack Mount Kit		

Comments:
Tuning steps: 10 Hz, 100 Hz, 1 kHz or 1 MHz. Filters are switchable independent of mode. This model was designed as the successor to the famous 51S-1. The Keypad Input Jack for the optional AC-3805 Wired Keypad was not introduced until 1982. The 451S-1 has a very similar appearance to the KWM-380 transceiver. Reportedly only ten 451S-1 radios were produced.

Mfg. In:	United States 1982-1983
Voltages:	105/115/125/210/220/230 240/250 VAC 47-63 Hz or 12-15 VDC
Readout:	00000.01 Digital LED
Physical:	15.5x6.5x18" 28 Lbs. 394x165x457mm 12.7 kg
Status:	Active Manufacturer Discontinued Model
Rarity:	Typically Unavailable.
Reviews:	*WRTH* 1982

	New	Used
Price:	$4800	$5000+
Rating:	★★★★★	★★★★

COLLINS

Rockwell International

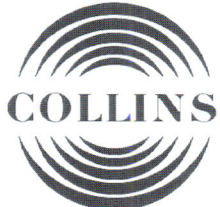

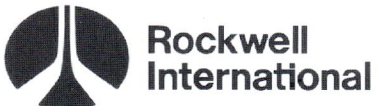

651S-1

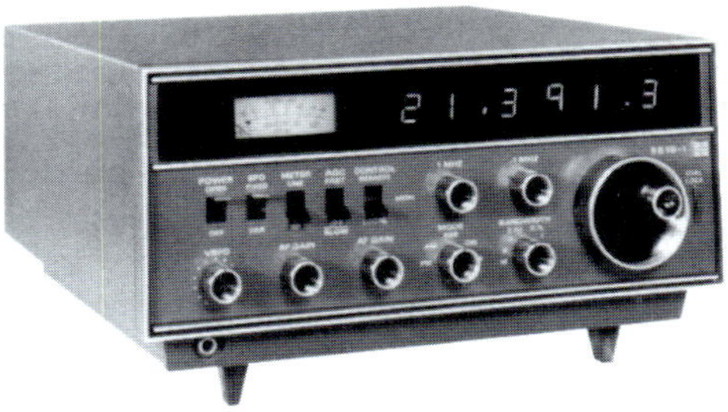

General Coverage Communications Receiver
Double Conversion Superheterodyne. Solid State.

Features:

• ¼" Head. Jack	• S/AF-Meter	• BFO & VBFO	• Modular Construction
• RF Gain	• Speaker	• Mute Line	• AGC SLO/FST
• IF Out Jack	• Dial Lock	• Ext. Standard	• Speaker Jack
• Line Out Jack	• Cooling Fan	• Spinner Knob	• Local/Remote

Specifications:

Coverage	250 - 30000 kHz	Modes	AM/SSB/CW
Selectivity	16/6/2.7/_ kHz -3dB.	Stability	5 x 10^7/30 days
Sensitivity	<.7µV 2-30 MHz SSB	IF Rejection	>80 dB
Image Rej	>80 dB	Antenna Input	BNC 50 ohms
Audio Out	2 W 8 ohms	Environment	0° to +55° C. 95% Humid.

Accessories:

	ISB Option	Sweep Feature	NBFM
635U-2 HF Preselector	514S-1 ASCII Remote	RS-232 Interface	Rack Kit
200 Hz Filter	370 Hz Filter	1 kHz Filter	
1.1 kHz Filter	3 kHz Filter	VLF Converter (12-250 kHz)	

Comments: Two rotary switches provide coarse frequency adjustment in 1 and .1 MHz steps while the main tuning knob tunes in 100 Hz increments (10 kHz/rev.). Early production had Nixie tubes, middle production had LED assemblies in the Nixie sockets. Final production had LEDs conventionally mounted. LED units more valuable. Has a "chuffy" VFO. Most units sold to gov't. agencies. [¥200,000].
Variants: Model **651S-1B** has the BFO tied to an internal synthesizer standard and is tunable ±9990 Hz. Model **651S-1A** exclusively uses an LED frequency display.

Mfg. In: United States 1971-1973

Voltages: 115/230 VAC 47-63 Hz 70W

Readout: **00000.1** Digital Nixie or later LED

Physical: 13.2x6.25x15.8" 28 Lbs. 335x159x401mm 13.7 kg

Status: Active Manufacturer Discontinued Model

Rarity: Very Scarce

Reviews: *Ham Radio* January 1971; *WRTH* 1976

	New	Used
Price:	$9000-40000	$800-1600
Rating:	★★★★★	★★★★

Rockwell International

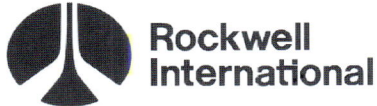

851S-1A

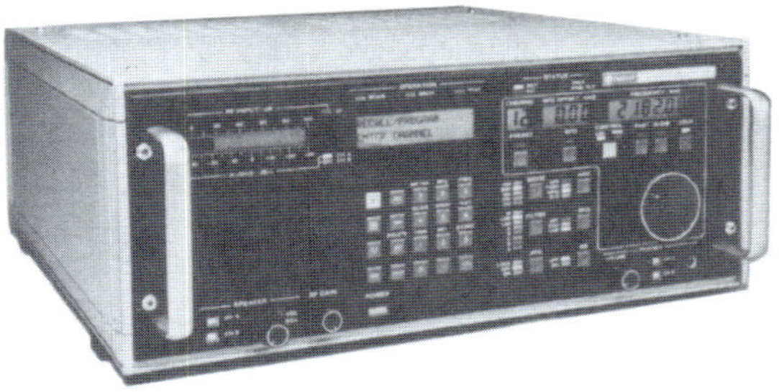

General Coverage Communications Receiver
Double? Conversion Superheterodyne. Solid State.

Features:

• ¼" Head. Jack	• S-Indicator	• Squelch	• Modular Construction
• RF Gain	• Speaker	• Speaker Jack	• AGC SLO/MED/FST/OFF
• IF Out Jack	• Keypad	• B.I.T.E.	• Line Out 600 ohm
• BFO ±9.9 kHz	• Local/Remote	• Dial Lock	• Two Tuning Speeds
• AFC	• Rack Handles	• 99 Memories	• Speaker Switch

Specifications:

Coverage	250 - 30000 kHz	Modes	AM/LSB/USB/CW
Selectivity	16/6/2.7/1/.5 kHz	Stability	5x10^{-7}
Sensitivity	<.7µV 10 dB (1.6-29.9)	Environment	0° to 50°C. 95% Humid.
3IP	+25 dBm	Antenna Input	BNC 50 ohms
Audio Out	2 W 8 / 600 ohms		

Accessories:

FM Mode Option	ISB Option	Oven Standard
Filters	AFC Option	VLF (12-249 kHz)
Table Top Case	+35 dBm 3IP	

Comments:
Designed for rack mounting. Shown with optional cabinet. There is a switch for the speaker and a switch for the headphone to select ISB channel A or ISB channel B if optional ISB board is installed.
Variants:
Model **851S-1** earlier production, fewer features and cannot be operated remotely.

Mfg. In: United States 1982-1990

Voltages: 100/115/215/230 VAC 47-63 Hz 100W

Readout: **00000.001** Digital LCD

Physical: 19x7x19" 38 Lbs. 483x178x483mm 17.2 kg

Status: Active Manufacturer Discontinued Model

Rarity: Extremely Scarce

	New	Used
Price:	$8000?	[5]
Rating:	[6]	[6]

Rockwell International

HF-2050

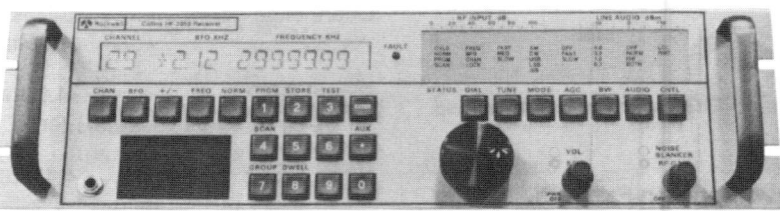

General Coverage Communications Receiver
Double Conversion Superheterodyne. Solid State.

Features:
- ¼" Head. Jack
- RF Gain
- IF Out Jack
- Line Out Jack
- BFO ±4 kHz
- S & AF Indicator
- Speaker
- Dial Lock
- Keypad
- Scan
- PBT
- Speaker Jack
- Atten. -20 dB
- B.I.T.E.
- RS232C
- Modular Construction
- AGC SLO/FST
- 30 Memories
- Rate Sensitive Tuning
- Rack Handles

Specifications:
Coverage ...... 14 - 30000 kHz
Selectivity ..... 6/3.2/2.8/1/.3 kHz -3dB.
Sensitivity ..... <.42µV 10 dB S/N
Image Rej ..... >100 dB
Audio Out ...... 2.5 W 4 ohms

Modes .... AM/LSB/USB/CW
Stability .... 5x10⁻⁷ per day
IF Rejection .... >100 dB
Antenna Input . BNC 50 ohms
Environment ... 0° to +50° C. 95% Hum.

Accessories:
Filters
Variable NB
Table Top Case
1 Hz Readout

100 Memories
Squelch
Display Light
NBFM Option

ISB Option
RS232/RS422 Interface
Scan/Sweep Option

Comments: Designed for rack mounting. Memories store: frequency, mode, bandwidth and AGC settings. Later production featured 100 memories. The first production receiver with DSP. Virtually the entire run was made in Canada for a single Canadian customer. Only few have found their way to the surplus market. Very scarce and very desirable. Less than 100 were made.

Mfg. In:	Canada	1984-1991

Voltages: 100/115/215/230 VAC
47-63 Hz 100W

Readout: `00000.01` Digital LCD

Physical: 19x5.25x18" 38 Lbs.
483x133x458mm 17.2 kg

Status: Active Manufacturer
Inactive Model

Rarity: Extremely Scarce

	New	Used
Price:	①	$4000A
Rating:	⑥	⑥

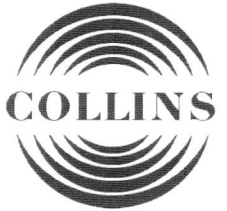

COLLINS

Rockwell International

HF-8050

General Coverage Communications Receiver
Double Conversion Superheterodyne. Solid State.

Features:
- ¼" Head. Jack
- RF Gain
- BFO
- S/AF-Meter
- Speaker
- Squelch
- Rack Handles
- IF Out Jack
- Modular Construction
- Dial Lock

Specifications:
Coverage .... .25 - 30 MHz
Selectivity ...... 19/2.7/_/_/_ kHz -3dB.
Sensitivity ...... <.7µV 10 dB S+S/N
Image Rej ...... >100 dB

Modes .... AM/LSB/USB/CW
Stability .... 5x10⁻⁷/day
IF Rejection .... >100 dB
Environment ... 0° to 50° C. 0-95% Hum.

Comments:
Designed for rack mounting. Uses decadic thumb wheel switches for frequency selection. Tunes to 10 Hz. [¥350,000-500,000].

Variants:
The models **HF-8054** (shown) and **HF-8054A** are a four channel ISB versions. [¥800,000]. The HF-8054A can be operated remotely.

Mfg. In: United States 1982-1997

Voltages: 100-127/207-253 VAC
47-63 Hz 80W

Readout: `00000.01` Digital Switch

Physical: 19x7x20.9" 42 Lbs.
483x133x458mm 17.2 kg

Status: Active Manufacturer
Active Model

Rarity: Extremely Scarce

Reviews: *RadCom* May 1985

	New	Used
Price:	①	$2900-3600
Rating:	⑥	⑥

R-388/URR

General Coverage Communications Receiver
Triple Conversion Superheterodyne. 18 Tubes.

Features:
- ¼" Head. Jack
- RF Gain
- Dial Set Adjust
- ANL
- S/AF Meter
- AVC ON/OFF
- IF Out Jack
- Mute
- Antenna Trimmer
- Dial Lamp
- Rack Handles
- Fiduciary Adj.
- Dial Lamp
- BFO
- Crystal Cal. 100 kHz.
- 600 Ohm Line Out

Specifications:
Coverage ...... 540 - 30500 kHz
Selectivity ...... 6/2-.2 kHz 5 steps
Sensitivity .... 3 V 10 dB S/N CW
IF Rejection ... >50 dB
Audio Out ...... 1.5 W 4/600 ohms

Modes .... AM/CW/MCW
Stability .... 1 kHz (After warm-up)
Image Rej .... >40 dB
Antenna Input . SO239 50 ohms

Circuit Complement: 6AK5 RF Amp, 6BE6 1st Mixer, 6BE6 Band 1 Mix, 6BA6 Calibration Osc, 6AK5 Crystal HF Osc, 6BE6 2nd Mixer, 6BA6 1st 500 kHz IF Amp, 6BA6 2nd 500 kHz IF Amp, 6BA6 3rd 500 kHz IF Amp, 12AX7 Detector/AVC Rect., 12AU7 AVC Amp/IF Out Cath. Follow, 12AX7 NL/1st Audio Amp, 6AQ5 Audio Power Amp, 6BA6 BFO, 5V4 Power Rect., 0A2 Volt.Reg., 6BA6 VFO Osc and 6BA6 Osc Isolation Amp. Double conversion from 3.5-30.5 MHz. Uses 70E-15 VFO.

Accessories: CV-182 FSK Converter

Comments: Designed for 19" rack mount. Ranges: thirty 1 MHz bands starting at .5 MHz. The military version of the Collins 51J-3. Some can be found labeled by Hallicrafters and Barker & Williamson. It is likely they were actually made by Collins.

Variants: An **R-388A** with mechanical filters, was built for the U.S. Navy (51J-4).

Mfg. In:	United States 1951-1962
Voltages:	115/230 VAC 45-70 Hz 120 W
Readout:	Analog Lin.
Physical:	19x10.5x13" 35 Lbs 483x267x330mm 15.8 kg
Status:	Active Manufacturer Discontinued Model
Rarity:	Common

	New	Used
Price:	①	$210-380
Rating:	⑦	⑦

R-389/URR

LF/MF Communications Receiver
Double Conversion Superheterodyne. 36 Tubes.

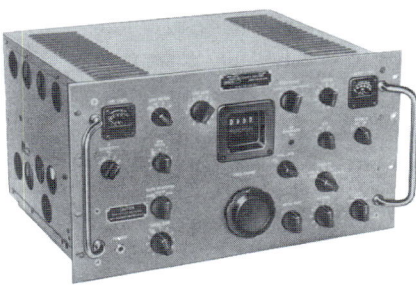

Features:
- ¼" Head. Jack
- RF Gain
- Dial Lock
- Squelch
- AGC/MGC
- S-Meter
- BFO
- Rack Handles
- Tone 3 Pos.
- Standby
- AF Meter
- Limiter
- Line Out
- Mute Line
- Antenna Trimmer
- AGC SLO/MED/FST
- IF Out Jack
- Break-In Switch

Specifications:
Coverage ...... 15 - 1500 kHz
Selectivity ...... 8/4/2/1/.1 kHz
Audio Out ...... 600 ohms

Modes .... AM/CW/MCW/FSK
IF .... 10 MHz, 455 kHz
Environment ... -40° to +65°C. 95% Humid

Circuit Complement:
Same IF strip as the R-390.

Comments:
Ranges: .015-.5 and .5-1.5 kHz. Please note the very limited frequency coverage of this receiver. The R-389A/URR is not an HF receiver. This receiver will have strong appeal to longwave and medium DX'ers and possibly collectors. Has motor assisted tuning for rapid frequency changes. Excellent dial accuracy is achieved. There are calibration marks every 1 kHz on medium wave and every 100 Hz on longwave. Should be supplied with CX-1358/U AC power cable. To convert a PL-259 to the balanced antenna input jack you need a UG-970/U connector. The meters may be missing. (See R-390A/URR Comments). Reportedly less than 700 of this model were produced.

Mfg. In:	United States 1951-1955
Voltages:	115/230VAC 48-62 Hz 250W
Readout:	00001. Digital Mech.
Physical:	19x10.5x17.25" 82 Lbs. 483x267x438mm 37.2 kg
Status:	Active Manufacturer Discontinued Model
Rarity:	Very Scarce
Reviews:	ODXA August 1987 Electric Radio May 1996

	New	Used
Price:	①	$300-450
Rating:	⑦	★★★★

R-390/URR

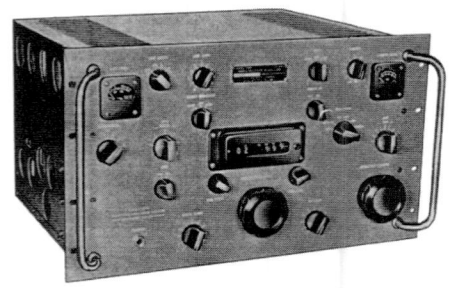

General Coverage Communications Receiver
Triple Conversion Superheterodyne. 32 Tubes.

Features:
- ¼" Head. Jack
- RF Gain
- Dial Lock
- Squelch
- S-Meter
- BFO
- Rack Handles
- Dial Adjust
- AF Line Meter
- Limiter
- IF Out Jack
- Antenna Trimmer
- Line Out 600 ohm
- AGC SLO/MED/FST
- Calibrator 100 kHz

Specifications:
Coverage ...... 500 - 32000 kHz	Modes .... AM/CW/MCW
Selectivity ...... 16/8/4/2/1/.1 kHz	IF .... 17.5-25, 3-2 MHz, 455 kHz
Stability .... ±300 Hz	Environment ... -40° to +65° C.
Audio Out ...... 600 ohm .5 W	Image Rej. ...... >55 dB

Accessories:
CV-1982/TSC26 SSB Conv.   CV-89/URA-8 AFSK Loop Conv.
AM-4823/U HF Preselector   CV-157/URR SSB Converter

Comments:
The type of frequency readout employed is often termed "odometer". The R-390 was designed and manufactured exclusively by Collins and Motorola. An estimated 16,900 were manufactured. To some collectors they are more desirable than the R-390A. [¥150,000-200,000].

Mfg. In:	United States 1951-1954
Voltages:	110/220 VAC 48-62 Hz 270W
Readout:	`00001.` Digital Mech. & Ana. Linear
Physical:	19x10.5x17.25" 65 Lbs. 483x267x438mm 29.5 kg
Status:	Active Manufacturer Discontinued Model
Rarity:	Scarce
Reviews:	CQ January 1971 Proceedings 1988

	New	Used
Price:	①	$220-450
Rating:	★★★★★	★★★★

R-390A/URR

General Coverage Communications Receiver
Triple Conversion Superheterodyne. 26 Tubes.

Features:
- ¼" Head. Jack
- RF Gain
- Dial Lock
- IF Out Jack
- S-Meter
- BFO
- Rack Handles
- Standby
- AF Line Meter
- Limiter
- Line Out 600 ohm
- Dial Adjust
- Antenna Trimmer
- AGC SLO/MED/FST
- Calibrator 100 kHz

Specifications:
Coverage ...... 500 - 32000 kHz	Modes .... AM/CW/MCW
Selectivity ...... 16/8/4/2/1/.1 kHz	IF .... 17.5-25, 3-2 MHz, 455 kHz
Audio Out ...... 600 ohm .5 W	Environment ... -40° to +65° C.

Accessories: CV-591A & 1982 SSB Convs. CV-89/URA-8 AFSK Loop Conv.
Comments: Over 54,000 "A" models were made by Collins **and** many other manufacturers including: Motorola, Stewart-Warner, Capehart, Amelco, Teledyne, EAC, Fowler Ind., Dittmore-Freimuth and Helena Rubenstein. Most, but not all "A" production has mechanical filters and is therefore more desirable to DXer's. Note the antenna trimmer knob directly above the frequency display that is not found on the earlier non-A model. This robustly built receiver is still in use worldwide. It has a hearty front-end, is unusually stable and accurate for a tube receiver. The R-390's and R-390A's are often found missing meters which allegedly gave off excessive radiation if meter case opened. [¥150,000].
Variants: The **R-725** is a DF variant prototype by Motorola, made by Servo and Arvin with an IF strip similar to the R-390's. Less than 300 produced. The **R-1274/GRC-129** is a high stability model used by NASA for the Apollo program.

Mfg. In:	United States 1954-1985
Voltages:	110/220 VAC 48-62 Hz 225W
Readout:	`00001.` Digital Mech. & Ana. Linear
Physical:	19x10.5x17.25" 65 Lbs. 483x267x438mm 29.5 kg
Status:	Active Manufacturer Discontinued Model
Rarity:	Common
Reviews:	CQ January 1971 Electric Radio April 1991 Electric Radio April 1996[1] Radio Bygones #4

	New	Used
Price:	$2500-35500	$250-570
Rating:	★★★★★	★★★★★

R-391/URR

General Coverage Communications Receiver
Triple Conversion Superheterodyne. 33 Tubes.

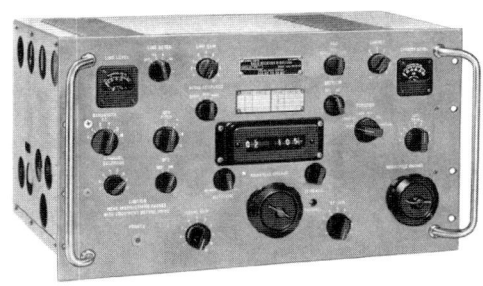

Features:
- ¼" Head. Jack
- RF Gain
- Dial Lock
- Squelch
- S-Meter
- BFO
- Rack Handles
- Channel Select
- AF Line Meter
- Limiter
- IF Out Jack
- Dial Adjust
- Line Out 600 ohm
- AGC SLO/MED/FST
- Calibrator 100 kHz
- Antenna Trimmer

Specifications:
Coverage ......500 - 32000 kHz
Selectivity ...... 16/8/4/2/1/.1 kHz
Stability ....±300 Hz
Audio Out ......600 ohm 0.5 W

Modes .... AM/CW/MCW
IF .... 9-18, 2-2.5 MHz, 455 kHz
Environment ... -40° to +55° C.
Image Rej. ...... >55 dB

Accessories:
CV-1982/TSC26 SSB Conv.
AM-4823/U HF Preselector
CV-89/URA-8 AFSK Loop Conv.
CV-157/URR SSB Converter

Comments:
The model R-391 is similar to the R-390, but supports 8 autotune channels. This is a mechanical preset scheme. In comparing the front panel to the R-390 you will note one additional knob on the left side of the radio directly below the bandwidth knob. This is for channel selection. A channel card will also be located on the front panel directly above the odometer frequency display. [¥165,000-250,00].

Variants:
The model **R-391A** (also referred to as the **R-391(XC-2)**), reached the prototype stage. Production units have not been confirmed.

Mfg. In:	United States	
Voltages:	110/220 VAC 48-62 Hz 270W	
Readout:	00001. ▤	Digital Mech. & Ana. Linear
Physical:	19x10.5x17.25" 65 Lbs. 483x267x438mm 29.5 kg	
Status:	Active Manufacturer Discontinued Model	
Rarity:	Scarce	

	New	Used
Price:	①	$300-450
Rating:	★★★★★	★★★★

R-392/URR

General Coverage Communications Receiver
Triple Conversion Superheterodyne. 25 Tubes.

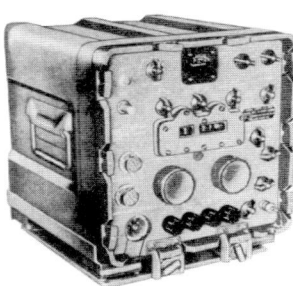

Features:
- ¼" Head. Jack
- RF Gain
- Dial Lock
- AGC ON/OFF
- S-Meter
- BFO ±3 kHz
- Dial Adjust
- Dimmer
- AF Meter
- Limiter
- Standby
- Squelch
- Antenna Trimmer
- IF Jack 455 kHz
- Calibrator 100 kHz
- Carry Handles

Specifications:
Coverage ......500 - 32000 kHz
Selectivity ......8/4/2 kHz -6 dB
Sensitivity ......<5µV 2-32 MHz.
Audio Out ......200 mw 600 ohms

Modes .... AM/CW/MCW
IF .... 9-18, 3-2 MHz, 455 kHz.
Antenna Input . Military BNC

Circuit Complement:
Two RF and six IF stages.

Accessories: LS-166/U Speaker

Comments: The radio has 32 one MHz bands. MHz is selected by the large left knob, and kHz is selected by the large right knob. The R-392 is the smaller, vehicular version of the R-390. Like the full size R-390, it does not have mechanical filters. Frequently mounted in jeeps. Immersion proof. Olive colored. Not a pretty radio. Try to obtain a radio with the power cord, speaker cord and antenna connectors. At least 25,000 were made. [¥90,000-150,000]

Variants: Designed by Collins. Manufactured by Collins, Western Electric, Stromberg Carlson, Dubrow Electronics, Colonial Mfg. Co., Philco, Stewart Warner and Motorola.

Mfg. In:	United States 1950-1959	
Voltages:	24-28 VDC 3.5 Amps	
Readout:	00001.	Digital Mech.
Physical:	11.5x11.5x14.375" 52 Lbs. 292x365x292mm 23.6 kg	
Status:	Active Manufacturer Discontinued Model	
Rarity:	Common	
Reviews:	*Electric Radio* Dec. 1990	

	New	Used
Price:	①	$60-250
Rating:	★★★★★	⑦

16 | ComFocus

The **ComFocus Corporation** abruptly entered the high end receiver market in 1994 with the very ambitious SoftWave receiver. This was the first fully integrated digital communications receiver for Microsoft Windows. The receiver was shown at the *Dayton Hamvention* in April 1994.

The receiver consists of two components: a plug-in DSP PC-card and external metal chassis. Complete control is exclusively by the Windows based graphical interface. Coverage is from .5-30 and 108-174 MHz. The radio has several personalities. As an AM radio, the display emulated a digital car radio. In the communications mode, 1 Hz tuning resolution is featured along with a spectrum display. The VHF radio personality tunes from 108-174 MHz. Various teletype demodulator options were announced, but never introduced.

It is evident that a great deal of research and development went into this product, especially on the digital and graphical user interface (GUI) side. However, the RF performance of the receiver, despite its many features, was rather modest. The graphical interface was interesting, but somewhat tiresome and slow to use. With just a bit more work on the RF side, this interesting receiver could have been a winner, but the company faded from the scene in 1995.

ComFocus Corporation
6160 Lusk Blvd.
San Diego, CA 92121   1994-1995

SoftWave

General Coverage Communications Computer Receiver
Double Conversion Superheterodyne. Solid State.

Features:

• Speaker Out	• Line Out	• Sync. Detection	• Attenuator
• Scan	• Sweep	• NB	• Notch Filter
• BFO	• Squelch	• AGC	• Spectrum Display
• RF Gain	• Dual VFOs	• VHF Audio Filter	• Database Storage
• Base	• Treble	• WWV Time Sync.	

Specifications:
Coverage ......500 - 30000 kHz +VHF[1]
Selectivity ......48 bandwidths.
Sensitivity ......-130/142 dBm/V
Dyn. Range ...97 dB
Audio Out ......1 W 8 ohms
VFO Resol .....18 Hz.
Noise Fig ....9 dB HF, 11 dB VHF

Modes .... AM/CW/USB/LSB/FM
3rd Ord Interc, 2.5 dBm 20 kHz.
IF Rejection .... >90 dB
Image Reject .. >90 dB HF, 50 dB VHF
Notch Filter ..... Variable -60 dB
Display Res .... 1 Hz
Antenna Jack . BNC

Comments:
The Softwave receiver by ComFocus was the first Windows-based receiver. VHF coverage: 108 - 174 MHz. Minimum PC requirement: 286/287 CPU, 1M RAM, DOS 3.2, Windows 3.1, EGA. Suggested PC configuration: 486, 4M RAM, DOS 5.0, Windows 3.1, VGA. This receiver offers many unique capabilities, but has only average shortwave performance. The Windows based interface was rather sluggish on the 286/386 computers of the time. Perhaps it was a radio ahead of its time. Obtaining service may be difficult.

Made In:	United States 1994-1995
Voltages:	(derived from the PC). 8 W
Readout:	**1 Hz.** Digital PC
Physical:	7.3"x2.4"x10.4" 2.5 Lbs. 185x60x264mm 1.1 Lbs
Status:	Inactive Manufacturer Discontinued Model
Rarity:	Very Scarce
Reviews:	*Passport* 1995 *Mon. Times* June 1994 *Mon. Times* October 1994

	New	Used
Price:	$1400-1500	②
Rating:	★★	★

17 Communications Product

The **Communications Product** HF1030 bears a striking resemblance to a Cubic receiver. The author suspects that this receiver was an OEM version or produced for a particular marketing channel. Another reported possibility is that the receiver was designed and developed by Communications Products and the rights then sold to Cubic.

Communications Product Corp.
Rohde & Schwarz Sales Co.
14 Gloria Lane
Fairfield, NJ 07006

HF 1030

General Coverage Communications Receiver
Double Conversion Superheterodyne. Solid State.

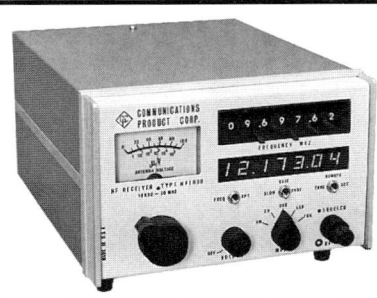

Features:
- NB SSB/CW
- RF Gain
- 455 IF Out Jack
- S-Meter
- 10 Hz VFO
- 1 MHz Ref. Osc.
- Squelch
- 100 Hz display
- Speaker Out
- AGC (2 Pos.)
- 2 Tuning Speeds
- Line Out 600 Ohms

Specifications:
Coverage ...... 10 - 30000 kHz
Selectivity ...... 5.8/2.2/1.9/.4 kHz -3dB.
Sensitivity ...... <.5µV SSB/CW
I.F.s .... 40.455 MHz, 455 kHz
Audio Out ...... 2 W 4 ohm 10% dist.
Antenna Input BNC

Modes .... AM/CW/LSB/USB/FSK
Stability .... 1 PPM/month
IF Rejection .... >80 dB
Image Rej. ...... >80 dB
Environment ... -25° to +50° C 95% Humid

Accessories:
NBFM
Single Rack Mount

ISB
Double Rack Mount

FSK to ASCII RS232 port
CW to ASCII RS232 port

Comments:
Three tuning methods: thumbwheel decadic switches, main tuning knob (1.8 or 18 kHz per rev), IEEE BCD bus. Filters are Collins. This product was marketed to the commercial sector by the Rohde & Schwarz Sales Company U.S.A. Inc. It was sold to amateur radio operators by Ehrhorn Technological Operations, Inc.

Made In:	United States 1978-1982
Voltages:	110/220 VAC 47-400 Hz 25W or 13-30 VDC 1.2A
Readout:	`00000.01` Digital LED
Physical:	8.25x5.25x14.5" 18 Lbs. 210x133x368mm 8.2 kg
Status:	Inactive Manufacturer Discontinued Model
Rarity:	Typically Unavailable

	New	Used
Price:	$4500-6500	⑤
Rating:	⑥	⑥

18 Cubic Communications

The **Cubic Corporation** was established in 1951 and is one of the world's leading suppliers of communications equipment and systems. They have expanded into other high technology fields including computer based fare collection, electronic countermeasures, defense and space systems plus electronic positioning devices.

Cubic Communications is a wholly owned subsidiary of Cubic Corporation. The Corporation has over 3,500 employees worldwide including more than 1,200 scientists, engineers and technicians.

Cubic Communications became better known to the hobbyist community when they acquired Swan in the seventies. The Swan Astro-150 is still a respected 80 to 10 meter amateur transceiver.

More recently, Cubic Communications has exclusively focused on the military, government and commercial markets offering high quality, leading-edge communications equipment.

Cubic Communications
305 Airport Rd.
Oceanside, CA 92054 — 1979-1981

Cubic Communications
9535 Waples St.
San Diego, CA 92121 — -1998

Cubic Communications
4285 Ponderosa Av.
San Diego, CA 92123 — 1993-1998

CUBIC

CDR-3130

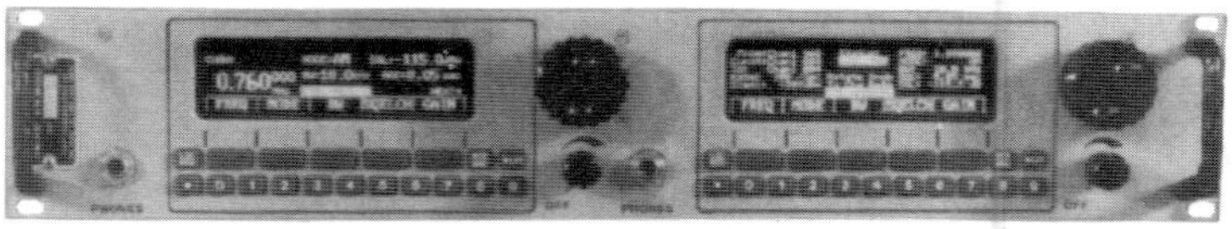

General Coverage Communications Dual Receiver
Double Conversion Superheterodyne. Solid State.

Features:
- ¼" Head. Jack
- BFO ±10 kHz
- Local/Remote
- 250 Memories
- DSP
- S/AF-Indicator
- AGC (variable)
- Scan
- Keypad
- Computer Port[1]
- BITE
- Rack Handles
- Sweep
- IF Shift
- Line Out 600 ohm
- Modular Construction
- Squelch
- Ext. Ref. 10/5/1 MHz

Specifications:
Coverage ...... 10 - 30000 kHz
Selectivity ...... 51 Bandwidths
Sensitivity ...... -124 dBm 10dB SINAD CW
Image Rej ...... >90 dB
Third Ord. I. ... +30 dBm
Audio Out ...... 15 ohms
Modes .... AM/LSB/USB/CW/FM/ISB
Stability .... ±1 PPM
IF .... 40.455 MHz, 455 kHz
IF Rejection .... >100 dB
Antenna Input . TNC 50 ohm
Environment ... 0° to +50° C.

Accessories: FSK Demod. RS-232 Bus RS-422 Bus IEEE-488 Bus RCU-3100 Controller

Comments: This is dual rack version of the CDR-3150 shown on the following page. Please see CDR-3150 Comments. [1]Customer specifies type of bus desired.

Variants:
Model **CDR-3180** is a half-rack receiver. It is shown to the right. Model **CDR-3120** is a remote version of the CDR-3130 and does not have a front panel.

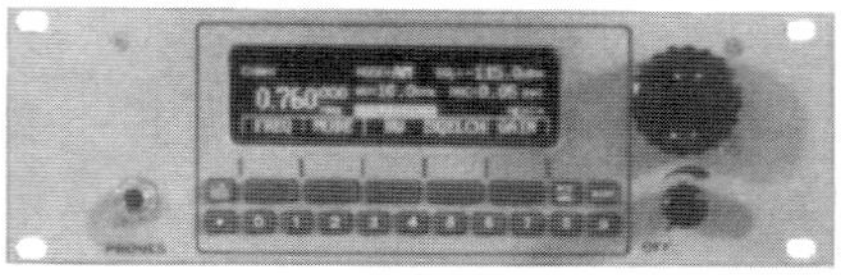

Made In: United States 1992-1998

Voltages: 90-260 VAC
47-400 Hz 60W

Readout: 00000.001 Digital Fluor.

Physical: 19x3.5x21" 25 Lbs.
483x89x538mm 11.4 kg

Status: Active Manufacturer
Active Model

Rarity: Typically Unavailable

	New	Used
Price:	①	⑤
Rating:	⑥	⑥

CUBIC

CDR-3150

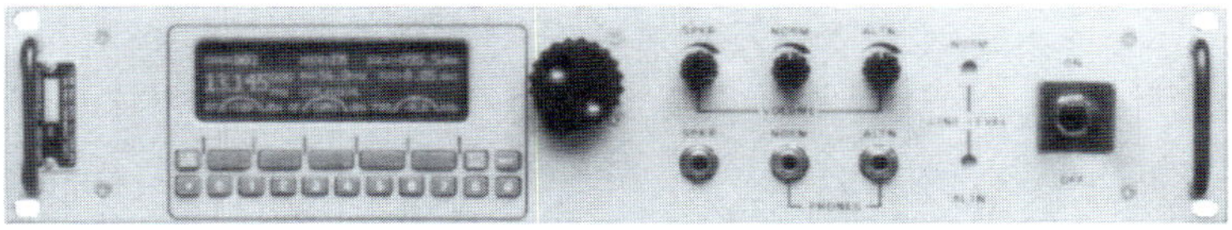

General Coverage Communications Receiver
Double Conversion Superheterodyne. Solid State.

Features:

• ¼" Head. Jack (2)	• S/AF-Indicator	• BITE	• Line Out 600 ohm
• BFO ±10 kHz	• AGC (variable)	• Rack Handles	• Modular Construction
• Local/Remote	• Scan	• Sweep	• Squelch
• 250 Memories	• Keypad	• IF Shift	• Ext. Ref. 10/5/1 MHz
• DSP	• Computer Port[1]		

Specifications:

Coverage ...... 10 - 30000 kHz	Modes .... AM/LSB/USB/CW/FM/ISB
Selectivity ...... 51 Bandwidths	Stability .... ±1 PPM
Sensitivity ...... -124 dBm 10dB SINAD CW	IF .... 40.455 MHz, 455 kHz
Image Rej ...... >90 dB	IF Rejection .... >100 dB
Third Ord. I. ... +30 dBm	Antenna Input . TNC 50 ohm
Audio Out ...... 15 ohms	Environment ... 0° to +50° C.

Accessories:
FSK Demod. RS-232 Bus RS-422 Bus IEEE-488 Bus
RCU-3100 Controller

Comments: The fully integrated vacuum fluorescent display coupled with menus and soft-keys provides maximum flexibility. The rotary knob, as well as keypad entry can be used for: frequency tuning, bandwidth selection, BFO offset, IF shift, manual gain, squelch level, scan-sweep speed, scan-sweep dwell and scan-sweep level offset. Controlled by an 80C188 Intel microprocessor. Memory is nonvolatile. [1]The customer, at time of order, must specify the type of computer bus desired.

Made In: United States 1992-1998

Voltages: 90-260 VAC
47-400 Hz 60W

Readout: 00000.001 Digital Fluor.

Physical: 19x3.5x21" 25 Lbs.
483x89x538mm 11.4 kg

Status: Active Manufacturer
Active Model

Rarity: Typically Unavailable

	New	Used
Price:	①	⑤
Rating:	⑥	⑥

CUBIC

CDR-3250

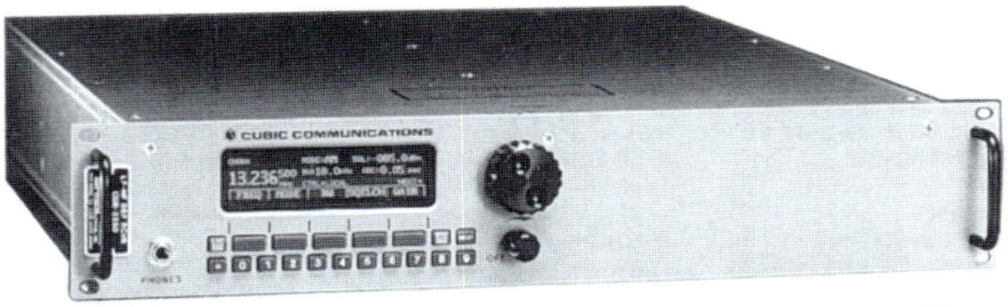

General Coverage Communications Receiver
Triple Conversion Superheterodyne. Solid State.

Features:

• ¼" Head. Jack	• S/AF-Indicator	• BITE	• Line Out 600 ohm
• BFO ±10 kHz	• AGC (variable)	• Rack Handles	• Modular Construction
• Local/Remote	• Scan	• Sweep	• Squelch
• 250 Memories	• Keypad	• WB IF Output	• Ext. Reference Input
• DSP	• Computer Port	• NB IF Output	• ISB Output

Specifications:

Coverage ...... 10 - 30000 kHz	Modes .... AM/LSB/USB/CW/FM/ISB
Selectivity ...... 51 Bandwidths	Stability .... ±1 PPM
Sensitivity ...... -122 dBm 10dB SINAD CW	IF .... 40.456 MHz, 456, 24 kHz
Image Rej ...... >100 dB	IF Rejection .... >100 dB
Third Ord. I. ... +30 dBm	Antenna Input . BNC 50 ohm
Audio Out ...... 8 ohms	Environment ... 0° to +50° C.

Accessories:
RS-232 Bus RS-422 Bus IEEE-488 Bus

Comments:
The fully integrated vacuum fluorescent display coupled with menus and soft-keys provides maximum flexibility. The rotary knob, as well as keypad entry can be used for: frequency tuning, bandwidth selection, BFO offset, IF shift, manual gain, squelch level, scan-sweep speed, scan-sweep dwell and scan-sweep level offset. Memory is nonvolatile. The 51 DSP bandwidths range from 0.1 to 16 kHz. Quadrature Digital IF output (I&Q). Gray semigloss finish.

Made In: United States 1995-1998

Voltages: 90-260 VAC
47-440 Hz 50W

Readout: 00000.001 Digital Fluor.

Physical: 19x3.5x22.25" 16 Lbs.
483x89x570mm 7.25 kg

Status: Active Manufacturer
Active Model

Rarity: Typically Unavailable

	New	Used
Price:	①	⑤
Rating:	⑥	⑥

CUBIC

CDR-3280

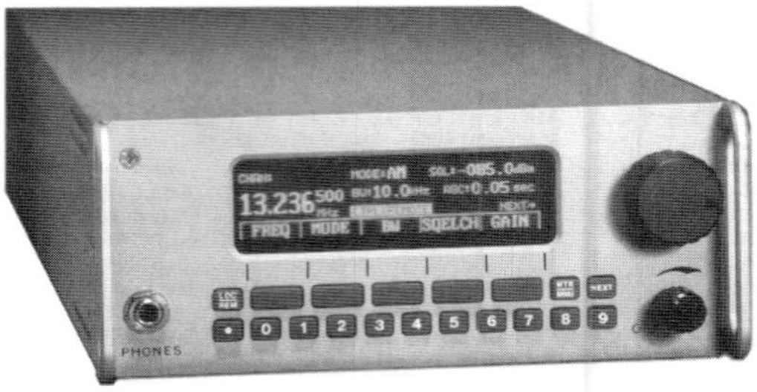

General Coverage Communications Receiver
Triple Conversion Superheterodyne. Solid State.

Features:

- ¼" Head. Jack
- BFO ±10 kHz
- Local/Remote
- 250 Memories
- DSP
- S/AF-Indicator
- AGC (variable)
- Scan
- Keypad
- Computer Port
- BITE
- Rack Handles
- Sweep
- WB IF Output
- NB IF Output
- Line Out 600 ohm
- Modular Construction
- Squelch
- Ext. Reference Input
- ISB Output

Specifications:

Coverage ...... 10 - 30000 kHz
Selectivity ...... 51 Bandwidths
Sensitivity ...... -122 dBm 10dB SINAD CW
Image Rej ...... >100 dB
Third Ord. I. ... +30 dBm
Audio Out ...... 8 ohms
Modes .... AM/LSB/USB/CW/FM/ISB
Stability .... ±1 PPM
IF .... 40.456 MHz, 456, 24 kHz
IF Rejection .... >100 dB
Antenna Input . BNC 50 ohm
Environment ... 0° to +50° C.

Accessories:

RS-232 Bus   RS-422 Bus   IEEE-488 Bus

Comments:

The CDR-3280 is a half-rack version of the CDR-3250. Gray semigloss finish. Scan and sweep rates are adjustable from 1 to 100 per second. AGC decay is selectable from 20 milliseconds to 4 seconds.

Made In: United States 1995-1998

Voltages: 90-260 VAC
47-440 Hz 50W

Readout: 00000.001 Digital Fluor.

Physical: 8.45x3.5x22.25" 16 Lbs.
222x89x570mm 7.25 kg

Status: Active Manufacturer
Active Model

Rarity: Typically Unavailable

	New	Used
Price:	①	⑤
Rating:	⑥	⑥

CUBIC

HF-1030

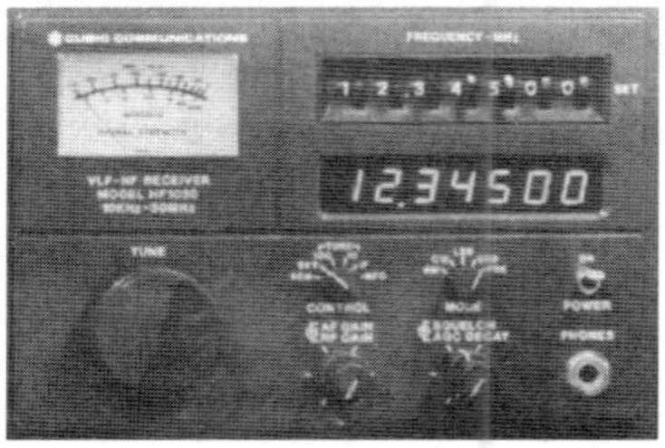

General Coverage Communications Receiver
Double Conversion Superheterodyne. Solid State.

Features:

- ¼" Head. Jack
- RF Gain
- IF Out 455 kHz
- S-Meter
- AGC
- AGC Out Jack
- IF Shift ±5 kHz
- Squelch
- IEEE488 Port
- Line Out 600 ohm
- Modular Construction
- BCD Control Port

Specifications:

Coverage ...... 10 - 30000 kHz
Selectivity ...... 5.8/2.2/1.2/.375 kHz -3dB.
Sensitivity ...... <.1-.3µV 10 dB S/N CW
Image Rej ...... >70 dB
IF .... 40.455 MHz, 455 kHz
Audio Out ...... 2 W 4 ohms & 600 ohm
Modes .... AM/LSB/USB/CW/FSK
Stability .... ±1 PPM/month
IF Rejection .... >100 dB
Antenna Input . BNC 50 ohm
Environment ... 0° to +50° C. 95% Humid

Accessories:

HF1030R Rack Kit   BA1030 Bus Adapter   DA1030 Diversity Adapter
HF1030R-2 Dual Rack Kit (shown in use below).

Comments: Tuning may be accomplished by rotating the main tuning knob, setting the frequency via the BCD decade switches above the LED display or remotely via the BCD parallel port. This receiver is robustly built to meet the most demanding commercial applications. See *Chapter 17 Communications Prod.*

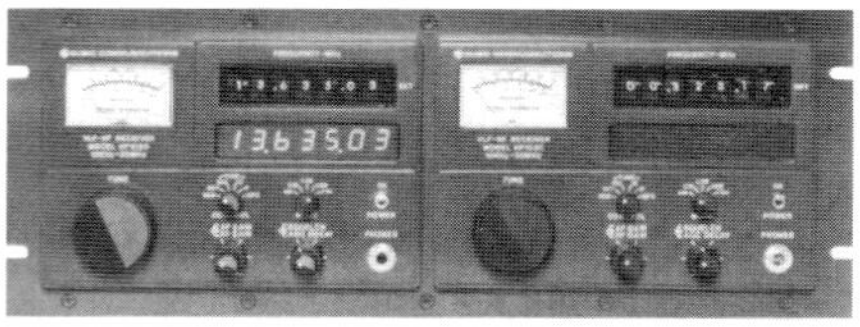

Made In: United States 1981-1986

Voltages: 115/230 VAC 47-400Hz 25W
or 12-30 VDC 1.2 A

Readout: 00000.01 Digital LED

Physical: 8.75x5.25x14.5" 18 Lbs.
210x133x368mm 8.2 kg

Status: Active Manufacturer
Inactive Model

Rarity: Extremely Scarce

	New	Used
Price:	①	$730-800
Rating:	⑥	⑥

CUBIC

R-3020

General Coverage Communications Receiver [2]
Double Conversion Superheterodyne. Solid State.

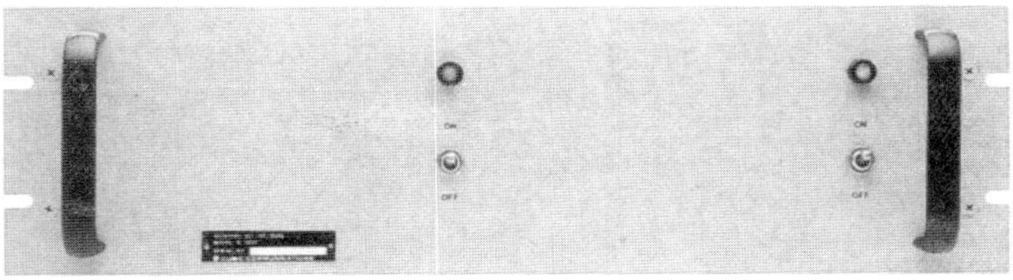

Features:

- Computer Port[1]
- BFO ±10 kHz
- IF Out 455 kHz
- 100 Memories
- AGC (5 pos)
- Scan
- IF Shift ±10 kHz
- Rack Handles
- Sweep
- Line Out 600 ohm
- Modular Construction
- 3 Tuning Steps

Specifications:

Coverage ...... 5 - 30000 kHz
Selectivity ...... 8/4/2/1/.5/_/_ kHz -3dB.
Sensitivity ...... -124 dBm 10dB SINAD CW
Image Rej ...... >90 dB
AGC .... 15, 50, 250 ms, 3 sec.
Audio Out ...... 15 ohms
Modes .... AM/LSB/USB/CW/FM
Stability .... ±1 PPM
IF .... 40.455 MHz, 455 kHz
IF Rejection .... >100 dB
Antenna Input . BNC 50 ohm
Environment ... -20° to +60° C.

Accessories:

1 MHz Ref. Module
Filters

Comments:

This model has only a sheet of solid metal for the front panel. The panel has one power switch and one pilot lamp for each of the two receivers. All functions must be controlled exclusively through the computer bus.
[1]The customer, at time of order, must specify the type of computer bus desired: RS-232, RS-422, RS-485, MIL-STD-188-114 or special parallel bus.

Made In: United States 1981-1998

Voltages: 95-135 or 190-270 VAC
47-420Hz 35W

Readout: 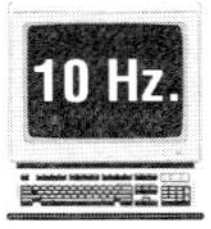

Digital on PC

Physical: 19x5.22x19.16" 40 Lbs.
483x132x486mm 18.1 kg

Status: Active Manufacturer
Active Model

Rarity: Typically Unavailable

	New	Used
Price:	①	⑤
Rating:	⑥	⑥

CUBIC

R-3030

General Coverage Communications Dual Receiver
Double Conversion Superheterodyne. Solid State.

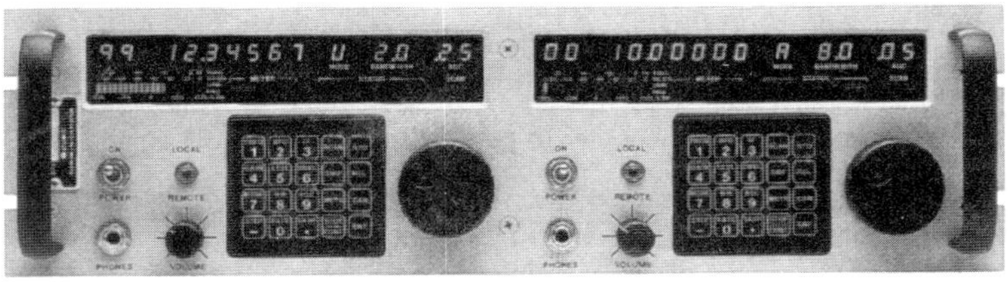

Features:

- ¼" Head. Jack
- BFO ±10 kHz
- IF Out 455 kHz
- 100 Memories
- S-Indicator
- AGC (5 Pos)
- Scan
- Keypad
- IF Shift ±10 kHz
- Rack Handles
- Sweep
- 10 MHz Ref.
- Line Out 600 ohm
- Modular Construction
- 3 Tuning Steps
- Computer Port[1]

Specifications:

Coverage ...... 5 - 30000 kHz
Selectivity ...... 8/4/2/1/.5/_/_ kHz -3dB.
Sensitivity ...... -124 dBm 10dB SINAD CW
Image Rej ...... >90 dB
AGC .... 15, 50, 250 ms, 3 sec.
Audio Out ...... 15 ohms
Modes .... AM/LSB/USB/CW/FM
Stability .... ±1 PPM
IF .... 40.455 MHz, 455 kHz
IF Rejection .... >100 dB
Antenna Input . (2) TNC 50 ohm
Environment ... -20° to +60° C.

Accessories:

Other IF Out Jack
Filters
1 MHz Ref. Module

Comments:

All receivers in the R-3000 series have 3 tuning step rates: 10, 100 and 1000 Hz.
[1]The customer, at time of order, must specify the type of computer bus desired: RS-232, RS-422, RS-485, MIL-STD-188-114 or special parallel bus. The R-3030 can also be found with a BNC antenna jack. [¥440,000].

Variants:

The **R-2307/U** model is the military version and is available in one configuration.

Made In: United States 1981-1998

Voltages: 95-135 or 190-270 VAC
47-420Hz 35W

Readout: 00000.01 Digital LED

Physical: 19x5.22x19.16" 47 Lbs.
483x132x486mm 21.3 kg

Status: Active Manufacturer
Active Model

Rarity: Very Scarce

	New	Used
Price:	①	$1875-2700
Rating:	⑥	⑥

 **CUBIC**

R-3050

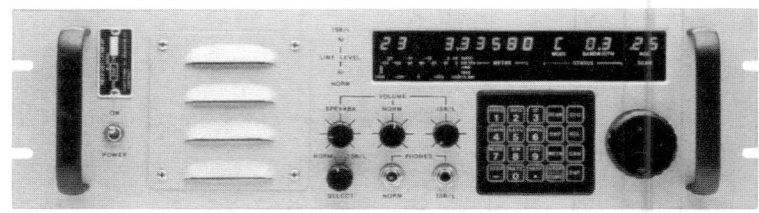

General Coverage Communications Receiver
Double Conversion Superheterodyne. Solid State.

Features:
- ¼" Head. Jack (2)
- S-Indicator
- IF Shift ±10 kHz
- Line Out 600 ohm
- BFO ±10 kHz
- AGC (5 Pos)
- Rack Handles
- Modular Construction
- IF Out 455 kHz
- Scan
- Sweep
- 3 Tuning Steps
- 100 Memories
- Keypad
- Speaker
- 1 MHz Ref. Module
- Computer Port[1]

Specifications:

Coverage ......5 - 30000 kHz	Modes .... AM/LSB/USB/CW/FM/ISB
Selectivity ......16/8/3/.5/_/_ kHz -3dB.	Stability .... ±1 PPM
Sensitivity ......-124 dBm 10dB SINAD CW	IF .... 40.455 MHz, 455 kHz
Image Rej ......>90 dB	IF Rejection .... >100 dB
AGC ....15, 50, 250 ms, 3 sec.	Antenna Input . N 50 ohm
Audio Out ......15 ohms	Environment ... -20° to +60° C.

Accessories:
10 MHz Ref. Module   Filters

Comments: Unlike the other models in the 3000 series, the R-3050 includes ISB and a 1 MHz Reference Module as standard plus has different bandwidths and antenna input jack. In the ISB position, the R-3050 will receive both the upper and lower sidebands using two separate detectors and sideband filters. AGC may be set for Zero, Short, Medium, Long or Off. [1]The customer, at time of order, must specify the type of computer bus desired: RS-232, RS-422, RS-485, MIL-STD-188-114 or special parallel bus.

Made In: United States 1981-1998

Voltages: 95-135 or 190-270 VAC
47-420Hz 40W

Readout: `00000.01` Digital LED

Physical: 19x5.22x19.16" 35 Lbs.
483x132x486mm 15.8 kg

Status: Active Manufacturer
Active Model

Rarity: Extremely Scarce

	New	Used
Price:	①	$1500
Rating:	⑥	⑥

 **CUBIC**

R-3080

General Coverage Communications Receiver
Double Conversion Superheterodyne. Solid State.

Features:
- ¼" Head. Jack
- S-Indicator
- IF Shift ±10 kHz
- Line Out 600 ohm
- BFO ±10 kHz
- AGC (5 Pos)
- Keypad
- Modular Construction
- IF Out 455 kHz
- Scan
- Sweep
- 3 Tuning Steps
- 100 Memories
- Computer Port[1]
- Spinner Knob

Specifications:

Coverage ......5 - 30000 kHz	Modes .... AM/LSB/USB/CW/FM
Selectivity ......8/4/2/1/.5/_/_ kHz -3dB.	Stability .... ±1 PPM
Sensitivity ......-124 dBm 10dB SINAD CW	IF .... 40.455 MHz, 455 kHz
Image Rej ......>90 dB	IF Rejection .... >100 dB
AGC ....15, 50, 250 ms, 3 sec.	Antenna Input . BNC 50 ohm
Audio Out ......15 ohms	Environment ... -20° to +60° C.

Accessories:
1 MHz Ref. Module   Filters

Comments:
The R-3080 is the half rack, single receiver, version of the R-3030. [1]The customer, at time of order, must specify the type of computer bus desired: RS-232, RS-422, RS-485, MIL-STD-188-114 or special parallel bus.

Made In: United States 1981-1998

Voltages: 95-135 or 190-270 VAC
47-420Hz 35W

Readout: `00000.01` Digital LED

Physical: 25 Lbs.

Status: Active Manufacturer
Active Model

Rarity: Extremely Scarce

	New	Used
Price:	①	$1000-1500
Rating:	⑥	⑥

19 Dansk Radio

The **Dansk Radio Company** produced a complete line of communications equipment including H.F. and VHF receivers. They also produce H.F. SSB/ISB transmitters.

Dansk Radio AS was acquired by TERMA Elektronik AS. In 1994 the Company ceased the production of H.F. communications equipment. TERMA is still producing military communications equipment.

Dansk Radio AS
Mårkærvej 2
DK-26320 Tåstrup,
Denmark -1992

TERMA Electronik AS
Hovmarken 4
DK-8520 Lystrup
Denmark 1994-1998

Elektromekano

M97S

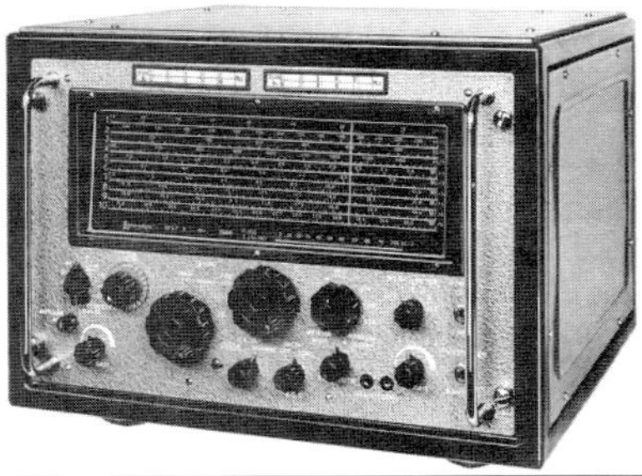

General Coverage Communications Receiver
Double Conversion Superheterodyne. 18 Tubes.

Features:
- ¼" Head. Jk. (2)
- Standby
- Dial Lamp
- Rack Handles
- BFO
- Speaker
- Spinner Knobs
- Fine Tuning
- Speaker Switch
- AGC OFF/SHT/LNG
- Calibrator 100 kHz
- Antenna Trimmer

Specifications:
Coverage ...... See Comments
Selectivity ...... 8.2/4/2/.6 kHz -3 dB
Sensitivity ...... <3 to 5 µV 20 dB S/N
Audio Out ...... 3.23 ohms 1.5 W
Modes .... AM/CW/MCW/SSB
IF .... 1300-1200, 560 kHz.
Image Rej. .... >65 dB (7-26 MHz).
IF Rej. .... >90 dB typical

Circuit Complement:
(2) EF85 {6BY7}, (7) EBF80 {6N8}, (4) ECH81 {6AJ8}, ECF80 {6BL8}, EAA91 {6AL5}, EL95 {6DL5} and (2) OB2. Single conversion below 3.7 MHz.

Accessories:
DC Power Supply

Comments:
Ranges: .014-.0215, .096-.23, .24-.53, .6-1.5, 1.5-3.8, 3.7-7.4, 7-11, 11-15, 15-19 and 19-26 MHz. In addition to this traditional general coverage, the M97S is supplied with nine built-in crystals giver a complete coverage of the marine coastal station frequencies in the 4, 8, 12, 16 and 22 MHz bands. Furthermore, sockets are provided for an additional ten crystals each giving coverage of 100 kHz band. These crystal controlled band segments provide extremely stable (better than 20 Hz) and accurate tuning. Designed for shipboard use.

Made In: Denmark

Voltages: 220 VAC 50/60 Hz

Readout: Analog

Physical: Cabinet Version:
22.2x16x17" 97 Lbs.
564x408x434mm 44 kg
Rack Version:
19x12.2x18.9" 70.5 Lbs.
483x310x480mm 32 kg

Status: Active Manufacturer
Discontinued Model

Rarity: Typically Unavailable

	New	Used
Price:	①	⑤
Rating:	⑥	⑥

Elektromekano

M114

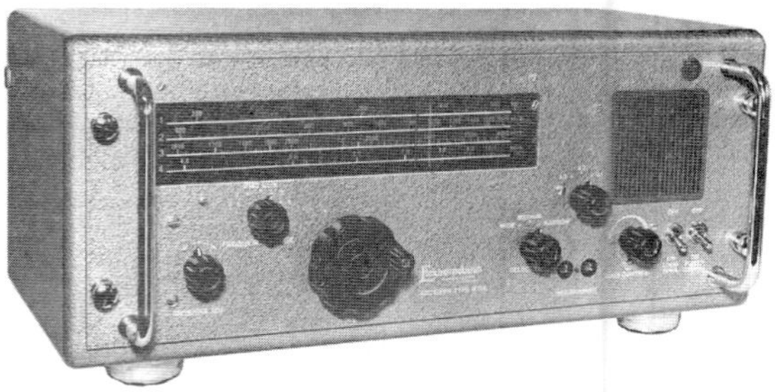

General Coverage Communications Receiver
Double Conversion Superheterodyne. Solid State.

Features:
- ¼" Head. Jk. (2)
- Speaker
- Dial Lamp
- Rack Handles
- Speaker Switch
- 2182 Preset (2nd version)
- AVC/MVC
- Standby
- Spinner Knob
- Antenna Trimmer

Specifications:

Coverage ...... See Comments
Selectivity ...... ±6/3/1.5 kHz -3 dB
Audio Out ...... 1.5 W
Modes .... AM/CW/MCW
Sensitivity .... <5 to 15 µV

Accessories:

PF114 Direction Finding Adapter, shown below, is for the 195-520 and 1580-4030 kHz bands only.

Comments:

Two versions of the M114 were produced, each with different coverage. The first version covers: .195-.52, .59-1.62, 1.58-4.03, 3.9-10.5 and 10-26 MHz. The second version does not cover the 10-26 MHz band, but rather features a 2182 kHz preset. The receiver may be supplied with either an "A" power supply of 110/220 VAC or a "B" power supply of 24 VDC. The M114 is primarily designed for use as a radiotelephone receiver for small vessels and as an emergency and reserve receiver for larger ships.

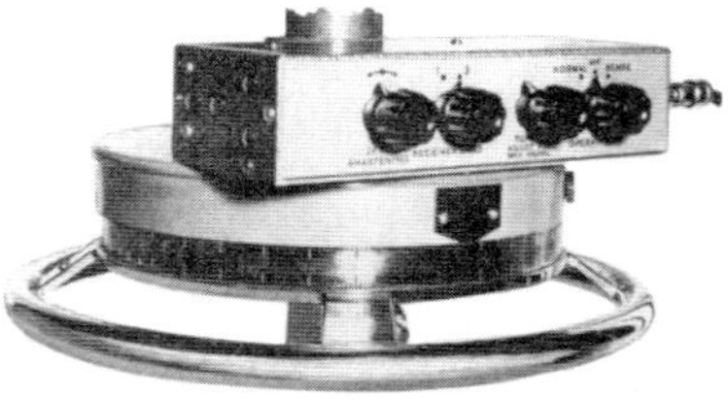

Made In: Denmark

Voltages: See Comments.

Readout: Analog

Physical: 20x9.3x12" 30 Lbs
508x237x300mm 13.5 kg

Status: Active Manufacturer
Discontinued Model

Rarity: Typically Unavailable

	New	Used
Price:	①	⑤
Rating:	⑥	⑥

Dansk Radio AS dra

M3000

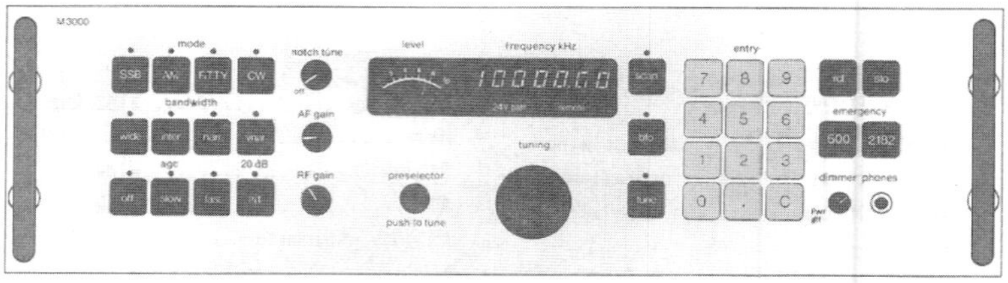

General Coverage Communications Receiver
Double Conversion Superheterodyne. Solid State.

Features:
- ¼" Head. Jack
- RF Gain
- Mute Line
- Ext. Spkr Out
- Scan & Sweep
- S-Meter
- IF Out Jack
- 75 Memories
- Preselector
- Rack Handles
- Keypad
- Notch
- BFO ±7 kHz
- BITE
- Dimmer
- AGC FST/SLO/OFF
- Line Out 600 ohm
- Presets 500/2182 kHz
- Attenuator -20 dB
- IF Out Jack

Specifications:

Coverage ...... 100 - 30000 kHz
Selectivity ...... 5.4/2.4/1/.2 kHz -6dB.
Sensitivity ...... <2µV 4-30 MHz 20 dB
Image Rej ...... >90 dB
Audio Out ...... 4 W 4 ohms
Modes .... AM/SSB/CW/RTTY
Stability .... ±1 PPM 0° to 40° C.
IF Rejection .... >90 dB
Antenna Input . BNC 50 ohm (>4 MHz)
Environment ... 0° to +50° C.

Accessories:

001 Suboctave Filters
003 Duplex Filters
004 RTTY Demodulator
005 Oven TCXO
006 IEEE-488 Remote
007 RS-232 Remote
Cabinet
Spare Parts Kit

Comments:

The memories store: frequency, mode, bandwidth, AGC setting, RF attenuation and BFO frequency and have battery backup. During scanning or sweeping, the dwell time can be set from .1 to 9 seconds. This high quality receiver was designed for the maritime user. The Dansk M3000 looks identical to the Marconi M3000. Please see *Chapter 105 Briefly Mentioned* where the Marconi M3000 is shown.

Made In: Denmark 1984

Voltages: 110/220 VAC 50/60 Hz 60W
or 24 VDC 85 W

Readout: 00000.01 Digital

Physical: 20x6.25x18.25" 33 Lbs.
509x159x463mm 15 kg

Status: Active Manufacturer
Discontinued Model

Rarity: Typically Unavailable

	New	Used
Price:	①	②
Rating:	⑥	⑥

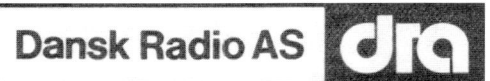

RX4000

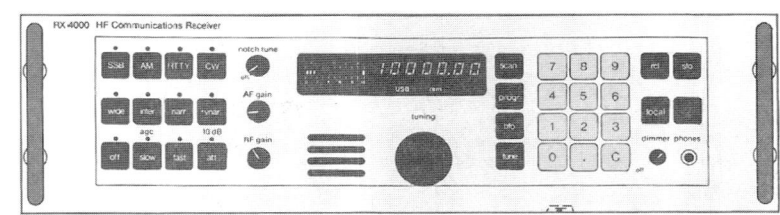

General Coverage Communications Receiver
Double Conversion Superheterodyne. Solid State.

Features:
- ¼" Head. Jack
- RF Gain
- Mute Line
- AGC
- Squelch
- Dial Lock
- S-Indicator
- IF Out Jack
- 75 Memories
- Ext. Spkr Out
- Clock
- Speaker
- Keypad
- Notch
- BFO ±7 kHz
- Atten. -10 dB
- Dimmer
- Speaker Switch
- Line Out 600 ohm
- Scan & Sweep
- 1.4 MHz IF Out Jack
- Rack Handles
- Sidetone Input
- 3 Tuning Rates

Specifications:

Coverage ...... 15 - 30000 kHz
Selectivity ...... 5.4/2.4/1/.2 kHz -6 dB.
Sensitivity ...... <2μV 20 dB SINAD SSB
Image Rej ...... >90 dB
Audio Out ...... 4 W 4 ohms

Modes .... AM/SSB/CW/RTTY
Stability .... ±.1 PPM -15° to 45° C.
IF Rejection .... >90 dB
Environment ... 0° to +50° C.
Antenna Input . BNC 50 ohm

Accessories:

471666 RS232 600/1200 bps
471577 RTTY Demod.
463302 RS232 9600 bps
475246 Cabinet

Comments:
Operation from 15 to 100 kHz with reduced performance. The RX4000 has 175 pre-programmed CCIR SSB channels and 257 pre-programmed CCIR-RTTY channels. The memories store frequency, mode, bandwidth, AGC setting, RF attenuation and BFO frequency and have battery backup. During scanning or sweeping, the dwell time can be set from .1 to 9 seconds. The clock can set for up to 24 alarms per day.

Made In:	Denmark	1985-1991

Voltages: 110-125, 220-250 VAC 50/60Hz 70W

Readout: `00000.01` Digital

Physical: 19x5.25x18.25" 33 Lbs. 483x132x463mm 15 kg

Status: Active Manufacturer Discontinued Model

Rarity: Typically Unavailable

	New	Used
Price:	①	②
Rating:	⑥	⑥

RX4010

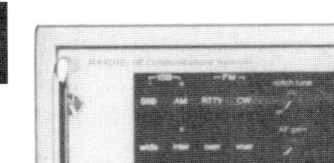

General Coverage Communications Receiver
Double Conversion Superheterodyne. Solid State.

Features:
- ¼" Head. Jack
- RF Gain
- Mute Line
- AGC
- Squelch
- Speaker
- S-Indicator
- IF Out Jack
- Memories
- Rack Handles
- Clock
- Keypad
- Notch
- BFO
- Ext. Spkr Out
- Dimmer
- Line Out 600 ohm
- Scan
- Sweep
- Attenuator -10 dB
- Dial Lock

Specifications:

Coverage ...... 10 - 30000 kHz
Environment .. 0° to +50° C.

Modes .... AM/SSB/ISB/CW/RTTY
Antenna Input . BNC 50 ohm

Accessories:

RC4010 Controller   Cabinet

Comments:
The author is seeking further information on this receiver.

Made In:	Denmark	1988-1998

Voltages: 110-125, 220-250 VAC 50/60 Hz

Readout: `00000.01` Digital

Physical: 19x5.25x18.25" 483x132x463mm

Status: Active Manufacturer Active Model

Rarity: Typically Unavailable

	New	Used
Price:	①	②
Rating:	⑥	⑥

20 Davco Electronics

Davco Electronics offered a most interesting amateur receiver in the mid-1960's called the DR-30, along with the less heavily advertised DT-20 matching transmitter. The DR-30 was heavily advertised with full page ads in the major American amateur radio magazines. Despite this promotion, the author does not think a great number were manufactured as this model rarely appears on the used market. The highest reported serial numbers for this radio are only in the 1100's.

The author believes the matching DT-20 transmitter was never put into production.

Davco Electronics Co.
113 Norwood Avenue
Asheville, NC   1962-1963

Davco Electronics, Inc.
2024 South Monroe St.
Tallahassee, FL 32304   1964-1968

DR-30

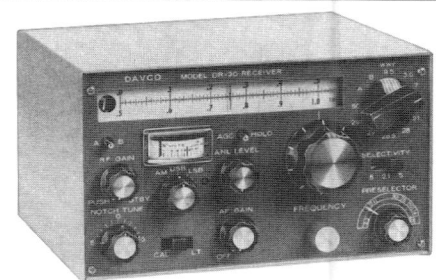

Amateur Band Communications Receiver
Double Conversion Superheterodyne. Solid State.

Features:
- ¼" Head. Jack
- RF Gain
- Mute
- Standby
- S-Meter
- ANL
- Notch Filter
- Dial Lamp Switch
- Preselector
- Dial Light
- Tone (2 Pos.)
- AGC (2 Pos.)
- Calibrator 100 kHz
- Modular Construction

Specifications:
Coverage ......Ham. See Comments.
Selectivity ......5/2.1/.5 kHz -6dB.
Sensitivity ......<.6 µV
Image Rej. .....>60 dB below 22 MHz
Audio ....0.6 W at 5% Distortion

Modes .... AM/LSB/USB
Stability .... <100 Hz/hour
IFs .... 2.405-2.955 MHz, 455 kHz
Antenna Input . RCA

Circuit Complement:
23 Transistors, 2 FET's, 15 diodes and 14 crystals. Separate AM and product detectors.

Accessories:
DR-30-S Power Supply, Speaker, Battery Holder (8xD). Please see *Chapter 107.*
Comments: Coverage: 3.5-4.05, 7-7.55, 9.5-10.05, 14-14.55, 21-21.55, 28-28.55, 28.5-29.05, 29.5-30.5 and 50-50.55 MHz plus two extra optional 550 kHz positions. The supplied 2.1 kHz bandwidth is a Collins mechanical filter. Companion to the DT-20 amateur 200 watt transmitter. Two tone gray metal cabinet. This receiver is highly coveted by collectors and consistently sells used for well over $1000.
Variants: The **CV-30** was an advertised commercial version. Ever produced?

Made In: United States 1964-1966

Voltages: 11.5-16 VDC 300 ma

Readout: Analog Lin.

Physical: 7.125x4x6" 7 Lbs.
181x101x152mm 3.2 kg

Status: Inactive Manufacturer
Discontinued Model

Rarity: Very Scarce
Reviews: *CQ* December 1964
CQ December 1966
QST January 1967
73 May 1965

	New	Used
Price:	$338-390	$700-1300
Rating:	★★★★	⑦

The **Debeg Company** of Germany manufactures commercial and marine grade communications equipment.

Debeg, from time to time, has also utilized other manufacturers to supplement its line of receivers. The Debeg MSR-2 was really a Drake MSR-2. The Debeg E755 was a modified Eddystone EC-958.

The Debeg company is now Telefunken Sytemtechnik GmbH.

Debeg GmbH
Behringstraße 120
D-22763 Hamburg 50
Germany

 **DEBEG**

2000

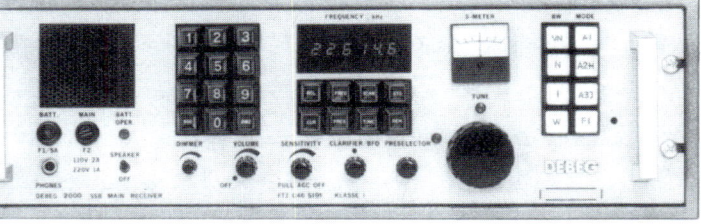

General Coverage Communications Receiver
Double Conversion Superheterodyne. Solid State.

Features:
- ¼" Head. Jack
- RF Gain
- Mute Line
- Line Out 600 ohm
- Scan
- S-Meter
- Rack Handles
- 29 Memories
- Ext. Spkr Out
- IF Out
- Keypad
- Dimmer
- BFO ±1.5 kHz
- Speaker
- Preselector
- AGC ON/OFF
- Presets 500/2182 kHz
- Modular Construction
- Speaker Switch

Specifications:
Coverage ...... 10 - 30000 kHz
Sensitivity ...... 0.13µV CW 4-30 MHz
Selectivity ...... 5.4/2.4/1/.3 kHz -6dB.
Spur. Rej. ...... >60 dB
Image Rej ...... >50 dB
Audio Out ...... 3 W 4 ohms

Modes .... AM/SSB/CW/RTTY
IF .... 38, 1.4 MHz
Accuracy .... ±50 Hz
IF Rejection .... >60 dB
Antenna Input . BNC 50 ohm
Environment ... 0° to +50° C.

Accessories:
FSK Demodulator

Comments:
The memories store frequency, mode and bandwidth. There is a provision for automatic switch-over to 24 VDC in the event of a mains failure. Sold for 12000 DM new and now sells for 2000 DM used.

Made In:	Germany	1979-1982

Voltages: 110/220 VAC 47-63Hz 50W or 24VDC 50W

Readout: `00000.1` Digital LED

Physical: Cabinet Version:
19.8x6.8x16.5" 30 Lbs.
502x172x420mm 13.6 kg

Status: Active Manufacturer
Discontinued Model

Rarity: Typically Unavailable

	New	**Used**
Price:	$5000A	$1500
Rating:	⑥	⑥

 **DEBEG**

2800

General Coverage Communications Receiver[1]
Double Conversion Superheterodyne. Solid State.

Features:
- ¼" Head. Jack
- RF Gain
- Mute Line
- S-Meter
- Rack Handles
- Ext. Spkr Out
- Keypad
- BFO ±1.5 kHz
- Speaker
- AGC ON/OFF
- Clock-Timer

Specifications:
Coverage ...... 100 - 10000 kHz +FM	Modes .... AM/SSB/CW
Selectivity ...... 7.5/2.4 kHz -6dB.	Accuracy .... ±50 Hz
Sensitivity ...... 0.5 µV S/N 10 dB CW	Image Rej .... >65 dB
Spur. Rej. ...... >60 dB	IF Rejection .... >65 dB
Audio Out ...... 2 W 4 ohms	Antenna Input . 50 ohm

Accessories:
7480 DF Unit

Comments:
[1]Please note that this receiver only covers H.F. up to 10 MHz. The WBFM bandwidth for the FM broadcast band (87.5-107.9 MHz) is 230 kHz. The Debeg 2800 is designed for operation on ocean-going yachts.

Made In:	Germany	1980-1982
Voltages:	10-32 VDC 21W	
Readout:	`00000.1`	Digital LED
Physical:	13.6x4.73x7" 14.3 Lbs	
	345x120x180mm 6.5 kg.	
Status:	Active Manufacturer	
	Discontinued Model	
Rarity:	Typically Unavailable	

	New	**Used**
Price:	①	②
Rating:	⑥	⑥

22 Del Mar

The **Del Mar** label was only found on this one interesting receiver. This model was manufactured in Japan and imported to the United States. It is unknown whether the receiver was designed in Japan or in America. The author would like to hear from any reader who has information on the history of this company.

Del Mar Ltd.
2201-C West Olive Street
Burbank, California   1959-1960

SW-59

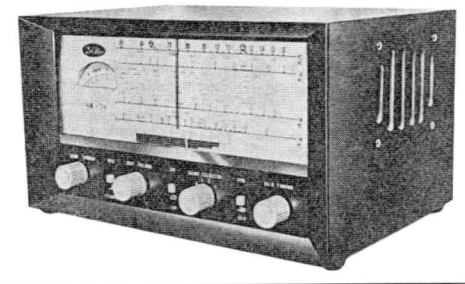

General Coverage Communications Receiver
Single Conversion Superheterodyne. 5 Tubes.

| | Made In: | Japan | 1959-1960 |

Features:
- ¼" Head. Jack
- S-Meter
- ANL
- Bandspread
- Standby
- BFO
- Dial Lamp
- Speaker 4x6"

Specifications:
Coverage ......540 - 35000 kHz
Sensitivity ......7µV average
Audio Out ......1.5W 3.2 ohms

Modes .... AM/CW
IF .... 455 kHz
Antenna Input . Terminals

Circuit Complement:
12BE6 Converter, 12BA6 IF Amp/S-Meter/BFO, 12AV6 Detector/ANL/1st Audio Amp, 35C5 Power Amp and 35W4 Rectifier.

Comments:
Ranges: .55-1.6, 1.6-5, 5-14 and 15-35 MHz. The main dial features a long black analog pointer against a white background. The bandspread is at the bottom of the tuning dial and features a small white pointer against black background. The bandspread scale covers the 80 through 10 meter amateur bands plus a 0-100 scale.

Made In: Japan   1959-1960

Voltages: 105-125 VAC 50-60 Hz
35 W

Readout: ⊔⊥⊥⊥⊥ Analog

Physical: 13.75x7.75x10" 11 Lbs
350x197x254mm 5 kg

Status: Inactive Manufacturer
Discontinued Model

Rarity: Very Scarce

	New	**Used**
Price:	$50	$100-125
Rating:	⑥	⑥

The **R.L. Drake Company**, established in 1943, has been an innovative leader in the engineering and design of amateur and shortwave radio products. The 1-A receiver was the first commercially built single sideband receiver for the radio amateur. The somewhat radical 1-A represented a classic paradigm shift in amateur receivers focusing on performance, economy and functionality rather than size! Bob Drake initially offered his design to Hallicrafters and National. When these firms showed no interest, Drake decided to build the radio under his own name. The project finally moved forward when Universal Service of Columbus, Ohio (now Universal Radio, Inc.) agreed to purchase the first 100 units.

The company was headquartered in Miamisburg, Ohio until November of 1996. The Corporate headquarters is now part of the 90,000 square foot plant in Franklin, Ohio. Drake also has international offices in Peterborough, Canada and Barcelona, Spain.

The company discontinued production of amateur and shortwave equipment in the early 1980's to concentrate all efforts on the home satellite television market. The Service Department was maintained for those in need of parts or service. The introduction of the R-8 in April of 1991 announced their return to the shortwave market. The success of the R-8 was followed by the SW-8, and the R-8A and B. Drake has once again positioned itself as one of the strongest shortwave receiver manufacturers in the world.

R.L. Drake Co.
P.O. Box 3006
Miamisburg, OH 45342   1943-1996

R.L. Drake Co.
230 Industrial Dr.
Franklin, OH 45005-4496   1978-1998

Drake Canada Sales & Service
Ctr. 655 The Queensway, Unit 16
Peterborough, Ontario
Canada   1985-1998

R.L. Drake Co.
Sucuesal Espana
C/Doctor Trueta, 1Y3
8860 Casteldefels, Spain   1990-1998

R.L. DRAKE COMPANY

1-A

Amateur Band Communications Receiver
Triple Conversion Superheterodyne. 12 or 13 Tubes.

Features:
- ¼" Head. Jack
- S-Meter
- RF Gain
- AVC SLOW/FAST
- Mute Jack
- Atten. 30 dB
- Speaker Jack
- Antenna Trimmer
- Standby
- Speaker (late)
- Calibrator 100 kHz. (late)

Specifications:
Coverage ...... Ham. See Comments.
Selectivity ......2.5 kHz -6dB.
Sensitivity ......<1µV 20 dB S/N
Image Reject .>60 dB
Audio Out ......4 ohms

Modes .... (AM)/LSB/USB(CW)
Stability .... <300 Hz After warmup
IF .... 2900-3500, 1100, 50 kHz
Antenna Input . SO-239 50-75 ohms

Circuit Complement:
6BZ6 RF Amp, 6BE6 1st Mixer, 6BE6 2nd Mixer, 6BY6 3rd Converter, 6BZ6 IF, 12AU7 Product Detector, 6BF6 Amp/Rectifier, 6AB4 Crystal Osc, 6BQ7A VF Osc, 12AU7 LF Osc/1st AF, 12AQ5 AF Out and 12X4 Rectifier.

Accessories: Speaker (Please see *Chapter 107 Additional Information*)

Comments: Tuning ranges: 3.5-4.1, 7-7.6, 14-14.6, 21-21.6, 28-28.6, 28.5-29.1 and 29.1-29.6 MHz plus an optional shortwave band. The speaker is on the rear panel. Very early production did not include the Crystal Calibrator function switched from the Antenna Trimmer control or the built-in speaker. This pace setting radio broke the long standing trend of building larger, ever heavier receivers. This receiver was superior in performance to its contemporaries at a fraction of the size, weight and price. There is strong collector demand for this historically significant model.

Made In: United States 1957-1959

Voltages: 115 VAC 60 Hz 50W

Readout: Analog Lin.

Physical: 6.75x11x15" 18 Lbs.
171x279x381mm 8.2 kg

Status: Active Manufacturer
Discontinued Model

Rarity: Very Scarce

Review: *QST* November 1957
CQ September 1986

	New	Used
Price:	$259-299	$180-300
Rating:	★★★★	⑦

R.L. DRAKE COMPANY

2-A

Amateur Band Communications Receiver
Triple Conversion Superheterodyne. 10 Tubes plus Semiconductors.

Features:
- ¼" Head. Jack
- S-Meter
- RF Gain
- AVC SLOW/FAST
- Preselector
- Mute Jack
- Sidetone Jack
- Speaker Jack
- Dial Lamp
- BFO
- Standby
- Diode/Prod. Detect.
- NL

Specifications:

Coverage ......Ham. See Comments.	Modes .... AM/LSB/USB(CW)
Selectivity ......4.8/2.4 kHz -6dB.	Stability .... <100 Hz After warmup
Sensitivity ......<.5µV 10 dB	Image Reject .. >60 dB
Audio Out ......1 W 5% distortion	Antenna Input . Terminals 50-75 ohms

Circuit Complement:
6BZ6 RF Amp, 6U8 1st Mixer, 6BE6 VFO 2nd Converter, 6BE6 3rd Converter, 6BA6 50 kHz IF Amp, 6BE6 Product Detector/BFO, 6AV6 1st Audio Amp, 6AQ5 Audio Out, 6BF6 AVC, 6X4 Rectifier and 1N34 Bias Rectifier.

Accessories:
2-AS Speaker   2-AC Calibrator 100 kHz   2-AQ Q-Multiplier/Speaker
Q-Xer Q-Multiplier

Comments:
Tuning ranges: 3.5-4.1, 6.9-7.5, 13.9-14.5, 20.9-21.5 and 28.5-29.1 MHz plus 7 additional empty ranges. The 2-A has a predrilled hole for an SO-239 jack. Requires a speaker. [¥51000].

Made In:	United States 1960-1961
Voltages:	120/240 VAC 50/60 40W
Readout:	[scale] Analog Lin.
Physical:	12x7x9" 14.5 Lbs. 304x177x228mm 6.6 kg
Status:	Active Manufacturer Discontinued Model
Rarity:	Scarce
Review:	*QST* July 1960

	New	Used
Price:	$270	$130-200
Rating:	★★★★	⑦

R.L. DRAKE COMPANY

2-B

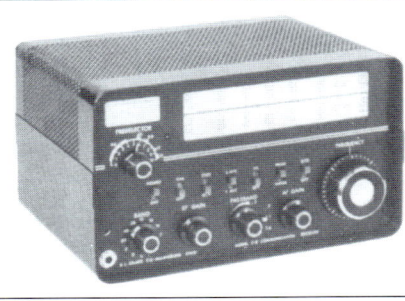

Amateur Band Communications Receiver
Triple Conversion Superheterodyne. 10 Tubes plus Semiconductors.

Features:
- ¼" Head. Jack
- S-Meter
- RF Gain
- AVC SLOW/FAST
- Preselector
- Mute Line
- Sidetone Jack
- Speaker Terminals
- BFO
- PBT
- Dial Set
- Diode/Prod. Detect.
- Dial Lamp
- ANL
- Socket for Q-Mult.

Specifications:

Coverage ......Ham. See Comments.	Modes .... AM/LSB/USB(CW)
Selectivity ......3.6/2.1/.5 kHz -6dB.	Stability .... <100 Hz After warmup
Sensitivity ......<.5µV 10 dB	Image Reject .. >60 dB
IF Reject ....>60 dB	Dial Accuracy . ±1 kHz
Audio Out ......1 W 5% distortion	Antenna Input . Terminals 50-75 ohms

Circuit Complement:
6BZ6 RF Amp, 6U8 1st Mixer, 6BE6 VFO 2nd Converter, 6BE6 3rd Converter, 6BA6 50 kHz IF Amp, 6BE6 Product Detector/BFO, 8BN8 1st Audio Amp/Bias Rectifier/ANL, 6AQ5 Audio Out, 6BF6 AVC, 6X4 Rectifier. Diode detect for AM with BFO off and product detect for SSB/CW with BFO on.

Accessories: 2-AC Calibrator 100 kHz   2-BQ Q-Multiplier/Speaker
2-LF LF Converter   2-BS Speaker

Comments: Supplied coverage: 3.5-4.1, 6.9-7.5, 13.9-14.5, 20.9-21.5 and 28.5-29.1 MHz plus 7 additional empty ranges. Dial skirt has 40 divisions representing 1 kHz each. The 2-B has a predrilled hole for an SO-239 jack. Requires a speaker. Reportedly, some units were produced with 4.8/2.4/12/.6 kHz selectivity.

Made In:	United States 1961-1965
Voltages:	120/240 VAC 50/60 Hz 40W
Readout:	[scale] Analog Lin.
Physical:	12x7x9" 14.5 Lbs. 305x179x228mm 6.6 kg
Status:	Active Manufacturer Discontinued Model
Rarity:	Common
Review:	*73* August 1961 *Electric Radio* July 1993 *Buyer's Guide to Amateur R.*

	New	Used
Price:	$230-280	$90-200
Rating:	★★★★	⑦

R.L. DRAKE COMPANY

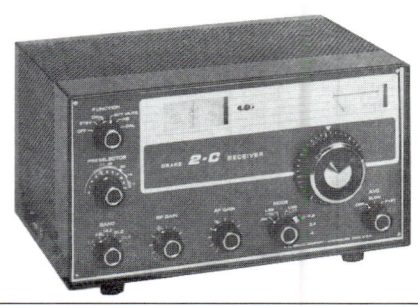

2-C

Amateur Band Communications Receiver
Triple Conversion Superheterodyne. 5 Tubes plus Semiconductors.

Features:
- ¼" Head. Jack
- S-Meter
- RF Gain
- AGC OFF/SLOW/FAST
- Preselector
- Mute Line
- Socket for Q-Mult
- Speaker Terminals
- Sidetone Jack
- Dial Lamp
- Fiduciary Adjust
- Standby
- PBT

Specifications:

Coverage ...... Ham. See Comments.
Selectivity ...... 4.8/2.4/.4 kHz -6dB.
Sensitivity ...... <.5µV 10 dB S/N
IF Rejection ... >60 dB
Audio Out ...... 1.8W 4 ohm 5% dis.

Modes .... AM/LSB/USB(CW)
Stability .... <100 Hz After warmup
Image Reject .. >60 dB
Antenna Input . RCA 52 ohm

Circuit Complement:
12BZ6 RF Amp, 12AU6 1st Mixer, 12BE6 VFO 2nd Mixer, 12BE6 3rd Mixer, 12BA6 50 kHz IF Amp, 2N3394 Crystal Osc, 2N3394 AGC Amp, 2N3394 50 kHz Osc, (2) 1N270 Product Detector, 1N270 & 2N3394 AM Detector & Amp, 2N3394 Audio Amp, 40310 Audio Out (2) 1N3194 High Voltage Rectifier, (2) 1N3194 Low Voltage Rectifier, 1N3194 Bias Rectifier. Total circuit: 5 Tubes, 7 transistors and 8 diodes.

Accessories:

2-CS Speaker   2-AC Calibrator 100 kHz   2-NB Noise Blanker
2-LF LF Converter   2-CQ Spkr/QMult/Notch

Comments: Coverage: 3.5-4, 7-7.5, 14-14.5, 21-21.5 and 28.5-29 MHz plus one 500 kHz position. Tunes at 45 kHz per revolution. Requires an external speaker. Matches the 2-NT CW transmitter.

Made:	United States 1966-1974	
Voltages:	120 VAC 50/60 Hz 30W	
Readout:	⊞⊟⊟⊟⊟ Analog Lin.	
Physical:	11.37x6.4x9.4" 13.5 Lbs. 289x162x239mm 6.1 kg	
Status:	Active Manufacturer Discontinued Model	
Rarity:	Common	
Review:	QST December 1966 CQ June 1967 73 January 1967	

	New	Used
Price:	$229-295	$80-180
Rating:	★★★★	⑦

DSR-1

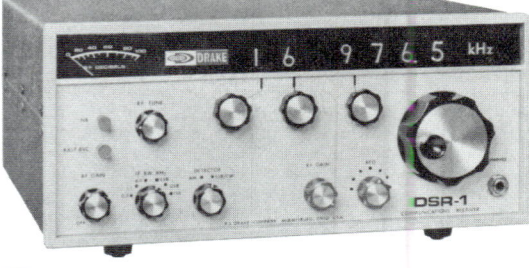

General Coverage Communications Receiver
Double Conversion Superheterodyne. Solid State.

Features:
- ¼" Head. Jack
- S-Meter
- BFO ±3 kHz.
- AVC (2 Pos.)
- RF Gain
- NB
- ISB
- Preselector
- Mute
- Speaker
- Line Out 600 ohm
- Modular Construction
- 1st IF Out Jack
- 2nd IF Out Jack
- Spinner Knob

Specifications:

Coverage ...... 10 - 30000 kHz
Selectivity ...... 6/2.4/1.2/.4 kHz -6dB.
Sensitivity ...... <.5µV 10 dB S/N SSB
IF Reject .... >60 dB
Audio Out ...... 3 W 4 ohms 5% dist
IF .... 50.5 MHz, 50 kHz

Modes .... AM/USB/LSB/CW/ISB
Stability .... < 150 Hz in 15 min. period
Image Reject .. >70 dB relative to 1µV
RF Blocking .... >100 dB relative to 1µV
Antenna Input . SO-239 50/1000 ohm

Comments:
The first tuning knob selects 0, 10 or 20 MHz. The second knob selects the frequency in 1 MHz increments. The third knob selects the .1 MHz position. The fourth (large) knob selects the range of 0 to .1 MHz and displays it on the last 3 positions of the digital readout. Display is by digital Nixie tubes. ISB output is to 600 ohm balanced line. A lab quality receiver. A rack mount version was shown in Drake advertising brochures, but few, if any, were made.

Variants:
It has also been reported that receiver was sold by **RCA Radio Marine** as model **6501**. Also see Drake MSR-1.

Made In:	United States 1969-1974	
Voltages:	115/230 AC 50-420 Hz 35 W	
Readout:	00000.1 Digital Nixie	
Physical:	19x5.25x15" 17 Lbs. 480x133x380mm 7.7 kg	
Status:	Active Manufacturer Discontinued Model	
Rarity:	Very Scarce	
Reviews:	Ham Radio October 1971	

	New	Used
Price:	$2195-2700	$800-950
Rating:	★★★★★	★★★★

DRAKE

DSR-2

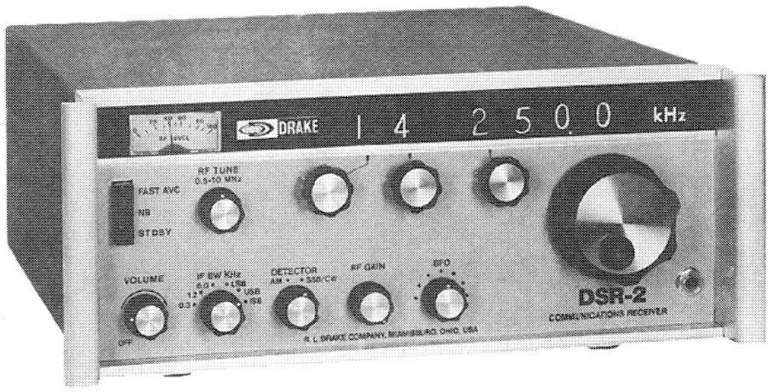

General Coverage Communications Receiver
Double Conversion Superheterodyne. Solid State.

Features:

• ¼" Head. Jack	• S-Meter	• BFO ±3 kHz.	• AVC (2 pos.)
• RF Gain	• NB	• ISB	• Preselector .5-10 MHz
• Mute	• Spinner Knob	• Modular Construction	

Specifications:

Coverage ...... 10 - 30000 kHz
Selectivity ...... 6/2.4/1.2/.3 kHz -6dB.
Sensitivity ...... <.3µV for 10 dB SSB
IF Reject .... >60 dB
Audio Out ...... 2 W 5% distortion
RF Blocking ... >100 dB (@1µV)

Modes .... AM/SSB-CW
Stability .... < 200 Hz in 8 hour period
Image Reject .. >60 dB below 10 MHz.
Image Reject .. >50 dB above 10 MHz.
Antenna Input . SO-239 50 ohms

Accessories:
Filters

Comments:
The first tuning knob selects 0, 10 or 20 MHz. The second knob selects the frequency in 1 MHz increments. The third knob selects the .1 MHz position. The fourth (large) knob, which is continuously adjustable, selects the range of 0 to .1 MHz and displays it on the last 3 positions of the digital readout. ISB output is to 600 ohm balanced line. The preselector is marked RF TUNE and operates only from .5 to 10 MHz. With the RF GAIN control pulled out, the AVC is deactivated. A lab quality receiver. Drake model #1242. [¥150,00-250,000].

Made In: United States 1974-1980

Voltages: 120/220/240 AC
50-420 Hz 35 W

Readout: 00000.1 Digital Nixie

Physical: 13.375x5.5x15" 17 Lbs.
340x140x380mm 7.7 kg

Status: Active Manufacturer
Discontinued Model

Rarity: Very Scarce

Reviews: *WRTH* 1976

	New	Used
Price:	$2750-3400	$900-1200
Rating:	★★★★★	⑥

DRAKE

MSR-1

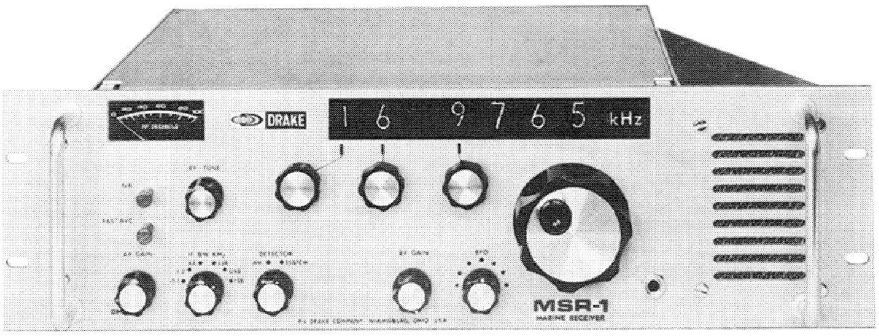

General Coverage Communications Receiver
Double Conversion Superheterodyne. Solid State.

Features:

• ¼" Head. Jack	• S-Meter	• BFO ±3 kHz.	• AGC FAST/SLOW
• RF Gain	• NB	• ISB	• Preselector
• Mute	• Speaker	• Spinner Knob	• Modular Construction
• 1st IF Out Jack	• 2nd IF Out Jack	• Line Out 600 ohm	

Specifications:

Coverage ...... 10 - 30000 kHz
Selectivity ...... 6/2.4/1.2/.4 kHz -6dB.
Sensitivity ...... <.5µV for 10 dB SSB
IF Reject .... >60 dB
Audio Out ...... 3 W 5% distortion
IF .... 50.5 MHz, 50 kHz

Modes .... AM/SSB/CW/ISB
Stability .... <40 Hz in 15 min. period
Image Reject .. >70 dB relative to 1µV
RF Blocking .... >100 dB relative to 1µV
Antenna Input . SO-239 & Terminals

Comments:
The first tuning knob selects 0, 10 or 20 MHz. The second knob selects the frequency in 1 MHz increments. The third knob selects the .1 MHz position. The fourth (large) knob selects the range of 0 to .1 MHz and displays it on the last 3 positions of the digital readout. ISB output is to 600 ohm balanced line. Available in desk or rack mount version. Rack mount version shown above. Marketed for the marine radio market. Virtually identical to the DSR-1. 12 or 24 VDC supply available as an option.

Variants:
This model was also sold as the **S.A.I.T. MR 1414**.

Made In: United States 1971-1974

Voltages: 115/230 AC
50-420 Hz 15 W

Readout: 00000.1 Digital Nixie

Physical: 19x5.25x15" 17 Lbs.
480x133x380 mm 7.7 kg

Status: Active Manufacturer
Discontinued Model

Rarity: Very Scarce

	New	Used
Price:	①	⑤
Rating:	★★★★	⑥

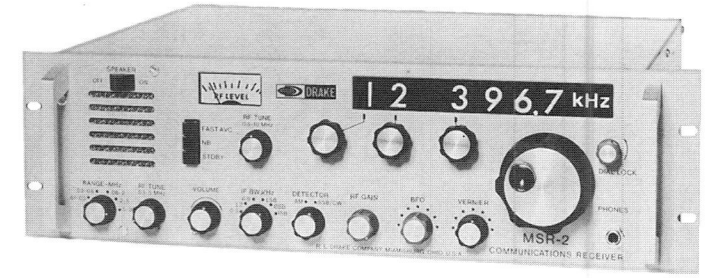

MSR-2

General Coverage Communications Receiver
Double Conversion Superheterodyne. Solid State.

Features:
- ¼" Head. Jack
- S-Meter
- BFO
- AVC (2 Pos.)
- RF Gain
- NB
- Dial Lock
- Speaker 3x5"
- Mute
- 2 Preselectors
- ISB
- Speaker Switch
- Spinner Knob
- Standby
- IF Out Jack
- Line Out 600 ohm
- Rack Handles

Specifications:

Coverage ...... 10 - 30000 kHz
Selectivity ...... 6/2.4/1.2/.3 kHz -6dB.
Sensitivity ...... <.5µV for 10 dB SSB
IF Reject .... >60 dB
Audio Out ...... 2W 4 ohms 5% dist.
IF .... 5.6 MHz, 50 kHz

Modes .... AM/SSB/CW/ISB
Stability .... <20 Hz 15 min period
Image Reject .. >60 dB
RF Blocking .... >100 dB (@1µV)
Antenna Input . 50 ohms SO-239

Accessories: Filters

Comments: Designed for rack mounted marine use. The first tuning knob selects 0, 10 or 20 MHz. The second knob selects the frequency in 1 MHz increments. The third knob selects the .1 MHz position. The fourth (large) knob selects the range of 0 to .1 MHz and displays it on the last 3 positions of the digital readout. A separate vernier knob provides fine tuning over ±200 Hz. There are two RF Tune (Preselector) knobs. One is for 30-500 kHz and the other for .5-10 MHz. With the RF GAIN control pulled out, the AVC is deactivated. Front-end protection can withstand 350 Volt emf!
Variants: This model was also sold as the **Debeg MSR-2** and the **S.A.I.T. MR 1415.**

Made In:	United States 1977-1978	
Voltages:	120/220/240 AC 50-420 Hz 35 W	
Readout:	00000.1 Digital Nixie	
Physical:	5.25x19x15" 25 Lbs. 133x482x381mm 11.3 kg	
Status:	Active Manufacturer Discontinued Model	
Rarity:	Very Scarce	

	New	Used
Price:	$5000A	$900-1300
Rating:	★★★★	⑥

R.L. DRAKE COMPANY

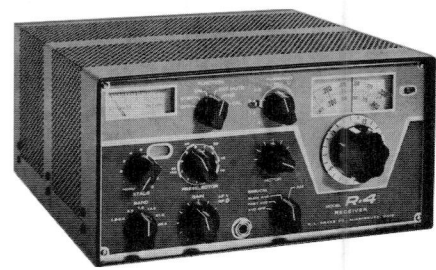

R-4

Amateur Band Communications Receiver
Double Conversion Superheterodyne. 13 Tubes plus Semiconductors.

Features:
- ¼" Head. Jack
- S-Meter
- Calibrator
- AVC FAST/SLOW/OFF
- Preselector
- IF Notch
- PBT
- Dial Lamp
- RF Gain
- Mute Jack
- Speaker Jack
- Anti-Vox Jack
- Standby
- NB

Specifications:

Coverage ...... Ham. See Comments.
Selectivity ...... 4.8/2.4/1.2/.4 kHz -6dB.
Sensitivity ...... <.5µV 10 dB S/N
I.F. .... 5645 kHz, 50 kHz
Audio Out ...... 1.4W 4 ohms 5% dist.

Modes .... AM/LSB/USB(CW)
Stability .... <100 Hz After warmup
Image Reject .. >60 dB
IF Reject .... >60 dB
Antenna Input . RCA 50 ohm

Circuit Complement:
12BZ6 RF Amp, 12BA6 Calibrator, 6HS6 1st Mixer, 6HS6 Premixer, 12BE6 2nd Mixer & Crystal Osc, 12BA6 50 kHz IF, 12BA6 50 kHz IF, 12AV6 AVC Amp/AVC Detector, 6GX6 Product Detector/Audio Amp, 6EH6 Audio Output, 12BA6 NB Amp, 12AX7 NB Pulse Amp/Shaper, 1N625 NB Gate, (2) 1N625 AM Detector, ED-3004 Bias Rectifier, (2) 1N3756 PS Rectifiers, 0B2 and 1N625 Voltage Regulator. Product detector for SSB/CW and diode detector for AM.

Accessories: MS-4 Speaker
Comments: Supplied coverage: 3.5-4, 7-7.5, 14-14.5, 21-21.5 and 28.5-29 MHz plus ten 500 kHz crystal positions. Dial accuracy is ±1 kHz. Requires an external speaker. Matches T4 or T4X Transmitter. [¥90,000].

Made In:	United States 1964-1965	
Voltages:	120/240 VAC 50/60 60W	
Readout:	Analog Lin.	
Physical:	10.75x5.5x11.7" 16 Lbs. 140x273x297mm 7.3 kg	
Status:	Active Manufacturer Discontinued Model	
Rarity:	Common	
Reviews:	*73* January 1966	

	New	Used
Price:	$379	$100-160
Rating:	★★★★	★★★

R.L. DRAKE COMPANY

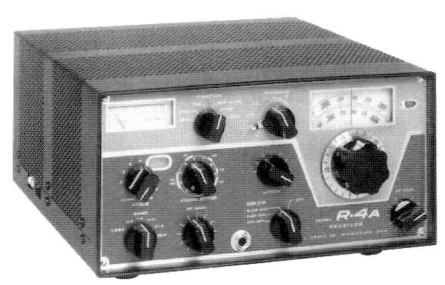

R-4A

Amateur Band Communications Receiver
Double Conversion Superheterodyne. 12 Tubes plus Semiconductors.

Features:
- ¼" Head. Jack
- S-Meter
- Anti-Vox Jack
- AVC FAST/SLOW/OFF
- Preselector
- IF Notch
- PBT
- Calibrator 100 kHz
- RF Gain
- Mute Jack
- Speaker Jack
- Dial Lamp
- Standby
- NB

Specifications:
Coverage ...... Ham. See Comments.
Modes .... AM/LSB/USB(CW)
Selectivity ...... 4.8/2.4/1.2/.4 kHz -6dB.
Stability .... <100 Hz After warmup
Sensitivity ...... <.5µV 10 dB
Image Reject .. >60 dB
IF Reject .... >60 dB
I.F. .... 5645 kHz, 50 kHz
Audio Out ...... 1.4 W 4 ohm 5% dist.
Antenna Input . RCA 52 ohms

Circuit Complement: 12BZ6 RF Amp, 12BA6 Calibrator, 6HS6 1st Mixer, 6HS6 Pre-mixer, 12BE6 2nd Mixer/Crystal Osc, 12BA6 50 kHz IF, 12BA6 50 kHz IF, 12AV6 AVC Amp/AVC Det., 6GX6 Product Detector/Audio Amp, 6EH6 Audio Output, 12BA6 NB Amp, 12AX7 NB Pulse Amp/Shaper, 2N3858 VFO, 2N3858 VFO Buffer, 2N3394 Crystal Oscillator, 1N714 Voltage Regulator, 0B2 & 1N625 Voltage Regulator, (2) 1N3194 P.S. Rectifiers, 1N3194 Bias Rectifier, 1N625 AM Detector, (2) 1N625 NB Pulse Clippers, 1N625 NB Gate and 1N483 Switching.

Accessories: MS-4 Speaker   FS-4 Frequency Synthesizer

Comments: Supplied coverage: 3.5-4, 7-7.5, 14-14.5, 21-21.5 and 28.5-29 MHz plus ten additional 500 kHz crystal positions. Dial accuracy is 1 kHz. Transceives with the T-4 to T-4X. Requires a speaker.

Made In:	United States 1966-1967
Voltages:	120/240 VAC 50/60 Hz 60W
Readout:	Analog Lin.
Physical:	10.75x5.5x11.7" 16 Lbs. 140x273x297mm 7.3 kg
Status:	Active Manufacturer Discontinued Model
Rarity:	Common
Reviews:	*73* January 1969

	New	Used
Price:	$379-399	$140-170
Rating:	★★★★★	★★★

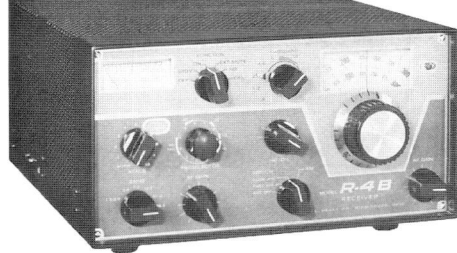

R-4B

Amateur Band Communications Receiver
Double Conversion Superheterodyne. 10 Tubes plus Semiconductors.

Features:
- ¼" Head. Jack
- S-Meter
- IF Notch
- Calibrator 25 kHz
- AVC 3 pos.
- NB
- Preselector
- PBT AM/CW/SSB
- RF Gain
- Mute Jack
- Speaker Jack
- Anti-Vox Jack
- Dial Lamp
- Standby

Specifications:
Coverage ...... Ham. See Comments.
Modes .... AM/SSB-CW
Selectivity ...... 4.8/2.4/1.2/.4 kHz -6dB.
Stability .... <100 Hz After warmup
Sensitivity ...... <.25 µV 10 dB
Image Reject .. >60 dB
IF .... 5645, 50 kHz.
Antenna Input . RCA 52 ohms
Audio Out ...... 1.5 W 4 ohms 5% dist

Circuit Complement:
6BZ6 RF Amp, 6HS6 1st Mixer, 12BE6 2nd Mixer/Crystal Osc, 12BA6 1st IF Amp, 12BA6 2nd IF Amp, 6EH5 Audio Out, 6HS6 Pre-Mix, 12BA6 NB plus 10 transistors, 17 diodes and 2 ICs. Product detector for SSB/CW and diode detector for AM.

Accessories:
MS-4 Speaker   4-NB Noise Blanker   FS-4 Frequency Synthesizer
SC-6 Converter 6M   SC-2 Converter 2M

Comments: Supplied coverage: 3.5-4, 7-7.5, 14-14.5, 21-21.5 and 28.5-29 MHz plus ten additional 500 kHz crystal positions. Dial accuracy is ±1 kHz. Bandwidth may be selected independently of mode. Requires external speaker. Matches T4XB Transmitter. An outstanding amateur receiver. [¥90,000].

Made In:	United States 1967-1973
Voltages:	120/240 VAC 50/60 Hz 60W
Readout:	Analog Lin.
Physical:	10.8x5.5x11.7" 16 Lbs. 274x140x297mm 7.3 kg
Status:	Active Manufacturer Discontinued Model
Rarity:	Common
Reviews:	*73* January 1966 *ODXA* November 1987 *Proceedings* 1989

	New	Used
Price:	$400-475	$170-250
Rating:	★★★★★	★★★★

R-4C

Amateur Band Communications Receiver
Triple Conversion Superheterodyne. 6 Tubes plus Semiconductors.

Features:
- ¼" Head. Jack
- PBT CW/SSB
- RF Gain
- Dial Lamp
- S-Meter
- Line out
- Mute Jack
- Spinner Knob
- IF Notch
- NB
- Speaker Jack
- Standby
- AGC OFF/FST/MED/SLO
- Calibrator 25 kHz
- Anti-Vox Jack
- Inj Jack

Specifications:

Coverage ......Ham. See Comments.	Modes .... AM/LSB/USB(CW)
Selectivity ......8/2.4/_/_ kHz -6dB.	Stability .... <100 Hz After warmup
Sensitivity ......<.25 V 10 dB	Image Reject .. >70 dB
I.F. ....5645, 5695, 50 kHz	Antenna Input . RCA 52 ohm
Audio Out ......2 W 3.2 ohms 5% dist.	

Circuit Complement: 6 Tubes, 15 transistors, 23 diodes and 1 I.C.

Accessories: 4-NB Noise Blanker   FS-4 Frequency Synthesizer
FL-6000 6 kHz Filter   FL-1500 1.5 kHz Filter   FL-500 .5 kHz Filter
FL-250 .25 kHz Filter   MS-4 Speaker

Comments:
Supplied coverage: 3.5-4, 7-7.5, 14-14.5, 21-21.5 and 28.5-29 MHz plus fifteen additional 500 kHz crystal positions. Dial accuracy is ±1 kHz. Requires external speaker. An outstanding amateur receiver. The optional (and very scarce!) FS-4 shown right, provides continuous coverage.

Made In:	United States 1973-1979	
Voltages:	120/240 VAC 50-420 Hz 60 W	
Readout:	⊔⊔⊔⊔ Analog Lin.	
Physical:	10.75x5.5x11.7" 17 Lbs. 273x140x324mm 7.7 kg	
Status:	Active Manufacturer Discontinued Model	
Rarity:	Common	
Reviews:	WRTH 1976 and 1978 QST January 1974 Buyer's Guide to Amateur R. SW Magazine April 1989	

	New	Used
Price:	$500-699	$190-270
Rating:	★★★★★	★★★★

R7/DR7

General Coverage Communications Receiver
Triple Conversion Superheterodyne. Solid State.

Features:
- ¼" Head. Jack
- RF Gain
- IF Notch Filter
- Speaker
- S-Meter
- RIT
- Calibrator 25 kHz
- Record Jack
- Preamp 10 dB
- PBT
- Speaker Jack
- Analog Sub-Dial
- AGC SLO/MED/FST/OFF
- Ant Switch (6 Pos.)
- Counter Input Jack

Specifications:

Coverage ......10 - 30000 kHz[1]	Modes .... AM/SSB/CW/RTTY
Selectivity ......2.3/_/_/_ kHz -6dB	Stability .... <100 Hz ±10% voltage
Sensitivity ......<.2µV SSB 1.8-30 MHz	Image Reject .. >80 dB
IF Reject ....>80 dB	Freq. Accur. .... 15 PPM ±100 Hz[1]
Ultimate Sel...>100 dB	IFs .... 48050, 5645, 50 kHz.
Audio Out ......2.5 W 4 ohm 10% dist.	Antenna Input . RCA

Accessories:
MS-7 Speaker   NB-7A Noise Blanker   AUX-7 Fixed Freq. Board
SL-6000 6 kHz Filter   SL-4000 4 kHz Filter   SL-1800 1.8 kHz Filter
SL-500 .5 kHz Filter   SL-300 .3 kHz Filter   RV-75 Extl. Synthz. VFO

Comments: [1]Very few, if any, units were manufactured without the DR7 digital display / general coverage board. These units were not general coverage and only with the analog sub-display (accurate to ±1 kHz). The digital display may be used as a digital counter to 150 MHz with external signal source. An excellent receiver for tropical band DXing. Also superb on longwave. Somewhat cumbersome to change frequency (sixty 500 kHz bands) and a bit drifty for radioteletype applications.

Made In:	United States 1978-1981	
Voltages:	100/120/200/240 AC 50/60 Hz 60 W or 11-16 VDC @ 3A	
Readout:	00000.1 Digital LED	
Physical:	13.6x4.6x13" 19 Lbs. 346x116x330mm 8.3 kg	
Status:	Active Manufacturer Discontinued Model	
Rarity:	Scarce	
Reviews:	Passport 1987 WRTH 1980 73 November 1980 QST Jan. & Feb. 1980 Ham Radio October 1978 Proceedings 1988	

	New	Used
Price:	$1295-1549	$650-770
Rating:	★★★★★	★★★★

R7A

General Coverage Communications Receiver
Triple Conversion Superheterodyne. Solid State.

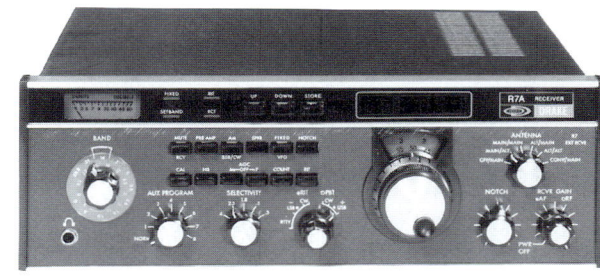

Features:
- ¼" Head. Jack
- RF Gain
- IF Notch Filter
- Speaker
- S-Meter
- RIT
- Calibrator
- Record Jack
- Preamp
- PBT
- NB
- Analog Sub-Dial
- AGC SLO/MED/FST/OFF
- Ant Switch (6 Pos.)
- Counter Input Jack
- Speaker Switch

Specifications:
Coverage ......10 - 30000 kHz
Selectivity ......9/2.3/.5/_/_ kHz -6dB
Sensitivity ......<.2µV SSB 1.8-30 MHz
IF Reject ....>80 dB
Ultimate Sel...>100 dB
Audio Out ......2.5 W 4 ohm 10% dist.

Modes .... AM/SSB/CW/RTTY
Stability .... <100 Hz ±10 voltage
Image Reject .. >80 dB
Freq. Accur. ..... 15 PPM ±100 Hz
IFs .... 48050, 5645, 50 kHz
Antenna Input . RCA

Accessories:
MS-7 Speaker   AUX-7 Fixed Freq. Board   RV-75 Extl. Synthz. VFO
SL-6000 6 kHz Filter   SL-4000 4 kHz Filter   SL-1800 1.8 kHz Filter
SL-500 .5 kHz Filter   SL-300 .3 kHz Filter

Comments: This enhanced version included the NB-7A Noise Blanker and 500 Hz filter as standard. This receiver exhibits excellent sensitivity, especially on the lower frequencies. Unlike many receivers, the RIT control does change the displayed frequency. Utilizes sixty 500 kHz bands. It is still a receiver of choice for tropical band DXer's. The optional RV-75 External VFO (1982) provides improved stability, 10 Hz tuning and two fixed channels. The RV-75 is very scarce and desirable (used $380).

Made In: United States 1981-1983

Voltages: 100/120/200/240 AC
50/60 Hz 60 W or
11-16 VDC @ 3A

Readout: `00000.1` Digital LED

Physical: 13.6x4.6x13" 19 Lbs.
346x116x330mm 8.3 kg

Status: Active Manufacturer
Discontinued Model

Rarity: Scarce

Reviews: 📖*Passport* 1987
WRTH 1982
Proceedings 1994-95

	New	Used
Price:	$1489-1649	$680-870
Rating:	★★★★★	★★★★

R8

General Coverage Communications Receiver
Double Conversion Superheterodyne. Solid State.

Features:
- ¼" Head. Jack
- RF Gain
- Notch Filter
- Scan/Sweep
- Speaker
- Mute Jack
- S-Meter
- Keypad
- RS232 port DB9
- Preamp
- Speaker Switch
- Squelch
- BFO
- 100 Memories
- Sync. Detection
- Record Jack
- Speaker Jack
- Two VFOs
- AGC OFF/SLOW/FAST
- Dual Clock-Timers
- Attenuator - 10 dB
- Recorder Activation
- Line Out Jacks (2)
- PBT   • Tone

Specifications:
Coverage ......100 - 30000 kHz
Selectivity ......6/4/2.3/1.8/.5 kHz -6dB
Sensitivity ......<.25 V 2-30MHz 10dBCW
IF Reject ....>80 dB
IFs ....45 MHz, 50 kHz.
Audio Out ......2.5 W 4 ohms 10% dist.

Modes .... AM/SSB/CW/RTTY/FM
Stability .... ±10 PPM -10° to 50°C.
Image Reject .. >80 dB .1-30 MHz
Freq. Accur. ..... ±100 Hz -10° to 50°C
Environment ... -10° to +50°C
Antenna Input . SO-239 and Terminals

Accessories:
MS-8 Speaker   VHF Converter

Comments: After a nearly ten year hiatus, Drake reentered the hobbyist receiver market with this very capable and hugely successful model. A 12 kHz FM bandwidth is also supplied. Memories store: Frequency, Mode, Bandwidth, AGC, RF, Antenna, Notch, NB and Synchro settings. The optional internal VHF converter covers 35-55 and 108 to 174 MHz in all modes. Also reviewed in *Proceedings* 1992-93.

Variants: Model **R8E** is an export version that has a different mains plug.

Made In: United States 1991-1995
Voltages: 100/120/200/240 AC
50/60 Hz 40 W or
11 - 16 VDC @ 2A

Readout: `00000.1` Digital

Physical: 13.2x5.25x13" 13 Lbs.
334x134x330 mm 6 kg

Status: Active Manufacturer
Discontinued Model

Rarity: Abundant

Reviews: 📖IBS-RDI Whitepaper
📖*Passport* 1992-1995
WRTH 1992, *BuyGuide* 1993
S.W. Mag. February 1992
Pop. Comm. October 1991
ODXA November 1991
QST March 1992

	New	Used
Price:	$980-999	$650-720
Rating:	★★★★★	★★★★★

R8A

General Coverage Communications Receiver
Double Conversion Superheterodyne. Solid State.

Features:

- ¼" Head. Jack
- RF Gain
- Notch Filter
- Scan/Sweep
- Speaker
- Mute Jack
- Dual VFOs

- S-Meter
- Keypad
- RS232 port DB9
- Tilt Stand
- Speaker Switch
- Squelch

- BFO
- PBT
- Sync. Detection
- Preamp
- Speaker Jack
- Tone

- AGC OFF/SLOW/FAST
- 440 Alpha Memories
- Recorder Activation
- Dual Clock-Timers
- Attenuator -10 dB
- Line Out Jacks (2)

Specifications:

Coverage	100 - 30000 kHz	Modes	AM/SSB/CW/RTTY/FM
Selectivity	6/4/2.3/1.8/.5 kHz -6dB.	Stability	±5 PPM -10° to 50°C.
Sensitivity	<.25 V 5-30 MHz 10dB	Image Reject ..	>80 dB
IF Reject	>80 dB	Freq. Accur.....	±100 Hz -10° to 50°C
IFs	45 MHz, 50 kHz.	Line Output	300 mV
Audio Out	2.5 W 10% distortion	Antenna Input .	SO-239 & Terminals

Accessories:

MS-8 Speaker   VHF Converter

Comments: All R8 features and functions are retained. Several ergonomic and performance improvements were added including: Alphanumeric memories, faster scanning, improved AGC, improved Notch, improved display, easier selection of mode and bandwidth, tilt-bar, enhanced tone control, detachable line cord and expanded RS-232 command set.

Made In: United States 1995-1997

Voltages: 100/120/200/240 AC
50/60 Hz 40 W or
11 - 16 VDC @ 2A

Readout: `00000.1` Digital

Physical: 13.2x5.25x13" 13 Lbs.
334x134x330mm 6 kg

Status: Active Manufacturer
Discontinued Model

Rarity: Abundant

Reviews: IBS-RDI Whitepaper
Passport 1996-1997
WRTH 1996
Monitoring Times May 1996
NASWA June 1996
SW Magazine Dec. 1995

	New	Used
Price:	$1050-1099	$730-810
Rating:	★★★★★	★★★★★

R8B

General Coverage Communications Receiver
Double Conversion Superheterodyne. Solid State.

Features:

- ¼" Head. Jack
- RF Gain
- Notch Filter
- Scan/Sweep
- Speaker
- Mute Jack
- Dual VFOs

- S-Meter
- Keypad
- RS232 port DB9
- Tilt Stand
- Speaker Switch
- Squelch

- BFO
- PBT
- Sync. Detection
- Preamp
- Speaker Jack
- Tone

- AGC OFF/SLOW/FAST
- 1000 Alpha Memories
- Recorder Activation
- Dual Clock-Timers
- Attenuator -10 dB
- Line Out Jacks (2)

Specifications:

Coverage	100 - 30000 kHz[1]	Modes	AM/SSB/CW/RTTY/FM
Selectivity	6/4/2.3/1.8/.5 kHz -6dB.	Stability	±5 PPM -10° to 50°C.
Sensitivity	<.25 V .1-30 MHz 10dB	Image Reject ..	>80 dB
IF Reject	>80 dB	Freq. Accur.....	±100 Hz -10° to 50°C
IFs	45 MHz, 50 kHz.	Line Output	300 mV
Audio Out	4 ohms 2.5 W 10% dist.	Antenna Input .	SO-239 & Terminals

Accessories:

MS-8 Speaker   VHF Converter

Comments: The R8B retains all of the model R8A features and functions. The three enhancements over the R8A include: faster scanning, *selectable* sideband synchronous AM detection and 1000 memories. (The FM mode bandwidth is 12 kHz). [1]Units produced with serial numbers greater than 7L1294 tune from 10-30000 kHz. Sensitivity below 100 kHz is not specified.

Made In: United States 1997-1998

Voltages: 100/120/200/240 AC
50/60 Hz 40 W or
11 - 16 VDC @ 2A

Readout: `00000.1` Digital

Physical: 13.2x5.25x13" 13 Lbs.
334x134x330mm 6 kg

Status: Active Manufacturer
Active Model

Rarity: (Too new)

Reviews: IBS-RDI Whitepaper
Passport 1998
NASWA January 1998

	New	Used
Price:	$1159-1199	$850-860
Rating:	★★★★★	★★★★★

R4245

General Coverage Communications Receiver
Triple Conversion Superheterodyne. Solid State.

Features:
- ¼" Head. Jack
- RF Gain
- IF Notch Filter
- Rack Handles
- S-Meter
- RIT
- Calibrator
- Speaker
- Preamp
- PBT
- NB
- Spinner Knob
- AGC SLO/MED/FST/OFF
- Ant Switch (6 Pos.)
- Counter Input Jack
- 8 Fixed Channels

Specifications:
Coverage ...... 10 - 30000 kHz
Selectivity ...... 4/2.3/1.8/.5/.3 kHz -6dB
Sensitivity ...... <.2µV SSB1.8-30 MHz
IF Reject .... >80 dB
Blocking .... >145 dB above floor
Ultimate Sel... >100 dB
Audio Out ...... 2.5 W 10% distortion

Modes .... AM/SSB/CW/RTTY
Stability .... <100 Hz ±10 voltage
Image Reject .. >80 dB
Freq. Accur. .... 15 PPM ±100 Hz
Third Ord Int. .. +10 dBm with preamp on
IFs .... 4805, 5645, 50 kHz
Antenna Input . RCA ohm

Accessories:
R577 Cabinet

Comments:
The R4245 is the military-commercial version of the R-7A featuring full frequency synthesis rather than an analog VFO. This synthesized VFO provides more stability than the R-7/A. The front panel is light beige, has built-in speaker and is ready for 19" rack mounting. This premium receiver is scarce. Matches TR4310 transceiver.

Variants:
The Drake model **R77** seems to be the identical unit.

Made In:	United States 1982-1985	
Voltages:	100/120/200/240 VAC 50/60 Hz 60 W or 11-16 VDC @3A	
Readout:	`00000.1` Digital LED	
Physical:	Rack Version: 19x5.25x12.5" 19 Lbs. 483x133x318mm 8.6 kg Cabinet Version: 20.75x7.5x14.5" 514x190x368mm	
Status:	Active Manufacturer Discontinued Model	
Rarity:	Very Scarce	
Reviews:	📖*Passport* 1985/86, 1987 *WRTH* 1982	

	New	Used
Price:	$2795-3795	$1200-2400
Rating:	★★★★★	★★★★

R8000

General Coverage Communications Receiver
Double Conversion Superheterodyne. Solid State.

Features:
- ¼" Head. Jack
- RF Gain
- Notch Filter
- Scan/Sweep
- Speaker
- Mute Jack
- S-Meter
- PBT
- RS232 port DB9
- Preamp
- Speaker Switch
- Squelch
- BFO
- Keypad
- Sync. Detection
- Record Jack
- Speaker Jack
- Tone
- AGC OFF/SLOW/FAST
- 100 Memories
- Two VFOs
- Recorder Activation
- Line Out Jacks (2)
- Attenuator - 10 dB

Specifications:
Coverage ...... 100-30000 kHz +VHF[1]
Selectivity ...... 6/4/2.3/1.8/.5 kHz -6dB
Sensitivity ...... <.25 V 1.8-30 MHz 10dB[2]
IF Reject .... >80 dB
IFs .... 45 MHz, 50 kHz.
Audio Out ...... 2.5W 4 ohm 10% dist.
Ult. Select. ...... >95 dB

Modes .... AM/SSB/CW/RTTY/FM
Stability .... <±10 PPM -10° to 50°C.
Image Reject .. >80 dB
Freq. Accur. .... <±100 Hz -10° to 50°C
Environmental -10° to 50° C.
Antenna Input . SO-239 and Terminals
Dyn. Range .... >90 dB .1-30 MHz

Comments:
The R8000 is a military-commercial version of the R8. It is designed for 19" rack mounting. [1]The VHF Converter is installed as standard and provides coverage of 35-55 and 108-174 MHz in all modes, with full readout. Drake PC control software also supplied as standard. [2] With preamp.

Made In:	United States 1992-1995	
Voltages:	100/120/200/240 AC 50/60 Hz 42 W or 11-16 VDC @ 2A	
Readout:	`00000.1` Digital	
Physical:	19x5.25x13" 13 Lbs. 483x133x330mm 5.9 kg	
Status:	Active Manufacturer Discontinued Model	
Rarity:	Extremely Scarce	

	New	Used
Price:	$2000	②
Rating:	★★★★	⑥

RR-1

Marine Band Communications Receiver
Double Conversion Superheterodyne. Solid State.

Features:
- ¼" Head. Jack
- RF Gain
- Speaker
- Spinner Knob
- S-Meter
- Notch Filter
- Speaker Switch
- Mute Jack
- Atten -20 dB
- Rack Handles
- Preselector
- Dial Lamp
- AGC (2 Pos.)
- Calibrator 100 kHz
- Audio In-Out Jack

Specifications:
Coverage ...... See Comments.	Modes .... AM/CW/LSB/USB
Selectivity ...... 4.8/2.4/.4 kHz -6dB.	Stability .... <100 Hz After warmup
Sensitivity ...... <1µV 4-26 MHz 20dB SSB	Image Reject .. >60 dB <15 MHz.
IFs .... 5.645 MHz, 50 kHz.	Image Reject .. >50 dB >15 MHz.
Audio Out ...... 2 W 10% distortion	Freq. Accur. .... 3 kHz. with calibrator
AGC .... <6 dB AF for 100 dB RF	Antenna Input . BNC 50 ohms

Accessories:
Line Amplifier   RY-4 RTTY Adapter   5-NB Noise Blanker
RP-500 Protector

Comments:
Ranges: .15-.535, 1.5-4, 6-6.5, 8-9, 12-13.5, 16-17.5, 22-23 and 25-26 MHz. Four additional 500 kHz crystal ranges are available. The RR-1 is basically an SPR-4 that has been optimized, rack mounted and fitted with crystals to serve the marine user as a maritime reserve receiver. [¥80,000].

Variants:
This model was also sold as the **S.A.I.T. MR 1541**.

Made In:	United States 1971-1977
Voltages:	120/240 AC 50-60 18W or 24 VDC 24W
Readout:	Analog Lin.
Physical:	19x7x11" 14 Lbs. 490x180x280mm 6.3 kg
Status:	Active Manufacturer Discontinued Model
Rarity:	Very Scarce

	New	Used
Price:	$1450	$500-650
Rating:	★★★★★	★★★★★

RR-2

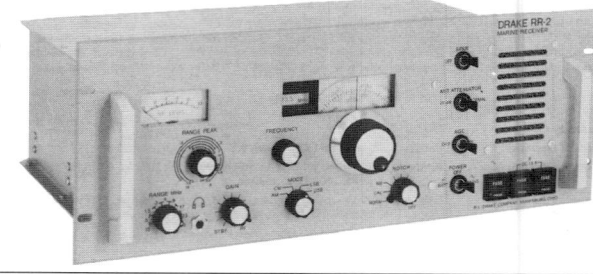

General Coverage Communications Receiver
Double Conversion Superheterodyne. Solid State.

Features:
- ¼" Head. Jack
- RF Gain
- Speaker
- Dimmer Jack
- S-Meter
- Notch Filter
- Speaker Switch
- Dial Lamp
- Atten. -20 dB
- Rack Handles
- Preselector
- Mute Jack
- AGC OFF/ON
- Calibrator 100 kHz
- Audio In-Out Jack
- Spinner Knob

Specifications:
Coverage ...... 150 - 30000 kHz	Modes .... AM/CW/LSB/USB
Selectivity ...... 4.8/2.4/.4 kHz -6dB.	Stability .... <100 Hz After warmup
Sensitivity ...... <1µV 4-26 MHz 20dB SSB	Image Reject .. >60 dB <15 MHz.
IFs .... 5.645 MHz, 50 kHz.	Image Reject .. >50 dB >15 MHz.
Audio Out ...... 2 W 3.2 ohm 10% dist.	Freq. Accur. .... 3 kHz. with calibrator
AGC .... <6 dB AF for 100 dB RF	Antenna Input . SO-239 50 ohms

Accessories:
Line Amplifier   RY-4 RTTY Adapter   5-NB Noise Blanker
RP-500 Protector

Comments:
Similar to the RR-1 but with continuous coverage provided by a frequency synthesizer which generates an injection signal every 500 kHz and PTO which tunes a 500 kHz range. PTO over-travel allows an additional 50 kHz off each band edge. The RR-2 is classified as a marine "reserve" (i.e. backup) receiver.

Variants:
This model was also sold as the **S.A.I.T. MR 1542**.

Made In:	United States 1974-1980
Voltages:	120/240 AC 50-60 40W or 24 VDC 24W
Readout:	Analog Lin.
Physical:	19x7x11" 14 Lbs. 483x178x279mm 6.35 kg
Status:	Active Manufacturer Discontinued Model
Rarity:	Very Scarce

	New	Used
Price:	$1795	$700-750
Rating:	★★★★★	★★★★★

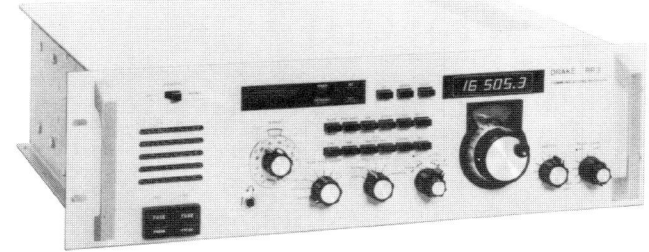

RR-3

General Coverage Communications Receiver
Triple Conversion Superheterodyne. Solid State.

Features:
- ¼" Head. Jack
- S-Meter
- Preamp
- AGC SLO/MED/FST/OFF
- RF Gain
- RIT
- PBT
- Preset 500, 2182 kHz
- IF Notch Filter
- Calibrator
- Analog Sub-dial
- AC/DC Switches
- Rack Handles
- Speaker
- Spinner Knob
- 8 Fixed Channels

Specifications:
Coverage ......10-30000 kHz	Modes .... AM/SSB/CW/RTTY
Selectivity ......10/2.3/_/_/_ kHz	Stability .... <150 Hz/hour after 1 hour
Sensitivity ......<.2µV 2-30 MHz SSB	Image Reject .. >80 dB
IF Reject ....>80 dB	Blocking .... >100 dB relative to 1µV
Ultimate Sel. ...>100 dB	IFs .... 48.05, 5.645 MHz, 50 kHz.
Audio Out ......2.5 W 10% distortion	Antenna Input . 50 ohms

Accessories:
SL-6000 6 kHz Filter   SL-4000 4 kHz Filter   SL-1800 1.8 kHz Filter
SL-500 .5 kHz Filter   SL-300 .3 kHz Filter   NB7A Noise Blanker
R577 Cabinet   IN7 DC to DC Isolation PS

Comments:
Similar to the R-7, but with different filters, without the antenna switch and rack mounted. Additionally, presets for 500 and 2182 kHz distress channels and 24 VDC operation make the RR-3 I.T.U. approved for use as ship's main or reserve receiver. AC and DC fuses are accessible from the front panel. The RR-3 features a built-in Drake RP-700 receiver front-end protector. White metal cabinet. [¥180,000].

Made In: United States 1982-1983

Voltages: 100/120/200/240 VAC
50/60 Hz 60 W or
24 VDC ±15% 65W

Readout: 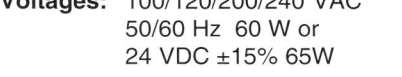 `00000.1` Digital LED

Physical: 19x5.25x14" 19 Lbs.
483x133x356mm 8.6 kg
514x190x368mm

Status: Active Manufacturer
Discontinued Model

Rarity: Extremely Scarce

	New	Used
Price:	$2395	$1000-1400
Rating:	★★★★★	★★★★

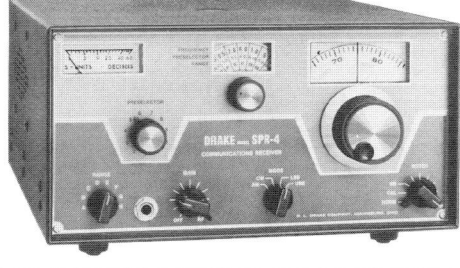

SPR-4

Shortwave Broadcast Band Communications Receiver
Double Conversion Superheterodyne. Solid State.

Features:
- ¼" Head. Jack
- S-Meter
- IF Notch
- Speaker
- RF Gain
- Mute Jack
- Speaker Jack
- Anti-Vox Jack
- Preselector
- Dial Lamp
- Dial Lamp Switch
- Audio In-Out Jack
- Spinner Knob

Specifications:
Coverage ......See Comments.	Modes .... AM/LSB/USB/CW
Selectivity ......4.8/2.4/.4 kHz -6dB.	Stability .... <100 Hz After warmup
Sensitivity ......<.25 V 10 dB CW/SSB	Image Reject .. >60 dB below 15 MHz
I.F. ....5645, 50 kHz	Image Reject .. >50 dB above 15 MHz
Audio Out ......3 W 4 ohms 5% dist.	Antenna Input . RCA 50 ohm

Accessories:
MS-4 Speaker   5-NB Noise Blanker   FS-4 Frequency Synthesizer
AL-4 LW/MW Loop   RY-4 RTTY Unit   TA-4 Transceive Adapter
DC-PC 12 VDC Cord   SCC-4 Calibrator 100 kHz

Comments:
Ranges: .15-.5, .5-1, 1-1.5, 6-6.5, 7-7.5, 9.5-10, 11.5-12, 15-15.5, 17.5-18 and 21.5-22 MHz. Fourteen additional 500 kHz crystal positions are available. Dial accuracy is ±1 kHz. Drake's first completely solid-state receiver. May be used for transceive with the T-4 or T-4XB/C with TA-4 adapter.

Made In: United States 1969-1978

Voltages: 120/240 VAC 18W
or 12 VDC 6W

Readout: Analog Lin.

Physical: 10.75x5.5x12.25" 14 Lbs.
274x140x324mm 8.2 kg

Status: Active Manufacturer
Discontinued Model

Rarity: Common

Reviews: *Popular Elec.* Nov. 1974
QST December 1970
WRTH 1976
CQ November 1970

	New	Used
Price:	$379-699	$250-360
Rating:	★★★★★	★★★

SSR-1

General Coverage Communications Receiver
Double Conversion Superheterodyne. Solid State.

Features:
- ¼" Head. Jack
- S-Meter
- Preselector
- Speaker
- RF Gain
- Mute Jack
- Record Jack
- Dial Light
- Clarify ±3 kHz
- Telescopic Ant.
- Dial Light Switch
- Atten. -20 dB
- Spinner Knob

Specifications:

Coverage ......500 - 30000 kHz	Modes .... AM/LSB/USB/CW
Selectivity ......5.5/3 kHz -6dB.	Image Reject .. >50 dB
Sensitivity ......<.5µV 10 dB 2-30 MHz CW	IF Reject .... >50 dB below 20 MHz
Dial Accur ......5 kHz.	IF Reject .... >40 dB above 20 MHz
Audio Out ......2 W 8 ohms	Antenna Input . Single Terminal 75 ohm
I.F. ....44.5-45.5 MHz, 455 kHz	

Accessories: DC-PC 12 VDC Cord

Comments: The MHz knob (top left) tunes inner MHz dial on display. The main tuning knob (large) rotates the outer dial (10 kHz graduations). If the AC line voltage fails an automatic circuit switches to the internal D cells. The dial lamp is always lit when operating on AC power. The dial light button must be pressed when the receiver is operating on batteries. This model did not live up to the usual high performance level of Drake receivers but represented a decent value for the money. This SSR-1 was made in Japan and was the only model not manufactured by Drake in Ohio. It appears that the Lowe SRX-30 is a close relative of the SSR-1. Please see the Lowe Chapter.

Made In:	Japan	1975-1978
Voltages:	117/240 VAC 50-60 or 8 D cells	
Readout:	⊔⊔⊔⊔⊔**▮** Analog Lin.	
Physical:	13x5.5x11" 14 Lbs. 330x140x280 mm 6.4 kg	
Status:	Active Manufacturer Discontinued Model	
Rarity:	Scarce	
Reviews:	*WRTH* 1978 *Popular Elec.* January 1977 *73* July 1977	

	New	Used
Price:	$299-350	$130-200
Rating:	★★	★★

SW1

General Coverage Broadcast Receiver
Double Conversion Superheterodyne. Solid State.

Features:
- Mini Head. Jack
- RF Gain
- Keypad
- 32 Memories
- Speaker
- Dimmer
- 2 Tuning Speeds

Specifications:

Coverage ......100 - 30000 kHz	Modes .... AM
Selectivity ......5.5 kHz -6dB	Stability .... ±10 PPM 0° to 50°C.
Sensitivity ......<2µV 10dB	Environment ... 0° to +50°C
IFs ....45 MHz, 455 kHz.	Antenna Input . SO-239 & Terminal
Audio Out ....7 W 10% distortion	

Accessories:
Carrying Handle   Automotive Bracket

Comments:
Tuning may be accomplished by the manual tuning dial, the Up/Down buttons or via direct keypad entry. This receiver is designed with simplicity in mind, primarily for the general consumer market. It does not have SSB/CW capability. A huge, easy-to-read 3/4 inch LED frequency display is featured.

Variants:
This model is similar to the model **PRN-1000** marketed on shortwave radio by the *People's Radio Network*. However, the newer SW1 has 32 memories instead of one and an RF Gain instead of a Tone control. It also has improved signal handling capability and adjacent interference rejection capability.

Made In:	United States 1996-1997	
Voltages:	12 VDC @.4A via supplied 120 VAC adapter.	
Readout:	**00001.** Digital LED	
Physical:	10.875x4.4x7.675" 4.7Lbs. 276x112x195mm 2.1 kg	
Status:	Active Manufacturer Active Model	
Rarity:	Scarce	
Reviews:	*Passport* 1997-1998 *Pop Comm.* December 1996 *Mon. Times* August 1996 *NASWA* March 1997 *QST* October 1997	

	New	Used
Price:	$200-300	$140-150
Rating:	★★★	★★★

SW2

General Coverage Communications Receiver
Double Conversion Superheterodyne. Solid State.

Features:
- Mini Head. Jack
- S Indicator
- Keypad
- Ext Speaker Jack
- Speaker
- 100 Memories
- Dimmer
- 2 Tuning Speeds
- Sync. Detection
- RF Gain

Specifications:

Coverage ...... 100 - 30000 kHz	Modes .... AM/LSB/USB
Selectivity ...... 6/2.3 kHz -6dB	IFs .... 55.845 MHz, 455 kHz
Sensitivity ...... <.5µV 10dB SSB	Environment ... 0° to +50°C
Audio Out ...... 0.7 W 5% distortion.	Antenna Input . SO-239 & Terminal

Accessories:

Infrared Remote   Tilt Carry Handle   MMK-1 Automotive Bracket

Comments:
Tuning may be accomplished by the manual tuning dial, the Up/Down buttons (5 kHz steps), manual tuning (50 Hz steps) or via direct keypad entry. A huge, easy-to-read 3/4 inch LED frequency display is featured. Sideband selectable synchronous detection is featured. Memories store frequency, mode and synchronous detection. The optional remote can be used to tune the radio up and down, mute audio, enter frequencies, change modes and even dim the display

Made In: United States 1997-1998

Voltages: 12 VDC @1.5A via supplied 120 VAC adapter.

Readout: `00000.1` Digital LED

Physical: 10.875x4.4x7.675" 5.8 Lbs. 276x112x195mm 2.6 kg

Status: Active Manufacturer Active Model

Rarity: (Too new)

Reviews: 📖 *Passport* 1998
Pop Comm. May 1997
Pop Comm. August 1997
NASWA July 1997
WRTH 1997, 1998
Mon. Times August 1997

	New	Used
Price:	$490-500	$290-310
Rating:	★★★★★	★★★★★

SW-4A

Shortwave Broadcast Band Receiver
Double Conversion Superheterodyne. 6 Tubes plus Semiconductors.

Features:
- ¼" Head. Jack
- S-Meter
- Preselector
- Speaker
- Tone
- Speaker Jack
- Dial Adjust
- Dial Lamp

Specifications:

Coverage ...... See Comments.	Modes .... AM
Selectivity ...... 5 kHz -6dB.	Stability .... <100 Hz After warmup
Sensitivity ...... <2 µV AM 10 dB	I.F. .... 5645 kHz, 455 kHz
Audio Out ...... 4/8 ohm 2 W 5% dis.	Antenna Input . 52 ohm Terminals only.

Circuit Complement:
12BZ6 RF Amp, 6HS6 1st Mixer, 12BE6 2nd Mixer, 12BA6 1st IF Amp, 12BA6 2nd IF Amp, 6HS6 Premix, 2N3394 Crystal Osc, 2N3858 VFO Buffer, 2N706 VFO, 2N3858 AGC Amp, (2) 2N3394 AF AMP, 40310 RCA Audio Out, 1N483 AVC Clamp, 1N714 Voltage Reg., 1N270 Detector, 1N3194 Bias Rect. and (4) 1N3194 PS Rect.

Accessories: MS-4 Speaker   AL-4 LW/MW Loop Antenna

Comments: Ranges: .15-.5, .485-1.05, .95-1.55, 5.95-6.55, 6.95-7.55, 9.45-10.05, 11.45-12.05, 14.95-15.45, 17.45-18.05, 21.45-22.05 and 25.45-26.05 MHz. Dial accuracy is ±3 kHz. Note that this model does not receive CW/SSB. The front is marked with a logo for the defunct shortwave station, *WNYW Radio New York Worldwide*. It was designed and promoted as an affordable, easy-to-use shortwave receiver. Requires a speaker. The AL-4 Loop is scarce. Becoming collectable.

Variants: The earlier model **SW-4** is all tubes and has a band *and* a range switch plus 3 aux. crystal sockets. The SW-4 is very scarce.

Made In: United States 1967-1974

Voltages: 120/240 VAC 50/60 Hz 30 W

Readout: `|....|..|` Analog Lin.

Physical: 10.75x5.5x12.25" 16 Lbs. 273x139x311mm 7.3 kg

Status: Active Manufacturer Discontinued Model

Rarity: Scarce

Reviews: *Popular Elec.* Nov. 1972

	New	Used
Price:	$289-335	$250-350
Rating:	★★★★	★★★

SW8

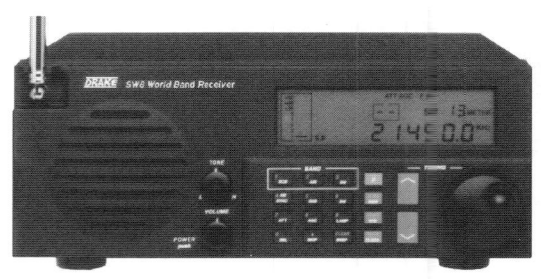

General Coverage Communications Receiver
Double Conversion Superheterodyne. Solid State.

Features:
- Mini Head. Jack
- 70 Memories
- Tone
- Keypad
- Backlit Display
- S-Indicator
- Memory Scan
- Speaker
- Whip Preamp
- Record Jack
- BFO
- Scan/Sweep
- Sync. Detection
- Attenuator
- Ant Select Switch
- AGC FAST/SLOW
- Dual Clock-Timer
- Tels. Whip Antenna
- Squelch (VHF air)
- Dial-Keypad Lock

Specifications:
Coverage ...... See Comments.
Selectivity ...... 6/4/2.3 kHz -6dB
Sensitivity ...... <.5µV .5-30 MHz 10dB
IF Reject .... >80 dB
IFs .... 55.845 MHz, 455 kHz.
Dyn. Range ... >95 dB .5-30 MHz SSB
Audio Out ...... 2.5 W 10% distortion

Modes .... AM/SSB/CW
Stability .... ±10 PPM 0° to 50°C.
Image Reject .. >80 dB .5-30 MHz
Freq. Accur..... ±100 Hz @50°C
Environment ... 0° to +50°C
Antenna Input . SO-239 or Terminals

Accessories:
MS-8 Speaker   Canvas Carrying Case

Comments: Coverage: 500 - 30000 kHz, FM band 88-108 MHz and VHF Air band 118-137 MHz. Memories store: Frequency, Mode, Bandwidth, AGC, Attenuator and Sync. settings. The carrying handle is removable. Units produced after 03/31/95 (S/N >= 5C) feature a preamp circuit for the telescopic whip antenna. Units produced after 01/01/96 (S/N >= 6A) include coverage down to 100 kHz (long wave) and have selectable sideband syncro.

Made In: United States 1994-1998
Voltages: 7-10 VDC @1A via supplied AC wall adapter. Or six D cells mounted internally.
Readout: `00000.1` Digital LCD
Physical: 11.5x5.25x13" 10 Lbs. 292x133x330mm 4.5 kg
Status: Active Manufacturer Active Model
Rarity: Abundant
Reviews: IBS-RDI White paper
Passport 1995 1998
Mon. Times August 1994
Mon. Times November 1996
NASWA July 1994
SW Magazine May 1994
QST October 1994

	New	**Used**
Price:	$589-699	$350-435
Rating:	★★★★★	★★★★★

24 Echophone

The **Echophone** company produced broadcast receivers during the 1930's. The company ran into fiscal difficulties in 1936 and Hallicrafters acquired their Chicago plant. In 1940 Hallicrafters revived the Echophone name to label their low-end models.

Chuck Dachis' book *Radios By Hallicrafters* is a must-read for anyone interested the radios and the history of Echophone or Hallicrafters.

It is suggested that Echophone and other such AC-DC style radios be operated from an isolation transformer for safety reasons.

Echophone Radio Co.
201 East 26th Street
Chicago, Illinois 1943-1945

Echophone Division
The Hallicrafters Co.
2611 Indiana Ave.
Chicago 16, Illinois 1946-1952

ECHOPHONE

EC-1

General Coverage Communications Receiver
Single Conversion Superheterodyne. 6 Tubes.

Features:
- Head. Jack
- Speaker
- Logging Scale
- Standby
- BFO
- Bandspread 0-100
- Spkr.-Head. Switch

Specifications:
Coverage ...... 550 - 30500 kHz
Audio .... 1.7 W
Modes .... AM/CW
Antenna Input . Terminals

Circuit Complement:
12K8 Oscillator/Mixer 12SK7 IF, 12SQ7 Detector/AVC/1st Audio, 12J5 BFO, 35L6 AF Output and 35Z5 Rectifier.

Comments:
Ranges: .55-2, 2-8 and 8- 30.5 MHz. The speaker is on the top panel. The Headphone Tip Jack is on the rear panel. This model is from the Echophone Commercial series. Black enameled metal cabinet.

Made In: United States 1941-1943

Voltages: 115-125 VAC 50/60 Hz or 115-125 VDC

Readout: Analog

Physical: 10.785x8x7.75" 12 Lbs. 276x203x197mm 5.5 kg

Status: Inactive Manufacturer
Discontinued Model

Rarity: Scarce

	New	Used
Price:	$20-29	⑤
Rating:	⑥	⑥

ECHOPHONE

EC-1A

General Coverage Communications Receiver
Single Conversion Superheterodyne. 6 Tubes.

Features:
- Head. Jack
- Speaker
- Standby
- BFO
- ANL
- Spkr.-Head. Switch
- Bandspread 0-100

Specifications:
Coverage ...... 550 - 30500 kHz
Modes .... AM/CW
Antenna Input Terminals

Circuit Complement:
12SA7 Mixer, 12SK7 IF Amp, 12SQ7 2nd Detector/1st Audio, 35L6GT 2nd Audio Amp, 12SQ7 BFO/ANL and 35Z5GT Rectifier.

Comments:
Ranges: .55-2, 2-8 and 8- 30 MHz. The speaker is on the top panel. The Headphone Tip Jack is on the rear panel. This model is from the Echophone Commercial series. This is an updated version of the EC-1.

Variants:
The model **EC-1B** (1945) is similar. This is the functional predecessor to the Hallicrafters S-38. It is virtually the physical twin of the Hallicrafters S-41G/S-41W.

Made In: United States 1942-1946

Voltages: 120 VAC
50/60 Hz

Readout: Analog

Physical: 11.75x8x8.125" 11 Lbs.
295x203x206mm 5 kg

Status: Inactive Manufacturer
Discontinued Model

Rarity: Scarce

	New	Used
Price:	$30	[5]
Rating:	[6]	[6]

25 Eddystone

Eddystone Radio was established in 1923. Initially they produced components, but quickly added assembled receivers, starting with a battery operated, two tube medium wave set.

The first H.F. receiver, simply called the *Eddystone Shortwave,* was also a two tubed affair covering 3.5 to 20 MHz. In 1928 the *Scientifics* series was introduced. In the thirties, Eddystone focused on robustly built export receivers for commercial enterprises through the Empire. Eddystone customers included rubber, coffee, tea, sugar and cocoa plantations. For the domestic hobby market models such as: *The Kilodyne Four, The Sphinx, The Homeland* and *The Overseas Four* were introduced. Ground breaking engineering in the VHF area also occurred.

The early sixties saw the acquisition of Eddystone by the Marconi company. In the 1990's Eddystone became a major provider of FM broadcast transmitters.

Model number prefixes are as follows: EA stands for Eddystone Amateur receiver, EB stands for Eddystone Broadcast receiver and EC stands for Eddystone Communications receiver. Like Hallicrafters in the U.S., Eddystone produced a staggering number of different models. Used models are common in the U.K., but usually scarce in America. Eddystone enthusiasts should consider joining the *Eddystone User Group* referenced in *Chapter 4.*

Stratton & Company Ltd.
Eddystone Works
West Heath
Birmingham 31
England   1947-1963

Eddystone Radio Ltd.
Alvechurch Road
Birmingham 31
England   1959-1972

Eddystone Radio Ltd.
Marconi Communications Ltd.
Alvechurch Road
Birmingham B31 3PP
England   1972-1996

Eddystone Radio Ltd.
Unit 8/9, Birkdale Avenue
Heeley Road, Selly Oak
Birmingham B29 6U8
England   1996-1998

504

General Coverage Communications Receiver
Single Conversion Superheterodyne. 10 Tubes.

Features:
- ¼" Head. Jack
- RF Gain
- NL
- BFO
- Bandspread
- Tone
- Mute
- Dial Lamp
- Rack Handles
- Flywheel Tuning
- Speaker Output

Specifications:
Coverage ...... 600 - 30000 kHz.
IF .... 450 kHz
Audio .... 3 ohms 3 W

Modes .... AM/CW
Antenna Input . Terminals

Circuit Complement:
EF39 RF Amp, EF39 RF Amp, ECH35 Freq. Changer, EF39 IF Amp, EF39 IF Amp, EBC33 Detector/AF Amp/AVC, 6V6GT Audio Output, 5Z4G Rectifier, EB34 S Meter/NL and EF39 BFO. Two RF stages. Crystal filter.

Comments:
This radio has five bands. This model is also referred to as the S.504. The "S" refers to Stratton. Finished in battleship gray with blue front panel and chromium fittings.

Variants:
Model **556** is a broadcast version. The 556 came out prior to the 504. The BFO has been removed and a tuning eye added. Model **556/B** is a battery operated version of the 556.

Made In:	England	1946

Voltages: 110 or 210/250 VAC
40-60 Hz

Readout: ⌊ı ı ı ■ ı⌋ Analog

Physical: 16.75x8.75x10"
426x222x254 mm

Status: Active Manufacturer
Discontinued Model

Rarity: Extremely Scarce

	New	Used
Price:	①	£90-100
Rating:	⑥	⑥

640

General Coverage Communications Receiver
Single Conversion Superheterodyne. 9 Tubes.

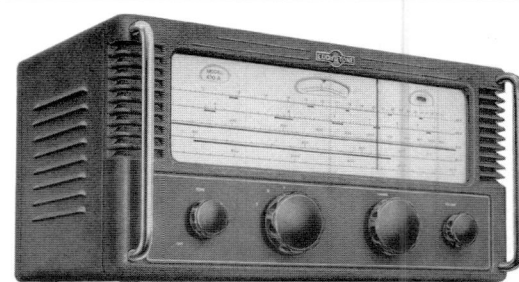

Features:
- ¼" Head. Jack
- BFO
- Mute
- Flywheel Tuning
- RF Gain
- Bandspread
- Dial Lamp
- Speaker Output
- Standby
- NL

Specifications:
Coverage ...... 1700 - 30000 kHz.
Sensitivity ...... 2 µV
IF .... 1620 kHz
Modes .... AM/CW
Antenna Input . 400 ohms Terminals

Circuit Complement:
One RF and two IF stages.

Accessories:
687 6V Accumulator   669 External S-Meter

Comments:
Three ranges. Features a ripple-black metal cabinet.

Variants:
Model **659** is a broadcast version covering 400 - 30000 kHz. Model **659B** is a 6 VDC battery version of the 659. The model 640 was also sold under the Marconi Label.

| **Made In:** | England | 1947-1949 |

Voltages: 110 or 200/250 VAC 40-60 Hz

Readout: Analog

Physical: 16.75x8.75x10"
426x222x254 mm

Status: Active Manufacturer
Discontinued Model

Rarity: Extremely Scarce

	New	**Used**
Price:	£27-42	£45-80
Rating:	⑥	⑥

670A

General Coverage Broadcast Receiver
Single Conversion Superheterodyne. 6 Tubes plus Semiconductors.

Features:
- Speaker Output
- Dial Lamp
- Rack Handles
- Tuning Indicator
- Speaker
- Phono Input
- Vernier Tuning
- Log Scale 0-2500
- Tone
- Flywheel Tuning

Specifications:
Coverage ...... 480 - 30000 kHz.
Selectivity ......
Audio Out ...... 3 ohms
Modes .... AM
IF .... 450 kHz
Antenna Input . 400 ohms Terminals

Circuit Complement:
UAF42 RF Amplifier, UCH42 Frequency Changer, UAF42IF Amp/AGC, UAF42 Detector/Audio, UL41 Audio Output, DM70 Tuning Indicator plus a diode rectifier.

Accessories:
732 Mains Filter

Comments: The 670 series was expressly designed for ship board cabin use. The 670A has five ranges: .15-.38, .54-1.5 3.7-10.6 and 10.5-30 MHz. The fifth "range" is marked "G" which is for a gramophone (phonograph) input. The broadcast bands are marked on the dial. The vernier tuning scale is a mechanical bandspread scheme that provides a quick and accurate means to return to a previously logged frequency. The tuning indicator (DM70) is a green fluorescent tube. With no signal the glow will extend to its full length. When a signal is received the glow contracts.

Variants: The earlier model **670** was introduced in 1948 (£40 new) and had 7 tubes. The original 670 had the medium wave scale marked in meters.

Made In: England   1953-1959

Voltages: 110/ 200/230 VAC or 110/ 200/230 VDC

Readout: Analog

Physical:

Status: Active Manufacturer
Discontinued Model

Rarity: Typically Unavailable

	New	**Used**
Price:	①	£80-100
Rating:	⑥	⑥

670C

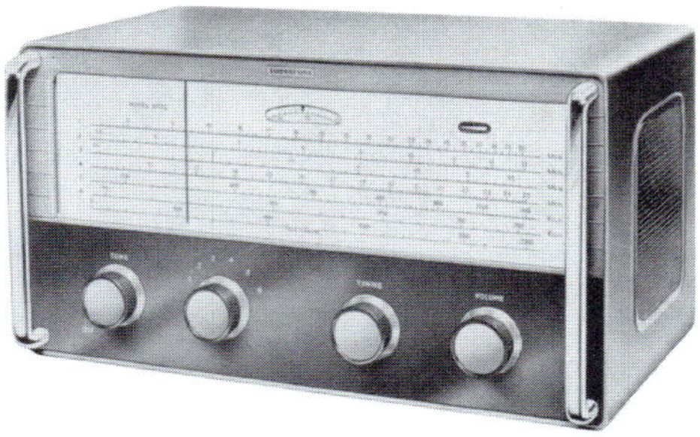

General Coverage Broadcast Receiver
Single Conversion Superheterodyne. 6 Tubes plus Semiconductors.

Features:
- Speaker Output
- Speaker
- Flywheel Tuning
- Dial Lamp
- Tone
- 140:1 Tune Ratio
- Rack Handles
- Vernier Tuning
- Phono Input
- Tuning Indicator

Specifications:
Coverage ...... 150 - 30000 kHz.
Selectivity ......
Audio Out ...... 2.5 ohms 2W
Modes .... AM
IF .... 450 kHz
Antenna Input . 400 ohms Terminals

Circuit Complement:
UAF42 RF Amplifier, UCH42 Frequency Changer, UAF42 IF Amp/AGC, UAF42 Detector/Audio, UL41 Audio Output, DM70 Tuning Indicator plus a diode rectifier.

Comments:
Ranges: .15-.35, .5-1.111, 1.1-2.5, 2.5-5.5, 5.5-13 and 13-30 MHz. Reception gap from 350-550 kHz. The 670C distinguishes itself from the earlier 670A by having six bands instead of four and having four front knobs of equal size.

Variants:
The similar model **670C/1** was sold at the **MIMCO 2232B** (1963).

Made In: England 1962-1964

Voltages: 110/200/230 VAC or 110/200/230 VDC

Readout: Analog

Physical:

Status: Active Manufacturer
Discontinued Model

Rarity: Typically Unavailable

	New	Used
Price:	①	⑤
Rating:	⑥	⑥

680

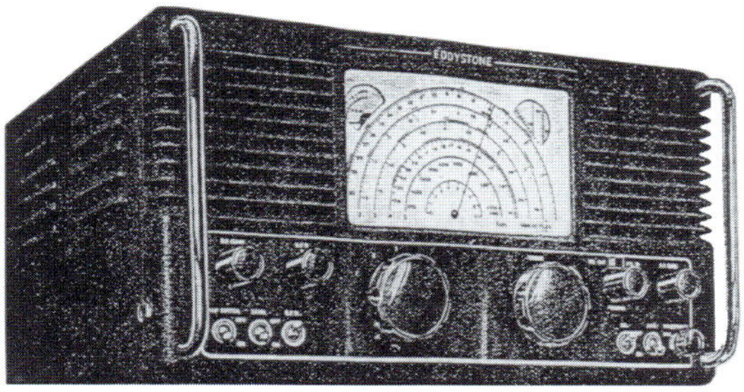

General Coverage Communications Receiver
Single Conversion Superheterodyne. 15 Tubes.

Features:
- ¼" Head. Jack
- RF Gain
- Standby
- Speaker Output
- S-Meter
- BFO
- NL
- Phono Input
- Rack Handles
- Mute
- Dial Lamp
- Flywheel Tuning
- Hinged Top Cover
- AVC ON/OFF
- Dimmer (back panel)

Specifications:
Coverage ...... 480 - 30000 kHz.
IF .... 450 kHz
Modes .... AM/CW
Antenna Input . 400 ohms Terminals

Circuit Complement:
6BA6 RF Amp, 6BA6 RF Amp, 7S7 Frequency Changer, 6AM6 Local Osc, 6BA6 IF, 6BA6 IF, 6AU6 Audio Amp, 6AU6 Phase Inverter, EL91 Push-Pull, 6BA6 BFO, 6AL5 NL, 6AL5 Detector/AVC, 5Z4 Rectifier and VR150 Voltage Stabilizer. Two RF stages and two IF stages. (Later production substituted the 6BE6 for the 7S7).

Accessories: Rack Mount

Comments:
The 680 series was introduced in 1949 and represented Eddystone's top of the line for many years. The headphone jack is on the left side panel. An early pre-production 9 tube 680 is reported to have the headphone jack on the front. Gray metal case. This receiver requires a speaker. Prototypes show headphones on the front panel. The 680 may also be referred to as the S.680.

Variants:
Model **680/2** and **680/2A** were produced for the New Zealand post office.

Made In: England 1949-1952

Voltages: 110 or 200/250 VAC 40-60 Hz

Readout: Analog

Physical: 16.75x8.75x14" 41 Lbs.
426x222x355 mm 18.6 kg

Status: Active Manufacturer
Discontinued Model

Rarity: Typically Unavailable

	New	Used
Price:	£85-89	£90-110
Rating:	⑥	⑥

680X

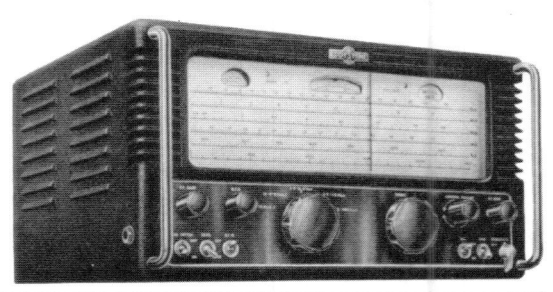

General Coverage Communications Receiver
Single Conversion Superheterodyne. 15 Tubes.

Features:
- ¼" Head. Jack
- S-Meter
- Rack Handles
- Calibrator 100 kHz
- RF Gain
- BFO ±3 kHz
- Mute
- 140:1 Tune Ratio
- Standby
- NL
- Dial Lamp
- Dimmer (back panel)
- AVC ON/OFF
- Flywheel Tuning

Specifications:
Coverage ...... 480 - 30000[1] kHz.
Selectivity ...... 6/3/1.3 kHz -6dB.
Sensitivity ...... 5 µV 15 dB S/N
Audio Out ...... 2.5 ohms 1.8W

Modes .... AM/CW
IF .... 450 kHz
Image Ratio .... 40 dB @25 MHz
Antenna Input . 400 ohms

Circuit Complement:
6BA6 RF Amp, 6BA6 RF Amp, 6BE6 Mixer, 8D3 Local Osc, 6BA6 IF, 6BA6 IF, 6AL5 Detector/AVC, 6BR7 AF, 6BR7 AF, 7D9 Audio Amp, 7D9 Audio Amp, 6BA6 BFO, 6AL5 S-Meter/NL, 5Z4QT Rectifier and VR150 Voltage Stabilizer. Two RF stages and two IF stages. Crystal filter.

Accessories:
812 Mounting Feet   811 Diecast Speaker   688 Speaker Black

Comments:
The 680X has features a 6BE6 for V3 rather than the 7S7 found in the 680. It also adds a crystal filter and product detector. Ranges: .48-1.12, 1.11-2.5, 2.5-5.7, 5.3-12.5 and 12.3-30 MHz.

Variants: Model **680X/RM** is for racking mounting.

Made In:	England	1952-1962

Voltages:	110/125 200/250 VAC 40-60 Hz 80W

Readout:	Analog

Physical:	16.75x8.75x15" 47 Lbs. 426x222x381 mm 21.3 kg

Status:	Active Manufacturer Discontinued Model

Rarity:	Typically Unavailable

	New	Used
Price:	£106-140	£90-160
Rating:	⑥	⑥

710/B
All World Six

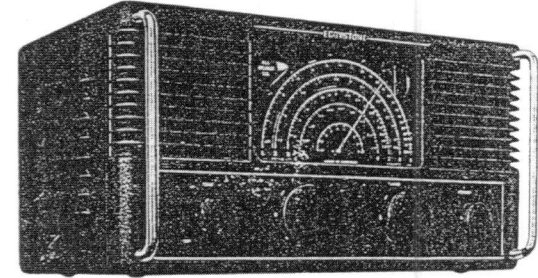

General Coverage Broadcast Receiver
Single Conversion Superheterodyne. 6 Tubes.

Features:
- ¼" Head. Jack
- BFO
- Phono Input
- Flywheel Tuning
- RF Gain
- Speaker
- Speaker Output
- 140:1 Ratio Tune
- Tone
- Dial Lamp
- Carrying Handles

Specifications:
Coverage ...... 500 - 30600 kHz.
IF .... 450 kHz

Modes .... AM
Antenna Input . 400 ohms Terminals

Circuit Complement:
EAF42 RF Amp, ECH42 Frequency Changer, EAF42 IF Amp, EAF42 AF Amp, EL42 Audio Amp and EL42 Audio Amp.

Accessories:
687 6V Accumulator

Comments:
This receiver has four tuning ranges. This receivers is also referred to as the S.710 (for Stratton 710). The name "All World Six" was also used on a different model that was introduced in 1939. The 710/B is similar to the 740 but without a BFO or NL.

Made In:	England	1950

Voltages:	Battery 6 VDC 2.5 A

Readout:	Analog

Physical:	16.75x8.75x10" 34 Lbs. 426x222x254 mm 15.4 kg

Status:	Active Manufacturer Discontinued Model

Rarity:	Extremely Scarce

	New	Used
Price:	①	⑤
Rating:	⑥	⑥

730/4

General Coverage Communications Receiver
Single Conversion Superheterodyne. 15 Tubes.

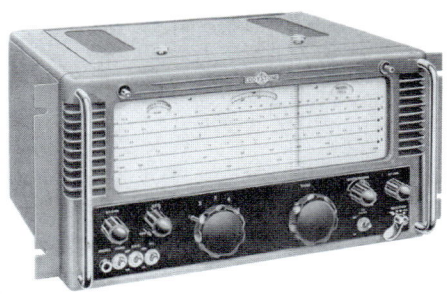

Features:

• ¼" Head. Jack	• S-Meter	• Dial Lamp	• Calibrator 500 kHz
• RF Gain	• BFO ±5 kHz.	• AF Filter 1000 Hz	• 140:1 Tune Ratio
• Standby	• AVC ON/OFF	• Rack Handles	• Dimmer • Mute
• Hinged Cover	• Flywheel Tuning	• IF Out Jack	• Line Out • NL

Specifications:

Coverage ...... 480 - 30000 kHz. Modes .... AM/CW
Selectivity ...... See Comments IF .... 450 kHz
Sensitivity ...... <2µV 15 dB S/N Image Rej. ...... >45 dB @18 MHz
Audio Out ...... 2.5 ohm 1W Antenna Input . 75 ohms

Circuit Complement:
6BA6 RF Amp, 6BA6 RF Amp, 6BE6 Mixer, 6AM6 Oscillator, 6BA6 IF Amp, 6BA6 IF Amp, 6AL5 AGC/Demodulator, 12AU7 Audio Amp, 6AL5 NL/S-Meter, 6AM6 Crystal Calibrator, 6AU6 Cathode Follower, 6BA6 BFO, 5Z4G Power Rectifier, VR150 Voltage Stabilizer and 6AM5 Output. Two RF stages and two IF stages.

Comments: Ranges: 12.3-30, 5.3-12.5, 2.5-5.7, 1.11-2.5 and .48-1.12 MHz. This radio requires a speaker (such as 688D). The vernier tuning has 2500 division and gives an equivalent scale of 34 feet. Small lamps on the right of the scale indicate band in use. Selectivity: 12/6/3/1.5/.5 kHz (.1 with audio filter). Light gray finish.

Variants: Model **730/6** has four crystal control fixed frequencies in the range of 2 - 20 MHz. Model **730/8** also has the four fixed positions and the lowest tuning range is changed to 200 - 410 kHz rather than the 480 - 1120 kHz.

Made In:	England	1957-1961
Voltages:	110/125 or 200/250 VAC 40-60 Hz 70W	
Readout:	Analog	
Physical:	16.75x8.75x13.75" 57 Lbs. 426x222x350 mm 25.8 kg	
Status:	Active Manufacturer Discontinued Model	
Rarity:	Typically Unavailable	
Reviews:	*Radio Bygones* #37	

	New	Used
Price:	①	£65-85
Rating:	⑥	⑥

740

General Coverage Communications Receiver
Single Conversion Superheterodyne. 8 Tubes.

Features:

• ¼" Head. Jack	• BFO	• Mute	• AVC ON/OFF
• RF Gain	• Standby	• NL	• Dial Lamp
• Speaker Output	• Flywheel Tuning		

Specifications:

Coverage ...... 480 - 30600 kHz. Modes .... AM/CW
Sensitivity ...... <10µV 15 dB S/N IF .... 450 kHz
Audio .... 2.5 ohms 3 W Antenna Input . 400 ohms Terminals

Circuit Complement:
EAF42, ECH42, EAF42, EAF42, EL42, EAF42, EB41 and EZ40.

Accessories: 687 6V Vibrator

Comments:
This receiver requires a speaker.

Variants:
Model 710/B is a broadcast version with no BFO or NL. The 710/B is shown earlier in this chapter.

Made In:	England	1950-1954
Voltages:	110 or 220/250 VAC 40-60 Hz 45 W	
Readout:	Analog	
Physical:	16.75x8.75x10" 34 Lbs. 426x222x254 mm 15.4 kg	
Status:	Active Manufacturer Discontinued Model	
Rarity:	Typically Unavailable	

	New	Used
Price:	£29.1-32.1	
Rating:	⑥	⑥

750

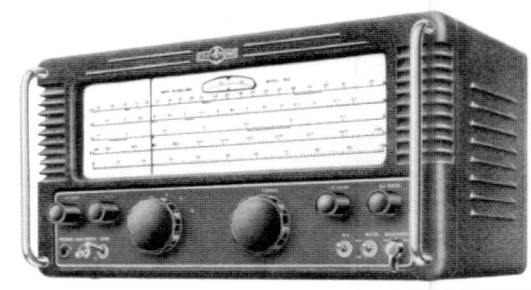

General Coverage Communications Receiver
Double Conversion Superheterodyne. 11 Tubes.

Features:
- ¼" Head. Jack
- RF Gain
- Standby
- Dial Lamp
- BFO ±3 kHz
- NL
- Rack Handles
- Mute
- Jack for S-Meter
- Flywheel Tuning
- AVC ON/OFF
- 200:1 Tune Ratio

Specifications:

Coverage ......480 - 30000 kHz.		Modes .... AM/CW	
Selectivity ......Variable		IF .... 1620, 85 kHz	
Sensitivity ......<5µV 15 dB S/N		Image Ratio .... >40 dB @30 MHz	
Audio Out ......3.5 W 2.5 ohm		Antenna Input . 400 ohms	

Circuit Complement:
6BA6 RF Amp, ECH42 Mixer, 6AM6 Osc., ECH42 Frequency Changer, 6BA6 IF Amp, DH77 Detector/AGC/AF, 6AL5 NL/S-Meter Diodes, N78 Output, 6BA6 BFO, 5Z4G Rectifier and VR150 Voltage Stabilizer.

Accessories:
811 Speaker Gray   687/1 Vibrator Power Unit 6VDC
669 External S-Meter

Comments:
Ranges: 12.3-30, 5.3-12.5, 2.5-5.7, 1.11-2.5, and .48-1.12 MHz. Fine black ripple finish. This radio requires a speaker.

Variants:
Model **750A** is similar, but with an S-Meter.

Made In:	England	1950-1957
Voltages:	110/125 200/250 VAC 40-60 Hz 70W	
Readout:	Analog	
Physical:	16.75x8.75x10" 40 Lbs. 426x222x254 mm 18.1 kg	
Status:	Active Manufacturer Discontinued Model	
Rarity:	Typically Unavailable	

	New	Used
Price:	£50-78	£70-95
Rating:	⑥	⑥

830/2

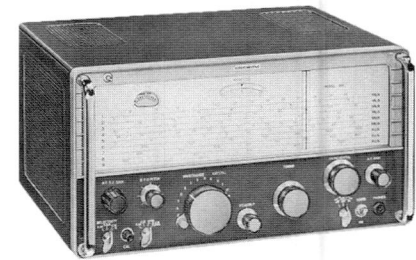

General Coverage Communications Receiver
Double Conversion Superheterodyne. 13 Tubes.

Features:
- ¼" Head. Jack
- RF Gain
- Dial Light
- S-Meter
- BFO
- Speaker Out
- Rack Handles
- Mute
- Line Out
- Calibrator 100 kHz
- 8 Crystal Positions
- IF Out

Specifications:

Coverage ......300 - 30000 kHz.		Modes .... AM/LSB/USB	
Selectivity ......6/3/1.3 kHz -6dB.		Stability .... ±2 kHz. after 10 mins.	
Sensitivity ......<3µV for 15 dB CW.		Image Reject .. >70 dB 1.5-10 MHz	
Audio Out ......2.5/600 ohm 2.5 W		Image Reject .. >50 dB 10 -30 MHz	
Dial Accuracy ±1 kHz.		IF .... 1250-1450 kHz, 100 kHz	

Circuit Complement: 6ES8 RF Amp, 6AK5 1st Mixer, 6U8 1st Local Osc, 6AK5 2nd Mixer, 6C4 2nd Local Osc, 6BA6 1st IF, 6BA6 2nd IF, 6AL5 NL, 6AU6 IF Output Cathode Follower, 6AT6 AM Detector/AGC, 6AQ5 Audio Output, 6AU6 Crystal Calibrator, 6BE6 CW/SSB Detector and (2) 0A2 HT Stabilizer.

Accessories: Rack Mount   EP20 Panoramic Adapter

Comments: 8 Internal crystal positions are available for fixed operation.

Variants: The 830/2 is the general production version. Model **830/1** is identical to the 830/2 but with smaller "pointer" knobs for Selectivity, Mode switch and AGC/NL switch. Model **830/3** uses a different RF Gain knob than the 830/2. Model **830/4** Canadian version, covers 120-560 and 1500-30000 kHz. and features a different mains connector, cabinet screws and incremental scale. Model **830/5** is Swedish version with a different antenna jack. Models ending in /RM are for rack mounting.

Made In:	England	1965-1969
Voltages:	110/125 200/250 VAC 40-60 Hz	
Readout:	Analog	
Physical:	16.75x8.75xx15" 49 Lbs 426x222x381mm 22 kg	
Status:	Active Manufacturer Discontinued Model	
Rarity:	Very Scarce	

	New	Used
Price:	$995	£110-180
Rating:	⑥	⑥

830/7

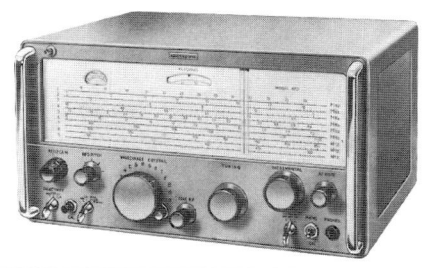

General Coverage Communications Receiver
Double Conversion Superheterodyne. 15 Tubes plus Semiconductors.

Features:
- ¼" Head. Jack
- S-Meter
- Rack Handles
- Calibrator 100 kHz
- RF Gain
- BFO
- Mute Line
- 8 Crystal Positions
- Dial Light
- AGC
- ANL
- 140:1 Reduction
- Speaker Out
- F Output
- Line Out 600 Ohm
- Line Out 2000 Ohm

Specifications:
Coverage ......300 - 30000 kHz.
Selectivity ......6/3/1.3 kHz -6dB.
Sensitivity ......<3µV for 15 dB CW.
Audio Out ......2.5W 3 ohm 10% dist.
Inter. Freq. ....1250-1450, 100 kHz.
Modes .... AM/LSB/USB/CW
Dial Accuracy . ±1 kHz.
Image Reject .. >70 dB 1.5-10 MHz
Image Reject .. >50 dB 10 -30 MHz
Antenna Input . 75 ohms

Circuit Complement: 6ES8 RF Amp, 6AK5 1st Mixer, 6AJ8 2nd Mixer/2nd Osc Iso. Amp, 6C4 2nd Local Osc, 6BA6 1st 100 kHz IF Amp, 6BA6 2nd 100 kHz IF Amp, 6AL5 AM NL, 6AU6 Cathode Follow, 6AT6 AM Detector AGC Rectifier, 6AQ5 Audio Out, 6AU6 Calibrator, 6U8 1st Local Osc, 6BE6 CW/SSB Detector and (2) 0A2 HT Stabilizer plus 4 diodes. Single conversion below 1.5 MHz.

Accessories: Rack Mount   EP20 Panoramic Adapter

Comments: Ranges: .30-.5, .5-.85, .58-1.5, 1.5-2.5, 2.5-4, 4-6.8, 6.8-11, 11-18 and 18-30 MHz. Replaces the 830/2. A great receiver.

Variants: Model **830/7/RM** is rack mount version. Model **830/8** is Dipl.Wire Service version. Model **830/9** has synthesized osc. input. Model **830/10** is the Canadian version. Model **830/11** was sold under the STC label. Model **830/12** is Hagenuk E81.

Made In:	England 1969-1973
Voltages:	110/125 200/250 VAC 40-60 Hz 85W
Readout:	Analog Lin.
Physical:	Cabinet Version: 16.75x8.75x15" 49 Lbs 426x222x381 mm 22.2 Kg Rack Version: 19x8.75x13" 50 Lbs 483x222x330mm 22.7 kg
Status:	Active Manufacturer Discontinued Model
Rarity:	Very Scarce

	New	Used
Price:	£309	£150-240
Rating:	⑥	⑥

840

General Coverage Communications Receiver
Single Conversion Superheterodyne. 7 Tubes.

Features:
- ¼" Head. Jack
- Standby
- Rack Handles
- 140:1 Ratio Tuning
- RF Gain
- Tone
- BFO ±3 kHz
- Phono Input
- Mute
- AGC
- NL
- Speaker

Specifications:
Coverage ......480 - 30600 kHz
Selectivity ......10 kHz -30dB.
Audio Out ......2.5 ohms 1.2 W
Modes .... AM/CW
Sensitivity .... <10 µV 15 dB S/N
Antenna Input . 400 Ohm

Circuit Complement:
UAF42 RF Amp, UCH42 Frequency Changer, UAF42 IF Amp/AVC, UAF42 AM Detector/Audio Amp, UL41 Audio Output, UAF42 BFO and UY41 Rectifier.

Accessories:
688 Speaker

Comments:
Ranges: .5-1.45, 1.4-3.8, 3.7-10.6 and 10.5-30 MHz. The medium wave range (range #4) is marked in meters. Amateur and broadcast bands are marked in color on the tuning scale.

Made In:	England 1953-1954
Voltages:	110/125 200/250 VAC 40-60 Hz 60W or DC
Readout:	Analog
Physical:	16.75x8.75x10.5" 30 Lbs 425x222x266mm 13.6 kg
Status:	Active Manufacturer Discontinued Model
Rarity:	Very Scarce

	New	Used
Price:	£45	£90-100
Rating:	⑥	⑥

840A

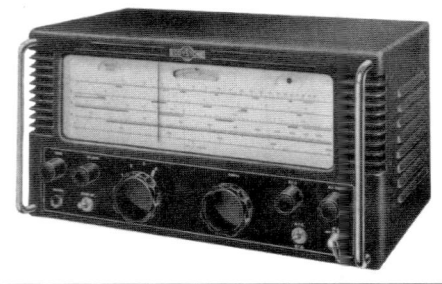

General Coverage Communications Receiver
Single Conversion Superheterodyne. 7 Tubes.

Features:
- ¼" Head. Jack
- RF Gain
- Mute
- NL
- Standby
- Speaker
- AGC
- Rack Handles
- Tone
- Phono Input
- Logging Scale
- BFO ±3 kHz
- 140:1 Ratio Tuning

Specifications:
Coverage ...... 480 - 30600 kHz
Selectivity ...... 20 kHz -30dB.
Sensitivity ...... <10µV 15 dB S/N
Audio Out ...... 2.5 ohms 0.75 W

Modes .... AM/CW/SSB
IF .... 450 kHz
Image Rej .... >15 dB
Antenna Input . 400 Ohms

Circuit Complement:
UAF42 RF Amp, UCH42 Frequency Changer, UAF42 IF Amp/AGC, UAF42 AM Detector/Audio Amp, UL41 Audio Output, UAF42 BFO and UY41 Rectifier.

Accessories:
688 Speaker   732 Mains Filter Unit   774 Receiver Mounting Blocks

Comments:
Ranges: .48-1.14, 1.4-3.8, 3.7-10.6 and 10.5-30.6 MHz. The dial frequency readout is not too accurate. However, the logging scale allows you to quickly and accurately return to a previously heard frequency. Audio quality is good. Features and oyster gray hammer finish metal cabinet. The amateur bands are marked in blue and the broadcast bands in red.

Made In:	England 1955-1961
Voltages:	110/125 200/250 VAC 40-60 Hz 60W or DC
Readout:	Analog
Physical:	16.75x8.75x10" 30 Lbs 425x222x254mm 13.6 kg
Status:	Active Manufacturer Discontinued Model
Rarity:	Very Scarce

	New	Used
Price:	£49-58	£50-75
Rating:	⑥	⑥

840C

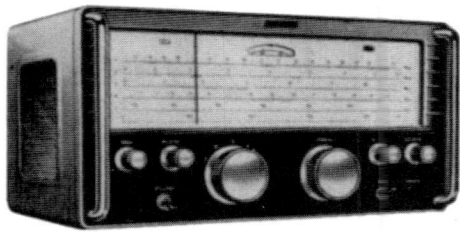

General Coverage Communications Receiver
Single Conversion Superheterodyne. 8 Tubes.

Features:
- ¼" Head. Jack
- RF Gain
- Mute
- Tune Indicator
- Speaker
- Standby
- Rack Handles
- Tone
- AGC
- Logging Scale
- BFO ±3 kHz
- 130:1 Tune Ratio

Specifications:
Coverage ...... 480 - 30000 kHz
Selectivity ...... 20 kHz -30dB.
Sensitivity ...... <10µV 15 dB S/N
Audio Out ...... 1.2 W

Modes .... AM/CW/SSB
IF .... 450 kHz
Image Rej .... >15 dB

Circuit Complement:
UAF42 RF Amp, UCH42 Frequency Changer, UAF42 IF Amp/AGC, UAF42 AM Detector/Audio Amp, UL41 Audio Output, UAF42 BFO, UY41 Rectifier and DM70 Tuning Eye.

Comments:
Ranges: .48-1.15, 1.15-2.58, 2.5-6.1, 5.2-12.9 and 12.4-30 MHz. Two tone gray metal cabinet. The 840C has a low impedance antenna input, while the 840 and 840A have high impedance antenna input. Also note that this is a five band receiver and has a newer style cabinet than its predecessor.

Variants:
Please see *Chapter 105 Briefly Mentioned* for the 850/2, 850/3, 850/4 and 850/5. The 850 series has a similar appearance, but only covers longwave 10 to 600 kHz.

Made In:	England 1962-1968
Voltages:	110/125 or 200/250 VAC 40-60 Hz 60W or DC
Readout:	Analog
Physical:	17x9x11" 28 Lbs 430x230x280mm 12.7 kg
Status:	Active Manufacturer Discontinued Model
Rarity:	Very Scarce

	New	Used
Price:	£58	£85-120
Rating:	⑥	⑥

870

General Coverage Broadcast Receiver
Single Conversion Superheterodyne. 5 Tubes.

Features:
- Speaker
- RF Gain
- Flywheel Tuning
- Logging Scale
- Dial Lamps
- Speaker
- Rack Handles

Specifications:
Coverage ...... 150 - 18000 kHz   Modes .... AM
IF .... 465 kHz

Circuit Complement:
12BE6 Detector, 12BA6 IF Amp, 12AT6 AF Amp/2nd Detector/AVC, 19AQ5 AF Output and 35W4 Rectifier.

Comments:
Ranges: .15-.38, .54-1.5, 1.95-6.3 and 5.9-18 MHz. Note reception gaps from .38-.54 and 1.5-1.95 MHz. This simple receiver was designed for home or shipboard cabin use. It was also advertised as a *semi-portable*. It has a two toned metal cabinet.

Accessories:
732 Mains Filter

Variants:
The model **870A** (1962-1963, new £33-35) is similar but with a fifth range extending coverage up to 24 MHz. The tube lineup is the same. The 870A ranges are: .15-.38, .51-1.4, 1.3-3.5, 3.2-7.5 and 7.5-24 MHz.

	Made In:	England	1957-1959

Made In: England   1957-1959

Voltages: 110/120 200/250 VAC 40-60 Hz 60W or DC

Readout: Analog

Physical: 11x6.375x8.22" 11.5 Lbs 279x162x208mm 5.2 kg

Status: Active Manufacturer Discontinued Model

Rarity: Very Scarce

	New	Used
Price:	£30-35	£35
Rating:	⑥	⑥

880/2

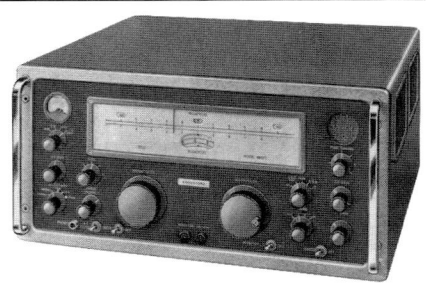

General Coverage Communications Receiver
Double Conversion Superheterodyne. 22 (or 23?) Tubes.

Features:
- ¼" Head. Jack
- RF Gain
- Mute
- Audio In Jack
- Osc. Output
- S-Meter
- IF Gain
- Rack Handles
- Fine Tuning
- Speaker Switch
- Speaker
- ANL
- AF Filter 400 Hz
- IF Out 500 kHz
- Bass Control
- AGC FST/SLO/SSB
- Line Out 600 ohms
- Calibrator 100 kHz
- Antenna Trimmer
- 140:1 Tune Ratio

Specifications:
Coverage ...... 500 - 30500 kHz   Modes .... AM/LSB/USB/CW
Selectivity ...... 14/7/3/_/_ kHz -6dB.   Stability .... ±20 Hz. after 4 hours.
Sensitivity ...... <1µV for 15 dB CW   IF Rejection .... 60 dB
Audio Out ...... 0.75W 2.5 ohm 10% dis.   Image Reject .. >15 dB
IF .... 2.5-3.5/3.5-4.5,500 kHz   Antenna Input . 75 ohms

Accessories:
A Filter 400 Hz   B Filter 1.2 Hz   C Filter 3 kHz
LP2773 SSB Filter   Rack Mount

Comments: The 880/2 tunes in 30 1 MHz switched ranges (24 revolutions per range) yield a tuning accuracy of ±1kHz. The Oscillator Output is for diversity with 2 or more receivers. A quiet receiver and great performer.

Variants: The initial model **880** was sold from 1959 to 1961 (£390 new). The model **880/3** (1964) had 23 tubes and was sold as the **Marconi H2301**. Model **880/4** is different from the 880/2: uses a combined RF/IF control, calibrator is 500 kHz, bandwidths are 14/7/4/1.2/.2 kHz, has no AF filter, mains filter & dimmer added, etc.

Made In: England   1961-1969

Voltages: 110/125 or 200/250 VAC 40-60 Hz

Readout: Analog Lin.

Physical: Cabinet Version:
19.5x9.5x20.5" 99 Lbs.
495x239x521mm 44.9 kg
Rack Version:
19x8.75x20.5" 87 Lbs.
483x222x521mm 39.5 kg

Status: Active Manufacturer Discontinued Model

Rarity: Very Scarce

	New	Used
Price:	£380	£180-250
Rating:	★★★★★	⑥

888A

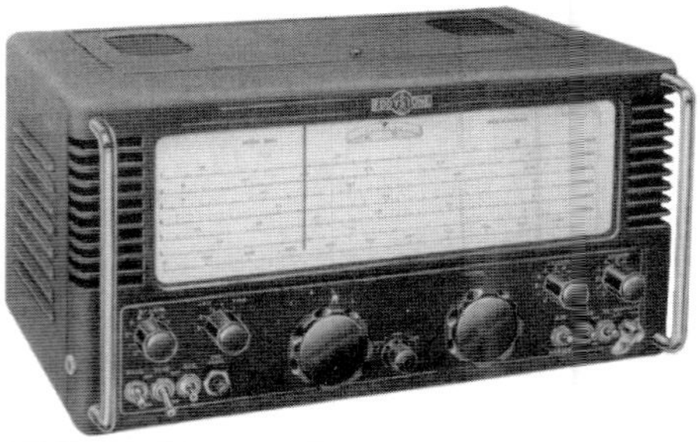

Amateur Band Communications Receiver
Double Conversion Superheterodyne. 12 Tubes.

Features:
- ¼" Head. Jack
- RF Gain
- Mute
- Dial Adjust
- AGC ON/OFF
- IF Gain
- Rack Handles
- Standby
- AF Filter 1000 Hz
- ANL
- Antenna Trimmer
- BFO ±3 kHz
- Calibrator 100 kHz
- Hinged Top Cover
- 40:1 Tuning Ratio
- Vernier Scale 0-1000

Specifications:
Coverage ...... Ham. See Comments.
Selectivity ...... 1 - 5 kHz Variable.
Sensitivity ...... <3μV for 20 dB S/N
Audio Out ...... 2.5W 2.5 ohm 10% dis.
Modes .... AM/LSB/USB/CW
Image Reject .. >35 dB @30 MHz
IF .... 1620, 85 kHz
Antenna Input . 75 ohms Terminals

Circuit Complement:
6BA6 RF Amp, 6AJ8 Mixer, 6C4 Oscillator, 6AJ8 Freq. Changer, 6BA6 IF Amp, 6AT6 Demodulator AGC, 6AL5 NL (S-Meter diodes), 6AQ5 Output, 6BE6 CW/SSB Converter, 5Z4G Rectifier, 0D3 Stabilizer and 6AU6 Crystal Calibrator.

Accessories:
669/E S-Meter Oyster
774/E Oyster Tilt Blocks
688 Speaker Black
688/A Speaker Gray
688/B Speaker Polychromatic
774 Black Tilt Blocks
688/E Speaker Oyster Gray
774/B Polychrom. Tilt Blocks
732 Line Filter
669 S-Meter Black (shown below)

Comments:
Ranges: 1.8-2, 3.5-4, 7-7.3, 14-14.35, 21-21.5 and 28-30 MHz. The headphone jack is on the left panel. Light gray.

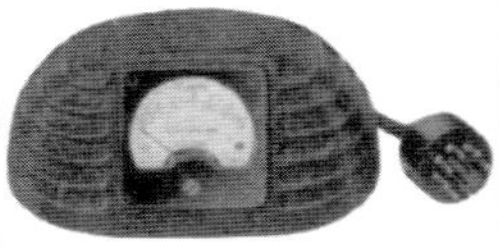

Variants: Earlier model **888** (1956) used nonstandard tubes.

Made In: England 1957-1962

Voltages: 110 and 200/250 VAC
40-60 Hz 80W

Readout: Analog

Physical: 16.75x8.75x10" 44 Lbs.
425x222x254mm 20 kg

Status: Active Manufacturer
Discontinued Model

Rarity: Extremely Scarce

Reviews: *Radio Bygones* #16

	New	Used
Price:	£110	£85-110
Rating:	⑥	⑥

910

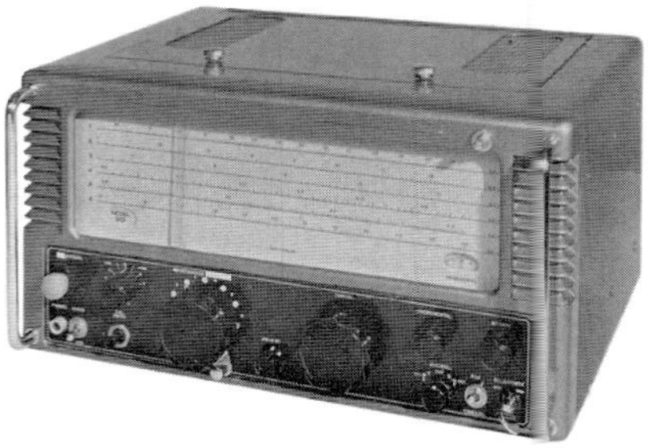

General Coverage Communications Receiver
Double Conversion Superheterodyne. 13 Tubes.

Features:
- ¼" Head. Jack
- RF Gain
- Standby
- Mute
- BFO ±3 kHz
- ANL
- Line Out
- Logging Scale
- Rack Handles
- AVC
- RIT ±50 kHz
- Vibrator Input
- Hinged Top Cover
- Line Out 600 Ohms
- 4 Fixed Crystal Pos.
- Calibrator 500 kHz

Specifications:
Coverage 375-525, 1500-30000kHz
Selectivity ...... Variable 5-1.5 kHz -6 dB
Sensitivity ...... <3.5 μV 20 dB S/N CW
IF .... 1350-1450, 85 kHz
Audio .... 2.5 ohms 3 W
Modes .... AM/CW/SSB
Stability .... <600 Hz/hr after warmup
Image Rej. ...... >32 dB 10-30 MHz
Image Rej. ...... >55 dB <10 MHz
Antenna Input . 75 ohms

Circuit Complement: 6AK5 RF Amp, 6BE6 Mixer, 6C4 2nd Local Osc., 6AK5 2nd Mixer, 6BA6 85 kHz IF Amp, 6AL5 NL/Meter Prot., 6AT6 AM Detector/AVC/RF Amp, 5763 AF Output, 6BA6 Calibrator, 6C4 Local Osc., 6BE6 CW Detector, 0A2 Voltage Stabilizer and 5Z4G HT Rectifier.

Comments: Ranges: .375-.525, 1.5-2.5, 2.5-4, 4-6.7, 6.7-11 and 18-30 MHz. Four crystal positions are provided, harmonics of fundamental crystal frequency permit up to 6 frequencies to be received from any one crystal. Gray oyster hammer finish metal cabinet. May be rack mounted with the addition of two steel plates. A rear panel socket enables the receiver to be operated from a 6 VDC vibrator unit.

Variants: This model is also referred to as the Eddystone **S.910**. It is even more commonly found as the **Marconi HR 101**.

Made In: England 1962

Voltages: 110 or 200-250 VAC
40-60 Hz

Readout: Analog

Physical: 16.75x8.75x15" 50 Lbs
425x222x381mm 22.7 kg

Status: Active Manufacturer
Discontinued Model

Rarity: Typically Not Available

	New	Used
Price:	①	⑤
Rating:	⑥	⑥

940

General Coverage Communications Receiver
Single Conversion Superheterodyne. 13 Tubes plus Semiconductors.

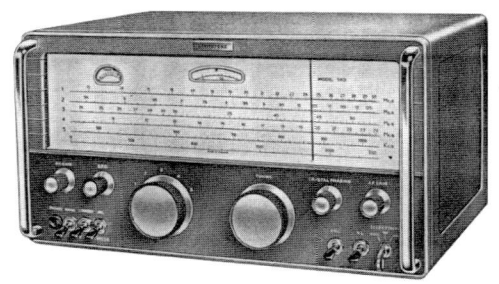

Features:
- ¼" Head. Jack
- S-Meter
- Rack Handles
- BFO ±3 kHz
- RF Gain
- ANL
- Standby
- Calibrator
- Mute
- Logging Scale
- Line Out
- 140:1 Tuning

Specifications:

Coverage ...... 480 - 30000 kHz
Selectivity ...... 10/4/.4 kHz -6 dB
Sensitivity ...... <3µV 15 dB S/N
Audio .... 2.5/600 ohms

Modes .... AM/CW/SSB
Image Rej. ...... 75dB @ 8, 40dB @ 20 MHz
IF .... 450 kHz
Antenna Input . 75 ohms

Circuit Complement:
ECC189 {6ES8} 1st RF Amp, 6BA6 2nd RF Amp, 6AJ8 Mixer Stage, 6C4 Local Osc., 6BA6 1st IF Amp, 6BA6 2nd IF Amp, 6AL5 Am Detector/AGC Rectifier, 6BE6 CW/SSB Detector, 12AU7 Audio Amp/Phase Splitter, (2) 6AM5 Push-pull Audio Output, GZ34 HT Rectifier, VR150 HT Stabilizer and 2E1 silicon diode NL Two RF stages.

Accessories: 688 Speaker   899F Speaker

Comments: One of Eddystone's last tube receivers. Ranges: .48-1.03, 1.03-2.4, 2.4-5.4, 5.4-12.7 and 12.7-30 MHz.

Variants: Model **940/RM** is a rack mount version. Model **940/1** has one fixed crystal position. Model **940/2** is configured for vehicular use. Model **940/3** produced for Her Majesty's Coast Guard. Eddystone's first solid state receiver was the **960** which had 19 semiconductors and physically resembled the 940. The 960 operated from 8 D cells or 12 VDC. Model **962** is similar to the 960, but with 8 fixed crystal positions.

Made In:	England 1962-1969
Voltages:	110/125 or 200/250 VAC 40/60 Hz 80W
Readout:	Analog
Physical:	16.75x8.75x15" 43 Lbs 425x222x381mm 19.5 kg
Status:	Active Manufacturer Discontinued Model
Rarity:	Very Scarce

	New	Used
Price:	£125	£120-190
Rating:	⑥	⑥

1001

General Coverage Communications Receiver
Single Conversion Superheterodyne. Solid State.

Features:
- ¼" Head. Jack
- S/Battery Meter
- Dual Speakers
- 10 Crystal Positions
- RF Gain
- Rack Handles
- Mute Line
- Fine Tuning
- Dial Light
- Standby
- BFO
- AGC MAN/LNG/SHT
- Logging Scale
- Flywheel Drive
- Dial Lamp Switch

Specifications:

Coverage ...... 550 - 30000 kHz
Selectivity ...... 10/4 kHz -6dB.
Sensitivity ...... <5 µV 1.5-30 MHz AM
Audio Out ...... 0.6 W 5% dist.
Environment .. 0° to 50° C.
I.F. .... 455 kHz
Antenna Input 75 ohm / 400 ohm

Modes .... AM/SSB/CW
Stability .... 1 part in 10⁴
IF Rejection .... >85 dB 1.5-30 MHz
IF Rejection .... >70 dB .55-1.5 MHz
Image Rej .... >50 dB at 2 MHz
Image Rej .... >35 dB at 18 MHz

Accessories:
Rack Mounting Kit

Comments:
Ranges: Band 1: 18-30, Band 2: 8.5-18, Band 3: 3.6-8.5, Band 4: 1.5-3.8, Band 5: .55-1.5 MHz. Crystals can be fitted without removing the outer covers. Internal batteries automatically take over in the event of a power loss at the mains.

Variants: Model **1000** (1972) is the same as the 1001, less the 10 crystal positions 1971-1974. The model 1000 uses 2 integrated circuits, 20 transistors and 30 diodes. Model **1001/1** is a rack version of the 1001.

Made In:	England 1972-1974
Voltages:	100/130 or 200/260 VAC 40-60 Hz and 12VDC and NiCad batteries
Readout:	Analog
Physical:	13.2x5.375x9.5" 18 Lbs. 335x137x242mm 8.2 kg
Status:	Active Manufacturer Discontinued Model
Rarity:	Very Scarce
Reviews:	*Ham Radio* January 1974

	New	Used
Price:	£200-300	£160
Rating:	⑥	⑥

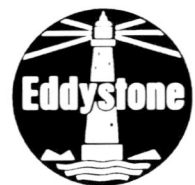

1002

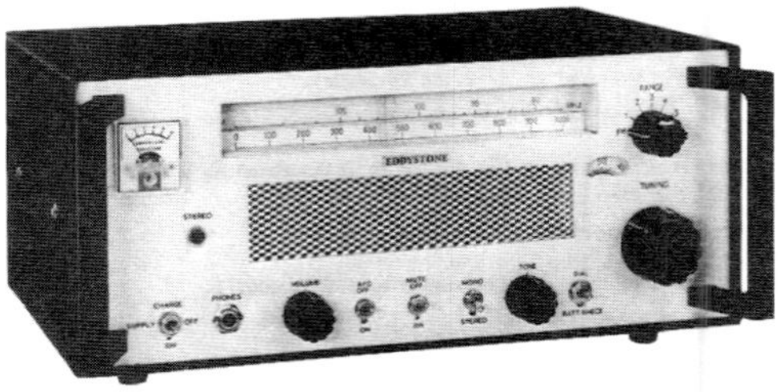

General Coverage Broadcast Receiver
Single Conversion Superheterodyne. Solid State.

Features:
- ¼" Head. Jack
- Tone
- Dial Light
- Telescopic Whip
- Dial Lamp Switch
- S/Battery Meter
- Rack Handles
- AFC (2 Pos.)
- Line Out
- Dual Speakers
- Mute Line
- Standby
- Flywheel reduction drive
- Stereo Lamp
- Fine Tuning
- Logging Scale

Specifications:
Coverage ...... 150 - 30000 kHz[1] +FM
Selectivity ...... 1 kHz -6dB. [250 FM]
Sensitivity ...... <5 µV 1.5-30 MHz AM
Audio Out ...... 0.6 W 5% dist.
Environment .. 0° to 50° C.
I.F. .... 455 kHz
Antenna Input 75 ohm / 400 ohm
Modes .... AM/SSB/CW [FM-W]
Stability .... 1part in 10^4
IF Rejection .... >85 dB 1.5-30 MHz
IF Rejection .... >60 dB .55-1.5 MHz
Image Rej .... >50 dB at 2 MHz
Image Rej .... >35 dB at 18 MHz

Comments:
Band ranges: Band 1: 18-30, Band 2: 8.5-18, Band 3: 3.6-8.5, Band 4: 1.5-3.8, Band 5: .55-1.5, Band 6: .150-.350 MHz. [1]Gap from 350-550 kHz. Internal batteries automatically take over in the event of a power loss at the mains. A telescopic rod aerial for MW and FM reception is mounted at the rear of the case. FM band reception is stereo.

Variants:
Model **1002/1** is the same but without Nicad batteries and telescopic whip antenna.

Made In: England 1972-1977

Voltages: 100/130 or 200/260 VAC
40-60 Hz or 12VDC
or NiCad batteries

Readout: Analog

Physical: 13.2x5.375x9.5" 18 Lbs.
335x137x242mm 8.2 kg

Status: Active Manufacturer
Discontinued Model

Rarity: Very Scarce

	New	Used
Price:	£250	£180
Rating:	[6]	[6]

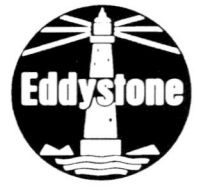

1004

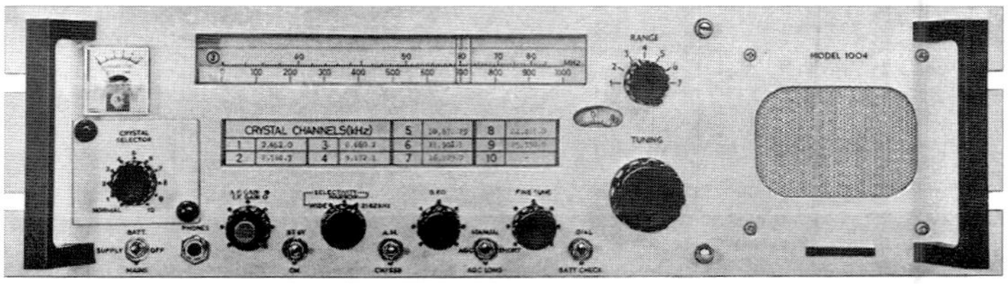

General Coverage Communications Receiver
Single Conversion Superheterodyne. Solid State.

Features:
- ¼" Head. Jack
- RF Gain
- Dial Light
- Line Out
- S/Battery Meter
- Rack Handles
- Standby
- Logging Scale
- Speaker
- Mute Line
- BFO ±3 kHz
- 2182 kHz Fixed
- 10 Crystal Positions
- Fine Tuning
- AGC MAN/LNG/SHT
- Flywheel Reduction

Specifications:
Coverage ...... See Comments.
Selectivity ...... 8/3 kHz -6dB.
Sensitivity ...... <5 µV 1.6-30 MHz AM
Audio Out ...... 0.6 W 5% dist.
Environment .. -15° to +55° C.
I.F. .... 720 kHz
Antenna Input 50/75 ohm
Modes .... AM/SSB/CW
Stability .... 1 part in 10^4/°C
IF Rejection .... >90 dB 12-30 MHz
IF Rejection .... >75 dB .55-12 MHz
Image Rej .... >60 dB at 1.6 MHz
Image Rej .... >40 dB at 18 MHz

Accessories: Cabinet 3 part in 10^5/°C Osc.

Comments: Band ranges: Band 1: 18-30, Band 2: 8.5-18, Band 3: 3.6-8.5, Band 4: 2.6-3.8, Band 5: 1.6-2.65, Band 6: .380-535, Band 7: .15-.385 MHz. Note that there is no MW reception. Crystals can be fitted without removing the outer covers. Fixed 2182 kHz reception for monitoring the international distress channel is standard. Designed for 19" rack mounting. Cabinet is optional. Internal batteries automatically take over in the event of a power loss at the mains.

Variants: This model was also sold as the **Marconi Sentinel** and the **Hagenuk E92.** It was also produced for Redifon and ITT.

Made In: England 1978-1980

Voltages: 100/130 or 200/260 VAC
40-60 Hz or 12 VDC
or NiCad batteries

Readout: Analog

Physical: 19x5.25x10" 17 Lbs.
488x133x250mm 8.2 kg

Status: Active Manufacturer
Discontinued Model

Rarity: Very Scarce

	New	Used
Price:	£300	[5]
Rating:	[6]	[6]

1570

General Coverage Communications Receiver
Single Conversion Superheterodyne. Solid State.

Features:
- ¼" Head. Jack
- S/Battery Meter
- Speaker
- Tilt bar
- RF Gain
- Rack Handles
- Mute Line
- Fine Tuning
- Dial Light
- Standby
- Tone
- BFO ± 3 kHz
- Record Jack
- Audio In Jack

Specifications:
Coverage ...... 150 - 30000 kHz +FM
Selectivity ...... 10/4 kHz -6dB.[250 FM]
Sensitivity ......<2 µV -12 dB S/N CW
Audio Out ......2.5 W 4 ohm 5% dist.
BFO Range ...±3 kHz.
I.F. ....455 kHz [FM 10.7 MHz]

Modes .... AM
Environment ... 0 to 40° C. 95% Humid.
IF Rejection .... >70 dB 2 MHz
Image Rej .... >50 dB at 2 MHz
Image Rej .... >25 dB at 22 MHz

Comments:
Coverage gap from 350-550 kHz. FM display resolution is 10 kHz. The Tone and BFO share the same knob. Ranges: .15-.35, .55-1.5, 1.5-3.5, 3.5-8.5, 8.5-18 and 14-30 MHz plus 88-108 MHz FM.

Variants:
Model **1570/3** (1930) has 100 kHz display resolution. Model **1570/2** has 1 kHz resolution, except 100 kHz on the FM broadcast band.

Made In:	England	1979-1981
Voltages:	100-230 or 200-250 VAC 40-60 Hz or 12 VDC or optional NiCad battery	
Readout:	**00001.** Digital LED	
Physical:	16x5x13" 20 Lbs 410x133x330mm 9 kg	
Status:	Active Manufacturer Discontinued Model	
Rarity:	Very Scarce	

	New	Used
Price:	①	£140
Rating:	⑥	⑥

1590

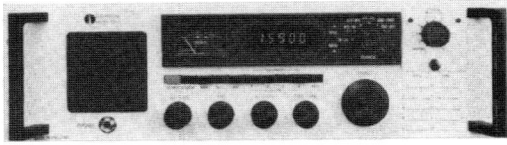

General Coverage Communications Receiver
Single Conversion Superheterodyne. Solid State.

Features:
- ¼" Head. Jack
- S/Battery Meter
- Speaker
- 10 Crystal Positions
- IF Gain
- Rack Handles
- Mute Line
- Fine Tuning
- AGC (2 pos)
- BFO
- IF Output

Specifications:
Coverage ...... 150 - 30000 kHz +FM
Selectivity ...... 10/4/2.5/.3 kHz -6dB
Sensitivity ......<1 µV -12 dB S/N CW
Audio Out ......2.5 W 4 ohm 5% dist.
BFO Range ... ±3 kHz.
Environment ..0 to 40° C. 95% Humid.

Modes .... AM/LSB/USB
Stability .... 1 part in 10⁴ /°C.
I.F. .... 455 kHz
IF Rejection .... >70 dB 2 MHz
Image Rej .... >70 dB at 2 MHz
Image Rej .... >25 dB at 22 MHz

Accessories:
Cabinet

Comments:
Coverage gap from 350-550 kHz. Designed for 19" rack mounting. The ten crystals can be changed from the front panel.

Made In:	England	1980-1983
Voltages:	100/120 200/250 VAC 40-60 Hz or 12 VDC	
Readout:	**00001.** Digital LED	
Physical:	Rack Version: 19x5.25x13" 22 Lbs 483x133x330mm 10 kg	
Status:	Active Manufacturer Discontinued Model	
Rarity:	Typically Unavailable	

	New	Used
Price:	①	£150
Rating:	⑥	⑥

1650/1

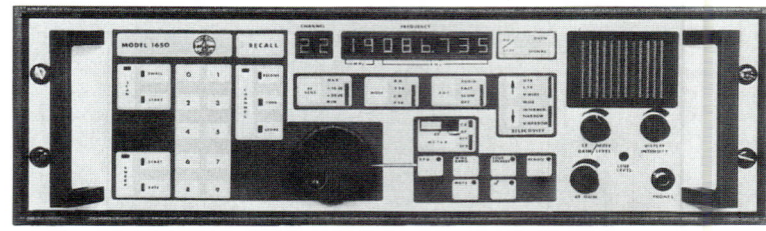

General Coverage Communications Receiver
Double Conversion Superheterodyne. Solid State.

Features:
- ¼" Head. Jack
- RF Gain
- Mute Line
- Line Out Jack
- Scan & Sweep
- Speaker Switch
- BFO ±3.9 kHz
- Speaker
- Ext. Spkr Jack
- IF Out 1.4 MHz
- Dimmer
- Standby
- 99 Memories
- RS232C/422A
- Rack Handles
- Self Diagnostics
- Keypad
- Atten. -10/20/40 dB
- AGC OFF/FST/SLO/AUD
- Ant. Switch (4 Pos).
- Membrane Front
- Ext Ref. Input

Specifications:

Coverage ...... 10 - 30000 kHz	Modes .... AM/USB/CW/MCW/FSK
Selectivity ...... 14/8/3/2.4/1 kHz -6dB.	Stability .... 10 Hz over entire range
Sensitivity ...... <1µV 30 dB S/N SSB	IF Rejection .... >90 dB
Image Rej ...... >90 dB	Environment ... -15 to 55° C 95% humid
Audio Out ...... 1 W 4 ohms	I.F. .... 46.205 & 1.4 MHz
Line Out .... 200 mw 600 ohms	I.F. Output ...... 20 mV into 50 ohms

Accessories:
Cabinet   1161 Panoramic Display

Comments: Memories store: Frequency, Mode, BFO, AGC, RF setting and remote antenna or preamp selection. Front panel utilizes membrane switches for most functions. LSB reception via BFO only.

Variants: Model **1650/2** also includes LSB and a .4 kHz filter. Model **1650/3** also includes LSB and ISB and looks like the 1650/10. Model **1650/6** was designed solely for operation via a computer.

Made In: England   1982–1998

Voltages: 100/150 & 220/260 VAC
40-60 Hz or 24VDC

Readout: `00000.005` Digital LED

Physical: Cabinet Version:
19.75x6.5x20.8" 51 Lbs.
502x164x528mm 23 kg
Rack Version:
19x5.25x20.8" 42 Lbs.
483x133x528mm 19 kg

Status: Active Manufacturer
Active Model

Rarity: Typically Unavailable

	New	Used
Price:	①	⑤
Rating:	⑥	⑥

1650/9

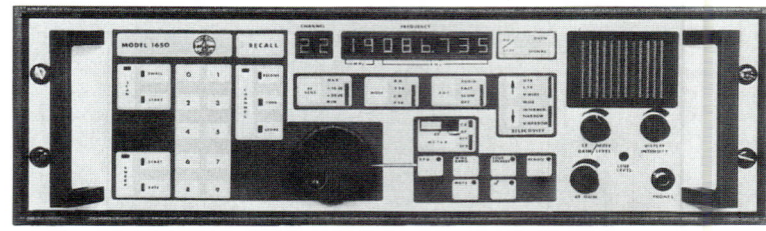

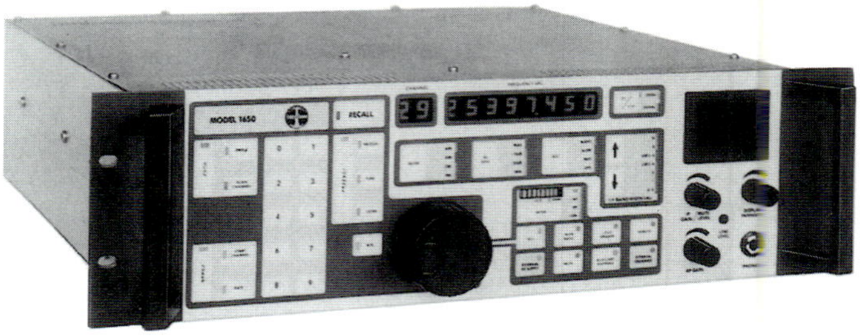

General Coverage Communications Receiver
Double Conversion Superheterodyne. Solid State.

Features:
- ¼" Head. Jack
- RF Gain
- Mute Line
- Line Out Jack
- Scan & Sweep
- BFO ±3.9 kHz
- Speaker
- Ext. Spkr Jack
- IF Out 1.4 MHz
- Dimmer
- 99 Memories
- RS232C/422A
- Rack Handles
- Self Diagnostics
- Keypad
- Atten. -10/20/40 dB
- AGC OFF/FST/SLO/AUD
- Ant. Switch (4 pos).
- Membrane front.
- Ext Ref. Input

Specifications:

Coverage ...... 10 - 30000 kHz	Modes .... AM/USB/CW/MCW
Selectivity ...... 16/8/3/2.4/1/.3 kHz -6dB.	Stability .... 10 Hz over entire range
Sensitivity ...... <1µV 12dB S/N SSB	IF Rejection .... >100 dB
Image Rej ...... >100 dB	Environment ... -15 to +55° C 95% humid
Audio Out ...... 1 W 4/8 ohms	I.F. .... 46.205 & 1.4 MHz
Line Out .... 200 mw 600 ohms	I.F. Output ...... 20 mV into 50 ohms
Antenna Input 50 ohms	

Accessories:
Cabinet   1529 FSK Unit   1855 Diversity Switch

Comments: Memories store: Freq., Mode, BFO, AGC, RF setting, and remote antenna or preamp selection. Options: **/A** with motor driven RF preselector, **/S** with external standard input 1/5 MHz, **/N** with NBFM reception, **/K** with FSK reception. All options are available together (apart from NBFM and FSK).

Variants: Model **1650/9H** features 1.6 MHz High pass filter. Model **1650/9AH** features 14 range sub-octave preselector.

Made In: England   1992–1998

Voltages: 100/150 & 200/260 VAC
40-60 Hz or 24VDC

Readout: `00000.005` Digital LED

Physical: Cabinet Version:
19.75x6.5x20.8" 51 Lbs.
502x164x528mm 23 kg
Rack Version:
19x5.25x20.8' 42 Lbs.
483x133x528mm 19 kg

Status: Active Manufacturer
Active Model

Rarity: Typically Unavailable

	New	Used
Price:	①	⑤
Rating:	⑥	⑥

1650/10

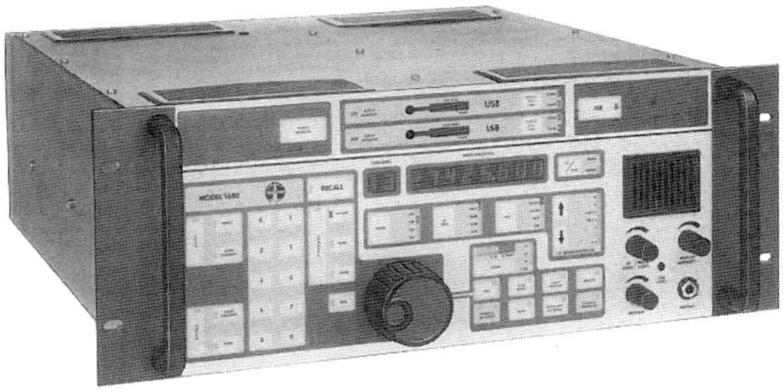

General Coverage Communications Receiver
Double Conversion Superheterodyne. Solid State.

Features:

• ¼" Head. Jack	• I.S.B.	• 99 Memories	• Atten. -10/20/40dB
• RF Gain	• Speaker	• RS232C/422A	• AGC OFF/FST/SLO/AUD
• Mute Line	• Ext. Spkr Jack	• Rack Handles	• Ant. Switch (4 pos).
• Line Out Jack	• IF Out 1.4 MHz	• Self Diagnostics	• Membrane front.
• Scan & Sweep	• Dimmer	• Keypad	• Ext Ref. Input
• Speaker Switch			

Specifications:

Coverage	10 - 30000 kHz	Modes	AM/LSB/USB/CW/ISB
Selectivity	16/8/3/2.4/1/.3 kHz -6dB.	Stability	±1 PPM
Sensitivity	-113 dBm 1 kHz CW 118dB	IF Rejection	100 dB
I.F.	46.205 & 1.4 MHz	Environment	-15 to 55° C 95% humid
Audio Out	1 W 4 ohms	I.F. Output	20 mV into 50 ohms
Line Out	200 mw 600 ohms		

Accessories:
Cabinet   1161 Panoramic Display   1775 Remote Control Unit

Comments: Memories store: Freq., Mode, BFO, AGC, RF setting and remote antenna or preamp selection. Options: **/A** with motor driven multirange RF preselector, **/S** with external standard input 1 or 5 MHz, **/N** with NBFM reception, **/E** with FSK reception **/F** with panoramic output. All options are available together except the NBFM and FSK options.

Variants: Model **1650/8** is the VLF version covering 10 to 160 kHz $8650

Made In: England 1989-1998

Voltages: 100/150 & 220/260 VAC 40-60 Hz or 24 VDC

Readout: 00000.005 Digital LED

Physical: Cabinet Version
19.75x8.25x21.4" 61 Lbs
502x209x543mm 28 kg
Rack Version
19x8.25x21.4" 52 Lbs
483x209x555mm 24 kg

Status: Active Manufacturer
Active Model

Rarity: Typically Unavailable

	New	Used
Price:	$11000	⑤
Rating:	⑥	⑥

1680/3

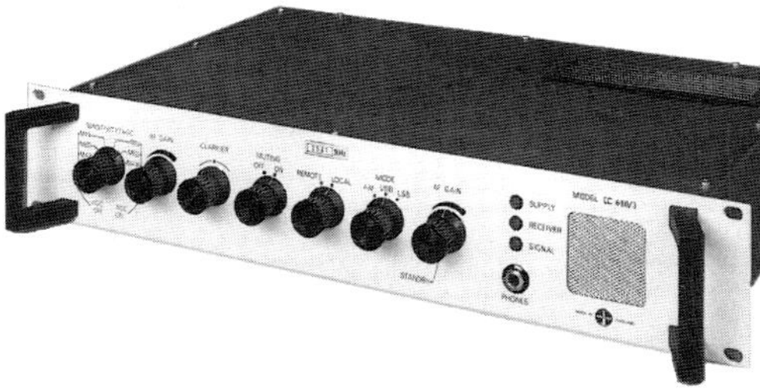

Fixed Channel Communications Receiver
Single Conversion Superheterodyne. Solid State.

Features:

• ¼" Head. Jack	• BFO ±3 kHz	• RF Gain	• Speaker
• Mute Line	• AGC (2 Pos.)	• Atten. (3 Pos.)	• Remote Operation
• Rack Handles			

Specifications:

Coverage	1 Ch. 1600- 30000 kHz	Modes	AM/USB
Selectivity	7/2.7 kHz -6dB.	Stability	20 Hz 0° - 40° C.
Sensitivity	1μV for 12 dB SINAD	IF Rejection	> 90 dB
Image Rej	70 dB <20 MHz	I.F.	1.4 MHz
Audio Out	2 W 8 ohms	Environment	-10 to 55° C 95% humid
Line Out	600 ohms		

Accessories:
LSB Mode   CW Mode

Comments:
Designed for rack mounting and commercial applications.

Variants:
Model **1680/4** is the same as above but with 2 channels. Model **1680/1** is fixed at 500 kHz, Model **1680/2** covers 7 channels in the range of 400 to 535 kHz. Variants have slightly different performance specifications.

Made In: England 1985-1989

Voltages: 100/130 & 220/250 VAC 40-60 Hz 25W or 24 VDC

Readout: No Display

Physical: 19x3.5x11" 14 Lbs
483x88x282mm 6.5 kg

Status: Active Manufacturer
Discontinued Model

Rarity: Typically Unavailable

	New	Used
Price:	①	⑤
Rating:	⑥	⑥

1830/1

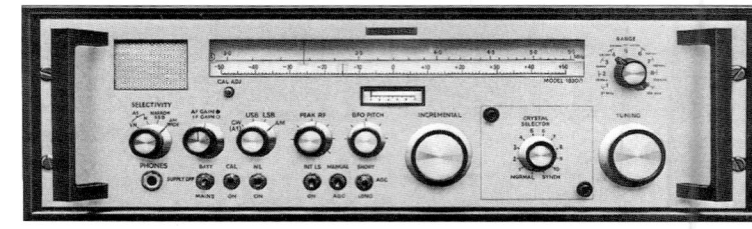

General Coverage Communications Receiver
Double Conversion Superheterodyne. Solid State.

Features:
- ¼" Head. Jack
- S-Meter
- Speaker
- 10 Crystal Positions
- IF Gain
- Rack Handles
- Mute Line
- Line Out
- Dial Light
- Standby
- BFO ±5 kHz
- AGC MAN/SLO/FST
- IF Out 100 kHz
- NL

Specifications:
Coverage ...... 120 - 31000 kHz
Selectivity ...... 8/3/1.3 kHz -6dB.
Sensitivity ...... <3 µV 1.6-30 MHz AM
Audio Out ...... 0.5 W 3 ohms 5% dist.
Environment .. -15° to 55° C.
I.F. .... 1300-1400, 100 kHz
Antenna Input BNC 75 ohm unbal.

Modes .... AM/LSB/USB/CW
Stability .... 1part in 10^4/° C.
IF Rejection .... >60 dB .12-2.9 MHz
IF Rejection .... >85 dB 2.9-31 MHz
Image Rej .... >70 dB 1.5-18 MHz
Image Rej .... >50 dB 18-31 MHz

Accessories:
961A Panoramic Display
Comments: The 1830/1 replaced the 830/7 tube receiver.
Variants: Model **1830/2** and **1830/5** have 50 crystal positions and a deeper front panel. Model 1830/5 has a special SSB filter. Model **1830/7** has a special SSB filter and 50 crystal positions. Model **1830/3** has a gap in coverage from 535-920 kHz. Model **1830/4** has a revised circuit. Model **1830/6** has 535-920 kHz gap and special SSB filter. Model **1830/8** is the same as 1830/6 but with 50 crystal positions. The 1830/1 was also sold as the **S.A.I.T. MR 1431**.

Made In: England   1971-1974

Voltages: 100/130 or 200/260 VAC
40-60 Hz or 12 VDC

Readout: ⊥⊥⊥⊥ ▮ ⊥⊥ Analog

Physical: Cabinet Version:
19.75x6.5x14.8" 40 Lbs
502x164x376mm 18.1 kg
Rack Version:
19x5.25x13.125" 40 Lbs.
483x133x344mm 8.2 kg

Status: Active Manufacturer
Discontinued Model

Rarity: Typically Unavailable

Reviews: *WRTH* 1976

	New	Used
Price:	①	£200-350
Rating:	⑥	⑥

1837/1

General Coverage Communications Receiver
Double Conversion Superheterodyne. Solid State.

Features:
- ¼" Head. Jack
- S-Meter
- Speaker
- AGC OFF/LNG/SHT
- IF Gain
- Rack Handles
- Mute Line
- Fine Tuning ±10 kHz.
- Atten. -20/40 dB
- BFO
- IF Output
- High Stability Lock
- Dimmer
- Line Out
- Speaker Out

Specifications:
Coverage ...... 100 - 31000 kHz
Selectivity ...... 8/3/2.4/1.3/.4 kHz -6dB
Sensitivity ...... <1 µV -15 dB S/N CW
Audio Out ...... 0.5 W 3 ohm 5% dist.
I.F. .... 1340-1360, 100 kHz
Radiation .... <400pW
Environment .. -15 to +50° C 95% humid

Modes .... AM/CW/MCW/USB
Stability .... 1 part in 10^4/°C
IF Rejection .... 60 dB .1-2.9 MHz
IF Rejection .... 85 dB 2.9-30 MHz
Image Rej .... 50-40 dB (.5-30 MHz)
Antenna Input . 50/75 ohm unbal.

Accessories:
Cabinet   1061 Panoramic Unit   12/24 VDC PS
Comments:
Intended for maritime applications. Designed for 19" rack mounting. Single conversion 840-1600 kHz. The Lock button puts the receiver into a high stability mode and locks the received frequency.
Variants:
Model **1837/1B** has fine tuning ±5 kHz. The **Marconi "Pacific"** is a version of the 1837/15.

Made In: England   1978-1980

Voltages: 100/130 or 200/260 VAC
40-60 Hz

Readout: `00000.1` Digital LED

Physical: Cabinet Version:
19.75x6.5x14.8" 48 Lbs
502x164x376mm 21.8 kg
Rack Version:
19x5.25x13.125" 37 Lbs.
483x133x334mm 16.8 kg

Status: Active Manufacturer
Discontinued Model

Rarity: Typically Unavailable

	New	Used
Price:	①	£575-625
Rating:	⑥	⑥

1837/2

General Coverage Communications Receiver
Double Conversion Superheterodyne. Solid State.

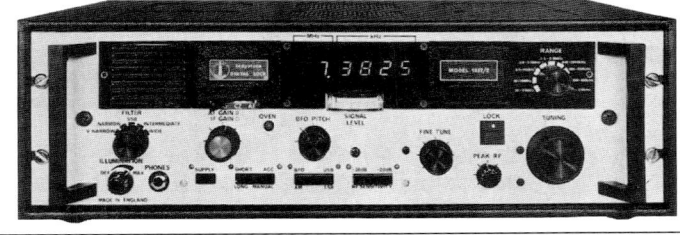

Features:
- ¼" Head. Jack
- IF Gain
- Atten. -20/40 dB
- Dimmer
- S/FSK-Meter
- Rack Handles
- BFO
- Line Out 600 ohm
- Speaker
- Mute Line
- IF Output
- Speaker Out
- AGC OFF/LNG/SHT
- Fine Tuning ±10 kHz.
- High Stability Lock

Specifications:
Coverage ......100 - 31000 kHz
Selectivity ......8/3/2.4/1.3/.4 kHz -6dB
Sensitivity ......<.5 µV -15 dB S/N CW
Audio Out ......0.5 W 3 ohm 5% dist.
I.F. ....1340-1360, 100 kHz
Radiation ....<400pW
Environment ..-15 to +50° C 95% humid

Modes .... AM/CW/MCW/LSB/USB
Stability .... 1 part in 10^4 per °C
IF Rejection .... 60 dB .1-2.9 MHz
IF Rejection .... 85 dB 2.9-30 MHz
Image Rej .... 50-40 dB (.5-30 MHz)
Antenna Input . 75 ohm unbal

Accessories:
Cabinet   EP1061 Panoramic Display   12/24 VDC PS
FSK Option

Comments:
Designed for 19" rack mounting. Single conversion 840-1600 kHz. The Lock button puts the receiver into a high stability mode and locks the received frequency. Unlike the 1837/1, the 1837/2 has provisions for the separate selection of LSB and USB rather than just USB.

Made In:	England 1977-1983

Voltages:	100/130 or 200/260 VAC 40-60 Hz

Readout:	**00000.1** Digital LED

Physical:	Cabinet Version: 19.75x6.5x14.8" 48 Lbs 502x164x376mm 21.8 kg Rack Version: 19x5.25x13.125" 37 Lbs. 483x133x334mm 16.8 kg

Status:	Active Manufacturer Discontinued Model
Rarity:	Very Scarce

	New	Used
Price:	①	£400
Rating:	⑥	⑥

1838/3

General Coverage Communications Receiver
Double Conversion Superheterodyne. Solid State.

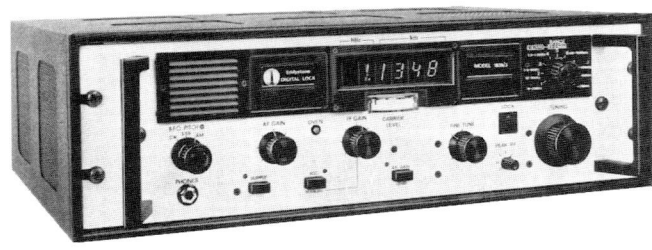

Features:
- ¼" Head. Jack
- IF Gain
- AGC OFF/ON
- Line Out
- S/FSK-Meter
- Rack Handles
- BFO
- Speaker
- Mute Line
- IF Output
- Attenuator -20 dB
- Fine Tuning ±10 kHz.
- High Stability Lock

Specifications:
Coverage ......100 - 30000 kHz
Selectivity ......8/3/2.4/1.3/.4 kHz -6dB
Sensitivity ......<.5 µV -15 dB S/N CW
Audio Out ......0.5 W 3 ohm 5% dist.
I.F. ....1340-1360, 100 kHz
Radiation ....<400pW
Environment ..-15 to 50° C 95% humid

Modes .... AM/CW/SSB
Stability .... 1 part in 10^4/°C
IF Rejection .... 60 dB .1-2.9 MHz
IF Rejection .... 85 dB 2.9-30 MHz
Image Rej .... 50-70 dB (.5-30 MHz)
Antenna Input . 75 ohm unbal.

Accessories:
Cabinet (shown)   EP1061 Panoramic Display   12/24 VDC PS

Comments:
Designed for 19" rack mounting. Single conversion 840-1600 kHz. The Lock button puts the receiver into a high stability mode and locks the received frequency.

Variants:
Model **1838/2** covers only .6-30 MHz. Model **1838/1** covers only 1.6-30 MHz and offers CW as an option.

Made In:	England 1976-1977

Voltages:	100/130 or 200/260 VAC 40-60 Hz

Readout:	**00000.1** Digital LED

Physical:	Cabinet Version: 19.75x7.5x14.8" 48 Lbs 502x191x376mm 21.8 kg Rack Version: 19x5.25x13.125" 37 Lbs. 483x159x334mm 16.8 kg

Status:	Active Manufacturer Discontinued Model
Rarity:	Very Scarce

	New	Used
Price:	①	£400
Rating:	⑥	⑥

6100/1

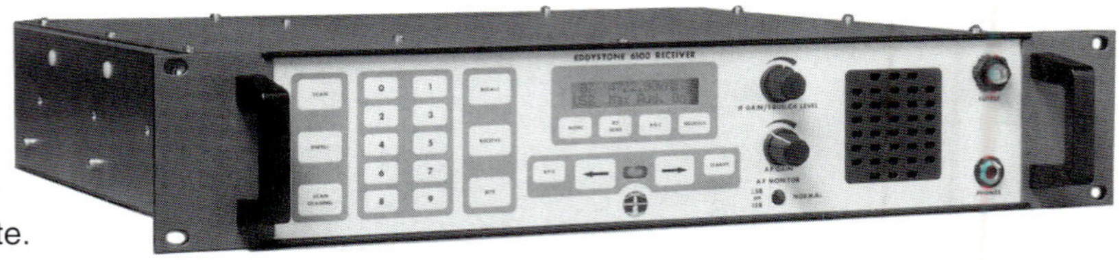

General Coverage Communications Receiver
Double Conversion Superheterodyne. Solid State.

Features:
- ¼" Head. Jack
- RF Gain
- Mute Line
- Line Out Jack
- AGC (4 Pos.)
- Speaker
- Ext. Spkr Jack
- Squelch
- 99 Memories
- Keypad
- Rack Handles
- Membrane front.
- Atten. -10 dB
- BFO ±2.4 kHz
- B.I.T.E.
- Scan & Sweep

Specifications:
Coverage ...... 1600 - 30000 kHz
Selectivity ...... 6/2.4/_/.3-2.7 kHz -6dB.
Sensitivity ...... <.4μV 10dB S/N CW
Image Rej ...... >100 dB 1st
Audio Out ...... 1 W 4/8 ohms
Line Out .... 200 mw 600 ohms
Modes .... AM/USB/LSB/CW
Stability .... 10 Hz over entire range
IF Rejection .... >100 dB
Environment ... -15 to +55° C 95% humid
I.F. .... 45 & 1.4 MHz
Antenna Input . BNC 50 ohm

Accessories (Options):
/A adds 10-1600 kHz coverage and preselector, **/B** adds external preselector drive option, **/F** adds 10-1600 kHz coverage, **/K** adds internal FSK, **/S** adds Ext. Standard Input, **/X** adds high stability. 6800 Remote Control Unit

Comments:
Memories store: Frequency, Mode, Bandwidth, BFO, AGC and Clarifier setting. Front panel utilizes membrane switches for most functions. Model 6100/1 has parallel remote control only.

Variants: Model **6100/2** has 50 memories with full serial remote control. Model **6100/3** has ISB, full serial remote control and 50 memories.

Made In: England 1991-1998

Voltages: 100/150 & 200/260 VAC
40-60 Hz or 24 VDC

Readout: 00000.01 Digital LCD

Physical: 19x3x17" 26.5 Lbs.
483x88x440mm 12 kg

Status: Active Manufacturer
Active Model

Rarity: Typically Unavailable

	New	Used
Price:	$6500	⑤
Rating:	⑥	⑤

6200/1

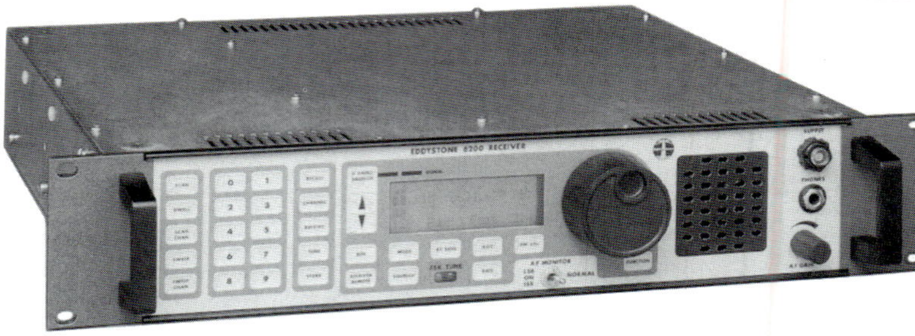

General Coverage Communications Receiver
Double Conversion Superheterodyne. Solid State

Features:
- ¼" Head. Jack
- RF Gain
- Mute Line
- Line Out Jack
- Dimmer
- AGC (4 Pos.)
- Speaker
- Ext. Spkr Jack
- Squelch
- Keypad
- 99 Memories
- RS232C/422A
- Rack Handles
- Membrane Front
- Atten. -10 dB
- BFO ±2.4 kHz
- B.I.T.E.
- Scan & Sweep

Specifications:
Coverage ...... 1600 - 30000 kHz
Selectivity ...... 14/8/3/2.4/1/.4 kHz -6dB.
Sensitivity ...... <1μV 12dB S/N SSB
Image Rej ...... >100 dB 1st
Audio Out ...... 1 W 4/8 ohms
Line Out .... 200 mw 600 ohms
Antenna Input BNC 50 ohm
Modes .... AM/USB/CW
Stability .... 10 Hz over entire range
IF Rejection .... >100 dB
Environment ... -15 to +55° C 95% humid
I.F. .... 45 & 1.4 MHz
I.F. Output ...... 20 mV into 50 ohms

Accessories (Options):
/A adds 100-1600 kHz coverage, **/B** adds 8 bit data out for antenna control, **/C** adds filters, **/F** adds wideband input, **/I** adds IF Output, **/K** adds internal FSK, **/P** adds panoramic output jack, **/S** adds Ext. Standard Input, **/T** adds clock, **/X** adds high stability. 6850 Remote Control Unit.

Comments: Memories store: Frequency, Mode, Bandwidth, BFO, AGC and RF setting. Front panel utilizes membrane switches for most functions.

Variants: Model **6200/2** also includes LSB. Model **6200/3** includes LSB and ISB.

Made In: England 1994-1998

Voltages: 100/130 & 200/260 VAC
40-60 Hz or 19-32 VDC

Readout: 00000.01 Digital LCD

Physical: 19x3.5x17.3" 26 Lbs.
483x88x440mm 12 kg

Status: Active Manufacturer
Active Model

Rarity: Typically Unavailable

	New	Used
Price:	①	⑤
Rating:	⑥	⑥

EA12

Amateur Band Communications Receiver
Double Conversion Superheterodyne. 13 Tubes.

Features:
- ¼" Head. Jack
- S-Meter
- Calibrator
- AGC (2 Pos.)
- RF Gain
- IF Gain
- Slot Filter
- Rack Handles
- Mute Line
- Twin NL
- FSK Output
- 140:1 Tune Ratio
- Standby
- BFO ±3.5 kHz
- Panoramic Output

Specifications:
Coverage ......Ham. See Comments.
Selectivity ......Variable to .5 kHz -6dB.
IF ....1.1-1.7 MHz, 100 kHz.
Sensitivity ......<.5µV 20 dB S/N CW
Modes .... AM/SSB/CW/MCW
Stability .... 20 Hz short term
Antenna Input . 75 ohm

Circuit Complement:
6ES8 RF Amp, 6AJ8 Mixer/Osc. Amp, 6C4 1st Osc., 6AJ8 2nd Mixer/Osc. Amp, 6C4 2nd Osc., 6BA6 1st IF Amp, 6BA6 2nd IF Amp, 6AL5 Detector/AVC, 12AX7 Cathode Follow, 6BE6 CW/SSB Detector, 6AQ5 AF Output, 0A2 HT Stabilizer and 6AU6 Calibrator.

Accessories: FSK Adapter Panoramic Display

Comments:
Ranges: 1.8-2.4, 3.4-4, 6.9-7.5, 13.9-14.5, 20.9-21.5, 27.9-28.5, 28.4-29, 28.9-29.5 and 29.4-30 MHz. Output jacks for FSK and panoramic adapter. Slow-motion drive (140 to 1 ratio). This is believed to be the last valve (tube) receiver that Eddystone manufactured. It replaced the model 888A and did provide improved performance.

Made In:	England	1964-1969

Voltages: 110 and 200/250 VAC 40-60 Hz

Readout: ⊥⊥⊥⊥⊥⊥ Analog Lin.

Status: Active Manufacturer Discontinued Model

Rarity: Extremely Scarce

	New	Used
Price:	£185	£190-350
Rating:	⑥	⑥

EB-35

General Coverage Broadcast Receiver
Single Conversion Superheterodyne. Solid State.

Features:
- ¼" Head. Jack
- Speaker 5"
- 110:1 Tune Ratio
- Aux. Tuning Scale
- Tone
- Dial Lamp
- Dial Lamp Switch
- Rack Handles
- Flywheel Tuning
- Record Jack
- Phono Input Jack
- 110:1 Reduction Tune

Specifications:
Coverage ......550-22000 kHz +LW/FM
Selectivity ......5 kHz -6dB [250 FM]
Sensitivity ......<5 µV -15 dB S/N 3.5-22MHz
IF ....465 kHz [10.7 MHz FM]
Audio Out ......0.75 W 10% distortion
Modes .... AM/[FM-W]
Image Reject .. 50 dB @2 MHz
Image Reject .. 15 dB @18 MHz
Antenna Input . Terminals (Jack for FM)

Circuit Complement: 12 Transistors and 6 diodes.

Accessories: 924 AC Power Supply 945 12/24VDC Power Supply

Comments: Ranges: 8.5-22, 3.5-8.5, 1.5-3.5, .55-1.5, .15-.35 MHz plus FM broadcast coverage from 88-108 MHz. FM mode reception is only on the FM broadcast band. Longwave coverage is 150-350 kHz.

Variants: Model **EB-35 Mark I** has an extra IF amp stage. Model **EB-35 Mark II** (1969) has a slightly different case and front-end diode protection. Model **EB-35 Mark IIS** has stereo FM with 2 speakers. Model **EB-35 Mark II/S** has stereo FM output and 1 speaker. Model **EB-35A** covers 155-175 MHz VHF marine instead of FM broadcast. Model **EB-36** (1966, £55 new) is the same as the EB-35, but without the FM band. Model **EB-36A** (1969), produced for the G.P.O, has SW coverage from 3.5-22 MHz, a 3.5 kHz filter. Model **EB-36 Mark II** (1970) has new circuit and case.

Made In:	England	1966-1969

Voltages: Batteries 6 x D cells

Readout: ⊥⊥⊥⊥⊥ Analog

Physical: 12.5x6.3x8" 13 Lbs. 317x162x203mm 5.8 kg

Status: Active Manufacturer Discontinued Model

Rarity: Very Scarce

	New	Used
Price:	£50	£45-85
Rating:	⑥	⑥

EB-35 Mark III
The Statesman

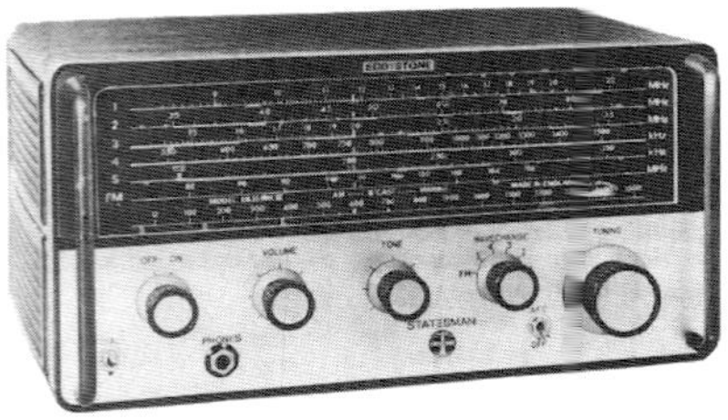

General Coverage Broadcast Receiver
Single Conversion Superheterodyne. Solid State.

Features:
- ¼" Head. Jack
- Tone
- AFC
- Speaker 5"
- Dial Light
- Phono Input Jack
- 110:1 Tune Ratio
- Record Jack
- Aux. Tuning Scale
- Rack Handles

Specifications:
Coverage ...... 550-22000 kHz +LW/FM
Selectivity ...... 5 kHz [250 FM] -6dB.
Sensitivity ...... <5 µV -15 dB S/N 3.5-22MHz
Audio Out .... .750 W 8 ohm 10% dist.
Modes .... AM/[FM-W]
Antenna Input . Terminals (Jack for FM)

Accessories:
924A AC Power Supply
945A 12/24VDC Power Supply

Comments:
The flywheel-loaded 110 to 1 reduction drive provides very smooth tuning. Ranges: 8.5-22, 3.5-8.5, 1.5-3.5, .55-1.5 and .15-.35 MHz plus FM coverage from 88-108 MHz. FM mode reception is only on the FM broadcast band. Longwave coverage is 150-350 kHz. Very similar to the EB-35 Mark II, but with an AFC switch and improved FM band performance.

Made In: England 1976

Voltages: Batteries 6 x D cells

Readout: Analog

Physical: 12.5x6.3x8" 11 Lbs.
317x162x203mm 5.8 kg

Status: Active Manufacturer
Discontinued Model

Rarity: Very Scarce

	New	Used
Price:	①	⑤
Rating:	⑥	⑥

EB-37

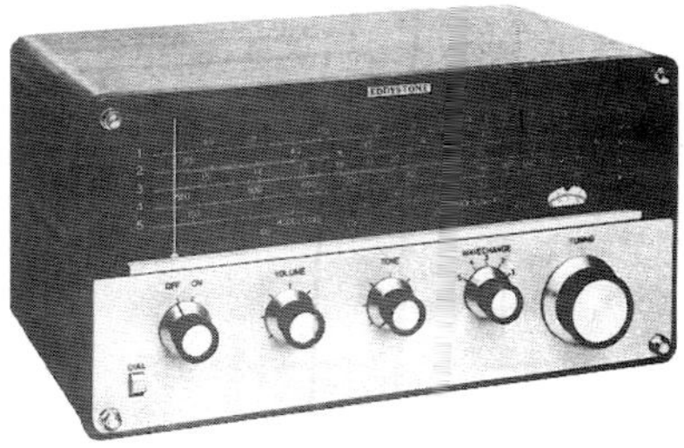

General Coverage Broadcast Receiver
Single Conversion Superheterodyne. Solid State.

Features:
- ¼" Head. Jack
- Tone
- Phono Input Jack
- Speaker 5"
- Dial Light
- 110:1 Tune Ratio
- Dial Lamp Switch
- Aux. Tuning Scale
- Record Jack

Specifications:
Coverage ...... 150 - 22000 kHz
Selectivity ...... 5 kHz -6dB.
IF .... 465 kHz
Stability .... 1 part in 10^4/°C
Audio Out .... .750 W 8 ohm 10% dist
Modes .... AM/FM
Image Reject .. 50 dB @ 2 MHz
Image Reject .. 15 dB @18 MHz
Sensitivity .... <5 µV -15 dB S/N 1.5-22MHz
Antenna Input . Screws 75 ohm

Circuit Complement:
10 Transistors and 5 diodes.

Accessories:
924 AC Power Supply
945 12/24 VDC Power Supply

Comments:
Ranges: 8.5-22, 3.5-8.5, 1.5-3.5, .55-1.5 and .15-.35 MHz. (Reception gap from 350-550 kHz). The "phones" jack is on the rear panel. Very similar to the EB-36 (and to the EB-35, but without the FM broadcast band). The EB-37 is designed primarily for operation from 6 D (U2) cells. However the model 924 optional AC power supply can be fitted directly in the battery cavity for operation from the mains.

Made In: England 1971-1976

Voltages: Batteries 6 x D cells

Readout: Analog

Physical: 12.5x6.3x8" 13 Lbs.
317x162x203mm 5.8 kg

Status: Active Manufacturer
Discontinued Model

Rarity: Extremely Scarce

	New	Used
Price:	①	⑤
Rating:	⑥	⑥

EC-10

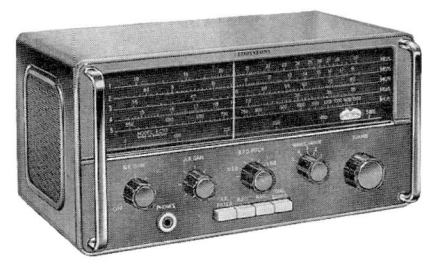

General Coverage Communications Receiver
Single Conversion Superheterodyne. Solid State.

Features:
- ¼" Head. Jack
- BFO
- CW Audio Filter
- AGC (2 Pos.)
- RF Gain
- Speaker 5"
- 110:1 Tune Ratio
- Aux. Vernier Scale
- Rack Handles
- Dial Light
- Logging Scale
- Flywheel Tuning
- Speaker Output

Specifications:
Coverage ...... 550 - 30000 kHz
Selectivity ...... 5 kHz -6dB.
IF .... 465 kHz
Audio Out ...... 0.8W 3 ohms 10% dist.
Antenna Input 75 ohms.

Modes .... AM/CW
Sensitivity .... <3 µV -15dB S/N 1.5-30MHz
Image Reject .. 50 dB @ 2 MHz
Image Reject .. 20 dB @ 18 MHz

Circuit Complement:
10 Transistors and 3 diodes.

Accessories:
924 AC Power Pack. 945 12/24 VDC Power Supply

Comments:
Ranges: 18-30, 8.5-18, 3.5-8, 1.5-3.5 and .55-1.5 MHz. The audio filter is fixed at 1000 Hz and provides 180 Hz bandwidth at 6 dB. Marginal SSB reception is possible with this radio.

Variants:
Model **EY11** (1969) is a marine version with ranges of: .15-.4, .48-1.25, 1.1-2.5 and 2.5-6.2 MHz. It also features a battery meter.

Made In: England 1964-1970

Voltages: Batteries 6 x D cells

Readout: Analog

Physical: 12.5x6.3x8" 12.75 Lbs.
317x162x203mm 5.8 kg

Status: Active Manufacturer
Discontinued Model

Rarity: Very Scarce

	New	Used
Price:	£53	£50-90
Rating:	⑥	⑥

EC-10 Mark II

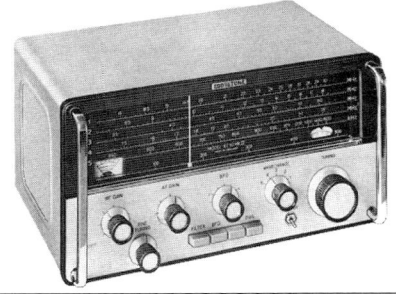

General Coverage Communications Receiver
Single Conversion Superheterodyne. Solid State.

Features:
- ¼" Head. Jack
- S-Meter
- Record-out Jack
- Audio CW Filter
- RF Gain
- AGC (2 Pos.)
- Dial Lamp
- 110:1 Tune Ratio
- BFO
- Fine Tuning
- Mute Line
- Rack Handles
- Speaker 5"
- Logging Scale
- Dial Lamp Switch
- Flywheel Tuning

Specifications:
Coverage ...... 550 - 30000 kHz.
Selectivity ...... 5 kHz -6dB.
Audio Out ...... 0.8 W 3 ohms 10% dist.
Inter. Freq. .... 465 kHz.

Modes .... AM/SSB/CW
Sensitivity .... <5µV 15dB S/N >1.5 MHz.
Stability .... 1 part in 10^4/°C.
Image Reject .. 50 dB @ 2, 20 dB @ 18 MHz

Accessories:
924 AC Power Supply   945 12/24 VDC PS

Comments:
The EC-10 MK.II is a restyled replacement of the EC-10 incorporating a number of additional features. These include an S-Meter, Fine Tuning Control and desensitizing facilities. An audio CW filter yields a bandwidth of 180 Hz @6 dB. The internal battery pack can be fitted with an AC power unit for operation from 100/125, 200/250 VAC 40-60 Hz, or external 12VDC or 24 VDC supplies. Ranges: 18-30, 8.5-18, 3.5-8, 1.5-3.5 and .55-1.5 MHz. The model EC-10A/2, shown later, is the same but with 2182 kHz preset facility.

Variants:
This model was also sold as the **MIMCO 6689**.

Made In: England 1969-1975

Voltages: 6 D Cells (internal pack). See comments.

Readout: Analog

Physical: 12.5x6.3x8" 14 Lbs.
320x162x203mm 6.3 kg

Status: Active Manufacturer
Discontinued Model

Rarity: Very Scarce

	New	Used
Price:	$200-210	£80-90
Rating:	⑥	⑥

EC-10A/2/RM

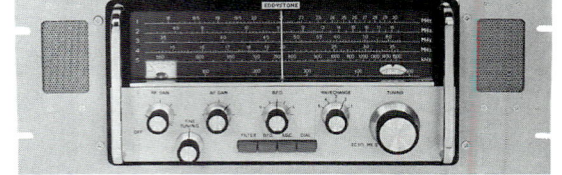

General Coverage Communications Receiver
Single Conversion Superheterodyne. Solid State.

Features:
- ¼" Head. Jack
- S-Meter
- Line Out Jack
- Audio CW Filter
- RF Gain
- AGC (2 Pos.)
- Dial Lamp
- 110:1 Tune Ratio
- BFO
- Fine Tuning
- Rack Handles

Specifications:
Coverage ...... 1500 - 30000 kHz.+LW
Selectivity ...... 7 kHz -6dB.
Audio Out ...... 0.8 W 10% distortion
Inter. Freq. .... 720 kHz.

Modes .... AM/SSB/CW
Sensitivity .... <5µV 15dB S/N >1.5 MHz.
Stability .... 1 part in 10^4/°C.

Comments:
The EC-10A series comprises four variants of the basic EC10 which have been tailored for maritime use. These receivers cover longwave 300 - 550 kHz instead of 550 - 1500 kHz. Also note the IF is now 720 kHz rather than 465 kHz.

Variants:
The **EC-10A/2** variants include crystal controlled receptions of 2182 kHz (international HF distress frequency). EC10A/2/RM (shown) is rack mount version with two panel speakers (one for the radio and one for the ship's intercom). The model **EC-10A/2/RM/S** is the same but less internal speaker and ship's intercom speaker.

Made In:	England	1976-1978
Voltages:	12 or 24 VDC	

Readout:	Analog
Physical:	19x7x8" 16.25 Lbs. 482x177x203mm 7.4 kg
Status:	Active Manufacturer Discontinued Model
Rarity:	Very Scarce

	New	Used
Price:	①	⑤
Rating:	⑥	⑥

EC-958

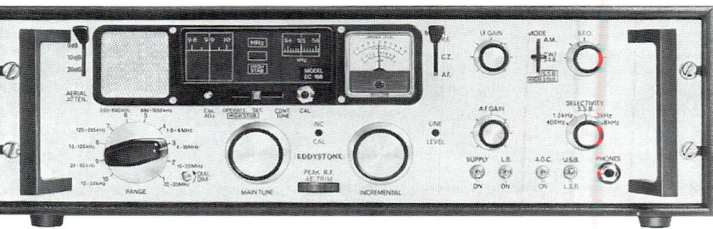

General Coverage Communications Receiver
Triple Conversion Superheterodyne. Solid State.

Features:
- ¼" Head. Jack
- S/AF/FSK Meter
- Line Out Jack
- Attenuator -10/20 dB
- RF Gain
- Dial Lamp
- Speaker Jack
- AGC ON/OFF
- BFO ± 5kHz
- Rack Handles
- Mute Line
- IF Out Jack
- Speaker

Specifications:
Coverage ...... 10 - 30000 kHz
Selectivity ...... 8/3/2.65/1.3/.4kHz -6dB.
Sensitivity ...... <1µV 10 dB S/N CW
Audio Out ...... 1 W 3 ohm 5% Dist.
Environment .. 0 to +50° C

Modes .... AM/SSB/CW
IF .... 1235-1335, 250, 100 kHz.
Image Rej. ...... >70 dB over 18 MHz
Stability .... 20 Hz after warmup
Antenna Input . BNC and Terminals

Accessories:
EP961 Panoramic Display   EP1061A/1 Panoramic Display

Comments: Ranges: 20-30, 10-20, 4-10, 1.6-4, .68-1.65, .29-.68, .125-.295, .53-.126, .023-.054 and .010-.023 MHz. One of the best receivers ever made by Eddystone. Over 20 versions were produced. Outstanding on longwave and VLF.
Variants: Model **EC-958/1** fitted with special filter for SSB, carrier controlled AGC and beat meter. Model **EC-958/2** is for specialized network monitoring and includes a 150 Hz CW bandwidth in lieu of SSB reception. Model **EC-958/3** is like the **/2** with additional 10 kHz scale check from 10 to 680 kHz. Model **EC-958/4** is a military variant. Model **EC-958/5** is a maritime variant. The **EC-958/D** version was produced for Debeg and sold as the **Debeg E755.** Model **EC-958/H** was sold as **Hagenuk E91.**

Made In:	England	1969-1973
Voltages:	100/130/200/260 VAC 40-60 Hz	

Readout:	Analog Lin.
Physical:	Cabinet Version: 19.75x6.5x18" 50 Lbs. 502x165x457mm 22.7 kg Rack Version: 19x5.25x16.2" 43.5 Lbs. 483x133x411mm 19.6 kg
Status:	Active Manufacturer Discontinued Model
Rarity:	Extremely Scarce

	New	Used
Price:	$3100	£370-450
Rating:	⑥	⑥

EC-958/7E

General Coverage Communications Receiver
Triple Conversion Superheterodyne. Solid State.

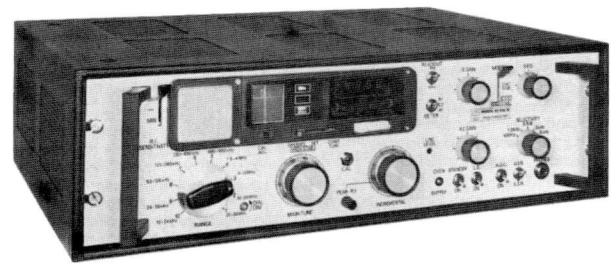

Features:
- ¼" Head. Jack
- S/AF/FSK Meter
- Speaker Jack
- Attenuator -10/20 dB
- RF Gain
- AGC ON/OFF
- Dial Lamp
- Line Out 600 ohms
- BFO ± 5 kHz
- Rack Handles
- Mute Line
- IF Out Jack
- Speaker

Specifications:
Coverage ...... 10 - 30000 kHz
Selectivity ...... 8/3/2.4/1.3/.4kHz -6dB.
Sensitivity ...... <1µV 20 dB SINAD CW
Audio Out ...... 1 W 3 ohm 5% Dist.
Environment .. 0 to +50° C

Modes .... AM/SSB/CW
IF .... 1235-1335, 250, 100 kHz.
Image Rej. ...... >60 dB over 18 MHz
Stability .... 20 Hz after warmup
Antenna Input . BNC and Terminals

Accessories:
989 Speaker

Comments:
Ranges: 20-30, 10-20, 4-10, 1.6-4, .68-1.65, .29-.68, .125-.295, .53-.126, .023-.054 and .010-.023 MHz. Similar to the EC-958, but with digital LED readout.

Variant:
Model **EC-958/7** (1977) features a -40 dB Attenuator. Model **EC-958/8** was produced for the Netherlands Navy. Model **EC-958/9** features ISB. Model **EC-958/10** was produced for the Canadian military.

Made In:	England	1973-1980

Voltages: 100/130/200/260 VAC
40-60 Hz

Readout: `00000.001` Digital LED

Physical: Cabinet Version:
19.75x6.5x18" 50 Lbs.
502x165x457mm 22.7 kg
Rack Version:
19x5.25x16.2" 43.5 Lbs.
483x133x411mm 19.6 kg

Status: Active Manufacturer
Discontinued Model

Rarity: Extremely Scarce

	New	Used
Price:	①	⑤
Rating:	⑥	⑥

EC-958/12

General Coverage Communications Receiver
Triple Conversion Superheterodyne. Solid State.

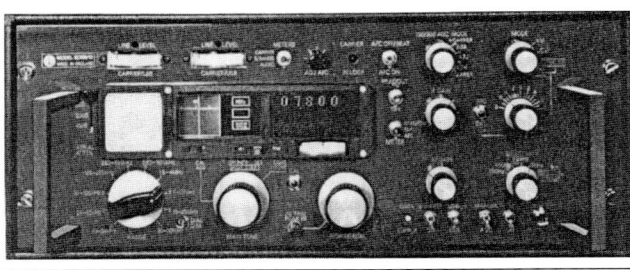

Features:
- ¼" Head. Jack
- S/AF Meters
- Line Out Jack
- Diversity AGC Out
- RF Gain
- AGC ON/OFF
- Dial Lamp
- Attenuator -10/20 dB
- BFO ± 5 kHz
- Rack Handles
- Mute Line
- IF Out Jack
- Speaker
- Speaker Jack
- (2) Line Out 600 ohms

Specifications:
Coverage ...... 10 - 30000 kHz
Selectivity ...... 8/3/2.4/1.3/.4 kHz -6dB.
Sensitivity ...... <1µV 10 dB S/N CW
Audio Out ...... 0.5 W 3 ohm 5% Dist.
IF Reject ....>100 dB 3-30 MHz.
Environment .. -15 to +55° C

Modes .... AM/SSB/CW/ISB
IF .... 1235-1335, 250, 100 kHz.
Image Rej. ...... >60 dB over 18 MHz
Stability .... 20 Hz after warmup
Antenna Input . BNC and Terminals

Accessories:
989 Speaker   LP3058 FSK Module

Comments:
Similar to the EC-958/7E, but with ISB capability. Note the two meters mounted above the speaker. They can display Carrier, Side Bands or Audio for each of two independent sidebands. Two independent 600 ohm (10mW) line out jacks are provided.

Made In:	England	1974-1981

Voltages: 100/130/200/260 VAC
40-60 Hz 90W

Readout: `00000.001` Digital

Physical: Cabinet Version:
19.75x8x18" 72 Lbs.
502x203x457mm 32.6 kg
Rack Version:
19x7x16.2" 60 Lbs.
483x177x411mm 27.2 kg

Status: Active Manufacturer
Discontinued Model

Rarity: Typically Unavailable

	New	Used
Price:	①	⑤
Rating:	⑥	⑥

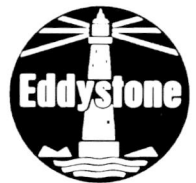

EC-964/1

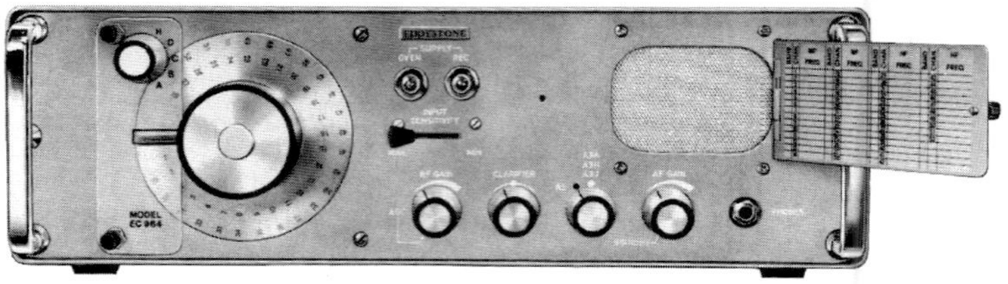

Fixed Channel Communications Receiver
Dual Conversion Superheterodyne. Solid State.

Features:

- ¼" Head. Jack
- Mute Line
- Rack Handles
- BFO ±3 kHz
- Atten (2 Pos.)
- RF Gain
- Speaker
- 52 Channels
- Modular Construction

Specifications:

Coverage ...... 52 Ch. 1600-27500 kHz
Selectivity ...... 6/2.4 kHz -6dB.
Sensitivity ...... 1μV for 12 dB S/N
Image Rej ...... 60 dB <15 MHz
Audio Out ...... 2 W 8 ohms
Line Out .... 600 ohms
Modes .... AM/USB/CW
Stability .... 20 Hz 0° - 40° C.
IF Rejection .... > 80 dB Below 4 MHz
IF Rejection .... > 90 dB Above 4 MHz
IF .... 1.2 MHz, 100 kHz

Comments:
Designed for the marine market.

Accessories:
978/12 12 VDC PS   978/24 24VDC PS

Variants:
Model **EC-964/3** (1969) covers 28 channels in the range of 1.6 - 4.5 MHz. Model **EC-964/2** (1969) has different channeling.

Made In: England 1969-1970

Voltages: 100/125 or 200/250 VAC
40-60 Hz 45W

Readout: [Channel] Channel

Physical: Cabinet Version:
16.75x5.25x17.375" 33 Lbs
426x134x441mm 15.2 kg
Rack Mount Version:
19x5.25x17.375" 33 Lbs
483x134x441mm 15.2 kg

Status: Active Manufacturer
Discontinued Model

Rarity: Typically Unavailable

	New	Used
Price:	①	⑤
Rating:	⑥	⑥

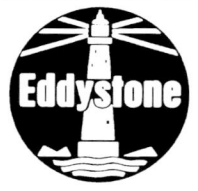

EC-964/4

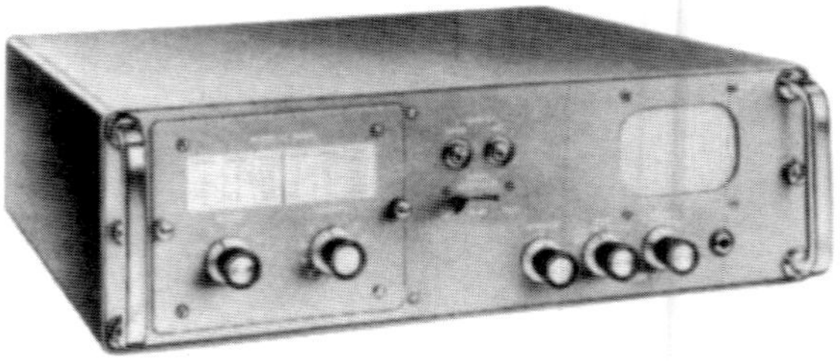

Fixed Channel Communications Receiver
Dual Conversion Superheterodyne. Solid State.

Features:

- ¼" Head. Jack
- Mute Line
- Rack Handles
- BFO ±3 kHz
- Atten. (2 Pos.)
- RF Gain
- Speaker
- 12 Channels
- Modular Construction

Specifications:

Coverage ...... 12 Ch. 1600-30000 kHz
Selectivity ...... 6/2.4 kHz -6dB.
Sensitivity ...... 1μV for 12 dB S/N
Image Rej ...... 60 dB <15 MHz
Audio Out ...... 2 W 8 ohms
Line Out .... 600 ohms
Modes .... AM/USB/CW
Stability .... 20 Hz 0° - 40° C.
IF Rejection .... > 80 dB Below 4 MHz
IF Rejection .... > 90 dB Above 4 MHz
IF .... 1.2 MHz, 100 kHz

Comments:
Designed for the marine market. Note that the 12 channel EC-964/4 will operate from AC or 24 VDC as standard.

Accessories:
ERC974/1 Remote Controller (shown right)

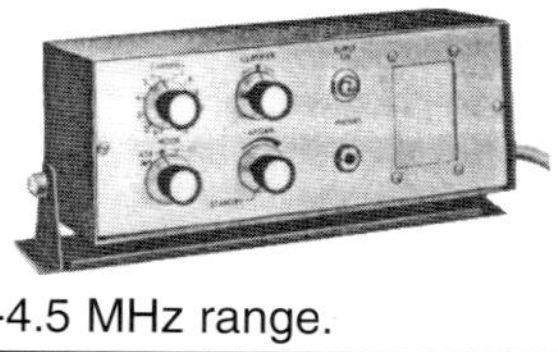

Variants:
Model **EC-964/5** (1969) covers 28 channels in the 1.6-4.5 MHz range.

Made In: England 1969

Voltages: 100/125 or 200/250 VAC
40-60 Hz 45W or 24VDC

Readout: [Channel] Channel

Physical: Cabinet Version:
16.75x5.25x17.375" 33 Lbs
426x134x441mm 15.2 kg
Rack Mount Version:
19x5.25x17.375" 33 Lbs
483x134x441mm 15.2 kg

Status: Active Manufacturer
Discontinued Model

Rarity: Typically Unavailable

	New	Used
Price:	①	⑤
Rating:	⑥	⑥

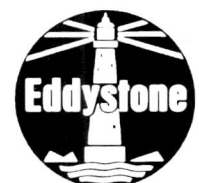

EC-964/7A

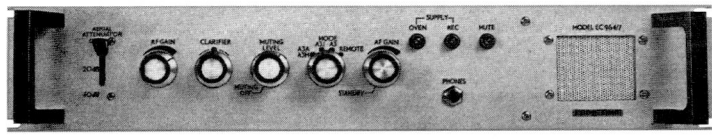

Fixed Channel Communications Receiver
Single Conversion Superheterodyne. Solid State.

Features:
- ¼" Head. Jack
- BFO ±300 Hz
- RF Gain
- Single Channel
- Standby
- Speaker
- Muting
- Atten. -20/40 dB
- Rack Handles

Specifications:

Coverage ...... 1 Chan. See Comments	Modes .... SSB (A3J)
Selectivity ...... Depends on version.	Stability .... 10 Hz -10° to +55° C.
Sensitivity ...... 1µV for 15 dB S/N SSB	IF Rejection .... > 80 dB Below 2 MHz
Image Rej ...... 70 dB <18 MHz	IF Rejection .... >100 dB Above 2 MHz
Audio Out ...... 1 W 8 ohms	IF .... 1400 kHz
Line Out .... 600 ohms	Antenna Input . 50 ohm

Comments:
Single channel operation in the range .4-.535 or 1.6-27.5 MHz.

Variants:
Model **EC-964/7B** (1980) is SSB (A3J/A3A). Model **EC-964/7C** (1980) is SSB (A3J/A3A) and AM. Model **EC-964/7D** is MCW. Models **EC-964/7F** and **EC-964/7G** (1980) are CW only.

Made In:	England	1973-1976
Voltages:	100/130 or 200/260 VAC 40-60 Hz 25W	
Readout:	▆▆▆ None	
Physical:	19x3.5x8.5" 17 Lbs 483x88x216mm 7.7 kg	
Status:	Active Manufacturer Discontinued Model	
Rarity:	Typically Unavailable	

	New	Used
Price:	①	⑤
Rating:	⑥	⑥

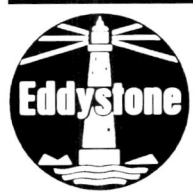

EC-1964/1

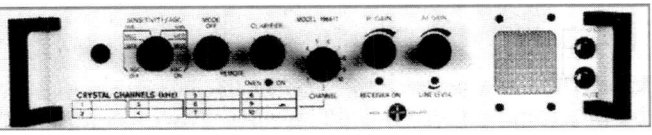

Fixed Channel Communications Receiver
Double Conversion Superheterodyne. Solid State.

Features:
- ¼" Head. Jack
- BFO ±300 Hz
- RF Gain
- Ten Channel
- Rack Handles
- Standby
- Speaker
- Atten. -20/40 dB
- Muting

Specifications:

Coverage ...... 10 Ch. See Comments	Modes .... AM/SSB/CW
Selectivity ...... 7/2.7 kHz.	Stability .... 10 Hz -10° to +55° C.
Sensitivity ...... 1µV for 15 dB S/N SSB	Image Reject .. See Comments
IF Rejection ... >100 dB	IF .... 400 kHz, 100 kHz
Audio Out ...... 1 W 8 ohms	Antenna Input . 50 ohm
Line Out .... 600 ohms	

Comments:
Ten channel operation in the range 1.6-20 MHz. Image rejection: 70 dB <9 MHz, 60 dB <15 MHz, 50 dB <20 MHz and 40 dB <27.5 MHz.

Variants:
Model **EC-1964/2** covers ten channels in the range of 1.6-27.5 MHz.

Made In:	England	1978-1980
Voltages:	100/130 or 200/260 VAC 40-60 Hz	
Readout:	[Channel] Channel	
Physical:	19x3.5x16.5" 22 Lbs. 483x88x420mm 10 kg	
Status:	Active Manufacturer Discontinued Model	
Rarity:	Typically Unavailable	

	New	Used
Price:	①	⑤
Rating:	⑥	⑥

S1670

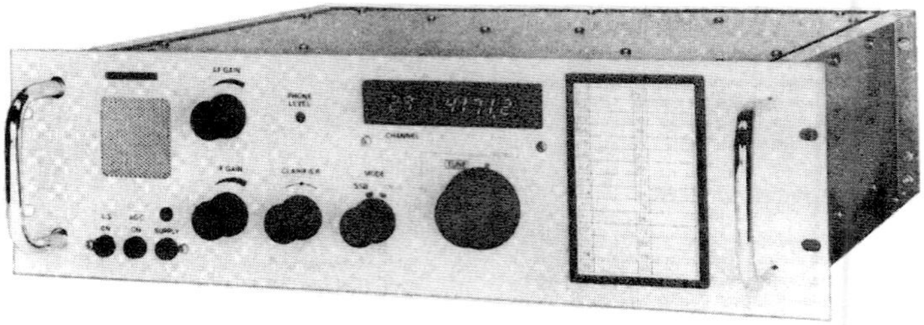

Fixed Channel Communications Receiver
Dual Conversion Superheterodyne. Solid State.

Features:
- ¼" Head. Jack
- Mute Line
- Rack Handles
- BFO ±150 Hz
- AGC
- RF Gain
- Speaker
- 30 Channels
- 2182 kHz Preset

Specifications:

Coverage	30 Ch. 1.6-4.2 MHz
Sensitivity	1μV for 20 dB S/N
Image Rej	>70 dB
Audio Out	2 W 8 ohms
Line Out	600 ohms
Modes	SSB/CW
Stability	50 Hz 0° - 40° C.
IF Rejection	>80 dB
IF	1400, 100 kHz
Environment	-10° to +55° C.

Comments:
Designed strictly for the marine market. The LED display shows the channel number and frequency to 100 Hz.

Made In: England

Voltages: 100-130 or 200-260 VAC
40-60 Hz or 24 VDC

Readout: 00000.1 Digital LED
[Channel] Chan. LED

Physical: Cabinet Version:
19.75x6.5x18.1" 39.5 Lbs.
502x165x460mm 17.9 kg
Rack Mount Version:
19x5.25x18.1' 28 Lbs.
483x134x460mm 12.7 kg

Status: Active Manufacturer
Discontinued Model

Rarity: Typically Unavailable

	New	Used
Price:	①	⑤
Rating:	⑥	⑥

26 Eldico

Eldico was a subsidiary of Radio Engineering Laboratories, Inc. of Long Island, New York. Radio Engineering Labs is best remembered for their line of commercial broadcast tuners. Their famous Precedent line of FM broadcast tuners is a professional 19 inch rack mounted monaural tuner that is still coveted by audiophiles to this day, and often bring a price of $1000 or more.

Quite a few stories or rumors, still circulate regarding the Eldico R-104 receiver. Reportedly the designer of the Collins 75S3 and the R-104 is one in the same. There are reports that this radio was a subject of a lawsuit between Collins and Radio Engineering Laboratories.

Radio Engineering Laboratories, Inc.
29-01 Borden Avenue
Long Island City 1, NY 1955-1961

R-104

Amateur Band Communications Receiver
Double Conversion Superheterodyne. 11 Tubes plus Semiconductors.

Features:
- ¼" Head. Jack
- RF Gain
- Mute
- S-Meter
- ANL
- Line Out 600 ohms
- Standby
- Dial Light
- Fiduciary Adjust
- Calibrator 100 kHz

Specifications:
Coverage ...... Ham. See Comments.
Selectivity ...... 2.1/.5 kHz -6dB.
Sensitivity ...... <1µV
Image Rej. ...... >50 dB
Audio .... 4 ohms 0.75 W
Modes .... AM/CW/LSB/USB
Stability .... <100 Hz after warm-up
IFs .... 2.95 MHz, 455 kHz
Antenna Input . RCA

Circuit Complement:
6DC6 RF Amp, 6AH6 HF Crystal Osc., 6BE6 1st Mixer, 6BE6 2nd Mixer, 6BA6 1st IF Amp, 6BA6 2nd IF Amp, 6AT6 AVC Rectifier/AM Detector/AF Amp, 6BE6 SSB-CW Product Detector/BFO, 6AQ5 AF Output plus semiconductors.

Comments:
Coverage: 3.4-3.6, 3.6-3.8, 3.8-4, 7-7.2, 7.2-7.4, 14-14.2, 14.2-14.4, 14.8-15 (WWV), 21-21.2, 21.2-21.4, 21.4-21.6, 28.5-28.7 and two additional positions for further 10 meter band coverage. Dial accuracy is 300 Hz electrical and 200 Hz visual. The R-104 is housed in a light gray metal cabinet with red logo.

Made In: United States 1960

Voltages: 115 VAC 60 Hz, 90W.

Readout: Analog Lin.

Physical: 14.875x6.875x12" 20 Lbs.
378x175x305mm 9 kg

Status: Inactive Manufacturer
Discontinued Model

Rarity: Very Scarce

	New	Used
Price:	①	⑤
Rating:	⑥	⑥

27 Elta

Elta Electronics Industries Ltd. is one of Israel's leading military electronics systems companies. Elta, a wholly-owned subsidiary of Israel Aircraft Industries, was established in 1960. Located in the port city of Ashdod, Elta's modern industrial complex covers an area of 576,000 square feet. Most of Elta's 2,455 employees are experienced scientists, engineers, programmers and technicians.

Elta has five divisions: Radar Systems, Electronic Warfare Systems, Communications Systems, Information Systems and Advanced Technologies. In 1994 sales were divided between the Israel Ministry of Defense (29%) and export markets (71%).

1994 Sales totaled $271 million of which over $200 million was for export.

Elta does not currently manufacture an H.F. receiver.

Elta Electronics Industries Ltd.
Israel Aircraft Industries Ltd.
P.O. Box 330
Ashdod 77102, Israel 1994-1998

IAI ELTA
ISRAEL AIRCRAFT INDUSTRIES LTD

EL/K-1160

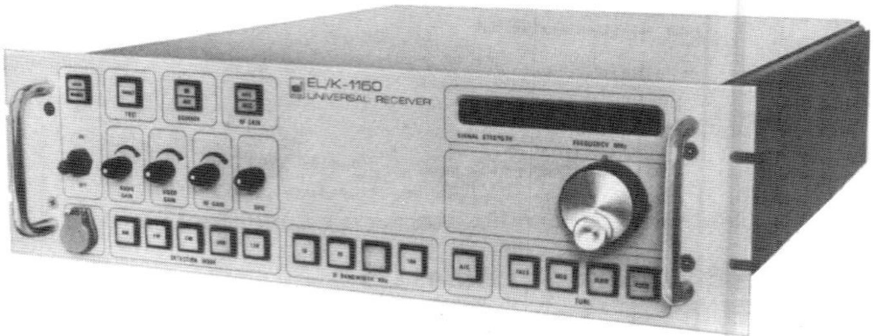

General Coverage Communications Receiver
Double Conversion Superheterodyne. Solid State.

Features:
- ¼" Head. Jack
- RF Gain
- BITE
- Spinner Knob
- S-Indicator
- Dial Lock
- Modular
- BFO
- AGC/MGC
- Rack Handles
- 3 Tuning Speeds
- Squelch
- Line Out 600 ohms

Specifications:
Coverage ...... 500 - 30000 kHz
Selectivity ...... 4 Positions
Sensitivity ...... -105dBm CW 16 dB
IF Rej .... >100 dB
Audio Out ...... 600 ohm
Modes .... AM/USB/LSB/CW/FM/ISB
Stability .... $6x10^{-8}$/day
Image Rej. ...... >100 dB
Antenna Input . 50 ohm
Environment ... -10° to +65° C.

Accessories:
Ext. Ref.

Comments:
Elta no longer makes this model, but does sell the very similar VHF/UHF version called the EL/K-1150 which covers 20 to 500 MHz. The EL/K-1150 has the same chassis, knob complement, etc.

Made In: Israel 1986-1994

Voltages: 220 VAC 50 Hz or 24 VDC

Readout: 00000.01 Digital LED

Physical: 19x5.25x21.7"
483x133x551mm

Status: Active Manufacturer
Discontinued Model

Rarity: Typically Unavailable

	New	Used
Price:	①	⑤
Rating:	⑥	⑥

28 Eska

The **Eska Company** has had an erratic history with a strong ability to comeback from the dead more than once.

Most of the receivers sold under the label were small, handheld crystal controlled devices (models RX12PL, RX12M and RX12S). These units provided good reception for those interested in monitoring fixed frequencies.

The Eska deluxe communications receiver was the dream of Eddy Visser. His ambitious plans for an uncompromising shortwave receiver for the hobbyist collided with economic reality. Having underestimated the cost of development, the target price of $800 crept up to over $3000. At this point the potential market shifted from the hobbyists to the commercial market. However, being too expensive for the listener market, and a bit too radical for the commercial market, the financial difficulties still remained. The company was acquired by Allan Eimert, the once business manager of rock star Rod Stewart. The August 1985 issue of *Monitoring Times* announced that Eska filed for bankruptcy.

Eska was not to totally leave the scene. For many years they continued to sell circuit upgrades, filters and ECSS adapters for other receivers. Later on the Eska<u>b</u> company emerged (note spelling). They sold an OEM version of the Sangean ATS-803A.

Eska Elektronik A/S
Møllestraede 5
DK-3400 Hillerød,
Denmark   1983-1985

RX-99M

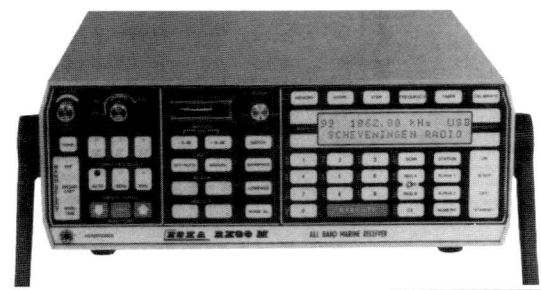

General Coverage Communications Receiver
Double Conversion Superheterodyne. Solid State.

Features:
• ¼" Head. Jack	• S-Meter	• PBT	• Tone (4 Pos.)
• RF Gain	• Squelch	• Keypad	• Ant.Gain +10/0/-15 dB
• 99 Memories	• Carry Handle	• AGC 4 Pos.	• 24 Hr. Clock/Timer
• Timer Terminals	• Phase Lock AM	• Sleep Function	• Scan & Sweep

Specifications:
Coverage ...... See Comments.	Modes .... AM/SSB/CW/FM/RTTY
Selectivity ...... 8/4/2.4 kHz -6dB.	3rd Ord. Int. .... +10 dBm
Sensitivity ...... <.4µV 10 dB S/N	Image Rej .... >80 dB
Dyn. Range ...95 dB	IF .... 70 MHz, 10.7 MHz
Audio Out ......1.5 W	

Accessories:
NBFM	12 kHz Filter	6 kHz Filter
1.8 kHz Filter	1 kHz Filter	.5 kHz Filter
Alpha Memory Display	I.F. NB	VHF/UHF Conv. (30-500 MHz)
Rack Mount Kit	External Speaker	External VFO

Comments:
Coverage: .04-30, 60-110 and 140-170 MHz. Memories store frequency, mode, bandwidth, antenna gain, sideband selection and AGC. All functions except, volume, PBT and (optional) audio filter, are controlled by front panel membrane switches. The RX-99M model was marketed primarily to the maritime industry. This receiver was not produced in large numbers.

Made In: Denmark   1984-1985

Voltages: 7.5 V 120 ma
6 UM2 and 2 UM2 cells,

Readout: `00000.01` Digital LCD

Physical: 10x3.5x11.2" 6.6 Lbs.
255x90x285mm 3 kg

Status: Active Manufacturer
Discontinued Model

Rarity: Typically Unavailable

Reviews: 📖 *Passport* 1985/86
Pop. Comm. February 1985

	New	Used
Price:	$3900	③
Rating:	⑥	⑥

RX-99PL

General Coverage Communications Receiver
Double Conversion Superheterodyne. Solid State.

Features:
- ¼" Head. Jack
- S-Meter
- PBT
- Tone (4 Pos.)
- RF Gain
- Squelch
- Keypad
- Ant.Gain +10/0/-15 dB.
- 99 Memories
- Carry Handle
- AGC 4 Pos.
- 24 Hr. Clock/Timer
- Timer Terminals
- Phase Lock AM
- Sleep Function
- Scan & Sweep

Specifications:
Coverage ...... 40 - 30000 kHz & FM
Selectivity ...... 4/2.4 kHz -6dB.
Sensitivity ...... <.4µV 10 dB S/N
Dyn. Range ... 95 dB
Audio Out ...... 1.5 W

Modes .... AM/SSB/CW/FM/RTTY
3rd Ord. Int. .... +10 dBm
Image Rej .... >80 dB
IF .... 70 MHz, 10.7 MHz

Accessories:
12 kHz Filter	8 kHz Filter	Alpha Memory Display
1.8 kHz Filter	1 kHz Filter	VHF/UHF Conv. 30-500 MHz
.5 kHz Filter	6 kHz Filter	VHF Converter 60-88 MHz
Rack Mount Kit	External Speaker	VHF Converter 140-170 MHz
I.F. NB		

Comments:
Memories store frequency, mode, bandwidth, antenna gain, sideband selection and AGC. All functions except, volume, PBT and (optional) audio filter, are controlled by front panel membrane switches. This receiver was not produced in large numbers.

Made In: Denmark   1984-1985

Voltages: 7.5 V 120 ma
6 UM2 and 2 UM2 cells,

Readout: `00000.01` Digital LCD

Physical: 10x3.5x11.2" 4.4 Lbs
255x90x285mm 2 kg

Status: Active Manufacturer
Discontinued Model

Rarity: Typically Unavailable

Reviews: 📖 *Passport* 1985/86
WRTH 1986

	New	Used
Price:	$3500	③
Rating:	⑥	⑥

29 Fairhaven

Fairhaven Electronics is a relatively new company, established in 1990. The company is best known in radio hobbyist's circles for their innovative new RD500 receiver. The company also produces flight altimeters and variometers.

Fairhaven Electronics Ltd.
47 Dale Rd.
Spondon, Derby DE21 7DG,
England   1997-1998

RD500

General Coverage Communications Receiver
Double Conversion Superheterodyne. Solid State.

Features:
- ¼" Head. Jack
- V.R.I.T.
- Scan
- Speaker
- P.B.T.
- Notch Filter
- S-Indicator
- NB
- Priority
- Clock
- Line Output
- AGC (6 settings)
- Squelch
- 14,200 Memories
- Sweep
- 4 Event Timer
- Record I/O Jack
- Noise Cancelling Antenna Input
- Digital Audio Record
- Attenuator -20 dB
- Synchronous Det.
- Recorder Activation
- PC Keyboard Input

Specifications:
Coverage ...... 30 - 40000 kHz
Selectivity ...... 6/2.4 kHz -6 dB [20 FM]
Sensitivity ...... <.1 µV .5-30 MHz SSB
Audio Out ...... 4 ohms 2 W

Modes .... AM/CW/LSB/USB/FM
IF .... 55 MHz, 455 kHz
IF Rejection .... >65 dB
Antenna Input . SO-239 50 Ohms

Accessories:
VHF/UHF Board   Mini PC Keyboard   Internal Battery Pack

Comments:
The RD500 tunes in 5 Hz in SSB/CW and AM synchronous modes and 100 Hz in AM modes. Step sizes: 1, 5, 9, 10, 12.5, 25 and 50 kHz. The memory system proves 234 groups (A1 to Z9) which can each hold between 1 and 999 records. Each group carries a 20 character plus each memories carries a 20 character label. The memory can also be partitioned to provide an area for digital sound recording. Recording up to thirty seconds may be accomplished using half the memory. Up to a maximum of 4 minutes recording time can be achieved with 2 MB or RAM.

Made In: England   1997-1998

Voltages: 12 VDC 0.5 Amps

Readout: `00000.01` Digital LCD

Physical: 8.1x2.6x7.6" 1.35 Lbs
205x65x193mm 0.6 kg

Status: Active Manufacturer
Active Model

Rarity: (Too New)

Reviews: *Shortwave Mag.* Sept. 1997

	New	Used
Price:	£799	④
Rating:	⑧	⑥

Galaxy Electronics was an active manufacturer of amateur radio equipment in the mid-1960's. Many low to mid priced transceivers were offered to the amateur community.

The Galaxy R-530 was their only receiver geared towards the hobbyist. It offered technology and performance that was cutting-edge for the time. The R-530 provided unprecedented value and frequency coverage.

Galaxy Electronics was acquired by Hygain Electronics, who in turn combined with Telex to become Telex Hygain. Telex Hygain continues to manufacture high quality amateur antennas and rotors. When Telex bought Hygain they did not acquire the Galaxy Division and it ceased to exist. Telex Hygain does not have parts for Galaxy.

Please see *Chapter 105 Briefly Mentioned,* for Galaxy fixed receivers.

Galaxy Electronics
10 South 34th St.
Council Bluffs, IA 51501   1968

Hygain Electronics Corp.
Galaxy Electronics Subsidiary
8601 N.E. Hwy. 6
Lincoln, NE 68507

Hygain Electronics Corp.
4900 Superior St.
Lincoln, NE 68504

 GALAXY

R-530

General Coverage Communications Receiver
Double Conversion Superheterodyne. Solid State.

Features:
- ¼" Head. Jack
- RF Gain
- Mute Line
- Speaker Jack

- S/AF-Meter
- BFO ±1 kHz.
- Preselector
- VFO In/Out Jack

- Attenuator -20 dB
- Calibrator 50 kHz
- Line Out Terminal
- AGC Output Jack

- AVC SLOW/FAST
- Noise Blanker
- Detector Out Jack
- Standby

Specifications:
Coverage ......500 - 30000 kHz.
Selectivity ......2.1/_/_ kHz -6dB.
Sensitivity ......<.1 μV 6 dB S/N SSB
Audio Out ......1 W 10% distortion
BFO Range ...±1 kHz.
Aud. Resp. ....250-3000 Hz -3 dB
IF ....41.625-42.125, 9 MHz

Modes .... AM/LSB/USB
Stability .... 100 Hz.
IF Rejection .... >50 dB
Dial Accuracy . ±1 kHz.
Intermod Dis. .. 3rd order > 50 dB
Antenna Input . SO-239 50 ohm

Accessories:
SPK530 Speaker   CL530 Clock (inside SPK530)   FL5306 6 kHz AM Filter
RPA530 Rack Kit   FL5305 500 Hz CW Filter   FL53015 1.5 kHz Filter

Comments:
A general coverage receiver with a ±1 kHz dial accuracy was exceptional at the time. Not often seen on the used market. Probably the finest piece of equipment ever produced by Galaxy. Possibly the first example of an American manufacturer using the Wadley loop design. The R-530 is a collector's item. Please also see the military version of this receiver called the model R-1530 on the following page.

Made In: United States 1967-1973
Voltages: 115/230 AC 50/60 Hz
24 W or 18 VDC .6 A

Readout: ⌊⌊⌊⌊▮⌋⌋ Analog Lin.

Physical: Cabinet Version:
17x6x14" 25 Lbs.
431x152x355mm 11.3 kg
Rack Version:
19x8.75x14" 25 Lbs
483x222x355mm 11.3 kg

Status: Inactive Manufacturer
Discontinued Model

Rarity: Very Scarce

Reviews: *CQ* Feb. 1969
QST May 1969
Ham Radio April 1968

	New	Used
Price:	$695-895	$490-750
Rating:	★★★★★	★★★★

R-1530

General Coverage Communications Receiver
Double Conversion Superheterodyne. Solid State.

Features:
- ¼" Head. Jack
- RF Gain
- Mute Line
- Speaker Jack
- VFO In Jack
- S/AF-Meter
- BFO ±1 kHz.
- Preselector
- VFO Out Jack
- Attenuator -20 dB
- Noise Blanker
- Line Out Terminal
- AGC Output Jack
- AVC FST/SLO/OFF
- Calibrator 50 kHz
- Standby
- Detector Out Jack

Specifications:
Coverage ...... 10 - 30000 kHz.
Selectivity ...... 2.1/1.8/_ kHz -6dB.
Sensitivity ...... <.25 V 6 dB SSB w/Pre
Audio Out ...... 1 W 10% distortion
BFO Range ... ±1 kHz.
Aud. Resp. .... 250-3000 Hz -3 dB
IF .... 41.625-42.125, 9 MHz

Modes .... AM/LSB/USB
Stability .... 100 Hz.
IF Rejection .... 50 dB
Dial Accuracy . ±1 kHz.
Intermod Dis. ... 3rd order > 50 dB
Antenna Input . SO-239 and Terminals

Circuit Complement:
Permeability tuned, phase lock oscillator, dual gate MOSFET RF amp and preselector, hot carrier diodes in first mixer, J-FET second mixer.

Accessories:
SC530 Speaker   CL530 Clock (inside SPK530)   FL5306 6 kHz AM Filter
RPA530 Rack Kit   FL5305 500 Hz CW Filter   FL53015 1.5 kHz Filter

Comments: This is the military version of the R-530. Hygain had purchased Galaxy by this time and the R-1530 was labeled and sold as the Hygain/Galaxy R-1530.

Made In:	United States 1967-1975	
Voltages:	115/230 AC 50/60 Hz	
	24 W or 18 VDC .6 A	
Readout:	Analog Lin.	
Physical:	Cabinet Version:	
	17x6x14" 25 Lbs.	
	431x152x355mm 11.3 kg	
	Rack Version:	
	19x8.75x14"	
	483x222x356mm	
Status:	Inactive Manufacturer	
	Discontinued Model	
Rarity:	Extremely Scarce	
Reviews:	*WRTH* 1976	

	New	Used
Price:	①	$470-800
Rating:	★★★★★	★★★★

31 Geloso

John Geloso was born in Argentina where his parents had temporarily moved from Italy. The entire family moved back to Savona, Italy in 1904 where John studied at the Nautical school. After finishing school John started an electromechanical workshop where he manufactured items that he had personally patented.

In 1920 he left Italy for America and worked for Pilot Electric Manufacturing in New York and attended Copper Square University. After graduation he became the chief engineer at Pilot. In 1928 he was involved with the first transmissions of television (using a mechanical spinning disk). In 1931 he returned to Italy and founded the John **Geloso S.A** company in Milano. The main production was home radios, televisions and tape recorders. In 1932 he began publication of *Bollettino tecnico Geloso*, a valuable technical bulletin of the time. During the 1950's and 1960's made semiprofessional receivers, transmitters and subassemblies for the amateur radio market.

Their most famous amateur band receiver was the G 209-R. Geloso also produced a G 222TR transmitter that had a very similar appearance. Geloso amateur equipment was very popular in Europe in the 1960's. John Geloso died in 1968. The Company went into receivership in the mid-1970's. Their former headquarters is now owned by a bank. Please see *Chapter 105 Briefly Mentioned* for the interesting Geloso G 512 and G 512-L models.

Geloso
Viale Brenta 29
Milano, Italy 1931-1973

American Geloso Electronics
251 Park Avenue
New York, NY

G 4/214

Amateur Band Communications Receiver
Double Conversion Superheterodyne. 12 Tubes.

Features:
- ¼" Head. Jack
- RF Gain
- NL
- S-Meter
- Standby
- Dial Lamp
- Antenna Trimmer
- Mute Line
- Dial Reset
- Calibrator
- 46:1 Tune Ratio
- Speaker Output

Specifications:
Coverage ...... Ham. See Comments
Selectivity ...... 5 Position
Sensitivity ...... <1µV 1W AF Output
Image Rej. ..... >50 dB
Audio Out ...... 1.5 W 3.2/500 ohms
Modes .... AM/CW/SSB
Stability .... ±1 kHz/MHz
IFs .... 4.6 MHz, 467 kHz
IF Rej. .... >70 dB
Antenna Input . Coaxial & Terminals

Circuit Complement:
6DC6 RF, 6BE6 1st Mixer, 12AT7 Crystal Oscillator/Buffer, 6BE6 2nd Mixer, 12AU7 Crystal Osc., 6BA6 IF Amp, 6AL5 Audio Detector/AVC/NL, 12AX7 BFO, 6BE6 Mixer, 12AX7 AF Amp/RF Osc., 6BQ5 AF Amp, 0A2 Voltage Stabilizer, VR625A Current Stabilizer plus selenium rectifiers.

Comments: Ranges: 3.5-4, 7-7.3, 14-14.4, 21-21.5, 26-28 and 28-30 MHz. The dial is also precalibrated (in red) for the 144-146, 146-148 MHz bands if the optional 2 meter converter is employed. The G 4/214 may be mounted in a standard 19 inch rack, taking up 8.75 inches of vertical space.

Variants: The model **G 4/218** is the continuous coverage, shortwave version covering .5-30 MHz. It has a speaker in the upper right corner of the front panel and has fewer knobs.

Made In: Italy 1964-1966

Voltages: 110/125/140/160/220 VAC
50/60 Hz 90W @160 VAC

Readout: Analog

Physical: 20x10x10.25" 38 Lbs
508x254x260mm 17.2 kg

Status: Inactive Manufacturer
Discontinued Model

Rarity: Very Scarce

	New	Used
Price:	①	⑤
Rating:	⑥	⑥

G 4/215

Amateur Band Communications Receiver
Single Conversion Superheterodyne. 11 Tubes.

Features:
- ¼" Head. Jack
- S-Meter
- Speaker Output
- Rack Handles
- RF Gain
- Standby
- Mute Line
- 46:1 Tune Ratio
- NL
- Dial Lamp
- BFO

Specifications:

Coverage ...... Ham. See Comments   Modes .... AM/CW/SSB
Selectivity ...... 5 Position   Stability .... ±50 Hz/MHz
Sensitivity ...... <1µV 1W AF Output   IF .... 467 kHz
Image Rej. ..... >50 dB   IF Rej. .... >70 dB
Audio Out ...... 1 W 3.2/500 ohms   Antenna Input . 50-100 ohms

Circuit Complement:
6BZ6 RF Amp, (2) 12AT7 Oscillator/Buffer, 6BE6 1st Mixer, ECH81 2nd Mixer, EF93 IF Amp, EF93 IF Amp, 12AX7 BFO, 6BE6 Product Detector, ECL86 Pre-Amplifier AF Amplifier, 0A2 Voltage Stabilizer, ZF10 Current Stabilizer plus selenium rectifiers.

Accessories:
4/161 2 Meter Converter

Comments:
Ranges: 3.5-4, 7-7.5, 14-14.5, 21-21.5 and 28-30 MHz. The dial is also precalibrated (in red) for the 144-146, 146-148 MHz bands if the optional 2 meter converter is employed. Matches the G 4/228 transmitter. The G 4/215 may be mounted in a standard 19 inch rack, taking up 8.75 inches of vertical space.

Made In:	Italy	1964-1966
Voltages:	110/125/140/160/220 VAC 50/60 Hz 90W @160 VAC	
Readout:	Analog	
Physical:	20x10x10.25" 28 Lbs 508x254x260mm 12.7 kg	
Status:	Inactive Manufacturer Discontinued Model	
Rarity:	Very Scarce	

	New	Used
Price:	①	⑤
Rating:	⑥	⑥

G 4/216

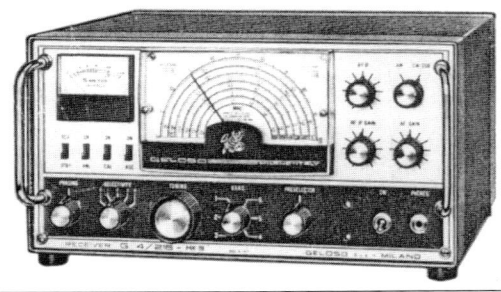

Amateur Band Communications Receiver
Double Conversion Superheterodyne. 11 Tubes.

Features:
- ¼" Head. Jack
- S-Meter
- Speaker Output
- Antenna Trimmer
- RF Gain
- Standby
- Mute Line
- 46:1 Tune Ratio
- ANL
- Dial Lamp
- Preselector
- Calibrator Reset
- Rack Handles
- BFO
- Calibrator

Specifications:

Coverage ...... Ham. See Comments   Modes .... AM/CW-SSB
Selectivity ...... 5 Position   Stability .... ±50 Hz/MHz
Sensitivity ...... <1µV 1W AF Output   IF .... 3.5-4 MHz, 467 kHz
Image Rej. ..... >50 dB   IF Rej. .... >70 dB
Audio Out ...... 1 W 3.2/500 ohms   Antenna Input . 50-100 ohms

Circuit Complement:
6BZ6 RF Amp, (2) 12AT7 Oscillator/Buffer, 6BE6 1st Mixer, ECH81 2nd Mixer, EF93 IF Amp, EF93 IF Amp, 12AX7 BFO, 6BE6 Product Detector, ECL86 Pre-Amplifier AF Amplifier, 0A2 Voltage Stabilizer, ZF10 Current Stabilizer plus selenium rectifiers.

Accessories:
4/161 2 Meter Converter   4/163 432-436 MHz Converter

Comments: Ranges: 3.5-4, 7-7.5, 14-14.5, 21-21.5 and 28-30 MHz. The dial is also precalibrated (in red) for the 144-146, 146-148 MHz bands if the optional 2 meter converter is employed. Matches the G 4/228 transmitter. The G 4/216 may be mounted in a standard 19 inch rack, taking up 8.75 inches of vertical space.

Variants: Model **G 4/216 MkII** and **G 4/216 MkIII** were later production.

Made In:	Italy	1964-1966
Voltages:	110/125/140/160/220 VAC 50/60 Hz 90W @160 VAC	
Readout:	Analog	
Physical:	20x10x10.25" 28 Lbs 508x254x260mm 12.7 kg	
Status:	Inactive Manufacturer Discontinued Model	
Rarity:	Very Scarce	

	New	Used
Price:	①	⑤
Rating:	⑥	⑥

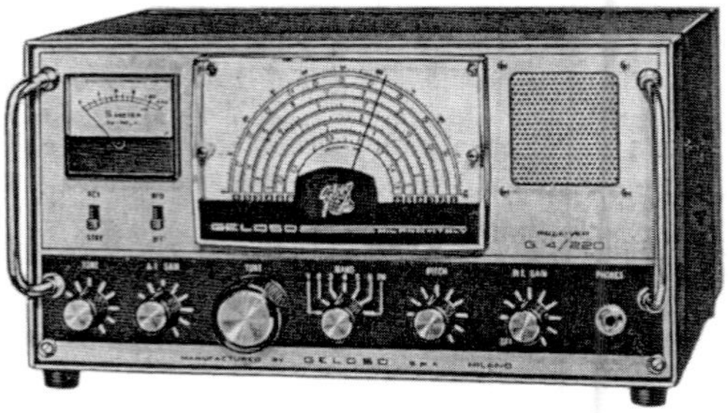

G 4/220

General Coverage Communications Receiver
Single Conversion Superheterodyne. 11 Tubes plus Semiconductors.

Features:
- ¼" Head. Jack
- RF Gain
- Dial Lamp
- S-Meter
- Standby
- BFO
- Speaker Output
- Mute Line
- Speaker
- Rack Handles
- 46:1 Tune Ratio
- Tone Control

Specifications:
Coverage ...... See Comments
Selectivity ...... 5 Position
Sensitivity ...... <1µV 1W AF Output
Audio Out ...... 1 W 500 ohm
Modes .... AM/CW
Stability .... ±1 kHz/MHz
IF .... 1900 kHz
Antenna Input . 50-100 ohms

Circuit Complement:
EF89 RF Amp, ECC82 Osc., ECH81, EF89 IF Amp, EF89 2nd IF Amp, 6AL5, ECC81, ECL86, 0A2 Voltage Stabilizer plus diodes.

Comments:
Ranges: .53-1.6, 2.2-6, 6-9, 9-13.8, 13.5-20.6 and 20.5-30.5 MHz.

Made In: Italy

Voltages: 110/240 VAC
50-60 Hz 55 VA

Readout: Analog

Physical: 15.3x7.5x10.6" 20 Lbs.
390x190x270mm 9 kg

Status: Inactive Manufacturer
Discontinued Model

Rarity: Very Scarce

	New	Used
Price:	①	⑤
Rating:	⑥	⑥

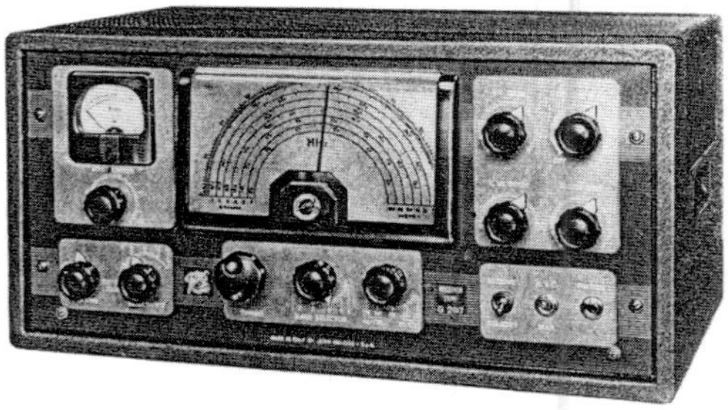

G 207-CR

Amateur Band Communications Receiver
Double Conversion Superheterodyne. 14 Tubes.

Features:
- ¼" Head. Jack
- RF Gain
- NL
- S-Meter
- Standby
- Dial Lamp
- Antenna Trimmer
- Mute Line
- Dial Reset
- Calibrator
- 46:1 Tune Ratio
- Speaker Output

Specifications:
Coverage ...... Ham. See Comments
Selectivity ...... 5 Position
Sensitivity ...... <1µV 1W AF Output
Image Rej. ..... >50 dB
Audio Out ...... 2.5 W 3.2/500 ohms
Modes .... AM/CW
Stability .... ±1 kHz/MHz
IFs .... 4.6 MHz, 467 kHz
IF Rej. .... >70 dB
Antenna Input . Coaxial & Terminals

Circuit Complement:
6CB6 RF, 6BE6 1st Mixer, 12AU7 Crystal Oscillator/Buffer, 6BA6 IF Amp, 6BE6 2nd Mixer, 6BA6 IF Amp, 6BA6 IF Amp, 6AU6 NBFM Limiter, 6AL5 NBFM Ratio Detector, 6AL5 Detector/AVC, 6SL7 BFO/Audio Amp, 6AL5 NL, 6V6 Power Output, 5V4 Rectifier and VR150 Voltage Stabilizer.

Comments:
Ranges: 3.5-4, 6.95-7.5, 13.8-14.6, 20.6-22, 26.4-28.1 and 28-29.8 MHz. The G 207-R may be mounted in a standard 19 inch rack, taking up 8.75 inches of vertical space.

Made In: Italy 1954-1956

Voltages: 110/125/140/160/220 VAC
50/60 Hz 100W @160 VAC

Readout: Analog

Physical: 20x10x10.25" 29 Lbs
508x254x260mm 13 kg

Status: Inactive Manufacturer
Discontinued Model

Rarity: Very Scarce

	New	Used
Price:	①	⑤
Rating:	⑥	⑥

G 208

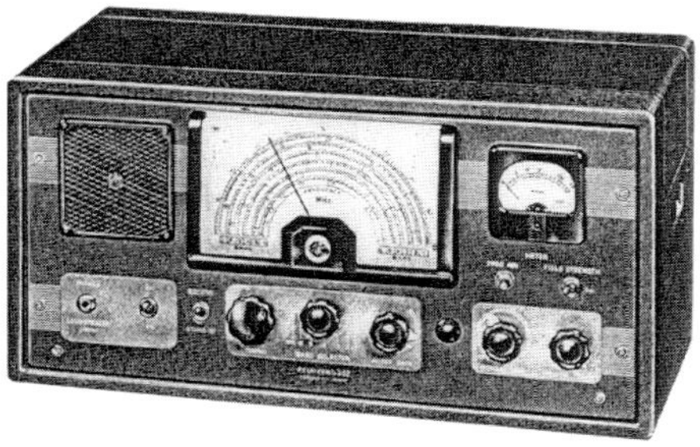

Amateur Band Communications Receiver
Single Conversion Superheterodyne. 8 Tubes.

Features:
- ¼" Head. Jack
- S-Meter
- Tone Control
- Speaker
- Standby
- Mute Line
- Dial Lamp
- Speaker Output
- Meter Zero Adj.

Specifications:
Coverage ...... Ham. See Comments
Sensitivity ...... <1µV .05W AF Output
Audio Out ...... 2.5 W 500 ohms
Modes .... AM/CW
IFs .... 467 kHz
Antenna Input . Terminals

Circuit Complement:
EF41 RF Amp, ECH42 Mixer, 6BA6 IF Amp, 6AL5 2nd Detector/AVC, 12AX7 AF Preamplifier/BFO, 6V6 Power Output and 5V4 Rectifier.

Accessories:
1481/6 6 Volt Vibrapack
1482/12 12 Volt Vibrapack

Comments:
This radio has six bands. May be either operated from main or optional battery supply.

Made In: Italy

Voltages: 110/125/140/160/220 VAC
42-60 Hz .45A @160 VAC

Readout: Analog

Physical: 20x10x10.25" 34 Lbs
508x254x260mm 15.4 kg

Status: Inactive Manufacturer
Discontinued Model

Rarity: Very Scarce

	New	Used
Price:	①	⑤
Rating:	⑥	⑥

G 209-R

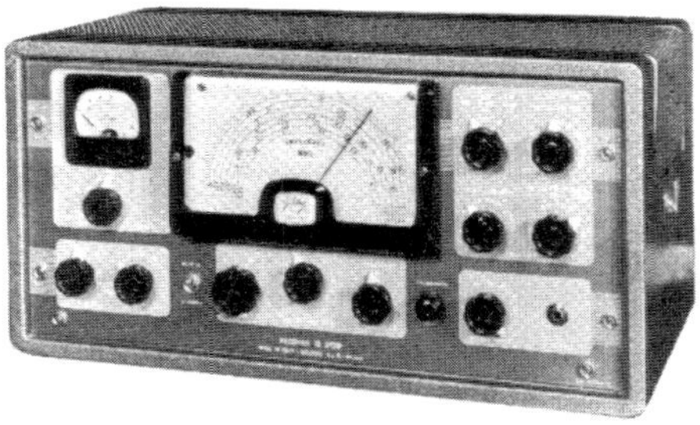

Amateur Band Communications Receiver
Double Conversion Superheterodyne. 13 Tubes.

Features:
- ¼" Head. Jack
- S-Meter
- Antenna Trimmer
- Calibrator 3500 kHz.
- RF Gain
- Standby
- Mute Line
- 46:1 Tune Ratio
- NL
- Dial Lamp
- Dial Reset
- Speaker Output

Specifications:
Coverage ...... Ham. See Comments
Selectivity ...... 5 Position
Sensitivity ...... <1µV 1W AF Output
Image Rej. ..... >50 dB
Audio Out ...... 2.5 W 3.2/500 ohms
Modes .... AM/SSB-CW
Stability .... 500 Hz/MHz
IFs .... 4.6 MHz, 467 kHz
Antenna Input . Coaxial & Terminals

Circuit Complement:
6BA6 RF Amp, 12AU7 Oscillator Buffer, 6BE6 Mixer, 6BE6 Mixer, 12AU7 Crystal Oscillator, 6BA6 IF Amp, 6BA6 IF Amp, 6T8 Audio/AVC Detector/BFO, 6BE6 SSB Mixer, 6AL5 Noise Limiter, 12AX7 AF Amp/Calibration Oscillator, 6AQ5 AF Amp, 0A2 Voltage Stabilizer, 6H6 Current Stabilizer and Selenium Rectifier.

Comments:
Ranges: 3.5-4, 6.95-7.5, 13.8-14.6, 20.6-22, 26.4-28.1 and 28-29.8 MHz. Matches the G 222-TR transmitter. The G 209-R may be mounted in a standard 19 inch rack, taking up 8.75 inches of vertical space.

Made In: Italy 1958-1961

Voltages: 110/125/140/160/220 VAC
50/60 Hz 90 W @160 VAC

Readout: Analog

Physical: 20x10x10.25" 38 Lbs
508x254x260mm 17.2 kg

Status: Inactive Manufacturer
Discontinued Model

Rarity: Very Scarce

Reviews: *QST* July 1959

	New	Used
Price:	$300	$150-210
Rating:	⑥	⑥

32 General Dynamics

General Dynamics is a huge American defense contractor. They are far better known for their submarines than their radios. This radio, designed by Bendix, replaced many R-390's in the 1960's. Many radios feature a Bendix label.

The R-1051/URR H.F. receiver does appear from time to time at hamfests, sometimes at very modest prices. You may be able to obtain one at 1 or 2 percent of what Uncle Sam paid for his.

General Dynamics Electronics Div.
177 Admiral Cochrane Dr.
Annapolis, MD 21401

GENERAL DYNAMICS

R-1051/URR

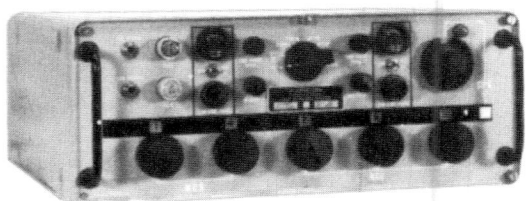

General Coverage Communications Receiver
Triple Conversion Superheterodyne. Tubes plus Semiconductors.

Features:
- ¼" Head Jack (2) • AF-Meters (2)   • Fine Tuning   • Rack Handles
- RF Gain   • ISB   • AGC FST/SLO   • Modular Construction
- Line Out 600 ohms

Specifications:

Coverage ...... 2000 - 30000 kHz	Modes .... AM/LSB/USB/CW/ISB/FSK	
Selectivity ...... 7/3 kHz -6dB.	Stability .... 1×10^8 per day	
Sensitivity ...... <.6µV CW 10 dB S/N	Environment ... 0° to +50° C.	
Audio Out ...... 600/1200 ohm	Ant. Jack .... 50 ohm	
Image Rej. ..... 90 dB		

Accessories: External LF Converter (14-2000 kHz).

Comments: The R-1051 was manufactured for the U.S. Navy and is optimized for Independent Sideband (ISB) reception. Double conversion on lower frequencies. The receiver is designed for rapid frequency change with a minimum of operator adjustments. The RF stages are tuned by a motor driven turret which operates with the MHz and 100 kHz knobs. Frequency selection is via decadic knobs. May be mounted in 19" rack or in supplied cabinet. A vernier knob is provided for tuning between 500 Hz synthesizer points. [¥95,000-120,000].

Variants: The model **R-1051A/URR** reportedly provided 500 Hz tuning increments. Model **R-1051B/URR** features a 100 Hz synthesizer. Model **R-1051F/URR** was manufactured by Stewart-Warner Corporation. Model **R-1051G/URR** was later production utilizing integrated circuits in the synthesizer (used value $600-625).

Made In: United States 1960-1981

Voltages: 115 VAC ±10%
48-450 Hz 70W

Readout: ●●●●●⓪. Digital Knobs

Physical: 17.375x7x18.5" 75 Lbs.
441x177x470mm 34 kg

Status: Active Manufacturer
Discontinued Model

Rarity: Scarce

Reviews: *Electric Radio* Nov. 1993

	New	Used
Price:	①	$300-550
Rating:	⑥	⑥

The **General Electric Co. Ltd. of England** can trace its origins to the Wireless Telegram and Signal Company founded in 1897. In 1900 the company name changed to Marconi Wireless Telegraph Company. The Marconi International Marine Communication Company Limited was later formed.

Please also see the Marconi Chapter in this book.

Eddystone is now a wholly owned subsidiary of G.E.C. Marconi Communications Limited. This company in turn is owned by the Italian arm of GEC Marconi, Marconi SpA.

General Electric Co. Ltd. of England
Magnet House
Kingsway, London WC2
England   1955

General Electric Co. Ltd. of England
Electronics Division
Ford Street
Coventry, England   1961

G.E.C.

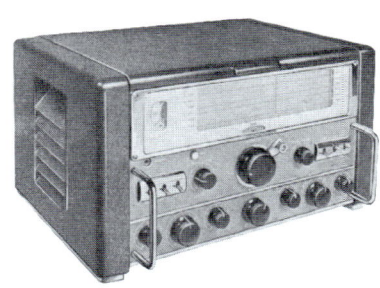

BRT 400D

General Coverage Communications Receiver
Single Conversion Superheterodyne. 15 Tubes.

Features:
- ¼" Head. Jack
- S-Meter
- Speech/Music
- Bandspread 0-3200
- RF Gain
- IF Gain
- Antenna Trimmer
- Flywheel Tuning
- Dial Lamp
- FSK Line Out
- Line Out Jack
- Dial Lock
- BFO
- Rack Handles
- AGC ON/OFF
- Audio Filter
- Standby
- NL

Specifications:
Coverage ...... 150 - 30000 kHz
Selectivity ...... 13/9/5.5/2/1/.5 kHz -6dB
IF .... 455 kHz
Audio Out ...... 2.5/15 ohms 2W 5%
Modes .... AM/CW
Image Rej. ...... 30-100 dB
Antenna Input . Terminals

Circuit Complement: 6BA6 1st RF Amp, 6BA6 2nd RF Amp, 6BE6 1st Detector, N77 Local Osc, 6BA6 1st IF Amp, 6BA6 2nd IF Amp, 6AT6 2nd Det./IF AGC Delay/ 1st Audio Amp, D63 NL/RF AGC Delay, N709 Output, Z77 AGC Amp, Z77 BFO, S130P Stabilizer, N709 Smoothing Valve, U52 Rect. and Z77 Calibrating Osc.
Accessories: Crystal Calibrator 500 kHz.   BRT401 PS 12VDC
Comments: Ranges: .15-.385, .51-1.3, 1.3-3.2, 3.2-8.5, 8.5-20 and 20-30 MHz. The bandspread is obtained by means of a 64:1 slow-motion flywheel drive.
Variants: Earlier model **BRT 400** (1948) 14 tubes, covers 150-33000 kHz. Model **BRT 402** is rack version of BRT 400. Model **BRT 400E** is same as BRT 400D, but with Calibrator. **BRT 402E** is rack version of BRT 400E. Model **BRT 400ES** was special production produced for the Swedish government (differences unknown). A localized version for the Swedish Army was **MT600** and the Navy **M50**.

Made In: England   1952-1957

Voltages: 95-130, 195-250 VAC
40-80 Hz 135 W

Readout: ⊥⊥⊥⊥■⊥ Analog

Physical: Cabinet Version:
20.2x11.7x17.25" 82 Lbs
513x297x438mm 37.2 kg
Rack Version:
19x10.5x17.25" 79 Lbs
483x267x438mm 35.8 kg

Status: Active Manufacturer
Discontinued Model

Rarity: Extremely Scarce

	New	Used
Price:	①	②
Rating:	⑥	⑥

BRT 402K

General Coverage Communications Receiver
Single Conversion Superheterodyne. 15 Tubes.

Features:
- ¼" Head. Jack
- RF Gain
- Dial Lamp
- NL
- Rack Handles
- S-Meter
- IF Gain
- FSK Line Out
- AGC ON/OFF
- Speech/Music
- BFO
- Standby
- Line Out Jack
- Dial Lock
- Crystal Calibrator
- Bandspread 0-3200
- Antenna Trimmer
- Audio Filter 1 kHz
- Flywheel Tuning

Specifications:
Coverage ......550 - 30000 kHz
Selectivity ...... 13/9/5.5/2/1/.5 kHz -6dB
IF ....455 kHz
Audio Out ......2.5/15 ohms 2W 5%

Modes .... AM/CW
Image Rej. ...... 30-100 dB
Antenna Input . 75 Ohms

Circuit Complement: Please see model BRT 400D
Accessories:
BRT401 PS 24 VDC
Comments: Ranges: .16-.41, .51-1.3, 1.3-3.2, 3.2-8.5, 8.5-20 and 20-30 MHz. The mechanical bandspread is obtained by means of a 64:1 slow-motion flywheel drive. A rack mount version was also produced. This model was used by the monitoring stations of the B.B.C. and Radio Australia.
Variants:
Model **BRT 400K** is the table top version. Model **BRT 400KN** is table version with a 9 kHz Audio Rejector Filter. Model **BRT 402KN** is the rack version of the BRT 400KN.

Made In: England   1961-1964

Voltages: 95-130, 195-250 VAC
40-80 Hz 135 W

Readout: Analog

Physical: Cabinet Version:
20.2x11.7x17.25" 82 Lbs
513x297x438mm 37.2 kg
Rack Version:
19x10.5x17.25" 79 Lbs
483x267x438mm 35.8 kg

Status: Active Manufacturer
Discontinued Model
Rarity: Extremely Scarce

	New	Used
Price:	①	$350-400
Rating:	⑥	⑥

It is believed that **Globe Electronics** produced only one shortwave communications receiver. The Globe 'Ceiver was a transformerless "all American five", and was not a stand-out. It is not know whether this model was produced at the main plant in Rockford, Illinois or the Western plant in Los Angeles, or elsewhere.

It is also not certain when Globe Electronics became the GC Electronics Company.

Globe is better known for their amateur transmitters such as the Globe Scott Deluxe SD-75A and Globe King 500C.

Globe Electronics Co.
Division of Textron Electronics Inc.
22-30 South 34th Street
Council Bluffs, IA   1960-1961

GC Electronics Co.
Division of Textron Electronics Inc.
400 S. Wyman St.
Rockford, Illinois   1962-1964

GC Electronics Co.
Division of Textron Electronics Inc.
3225 Exposition Place
Los Angeles 18, California 1962-1964

Globe 'ceiver

General Coverage Communications Receiver
Single Conversion Superheterodyne. 5 Tubes.

Features:
- ¼" Head. Jack
- Dial Lamp
- Speaker
- Telescopic Whip
- Standby
- Bandspread 0-100
- MW Ferrite Antenna

Specifications:
Coverage ......550 - 30000 kHz
IF ....455 kHz
Audio Out ......8 ohms

Modes .... AM/CW
Antenna Input . Terminals

Circuit Complement:
12BE6 Converter, 12BA6 IF Amp, 12AV6 Detector Amp, 50C5 Audio Output and 35W4 Rectifier. This radio does not have a power transformer.

Comments:
Ranges: .55-1.6, 1.6-4.5, 4.5-12 and 11.5-30 MHz. The 7½ inch dial has the amateur bands noted and the Conelrad indicators at 640 and 1240 kHz. This model is also referred to as the World-Wide Globe' Ceiver. It is also referred to as GC Electronics model #65-320.

Made In: United States 1962-1963

Voltages: 105-125 VAC 50-60 Hz 30 W

Readout: Analog

Physical:

Status: Inactive Manufacturer Discontinued Model

Rarity: Very Scarce

	New	Used
Price:	$52	$40-50
Rating:	⑨	⑨

35 Gonset

The **Gonset Radio Company** was established in 1955. They are best known for their VHF receivers and G-66 mobile amateur band receiver and matching G-77 transmitter.

Their table top amateur band receivers and general coverage receivers were not as widely produced.

Gonset was purchased by the Young Spring and Wire Corporation in 1957.

Hobbyists should be careful if they come upon a *Gonset Super-Ceiver* (model 3041). This device was designed to serve as the "tail end" for an amateur band converter; the combination of the two making a complete receiver. It will not function unless it is matched with a *Gonset Super-Six Converter*, or equivalent.

Gonset
801 S. Main St.
Burbank, CA   1955-1957

Gonset
Div. of Young Spring & Wire Corp.
801 S. Main St.
Burbank, CA   1957-1962

Gonset Inc.
LTV Ling Altec, Inc.
1515 S. Manchester Ave.
Anaheim, CA   1968

G-33

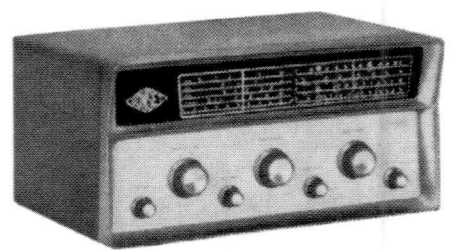

General Coverage Communications Receiver
Single Conversion Superheterodyne. 6 Tubes.

Features:
- ¼" Head. Jack
- Speaker 4"
- Bandspread
- Antenna Trimmer
- Dial Light
- Speaker Out

Specifications:
Coverage ......540 - 34000 kHz
IF ....1650 kHz.
Modes .... AM/CW
Antenna Input . Terminals

Circuit Complement:
6BE6 Converter, 6BA6 1st IF, 6BA6 2nd IF/BFO, 6AV6 Detector/AVC/1st Audio, 6CM6 2nd Audio and 6X4 Rectifier.

Accessories:
3285 Speaker   3269 Crystal Calibrator 100 kHz

Comments:
Ranges: .54-1.6, 1.8-6, 6-13 and 13-34 MHz. Drum type dial. An early example of the use of printed circuits. Gonset model #3222.

Made In: United States 1958-1961

Voltages: 110-120 VAC

Readout: Analog

Physical: 16.5x8x10" 22 Lbs.
419x203x254mm 10 kg

Status: Inactive Manufacturer
Discontinued Model

Rarity: Scarce

	New	Used
Price:	$90	$40-50
Rating:	★★★	⑦

G-43

General Coverage Communications Receiver
Single Conversion Superheterodyne. 8 Tubes.

Features:
- ¼" Head. Jack
- S-Meter
- Speaker
- Antenna Trimmer
- Sensitivity
- ANL
- Mute Line
- Bandspread
- Dial Light
- Standby
- BFO
- Flywheel Tuning

Specifications:

Coverage ......540 - 30000 kHz
Selectivity ......6 kHz -6dB.
Sensitivity ......<3 µV

Modes .... AM/CW-SSB
IF .... 1650 kHz
Antenna Input . Terminals

Circuit Complement:
6BE6 Converter, 6BA6 1st IF, 6BA6 2nd IF, 6AU6 3rd IF, 6AL5 Detector/ANL/AVC, 12AX7 1st Audio/BFO, 6CM6 2nd Audio and 6X4 Rectifier. Three IF stages and 12 tuned circuits.

Accessories:
3285 Speaker   3269 Crystal Calibrator 100 kHz

Comments:
Ranges: .54-1.6, 1.9-5.7, 5.7-13, 13-20, 20-25 and 25-30 MHz. The individual ranges are on a turret-type band display. Each range (except MW) also displays a bandspread amateur band. The ¼" headphone jack, which also serves as an external speaker jack is found on the back of the radio. The G-43 uses printed circuit board technology. Gonset #3241.

Made In:	United States 1958-1961
Voltages:	110-120 VAC
Readout:	Analog
Physical:	16.5x8x10.5" 26 Lbs. 419x203x267mm 11.8 kg
Status:	Inactive Manufacturer Discontinued Model
Rarity:	Scarce

	New	Used
Price:	$160	$50-80
Rating:	★★★	⑦

G-63

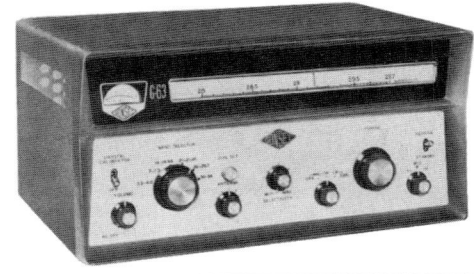

Amateur Band Communications Receiver
Double Conversion Superheterodyne. 11 Tubes.

Features:
- ¼" Head. Jack
- S-Meter
- Speaker
- Antenna Trimmer
- Sensitivity
- ANL
- Mute Line
- Bandspread
- Dial Light
- Standby
- Dial Set
- BFO
- Speaker Jack

Specifications:

Coverage ......Ham Bands 80-6M
Selectivity ......3.3-.2 kHz -6dB.
Sensitivity ......<1 µV -6 dB S/N
IF ....2065, 263 kHz
Audio Out ......3 W 3.2 ohm 5% dist.

Modes .... AM/CW-SSB
Stability .... .01% during warm-up
Image Rej .... >40-86 dB
Antenna Input . Terminals 50-100 ohms

Circuit Complement:
6BZ6 RF Amp, 6U8A Mixer/HF Osc, 6BE6 Converter, 12AX7 AF Amp/Q-Multiplier, 6BA6 IF Amp, 6BA6 IF Amp, 6AL5 AM Detector/AVC/ANL, 6BE6 Product Detector/BFO, 6AQ5 Audio Out, 5Y3GT Rectifier and 0B2 Voltage Regulator.

Accessories:
3285 Speaker   3269 Crystal Calibrator 100 kHz

Comments:
Ranges: 3.5-4, 7-7.3, 14-14.35, 21-21.45, 28-29.7 and 50-54 MHz. A turret-type band display shows one band at a time. The rear panel has two ¼" audio jacks. One can be used for phones and one for a speaker.

Made In:	United States 1960-1961
Voltages:	117 VAC 50-60 Hz 65W
Readout:	Analog
Physical:	16.5x8x10" 22 Lbs. 419x203x254mm 10 kg
Status:	Inactive Manufacturer Discontinued Model
Rarity:	Scarce

	New	Used
Price:	$240	$90-100
Rating:	★★★	⑦

G-66

Amateur Band Communications Receiver
Double Conversion Superheterodyne. 9 Tubes

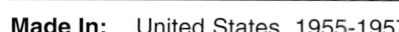

Features:
- ¼" Head. Jack
- S-Meter
- BFO
- Antenna Trimmer
- ANL
- Mute Line
- Dial Light

Specifications:

Coverage ...... Ham Bands 160-10 +MW	Modes .... AM/CW-SSB	
Selectivity ...... 4 kHz -6dB.	IF .... 2050, 265 kHz	
Sensitivity ...... <1.5 μV -10 dB S/N	Image Rej .... >60 dB	
Audio Out ...... 3 W 8% dist..	Antenna Input . Motorola	

Circuit Complement:
6DC6 RF Amp, 6U8 Mixer-Buffer, 6C4 Local Osc, 6BE6 Conv, 6AU6 IF Amp, 6AL5 Det/AVC/ANL, 6AW8 Audio Amp/BFO, 6AQ5 Audio Output and OB2 Voltage Reg.

Accessories:
3098 PS 12VDC
3069 PS/Spkr 115VAC/12VDC

Comments: Designed for mobile use. Headphone jack on the left side. Ranges: .54-2, 3.5-4, 7-7.3, 14-14.35 21-21.45 and 28-29.7 MHz. Requires speaker (unless used with 3069 Power Supply / Speaker).

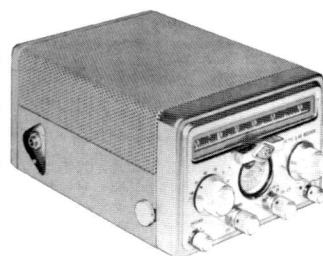

Variants: The more common model **G-66B** 1957 $189-210 has logo above S-meter and incorporates an additional stage of I.F. amplification (6BH6) which improves noise limiting on the higher frequencies.

Made In:	United States 1955-1957
Voltages:	Optional 6/12 VDC PS or optional 115 VAC PS
Readout:	Analog
Physical:	6.5x4.5x9" 8 Lbs. 165x114x229mm 3.6 kg
Status:	Inactive Manufacturer Discontinued Model
Rarity:	Scarce
Reviews:	*QST* June 1956 *CQ* July 1956

	New	Used
Price:	$170-210	$100-160
Rating:	★★★★	⑦

GONSET

GR-211

General Coverage Communications Receiver
Single Conversion Superheterodyne. 5 Tubes plus Semiconductors.

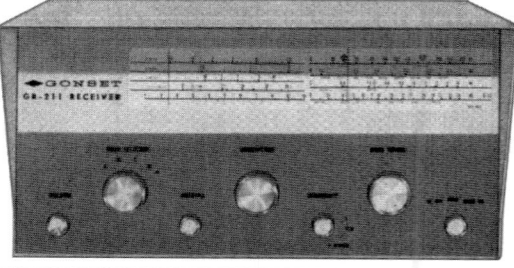

Features:
- Dial Lamp
- Speaker
- Antenna Trimmer
- Bandspread
- Flywheel Tuning

Specifications:

Coverage ...... 540 - 34000 kHz	Modes .... AM/CW
Antenna Input Terminals	

Circuit Complement:
5 Tubes and 2 solid state rectifiers.

Comments:
Bandspread for 80 to 10 meters plus CB.

Made In:	United States 1961-1963
Voltages:	115 VAC
Readout:	Analog
Physical:	16.25x8x10" 413x203x254mm
Status:	Inactive Manufacturer Discontinued Model
Rarity:	Scarce

	New	Used
Price:	$70	$35-50
Rating:	★★★	⑦

 # GONSET

GR-212

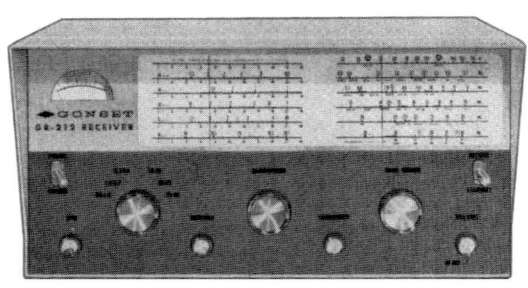

General Coverage Communications Receiver
Double Conversion Superheterodyne. 7 Tubes plus Semiconductors.

Features:
- ¼" Head. Jack
- S Meter
- Speaker
- Bandspread
- Sensitivity
- BFO
- Mute Line
- Antenna Trimmer
- Dial Light
- Standby
- Flywheel Tuning

Specifications:
Coverage ......550 - 34000 kHz   Modes .... AM/CW-SSB
IF ....1650, 455 kHz   Antenna Input . Terminals
Sensitivity ......<1 µV 6 dB S/N CW

Circuit Complement:
6U8A Mixer/HF Osc, 6BA6 IF Amp, 6BE6 2nd Mixer/Osc, 6BA6 2nd IF, 6AL5 Detector/ANL, 12AX7 1st Audio/BFO and 6AQ5 Audio Amp.

Comments:
Ranges: .55-1.8, 1.8-5.7, 5.7-13, 13-20, 20-25 and 25-30 MHz. Bandspread for 80 through 10 meters.

Made In:	United States 1961-1963
Voltages:	117 VAC 60 Hz
Readout:	Analog
Physical:	16.25x8x10.5" 20 Lbs. 413x203x267mm 9.1 kg
Status:	Inactive Manufacturer Discontinued Model
Rarity:	Scarce
Rarity:	*QST* May 1962

	New	**Used**
Price:	$100-110	$50-70
Rating:	★★★	⑦

36 Hagenuk

Hagenuk has a long and proud history in telecommunications. This German company was founded in 1899 by Neufeldt and Kuhnke and originally made telegraph equipment for naval ships.

In 1937 Neufeldt & Kuhnke merge with Hanseatische Apparatebaugesellschaft. The new name for the company became: Hagenuk (Hanseatische Apparatebaugesellschaft Newfeldt und Kuhne).

Radio and telephony transmission products were the company's main business until the 1980's. The focus then shifted to both fixed and mobile digital telephony. The Hagenuk Company is currently producing cellular telephones and satellite equipment.

Also please see Eddystone 830/7 entry in this book. The 830/12 variant was sold as the Hagenuk E81. Also please see the Eddystone 1004 entry in this book. The 1004 variant was sold as the Hagenuk E92. Also the Eddystone 958/H is the Hagenuk E91.

Hagenuk
Westring 431-451
Postfach 500
2300 Kiel, Germany

E 80

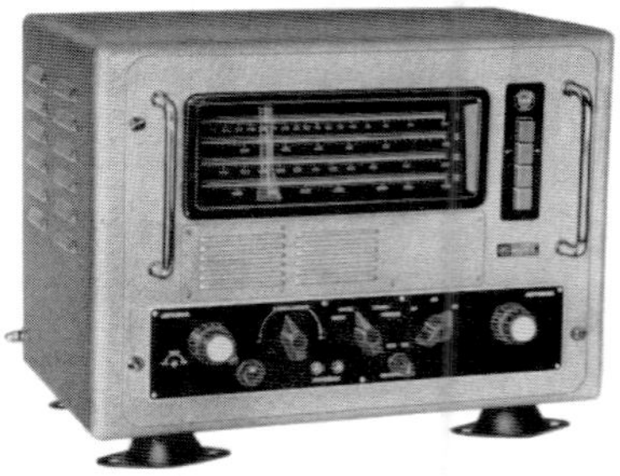

Marine Band Communications Receiver
Single Conversion Superheterodyne. 6 Tubes.

Features:
- ¼" Head. Jack
- Ext. Spkr Jack
- Line Out
- Rack Handles
- Standby
- Speaker
- AVC/MVC
- Dial Lamp
- BFO ±2.5 kHz
- 1:32 Fine Ratio Tuning

Specifications:
Coverage ...... See Comments.
Selectivity ...... 5.3/3.1/1.1/.2 kHz -3 dB
Sensitivity .... .5 to 3µV CW 3:1 S/N
Audio Out ...... 3 W
Modes .... AM/CW/MCW
IF .... 580 kHz
Antenna Input . 75 ohm unbalanced

Circuit Complement:
EF85, (3) EBF89, ECH81 and EL95

Accessories:
12 VDC PS
Fixed Frequency option for 6 channels.

Comments:
Ranges: .1-.264, .25-.53, .62-1.66 and 1.6-4.35 MHz. Please note that the E80 is more correctly classified as an LF/MF marine receiver. This receiver was commonly found on non-convention ships (ships less than 1600 tons). The E80KM variant listed below has more traditional coverage.

Variants:
Model **E 80KM** is a special type approved version for the Deutsche Bundespost. It has different tuning ranges. They are: .25-.53, 1.6-4.35, 4.2-10.15, 10-16.6 and 16.5-25.1 MHz.

Made In: Germany 1956-1968

Voltages: 220 VAC 50 Hz
40 W

Readout: Analog

Physical: 18.5x14.4x12" 55 Lbs
475x365x307mm 25 kg

Status: Active Manufacturer
Discontinued Model

Rarity: Typically Unavailable

	New	Used
Price:	①	⑤
Rating:	⑥	⑥

E 408R

General Coverage Communications Receiver
Double Conversion Superheterodyne. Solid State.

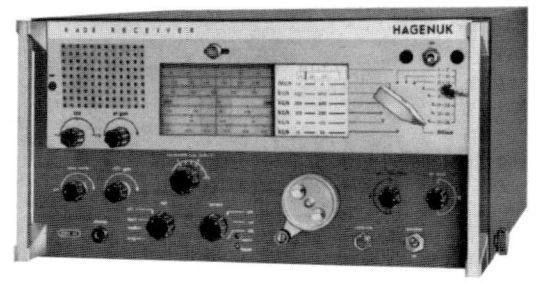

Features:
- ¼" Head. Jack
- BFO ±3 kHz
- Speaker
- 500 kHz Preset
- RF Gain
- ANL (Variable)
- Antenna Trimmer
- Calibrator 100 kHz
- Mute
- Fine Tuning
- Semi-Modular
- AGC OFF/SHT/MED/LNG
- IF Out Jack
- AGC Out Jack
- Dial Lamp
- Line Out 600 ohm
- Rack Handles
- Fiduciary Adjust
- Ext. Speaker Jack

Specifications:

Coverage ...... 13 - 28000 kHz	Modes .... AM/CW/MCW/LSB/USB
Selectivity ...... 8/3/1 kHz -6dB.	Stability .... +55° Hz after warmup
Sensitivity ...... <1µV above 650 kHz	IF Rejection .... >90 dB 3-28 MHz
Image Rej ...... >70 dB 3-28 MHz	IF .... See Comments
BFO Range ... ±3 kHz.	Environment ... -15° to +55° C. 95% Humid
Audio Out ...... 0.5 W 10 ohms	Ant. Jack .... 75 ohm

Comments:
Ranges: .013-.036, .036-.1, .1-.25, .25-.65, .65-1.6, 1.5-4, 4-7, 7-10, 10-13, 13-16, 16-19, 19-22, 22-25 and 25-28 MHz. Single conversion below 650 kHz. A 90 inch film tuning scale, in conjunction with the 100 kHz crystal calibrator, provides very accurate frequency readout. A 24 VDC and 100-125 VAC versions of this receiver were also produced. The first IF is 470, 1500 or 4500 kHz depending on frequency. The second IF is 80 kHz. Selectivity is continuously variable form 800 Hz to 8 kHz for AM and 800 Hz to 4 kHz for SSB. This receiver is designed for the marine market.
Variants: This model is nearly identical to the Redifon R408.

Made In: England? 1968

Voltages: 220-250 VAC
50 Hz 17W

Readout: |,,,,█,,| Analog

Physical: 17.5x8.75x19.75" 55 Lbs.
445x223x502mm 25 kg

Status: Active Manufacturer
Discontinued Model

Rarity: Typically Unavailable

	New	Used
Price:	①	⑤
Rating:	⑥	⑥

UE 12

General Coverage Communications Receiver
Double Conversion Superheterodyne. 9 Tubes.

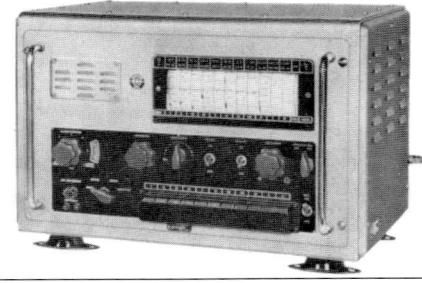

Features:
- ¼" Head. Jack
- Rack Handles
- AVC/MVC
- Two Tuning Speeds
- Ext. Spkr Jack
- Standby
- Dial Lamp
- 500 kHz Preset
- Line Out
- AF Filter
- RF Gain
- Calibrator 100 kHz
- BFO ±2.5 kHz
- Speaker
- 1:64 Fine Tune Ratio

Specifications:

Coverage ...... 95 - 28000 kHz.	Modes .... AM/CW/MCW
Selectivity ...... 5/1/.2 kHz -3 dB	Image Rej. ...... >90-60 dB 1.6-16.5 MHz
Sensitivity ...... <.1 to 1µV	Image Rej. ...... >60-34 dB 16.6-28 MHz
IF Reject. .... >60 dB	Antenna Input . 75 ohm unbalanced
Audio Out ...... 1 W	

Circuit Complement:
(2) UG85, (2) UCH81, (2) UBF80, (2) UCC85 and UCl82. Double conversion only above 1.6 MHz.

Accessories:
LF Converter
Fixed Frequency option for 5 channels.

Comments:
The UE12 has 11 ranges. From 95 - 1700 kHz the IF is 80 Hz. From 1.6-28 MHz the IFs are 1522 and 80 kHz. This model was often the standard receiver on Swedish flagships into the early 1980's.

Made In: Germany 1956-1968

Voltages: 110/220 VAC 50 Hz or
110/220 VDC

Readout: |,,,,█,,| Analog

Physical: 24.6x15.2x17.7" 88 Lbs
625x385x450mm 40 kg

Status: Active Manufacturer
Discontinued Model

Rarity: Typically Unavailable

	New	Used
Price:	①	⑤
Rating:	⑥	⑥

The **Hallicrafters** Company was founded by Bill Halligan in 1932. The first receivers were hand made (thus the name *Halli* for Halligan and *crafters* for hand crafted). To increase production and obtain the proper manufacturing licenses, Halligan purchased the Silver-Marshall Company. This business struggled to an end in 1934. Halligan then combined with the then anemic Echophone Company.

By the end of the 1930's the company was moving along well, bringing out new models to meet each new circuit advancement as well as making transmitters. Hallicrafters produced all types of equipment for the war effort. After World War II the company began producing consumer electronics items including clock radios, TVs and phonographs. The 1950's were very good for Hallicrafters, but the company was sold twice, ultimately to Northrop in 1966. Hallicrafters did not prosper under the wing of Northrop. By the mid 1970's sales were down sharply and Northrop sold the company in 1975. Efforts to successfully continue and revive the company have failed. Chuck Dachis' book *Radios By Hallicrafters* is a must-read for any Hallicrafters enthusiast or collector. Collectors should note that many low priced models of the 1940's did not feature power transformers. Caution should be followed when dealing with these models to avoid shock or electrocution. Also note that Hallicrafters shortwave receivers with an "X" in the model number usually indicated that a crystal filter or a fixed crystal position was used. This rule did not apply to their VHF radios.

Hallicrafters Co.
Chicago 16, Ill.   1945-1947

Hallicrafters Co.
4401 W. Fifth Ave.
Chicago 24, Ill.   1950-1966

Hallicrafters Co.
Subsidiary of Northrop Corp.
600 Hicks Rd.
Rolling Meadows, IL 60018 1968-1972

---

the hallicrafters co.

5R10

General Coverage Broadcast Receiver
Single Conversion Superheterodyne. 5 Tubes.

Features:
- Tip Head. Jack   • Standby   • Bandspread 0-100
- Speaker 5"   • Dial Lamp   • Speaker-Headphone Switch

Specifications:
Coverage ......540 - 31000 kHz   Modes .... AM
IF ....455 kHz   Antenna Input . Terminals
Audio Out ......1 W 3.2 ohms

Circuit complement:
12SA7 Converter, 12SG7 IF Amp, 12SQ7 Detector/Audio Amp, 50L6GT Audio Output and 35Z5GT Rectifier.

Comments:
Ranges: .55-1.6, 1.7-5, 5-14 and 13-31 MHz. Black cabinet. The speaker is on the top cover of the radio and the headphone tips are on the back panel. This series appears to be the predecessor to the S-38D. [¥35,000].

Variants:
Model **5R100** is the same but with a Hammertone grey cabinet. Models **5R10A** and **5R100A** were later production.

Made In:	United States 1952-1954
Voltages:	105-125 VAC 50/60 Hz 30W or 105-125 VDC
Readout:	Analog
Physical:	13x7.5x8.875" 11 Lbs. 330x191x225mm 5 kg
Status:	Inactive Manufacturer Discontinued Model
Rarity:	Scarce

	New	Used
Price:	$40	$45-65
Rating:	⑥	⑦

the hallicrafters co.

5R30
Continental

Shortwave Broadcast Receiver
Single Conversion Superheterodyne. 5 Tubes.

Features:
• Speaker 5"   • Dial Lamp

Specifications:
Coverage ......540 - 16000 kHz   Modes .... AM
IF ....455 kHz   Antenna Input . Terminals
Audio Out ......3.2 ohms

Circuit complement:
12SA7 Converter, 12SK7 IF Amp, 12SQ7 Detector/AVC/Audio Amp, 50L6GT Audio Output and 35Z5GT Rectifier.

Comments:
Ranges: .55-1.6 and 6-18 MHz. This model has a black cabinet and black knobs except the volume knob is red.

Variants:
Several identical models were produced in different colors. Model **5R31** has a teal cabinet and knobs except the volume knob is white. Model **5R32** has a brown cabinet and knobs except the volume knob is white. Model **5R33** has a beige cabinet and knobs except the volume knob is white. Model **5R34** has a powder blue cabinet and knobs except the volume knob is white. Models **5R30A, 5R31A, 5R32A, 5R33A** and **5R34A** were latter production using different tubes. The tube line up for this series was: 12BE6 Converter, 12BA6 IF Amp, 12AV6 Detector/AVC/Audio Amp, 50C5 Audio Output and 35W4 Rectifier. Models **5R50, 5R51** and **5R52** are similar but with a clock-timer feature.

Made In:	United States 1951-1953	
Voltages:	105-125 VAC 50/60 Hz 30W or 105-125 VDC	
Readout:	Analog	
Physical:		
Status:	Inactive Manufacturer Discontinued Model	
Rarity:	Very Scarce	

	New	**Used**
Price:	$30	$30-50
Rating:	⑥	⑦

the hallicrafters co.

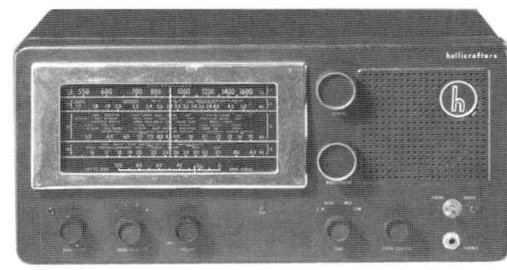

8R40

General Coverage Communications Receiver
Single Conversion Superheterodyne. 7 Tubes.

Features:
• ¼" Head. Jack   • BFO   • Standby   • Bandspread 0-100
• Sensitivity   • Speaker 5"   • BFO   • Flywheel Tuning
• Phono Input   • Tone (3 Pos.) • Hinged Top   • Speaker Terminals
• Dial Lamp

Specifications:
Coverage ......550 - 44000 kHz   Modes .... AM/CW
IF ....455 kHz.   Antenna Input . Terminals
Audio Out ......1 W 10% Dist.

Circuit Complement:
6SG7 RF Amp, 6SA7 Converter, (2) 6SK7 IF Amp, 6H6 Detector/AVC, 6SC7 AF Amp, 6K6GT Audio Output and either 5Y3 or selenium rectifiers.

Comments:
Ranges: .54-1.69, 1.7-5.5, 5.-15.6 and 15.4-44.2 MHz. Black cabinet with silver trim. [Photo by Joe Veras N4QB].

Variants:
Model **8R40C** was in a wooden console with record player.

Made In:	United States 1950-1954	
Voltages:	105-125 VAC 50/60 Hz	
Readout:	Analog	
Physical:		
Status:	Inactive Manufacturer Discontinued Model	
Rarity:	Scarce	

	New	**Used**
Price:	$90	$50-90
Rating:	⑥	⑦

the hallicrafters co.

S-20R
Sky Champion

General Coverage Communications Receiver
Single Conversion Superheterodyne. 9 Tubes.

Features:
- ¼" Head. Jack
- Mute Terminals
- Speaker 5"
- Dial Light
- Standby
- AVC ON/OFF
- RF Gain
- Flywheel Tuning
- ANL
- BFO
- Tone Hi/Med/Lo
- Hinged Top Cover
- Bandspread 0-100
- Speaker Terminals
- Jack for S-Meter

Specifications:

Coverage ......550 - 44000 kHz	Modes .... AM/CW
IF ....455 kHz.	Antenna Input . Terminals 400 ohm
Audio Out ......2.5 W	

Circuit Complement:
6SK7 RF Amp, 6K8 1st Detector-Mixer/HF Osc, 6SK7 1st IF Amp, 6SK7 2nd IF Amp, 6SQ7 2nd Detector/AVC/1st Audio Amp, 6F6G 2nd Audio Amp, 6H6 ANL, 6J5GT BFO and 80 Rectifier.

Accessories:
SM-20 Ext. S-Meter

Comments:
Ranges: .55-1.78, 1.74-5.4, 5.3-15.8 and 15.5-44 MHz. Gray metal cabinet with stainless steel trim.

Variants:
The S-20R replace the prior model **S-20** and had 8 tubes 1938 $50.

Made In: United States 1939-1945

Voltages: 117 VAC

Readout: Analog

Physical: 18.5x8.5x9.375" 32 Lbs. 470x216x238mm 14.5 kg

Status: Inactive Manufacturer Discontinued Model

Rarity: Common

Reviews: *Radio* February 1940

	New	Used
Price:	$50-60	$50-125
Rating:	★★★	⑦

---

the hallicrafters co.

S-22R
Skyrider Marine

Marine Communications Receiver
Single Conversion Superheterodyne. 8 Tubes.

Features:
- ¼" Head. Jack
- Mute Terminals
- Speaker 5"
- Hinged Cover
- Standby
- AVC
- RF Gain
- Dial Light
- ANL
- BFO
- Tone
- Bandspread 0-100
- Speaker Terminals
- Jack for S-Meter

Specifications:

Coverage ......110 - 18000 kHz	Modes .... AM/CW
IF ....1600 kHz.	Antenna Input . Terminals
Audio Out ......2.5 W	

Circuit Complement:
6SK7 RF Amp, 6K8 1st Detector-Mixer/HF Osc, 6SK7 1st IF Amp, 6SK7 2nd IF Amp, 6SQ7 2nd Detector/AVC/1st Audio Amp, 25L6 2nd Audio Amp, 6J5 BFO and 25Z5 Rectifier.

Comments:
Reception gap from 1500-1700 kHz. Ranges: .11-.41, .4-1.5, 1.7-5.9, and 5.3-18 MHz. Designed for marine use. Black wrinkle metal cabinet

Variants:
The earlier model **S-22** 1938 was similar, but the entire 360 degrees of the tuning dial was exposed rather than just a window as on the S-22R. The S-22 covered 140-18500 kHz and had a 465 kHz IF.

Made In: United States 1941-1945

Voltages: 110 VAC or 110 VDC 50W

Readout: Analog

Physical: 18.5x8.5x9.375" 31 Lbs. 470x216x238mm 14.1 kg

Status: Inactive Manufacturer Discontinued Model

Rarity: Common

	New	Used
Price:	$75	$30-90
Rating:	★★★	⑦

the hallicrafters co.

S-35

General Coverage Communications Receiver
Single Conversion Superheterodyne. 29 Tubes.

Features:
- ¼" Head. Jack
- S-Meter
- Phono Input Jack
- Tone
- NL
- Sensitivity
- Bandspread
- Dial Lock
- Speaker Terminals
- Flywheel Tuning
- AVC ON/OFF
- Dial Lamps
- Hinged Top Cover
- Mute Line
- Bass Switch
- Standby
- Antenna Trimmer
- Vertical Gain
- Horizontal Gain
- Sweep Width
- Panoramic Display

Specifications:
Coverage ...... 550 - 43000 kHz
Selectivity ...... 6 Positions
Audio Out ...... 8 W 500/5000 ohms
Modes .... AM/CW
IF .... 455 kHz
Antenna Input . Terminals

Circuit Complement:
6AB7 1st RF Amp, 6SK7 2nd RF Amp, 6SA7 Mix., 6SA7 HF Osc, 6L7 1st IF ANL, 6SK7 2nd IF Amp, 6B8 AVC Amp, 6B8 2nd Det./S Meter, 6AB7 Noise Amp, 6H6 Noise Rect., 6J5 BFO, 6SC7 1st Audio Amp, (2) 6V6GT Audio Amps and 5Z3 Rect. Tubes in the panoramic display: 6SG7 455 kHz Input Amp, 6SA7 1st Det., 6SK7 100 kHz IF Amp, 6SQ7 2nd Det./Vert. Amp, 6SN7GT Sawtooth Osc., 6SJ7 Return Trace Blanking, 6AC7 Reactance Modulator, 6J5 RF Osc., 6SC7 Horiz. Amp, 2X2/879 High Volt. Rect., 80 Low Volt. Rect., (2) VR150 Volt. Reg. and 5AP1 C.R.T.

Comments: Ranges: .55-1.6, 1.6-3, 3-5.8, 5.8-11, 11-21 and 21-43 MHz. Bandspread: 3.5-4, 7-7.3, 14-14.4 and 28-30 MHz. Up to 100 kHz segments can be viewed. The **S-35 system** consists of an SX-28A receiver and a spectrum display. The spectrum display was also sold separately and could be used on other receivers.

Made In: United States 1943-1945

Voltages: 105-125 VAC 50/60 Hz 120W

Readout: Analog

Physical: 20.5x18.675x18" 105 Lbs. 521x475x457mm 47.6 kg

Status: Inactive Manufacturer Discontinued Model

Rarity: Extremely Scarce

	New	Used
Price:	①	⑤
Rating:	⑥	⑦

the hallicrafters co.

S-38

General Coverage Communications Receiver
Single Conversion Superheterodyne. 6 Tubes.

Features:
- Tip Head. Jack
- ANL
- Standby
- Bandspread 0-100
- Mute Terminals
- Speaker 5"
- BFO
- Spkr.-Phone Switch
- Dial Light

Specifications:
Coverage ...... 540 - 32000 kHz
Selectivity ...... 7 kHz -6dB.
Sensitivity ...... <11µV @5MHz 50 mW
Audio Out ...... 1.6 W
Modes .... AM/CW
IF .... 455 kHz.
Antenna Input . Terminals

Circuit Complement:
12SA7 Converter, 12SK7 IF Amp, 12SQ7 BFO/NL, 12SQ7 Detector/AVC, 35L6GT Audio Output and 35Z5GT Rectifier.

Comments:
Ranges: .540-1.65, 1.65-5, 5-14.5 and 13.5-32 MHz. Black metal cabinet. The S-38 series did not feature a power transformer. The S-38 was produced in large numbers.

Made In: United States 1946-1949

Voltages: 105-125 VAC 50/60Hz 29W or 105-125 VDC

Readout: Analog

Physical: 12.875x7x7.25" 11 Lbs. 327x178x184mm 5 kg

Status: Inactive Manufacturer Discontinued Model

Rarity: Common

Reviews: *CQ* July 1946
CQ September 1986

	New	Used
Price:	$40-49	$30-60
Rating:	★★★	⑦

the hallicrafters co.

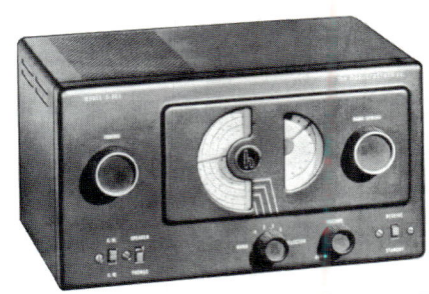

S-38A

General Coverage Communications Receiver
Single Conversion Superheterodyne. 5 Tubes.

Features:
- Tip Head. Jack
- Standby
- BFO
- Bandspread 0-100
- Mute Terminals
- Speaker 5"
- Dial Light
- Spkr.-Phone Switch

Specifications:
Coverage ......540 - 32000 kHz
IF ....455 kHz.
Audio Out ......1.6 W

Modes ....AM/CW
Antenna Input . Terminals

Circuit Complement:
12SA7 Converter, 12SK7 IF Amp/BFO, 12SQ7 Detector/AVC, 50L6 Audio Output and 35Z5GT Rectifier.

Comments:
Ranges: .540-1.65, 1.65-5, 5-14.5 and 13.5-32 MHz. Similar to model S-38 sans CW Pitch knob and NL switch. Black cabinet.

Made In:	United States 1947-1950	
Voltages:	105-125 VAC 50/60 Hz 105-125 VDC	
Readout:	Analog	
Physical:	12.875x7x7.25" 14 Lbs. 327x178x184mm 6.4 kg	
Status:	Inactive Manufacturer Discontinued Model	
Rarity:	Common	

	New	Used
Price:	$40	$35-50
Rating:	★★★	⑦

---

the hallicrafters co.

S-38B

General Coverage Communications Receiver
Single Conversion Superheterodyne. 5 Tubes.

Features:
- Tip Head. Jack
- Standby
- Mute Line
- Bandspread 0-100
- Speaker 5"
- Dial Lamp
- BFO
- Speaker Terminals
- Spkr/Phone Switch

Specifications:
Coverage ......540 - 32000 kHz
IF ....455 kHz.
Audio Out ......1.6 W 3.2 ohms

Modes ....AM/CW
Antenna Input . Terminals

Circuit Complement:
12SA7 Converter, 12SK7 IF Amp/BFO, 12SQ7 Detector/AVC, 50L6 Audio Out and 35Z5GT Rectifier.

Comments:
Ranges: .540-1.65, 1.65-5.1, 5-14.5 and 13-32 MHz. Black wrinkle steel cabinet. [¥35,000].

Made In:	United States 1948-1952	
Voltages:	105-125 VAC 50/60 Hz 30W or 105-125 VDC	
Readout:	Analog	
Physical:	12.875x7x7.25" 14 Lbs. 327x178x197mm 6.4 kg	
Status:	Inactive Manufacturer Discontinued Model	
Rarity:	Common	

	New	Used
Price:	$49	$35-55
Rating:	★★★	⑦

the hallicrafters co.

S-38C

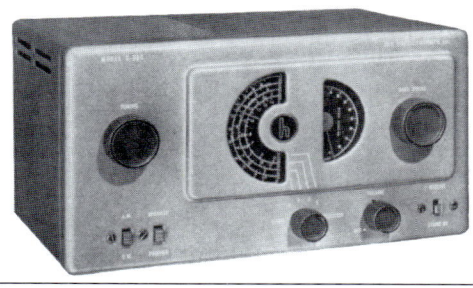

General Coverage Communications Receiver
Single Conversion Superheterodyne. 5 Tubes.

Features:
- Tip Head. Jack
- Standby
- Mute Line
- Bandspread 0-100
- Speaker 5"
- Dial Lamp
- Speaker Terminals
- Spkr/Phone Switch

Specifications:
Coverage ......540 - 32000 kHz
IF ....455 kHz.
Audio Out ......1.6 W

Modes .... AM/CW
Antenna Input . Terminals

Circuit complement:
12SA7 Converter, 12SK7 IF Amp/BFO, 12SQ7 Detector/AVC, 50L6GT Output and 35Z5GT Rectifier.

Comments:
Ranges: .54-1.65, 1.65-5, 5-14.5 and 13.5-32 MHz. Has black dial with white lettering. Gray steel cabinet. The headphone tip jack is on the back panel.

Made In:	United States 1952-1954
Voltages:	105-125 VAC 50/60 Hz 30W or 105-125 VDC
Readout:	Analog
Physical:	12.875x7x7.75" 12 Lbs. 327x178x197mm 5.4 kg
Status:	Inactive Manufacturer Discontinued Model
Rarity:	Common

	New	**Used**
Price:	$50	$35-55
Rating:	★★★	⑦

---

the hallicrafters co.

S-38D

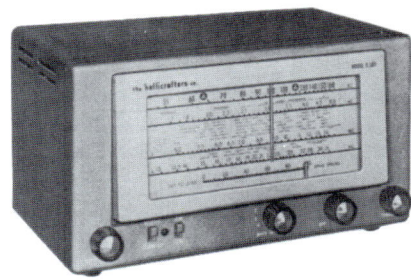

General Coverage Communications Receiver
Single Conversion Superheterodyne. 5 Tubes.

Features:
- Tip Head. Jack
- Standby
- Bandspread 0-100
- Speaker 5"
- Dial Lamp
- Speaker-Headphone Switch

Specifications:
Coverage ......540 - 32000 kHz
IF ....455 kHz
Audio Out ......1 W 3.2 ohms

Modes .... AM/CW
Antenna Input . Terminals

Circuit complement:
12SA7 Converter, 12SG7 IF Amp/BFO, 12SQ7 Detector/Audio Amp, 50L6GT Audio Out and 35Z5GT Rectifier.

Comments:
Ranges: .55-1.6, 1.7-5, 5-14 and 13-30 MHz. Gray cabinet. The headphone tip jack is on the back panel. See also earlier model 5R10A.

Made In:	United States 1954-1957
Voltages:	105-125 VAC 50/60 Hz 30W
Readout:	Analog
Physical:	13x7.5x8.875" 11 Lbs. 330x191x225mm 5 kg
Status:	Inactive Manufacturer Discontinued Model
Rarity:	Common

	New	**Used**
Price:	$50-60	$35-60
Rating:	★★★	⑦

the hallicrafters co.

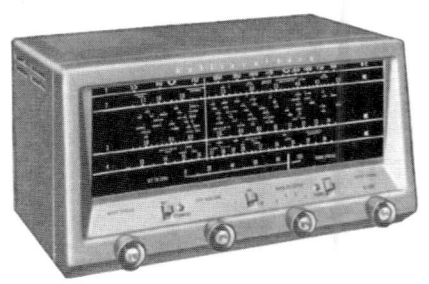

S-38E

General Coverage Communications Receiver
Single Conversion Superheterodyne. 5 Tubes.

Features:
- Head. Tip Jack
- Standby
- Mute Line
- Bandspread 0-100
- Speaker 5"
- Dial Lamp
- Speaker-Phone Switch

Specifications:
Coverage ...... 540 - 32000 kHz	Modes .... AM/CW
IF .... 455 kHz	Antenna Input . Terminals
Audio Out ...... 1W 3.2 ohms	

Circuit complement:
12BE6 Converter, 12BA6 IF Amp/BFO, 12AU6 Amp, 50C4 Audio Amp and 35W4 Rectifier.

Comments:
Ranges: .54-1.6, 1.6-5, 5-14 and 13-30 MHz. Gray metal cabinet with silver trim. The headphone tip jack is on the back panel.

Variants:
Model **S-38EM** has a mahogany colored painted steel case with gold trim. Model **S-38EB** has a "blond" colored metal case with gold trim.

Made In:	United States 1956-1961
Voltages:	105-125 VAC 50/60 Hz 30W
Readout:	Analog
Physical:	12.875x7x9.25" 13 Lbs. 327x178x235mm 5.9 kg
Status:	Inactive Manufacturer Discontinued Model
Rarity:	Common
Reviews:	*Radio Bygones* #40

	New	Used
Price:	$55-60	$25-50
Rating:	★★★	⑦

---

the hallicrafters co.

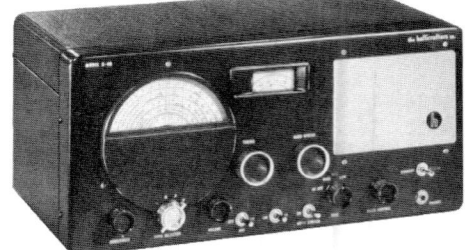

S-40

General Coverage Communications Receiver
Single Conversion Superheterodyne. 9 Tubes.

Features:
- ¼" Head. Jack
- Pitch
- Standby
- Bandspread 0-100
- Sensitivity
- Speaker
- BFO
- Flywheel Tuning
- Mute Terminals
- Tone (3 Pos.)
- Dial Lamp
- Speaker Terminals
- ANL
- AVC ON/OFF
- Hinged Top Cover

Specifications:
Coverage ...... 540 - 43000 kHz	Modes .... AM/CW
Selectivity ...... 6.8 kHz -6dB	IF .... 455 kHz.
Sensitivity ...... 8µV for 500 mw @40MHz	Antenna Input . Terminals
Audio Out ...... 1 W 10% Dist.	

Circuit Complement:
6SG7 RF Amp, 6SA7 Converter, (2) 6SK7 IF Amp, 6F6G 2nd Audio Amp, 6H6 ANL/AVC, 6J5GT BFO, 6SQ7 Detector/AF Amp and 80 Rectifier.

Accessories:
SM-40 External S-Meter (shown)   SP-44 Panoramic Unit

Comments:
Ranges: .54-1.7, 1.7-5.35, 5.35-15.7 and 15.7-43 MHz.

Variants:
Model **S-40U** 110/130/150/220/250 VAC.

Made In:	United States 1946-1947
Voltages:	105-125 VAC 50/60 Hz
Readout:	Analog
Physical:	18.5x9x11" 28 Lbs. 470x229x280mm 12.7 kg
Status:	Inactive Manufacturer Discontinued Model
Rarity:	Common
Reviews:	*QST* July 1946

	New	Used
Price:	$79	$30-80
Rating:	★★★	⑦

---

the hallicrafters co.

S-40A

General Coverage Communications Receiver
Single Conversion Superheterodyne. 9 Tubes.

Features:
- ¼" Head. Jack
- Pitch
- Standby
- Bandspread 0-100
- Sensitivity
- Speaker 5"
- BFO
- AVC ON/OFF
- Mute Terminals
- Tone (3 Pos.)
- Hinged Top
- Speaker Terminals
- ANL
- Dial Lamp

Specifications:
Coverage ......540 - 43000 kHz   Modes .... AM/CW
Selectivity ......6.8 kHz -6dB   IF .... 455 kHz.
Sensitivity ......8µV for 500mw @40MHz Antenna Input . Terminals
Audio Out ......1 W 3.2 ohms 10% Dist.

Circuit Complement:
6SG7 RF Amp, 6SA7 Converter, (2) 6SK7 IF Amp, 6H6 ANL/AVC, 6J5GT BFO, 6SQ7 Detector/AF Amp., 6FG6 Output and 5Y3GT Rectifier. One RF and two IF stages.

Accessories:
SM-40 Ext. S-Meter

Comments:
Ranges: .54-1.7, 1.7-5.3, 5.3-15.7 and 15.7-43 MHz. Black metal cabinet.

Variants:
Model **S-40AU** 110/130/150/220/250 VAC 25-60 Hz $100-111. Please see model S-52 which is an AC-DC version of the S-40A.

Made In:	United States 1947-1950
Voltages:	105-125 VAC 50/60 Hz 75W
Readout:	Analog
Physical:	18.5x9x11" 30 Lbs. 470x229x280mm 13.6 kg
Status:	Inactive Manufacturer Discontinued Model
Rarity:	Common

	New	**Used**
Price:	$80-110	$30-80
Rating:	★★★	⑦

the hallicrafters co.

S-40B

General Coverage Communications Receiver
Single Conversion Superheterodyne. 8 Tubes.

Features:
- ¼" Head. Jack
- Pitch
- AVC ON/OFF
- Bandspread 0-100
- Sensitivity
- Speaker 5"
- BFO
- ANL
- Mute Terminals
- Tone (3 Pos.)
- Dial Lamp
- Hinged Cover

Specifications:
Coverage ......540 - 43000 kHz   Modes .... AM/CW
Selectivity ......6.8 kHz -6dB.   IF .... 455 kHz.
Audio Out ......1 W 3.2 ohms   Antenna Input . Terminals 50-600 ohm

Circuit Complement:
6SG7 RF Amp, 6SA7 Converter, (2) 6SK7 IF Amp, 6H6 ANL/AVC, 6SL7 BFO/Detector, 6FG6 Output and 5Y3GT Rectifier. One RF and two IF stages.

Comments:
Ranges: .54-1.7, 1.7-5.3, 5.3-15.7 and 15.7-43 MHz. Black steel cabinet.

Variants:
Model **S-40BU** is an export version of the S-40B operating 115/250 VAC 25-60 Hz $100-130. Models **S-77** and **S-77A** are AC-DC versions of the S-40B $100-130 new.

Made In:	United States 1950-1955
Voltages:	105-125 VAC 50/60 Hz 75W
Readout:	Analog
Physical:	18.5x9x11" 33 Lbs. 470x229x279mm 15 kg
Status:	Inactive Manufacturer Discontinued Model
Rarity:	Common

	New	**Used**
Price:	$89-130	$30-80
Rating:	★★★	⑦

the hallicrafters co.

S-41G
Skyrider Jr.

General Coverage Communications Receiver
Single Conversion Superheterodyne. 6 Tubes.

Features:
- Tip Head. Jack
- Standby
- BFO
- Bandspread 0-100
- Sensitivity
- Speaker
- Dial Lamp
- Spkr.-Phone Switch
- ANL

Specifications:
Coverage ...... 550 - 30000 kHz   Modes .... AM/CW
IF .... 455 kHz.   Antenna Input . Terminals

Circuit Complement:
12SA7 Converter, 12SK7 IF Amp, 12SQ7GT Detector/AVC/AF, 12SQ7GT BFO/ANL, 35L6GT Output and 35Z5GT Rectifier.

Comments:
Ranges: .55-2.1, 2.1-7.7 and 7.7-30 MHz. Gray and black metal cabinet. This model is physically indentical to the Echophone EC-1A. Please see the Echophone chapter.

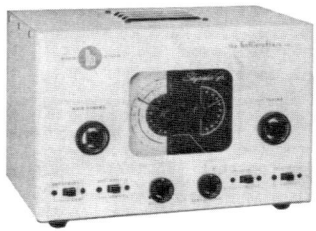

Variants:
The model **S-41W** was produced in a white cabinet (shown right). The S-41W is a very scarce and highly collectable model. In good condition it sells for $100 - $150. A version with a maroon and gray cabinet was also reportedly produced.

Made In:	United States 1945-1946
Voltages:	105-125 VAC 50/60 Hz or 105-125 VDC
Readout:	Analog
Physical:	11.75x8x8.125" 11 Lbs. 295x203x206mm 5 kg
Status:	Inactive Manufacturer Discontinued Model
Rarity:	Common

	New	Used
Price:	$34-37	$40-80
Rating:	★★★	⑦

---

the hallicrafters co.

S-47

General Coverage Broadcast Receiver
Single Conversion Superheterodyne. 15 Tubes.

Features:
- 5 AM Presets
- 5 FM Presets
- FM AFC
- Phono Input Jack
- Bass (3 Pos.)
- Treble (4 Pos.)
- Speaker Terminals
- 115 VAC Socket
- Dial Lamp
- Hinged Cover

Specifications:
Coverage ...... See Comments.   Modes .... AM/[FM]
IF .... 455 kHz. [10.7 MHz FM]   Antenna Input . Terminals
Audio Out ...... 10 W 500 ohm

Circuit Complement:
6BA6 RF Amp, 6BE6 RF Amp, 6J6 HFO/AFC, (2) 6SG7 IF amps, 6SG7 FM 3rd IF Amp/AM Detector, 6SH7 FM 4th IF, 6AL5 FM Detector, (2) 6J5 AF Amp, (2) 6SQ7 AF Amp, (2) 6V6GT Output and 5U4G Rectifier. Temperature compensated oscillator, one RF and three IF stages. 10 watt push-pull audio.

Accessories:
R-42 Speaker   R-44 Speaker   Rack Mount Kit

Comments: Ranges: .54-1.72, 5.9-18.2, 9-12 and 15-18 MHz plus FM broadcast band 88-108 MHz. Sold primarily for its virtues as an AM, FM hi-fidelity broadcast receiver. It was also built as an OEM radio, and therefore may be found under different labels (Sears, Concord, etc.). Requires a speaker.

Variants: Model S-47U is the same except that it will operate from 110-250 VAC 25-60 Hz $246. Model **S-47C** is a chassis-only version for custom or rack installations $189-209 new.

Made In:	United States 1947-1950
Voltages:	105-125 VAC 50/60 Hz 180W
Readout:	Analog
Physical:	Cabinet Version: 20x10.25x16" 63 Lbs. 508x260x406mm 28 kg Rack Version: 19x8.75x13.75" 483x222x350mm
Status:	Inactive Manufacturer Discontinued Model
Rarity:	Scarce

	New	Used
Price:	$190-230	$95-145
Rating:	★★★	⑦

the hallicrafters co.

S-51
Sea Farer

Marine Band Communications Receiver
Single Conversion Superheterodyne. 10 Tubes.

Features:
- ¼" Head. Jack
- Sensitivity
- Speaker
- BFO
- Mute Terminals
- Tone (3 Pos.)
- Flywheel Tuning
- 3 Pretuned Channels
- ANL
- Standby
- Dial Lamp
- Hinged Cover

Specifications:
Coverage ...... 132 - 13000 kHz
IF .... 455 kHz.
Audio Out ...... 1 W 3.2 ohms
Modes .... AM/CW
Antenna Input . Terminals

Circuit Complement:
6SS7 RF Amp, 7A8 Converter, (2) 6SS7 IF Amp, 7C6 Detector, 7A6 ANL, 6SS7 BFO, 35L6GT, 6V6GT Power Output, 35L6GT Power Output and 35Z5GT Rectifier.

Accessories:
1X629 6VDC Power Adapter   1X630 12 VDC Power Adapter
1X631 32 VDC Power Adapter

Comments:
Ranges: .132-.405, .485-1.53, 1.45-4.55 and 4.2-13 MHz. Also three pre-tuned channels (A/B/C) for fixed operation. Range A: .2-.3, B&C: 2-3 MHz. Fixed channels are set by lifting the hinged top cover and making internal adjustments.

Made In:	United States 1948-1950	
Voltages:	105-125 VAC 50/60 Hz or 105-125 VDC	
Readout:	Analog	
Physical:	18.5x9x9.5" 27 Lbs. 470x229x241mm 12.2 kg	
Status:	Inactive Manufacturer Discontinued Model	
Rarity:	Scarce	

	New	Used
Price:	$130-150	$70-90
Rating:	★★★	⑦

---

the hallicrafters co.

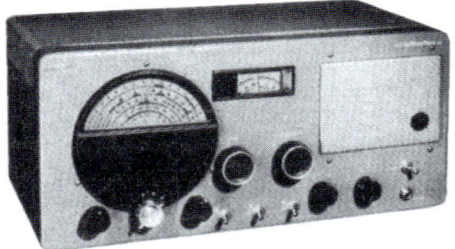

S-52

Marine Band Communications Receiver
Single Conversion Superheterodyne. 8 Tubes.

Features:
- ¼" Head. Jack
- Sensitivity
- Speaker 5"
- Bandspread 0-100
- Mute Terminals
- Tone (3 Pos.)
- BFO
- Standby
- AVC ON/OFF
- ANL
- Dial Lamp

Specifications:
Coverage ...... 530 - 43000 kHz
IF .... 455 kHz.
Audio Out ...... 1 W 3.2 ohms
Modes .... AM/CW
Antenna Input . Terminals

Circuit Complement:
6SG7 RF Amp, 6SA7 Osc./Mixer, 6SK7 1st IF, 6SK7 2nd IF, 6H6 2nd Detector/ANL, 6SC7 1st Audio/BFO, 25L6 Audio Output and 25Z5 Rectifier.

Comments:
Ranges: .54-1.68, 1.68-5.4, 5.3-15.5 and 15.5-44 MHz.
The S-52 is very similar to the S-40.

Made In:	United States 1948-1949	
Voltages:	105-125 VAC 60 Hz 40W or 105-125 VDC	
Readout:	Analog	
Physical:	18.5x9.675x8.25" 470x246x210mm	
Status:	Inactive Manufacturer Discontinued Model	
Rarity:	Scarce	

	New	Used
Price:	$100	$70-90
Rating:	★★★	⑦

the hallicrafters co.

S-53

General Coverage Communications Receiver
Single Conversion Superheterodyne. 8 Tubes.

Features:
- Tip Head. Jack
- S-Meter
- Standby
- Speaker 5"
- Mute Terminals
- Sensitivity
- Hinged Cover
- Tone (2 Pos.).
- ANL
- Phono Input Jack
- Bandspread 1-100
- Dial Lamp
- Speaker Terminals

Specifications:
Coverage ....54-31[1], 48-54.5 MHz
Audio Out ......1 W 3.2 ohms
Sensitivity ......<5µV
Modes .... AM/CW
IF .... 2075 kHz.
Antenna Input . Terminals 50-600 ohms

Circuit Complement:
6C4 HF Osc, 6BA6 Mixer, 6BA6 1st IF Amp, 6BA6 2nd IF Amp, 6H6 Detector/ANL, 6SC7 Audio Amp/BFO, 6K6GT Audio Out and 5Y3GT Rectifier. Has two IF stages.

Comments:
Ranges: .54-1.63, 2.5-6.3, 6.3-16, 14-30 and 48-54.5 MHz. [1] Reception gap from 1630-2500 kHz. Black metal case. Note the IF.

Variants:
Model **S-53U** 105-250 VAC 25-60 Hz $88.

Made In: United States 1948-1950

Voltages: 105/125 VAC 50/60 Hz

Readout: Analog

Physical: 12.875x7x7.75" 18 Lbs. 327x178x197mm 8.2 kg

Status: Inactive Manufacturer Discontinued Model

Rarity: Scarce

Reviews: *CQ* September 1986

	New	Used
Price:	$70-90	$30-70
Rating:	★★★	⑦

---

the hallicrafters co.

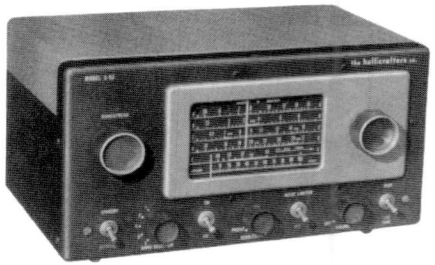

S-53A

General Coverage Communications Receiver
Single Conversion Superheterodyne. 8 Tubes.

Features:
- Tip Head. Jack
- S-Meter
- Standby
- Speaker 5"
- Mute Terminals
- Sensitivity
- Hinged Cover
- Tone (2 Pos.).
- ANL
- Phono Input Jack
- Bandspread 1-100
- Dial Lamp
- Speaker Terminals

Specifications:
Coverage ....54-31[1] and 48-54.5 MHz
Audio Out ......1 W 3.2 ohms
IF ....455 kHz
Modes .... AM/CW
Antenna Input . Terminals 52-600 ohms

Circuit Complement:
6C4 HF Osc, 6BA6 Mixer, 6BA6 1st IF Amp, 6BA6 2nd IF Amp, 6H6 Detector/AVC/ANL, 6SC7 Audio Amp/BFO, 6K6GT Audio Out and 5Y3GT Rectifier. Has two IF stages.

Comments:
Ranges: .54-1.63, 2.5-6.3, 6.3-16, 14-30 and 48-54.5 MHz. [1] Reception gap from 1630-2500 kHz. Black steel case.

Variants:
Model **S-53AU** operates on 105-250 VAC 25-60 Hz.

Made In: United States 1951-1957

Voltages: 105-125 VAC 50/60 Hz 50W

Readout: Analog

Physical: 12.875x7x7.75" 18 Lbs. 327x178x197mm 8.2 kg

Status: Inactive Manufacturer Discontinued Model

Rarity: Common

	New	Used
Price:	$80-100	$40-70
Rating:	★★★	⑦

the hallicrafters co.

S-76

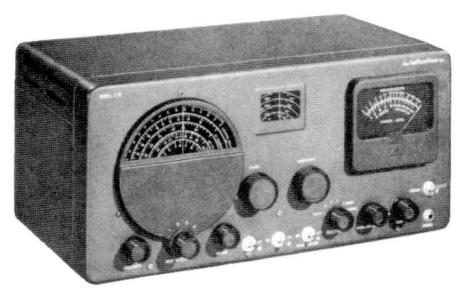

General Coverage Communications Receiver
Double Conversion Superheterodyne. 10 Tubes.

Features:
- ¼" Head. Jack
- S-Meter
- AVC ON/OFF
- BFO
- Sensitivity
- ANL
- Mute Line
- Bandspread
- Speaker Output
- Dial Lamp
- Hinged Cover
- Line Out 500 ohm
- Standby
- Tone
- Phono Input Jack
- AVC

Specifications:
Coverage ...... 538 - 34000 kHz[1]
Selectivity ...... 5 pos. 5.6-.5 kHz
Audio Out ...... 2.5 W 3.2 ohms
Modes .... AM/CW
IF .... 1650, 50 kHz
Antenna Input . Terminals and SO239 hole

Circuit Complement:
6CB6 RF Amp, 6AU6 Converter, 6C4 Osc., 6BA6 1st IF, 6BE6 2nd Converter, 6BA6 2nd IF, 6AL5 Detector/ANL, 6SC7 BFO, 6K6GT Output, VR150 Regulator and 5Y3GTY Rectifier. One RF, two conversion and two IF stages.

Accessories:
R-46 Speaker

Comments:
Ranges: .538-1.58, 1.72-4.9, 4.6-13, 12-34 MHz. [1] Reception gap from 1580 -1720 kHz because of the first IF frequency. Bandspread 80 to 10 meters. Gray metal cabinet with huge S-Meter.

Variants:
Model **S-76U** is an export version operating on 100-250 VAC 25-60 Hz

Made In:	United States 1951-1954
Voltages:	105-125 VAC 50/60 Hz 77W
Readout:	Analog
Physical:	18.5x9x9.5" 41 Lbs. 470x229x241mm 18.6 kg
Status:	Inactive Manufacturer Discontinued Model
Rarity:	Scarce

	New	Used
Price:	$170-200	$70-140
Rating:	★★★★	⑦

the hallicrafters co.

S-85

General Coverage Communications Receiver
Single Conversion Superheterodyne. 8 Tubes.

Features:
- ¼" Head. Jack
- BFO
- AVC ON/OFF
- Bandspread
- Sensitivity
- ANL
- Mute Line
- Tone (3 Pos.)
- Speaker 5"
- Dial Lamp
- Standby

Specifications:
Coverage ...... 540 - 34000 kHz
IF .... 455 kHz
Audio Out ...... 2W 3.2 ohms
Modes .... AM/CW
Antenna Input . Terminals

Circuit Complement:
6SG7 RF Amp, 6SA7 Converter, 6SK7 1st IF Amp, 6SK7 2nd IF Amp, 6SC7 BFO/Audio Amp, 6K6GT Audio Out, 6H6 Detector/ANL/AVC and 5Y3GTY Rectifier. Two IF stages.

Accessories:
R-46B Speaker

Comments:
Ranges: .538-1.6, 1.55-4.6, 4.6-13, 12-34 MHz. The bandspread covers the 80 to 10 meter amateur bands. Some S-85's featured a phono input jack on the back. Gray-black metal cabinet.

Variants:
Model **S-85U** is an export version operating on 100-250 VAC 25-60 Hz. Model **S-86** is an AC-DC version of the S-85 $119.95. The S-86 substitutes 25L6 for 6K6 and 25Z6 for 5Y3 and add ballast.

Made In:	United States 1954-1957
Voltages:	105/125 VAC 50/60 Hz 75W
Readout:	Analog
Physical:	18.5x9x10.625" 27.5 Lbs. 470x229x270mm 125 kg
Status:	Inactive Manufacturer Discontinued Model
Rarity:	Scarce

	New	Used
Price:	$120	$85-100
Rating:	★★★	⑦

the hallicrafters co.

S-107

General Coverage Communications Receiver
Single Conversion Superheterodyne. 8 Tubes.

Features:
- Head Tip Jack
- BFO
- Standby
- Sensitivity
- Speaker 5"
- Dial Lamp
- Phono Input Jack
- Bandspread 0-100
- Tone LOW/HI
- ANL

Specifications:

Coverage ...... See Comments	Modes .... AM/CW
IF .... 455 kHz.	Antenna Input . Terminals 50-300 ohms
Audio Out ...... 1 W 3.2 ohm	

Circuit Complement:
6C4 Osc, 6BA6 Mixer, 6BA6 IF Amp, 6BA6 IF Amp, 6H6 Detector/AVC/ANL, 6SC7 BFO/AF Amp, 6K6GT Output and 5Y3GT Rectifier.

Comments:
Ranges: .54-1.63, 2.5-6.3, 6.3-16, 14-31 and 48-54.5 MHz. The S-107 replaces the S-53A.

Variants:
Model **S-107 Mark I** has: 4x6" speaker, 2 prong phone jack and speaker-phone switch. Model **S-107 Mark II** has: 4x6" speaker, ¼" Head. Jack, no speaker-phone switch, no phono input jack and has external amp jack.

Made In:	United States 1958-1962
Voltages:	105/125 VAC 50/60 Hz 50W
Readout:	Analog
Physical:	13.375x7x8.875" 16 Lbs. 340x178x225mm 7.2 kg
Status:	Inactive Manufacturer Discontinued Model
Rarity:	Common

	New	Used
Price:	$60-129	$35-70
Rating:	★★★	⑦

allicrafters

S-108

General Coverage Communications Receiver
Single Conversion Superheterodyne. 8 Tubes.

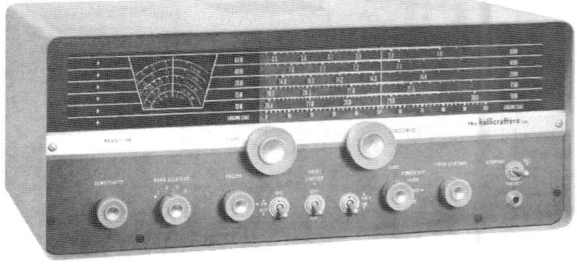

Features:
- ¼" Head. Jack
- AVC ON/OFF
- BFO
- Bandspread
- RF Gain
- ANL
- Mute Line
- Speaker Terminals
- Tone
- Standby
- Speaker

Specifications:

Coverage ...... 540 - 34000 kHz	Modes .... AM/CW
IF .... 455 kHz	Antenna Input . Terminals 52-600 ohms
Audio Out ...... 3.2/500 ohms 2W	

Circuit Complement:
6SG7 RF Amp, 6SA7 Converter, 6SK7 1st IF Amp, 6SK7 2nd IF Amp, 6SC7 Audio Amp/BFO, 6K6GT Audio Out, 6H6 Detector/ANL/AVC and 5Y3GT Rectifier.

Accessories:
R-46B Speaker   R-47 Speaker   R-48 Speaker

Comments:
Ranges: .538-1.6, 1.55-4.6, 4.6-13 and 12-34 MHz. Bandspread: 3.5-4, 7-7.3, 14-14.4, 20.1-21.5 and 26-30 MHz. Note that the bandspread ranges are on the long horizontal scale. Gray metal cabinet.

Variants:
Model **S-109** operates from AC or DC current. Also see model SX-110 which is the same but with: S-Meter, Antenna Trimmer and Crystal Filter.

Made In:	United States 1959-1963
Voltages:	105-125 VAC 50/60 Hz 75W
Readout:	Analog
Physical:	18.75x8.5x10.2" 32 Lbs. 476x216x259mm 14.5 kg
Status:	Inactive Manufacturer Discontinued Model
Rarity:	Common
Reviews:	ODXA May 1987

	New	Used
Price:	$130-150	$40-90
Rating:	★★★	⑦

 **allicrafters**

S-118

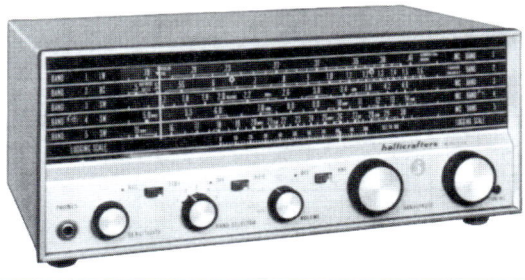

General Coverage Communications Receiver
Single Conversion Superheterodyne. 5 Tubes plus Semiconductors.

Features:
- ¼" Head. Jack
- BFO
- Standby
- Bandspread 0-100
- Speaker 4"
- Dial Lamp
- Record Out Jack
- MW Ferrite Antenna
- Phono Input
- Sensitivity
- ANL

Specifications:
Coverage ...... 185 - 31000 kHz[1]
IF .... 455 kHz.
Audio Out ...... 3.2 ohms
Modes .... AM/SSB-CW
Antenna Input . Terminals 50-75 ohms

Circuit Complement:
6BL8 Mixer/Osc., 12BA6 IF Amp, 6BL8 IF Amp/BFO, 6T8A 1st Audio/Detector/AVC/ANL, 6AQ5A Output plus 2 silicon rectifiers.

Comments:
Ranges: .185-.42, .495-1.62, 1.6-4.95, 4.85-15 and 14.8-31 MHz. [1] Reception gap from 420-495 kHz.

Variants:
Model **S-118 Mark II** was later production.

Made In:	United States 1961-1963	
Voltages:	105-125 VAC 50/60 Hz 33W	
Readout:	Analog	
Physical:	14.5x6.38x9.875" 15 Lbs. 368x162x251mm 6.8 kg	
Status:	Inactive Manufacturer Discontinued Model	
Rarity:	Common	

	New	Used
Price:	$100	$40-60
Rating:	★★★	⑦

 **allicrafters**

S-119
Sky Buddy II

Shortwave Broadcast Receiver
Single Conversion Superheterodyne. 3 Tubes plus Semiconductors.

Features:
- Head. Tip Jack
- Speaker
- Dial Lamp
- Voice/Code Switch
- Ferrite MW Ant.
- Speaker/Headphone Switch

Specifications:
Coverage ...... See Comments
IF .... 455 kHz.
Modes .... AM/CW
Antenna Input . Terminals

Circuit Complement:
6BE6 Converter, 6BA6 IF Amp/BFO, 6CM6 1st Audio/Audio Out plus a diode detector (1N295).

Comments: Ranges: .535-1640, 2.0-5.5 and 6-16.4 MHz. Has planetary drive tuning. Sold both as a kit and assembled. The S-119 was closed-out in 1964 at $30. The original *Sky Buddy* was the model **5-T** (shown right) which sold in 1935. The better known S-19 and S-19R were introduced in 1938 and 1939. The *Sky Buddy* series is very collectable and increasingly difficult to assemble.

Variants:
The **S-119K** unassebled kit version sold new for $40

Made In:	United States 1959-1964	
Voltages:	115 VAC 60 Hz 16W	
Readout:	Analog	
Physical:	10.5x5x7.5" 8 Lbs. 267x127x191mm 3.6 kg	
Status:	Inactive Manufacturer Discontinued Model	
Rarity:	Scarce	

	New	Used
Price:	$50	$45-90
Rating:	★★	⑦

S-120

General Coverage Communications Receiver
Single Conversion Superheterodyne. 4 Tubes plus Semiconductors.

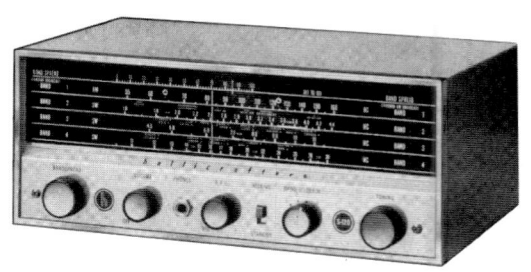

Features:
- ¼" Head. Jack
- Speaker 5"
- BFO
- Dial Lamp
- Standby
- Tels. Whip 45"
- Bandspread 0-100
- MW Ferrite Antenna

Specifications:
Coverage ......550 - 30000 kHz
IF ....455 kHz.
Modes .... AM/CW
Antenna Input . Terminals

Circuit Complement:
12BE6 Converter, 12BA6 IF Amp/BFO, 12AV6 Audio Amp/AVC, 50C5 Audio Power Amp plus 1 selenium rectifier.

Comments:
Ranges: .55-1.6, 1.6-4.4, 4.5-11 and 11-30 MHz. Gray steel cabinet with silver trim. Billed as "The world's most popular shortwave receiver." This entry-level model was produced in large numbers. Examples of the S-120 made in both Japan and America can be found.

Variants:
Model **SW-500** is virtually the same radio, but with a dark blue metallic cabinet. $70 1961.

Made In:	United States 1960-1963 and Japan
Voltages:	105-125 VAC 50/60 Hz 30W or 105-125 VDC
Readout:	Analog
Physical:	13.5x5.875x8.75" 10.2 Lbs. 343x149x222mm 4.6 kg
Status:	Inactive Manufacturer Discontinued Model
Rarity:	Abundant

	New	**Used**
Price:	$60-70	$30-50
Rating:	★★	⑦

S-120A
Star-Quest

General Coverage Communications Receiver
Single Conversion Superheterodyne. Solid State.

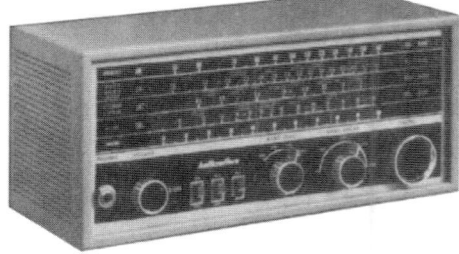

Features:
- ¼" Head. Jack
- Speaker 5"
- BFO
- Dial Lamp
- Standby
- Ferrite MW Antenna
- Bandspread 0-100

Specifications:
Coverage ......540 - 31000 kHz
IF ....455 kHz.
Modes .... AM/CW
Antenna Input . Terminals

Circuit Complement:
9 Transistors and 6 diodes.

Comments:
Ranges: .55-1.6, 1.6-4.4, 4.5-11 and 11-30 MHz. The S-120A is a solid state version of the popular S-120 and was manufactured in Japan. It was very different from the S-120. Based on the appearance and circuit of this radio, it may have come from the same factory that produced for Midland.

Made In:	Japan	1969-1972
Voltages:	105-125 VAC 50/60 Hz 30W or 12 VDC	
Readout:	Analog	
Physical:	12x5.5x5" 8 Lbs. 304x140x127mm 3.7 kg	
Status:	Inactive Manufacturer Discontinued Model	
Rarity:	Scarce	

	New	**Used**
Price:	$60	$40-60
Rating:	★	⑦

S-125
Star-Quest II

General Coverage Communications Receiver
Single Conversion Superheterodyne. Solid State.

Features:

- ¼" Head. Jack
- BFO
- Speaker
- Bandspread 0-100
- Dial Lamp
- AGC
- Ferrite MW Antenna

Specifications:
Coverage ......550 - 30000 kHz
Audio Out ......1 W

Modes .... AM-CW
Antenna Input . Terminals

Comments:
Virtually identical to the S-120A. Perhaps produced by the same factory that manufactured the Midland line?

Made In:	Japan	1970-1973
Voltages:	117 VAC or 12 VDC	
Readout:	Analog	
Physical:		
Status:	Inactive Manufacturer Discontinued Model	
Rarity:	Very Scarce	
Reviews:	*Ham Radio* May 1973	

	New	Used
Price:	$60	$60-80
Rating:	★★	⑦

S-129

General Coverage Communications Receiver
Single Conversion Superheterodyne. 7 Tubes plus Semiconductors.

Features:
- ¼" Head. Jack
- Bandspread
- Antenna Trimmer
- Product Det. SSB/CW
- RF Gain
- ANL
- BFO
- Speaker Terminals
- Mute Terminals

Specifications:
Coverage ......535 - 31500 kHz[1]
Selectivity ......Wide/Normal/Narrow
Audio Out ......2 W 3.2 ohms

Modes .... AM/SSB-CW
IF .... 1650 kHz.
Antenna Input . Terminals (50-600 ohms)

Circuit Complement:
6DC6 RF Amp, 6EA8 Mixer/Osc, 6BA6 1st IF Amp, 6BA6 2nd IF Amp, 6AL5 AM Detector/NL, 6BE6 BFO/Product Detector, 6GW8 AF Amp/Output plus a diode.

Accessories:
HA-7 Calibrator   R-47 Speaker   R-50 Speaker
R-51 Speaker/Clock

Comments:
Main ranges: .535-1.61, 1.725-4.7, 4.5-13 and 11.9-31.5 MHz. [1] Reception gap from 1610-1725 kHz. Bandspread ranges: 3.5-4, 7-7.4, 13.9-14.4, 20-22 and 26-30 MHz. Note that the bandspread occupies the long horizontal dial, and the main tuning occupies the smaller rotary dial. The S-129 is very similar to the model SX-130, but less the crystal filter, crystal phasing and S-Meter.

Made In:	United States 1965-1967	
Voltages:	105-125 VAC 50/60 Hz 48W	
Readout:	Analog	
Physical:	18.875x8x9.75" 22 Lbs. 479x203x248mm 10 kg	
Status:	Inactive Manufacturer Discontinued Model	
Rarity:	Scarce	

	New	Used
Price:	$165	$70-100
Rating:	★★★	⑦

 **allicrafters**

S-200
Legionnaire

Shortwave Broadcast Receiver
Single Conversion Superheterodyne. 4 Tubes plus Semiconductors.

Features:
- ¼" Head. Jack
- Speaker 4"
- Dial Lamp
- Ferrite MW Antenna
- Tone Control

Specifications:

Coverage ...... SWL. See Comments.		Modes .... AM	
IF .... 455 kHz.		Antenna Input . Terminals	

Circuit Complement:
12BE6 Converter, 12BA6 IF Amp, 12AV6 AVC/Audio Amp, 50C5 Power Output, plus diode.

Comments:
Ranges: .54-1.6, 5.9-6.25, 9.45-9.8, 11.65-12.05 and 15.05-15.55 MHz. Walnut grained vinyl-covered metal cabinet.

Made In:	United States 1965-1969
Voltages:	120 VAC 50/60 Hz 30W
Readout:	Analog
Physical:	13.5x6x8.375" 12 Lbs. 343x152x213mm 5.5 kg
Status:	Inactive Manufacturer Discontinued Model
Rarity:	Very Scarce

	New	Used
Price:	$60	$50-80
Rating:	★★★	⑦

 **allicrafters**

S-210

Shortwave Broadcast Receiver
Single Conversion Superheterodyne. 6 Tubes plus Semiconductors.

Features:
- ¼" Head. Jack
- Speaker 4"
- Tone
- Dial Lamp
- FM Band
- Ferrite MW Antenna

Specifications:

Coverage ...... SWBC +FM see Comments	Modes .... AM [WBFM]	
IF .... 455 kHz, 10.7 MHz	Antenna Input . Terminals	
Audio Out ...... 1 W		

Circuit Complement:
12DT8 FM RF Amp, 12BE6 AM Converter, 12BA6 FM/AM IF Amp, 12BA6 FM IF Amp/AM Detector, 12AX7 Audio Amp, 35C5 Audio Power Amp, plus 3 diodes. Double conversion on FM band only.

Comments:
Ranges: .54-1.62, 5.9-6.25, 9.45-9.8, 11.65-12.05, 15.05-15.55 MHz plus FM 88-108 MHz. The headphone jack is on the rear panel. Walnut grained vinyl-covered metal cabinet with silver trim. Not a strong performer on medium wave or shortwave.

Made In:	United States 1967-1969
Voltages:	105-125 VAC 50/60 Hz or 105-125 VDC
Readout:	 Analog
Physical:	13.5x5.875x8.375" 12 Lbs. 343x149x213mm 5.5 kg
Status:	Inactive Manufacturer Discontinued Model
Rarity:	Very Scarce

	New	Used
Price:	$80	$50-90
Rating:	★	⑦

S-214

Shortwave Broadcast Receiver
Single Conversion Superheterodyne. Solid State.

Features:
- ¼" Head. Jack
- Speaker 4"
- Dial Lamp
- Tone
- Fine Tuning
- AFC [FM]
- Line Out Jack
- Telescopic Antenna
- Ferrite MW Antenna

Specifications:
Coverage ......See Comments.
Audio Out ......1 W
Modes .... AM, [WBFM]
Antenna Input . Terminals

Circuit Complement:
Ten transistors and six diodes.

Accessories:
RSP-1 Speaker

Comments:
Ranges: .55-1.6, 5.9-6.25, 9.4-9.8, 11.5-12, 15.05-15.55 and FM 88-108 MHz. The headphone jack is on the rear panel. Vinyl-covered metal cabinet with silver trim.

Made In:	Japan	1967-1971

Voltages: 105-125 VAC 50/60 Hz

Readout: Analog

Physical: 13.92x5.9x8" 10 Lbs. 354x150x203mm 4.5 kg

Status: Inactive Manufacturer Discontinued Model

Rarity: Very Scarce

	New	Used
Price:	$90	$60-90
Rating:	★★	⑦

S-240

General Coverage Communications Receiver
Double Conversion Superheterodyne. Solid State.

Features:
- ¼" Head. Jack
- Speaker 4"
- BFO
- S-Meter
- Fine Tuning
- Tone
- AFC [FM]
- Dial Lamp
- Line Out Jack
- Telescopic Whip
- Ferrite MW Antenna

Specifications:
Coverage ....55-30 MHz + FM
IF ....455 kHz [10.7 MHz]
Audio Out ......1 W 8 ohms
Modes .... AM/CW, [WBFM]
Antenna Input . Terminals

Circuit Complement:
Ten transistors and six diodes.

Accessories:
RSP-1 Speaker

Comments:
Ranges: .55-1.6, 2-5, 4.8-11.5, 11-30 MHz and FM 88-108 MHz. Reception gap from 1.6-2 MHz. Headphone jack is on the back. A fine tuning knob is mounted concentrically on the main tuning knob. This is referred to as "Bandspread" in some ads. Walnut grained vinyl-covered metal cabinet with die-cast chromed front.

Made In:	Japan	1968-1971

Voltages: 105-125 VAC 50/60 Hz

Readout: Analog

Physical: 13.8x6.8x8" 10.5 Lbs. 343x173x203mm 4.8 kg

Status: Inactive Manufacturer Discontinued Model

Rarity: Very Scarce

	New	Used
Price:	$110-120	$80-110
Rating:	★★	⑦

---

the hallicrafters co.

SX-23
Super Skyrider 23

General Coverage Communications Receiver
Single Conversion Superheterodyne. 11 Tubes.

Features:
- ¼" Head. Jack
- S-Meter
- Tone Control
- Bandspread
- RF Gain
- Mute Line
- Standby
- Speaker Terminals
- ANL
- AVC ON/OFF
- Dial Lamp
- Hinged Top Cover
- BFO

Specifications:
Coverage ...... 540 - 34000 kHz
Selectivity ...... 5 positions
Audio Out ...... 5 W 500/5000 ohms
Modes .... AM/CW
IF .... 455 kHz
Antenna Input . Terminals 400 ohms

Circuit Complement:
6SK7 1st RF Amp, 6SA7 1st Detector/Mixer, 6SJ7 High Frequency Osc., 6SK7 1st IF, 6SK7 2nd IF, 6SQ7 Detector/1st Audio Amp, 6B8 AVC, 6SJ7 BFO, 6H6 ANL, 6F6 AF Output and 80 Rectifier. Crystal filter.

Accessories:
PM-23 Speaker

Comments:
Ranges: .54-1.7, 1.7-5.2, 5.2-16.5 and 11-34 MHz. Bandspread ranges: 3.5-4, 7-7.3, 14-14.4 and 28-30 MHz. This receiver requires a speaker. Mechanically complex. The bandspread bands are not directly calibrated. Rather, when the bandspread is engaged, one still tunes from the main dial noting the logging scale. Then the tuned frequency is interpolated from a chart. Gray crackle finish.

Made In: United States 1939-1940

Voltages: 110-125 VAC 60 Hz 110W

Readout: Analog

Physical: 19x9.25x12.5" 51 Lbs. 483x235x318mm 23 kg

Status: Inactive Manufacturer Discontinued Model

Rarity: Extremely Scarce

Reviews: *Electric Radio* March 1992

	New	Used
Price:	$116	⑤
Rating:	★★★★	⑦

the hallicrafters co.

SX-24
Skyrider Defiant

General Coverage Communications Receiver
Single Conversion Superheterodyne. 9 Tubes.

Features:
- ¼" Head. Jack
- S-Meter
- Tone Switch
- Bandspread
- RF Gain
- Mute Line
- Standby
- Speaker Terminals
- ANL
- AVC ON/OFF
- Dial Lamp
- Hinged Top Cover
- BFO

Specifications:
Coverage ...... 540 - 43500 kHz
Selectivity ...... 6 positions
Audio Out ...... 5 W 500/5000 ohms
Modes .... AM/CW
IF .... 455 kHz
Antenna Input . Terminals 400 ohm

Circuit Complement:
6SK7 RF Amp, 6K8 1st Detector/Converter, 6SK7 1st IF, 6SK7 2nd IF, 6SQ7 2nd Detector/AVC/1st Audio Amp, 6H6 ANL, 76 BFO, 6F6 AF Output and 80 Rectifier.

Accessories:
PM-23 Speaker

Comments:
Ranges: .54-1.73-1.7-5.1, 5-15.7 and 15.2-43.5 MHz. The true bandspread covers: 80, 40, 20 and 10 meters. This receiver requires a speaker.

Made In: United States 1939-1943

Voltages: 100-125 VAC 50-60 Hz 70W

Readout: Analog

Physical: 19.5x9.5x11.125" 495x241x283mm

Status: Inactive Manufacturer Discontinued Model

Rarity: Scarce

	New	Used
Price:	$70	⑤
Rating:	★★★★	⑦

the hallicrafters co.

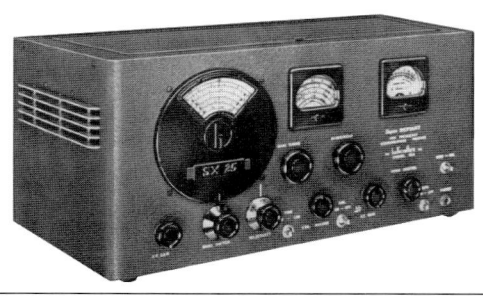

SX-25
Super Defiant

General Coverage Communications Receiver
Single Conversion Superheterodyne. 12 Tubes.

Features:
- ¼" Head. Jack
- S-Meter
- Tone (2 Pos.)
- Bandspread
- RF Gain
- Mute Line
- Standby
- Speaker Terminals
- ANL
- AVC
- Dial Lamp
- Hinged Top Cover
- BFO
- Flywheel Tuning

Specifications:
Coverage ...... 5540 - 42000 kHz
Selectivity ...... 6 positions
Audio Out ...... 8 W 500/5000 ohms
Modes .... AM/CW
IF .... 455 kHz
Antenna Input . Terminals

Circuit Complement:
6SK7 1st RF Amp, 6SK7 2nd RF Amp, 6K8 1st Detector-Mixer, 6SK7 1st IF ANL, 6SK7 2nd IF Amp, 6SQ7 2nd Detector/AVC/1st Audio Amp, 6SQ7 Phase Inverter, (2) 6F6 Audio Amp, 6H6 ANL, 6J5GT BFO and 80 Rectifier.

Accessories:
PM-23 Speaker

Comments:
Ranges: .54-1.7, 1.7-5.1, 5-15.7 and 15.2-42 MHz. The true bandspread covers: 80, 40, 20 and 10 meters. [¥59,000].

Made In:	United States 1940-1946	
Voltages:	100-125 VAC 50-60 Hz 120W	
Readout:	Analog	
Physical:	19.5x9.5x11.125" 46 Lbs. 495x241x283mm 21 kg	
Status:	Inactive Manufacturer Discontinued Model	
Rarity:	Common	
Reviews:	*Radio* May 1940	

	New	Used
Price:	$95-99	$100-210
Rating:	★★★★	⑦

the hallicrafters co.

SX-28A
Super Skyrider

General Coverage Communications Receiver
Single Conversion Superheterodyne. 15 Tubes.

Features:
- ¼" Head. Jack
- S-Meter
- Tone
- Phono Input Jack
- Sensitivity
- Bandspread
- Dial Lock
- Speaker Terminals
- Flywheel Tuning
- AVC ON/OFF
- Dial Lamps
- Hinged Top Cover
- Mute Line
- Bass Switch
- Standby
- Antenna Trimmer
- NL

Specifications:
Coverage ...... 550 - 43000 kHz
Selectivity ...... 6 positions
Audio Out ...... 8 W 500/5000 ohms
Modes .... AM/CW
IF .... 455 kHz
Antenna Input . Terminals

Circuit Complement:
6AB7 1st RF Amp, 6SK7 2nd RF Amp, 6SA7 Mixer, 6SA7 HF Osc, 6L7 1st IF ANL, 6SK7 2nd IF Amp, 6B8 AVC Amp, 6B8 2nd Detector/S Meter, 6AB7 Noise Amp, 6H6 Noise Rectifier, 6J5 BFO, 6SC7 1st Audio Amp, (2) 6V6GT Audio Amps and 5Z3 Rectifier.

Accessories: PM-23 Speaker

Comments: Ranges: .55-1.6, 1.6-3, 3-5.8, 5.8-11, 11-21 and 21-43 MHz. Bandspread: 3.5-4, 7-7.3, 14-14.4 and 28-30 MHz.

Variants: The SX-28A is an enhanced government model. The regular model **SX-28** $180 was made in 1941. A large number of SX-28s were sent to Russia as a part of the World War II Lend-Lease Act. When they reached Russia they were modified to accommodate Russian tubes. They are plentiful on the Russian used market.

Made In:	United States 1944-1945	
Voltages:	105-125 VAC 50/60 Hz 120W	
Readout:	Analog	
Physical:	20.5x10x14.75" 75 Lbs. 521x254x375mm 34 kg	
Status:	Inactive Manufacturer Discontinued Model	
Rarity:	Common	
Reviews:	*Electric Radio* June 1990	

	New	Used
Price:	$225-275	$180-290
Rating:	★★★★	⑦

the hallicrafters co.

SX-42

General Coverage Communications Receiver
Double Conversion Superheterodyne. 15 Tubes.

Features:
- ¼" Head. Jack
- S-Meter
- Tone (4 Pos.)
- Phono Input Jack
- Sensitivity
- Bandspread
- Dial Lock
- Speaker Terminals
- NL
- AVC
- Dial Lamp
- Hinged Top Cover
- Standby

Specifications:

Coverage ...... 540 kHz - 110 MHz
Selectivity ...... 6 Positions
Audio Out ...... 10 W 500/5000 ohms

Modes .... AM/CW/FM
IF .... 455 kHz, 10.7 MHz
Antenna Input . Terminals 300 ohms

Circuit Complement:
(2) 6AG5 RF Amp, 7F8 Converter, 6SK7 IF Amp, 6SG7 IF Amp, 7H7 IF Amp, 7H7 FM Limiter/AM/Detector, 6H6 FM Detector, 7A4 BFO, 6H6 ANL, 6SL7 AF Amp, (2) 6V6 Output, VR-150 Regulator and 5U4G Rectifier.

Accessories:

R-42 Speaker	R-44 Speaker	PM-23 Speaker
R-45 Speaker	B-42 Tilt Base	

Comments: Ranges: .54-1.62, 1.62-5, 5-15, 15-30, 27-55 and 55-109 MHz. Wideband FM detection available from 27 to 110 MHz. Bandspread for 80, 40, 20, 10 and 6 meters. Requires a speaker. One of the first receivers designed after World War II. Designed for excellent audio fidelity to appeal to the FM broadcast listener.

Variants:
Model **SX-42U** operates from 110/130/150/220/250 VAC $288.

Made In:	United States 1947-1950
Voltages:	105-125 VAC 50 60 Hz 120W
Readout:	Analog
Physical:	20x10.25x16" 52 Lbs. 508x260x203mm 23.6 kg
Status:	Inactive Manufacturer Discontinued Model
Rarity:	Scarce
Reviews:	*QST* May 1947 *Proceedings* 1994-95

	New	Used
Price:	$250-275	$250-450
Rating:	★★★★★	⑦

the hallicrafters co.

SX-43

General Coverage Communications Receiver
Double Conversion Superheterodyne. 11 Tubes.

Features:
- ¼" Head. Jack
- S-Meter
- Tone (2 Pos.)
- Bandspread 0-100
- Sensitivity
- BFO
- Phono Input Jack
- Speaker Terminals
- NL
- AVC
- Dial Lamps
- Mute Line
- Standby
- Hinged Top

Specifications:

Coverage .... .54-55 & 86-109 MHz
Selectivity ...... 4 Positions
Audio Out ...... 3 W

Modes .... AM/CW/FM
IF .... 455 kHz, 10.7 MHz
Antenna Input . Terminals 72-600 ohms

Circuit Complement:
6BA6 RF Amp, 7F8 Converter, 6SG7 IF Amp, (2) 6SH7 IF Amp, 6H6 AM Detector/ANL, 6AL5 FM Detector, 6J5 BFO, 6SQ7 AM Amp, 6V6 Output and 5Y3GT Rectifier. One RF and two IF stages.

Accessories:

R-42 Speaker	R-44 Speaker

Comments:
Ranges: .54-1.7, 1.7-5, 5-16, 14-14.4, 15.5-44, 44-54 and 86-109 MHz. Bandspread: 80, 40, 20, 10 and 6 meters.

Variants:
Model **SX-43U** operates from 110/130/150/220/250 VAC $178.

Made In:	United States 1947-1950
Voltages:	105-125 VAC 50/60 Hz 120W
Readout:	Analog
Physical:	18.5x8.875x12" 41 Lbs. 470x225x305mm 18.6 kg
Status:	Inactive Manufacturer Discontinued Model
Rarity:	Common

	New	Used
Price:	$160-190	$150-260
Rating:	★★★	⑦

the hallicrafters co.

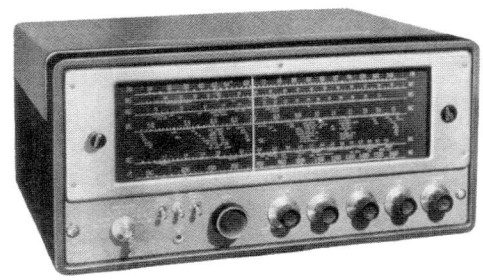

SX-62

General Coverage Communications Receiver
Double Conversion Superheterodyne. 16 Tubes.

Features:
- ¼" Head. Jack
- S-Meter
- Tone (4 Pos.)
- Dial Lamps • NL
- Sensitivity
- Standby
- Calibrator
- Speaker Terminals
- Hinged Cover
- Pointer Reset
- 60:1 Tuning Ratio • Phono Input Jack

Specifications:
Coverage ......540 kHz - 109 MHz   Modes .... AM/CW/FM
Selectivity ......6 Positions   IF .... 455 kHz, 10.7 MHz
Audio Out ......10W 500/5000 ohms   Antenna Input . Terminals

Circuit Complement:
6AG5 1st Amp, 6AG5 2nd RF, 7F8 Converter, 6SK7 IF Amp, 6SG7 IF Amp, 7H7 IF Amp, 7H7 FM Limiter/AM Detector, 6H6 FM Detector, 7A4 BFO, 6H6 ANL, 6SL7 AF Amp, (2) 6V6 Push-pull Audio Output, 6C4 Calibration Oscillators, VR-150 Regulator and 5U4G Rectifier. Two RF and three IF stages.

Accessories:
R-42 Speaker   R-44 Speaker   PM-23 Speaker
R-46 Speaker

Comments: Ranges: .55-1.62, 1.62-4.9, 4.9-15, 15-32, 27-56 and 54-109 MHz. The SX-62 is an SWL version of the SX-42. Again designed for outstanding audio fidelity. Only the selected band is illuminated. Two tone gray metal cabinet.

Variants: Model **SX-62U** operates 105-250 VAC 25/60 kHz. Model **SX-62A** seems to be the same, but with a different audio impedance 1956 $350-430. The **SX-62AU** operates 105-250 VAC 20/100 kHz. Model **SX-62B** 14 tubes 1965 $525.

Made In:	United States 1949-1953	
Voltages:	105-125 VAC 50/60 Hz 120W	
Readout:	⊞⊞ Analog	
Physical:	20x10.25x16" 64 Lbs. 508x260x406mm 29 kg	
Status:	Inactive Manufacturer Discontinued Model	
Rarity:	Scarce	
Reviews:	*Proceedings* 1994-95	

	New	Used
Price:	$269-299	$160-330
Rating:	★★★★	⑦

---

the hallicrafters co.

SX-71

General Coverage Communications Receiver
Double Conversion Superheterodyne. 14 Tubes.

Features:
- ¼" Head. Jack
- S-Meter
- Bandspread
- Antenna Trimmer
- Sensitivity
- Tone
- Dial Lamp
- Hinged Top Cover
- Mute Terminals
- Standby
- Phono Input
- BFO   • ANL

Specifications:
Coverage ......538-35000 kHz, 46-56 MHz   Modes .... AM/CW/NBFM
Selectivity ......3 Position   IF .... 2075, 455 kHz
Sensitivity ......<1µV   Antenna Input . Terminals
Audio Out ......3 W 3.2/500 ohms

Circuit Complement: 6BA6 RF Amp, 6C4 HF Osc, 6AU6 Mixer, 6BE6 2nd Conv., (3) 6SK7 IF Amp, 6SH7 IF Amp, 6H6 ANL/AVC, 6SC7 BFO/RF Amp, 6AL5 Det., 6K6GT Output, VR-150 Voltage Reg. and 5Y3GT Rectifier. One RF, two conv. and three IF stages. Single conv. < 4.7 MHz.

Accessories:
R-44B Speaker   R-46 Speaker

Comments: Ranges: .538-1.65, 1.6-4.8, 4.6-13.5, 12.5-35 and 46-56 MHz. Bandspread: 3.5-4, 7-7.3, 14-14.4, 27-30 and 47-55 MHz. Requires speaker. Also produced with white dials (shown right).

Variants: Model **SX-71U** operates from 115-230 VAC 25/60 Hz. $210.

Made In:	United States 1949-1954	
Voltages:	105-125 VAC 50/60 Hz 90W	
Readout:	⊞⊞ Analog	
Physical:	18.5x8.875x12" 33 Lbs. 330x130x280mm 15 kg	
Status:	Inactive Manufacturer Discontinued Model	
Rarity:	Common	

	New	Used
Price:	$180-250	$110-190
Rating:	★★★★	⑦

the hallicrafters co.

SX-73
R-274/FRR

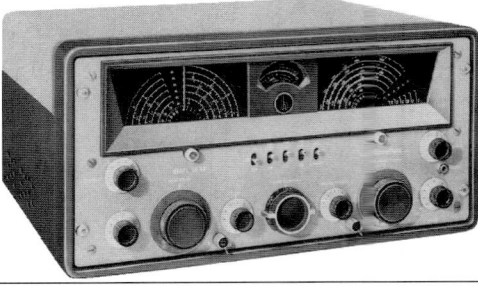

General Coverage Communications Receiver
Double Conversion Superheterodyne. 20 Tubes.

Features:
- ¼" Head. Jack
- S-Meter
- Dial Lock
- 6000° of Tuning
- RF Gain
- AGC
- BFO
- Antenna Trimmer
- Mute Terminals
- Standby
- ANL
- 6 Fixed Channels
- Crystal Vernier
- Rack Handles
- Dial Lamp
- Hinged Top
- 50:1 Tune Ratio
- IF Out Jack

Specifications:
Coverage ......540 - 54000 kHz
Selectivity ......6 Pos. .3-14.5 kHz
Sensitivity ......<1µV
Audio Out ......2W 600/50 ohms

Modes .... AM/CW/MCW
Image Rej. ...... 80-120 dB
IF .... 455 kHz, 6 MHz.
Antenna Input . 50-200 Ohms

Circuit Complement:
Two IF and two RF stages. Single conversion below 7 MHz. Coils are mounted on a turret and mechanically move into active circuit to reduce cross modulation.

Accessories: R-46 Speaker

Comments: Has 6 fixed crystal positions with crystal vernier tuning. Designed for rack mounting. Ranges: .54-1.35, 1.35-3.45, 3.45-7, 7-14.4, 14.4-29.7 and 29.7-54 MHz. Requires a speaker. Designed initially for the military-commercial market. Ready for mounting in a 19" rack. Two tone gray steel cabinet. Very collectable. Note that the Hammarlund SP-600 and this model both carry the "R-274" military nomenclature, but have mechanical and circuitry differences. It has also been alleged that Hammarlund made a few R-274s in this SX-73 configuration.

Made In: United States 1952-1954

Voltages: 75/105/117/130/130
210/234/260 VAC 50/60 Hz

Readout: Analog

Physical: 20x11x18.5" 58 Lbs.
508x280x470mm 26.3 kg

Status: Inactive Manufacturer
Discontinued Model

Rarity: Very Scarce

	New	Used
Price:	$975	$460-850
Rating:	★★★★	⑦

the hallicrafters co.

SX-88

General Coverage Communications Receiver
Double Conversion Superheterodyne. 20 Tubes.

Features:
- ¼" Head. Jack
- S-Meter
- Bandspread
- Calibrator 100 kHz
- Sensitivity
- Notch Filter
- BFO ±2.5 kHz
- Antenna Trimmer
- Mute Terminals
- Standby
- Phono Input Jack
- Speaker Terminals
- Flywheel Tuning
- AVC ON/OFF
- IF Out Jack
- Dial Locks
- Dial Lamp
- ANL

Specifications:
Coverage ......535 - 33000 kHz
Selectivity ......10/5/2.5/1.25/.5/.25 kHz
Sensitivity ......<2 µV A
Audio Out ......10 W 3.2/8/500 ohms.

Modes .... AM/CW/SSB
IF .... 50, 2075 kHz.
IF .... 50, 1550 kHz. 1.69-3 MHz
Antenna Input . Terminals & SO239 hole

Circuit Complement:
6CB6 RF Amp, 6BA6 RF Amp, 6U8 1st Mixer/1st Converter. Osc, 6BA6 2nd Mixer, 12AT7 2nd Converter Osc, 6BA6 IF Amp, 6BA6 IF Amp, 6BA6 IF Amp, 6AL5 Det/ANL, 6CB6 AVC Amp, 12AU7 AVC Rect/Cath Fol, 12AX7 Audio Amp/Phase Inverter, 6V6GT Audio Out, 6V6GT Audio Out, 6C4 BFO, 6BA6 BFO Amp, 5U4G Rectifier, 0D3 Voltage Reg, 4H4 Current Regulator and 6BA6 Crystal Calibrator.

Accessories: R-46 Speaker   R-46A Speaker

Comments: Ranges: .535-1.7, 1.69-3, 2.98-5.5, 5.4-10, 9.8-18.3 and 17.8-33 MHz. Bandspread: 160, 80, 40, 20, 15 and 11-10 meters. Illuminated dial-in-use indicator. Requires a speaker. Gear driven. A highly regarded, rare and collectable model.

Variants: Model **SX-88U** operates from 100-250V 25-60 Hz.

Made In: United States 1954-1955

Voltages: 105-125 VAC 50-60 Hz
138 W

Readout: Analog

Physical: 20x9.25x17.5"
508x235x444mm

Status: Inactive Manufacturer
Discontinued Model

Rarity: Extremely Scarce

Reviews: *QST* June 1954
Electric Radio April 1992

	New	Used
Price:	$595	$100-2500
Rating:	★★★★★	⑦

the hallicrafters co.

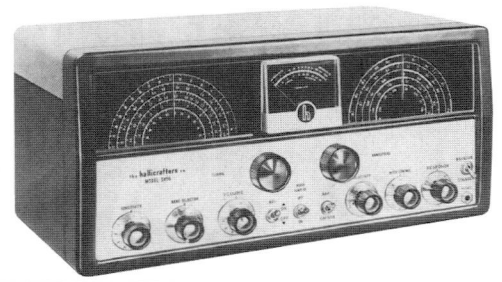

SX-96

General Coverage Communications Receiver
Double Conversion Superheterodyne. 12 Tubes.

Features:
- ¼" Head. Jack
- S-Meter
- Bandspread
- Tone (2 Pos.)
- Sensitivity
- Standby
- BFO
- Antenna Trimmer
- Mute Terminals
- Dial Lamp
- Phono Input Jack
- Speaker Terminals
- ANL
- AVC ON/OFF
- Hinged Top Cover
- Flywheel Tuning
- Line Out 500 ohms

Specifications:
Coverage ......538 - 34000 kHz[1] Modes .... AM/CW-SSB
Selectivity ......5/3/2/1/.5 kHz -6dB. IF .... 1650, 50.5 kHz.
Audio Out ......1.5W 3.2 ohms 10% dis. Antenna Input . Terminals 300 ohms

Circuit Complement:
6CB6 RF Amp, 6AU6 1st Mixer, 6C4 1st Converter Osc., 6BA6 1650 IF Amp, 6BA6 2nd Mixer, 6BA6 50.5 kHz IF Amp, 6BJ7 Detector/ANL/AVC, 6SC7 BFO/Audio Amp, 6K6GT, 0D3 Voltage Regulator and 5Y3GT Rectifier.

Accessories:
R-46A Speaker   R-46B Speaker

Comments:
Ranges: .538-1.58, 1.72-4.9, 4.6-13 and 12-34 MHz. Bandspread: 80, 40, 20, 15 and 10-11 meters. [1] Reception gap from 1580-1720 kHz. Gear driven tuning. Gray-black cabinet.

Made In:	United States 1954-1956
Voltages:	105-125 VAC 50/60 Hz 85W
Readout:	Analog
Physical:	18.75x8.875x11" 34.5 Lbs. 476x225x279mm 15.6 kg
Status:	Inactive Manufacturer Discontinued Model
Rarity:	Common
Reviews:	QST June 1955

	New	Used
Price:	$250	$100-150
Rating:	★★★★	⑦

the hallicrafters co.

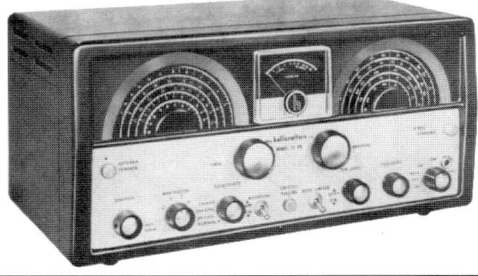

SX-99

General Coverage Communications Receiver
Single Conversion Superheterodyne. 8 Tubes.

Features:
- ¼" Head. Jack
- S-Meter
- Bandspread
- Hinged Cover
- Sensitivity
- Stand-By
- BFO
- Antenna Trimmer
- Dial Lamp
- NL
- AVC ON/OFF
- Speaker Terminals

Specifications:
Coverage ......538 - 34000 kHz Modes .... AM/CW-SSB
Selectivity ......4 Position. IF .... 455 kHz.
Audio Out ......2W 3.2/500 ohms Antenna Input . Terminals

Circuit Complement:
6SG7 RF Amp, 6SA7 Converter, 6SG7 1st IF Amp, 6SK7 2nd IF Amp, 6SC7 Audio Amp/BFO, 6K6GT Audio Output, 6H6 Detector/ANL/AVC and 5Y3 Rectifier. One RF and two IF stages.

Accessories:
R-46 Speaker   R-46A Speaker   R-46B Speaker

Comments:
Ranges: .538-1.6, 1.55-4.6, 4.6-13 and 12-34 MHz.
Bandspread: 80-10 Meters. Requires a speaker. Black metal cabinet with brushed chrome trim.

Variants:
The "universal" model, **SX-99U** operates 100-125 VAC 25-60 Hz.

Made In:	United States 1955-1958
Voltages:	105-125 VAC 50/60 Hz
Readout:	Analog
Physical:	18.75x9x10.75" 28.4 Lbs. 476x229x273mm 13 kg
Status:	Inactive Manufacturer Discontinued Model
Rarity:	Common

	New	Used
Price:	$150	$90-125
Rating:	★★★★	⑦

the hallicrafters co.

SX-100

General Coverage Communications Receiver
Double Conversion Superheterodyne. 14 Tubes.

Features:
- ¼" Head. Jack
- S-Meter
- Bandspread
- Calibrator 100 kHz
- RF Gain
- Notch Filter
- BFO
- Antenna Trimmer
- Mute Terminals
- Record Out Jack
- Phono Input Jack
- Speaker Terminals
- ANL
- AVC ON/OFF
- Standby
- Line Out 500 ohms
- Dial Lamp

Specifications:
Coverage ...... 538 - 34000 kHz[1]   Modes .... AM/CW/SSB
Selectivity ...... 5/3/2/1/.5 kHz -6 dB.   IF .... 1650, 50 kHz.
Sensitivity ...... <.1µV   Antenna Input . Terminals & SO239 Hole
Audio Out ...... 1.5W 3.2 ohms 10% dis.

Circuit Complement:
6CB6 RF Amp, 6AU6 1st Mixer, 6C4 HF Osc, 6BA6 1650 kHz IF Amp, 6BA6 2nd Mixer, 6BA6 50 kHz IF Amp, 6BJ7 Detector/ANL/AVC, 6SC7 Audio Amp/BFO, 6K6GT Audio Out, 0A2 Voltage Regulator, 5Y3GT Rectifier, 12AT7 2nd Converter Osc, 6C4 50 kHz IF Amp and 6AU6 Crystal Calibrator.

Accessories:
R-46 Speaker   R-46A Speaker   R-46B Speaker

Comments: Ranges: .538-1.58, 1.72-4.9, 4.6-13 and 12-34 MHz. [1]Reception gap from 1580-1720 kHz. Bandspread: 3.5-4, 7-7.3, 14-14.35, 21-21.45 and 28-29.7 MHz. Gear driven tuning. Requires a speaker. Gray-black cabinet.

Variants: Model **SX-100 Mark 1A** and **SX-100 Mark II** were later production.

Made In:	United States 1955-1962
Voltages:	105-125 VAC 50/60 Hz 88W
Readout:	Analog
Physical:	18.375x8.5x10.6" 42 Lbs. 467x216x269mm 19 kg
Status:	Inactive Manufacturer Discontinued Model
Rarity:	Abundant
Reviews:	*QST* December 1955

	New	Used
Price:	$290-325	$160-290
Rating:	★★★★	⑦

the hallicrafters co.

SX-101

Amateur Band Communications Receiver
Double Conversion Superheterodyne. 15 Tubes.

Features:
- ¼" Head. Jack
- S-Meter
- Pointer Reset
- Calibrator 100 kHz
- RF Gain
- Notch Filter
- BFO
- Antenna Trimmer
- Mute Terminals
- Record Out Jack
- Phono Input Jack
- Speaker Terminals
- ANL
- Conv. Input Jack
- AVC ON/OFF
- 40:1 Tune Ratio

Specifications:
Coverage ...... Ham. See comments.   Modes .... AM/CW/LSB/USB
Selectivity ...... 5/3/2/1/.5 kHz -6 dB.   IF .... 1650, 50 kHz.
Sensitivity ...... <.1µV CW   Antenna Input . SO239 & Terminals
Audio Out ...... 1W 3.2 ohms 10% dist.

Circuit Complement:
6CB6 RF Amp, 6BY6 1st Mixer, 12BY7 1st Converter Osc, 6BA6 1650 kHz IF Amp, 6BA6 2nd Mixer, 6BA6 50 kHz IF Amp, 6BJ7 Detector/ANL/AVC, 6SC7 Audio Amp/BFO, 6K6GT Audio Out, 0A2 Voltage Regulator, 5Y3GT Rectifier, 12AT7 2nd Converter Osc, 6C4 50 kHz IF Amp and 6AU6 Crystal Calibrator.

Accessories: R-46B Speaker   R-47 Speaker

Comments: Ranges: 3.5-4, 7-7.3, 14-14.4, 21-21.5, 27-29.7 and 10 MHz. The dial is also pre-calibrated for a 2 meter or 6 meter converter on the later A variant. The band-in-use is individually illuminated. Touted as the *New heavyweight champion - employs heaviest chassis in the industry*. Requires a speaker. [¥100,000].

Variants: Model **SX-101 Mark II**. Model **SX-101 Mark III** 1958. Model **SX-101 Mark IIIA** (shown) 1959 $400 *CQ* May 1959. Model **SX-101A** $399-445 1959-1963.

Made In:	United States 1956-1958
Voltages:	105-125 VAC 50/60 Hz 115W
Readout:	Analog
Physical:	20x10.5x16" 70 Lbs. 508x267x406mm 31.7 kg
Status:	Inactive Manufacturer Discontinued Model
Rarity:	Common
Reviews:	*QST* October 1957

	New	Used
Price:	$395	$170-360
Rating:	★★★★	⑦

SX-110

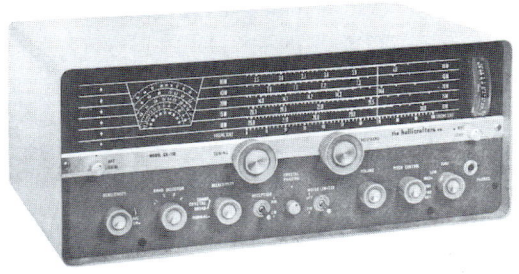

General Coverage Communications Receiver
Single Conversion Superheterodyne. 8 Tubes.

Features:
- ¼" Head. Jack
- S-Meter
- AVC ON/OFF
- Tone LO/MED/HI
- RF Gain
- ANL
- Mute Line
- Antenna Trimmer
- BFO
- Standby
- Dial Lamp
- Speaker Terminals

Specifications:
Coverage ...... 538 - 34000 kHz
Selectivity ...... 3 Position
Audio Out ...... 3.2/500 ohms 2W
Modes .... AM/CW
IF .... 455 kHz
Antenna Input . Terminals 52-600 ohms

Circuit Complement:
6SG7 RF Amp, 6SA7 Converter, 6SG7 1st IF Amp, 6SK7 2nd IF Amp, 6SC7 Audio Amp/BFO, 6K6GT Audio Output, 6H6 Detector/ANL/AVC and 5Y3GT Rectifier.

Accessories:
R-46B Speaker   R-47 Speaker   R-48 Speaker

Comments:
Ranges: .538-1.6, 1.55-4.6, 4.6-13 and 12-34 MHz. Bandspread: 3.5-4, 7-7.3, 14-14.4, 20.1-21.5 and 26-30 MHz. Note that the bandspread ranges are on the long horizontal scale. The model S-108 is the same sans: S-Meter, Antenna Trimmer and Crystal Filter. Gray steel cabinet.

Made In:	United States 1960-1963
Voltages:	105-125 VAC 50/60 Hz 75W
Readout:	Analog
Physical:	18.75x8.5x10.2" 28.25 Lbs. 476x216x259mm 13 kg
Status:	Inactive Manufacturer Discontinued Model
Rarity:	Common

	New	Used
Price:	$160-170	$80-120
Rating:	★★★★	⑦

SX-111

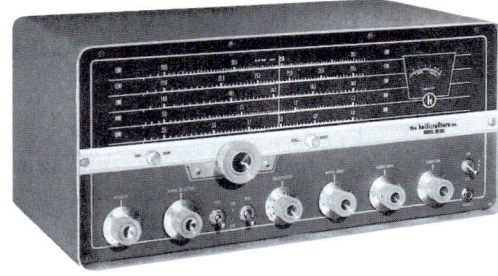

Amateur Band Communications Receiver
Double Conversion Superheterodyne. 13 Tubes.

Features:
- ¼" Head. Jack
- S-Meter
- AVC ON/OFF
- Calibrator 100 kHz
- RF Gain
- ANL
- Mute Line
- Antenna Trimmer
- Notch
- Pointer Reset
- BFO
- Speaker Terminals
- Standby
- 48:1 Tuning Ratio

Specifications:
Coverage ...... Ham. See Comments.
Selectivity ...... 5/3/2/1/.5 kHz
Sensitivity ...... <1µV 10 dB S/N
Audio Out ...... 3.2/500 ohms
Modes .... AM/CW/SSB
IF .... 50.75, 1650 kHz
Antenna Input . Terminals&RCA 50-70 ohm

Circuit Complement: 6DC6 RF Amp, 6BY6 1st Mixer, 6CB6 1650 kHz IF Amp, 6DC6 50.75 kHz IF Amp, 6BJ7 Detector/AVC/ANL, 6C4 1st Osc, 5Y3GT Rectifier, 0A2 Regulator, 12AX7 Audio Amp/BFO, 6AQ5A Audio Out, 6BA6 2nd Mixer, 6AU6 Crystal Calibrator and 12AT7 2nd Converter Osc.

Accessories:
R-46B Speaker   R-47 Speaker   R-48 Speaker

Comments: Ranges: 3.5-4, 7-7.3, 14-14.4, 21-21.5, 28-29.7 MHz and WWV 10 MHz. As is usually the convention, the headphones will disengage the 3.2 ohm speaker output, but the 500 ohm output remains active at all times. The Calibrator was not included in very early production.

Variants: Model **SX-111 Mark I** was later production and included a product detector for improved CW and SSB reception.

Made In:	United States 1960-1963
Voltages:	105-125 VAC 50/60 Hz 83W
Readout:	Analog
Physical:	18.7x8.81x10.3" 35.75 Lbs. 475x224x262mm 16.2 kg
Status:	Inactive Manufacturer Discontinued Model
Rarity:	Common
Reviews:	QST May 1960 CQ February 1961

	New	Used
Price:	$250-280	$90-150
Rating:	★★★★	⑦

SX-115

Amateur Band Communications Receiver
Triple Conversion Superheterodyne. 18 Tubes plus Semiconductors.

Features:

• ¼" Head. Jack	• S-Meter	• AVC ON/OFF	• Calibrator 100 kHz
• RF Gain	• ANL	• Mute Line	• Antenna Trimmer
• Notch	• Dial Lamp	• BFO	• Speaker Terminals
• Standby	• 50 kHz IF Out Jack		

Specifications:

Coverage ...... Ham. See Comments
Selectivity ...... 5/3/2/1/.5 kHz
Sensitivity ...... <.5µV 10 dB S/N SSB
IF Rej .... 1000x @ 60 dB down
Audio Out ...... 1.5 W 3.2/500 ohms

Modes .... AM/CW/SSB
Stability .... < 300 Hz /hr after warm-up
IF .... 6505-6005, 1005, 50.75 kHz
Image Rej. ...... >60 dB
Antenna Input . SO239

Circuit Complement:
6DC6 RF Amp, 6BA6 1st Mixer, 12AT7 Crystal Osc, 6DC6 1st IF Amp, 6BA6 2nd Mixer, 6CB6 VFO, 6DC6 2nd IF Amp, 6BA6 3rd Mixer, 12AT7 SSB Switch Osc., 6DC6 3rd IF Amp, 6BY6 Product Detector, 6BJ7 AM Detector/2nd AVC Rectifier/ANL, 12AX7 BFO/1st Audio Amp, 6AQ5 Audio Out, 6AU6 Crystal Calibrator, 6AU6 S-Meter, 6AU6 1st Loop AVC Amp, 0A2 Voltage Regulator and 5 silicon diodes.

Comments:
Ranges: 3.5-4, 7-7.5, 14-14.5, 21-21.5, 28-28.5, 28.5-29, 29-29.5, 29.5-30 and 9.6-10 MHz. (9.6-10 MHz not calibrated). 1 kHz dial accuracy between calibration points. Gear driven tuning with virtually no backlash. Requires a speaker. Matches the HT-33 and HT-32 transmitters. Designed to compete with the Collins 75A/S series.

Made In: United States 1961-1964

Voltages: 105-125 VAC 50/60 Hz 85W

Readout: Analog Lin.

Physical: 16x10.5x16" 44 Lbs. 406x267x406mm 20 kg

Status: Inactive Manufacture Discontinued Model

Rarity: Very Scarce

Reviews: QST March 1962
CQ March 1964

	New	Used
Price:	$595-600	$450-800
Rating:	★★★★	⑦

SX-117

Amateur Band Communications Receiver
Triple Conversion Superheterodyne. 13 Tubes plus Semiconductors.

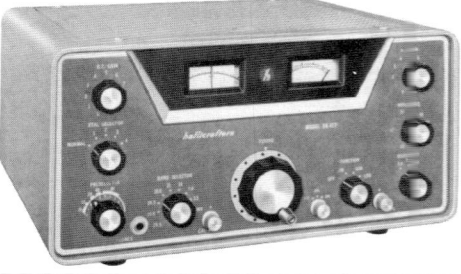

Features:

• ¼" Head. Jack	• S-Meter	• AVC ON/OFF	• Calibrator 100 kHz
• RF Gain	• ANL	• Mute Line	• Antenna Trimmer
• Notch	• Dial Lamp	• BFO	• Speaker Terminals
• Standby	• VFO Out Jack	• Cal Reset	• Crystal Osc Out Jack
• Preselector			

Specifications:

Coverage ...... Ham. See Comments
Selectivity ...... 5/2.5/.5 kHz -6 dB.
Sensitivity ...... <.5µV 10 dB S/N SSB
IF Rej .... >50 dB
Audio Out ...... 0.75 W 3.2/500 ohms

Modes .... AM/USB/LSB
Stability .... < 300 Hz /hr after warm-up
IF .... 6500-6000, 1650, 50.75 kHz
Antenna Input . RCA 50-70 ohms

Circuit Complement: 6DC6 RF Amp, 6EA8 1st Mixer/Cathode Follow, 12AT7 Crystal Osc, 6BA6 1st IF Amp, 6BE6 2nd Mixer, 6EA8 VFO/Cathode Follow, 6DC6 2nd IF Amp, 6EA8 3rd Mixer/SSB Swit Crystal Osc, 6BA6 3rd IF Amp, 6BE6 Product Detector/BFO, 6BN8 AM Detector/AVC Amp/AVC Rectifier, 6GW8 1st Audio Amp/Audio Out, 6AU6 Crystal Calibrator and four silicon diodes.

Accessories:
R-47 Speaker   R-48 Speaker   HA-10 LF/MF Converter (85-3000 kHz)
Comments: Ranges: 3.5-4, 7-7.5, 14-14.5, 21-21.5 and 28.5-29 MHz. Optional crystal positions for: 9.5-10, 28-28.5, 29-29.5 and 29.5-30 MHz. Product detector for CW/SSB. Requires a speaker. Matches the HT-44 Transmitter.

Made In: United States 1962-1965

Voltages: 105-125 VAC 50/60 Hz 70W

Readout: Analog Lin.

Physical: 15x7.75x14.75" 18.5 Lbs. 281x197x375mm 8.4 kg

Status: Inactive Manufacturer Discontinued Model

Rarity: Common

Reviews: QST May 1963
CQ August 1964

	New	Used
Price:	$380-400	$100-200
Rating:	★★★★	⑦

SX-122

General Coverage Communications Receiver
Double Conversion Superheterodyne. 11 Tubes plus Semiconductors.

Features:
- ¼" Head. Jack
- S-Meter
- Bandspread
- Antenna Trimmer
- RF Gain
- ANL
- BFO
- Speaker Terminals
- Dial Lamp
- Mute Terminals

Specifications:
Coverage ...... 538 - 34000 kHz[1] Modes .... AM/SSB-CW
Selectivity ...... 5/2.5/.5 kHz -6 dB. IF .... 1650, 50 kHz.
Sensitivity ...... <.5µV SSB Antenna Input . Terminals
Audio Out ...... 1 W 3.2 ohms 10% dist.

Circuit Complement:
6DC6 RF Amp, 6AU6 1st Mixer, 6C4 HF Osc, 6DC6 1650 kHz IF Amp, 6EA8 2nd Mixer/Crystal Osc, 6BA6 50 kHz IF Amp, 6BE6 BFO/Product Detector, 6BN8 AVC Amp/AVC Rect/AM Detector, 6GW8 1st AF Amp/Audio Out, 5Y3 Rectifier and 0A2 Voltage Regulator. Product detector for SSB/CW and envelope detector for AM.

Accessories:
R-47 Speaker R-50 Speaker HA-7 Calibrator 100 kHz
R-51 Speaker/Clock

Comments:
Ranges: .538-1.58, 1.72-4.9, 4.6-13 and 12-34 MHz. [1]Reception gap from 1580-1720 kHz. Mounting hole for SO-239.
Variants: Model **SX-122A** introduced in 1967 at $395. Same appearance and tube complement. Model **SX-122R** rack version not put into production.

Made In:	United States 1964-1968
Voltages:	105/125 VAC 50/60 Hz 85W
Readout:	[⊔⊔⊔∎∎] Analog
Physical:	18.75x8x9.75" 29 Lbs. 476x203x248mm 13.2 kg
Status:	Inactive Manufacturer Discontinued Model
Rarity:	Common
Reviews:	QST August 1970 Popular Elec. Sept. 1969 Electric Radio Dec. 1994

	New	Used
Price:	$290-295	$130-190
Rating:	★★★★	⑦

SX-130

General Coverage Communications Receiver
Single Conversion Superheterodyne. 7 Tubes plus Semiconductors.

Features:
- ¼" Head. Jack
- S-Meter
- Bandspread
- Product Det. SSB/CW
- RF Gain
- ANL
- BFO
- Antenna Trimmer
- Mute Terminals
- Crystal Phasing
- Dial Lamp
- Speaker Terminals

Specifications:
Coverage ...... 535 - 31500 kHz[1] Modes .... AM/SSB-CW
Selectivity ...... Wide/Normal/Narrow IF .... 1650 kHz.
Audio Out ...... 2 W 3.2 ohms Antenna Input . Terminals 50-600 ohms

Circuit Complement:
6DC6 RF Amp, 6EA8 Mixer/Osc, 6EA8 1st IF Amp/Crystal Sel., 6BA6 2nd IF Amp, 6AL5 AM Detector/NL, 6BE6 BFO/Product Detector, 6GW8 AF Amp/Output and diode.

Accessories:
HA-7 Calibrator R-47 Speaker R-50 Speaker
R-51 Speaker/Clock

Comments:
Ranges: .535-1.61, 1.725-4.7, 4.5-13 and 11.9-31.5 MHz. [1]Reception gap from 1610-1725 kHz. Bandspread ranges: 3.5-4, 7-7.4, 13.9-14.4, 20-22 and 26-30 MHz. Note that the bandspread occupies the long horizontal dial and the main tuning occupies the smaller rotary dial.
Variants: The model S-129, previously shown, is the same as the SX-130 less crystal filter, crystal phasing and S-Meter $165.

Made In:	United States 1965-1969
Voltages:	105-125 VAC 50/60 Hz 48W
Readout:	[⊔⊔⊔⊔∎] Analog
Physical:	18.875x8x9.75" 22 Lbs. 479x203x248mm 11.3 kg
Status:	Inactive Manufacturer Discontinued Model
Rarity:	Common

	New	Used
Price:	$170-200	$90-140
Rating:	★★★	⑦

SX-133

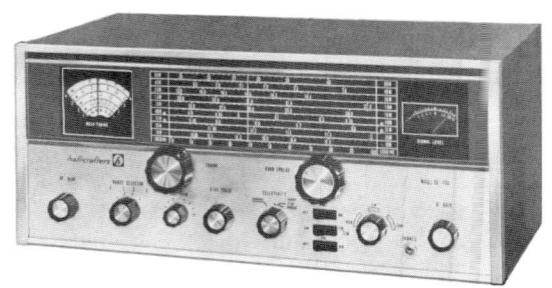

General Coverage Communications Receiver
Single Conversion Superheterodyne. 7 Tubes.

Features:
- ¼" Head. Jack
- S-Meter
- Bandspread
- Product Det. SSB/CW
- RF Gain
- ANL
- BFO
- Antenna Trimmer
- Mute Terminals
- Crystal Phasing
- Dial Lamp
- Speaker Terminals

Specifications:
Coverage ...... 535 - 31500 kHz[1]   Modes .... AM/LSB/USB-CW
Selectivity ...... 5/2.5/.5 kHz   IF .... 1650 kHz.
Audio Out ...... 2 W 3.2 ohms   Antenna Input . Terminals

Circuit Complement:
6DC6 RF Amp, 6EA8 Mixer/Osc, 6EA8 1st IF Amp/Crystal Sel., 6BA6 2nd IF Amp, 6AL5 AM Detector/NL, 6BE6 BFO/Product Detector, 6GW8 AF Amp/Output and diode.

Accessories:
R-47 Speaker   R-50 Speaker   HA-19 Calibrator 100 kHz
R-51 Speaker/Clock

Comments:
Main ranges: .535-1.61, 1.725-4.7, 4.5-13 and 11.9-31.5 MHz. [1] Reception gap from 1610-1725 kHz. Bandspread ranges: 3.5-4, 6.9-7.4, 13.5-14.4, 20-22, 26-30, 5.9-6.25, 9.45-9.8, 11.65-12 and 15.05-15.5 MHz. Note that the bandspread occupies the long horizontal dial and the main tuning occupies the smaller rotary dial. The bandspread of the 49, 31, 25 and 19 meter bands makes this radio more attractive to the shortwave listener than the SX-130.

Made In: United States 1968- 972

Voltages: 117 VAC 50/60 Hz
48W

Readout: Analog

Physical: 18.25x8x9.75" 31 Lbs.
464x203x248mm 14.1 kg

Status: Inactive Manufacturer
Discontinued Model

Rarity: Scarce

Reviews: *Popular Elec.* August 1970

	New	Used
Price:	$250-350	$90-160
Rating:	★★★★	⑦

SX-140

Amateur Band Communications Receiver
Single Conversion Superheterodyne. 5 Tubes plus Semiconductors.

Features:
- ¼" Head. Jack
- S-Meter
- Standby
- Calibrator 3500 kHz
- RF Gain
- ANL
- BFO
- Antenna Trimmer
- Mute Terminals
- Cal Reset
- 25:1 Tuning Ratio
- Speaker Terminals
- Dial Lamp

Specifications:
Coverage ...... See Comments.   Modes .... AM/SSB-CW
IF .... 1650 kHz.   Antenna Input . Terminals 50-75 ohms
Audio Out ...... 3.2 ohms

Circuit Complement:
6AZ8 RF Amp/Crystal Calibrator, 6U8 Mixer/Osc, 6BA6 IF Amp, 6T8 AVC Detector/ANL/1st Audio Amp, 6AW8A Audio Out/S-Meter plus 2 silicon rectifiers.

Accessories:
R-47 Speaker   R-48 Speaker

Comments:
Ranges: 3.5-4, 7-7.3, 14-14.4, 21-21.5, 28-29.9 and 50-54 MHz. Matches the model HT-40 transmitter. [¥70,000].

Variants:
Model **SX-140K** kit version sold for $95-115. Reviewed in *73* November 1963.

Made In: United States 1960-1965

Voltages: 117 VAC 50/60 Hz
47W

Readout: Analog

Physical: 13.4x7.4x8.25" 3.5 Lbs.
340x188x210mm 6.1 kg

Status: Inactive Manufacturer
Discontinued Model

Rarity: Common

Reviews: *QST* December 1961
Electric Radio April 1995

	New	Used
Price:	$95-140	$75-100
Rating:	★★★★	⑦

SX-146

Amateur Band Communications Receiver
Single Conversion Superheterodyne. 9 Tubes plus Semiconductors.

Features:
- ¼" Head. Jack
- S-Meter
- AVC ON/OFF
- Preselector
- RF Gain
- ANL
- Mute Line
- Extl. Osc. Input
- BFO
- Fiduciary Adjust
- Flywheel Tuning
- Speaker Terminals

Specifications:

Coverage ......Ham. See Comments.	Modes .... AM/CW-LSB/USB
Selectivity ......2.1/_/_ kHz	Stability .... <500 Hz/hr after warmup
Sensitivity ......<1μV 20 dB S/N	IF .... 9 MHz
Image Rej ......>50 dB	Antenna Input . RCA 50-70 ohms
Audio Out ......0.75 W 3.2 ohms 10% Dist.	

Circuit complement:
6JD6 RF Amp, 6GW8 Audio Amp & Output, 12AT7 Mixer, 6BA6 VFO, 6AU6A IF Amp, 6AU6 IF Amp, 6EA8 Osc. & Premix, 12AT7 AM Detector/AVC Rectifier, 12AT7 USB/LSB Crystal Osc. and 6AU6A Calibrator.

Accessories:
49-321 .5 kHz Filter	49-319 5 kHz Filter	HA-19 Calibrator 100 kHz
R-50 Speaker	R-51 Speaker/Clock	

Comments:
Ranges: 3.5-4, 7-7.5, 14-14.5, 21-21.5, 28-28.5, 29-29.5[1] and 29.5-30[1] MHz. [1]This range requires an optional crystal. Matches the HT-46 transmitter. Requires a speaker. [¥72,000].

Made In: United States 1965-1968

Voltages: 105-125 VAC 50/60 Hz
55W

Readout: Analog Lin

Physical: 13.125x5.875x11" 18 Lbs.
333x149x279mm 8.2 kg

Status: Inactive Manufacturer
Discontinued Model

Rarity: Common

Reviews: *CQ* June 1966
QST April 1966

	New	Used
Price:	$250-295	$140-200
Rating:	★★★★	⑦

WR-600G

General Coverage Communications Receiver
Single Conversion Superheterodyne. 4 Tubes plus Semiconductors.

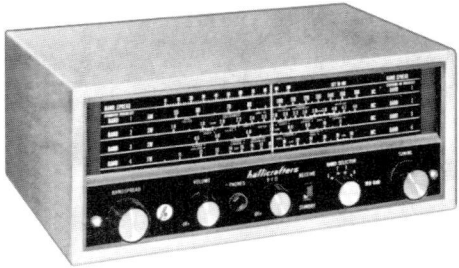

Features:
- ¼" Head. Jack
- BFO
- Speaker
- Bandspread 0-100
- Standby
- Dial Lamp
- Ferrite MW Antenna

Specifications:

Coverage ......550 - 30000 kHz	Modes .... AM/CW
IF ....455 kHz.	Antenna Input . Terminals
Audio Out ......8 ohms	

Circuit Complement:
12BE6 Converter, 12BE6 IF Amp/BFO, 12AV6 Audio Amp/AVC, 50C5 Audio Power Amp plus 1 selenium rectifier.

Comments:
Ranges: .55-1.6, 1.6-4.4, 4.5-12 and 11-30 MHz. This radio has a gray case and it is virtually identical to the S-120.

Variants;
Model **WR-600W** has a simulated wooden cabinet and a lighter front panel.

Made In: United States 1962-1967

Voltages: 105-125 VAC 50/60 Hz

Readout: Analog

Physical: 13.5x6x8.3" 12 Lbs.
243x152x210mm 5.4 kg

Status: Inactive Manufacturer
Discontinued Model

Rarity: Common

	New	Used
Price:	$70	$40-60
Rating:	★★★★	⑦

WR-1000

General Coverage Communications Receiver
Single Conversion Superheterodyne. 4 Tubes plus Semiconductors.

Features:
- ¼" Head. Jack
- BFO
- Speaker 5"
- Bandspread 0-100
- Standby
- Dial Lamp
- Ferrite MW Antenna

Specifications:

Coverage ......530 - 30000 kHz		Modes .... AM/CW	
IF ....455 kHz.		Antenna Input . Terminals	
Audio Out ......8 ohms			

Circuit Complement:
12BE6 Converter, 12BA6 IF Amp/BFO, 12AV6 Audio Amp/AVC, 50C5 Audio Power Amp plus 1 selenium rectifier.

Comments:
Ranges: .55-1.6, 1.6-4.4, 4.5-11 and 11-30 MHz. This radio has a wood case and is otherwise virtually identical to the S-120.

Made In:	United States 1964-1962
Voltages:	105-125 VAC 50/60 Hz 30W
Readout:	Analog
Physical:	14.5x6.75x9" 11.75 Lbs. 368x171x229mm 5.3 kg
Status:	Inactive Manufacturer Discontinued Model
Rarity:	Scarce

	New	Used
Price:	$70	$40-60
Rating:	★★★	⑦

WR-1500

General Coverage Communications Receiver
Single Conversion Superheterodyne. 5 Tubes plus Semiconductors.

Features:
- ¼" Head. Jack
- BFO
- Speaker 4"
- Bandspread 0-100
- Standby
- Dial Lamp
- ANL
- Ferrite MW Antenna
- Sensitivity

Specifications:

Coverage ......185 - 31000 kHz[1]		Modes .... AM/CW-SSB	
IF ....455 kHz.		Antenna Input . Terminals 52/600 ohms	
Audio Out ......8 ohms			

Circuit Complement:
6BL8 Mixer/Osc, 12BA6 IF Amp, 6BL8 IF Amp/BFO, 6T8 1st Audio Detector/ANL/AVC, 6AQ5 Output plus 2 selenium rectifiers.

Comments:
Ranges: .185-.42, .495-1.62, 1.6-4.95, 4.85-15 and 14.8-31 MHz. [1]Reception gap from 420-495 kHz. This radio is virtually identical to the S-118, except for having a wooden cabinet.

Made In:	United States 1962
Voltages:	105-125 VAC 50/60 Hz 33W
Readout:	Analog
Physical:	14.5x6.75x9" 4.5 Lbs. 368x171x229mm 6.6 kg
Status:	Inactive Manufacturer Discontinued Model
Rarity:	Very Scarce

	New	Used
Price:	$100	$40-65
Rating:	★★★	⑦

WR-2000

Limited General Coverage Broadcast Receiver
Single Conversion Superheterodyne. 5 Tubes plus Semiconductors.

Features:
- ¼" Head. Jack
- Speaker 5x7"
- Speaker Jack
- Tone
- AFC
- FM Bdcst. Band
- Dial Lamp
- Ferrite MW Antenna
- MUX/Phono Jack

Specifications:
Coverage ......See Comments.
IF ....455 kHz [10.7 MHz FM]
Audio Out ......0.5W 8 ohms 3% dist.
Modes .... AM
Antenna Input . Terminals 50-600 ohm

Circuit Complement:
6BL8 AM Mixer/FM IF Amp/AM Osc., 6BA6 IF Amp, 6KL8 FM Limiter/AM Detector/AVC, 6GW8A 1st Audio/Audio Output and 1N1764 diode.

Comments:
Ranges: .54-1.6, 2-6 and 6-18 MHz plus FM 88-108 MHz.

Made In: United States 1962

Voltages: 105-125 VAC 50/60 Hz 35W

Readout: Analog

Physical: 14.5x6.5x9.25" 15.25 Lbs. 365x165x238mm 6.9 kg

Status: Inactive Manufacturer Discontinued Model

Rarity: Scarce

	New	Used
Price:	①	$40-60
Rating:	★★★	⑦

WR-2500

General Coverage Broadcast Receiver
Single Conversion Superheterodyne. 5 Tubes plus Semiconductors.

Features:
- ¼" Head. Jack
- Speaker 5x7"
- Speaker Jack
- Tone
- AFC
- FM Bdcst. Band
- Dial Lamp
- Ferrite MW Antenna
- MUX/Phono Jack

Specifications:
Coverage ......540 - 18000 kHz +FM
IF ....455 kHz [10.7 MHz FM]
Audio Out ......0.5W 8 ohms 3% dist.
Modes .... AM
Antenna Input . Terminals 50-600 ohm

Circuit Complement:
6BL8 AM Mixer/FM IF Amp/AM Osc., 6BA6 IF Amp, 6KL8 FM Limiter/AM Detector/AVC, 6GW8A 1st Audio/Audio Output and 1N1764 diode.

Comments:
Ranges: .54-1.6, 2-6 and 6-18 MHz plus FM 88-108 MHz. This model is similar to the WR-2000, but with a walnut veneer wood cabinet.

Made In: United States 1962

Voltages: 105-125 VAC 50/60 Hz 35W

Readout: Analog

Physical: 15.5x7.32x9.25" 18 Lbs. 393x186x235mm 8.2 kg

Status: Inactive Manufacturer Discontinued Model

Rarity: Very Scarce

	New	Used
Price:	①	$40-65
Rating:	⑥	⑦

---

38 Hammarlund Mfg. Co.

The **Hammarlund Radio Company** was established in 1910. They produced some AM radios and components. Hammarlund got on the map with the introduction of the Comet and Comet Pro superheterodyne receivers in 1931. The famous (Comet) Super Pro followed. This would be the first in a series that would run over thirty years. The respected Super Pro series culminated with the famous SP-600 Super Pro still in active use today. Hammarlund is perhaps best known for their "HQ" series of affordable amateur and general coverage receivers. All resemble their original ancestor, the HQ-120, introduced in 1938. Unlike other manufacturers, Hammarlund made most of their own components, rather than using off the shelf components. Most agree that this and other reasons put Hammarlund ahead of its main competitor, Hallicrafters. Like Hallicrafters, Hammarlund ran into difficulties in the late 1960's.

In 1971 Cardwell Capacitor purchased the capacitor division of Hammarlund from the Electronic Assistance Corporation. Five years later Pax Manufacturing, which is owned by the same family as Cardwell acquired the remaining Hammarlund assets, mainly for military spare parts. Pax also has Gonset parts. Pax Mfg. did not manufacture any receivers under the Hammarlund name. They remain a source for Hammarlund parts.

Hammarlund Mfg. Co.
424-438 West 33rd St.
New York, NY   1941-1944

Hammarlund Mfg. Co.
460 West 34th St.
New York 1, NY   945-1959

Hammarlund Mfg. Co.
A Giannini Scientific Co.
53 West 23rd St.
New York 10, NY   1962-1966

Hammarlund Mfg. Co.
73-88 Hammarlund Dr.
Mars Hill, NC 28754   1968-1970

Hammarlund Mfg. Co.
Electronic Assistance Corp.
20 Bridge Av.
Red Bank, NJ 07701   1970-1971

Pax Mfg. Corp.
Hammarlund Radio Division
100 East Montauk Hwy.
Lindenhurst, NY 11757   1971-1997

HQ-88

Amateur Band Communications Receiver
Double Conversion Superheterodyne. 10 Tubes plus Semiconductors.

Features:
- ¼" Head. Jack
- BFO ±3 kHz
- Q-Multiplier
- Preselector
- S Meter
- Dial Cal Set
- AVC FAST/SLOW
- Standby
- Sensitivity
- NL
- Mute Contacts
- Spinner Knobs
- Antenna Trimmer
- Calibrator 100 kHz
- Speaker Terminals
- Dial Adjust

Specifications:
Coverage ...... Ham bands.
Selectivity ...... 5/2.2 kHz -6dB.
Sensitivity ...... <.4µV 10 dB S/N SSB
I.F. ....?, 262 kHz.
Audio .... 3.2/500 ohm 1W

Modes .... AM/(CW-SSB)
Stability .... <±50 Hz after warm-up
Image Rej .... >60 dB (45 dB on 10M)
Antenna Input . SO-239 and Terminals

Circuit Complement:
12BZ6 RF Amp, 12AT7 1st Converter (Crystal Osc.), 6BR8 2nd Converter (Tunable Osc.), 12BA6 1st IF Amp, 12BA6 2nd IF Amp, 6FM8 AVC Amp/AVC Detector, 6BE6 Product Detector, 12AX7 Q-Multiplier, 6GW8 Audio Amp and 12BV6 Crystal Calibrator. Separate diode AM detector and SSB product detector. Single conversion below 3 MHz. Tuning is 20 kHz per revolution. Dial accuracy is ±1 kHz.

Accessories:
S-88 Speaker   Noise Immunizer

Comments:
Ranges: 1.8-2.4, 3.5-4.1, 6.9-7.5, 20.9-21.5 and 28.5-29.1 MHz. Optional ranges: 27.9-28.5, 29.1-29.6 and three extra 600 Hz crystal positions.

Made In: United States 1964-1966

Voltages: 105-125 VAC 50-60 Hz
80 W

Readout: Analog Lin.

Physical: 15x7.75x10.5 16 Lbs.
381x197x267 mm 7.3 kg

Status: Inactive Manufacturer Discontinued Model

Rarity: Very Scarce

Reviews: *CQ* November 1964
How to Listen to W. 1965/66

	New	Used
Price:	$299-325	②
Rating:	⑥	⑥

HQ-100

General Coverage Communications Receiver
Single Conversion Superheterodyne. 10 Tubes.

Features:
- ¼" Head. Jack
- S Meter
- Sensitivity
- Antenna Trimmer
- Bandspread
- ANL
- AVC MAN/ON
- Speaker Terminals
- Q-Multiplier
- BFO ±4 kHz

Specifications:
Coverage ......540 - 30000 kHz   Modes .... AM/(CW-SSB)
Selectivity ......6kHz & var .1-3 kHz -6dB   I.F. .... 455 kHz
Sensitivity ......<1.75µV -10 dB S/N   Antenna Input . Terminals 100 ohm
Audio .... 1W 3.2 ohms

Circuit Complement:
6BZ6 RF Amp, 6BE6 Mixer, 6C4 HF Osc, 6BA6 1st IF Amp, 6BA6 2nd IF Amp, 6AL5 Detector/NL, 12AX7 1st AF Amp/Q-Multiplier/BFO, 6AQ5 Audio Out, 0B2 Regulator and 5Y3 Rectifier. The XC-455 option kit adds a 455 kHz crystal controlled BFO to the second detector. Permits CW reception with a variable bandwidth of 100 to 3000 kHz.

Accessories:
S-100 Speaker   XC-100 Calibrator   Clock/Timer (shown)
XC-455 BFO Kit
Comments: Ranges: .54-1.6, 1.6-4, 4-10 and 10-30 MHz. Requires a speaker. Shown with optional clock. [¥72,000].
Variants:
Model **HQ-100C** (shown) includes the clock. Model **HQ-100E** is export version.

Made In:	United States 1956-1961
Voltages:	105-125 VAC 50-60 Hz 68 W
Readout:	Analog
Physical:	16.25x9.5x9.2" 30 Lbs. 413x241x233mm 13.6 kg
Status:	Inactive Manufacturer Discontinued Model
Rarity:	Common
Reviews:	QST January 1957 CQ January 1958

	New	Used
Price:	$169-189	$75-100
Rating:	★★★	★★

HQ-100A

General Coverage Communications Receiver
Single Conversion Superheterodyne. 10 Tubes.

Features:
- ¼" Head. Jack
- S Meter
- Sensitivity
- Antenna Trimmer
- BFO ±4 kHz
- Bandspread
- ANL
- AVC MAN/ON
- Q-Multiplier
- Speaker Terminals

Specifications:
Coverage ......540 - 30000 kHz   Modes .... AM/(CW-SSB)
Selectivity ......6kHz & var .1-3 kHz -6dB   I.F. .... 455 kHz
Sensitivity ......<1.75µV -10 dB S/N   Antenna Input . Terminals 100 ohm
Audio .... 1W 3.2 ohms

Circuit Complement:
6BZ6 RF Amp, Mixer 6BE6, HF Osc 6C4, 6BA6 1st IF Amp, 6BA6 2nd IF Amp, 6BV8 Detector /NL/BFO, 12AX7 1st AF Amp/Q-Multiplier, 6AQ5 Audio Out, 0B2 Regulator and 5Y3 Rectifier.

Accessories:
S-100 Speaker   XC-100 Calibrator   Clock/Timer (shown)
Comments:
The HQ-100A is similar to the HQ-100, but with a separate BFO built-in (note the knob between the tuning dials), and the optional clock is in the 24 hour format. Ranges: .54-1.6, 1.6-4, 4-10 and 10-30 MHz. Requires a speaker.

Variants:
Model **HQ-100AC** (shown) includes the clock. Model **HQ-100AE** is export version.

Made In:	United States 1961-1966
Voltages:	105-125 VAC 50-60 Hz 68 W
Readout:	Analog
Physical:	16.25x9.5x9.2" 26 Lbs. 413x241x233mm 11.8 kg
Status:	Inactive Manufacturer Discontinued Model
Rarity:	Common
Reviews:	QST December 1961

	New	Used
Price:	$190-200	$80-130
Rating:	★★★★	★★★

  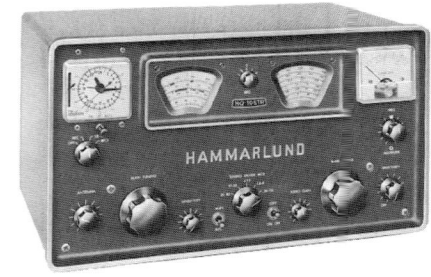

HQ-105-TR

Amateur Band Communications Receiver / Transceiver
Single Conversion Superheterodyne. 11 Tubes.

Features:
- ¼" Head. Jack
- S Meter
- Sensitivity
- Antenna Trimmer
- BFO ±3 kHz
- NL
- AVC MAN/ON
- Speaker Terminals
- Q-Multiplier
- Mute Contacts
- Speaker
- 5W CB/10M Xmit

Specifications:
Coverage ...... Ham See Comments.   Modes .... AM/(CW-SSB)
Selectivity ...... 6 +Var. .3-2.9 kHz -6dB.   I.F. .... 455 kHz
Sensitivity ...... <1.75µV -10 dB S/N   Antenna Input . Phono Jack 50-600 ohms
Audio .... 1W 3.2 ohm

Circuit Complement: 6BZ6 RF Amp, 6BE6 Mixer, 6C4 HF Osc, 12AX7 Q-Mult/ Preamp, 6BA6 1st IF Amp, 6BA6 2nd IF Amp, 6BV8 Det/AVC/NL/BFO, 6BM8 Audio Out/1st AF Amp, OB2 Volt Reg, 6CA4 Rectifier, 6CX8 Crystal Osc/RF Power Amp

Accessories: OCT-X External 8 Position Crystal Box
S-100 Speaker   XC-100 Calibrator 100 kHz   Telechron Clock/Timer

Comments: Ranges: .64-1.6, 1.6-4, 4-10, 10-30 MHz (20 BS). Bandspread: 3.5-4, 6.8-7.3, 20-21.6, 27.8-30 and 13.8-14.4 MHz. This radio may be more properly classified as a *transceiver* because it features a crystal controlled 5 watt CB (or 10 meter) transmitter in it. Crystal socket, headphone jack and 4 pin mic jack are on the rear panel of the radio. Requires a speaker.

Variants: Model **HQ-105-TR-C** (shown) includes the 24 hour analog clock. Model **HQ-105-TR-E** operates from 115-230 VAC 50-60 Hz. Model **HQ-105-TRS** features a speaker in place of the optional clock-timer.

Made In: United States 1951-1962

Voltages: 105-125 VAC 50 60 Hz 80 W

Readout: Analog

Physical: 16.5x9.5x9.25" 35 Lbs. 419x241x235mm 15.9 kg

Status: Inactive Manufacturer Discontinued Model

Rarity: Very Scarce

Reviews: *QST* December 961

	New	Used
Price:	$220	150-225
Rating:	★★★★	★★

HQ-110

Amateur Band Communications Receiver
Double Conversion Superheterodyne. 12 Tubes.

Features:
- ¼" Head. Jack
- S Meter
- Sensitivity
- Speaker Terminals
- BFO ±4 kHz.
- Dial Cal Set
- Antenna Trimmer
- Calibrator 100 kHz
- Q-Multiplier
- AVC MAN/ON
- Mute Contacts
- Standby
- NL

Specifications:
Coverage ...... Ham. See Comments.   Modes .... AM/(CW-SSB)
Selectivity ...... 6 +Var. .1-3 kHz -6dB.   I.F. .... 3045, 455 kHz
Sensitivity ...... <1.5µV 10 dB S/N   Antenna Input . Terminals
Audio .... 1W 3.2 ohm

Circuit Complement: 6BZ6 RF Amp, 6BE6 Mixer, 6BE6 Converter, 12AX7 Q-Mult/ 1st AF Amp, 6BA6 1st IF Amp, 6AZ8 Lin Detector/2nd IF Amp/BFO, 6BJ7 Detector/ NL/AVC, 6AQ5 AF Out, 6BZ6 Crystal Cal Osc., 6C4 HF Osc, 0B2 Regulator and 5U4GB Rectifier. Single conversion below 7 MHz. There were production changes after serial number 7000 (7 caps, 2 coils and 1 resistor were added).

Accessories:
S-100 Speaker   12 Hr. Clock/Timer (shown)

Comments: Ranges: 1.8-2, 3.5-4, 7-7.3, 14-14.4, 21-21.6, 28-30 and 50-54 MHz. Variable selectivity of .1 - 3 kHz provided by the Q-Multiplier. Both dials are controlled by the large left tuning knob. Requires a speaker. The headphone jack is on back.

Variants:
Model **HQ-110C** (shown) includes the 12 hour clock. The export model **HQ-110E** is 115-230 VAC 50/60 Hz.

Made In: United States 1957-1962

Voltages: 105-125 VAC 50-60 Hz 80 W

Readout: Analog

Physical: 16.2x9.5x9.2" 30 Lbs. 411x241x235mm 13.6 kg

Status: Inactive Manufacturer Discontinued Model

Rarity: Common

Reviews: *QST* August 1958 *CQ* September 1986 *Electric Radio* January 1994

	New	Used
Price:	$229-249	$70-140
Rating:	★★★★	★★

HQ-110A

Amateur Band Communications Receiver
Double Conversion Superheterodyne. 12 Tubes.

Features:
• ¼" Head. Jack	• S Meter	• Sensitivity	• Speaker Terminals
• BFO ±4 kHz.	• Dial Cal Set	• Antenna Trimmer	• Calibrator 100 kHz
• Q-Multiplier	• AVC MAN/ON	• Mute Contacts	• Standby • NL

Specifications:
Coverage ......Ham. See Comments.   Modes .... AM/(CW-SSB)
Selectivity ......6 +Var. .1-3 kHz -6dB.   Audio 1W
Sensitivity ......<1.5µV CW 10:1 S/N   I.F. .... 3045, 455 kHz
Audio ....1W 3.2 ohms   Antenna Input . Terminals

Circuit Complement: 6BZ6 RF Amp, 6BE6 Mixer, 6BE6 Converter, 12AX7 Q-Multiplier/1st AF Amp, 6BA6 1st IF Amp, 6AZ8 Lin Detector/2nd IF Amp/BFO, 6BJ7 Detector/NL/AVC, 6AQ5 AF Out, 6BZ6 Crystal Calibrator Osc., 6C4 HF Osc, 0B2 Regulator and 5U4GB Rectifier. Single conversion below 7 MHz.

Accessories:
S-100 Speaker   Clock/Timer (shown)   2 Meter Converter

Comments: Ranges: 1.8-2, 3.5-4, 7-7.3, 14-14.4, 21-21.6, 28-30 and 50-54 MHz. The "A" model has a dial that is precalibrated for the optional 2 meter converter, a 6 meter coax input and an accessory socket. Variable selectivity of .1 - 3 kHz is provided by the Q-Multiplier. Requires a speaker. The headphone jack is on back.

Variants: Model **HQ-110AC** (shown) includes the clock. Model **HQ-110AE** is the 115-230VAC export version. Model **HQ-110A-VHF** includes a preamp for improved 6M reception and a 2M converter. Model **HQ-110AC-VHF** is the same with clock.

Made In: United States 1961-1969

Voltages: 117/230 VAC 50-60 Hz 68 W

Readout: ⊔⊔⊔⊓⊔⊔ Analog

Physical: 16.2x9.5x9.2" 30 Lbs. 411x241x233mm 13.6 kg

Status: Inactive Manufacturer Discontinued Model

Rarity: Common

Reviews: *73* January 1967

	New	Used
Price:	$189-299	$80-170
Rating:	★★★★	★★

HQ-120-X

General Coverage Communications Receiver
Single Conversion Superheterodyne. 12 Tubes.

Features:
• ¼" Head. Jack	• S Meter	• RF Gain	• Antenna Trimmer
• BFO	• Bandspread	• AVC OFF/ON	• Standby
• Relay Contacts	• Dial Lamp	• NL	• Speaker Terminals
• Carry Handles			

Specifications:
Coverage ......540 - 31000 kHz   Modes .... AM/CW
Selectivity ......6/4/2/1/.5 kHz -6dB.   I.F. .... 455 kHz
Audio Out ......4 W   Aerial Input ..... Terminals 400 ohm

Circuit Complement:
6S7 RF Amp, 6K8 Converter, 6S7 1st IF Amp, 6S7 2nd IF Amp, 6F6 3rd IF Amp, 6H6 Detector/AVC, 6H6 NL, 6J7 BFO, 6V6G Audio, 6SF5 Meter, VR150 Voltage Regulator and 5Z4G Rectifier.

Accessories:
SC-10 Speaker Cabinet

Comments:
Ranges: .54-1.32, 1.32-3.2, 3.2-5.7, 5.7-10, 10-18 and 18-31 MHz. Bandspread: 80, 40, 20, 15 and 10 meters. Requires an external speaker which was supplied. Black or gray front panel. A great medium wave performer. The early ancestor of the famous Hammarlund "HQ" series.

Variants:
Model **HQ-120** 1938 does not have crystal filter.

Made In: United States 1938-1945

Voltages: 105-125 VAC 50-60 Hz

Readout: ⊔⊔⊔⊓⊔⊔ Analog

Physical: 17.125x10x12.25" 435x255x311mm

Status: Inactive Manufacturer Discontinued Model

Rarity: Scarce

Reviews: *QST* December 1941

	New	Used
Price:	$129-215	$65-100
Rating:	⑦	⑦

 HAMMARLUND

HQ-129-X

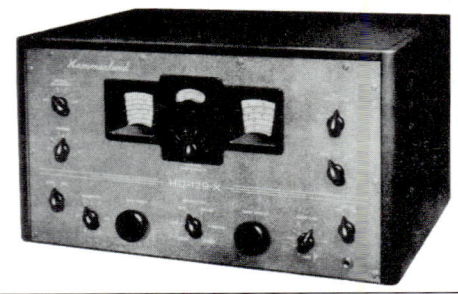

General Coverage Communications Receiver
Single Conversion Superheterodyne. 11 Tubes.

Features:
- ¼" Head. Jack
- S Meter
- RF Gain
- Antenna Trimmer
- BFO
- Bandspread
- AVC OFF/ON
- Standby
- Relay Contacts
- Dial Lamp
- NL
- Speaker Terminals
- Flywheel Tuning

Specifications:
Coverage ......540 - 31000 kHz   Modes .... AM/CW
Audio Out ......3 W 4 ohm   I.F. .... 455 kHz
Aerial Input .... Terminals 400 ohm

Circuit Complement:
6SS7 RF Amp, 6K8 Converter 1st Detector Osc, 6SS7 1st IF Amp, 6SS7 2nd IF Amp, 6SS7 3rd IF Amp, 6H6 Detector/NL, 6SN7GT/G 1st Audio Amp/S-Meter, 6V6GT/G Audio Power Amp, 6SJ7 BFO, 5U4G Rectifier and 0C3 Regulator.

Accessories:
SC-10 Speaker   FS-135-C Calibrator

Comments:
Ranges: .54-1.32, 1.32-3.2, 3.2-5.7, 5.7-10, 10-18 and 18-31 MHz. Bandspread: 80, 40, 20, 15 and 10 meters. Requires a speaker. This unit is often found with larger, user-supplied, tuning knobs. A great medium wave performer. Early production featured the Hammarlund name and model in red lettering. Most units had all the front panel labeling in white. Some late production units had black lettering for the name and model number. This is the post war successor the HQ-120-X.

Made In:	United States 1945-1953
Voltages:	105-125 VAC 50-60 Hz
Readout:	[analog scale] Analog
Physical:	20.125x11x13.5" 47 Lbs. 511x279x343mm 21.3 kg
Status:	Inactive Manufacturer Discontinued Model
Rarity:	Common
Reviews:	*QST* June 1946

	New	Used
Price:	$129-189	$50-140
Rating:	★★★★★	★★★

 HAMMARLUND

HQ-140-X

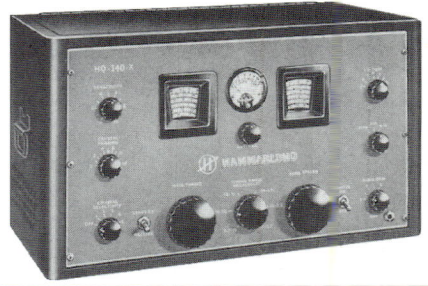

General Coverage Communications Receiver
Single Conversion Superheterodyne. 11 Tubes.

Features:
- ¼" Head. Jack
- S Meter
- RF Gain
- Antenna Trimmer
- BFO ±3 kHz
- Bandspread
- AVC OFF/ON
- Speaker Terminals
- Standby
- NL
- Mute Line
- Dial Lamp
- Carry Handles
- Flywheel Tuning

Specifications:
Coverage ......540 - 31000 kHz   Modes .... AM/CW
Selectivity ......6 position   I.F. .... 455 kHz
Audio Out ......3.5 W 6 ohms   Antenna Input . Terminals 400 ohm

Circuit Complement:
6C4 Osc, 6BA6 RF Amp, 6BA6 1st IF Amp, 6BA6 2nd IF Amp, 6BA6 3rd IF Amp, 6AL5 Detector/AVC/NL, 12AU7 1st AM Amp/BFO, 6V6GT/G Audio Power Amp, 0C3 Regulator and 5U4G Rectifier.

Accessories:
S-200 Speaker   Rack Mounting Kit   XC-100 Crystal Calibrator

Comments:
Ranges: .54-1.32, 1.32-3.2, 3.2-5.7, 5.7-10, 10-18 and 18-31 MHz. Bandspread: 80, 40, 20, 15 and 10 meters. Requires a speaker. Two tone gray metal cabinet. [¥83,000].

Variants:
Model **HQ-140-XA** 1956 $250 enhanced and has a square S-Meter instead of the round, giving the receiver a more modern appearance.

Made In:	United States 1953-1956
Voltages:	105-125 VAC 50/60 Hz 100 W
Readout:	[analog scale] Analog
Physical:	20.125x11x13.5" 47 Lbs. 511x279x343mm 21.3 kg
Status:	Inactive Manufacturer Discontinued Model
Rarity:	Common

	New	Used
Price:	$265	$120-250
Rating:	★★★★★	★★★

HQ-145

General Coverage Communications Receiver
Double Conversion Superheterodyne. 11 Tubes.

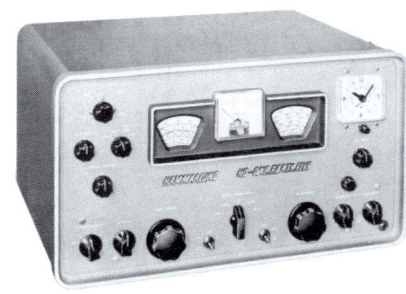

Features:
- ¼" Head. Jack
- S Meter
- Bandspread
- Slot Filter ±5 kHz
- Mute Line
- RF Gain
- Antenna Trimmer
- Speaker Terminals
- Dial Lamp
- BFO ±2 kHz.
- ANL
- AVC OFF/SLO/MED/FST
- Cal. Adjust
- Flywheel Tuning

Specifications:
Coverage ......540 - 30000 kHz   Modes .... AM/LSB/USB/CW
Selectivity ......6 Position   I.F. .... 3035, 455 kHz
Sensitivity ......<.6µV -10 dB S/N CW   Audio Out .... 2.5 W 4 ohm 5% dist.
Sensitivity ......<1.75µV -10 dB S/N AM   Antenna Input . Terminals 72 ohm
Image Rej ......>25 dB at 22 MHz

Circuit Complement: 6BZ6 RF Amp, 6BE6 1st Mixer, 6BE6 Converter/455 kHz IF Amp, 6BA6 455 kHz IF Amp, 6BA6 455 kHz IF Amp, 6AL5 Detector/NL, 12AX7 455 kHz BFO/Audio Amp, 6AQ5 Audio Power Output, 6C4 HF Oscillator, 0B2 Voltage Regulator, and 5U4GB Rectifier.

Accessories:
S-200 Speaker   Clock/Timer (shown)   XC-100 Crystal Calibrator

Comments: Ranges: .54-1.6, 1.6-4, 4-10, 10-30 MHz. Single conversion below 10 MHz. Requires a speaker. [¥103,000].

Variants: Model **HQ-145C** (shown) has built-in clock. Model **HQ-145X** has a fixed crystal position (1961 $269) QST December 1961. Model **HQ-145XC** has fixed crystal position and the clock. Model **HQ-145E** operates from 115/230 VAC 50/60 Hz.

Made In:	United States 1959-1963
Voltages:	105-125 VAC 50/60 Hz 80 W
Readout:	Analog
Physical:	19x10.5x13" 482x266x330mm
Status:	Inactive Manufacturer Discontinued Model
Rarity:	Common
Reviews:	QST June 1959 QST December 1961 CQ May 1959

	New	Used
Price:	$269-289	$130-190
Rating:	★★★★★	★★★★

HQ-145A

General Coverage Communications Receiver
Double Conversion Superheterodyne. 10 Tubes plus Semiconductors.

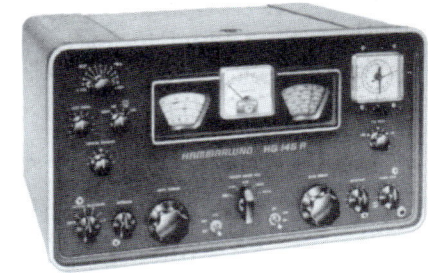

Features:
- ¼" Head. Jack
- S Meter
- Hinged Cover
- Mute Line
- RF Gain
- Slot Filter ±5 kHz
- Antenna Trimmer
- Speaker Terminals
- Dial Lamp
- BFO ±2 kHz.
- Bandspread
- AVC OFF/SLO/MED/FST
- Flywheel Tuning
- ANL

Specifications:
Coverage ......540 - 30000 kHz   Modes .... AM/LSB/USB/CW
Selectivity ......6 Position   I.F. .... 3035, 455 kHz
Sensitivity ......<.6µV -10 dB S/N CW   Sensitivity .... <1.75µV -10 dB S/N AM
Image Rej ......>25 dB at 22 MHz   Antenna Input . Terminals 72 ohm^
Audio Out ......2.5 W 4 ohm 5% dist.

Circuit Complement:
6BZ6 RF Amp, 6BE6 1st Mixer, 6BE6 Converter/455 kHz IF Amp, 6BA6 455 kHz IF Amp, 6BA6 455 kHz IF Amp, 6AL5 Detector/NL, 12AX7 455 kHz BFO/Audio Amp, 6AQ5 Audio Power Output, 6C4 HF Oscillator, 0B2 Voltage Regulator and (2) CER72C silicon diodes for rectification. Single conversion below 10 MHz.

Accessories:
S-200 Speaker   XC100 Crystal Calibrator   24 Hour Clock/Timer (shown)

Comments: Ranges: .54-1.6, 1.6-4, 4-10 and 10-30 MHz. The newer "A" model adds a hinged cover and ^SO-239 hole. Requires a speaker.

Variants: Model **HQ-145AX** ($349) has 11 fixed crystal positions. Model **HQ-145AC** (shown) has built-in clock.

Made In:	United States 1966-1969
Voltages:	117 or 220/230 VAC 50/60 Hz
Readout:	Analog
Physical:	19x10.5x13" 482x266x330mm
Status:	Inactive Manufacturer Discontinued Model
Rarity:	Common

	New	Used
Price:	$299	$140-220
Rating:	★★★★★	★★★★

 **HAMMARLUND**

HQ-150

General Coverage Communications Receiver
Single Conversion Superheterodyne. 13 Tubes.

Features:
- ¼" Head. Jack
- RF Gain
- BFO ±2 kHz.
- Dial Lamp
- S Meter
- ANL
- Bandspread
- Hinged Cover
- Q-Multiplier
- AVC ON/MAN
- Mute contacts
- Flywheel Tuning
- Antenna Trimmer
- Calibrator 100 kHz
- Speaker Terminals

Specifications:
Coverage ...... 540 - 31000 kHz
Selectivity ...... 6 Position
Audio Out 2W

Modes .... AM/CW-SSB
I.F. .... 455 kHz
Antenna Input . Terminals 100 ohm

Circuit Complement:
6BA6 RF Amp, 6BE6 Mixer, 6C4 HF Osc, 6BA6 1st IF Amp, 6BA6 2nd IF Amp, 6BA6 3rd IF Amp, 6AL5 Detector/AVC/NL, 12AX7 1st AF Amp/BFO, 6V6GT/G Audio Power Amp, 0C3 Voltage Regulator, 5U4GB Rectifier, 12AX7 Q-Multiplier and 6BZ6 Calibrator.

Accessories:
Speaker   Rack Speaker   Rack Mounting Kit

Comments:
Requires a speaker. Ranges: .54-1.32, 1.32-3.2, 3.2-5.7, 5.7-10, 10-18 and 18-31 MHz. Bandspread 80 to 10 meters. The HQ-150 features a true Q-Multiplier with both peak and reject. It has a very flexible filter system. Two toned gray metal cabinet. The HQ-150 was the last of the large single conversion Hammarlund general coverage receivers.

Made In:	United States 1956-1958
Voltages:	105-125 VAC 50/60 Hz
Readout:	Analog
Physical:	20.125x11x13.5" 70 Lbs. 511x279x343mm 31.7 kg
Status:	Inactive Manufacturer Discontinued Model
Rarity:	Scarce
Reviews:	QST December 1956 Proceedings 1958

	New	**Used**
Price:	$294-299	$150-210
Rating:	★★★★★	★★★

 **HAMMARLUND**

HQ-160

General Coverage Communications Receiver
Double Conversion Superheterodyne. 13 Tubes.

Features:
- ¼" Head. Jack
- Sensitivity
- Standby
- Dial Lamps
- S Meter
- Slot Filter ±5 kHz
- BFO
- Mute Line
- ANL
- Fiduciary Adj.
- Bandspread
- Flywheel Tuning
- Calibrator 100 kHz
- Speaker Terminals
- Antenna Trimmer

Specifications:
Coverage ...... 540 - 31000 kHz
Selectivity ...... 4/.1-2 kHz -6dB.
Sensitivity ...... <1.5µV -10 dB S/N

Modes .... AM/CW-SSB
I.F. .... 3035, 455 kHz
Antenna Input . Terminals

Circuit Complement:
6BA6 RF Amp, 6BE6 Mixer, 6C4 HF Osc, 6BE6 Converter, 6BA6 IF, 6BA6 IF, 6BJ7 Detector Limiter/AVC, 12AU7 AF, 6U8 Detector/BFO, 6AQ5 Output, 6BZ6 Calibrator, 5U4 Rectifier and 0B2 Regulator. 14 Tuned circuits in IF, crystal controlled second oscillator.

Accessories:
S-200 Speaker

Comments:
Ranges: .54-1.32, 1.32-3.2, 3.2-5.7, 5.7-10, 10-18 and 18-31 MHz. Bandspread: 80, 40, 20, 15 and 10 Meters. Requires a speaker.

Made In:	United States 1958-1960
Voltages:	105-125 VAC 50/60 Hz 100 W
Readout:	Analog
Physical:	19x10.5x13" 482x266x330mm
Status:	Inactive Manufacturer Discontinued Model
Rarity:	Common
Reviews:	QST October 1958 CQ March 1959

	New	**Used**
Price:	$379	$125-220
Rating:	★★★★★	★★★★

HQ-170

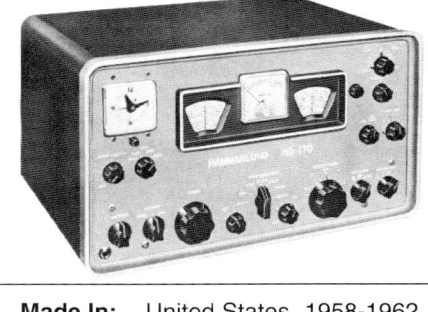

Amateur Band Communications Receiver
Triple Conversion Superheterodyne. 17 Tubes.

Features:
- ¼" Head. Jack
- BFO ±2 kHz.
- Slot Filter
- Fiduciary Adj.
- S Meter
- Vernier ±3 kHz.
- Mute Line
- Flywheel Tuning
- Sensitivity
- ANL
- Antenna Trimmer
- Speaker Terminals
- Calibrator 100 kHz
- AVC OFF/SLO/MED/FST

Specifications:
Coverage ......Ham. See Comments.
Selectivity ......6/4/3/2/1/.5 kHz
Sensitivity ......<.7µV -10 dB S/N CW
BFO Range ...±2 kHz.
Modes .... AM/(CW-SSB)
I.F. .... 3035, 455, 60 kHz
Audio .... 1W 3.2 ohms
Antenna Input . Terminals

Circuit Complement:
6BZ6 RF Amp, 6BE6 1st Mixer, 6BE6 Converter IF Amp, 6BA6 455 IF Amp, 6BE6 Converter, 6BA6 60 kHz IF Amp, 6BA6 60 kHz IF Amp, 6BV8 60 kHz IF Amp/AVC/ AM Detector, 12AU7 SSB Product Detector, 6AL5 NL, 6BZ6 Crystal Cal Osc. 6C4 HF Osc, 12AU7 60 kHz BFO/S Meter, 0B2 Voltage Regulator, 5U4GB Rectifier, 6AV6 1st AF Amp/AVC and 6AQ5 AF Out. Double conversion under 7 MHz. Some circuit changes were implemented at serial number 3300 and again at 3900.
Accessories: S-200 Speaker   Clock/Timer (shown)   IF Noise Silencer
Comments: Ranges: 1.8-2, 3.5-4, 7-7.3, 14-14.4, 21-21.6, 28-30 and 50-54 MHz. The large right knob is the ±3 kHz vernier tuning knob. Requires a speaker.
Variants: Model **HQ-170C** (shown) includes the clock. Model **HQ-170E** is for export 115-230 VAC 50-60 Hz.

Made In:	United States 1958-1962
Voltages:	105-125 VAC 50-60 Hz 120 W
Readout:	Analog
Physical:	19x10.5x13" 45 Lbs. 482x266x330mm 20.4 kg
Status:	Inactive Manufacturer Discontinued Model
Rarity:	Abundant
Reviews:	*QST* February 1959 *CQ* November 1958 *Electric Radio* May 1996

	New	**Used**
Price:	$359-379	$90-240
Rating:	★★★★	★★

HQ-170A

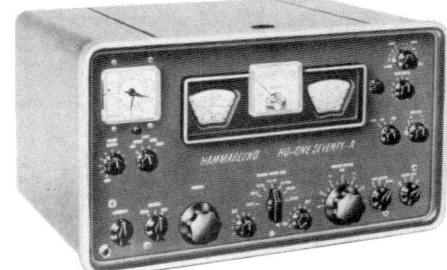

Amateur Band Communications Receiver
Triple Conversion Superheterodyne. 16 Tubes plus Semiconductors.

Features:
- ¼" Head. Jack
- BFO ±2 kHz.
- ANL
- Flywheel Tuning
- S Meter
- Vernier Tuning
- IF Out Jack
- Hinged Cover
- Sensitivity
- Slot Filter ±5 kHz.
- Antenna Trimmer
- Fiduciary Adj.
- Speaker Terminals
- Calibrator 100 kHz
- AVC OFF/SLO/MED/FST

Specifications:
Coverage ......Ham. See Comments.
Selectivity ......6/4/3/2/1/.5 kHz
Sensitivity ......<.7µV -10 dB S/N CW
BFO Range ...±2 kHz.
Modes .... AM/(CW-SSB)
I.F. .... 3035, 455, 60 kHz
Audio .... 1W 3.2 ohms, 500 ohms
Antenna Input . Terminals and SO-239

Circuit Complement: 6BZ6 RF Amp, 6BE6 1st Converter, 6C4 HF Osc., 6BE6 2nd Mix/Crystal Osc, 6BA6 455 IF Amp, 6BE6 3rd Mixer, 6BA6 60 kHz IF Amp, 6BA6 60 kHz IF Amp, 6BV8 60 kHz IF Amp/AVC/AM Det, 12AU7 SSB Product Detector, 6AL5 NL, 12AU7 BFO/S Meter, 6AV6 1st AF Amp/AVC, 6AQ5 AF Out, 0B2 Voltage Regulator, 6BZ6 Crystal Calibrator and silicon rectifier.
Accessories: S-200 Speaker   24 Hr. Clock/Timer
Comments: The A version features a dial scale for optional 2M Converter accessory jack and hinged top trap door. Requires a speaker. Ranges: 1.8-2, 3.5-4, 7-7.3, 14-14.4, 21-21.6, 28-30 and 50-54 MHz. Double conversion under 7 MHz.
Variants: Model **HQ-170A-RC** is the rack mounted version. Model **HQ-170A-VHF** includes a 6 meter preamp and a 2 meter converter for reception from 144-148 MHz $429. Model **HQ-170AC-VHF** includes the clock as well.

Made In:	United States 1962-1967
Voltages:	105-125 VAC 50-60 Hz 120 W
Readout:	Analog
Physical:	19x10.5x13" 38 Lbs. 482x266x330mm 17.2 kg
Status:	Inactive Manufacturer Discontinued Model
Rarity:	Abundant
Reviews:	*73* January 1967 *73* August 1964

	New	**Used**
Price:	$369-399	$100-250
Rating:	★★★★	★★

  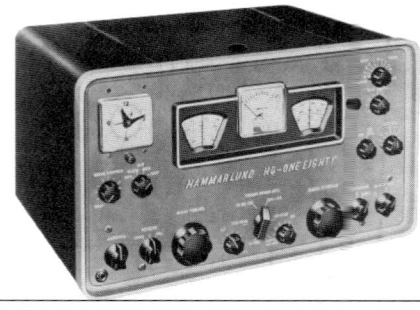

HQ-180

General Coverage Communications Receiver
Triple Conversion Superheterodyne. 18 Tubes.

Features:
- ¼" Head. Jack
- RF Gain
- Spkr. Terminals
- Mute Line
- S Meter
- ANL
- BFO ±2 kHz
- Flywheel Tuning
- Fiduciary Adjust
- Antenna Trimmer
- Bandspread
- Slot Filter ±5 kHz
- Calibrator 100 kHz
- Vernier Tune ±3 kHz
- AVC OFF/SLO/MED/FST

Specifications:
Coverage ......540 - 30000 kHz
Selectivity ......6/4/3/2/1/.5 kHz -6dB.
Sensitivity ......<.7µV -10 dB S/N CW
Audio Out ......2.5 W 4 ohm 5% dist.

Modes .... AM/LSB/USB/CW
I.F. .... 3035, 455, 60 kHz
Image Rej .... >25 dB at 22 MHz
Antenna Input . SO239 and Terminals

Circuit Complement:
6BZ6 RF Amp, 6BE6 1st Mixer, 6BE6 Converter, 6BA6 455 IF Amp, 6BE6 Converter, 6BA6 60 kHz IF Amp, 6BA6 60 kHz IF Amp, 6BV8 60 kHz IF Amp/AVC/AM Detector, 12AU7 SSB Product Detector, 6AL5 NL, 6BZ6 Crystal Cal, 6C4 HF Osc, 12AU7 60 kHz BFO/S Meter, 0A2 Voltage Regulator, 5U4GB Rectifier, 6AV6 1st AF Amp/AVC, 6AQ5 AF Output and 6BA6 455 kHz Gate. Double conversion below 7.85 MHz.

Accessories: S-200 Speaker   24 Hour Clock/Timer (shown)   IF Noise Silencer

Comments: Ranges: .54-1.05, 1.05-2.05, 2.05-4.04, 4-7.85, 7.85-15.35 and 15.35-30 MHz. Bandspread: 3.44-4.04, 6.81-7.3, 13.98-14.425, 20.925-21.6 and 27.89-29.7 MHz. Requires a speaker. An outstanding tube radio. [¥70,000]

Variants: Model **HQ-180C** (shown) has clock. Model **HQ-180RC** is the rack version with clock. Model **HQ-180XE** has 11 fixed crystal positions and adds 230 VAC.

Made In: United States 1959-1962

Voltages: 105-125 VAC 50/60 Hz
120 W

Readout: [⊞⊞⊞⊞] Analog

Physical: 19x10.5x13" 38 Lbs.
482x266x330mm 17.2 kg

Status: Inactive Manufacturer
Discontinued Model

Rarity: Common

Reviews: QST June 1960
CQ April 1960
QST July 1960
NASWA July 1995
Electric Radio January 1996

	New	Used
Price:	$429	$210-300
Rating:	★★★★★	★★★★

  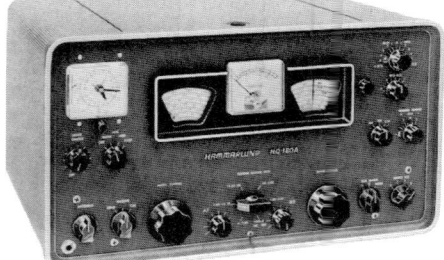

HQ-180A

General Coverage Communications Receiver
Triple Conversion Superheterodyne. 17 Tubes plus Semiconductors.

Features:
- ¼" Head. Jack
- RF Gain
- ANL
- Mute Line
- S Meter
- Bandspread
- BFO ±2 kHz.
- Fiduciary Adjust
- Slot Filter ±5 kHz
- Antenna Trimmer
- Calibrator 100 kHz
- Flywheel Tuning
- Speaker Terminals
- Vernier Tune ±3 kHz
- AVC OFF/SLO/MED/FST

Specifications:
Coverage ......540 - 30000 kHz
Selectivity ......6/4/3/2/1/.5 kHz -6dB.
Sensitivity ......<.8µV -10 dB S/N CW/SSB
Audio Out ......2.5 W 4 ohm 5% dist.

Modes .... AM/LSB/USB/CW
Image Rej .... >25 dB at 22 MHz
I.F. .... 3035, 455, 60 kHz
Antenna Input . SO-239 and Terminals

Circuit Complement: Very similar to the HQ-180 except the 5U4GB has been replaced by two silicon rectifiers. Double conversion below 7.85 MHz.

Accessories: S-200 Speaker   24 Hour Clock/Timer (shown)

Comments: Ranges: .54-1.05, 1.05-2.05, 2.05-4.04, 4-7.85, 7.85-15.35 and 15.35-30 MHz. This newer "A" model features a solid state PS, 230 VAC, 3 position BFO, accessory socket, less drift and 500 ohm line output. Requires a speaker. An excellent receiver. HQ-180A production started with serial number 6900.

Variants: Model **HQ-180AC** (shown) has the clock timer built-in. Model **HQ-180ARC** is rack and clock. Model **HQ-180AR** was rack mounted version. Export model **HQ-180AE** operates from 117 or 230 VAC 50 or 60 Hz. Model **HQ-180AX** accepts 11 fixed crystals (6 can be changed from the front) $625. (Note that the X models cannot have the clock). Model **HQ-180AXR** same but rack mounted $675.

Made In: United States 1963-1972

Voltages: 105-125 VAC 50/60 Hz
120 W

Readout: [⊞⊞⊞⊞] Analog

Physical: 19x10.5x13" 38 Lbs.
482x266x330mm 17.2 kg

Status: Inactive Manufacturer
Discontinued Model

Rarity: Common

Reviews: 73 January 1967
Pop. Elec. February 1964
Proceedings 988

	New	Used
Price:	$439-575	$250-360
Rating:	★★★★★	★★★★

HQ-200

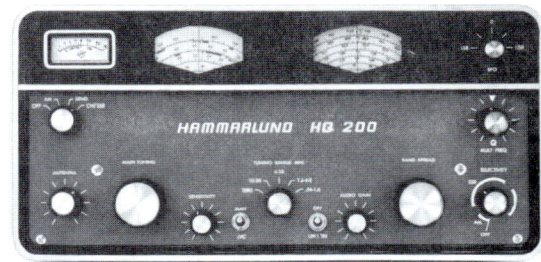

General Coverage Communications Receiver
Single Conversion Superheterodyne. 8 Tubes plus Semiconductors.

Features:
- ¼" Head. Jack
- S Meter
- NL
- Dial Lamp
- RF Gain
- Q-Multiplier
- Antenna Trimmer
- AVC OFF/ON
- Standby
- BFO ±2 kHz
- Bandspread
- Speaker Terminals

Specifications:
Coverage ......540 - 30000 kHz
Selectivity ....1-12 kHz Variable
Sensitivity ......<.5µV 10 dB S/N CW
Audio Out ......2.5 W 4 ohm 5% dist.

Modes .... AM/CW/SSB
IF .... 455 kHz
Antenna Input . Terminals 30-100 ohm

Circuit Complement:
6BZ6 RF Amp, 6BE6 Mixer, 6C4 HF Osc, 6BA6 1st IF Amp, 6BA6 2nd IF Amp, 6BE6 Product Detector/BFO, 12AX7 1st AF Amp, 6AQ5 Audio Power Output, Zener Voltage Regulator, (2) CER72C Rectifier, 1N34A AM Detector and 1N541A NL.

Accessories:
S-101 Speaker   XC-100 Calibrator 100 kHz

Comments:
Ranges: .54-16, 1.6-4, 4-10 and 10-30 MHz. The built-in Q-Multiplier provides continuously variable selectivity from .1 to 12 kHz. The headphone jack is on the back of the radio. Although not a great performer, this radio is becoming a collector's item due to limited production.

Made In:	United States 1968-1972
Voltages:	117/230 VAC 50-60 Hz 50W
Readout:	Analog
Physical:	16.5x9x9.2" 22 Lbs. 419x229x233mm 10 kg
Status:	Inactive Manufacturer Discontinued Model
Rarity:	Very Scarce
Reviews:	*Pop. Elec.* September 1969

	New	**Used**
Price:	$229-338	$150-250
Rating:	★★★	★★

HQ-205

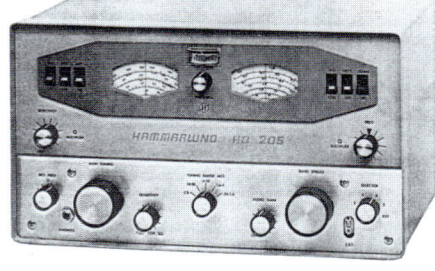

Amateur Band Communications Receiver / Transceiver
Single Conversion Superheterodyne. 11 Tubes plus Semiconductors.

Features:
- ¼" Head. Jack
- S Meter
- NL
- Dial Lamp
- RF Gain
- Q-Multiplier
- Antenna Trimmer
- AVC OFF/ON
- Standby
- BFO
- Bandspread
- Speaker Terminals
- Squelch
- 6 CB Crystal Positions

Specifications:
Coverage ......540 - 30000 kHz
Selectivity ......6 & .1-2.9 kHz Variable
IF ....455 kHz
Audio Out ......2 W 3.2 ohm 5% dist.

Modes .... AM/CW/SSB
Antenna Input . Terminals
CB Ant. Input .. SO-239

Circuit Complement:
6BZ6 RF Amp, 6BE6 Mixer, 6HF8 Tunable HF Osc, 6HF8 Crystal HF Osc., 12AX7 Q-Multiplier, 12AX7 1st Speech Amp, 6BA6 First IF Amp, 6BA6 2nd IF Amp, 6GW8 First Audio Amp/Second Speech Amp, 6GW8 Audio Output/Modulator, 6C4 BFO, 6BQ5 Final RF Amp (CB), 6BA6 Transmitter Osc. and 0B2 Voltage Regulator.

Accessories:
S-205 Speaker   XC-100 Crystal Calibrator

Comments: This radio is painted in various shades of brown. Ranges: .54-16, 1.6-4, 4-10 and 10-30 MHz. This radio may be more properly classified as a *transceiver* because it features a crystal controlled 5 watt CB (or 10 meter) transmitter in it. A 8 pin mic jack is on the rear panel of the radio. Five crystal sockets are internal and one is on the front panel. Requires a speaker.

Made In:	United States 1967
Voltages:	115 VAC 50-60 Hz
Readout:	Analog
Physical:	16.5x9x9.2" 419x229x233mm
Status:	Inactive Manufacturer Discontinued Model
Rarity:	Very Scarce

	New	**Used**
Price:	$259	$150-250
Rating:	★★★	★★

 HAMMARLUND

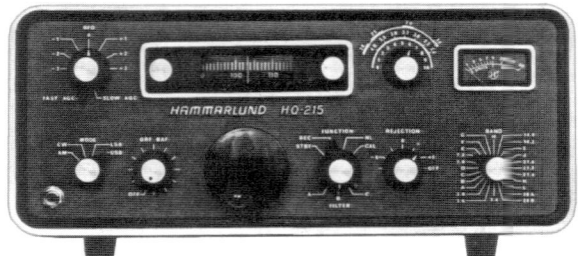

HQ-215

Amateur Band Communications Receiver
Double Conversion Superheterodyne. Solid State.

Features:
- ¼" Head. Jack
- S-Meter
- NL
- Calibrator 100 kHz
- RF Gain
- Dimmer
- Preselector
- AGC SLOW/FAST
- BFO ±3 kHz
- Dial Adjust
- Mute
- IF Notch
- Standby
- Speaker Jack
- VFO Output Jack
- Crystal HFO Output
- Fiduciary Adjust
- HF Osc. Jack
- Line Out Jack

Specifications:
Coverage ...... See Comments.
Selectivity ...... 2.1/_/_ kHz -6 dB
Sensitivity ...... <.5µV -10 dB S/N CW
Audio Out ...... 1.5 W 3.2 ohm 10% dist.
Environment .. 0° to +50° C

Modes .... AM/CW/LSB/USB
Stability .... <100 Hz per hour
Image Reject. . > 50 dB
IF .... 3055, 455 kHz
Antenna Input . SO-239 50-75 ohms

Circuit Complement:
26 transistors, 13 diodes and 2 zeners.

Accessories:
.5 kHz Filter   6 kHz Filter   S-215 Speaker

Comments: Ranges: 3.4-4, 7-7.4, 14-14.4, 21-21.6 and 28.5-28.7 MHz plus 13 optional 200 Hz positions. Dial accuracy is ±100 Hz. Main dial yields 10 kHz per revolution. The supplied 2.1 kHz bandwidth is a Collins mechanical filter. Closed-out in late 1970 at $295. Collectable.

Variants: Model **HQ-215 MK II** 1969-1970. Model **HQ-225** was an announced SWL version with SWL bands and 6 kHz filter as standard. Reportedly was not produced.

Made In:	United States 1968- 969	
Voltages:	117/234 VAC 50-60 Hz 19W or 12 VDC 460 ma	
Readout:	Analog Lin.	
Physical:	15.8x6.8x14" 18 Lbs 401x173x355mm 8.2 kg	
Status:	Inactive Manufacturer Discontinued Model	
Rarity:	Very Scarce	
Reviews:	QST December 1968 CQ October 1968	

	New	Used
Price:	$530	$300-400
Rating:	★★★★	★★★★

 HAMMARLUND

PRO-310

General Coverage Communications Receiver
Double Conversion Superheterodyne. 14 Tubes.

Features:
- ¼" Head. Jack
- S Meter
- Bandspread
- Speaker Terminals
- Sensitivity
- ANL
- BFO
- Antenna Trimmer
- Standby
- AGC
- Phono Input Jack
- Calibrator 100 kHz.
- IF Out Jack
- Spinner Knobs

Specifications:
Coverage ...... 550 - 35520 kHz
Selectivity ...... 4/2/.5 kHz -6dB.
Sensitivity ...... <1µV -10 dB S/N CW
Audio Out ...... 2 W 6 ohm

Modes .... AM/CW-SSB
Image Rej .... >70 dB (2.2-17.6 MHz)
I.F. .... 1802, 52 kHz
Antenna Input . RF Jack 75-300 ohm

Circuit Complement: 6BA6 RF Amp, 6BZ6 1st IF Amp, 6AN8 BFO/Buffer, 6BE6 1st Mixer, 6BZ6 2nd IF Amp, 6V6GT/G Audio Power Output, 6C4 Osc., 6AL5 Detector/Limiter, 5U4GB Rectifier, 6BE6 2nd Mixer, 12AX7 1st AF Amp, 6AL5 Bias Rectifier, 6C4 2nd Crystal/AGC and 0B2 Voltage Regulator. Single conversion below 2.2 MHz. Has five tuned circuits. Turret type band display.

Accessories:
Gray/Silver Speaker   Black/Gold Speaker   Rack Mounting Kit
Gray/Silver Rack Speaker Black/Gold Rack Speaker

Comments:
Ranges: .55-1.11, 1.1-2.22, 2.2-4.44, 4.4-8.88, 8.8-17.76 and 17.6-35.52 MHz. Requires a speaker. A somewhat radical design for the time. Cabinet available in gray and silver or black and gold. Reportedly only 1000 made. Very collectable.

Made In:	United States 1955-1957	
Voltages:	105-125 VAC 50-60 Hz 110 W	
Readout:	Analog	
Physical:	18x17.875x15.5" 65 Lbs. 457x454x394mm 29.4 kg	
Status:	Inactive Manufacturer Discontinued Model	
Rarity:	Very Scarce	
Reviews:	QST April 1956 QST May 1956	

	New	Used
Price:	$595	$550-1000A
Rating:	★★★★★	★★★★

SP-110
Super-Pro

General Coverage Communications Receiver
Single Conversion Superheterodyne. 16 Tubes.

Features:
- Head. Terminals
- S Meter
- Bandspread
- Speaker Terminals
- Sensitivity
- Dial Lamp
- BFO
- Phones-Spkr Switch
- Standby
- AVC ON/OFF

Specifications:

Coverage ......540 - 20000 kHz
Selectivity ......Variable 16-3 kHz.
Sensitivity ......<.85µV 6:1 S/N
Audio Out ......10 W

Modes .... AM/CW
I.F. .... 465 kHz
Antenna Input . Terminals 100 ohms

Circuit Complement:
6K7 1st RF Amp, 6K7 2nd RF Amp, 6L7 Mixer, 6J7 HF Osc, 6D6 1st IF Amp, 6D6 2nd IF Amp, 6D6 3rd IF Amp, 6B7 Detector, 6C6 BFO, 6B7 AVC, 6C5 1st Audio, 6F6 2nd Audio, (2) 6F6 AF Output, 5Z3 High Voltage Rectifier and 80 C-Bias Rectifier

Comments: Ranges: .54-1.6, 1.6-2.5, 2.5-5, 5-10 and 10-20 MHz. Supplied with external AC power supply and interconnecting cable. The power supply is separate and is approximately 13x8.5x17.5" 35 Lbs.

Accessories:
Rack Mounting Kit.

Variants:
Please see *Chapter 107 Additional Information* for a chart describing these Series 100 variants: **SP-110-X, SP-120-X, SP-110-S, SP-110-SX, SP-120-SX, SP-150** and **SP-150-S.** The SP-150 console measures 29.75x44.5x18"

| | | **Made In:** | United States 1937-1941 |

Made In: United States 1937-1941

Voltages: 110-115-125 VAC 50-60 Hz

Readout: ⊞⊞⊞ Analog

Physical: Cabinet Version:
18.5x10.5x14.75"
470x267x375mm

Status: Inactive Manufacturer
Discontinued Model

Rarity: Very Scarce

	New	Used
Price:	$238-405	$220-300
Rating:	⑦	⑦

SP-210-X
Super-Pro

General Coverage Communications Receiver
Single Conversion Superheterodyne. 18 Tubes.

Features:
- Head. Terminals
- S Meter
- Bandspread
- Speaker Terminals
- Sensitivity
- ANL
- BFO
- Phones-Spkr Switch
- Standby
- Dial Lamp
- AVC ON/OFF
- Hinged Top Cover
- Carry Handles
- Phono Terminals

Specifications:

Coverage ......540 - 20000 kHz
Selectivity ......6 Pos. 16-.1 kHz.
Sensitivity ......<1µV -10 dB S/N CW
Audio Out ......14 W

Modes .... AM/CW
I.F. .... 465 kHz
Antenna Input . Terminals 112 ohms

Circuit Complement:
6K7 1st RF Amp, 6K7 2nd RF Amp, 6L7 1st Detector/Mixer, 6J7 HF Osc, 6K7 1st IF Amp, 6SK7 2nd IF Amp, 6SK7 3rd IF Amp, 6H6 2nd Detector, 6N7 NL, 6SK7 AVC Driver, 6H6 AVC Diode, 6SJ7 BFO, 6C5 1st AF Amp, 6F6 2nd AF Amp, 6F6 3rd AF Amp, 6F6 3rd AF Amp, 5Z3 High Voltage Rectifier and 80 C-Bias Rectifier

Comments: Ranges: .54-1.16, 1.16-2.5, 2.5-5, 5-10 and 10-20 MHz. Supplied with separate external AC power supply (13x8.5x17.5" 35 Lbs.) and interconnecting cable. Initially supplied with an ext. speaker.

Variants: The military models **BC-779, BC-779A** and **BC-779B** are similar to the SP-210X. See *Chapter 107 Additional Information* for a chart describing these Series 200 variants: **SPR-210-X, SP-220-X, SPR-220-X, SP-210-SX, SPR-210-SX, SP-220-SX, SPR-220-SX, SP-210-LX, SPR-210-LX, SP-220-LX** and **SPR-220-LX.**

Made In: United States 1942-1950

Voltages: 110-115-125 VAC 50-60 Hz

Readout: ⊞⊞⊞ Analog

Physical: Cabinet Version:
21.5x12.25x15.25" 50 Lbs
546x311x387mm 23 kg
Rack Version:
19x10.6x14.7"
483x269x373mm

Status: Inactive Manufacturer
Discontinued Model

Rarity: Scarce

	New	Used
Price:	$308-507	$200-300
Rating:	⑦	⑦

SP-400-X
Super-Pro

General Coverage Communications Receiver
Single Conversion Superheterodyne. 18 Tubes.

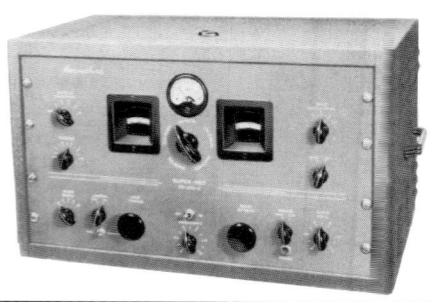

Features:
- ¼" Head. Jack
- S Meter
- Bandspread
- Speaker Terminals
- Sensitivity
- ANL
- BFO
- Phono Terminals
- Standby
- Dial Lamp
- AVC ON/OFF
- Hinged Top Cover
- Carry Handles

Specifications:
Coverage ......540 - 30000 kHz
Selectivity ......6/4/2/1/.5 kHz -6dB.
Audio Out ......8 W 500 ohm 10% dist.
Modes .... AM/CW
I.F. .... 455 kHz
Antenna Input . Terminals

Circuit Complement:
6K7 1st RF Amp, 6K7 2nd RF Amp, 6L7 1st Detector/Mixer, 6J7 HF Osc, 6K7 1st IF Amp, 6SK7 2nd IF Amp, 6SK7 3rd IF Amp, 6H6 2nd Detector, 6N7 NL, 6SJ7 BFO, 6SK7 AVC Amp, 6H6 AVC Rectifier, 6J5 1st AF Amp, 6F6 2nd AF Amp, 6F6 3rd AF Amp, 6F6 3rd AF Amp, 5U4G High Voltage Rectifier and 5Y3GT/G C-Bias Rectifier.

Accessories: SC-46 Speaker

Comments:
Ranges: .54-1.24, 1.24-2.86, 2.85-6.3, 6.3-14 and 13.4-30 MHz. Supplied with external AC power supply and interconnecting cable. The power supply is separate and is approximately 13.6x7.62x8.62" 28 Lbs. For battery operation use: quantity five 45V "B", one "A" and one "C". Requires a speaker.

Variants:
Model **SPR-400-X** is a rack version. Model **SP-400-SX** covers: 1.25-40 MHz $399.

Made In:	United States 1946-1948
Voltages:	105-125 VAC 50-60 Hz 180 W or batteries.
Readout:	Analog
Physical:	Cabinet Version: 21.5x12.25x15.25" 67 Lbs. 546x311x387mm 30.3 kg Rack Version: 19x10.6x14.7" 49 Lbs. 483x269x373mm 22.3 kg
Status:	Inactive Manufacturer Discontinued Mode
Rarity:	Scarce

	New	Used
Price:	$340-385	$160-270
Rating:	★★★★★	★★★★

SP-600-JX
Super-Pro 600

General Coverage Communications Receiver
Dual Conversion Superheterodyne. 20 Tubes.

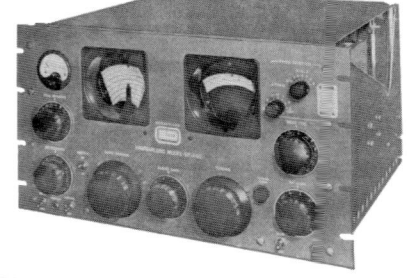

Features:
- ¼" Head. Jack
- S/AF Meter
- IF Output Jack
- AVC ON/MAN
- Standby
- Tune Lock
- 6 Fixed Positions
- Xtal Fine Tune ±3 kHz
- BFO ±3 kHz
- Internal P.S.
- Flywheel Tuning
- RF Gain
- ANL

Specifications:
Coverage ......540 - 54000 kHz
Selectivity ......13/8/3/1.3/.5/.2 kHz -6dB.
Sensitivity ......<.75µV -10 dB S/N CW
Audio Out ......2 W 600 ohm
Modes .... AM/CW
Image Rej .... >80 to 100 dB 15 min.
Drift .... .001% to .01%
Antenna Input . UG-103 72 ohm

Circuit Complement: (2) 6BA6 RF Amp, 6AC7 FFO, 6C4 VFO, (2) 6BE6 Mix, 6BA6 Gate, 6C4 FFO, (2) 6BA6 IF Amp, 6BA6 Driver, 6BA6 Buf, 6C4 BFO, 6AL5 Det./AVC, 6AL5 Lim/Meter, 12AU7 IF Out/AF Amp, 6V6GT Output, 0A2 Reg., 5R4GY Rect. and 6AL5 Rect. Single conv. under 7.4 MHz. Four IF and two RF stages.

Accessories: SP-300 Speaker SPC-10 SSB Converter

Comments: Ranges: .54-1.35, 1.35-3.45, 3.45-7.4, 7.4-14.8, 14.8-29.7 and 29.7-54 MHz. Built-in power supply. The single knob tuning is also geared to the logging scale. This extraordinary receiver had a long and illustrious career. [¥120,000].

Variants: SP-600-J has no fixed positions. **SP-600-JLX-2,15, 23** cover .1-.4, 1.35-29.7 MHz. **SP-600-JX-17, 30** are diversity. **SP-600-VLF-31, 38** is only 10-540 kHz. **SP-600-JX-21A** has 22 tubes, SSB. **SP-600-JLX-27** covers .2-.4, .54-29.7 MHz. **SP-600-JX-28** is R-620. **SP-600-JX-6** is R-274B. **SP-600-JX-12** is R-274A. **SP-600-JX-29** for CIA. **SP-600-JX-39** for FAA. Please see *Chapter 107 Additional Information.*

Made In:	United States 1951-1972
Voltages:	95/105/117/130/210/234/260 VAC 130 W 50-60 Hz
Readout:	Analog
Physical:	Cabinet Version 21.5x12.75x17.25" 87 Lbs. 546x324x438mm 39.4 kg Rack Version: 19x10.5x17.25" 66 Lbs 482x266x438mm 30 kg
Status:	Inactive Manufacturer Discontinued Model
Rarity:	Common
Reviews:	*Electric Radio* May 1990 *Proceedings* 1994-95

	New	Used
Price:	$985-1140	$190-450
Rating:	★★★★★	★★★★

39 Harvey Wells

The **Harvey-Wells Electronics** company was established in 1939 by Cliff Harvey (of *Harvey Radio* fame) and John Wells. The company produced amateur transmitters, tuners, power supplies and receivers. The Harvey-Wells factory was flooded two successive summers, possibly contributing to the company's demise.

A modest number of R-9 and R-9A receivers were produced. They are sometimes sold in conjunction with the matching T-90 transmitter. They trade at modest prices now, but will likely go up in value as collector items.

Harvey-Wells also considered a general coverage version of the R-9A. This would have been a six band version covering 540 to 32000 kHz. Apparently they were going to sell it either as the Harvey-Wells RG-9A or produce it for R.C.A. A photograph of the R.C.A. prototype is shown in *Chapter 106* of this book. It is uncertain whether this general coverage receiver was ever produced under either the Harvey-Wells label or the R.C.A. label.

Harvey-Wells Electronics, Inc.
Southbridge, Mass.   1955

R-9

Amateur Band Communications Receiver
Double Conversion Superheterodyne. 10 Tubes.

Features:
- ¼" Head. Jack
- S-Meter
- BFO
- AVC
- Mute
- RF Gain
- ANL
- Standby
- Dial Lamp

Specifications:

Coverage ......Ham. See Comments.	Modes .... AM/CW
Selectivity ......4 kHz -6dB.	Image Rej .... >25 dB
Sensitivity ......<2µV 10 dB S/N	IF .... 1600, 260 kHz.
Audio Out ......5W 3.2/600 ohms	Ant. Jack .... Phono Jack 50 ohms

Accessories:
FS-1 Speaker   VPS-R9 6/12 VDC PS   MS-1 Mobile Speaker

Circuit Complement:
6BJ6 RF Amp, 6U8 Mixer/Osc, 0A2 Regulator, 6U8 2nd Mixer/Osc., (2) 6BJ6 IF Amp, 6AL5 Detector/ANL, 12AX7 1st Audio/BFO, 6CM6 Audio Output and 5Y3GT Rectifier. The circuit consists of one stage of RF amplification, a first detector and stabilized high frequency variable oscillator. This is followed by a second detector and a second high frequency (fixed) osc. providing dual conversion on all bands.

Comments: Ranges: 3.5-4, 7-7.3, 14-14.4, 21-21.45 and 26.96-30 MHz. Matches the T-90 Transmitter. Designed for mobile or base operation.

Variants: Model **R-9A** 1958 $160. The R-9 was manufactured with point-to-point wiring. The later model R-9A featured half point-to-point wiring and half printed circuit board construction. This model later was closed-out for $90 in December of 1958. Model **RG-9A** general coverage version (please see Chapter introduction).

Mfg. In:	United States 1954-1956
Voltages:	100/120 VAC 50/60 Hz 70 W
Readout:	Analog
Physical:	12.375x6.75x10.5" 18 Lbs. 314x171x267mm 8.2 kg
Status:	Inactive Manufacturer Discontinued Model
Rarity:	Very Scarce

	New	**Used**
Price:	$150	$80-130
Rating:	★★★★	⑦

The fascinating story of Heath has been told many times on the inside cover of Heathkit catalogs. Edward Bayard Heath founded the Heath Aeroplane Company during the early 1900's. The first Heath "kit" was in fact an airplane introduced in 1926. In 1931 the founder was killed in a flight test. Heath remained an aircraft and aircraft parts company through World War II. Howard E. Anthony, who had purchased Heath in 1935 gave it a different direction after the war. With surplus electronic parts, he marketed the "O-1" oscilloscope kit for $39.50. A line of test instruments, amateur radios and hi-fi equipment kits followed.

Howard Anthony also died in a plane crash in 1954. Daystrom Inc. then acquired the Heath Company. In 1962 Daystrom was purchased by Schlumberger Limited, a leader in the development of electronic techniques for oil exploration.

Zenith purchased Heath from Schlumberger in 1979. They were interested in Heath's computer production capacity more than amateur radio gear. Budget cuts, difficulties with new models and increased offshore competition led to increased problems in the mid 1980s. In the late 80's and early 90's the amateur radio segment consisted primarily of assembled, private labeled equipment from Standard, Yaesu and Ameritron. Heath now focuses on educational videos and workbooks and is not in the radio market. Please see *Chapter 105 Briefly Mentioned* for more Heathkit models.

Heath Company
Benton Harbor, MI   1945- 961

The HEATH COMPANY

AR-1

General Coverage Broadcast Receiver Kit
Single Conversion Superheterodyne. 6 Tubes.

Features:
- RF Gain
- Tone
- Standby
- Phono/Radio Switch
- Mute
- Phono Input Jack

Specifications:
Coverage ......550 - 20000 kHz	Modes .... AM
Audio Out ......8 ohms	IF .... 455 kHz
Sensitivity ......15 µV	Antenna Input . Terminal 50 ohm

Accessories:
Metal Cabinet

Comments:
Ranges: .55-1.6, 1.7-5.5, 6-20 and 20-35 MHz. Does not have a dial lamp. Matches the AT-1 Transmitter. Requires a speaker. May be modified for CW. See *QST* May 1953.

Mfg. In:	United States 1949-1953
Voltages:	105-125 VAC 50/60 Hz 40W
Readout:	Analog
Physical:	11.5.x5.75x6.75" 11 Lbs. 292x146x171mm 5.5 kg
Status:	Inactive Manufacturer Discontinued Model
Rarity:	Very Scarce

	New	Used
Price:	$23 Kit	$30-50
Rating:	⑨	⑨

AR-2

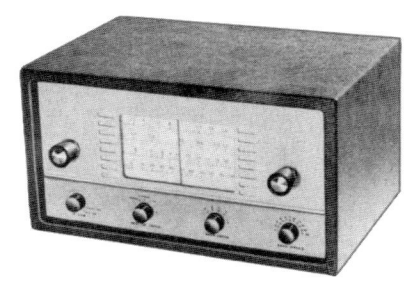

General Coverage Communications Receiver Kit
Single Conversion Superheterodyne. 5 Tubes.

Features:
- ¼" Head. Jack
- BFO
- Standby
- Bandspread 0-100
- RF Gain
- Mute
- Speaker 5.5"
- Accessory Socket
- AVC/MVC
- NL

Specifications:
Coverage ...... 550 - 35000 kHz	Modes .... AM/CW
Audio Out ...... 8 ohms	Antenna Input . Terminal 50 ohm
IF .... 455 kHz	

Circuit Complement:
12BE6 Osc & Mixer, 12BA6 IF Amp, 12AV6 2nd Detector/AVC/1st Audio Amp/Reflex BFO, 12A6 Beam Power Output and 5Y3 Rectifier.

Accessories:
QFQ-1 Q Multiplier   91-10 Cabinet (shown)

Comments:
Ranges: .55-1.5, 1.5-3.9, 3.9-10.5, 10.5-35 MHz. Does not have a dial lamp. Matches the AT-1 Transmitter. The headphone jack is on the rear panel. Shown in optional cabinet.

Mfg. In:	United States 1953-1956
Voltages:	105-125 VAC 50/60 Hz 45W
Readout:	Analog
Physical:	11.5.x5.75x6.75" 12 Lbs. 292x146x171mm 5.5 kg
Status:	Inactive Manufacturer Discontinued Model
Rarity:	Scarce

	New	Used
Price:	$25 Kit	$30-35
Rating:	⑨	⑨

AR-3

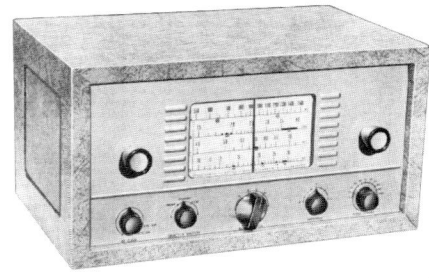

General Coverage Communications Receiver Kit
Single Conversion Superheterodyne. 5 Tubes.

Features:
- ¼" Head. Jack
- BFO
- Dial Lamp
- Bandspread
- RF Gain
- Mute
- Speaker 5.25"
- Antenna Trimmer
- Standby
- Q-Mult. Jack
- NL
- Accessory Socket
- AVC/MVC

Specifications:
Coverage ...... 550 - 30000 kHz	Modes .... AM/CW
Audio Out ...... 8 ohms	Antenna Input . Terminal 50 ohm

Circuit Complement:
12BE6 Osc & Mixer, 12BA6 IF Amp, 12AV6 2nd Detector/AVC/1st Audio Amp/Reflex BFO, 12A6 Beam Power Output and 5Y3 Rectifier.

Accessories:
QFQ-1 Q Multiplier   91-15A Cabinet (shown)

Comments:
Ranges: .55-1.5, 1.5-4.5, 4.5-10, 10-30 MHz. Similar to the AR-2. The headphone jack is on the rear panel. The rear panel of the AR-3 adds a Q-Multiplier jack and an Accessory jack. Shown in optional cabinet.

Mfg. In:	United States 1955-1961 40W
Voltages:	105-125 VAC 50/60 Hz 40W
Readout:	Analog
Physical:	11.5.x5.75x6.75" 12 Lbs. 292x146x171mm 5.5 kg
Status:	Inactive Manufacturer Discontinued Model
Rarity:	Common

	New	Used
Price:	$28-30 Kit	$30-35
Rating:	⑨	⑨

GC-1A
Mohican

General Coverage Communications Receiver Kit
Single Conversion Superheterodyne. Solid State.

Features:

• ¼" Head. Jack	• S-Meter	• Antenna Trimmer	• Telescopic Antenna
• RF Gain	• ANL	• Carry Handle	• Bandspread
• Mute Line	• Speaker	• Dial Lamp	• Dial Lamp Switch
• Standby	• BFO	• Flywheel Tuning	• AVC ON/OFF

Specifications:

Coverage ...... 550 - 32000 kHz
Selectivity ...... 3 kHz -6dB.
IF .... 455 kHz.

Modes .... AM/CW-SSB
Sensitivity .... <2µV CW 10 dB S/N
Antenna Input . Terminals 50 ohm

Circuit Complement:
The GC-1A was one of the first fully transistorized shortwave receivers. It incorporates 10 transistors and 6 diodes. The filter is a ceramic Clevite. The GC-1A uses printed circuits and conventionally wired boards.

Accessories: XP-2 117 VAC Power Supply

Comments: This receiver is properly classified as a portable or at least a "porta-top". Green metal case with carry handle. The telescopic whip is 54" long. Ranges: .55-1.6, 1.6-4, 4-9, 9-20 and 20-32 MHz. Bandspread: 3.5-4, 7-7.3, 14-14.35, 21-21.4 and 26-29.7 MHz. The dial lamp switch is spring loaded. The headphone jack is on the rear panel. This receiver has received unusually mixed reviews. It is becoming a collector's item.

Variants: GC-1 was earlier model with a similar appearance (1960-1962). Model **GCW-1A** was wired version $193 new.

Mfg. In:	United States 1965-1968
Voltages:	8 C cells or 12 VDC 50 ma or AC via optional XP-2 PS.
Readout:	Analog
Physical:	12x7x10" 18 Lbs. 305x275x254mm 3.2 kg
Status:	Inactive Manufacturer Discontinued Model
Rarity:	Scarce
Reviews:	QST December 1960 73 October 1960

	New	Used
Price:	$100-110 Kit	$50-70
Rating:	⑨	⑨

GR-54

General Coverage Communications Receiver Kit
Single Conversion Superheterodyne. 6 Tubes plus Semiconductors.

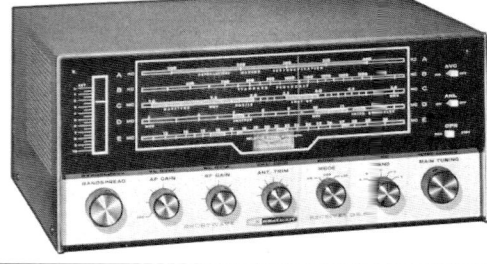

Features:

• Spkr/Head. Jack	• S-Meter	• Antenna Trimmer	• Ferrite MW Antenna
• RF Gain	• ANL	• Morse Key Input	• Bandspread 0-10
• Mute	• Speaker 4x6"	• Standby	• AVC ON/OFF

Specifications:

Coverage ...... See Comments
Selectivity ...... 3 kHz -6dB.
IF .... 1682 kHz.
Audio Out ...... 8 ohms

Modes .... AM/LSB/USB
Sensitivity .... <.4-4µV CW
Antenna Input . Terminals

Circuit Complement:
6BH6 RF Amp, 6EA8 Osc Mixer, 6BA6 IF Amp, 6BA6 IF Amp, 12AT7 BFO Product Detector, 6HF8 AF Amp and diodes.

Comments:
Ranges: .18-.42, .55-1.55, 2-5, 5-12.5 and 12.5-30 MHz.

Variants:
An export model was also available for 115/230 VAC 50-60 Hz operation.

Mfg. In:	United States 1966-1971
Voltages:	120/240 VAC 50/60 Hz 45W
Readout:	Analog
Physical:	14.58.x6.5x11" 24 Lbs. 370x160x280mm 8.5 kg
Status:	Inactive Manufacturer Discontinued Model
Rarity:	Common
Reviews:	73 November 1966

	New	Used
Price:	$85-135 Kit	$40-50
Rating:	⑨	⑨

GR-64

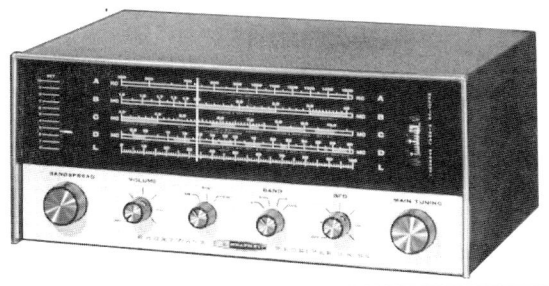

General Coverage Communications Receiver Kit
Single Conversion Superheterodyne. 4 Tubes plus Semiconductors.

Features:
- Spkr/Head. Jack
- S-Meter
- RF Gain
- ANL
- Mute
- Speaker 5"
- BFO
- Antenna Trimmer
- Morse Key Input
- Standby
- Ferrite MW Antenna
- Bandspread 0-10
- Dial Lamp

Specifications:
Coverage ...... 550 - 30000 kHz
IF .... 455 kHz.
Modes .... AM/CW-SSB
Antenna Input . Terminals 50 ohm

Circuit Complement:
12BE6 Osc/Mixer, 12BA6 IF Amp/BFO, 12AV6 Detector/Audio Amp, 12AQ5 Audio Output and diodes.

Comments:
Ranges: .55-1.5, 1.5-4, 4-10 and 9.5-30 MHz.

Variants:
An export model was also available for 115/230 VAC 50-60 Hz operation.

Mfg. In:	United States 1964-1971	
Voltages:	105-120 VAC 50/60 Hz 30W	
Readout:	Analog	
Physical:	13.5.x6x9" 11.5 Lbs. 342x152x228mm 5.2 kg	
Status:	Inactive Manufacturer Discontinued Model	
Rarity:	Common	
Reviews:	*73* March 1965	

	New	Used
Price:	$38-43 Kit	$25-40
Rating:	⑨	⑨

GR-78

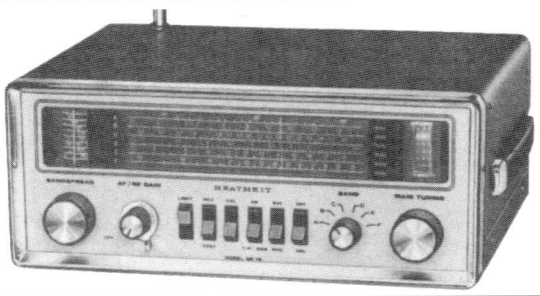

General Coverage Communications Receiver Kit
Double Conversion Superheterodyne. Solid State.

Features:
- ¼" Head. Jack
- S-Meter
- RF Gain
- ANL
- Mute
- Speaker
- AVC ON/OFF
- Standby
- Bandspread
- Carry Handle
- Dial Lamp
- Telescopic Whip
- Ferrite MW Antenna
- Calibrator 500 kHz
- Dial Lamp Switch

Specifications:
Coverage ...... See Comments
Selectivity ...... 7.5 kHz -6dB.
Sensitivity ...... <2 to 10 µV
Audio Out ...... 8 ohms
Modes .... AM/CW-SSB
IF .... 4034 kHz, 455 kHz.
Antenna Input . Terminals 50 ohm

Circuit Complement:
11 Transistors, 5 FETs and 7 diodes. Single conversion below 18 MHz.

Comments:
Ranges: .19-.41, .55-1.3, 1.3-3, 3-7.5, 7.5-18 and 18-30 MHz. Configured for portable operation. The GR-78 was supplied with two bandspread dials; one for the shortwave broadcast bands and one for the amateur bands. Selection had to be made during the assembly of this kit. With a built-in NiCad pack and charger, plus a telescopic antenna, the GR-78 may be properly classified as portable.

Mfg. In:	United States 1970-1977	
Voltages:	120/240 VAC 50/60 Hz or Internal NiCad 9.6 VDC	
Readout:	Analog	
Physical:	11.5x6.25x9" 14 Lbs. 292x158x228mm 6.3 kg	
Status:	Inactive Manufacturer Discontinued Model	
Rarity:	Scarce	
Reviews:	*QST* October 1970 *73* December 1974	

	New	Used
Price:	$130-170 Kit	$55-80
Rating:	⑨	⑨

GR-81

General Coverage Broadcast Receiver Kit
Regenerative. 3 Tubes.

Features:
• ¼" Head. Jack   • Speaker   • Bandspread

Specifications:
Coverage ...... 140 - 18000 kHz   Modes .... AM
Antenna Input Terminals

Circuit Complement:
12AT7 Detector/1st Audio, 50C5 Audio Output and 35W4 Rectifier.

Accessories:
GD125 Q-Multiplier

Comments:
Ranges: .55-1.6, 1.6-4, 4-10 and 10-30 MHz. Green and beige metal cabinet. The GR-81 was a popular introductory set. It features a regenerative-detector circuit and two stages of audio amplification.

Mfg. In:	United States 1962-1971
Voltages:	105-125 VAC 50/60 Hz 30W
Readout:	Analog
Physical:	10x7x7" 12 Lbs 254x178x178mm 5.4 kg
Status:	Inactive Manufacturer Discontinued Model
Rarity:	Scarce

	New	**Used**
Price:	$23-30 Kit	$20-40
Rating:	⑨	⑨

GR-91

General Coverage Communications Receiver Kit
Single Conversion Superheterodyne. 4 Tubes plus Semiconductors.

Features:
• ¼" Head. Jack   • S-Meter   • Dial Lamp   • Antenna Trimmer
• BFO   • ANL   • Speaker 3x5"   • Bandspread 0-10
• Mute   • Standby   • Q-Mult. Jack

Specifications:
Coverage ...... 550 - 30000 kHz   Modes .... AM/CW
IF .... 455 kHz.   Antenna Input . Terminals 75/300 ohms
Audio Out ...... 8 ohms

Circuit Complement:
12BE6 Osc/Mixer, 12BA6 IF Amp/BFO, 12AV6 Detector/Audio Amp, 50C5 Audio Output and silicon diode rectifier.

Accessories:
GD125 Q-Multiplier

Comments:
Ranges: .55-1.5, 1.5-4, 4-10 and 10-30 MHz. The headphone jack and ANL switch are on the rear panel. Beige and aquamarine cabinet.

Variants:
Model **GR-91E** is 115/230 VAC.

Mfg. In:	United States 1961-1963
Voltages:	105-125 VAC 50/60 Hz 30W
Readout:	Analog
Physical:	12.25x5.25x8.25" 9 Lbs. 311x133x209mm 4 kg
Status:	Inactive Manufacturer Discontinued Model
Rarity:	Scarce

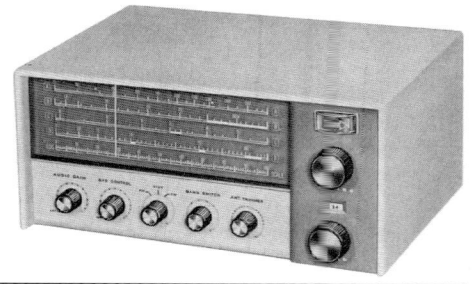

	New	**Used**
Price:	$40-50 Kit	$20-40
Rating:	⑨	⑨

HR-10

Amateur Band Communications Receiver Kit
Single Conversion Superheterodyne. 7 Tubes.

Features:
- ¼" Head. Jack
- S-Meter
- RF Gain
- Antenna Trimmer
- BFO
- ANL
- AVC ON/OFF
- Speaker
- Mute
- Standby
- Dial Lamp
- Speaker Jack

Specifications:
Coverage ......Ham. See Comments. Modes .... AM/SSB/CW
Selectivity ......3 kHz -6dB. Image Rej .... >40 dB
Sensitivity ......<1µV CW 10 dB S/N IF .... 1681 kHz.
Audio Out ......8 ohms Antenna Input . 50-75 ohms RCA

Circuit Complement:
6BZ6 RF Amp, 6EA8 Mixer Oscillator, 6BA6 1st IF Amp, 6EA8 2nd IF Amp/BFO, 6BJ7 Detector/AVC/ANL, 6EB8 1st Audio/Audio Output and 6X4 Rectifier.

Accessories:
HS-24 Speaker HRA-10-1 Calibrator 100 kHz

Comments:
Ranges: 3.5-4, 7-7.3, 14-14.35, 21-21.5 and 28-29.7 MHz. Matches the DX-60 transmitter.

Variants:
Model **HR-10B** 1968-74 $75-$90. It is a different color and matches the DX-60B. An export model operating from 115/230 VAC was also produced.

Mfg. In:	United States 1961-1967	
Voltages:	117 VAC 50/60 Hz 50W	
Readout:	Analog	
Physical:	13.75x6.5x11.5" 18 Lbs. 349x165x292mm 8.1 kg	
Status:	Inactive Manufacturer Discontinued Model	
Rarity:	Common	
Reviews:	*QST* July 1963	

	New	**Used**
Price:	$75-83 Kit	$40-70
Rating:	⑨	⑨

HR-20

Amateur Band Communications Receiver Kit
Single Conversion Superheterodyne. 8 Tubes.

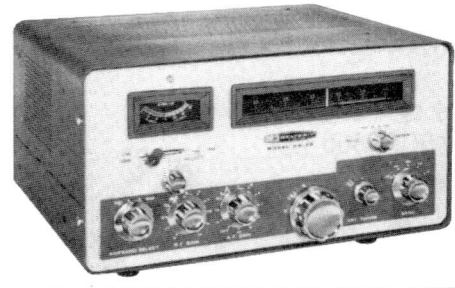

Features:
- ¼" Head. Jack
- S-Meter
- RF Gain
- Antenna Trimmer
- BFO
- ANL
- 30:1 Ratio Tuning
- AVC OFF/FST/SLO
- Mute
- Dial Lamp
- Line Out 500 ohms

Specifications:
Coverage ......Ham. See Comments. Modes .... AM/SSB-CW
Sensitivity ......<1µV 10 dB S/N IF .... 3000 kHz.
Selectivity ......3 kHz -6dB. Antenna Input . RCA 50-75 ohms
Audio Out ......8 ohms

Circuit Complement:
6BZ6 RF Amp, 6EA8 Mixer/Oscillator, 6BZ6 1st IF Amp, 6EA8 2nd IF Amp/S-Meter Amp, 6BE6 Product Detector/BFO, 6BJ7 AM Detector/AVC/ANL, 6EB8 1st Audio/Audio Output and 0A2 Voltage Regulator.

Accessories:
HP-10 Mobile PS HP-12 Mobile PS MP-1 AC PS
HP-20 AC PS HP-23 AC PS AK-6 Mobile Mount
AK-7 Mobile Speaker

Comments:
Ranges: 3.5-4, 7-7.3, 14-14.35, 21-21.5 and 28-29.7 MHz. Designed for mobile use. Matches the HX-20 transmitter. Requires an external power supply.

Mfg. In:	United States 1962-1964	
Voltages:	275-300 VDC @120 ma, 6/12 VDC @2.5 amps.	
Readout:	Analog	
Physical:	12.125x6.125x10" 16 Lbs. 307x155x254mm 8.5 kg	
Status:	Inactive Manufacturer Discontinued Model	
Rarity:	Scarce	
Reviews:	*QST* March 1964 *73* December 1962	

	New	**Used**
Price:	$135 Kit	$40-50
Rating:	⑨	⑨

HR-1680

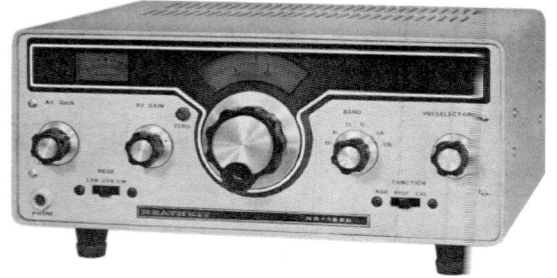

Amateur Band Communications Receiver Kit
Double Conversion Superheterodyne. Solid State.

Features:
- ¼" Head. Jack
- RF Gain
- Mute
- S-Meter
- Dial Set
- Dial Lamp
- Spinner Knob
- Preselector
- Calibrator 100 kHz

Specifications:

Coverage ...... Ham. See Comments	Modes .... LSB/USB/CW
Selectivity ...... 2.1/.25 kHz -6dB.	Stability .... <100 Hz after 30 mins.
Sensitivity ...... <1µV 10 dB S/N	IF .... 8.395-8.895 MHz, 3.395 MHz
IF Rejection ... >60 dB	Image Rej. ...... >50 dB
Audio Out ...... 1.2W 4 ohms	Antenna Input . 50 ohm

Accessories:
HS-1661 Speaker

Comments:
Ranges: 3.5-4, 7-7.5, 14-14.5, 21-21.5, 28-28.5 and 28.5-29 MHz. Dial calibration accuracy is <2 kHz. AGC slow for SSB and AGC fast for CW is automatically selected. Note there is no AM mode. Light green cabinet with red dial window. Matches the HTX-1681 Transmitter.

Mfg. In:	United States 197?-1981
Voltages:	120/240 VAC 50/6?Hz 27W or 11.5-15 VDC @ 75 Amps
Readout:	Analog Lin
Physical:	12.75x6.75x12" 1? Lbs. 324x171x304mm 3.3 kg
Status:	Inactive Manufacturer Discontinued Model
Rarity:	Common
Reviews:	QST January 197? CQ October 1976 73 June 1977

	New	Used
Price:	$200-225 Kit	$?0-140
Rating:	⑨	⑨

MR-1
Comanche

Amateur Communications Receiver Kit
Single Conversion Superheterodyne. 8 Tubes.

Features:
- ¼" Head. Jack
- RF Gain
- Mute Line
- S-Meter
- AVC ON/OFF
- Dial Lamp
- ANL
- BFO
- Speaker Out
- Antenna Trimmer
- AVC (On/Off)
- 30:1 Tune Ratio

Specifications:

Coverage ...... Ham. See Comments.	Modes .... AM/SSB/CW
Selectivity ...... 3 kHz -6dB.	IF .... 3000 kHz.
Sensitivity ...... <1µV SSB 10 dB S/N	Antenna Input . SO-239 50 ohms
Audio Out ...... 2 W 8/500 ohms	

Circuit Complement:
6BZ6 RF Amp, 6EA8 Mixer-Oscillator, 6BZ6 1st IF Amp, 6EA8 2nd IF Amp/S-Meter Amp, 6BE6 Product Detector, 6T8 1st Audio/Det/AVC/NL, 6AQ5 Audio Output and 0A2 Voltage Regulator. The circuit includes an RF stage, converter, two IF stages, two detectors, noise limiter, two audio stages and a voltage regulator.

Accessories:
HP-10 Mobile PS   HP-20 AC Power Supply   AK-7 Speaker

Comments:
Ranges: 3.5-4, 7-7.3, 14-14.35, 21- 21.5 and 28-29.7 MHz. Designed for mobile use. The MR-1 matches the MT-1 Cheyenne transmitter. Requires an external power supply.

Mfg. In:	United States 1959-1962
Voltages:	6/12/350 VDC 64W
Readout:	Analog Lin.
Physical:	12.125x6.125x10" 15 Lbs. 308x155x254mm 7 kg
Status:	Inactive Manufacturer Discontinued Model
Rarity:	Scarce
Reviews:	QST April 1960 CQ October 1959

	New	Used
Price:	$120-130 Kit	$30-75
Rating:	⑨	⑨

RX-1
Mohawk

Amateur Communications Receiver Kit
Double Conversion Superheterodyne. 15 Tubes.

Features:
- ¼" Head. Jack
- S-Meter
- ANL
- Antenna Trimmer
- RF Gain
- AVC ON/OFF
- BFO`
- Calibrator 100 kHz
- Mute Line
- Standby
- Dial Lamp
- Hinged Top Cover
- Notch

Specifications:
Coverage ......Ham. See Comments.
Selectivity ......5/3/2/1/.5 kHz -6dB.
Sensitivity ......<1µV SSB 10 dB S/N
Audio Out ......2 W 8/500 ohms
Modes .... AM/LSB/USB/CW
IF .... 1682, 50 kHz.
Antenna Input . SO-239 & Terminals

Circuit Complement:
6BZ6 RF Amp, 6CS6 1st Mixer, 12AT7 Osc-Cathode Follower, 6BA6 1st IF Amp, 6CS6 2nd Mixer, 12AT7 1632-1732 kHz Crystal Osc, 6BA6 2nd IF Amp, 6BA6 2nd IF Amp, 6BJ7 Detector/AVC/ANL, 6CS6 Product Detector, 12AT7 1st Audio/S-Meter Amp, 6AQ5 Audio Output, 0A2 Voltage Regulator and 5V4 Rectifier.

Accessories:
AK-5 Speaker   XC-6 6 Meter Converter   XC-2 2 Meter Converter

Comments:
Ranges: 1.8-2, 3.5-4, 7-7.3, 14-14.35, 21- 21.45, 26.96-27.23 and 28-29.7 MHz. The RX-1 is pre-calibrated for 6 and 2 meter converters. Green metal case. The RX-1 matches the TX-1 Apache transmitter.

Mfg. In:	United States 1958-1964	
Voltages:	117 VAC	
	50/60 Hz 75W	
Readout:	Analog Lin.	
Physical:	19.5x11.675x16" 52 Lbs.	
	495x296x406mm 23 kg	
Status:	Inactive Manufacturer	
	Discontinued Model	
Rarity:	Scarce	
Reviews:	*QST* December 1958	
	Electric Radio Nov. 1997	

	New	Used
Price:	$275-300 Kit	$50-70
Rating:	⑨	⑨

SB-300

Amateur Communications Receiver Kit
Double Conversion Superheterodyne. 10 Tubes plus Semiconductors.

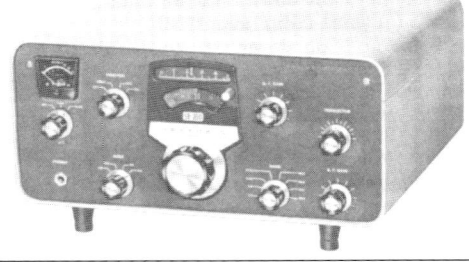

Features:
- ¼" Head. Jack
- S-Meter
- Preselector
- Standby
- ANL
- RF Gain
- LMO Out Jack
- BFO Jack
- AGC OFF/SLO/FST
- Mute Line
- Anti-Vox
- Fiduciary Adj.
- Calibrator 100 kHz

Specifications:
Coverage ......Ham. See Comments.
Selectivity ......2.1/_/_ kHz -6dB.
Sensitivity ......<1µV SSB 15 dB S/N
Image Rej ......>60 dB
Audio Out ......1 W 8 ohms 8% dist.
Modes .... AM/LSB/USB/CW
Stability .... <100 Hz after 20 mins.
IF .... 8400-8900, 3395 kHz.
IF Rej .... 50 dB
Antenna Input . RCA 50 ohms

Circuit Complement:
6BZ6 RF Amp, 6AU6 Crystal Calibrator, 6AU6 1st Mixer, 6AU6 2nd Mixer, 6AB4 Het Osc., 6BA6 IF Amp, 6BA6 IF Amp, 6HF8 AF Amp and 6AS11 Product Det./BFO.

Accessories:
SBA-300-1 .4 kHz CW Filter
SBA-300-2 3.75 kHz AM Filter
Crystal Calibrator 100 kHz
SBA-300-4 2M Converter
SBA-300-3 6M Converter
HS-24 Speaker

Comments:
Ranges: 3.5-4, 7-7.5, 9.5-10, 14-14.5, 21-21.5, 28-28.5, 29-29.5 and 29.5-30 MHz. Matches the SB-400 transmitter. 25 kHz per revolution. Dial accuracy is 400 Hz. Requires a speaker.
Variants: An export model operating from 115/230 VAC was also produced.

Mfg. In:	United States 1963-1966	
Voltages:	105-125 VAC	
	50/60 Hz 50W	
Readout:	Analog Lin.	
Physical:	14.875x6.68x13.4" 17 Lbs.	
	378x170x340mm 7.7 kg	
Status:	Inactive Manufacturer	
	Discontinued Model	
Rarity:	Common	
Reviews:	*QST* July 1964	
	CQ September 1964	
	73 August 1964	

	New	Used
Price:	$250-265 Kit	$75-175
Rating:	⑨	⑨

SB-301

Amateur Communications Receiver Kit
Double Conversion Superheterodyne. 10 Tubes plus Semiconductors.

Features:
- ¼" Head. Jack
- S-Meter
- Preselector
- Standby
- ANL
- RF Gain
- Anti-Vox
- Fiduciary Adj.
- AGC OFF/SLO/FST
- Mute Line
- LMO Jack
- BFO Out Jack
- HFO Jack

Specifications:

Coverage ...... Ham. See Comments.
Selectivity ...... 2.1/_/_ kHz -6dB.
Sensitivity ...... <.25µV SSB 10 dB S/N
Image Rej ...... >60 dB
Audio Out ...... 1W 8 ohms 8% dist.

Modes .... AM/LSB/USB/CW/RTTY
Stability .... <100 Hz after 20 mins.
IF .... 8395-8895, 3395 kHz.
IF Rej .... >50 dB
Antenna Input . RCA 50 ohms

Circuit Complement:
6CB6 Master Osc, 6BZ6 RF Amp, 6AU6 Crystal Calibrator, 6AU6 1st Mixer, 6AU6 2nd Mixer, 6AB4 Het Oscillator, 6BA6 IF Amp, 6BA6 IF Amp, 6HF8 AF Amp, 6AS11 Product Detector/BFO, 1N191 AM Detector, (2) 1N458 AGC, (2) S187 NL and (3) 1N2079 Rectifier.

Accessories:

SB-600 Speaker	Crystal Calibrator 100 kHz
SBA-301-2 .4 kHz CW Filter	SBA-301-1 3.75 kHz AM Filter
SBA-300-3 6M Converter	SBA-300-4 2M Converter

Comments: Ranges: 3.5-4, 7-7.5, 9.5-10, 14-14.5, 15-15.5, 21-21.5, 28-28.5, 29-29.5 and 29.5-30 MHz. The SB-301 features 15-15.5 MHz and a RTTY mode position not found on the SB-300. The SB-301 matches the SB-401 transmitter.

Mfg. In:	United States 1966-1970
Voltages:	105-125 VAC 50/60 Hz 50W
Readout:	Analog Lin.
Physical:	14.875x6.68x13.4' 17 Lbs. 377x169x340mm 7.7 kg
Status:	Inactive Manufacturer Discontinued Model
Rarity:	Common
Reviews:	QST March 1967 73 August 1967 Radio Bygones #42

	New	Used
Price:	$260-270 Kit	$90-170
Rating:	⑨	⑨

SB-303

Amateur Communications Receiver Kit
Single Conversion Superheterodyne. Solid State.

Features:
- ¼" Head. Jack
- S-Meter
- ANL
- Preselector
- RF Gain
- HFO Out Jack
- LMO Out Jack
- Calibrator 25/100 kHz
- Mute Line
- Anti-Vox
- Standby
- AGC OFF/SLO/FST
- Speaker Switch
- Attenuator
- Fiduciary Adj.

Specifications:

Coverage ...... Ham. See Comments.
Selectivity ...... 2.1 kHz -6dB.
Sensitivity ...... <.25µV SSB 10 dB S/N
Image Rej ...... >60 dB
Audio Out ...... 4 W 8 ohms 10% dist.

Modes .... AM/LSB/USB/CW/RTTY
Stability .... <100 Hz after 10 mins.
IF .... 3395 kHz.
IF Rej .... >50 dB
Antenna Input . RCA 50 ohms

Circuit Complement:
1 Integrated Circuit, 27 transistors with MOSFET front-end.

Accessories:

SBA-301-2 .4 kHz CW Filter	SB-650 Digital Display
SBA-301-1 3.75 kHz AM Filter	VHF Converters
SB-600 Speaker	

Comments:
Ranges: 3.5-4, 7-7.5, 9.5-10, 14-14.5, 15-15.5, 21-21.5 and 28-30 MHz. Dial readout is 1 kHz with dial accuracy at 400 Hz. The headphone jack is on the rear panel. The SB-303 matches SB-401 transmitter.

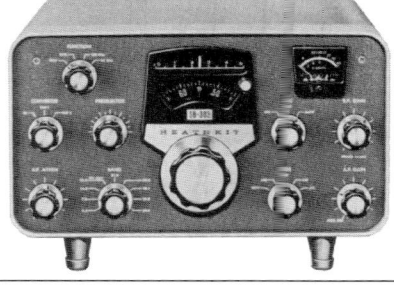

Mfg. In:	United States 1970-1976
Voltages:	105-125 & 210-250 VAC 50/60 Hz 40W
Readout:	Analog Lin.
Physical:	12.25x8x14" 15.75 Lbs. 311x203x355mm 7.2 kg
Status:	Inactive Manufacturer Discontinued Model
Rarity:	Common
Reviews:	QST July 197 CQ April 197

	New	Used
Price:	$320-345 Kit	$110-175
Rating:	⑨	⑨

SB-310

Shortwave Broadcast Communications Receiver Kit
Double Conversion Superheterodyne. 10 Tubes plus Semiconductors.

Features:
- ¼" Head. Jack
- S-Meter
- ANL
- AGC SLO/MED/FST
- RF Gain
- Preselector
- Standby
- Fiduciary Adj.
- Mute

Specifications:

Coverage ......See Comments.	Modes .... AM/LSB/USB/CW
Selectivity ......5/_/_ kHz -6dB.	IF .... 8395-8895, 3395 kHz.
Sensitivity ......<.3µV SSB 10 dB S/N	Audio Out .... 1 W 8 ohms
Image Rej......>60 dB	Drift .... <100 Hz/hour

Circuit Complement:
6CB6 Master Osc, 6BZ6 RF Amp, 6AU6 Crystal Calibrator, 6AU6 1st Mixer, 6AU6 2nd Mixer, 6AB4 Het Oscillator, 6BA6 IF Amp, 6BA6 IF Amp, 6HF8 AF Amp, 6AV11 Product Detector/BFO, 1N191 AM Detector, (2) 1N458 AGC, 2S187 NL and 1N2079 Rectifier.

Accessories:
.4 kHz CW Filter   2.1 kHz SSB Filter   2.1 kHz Deluxe SSB Filter
SB-600 Speaker

Comments:
This is the shortwave listener's version of the SB-300. Dial accuracy is rated at ±400 Hz. Ranges: 3.5-4, 5.7-6.2, 7-7.5, 9.5-10, 11.5-12, 14-14.5, 15-15.5, 17.5-18 and 26.9-27.4 MHz. Note the lack of coverage of the 21 MHz band.

Mfg. In:	United States 1968-1972
Voltages:	120/240 VAC 50/60 Hz 50W
Readout:	Analog Lin.
Physical:	15x6.75x14" 17 Lbs. 380x170x350mm 8 kg
Status:	Inactive Manufacturer Discontinued Model
Rarity:	Scarce

	New	Used
Price:	$259-268 Kit	$100-150
Rating:	⑨	⑨

SB-313

Shortwave Broadcast Communications Receiver Kit
Double Conversion Superheterodyne. Solid State.

Features:
- ¼" Head. Jack
- S-Meter
- ANL
- Calibrator 100/25 kHz
- RF Gain
- Standby
- Preselector
- AGC SLO/MED/FST
- Mute Line
- Attenuator
- Fiduciary Adj.

Specifications:

Coverage ......See Comments	Modes .... AM/LSB/USB/CW
Selectivity ......5/_/_ kHz -6dB.	Image Rej .... >60 dB
Sensitivity ......<.5µV SSB 10 dB S/N	IF .... 8395-8895, 3395 kHz.
Audio Out ......4 W 8 ohms	Stability .... <10 Hz/hour after 10 mins.

Circuit Complement:
One integrated circuit, 4 MOSFETs, 11 transistors and 11 crystals.

Accessories:
SB-600 Speaker   SBA-301-2 .4 kHz CW Filter
SBA-310-2 2.1 kHz SSB Filter

Comments:
Ranges: 3.5-4, 5.7-6.2, 7-7.5, 9.5-10, 11.5-12, 14-14.5, 15-15.5, 17.5-18 and 21.3-21.8 MHz. This model is the SWL version of the SB-303. It is also the solid state successor to the SB-310. Note the difference coverage on the last band versus the SB-310. The dial accuracy is ±1 kHz.

Mfg. In:	United States 1972-1975
Voltages:	105-130/210-260 VAC 50/60 Hz 40W
Readout:	Analog Lin.
Physical:	12.25x8x14" 18 Lbs. 380x170x350mm 8 kg
Status:	Inactive Manufacturer Discontinued Model
Rarity:	Very Scarce
Reviews:	*Popular El.* November 1972

	New	Used
Price:	$340-360 Kit	$120-225
Rating:	⑨	⑨

SW-717

General Coverage Communications Receiver Kit
Single Conversion Superheterodyne. Solid State.

Features:
- ¼" Head. Jack
- S-Meter
- BFO
- Ferrite MW Antenna
- RF Gain
- ANL
- Standby
- Bandspread ±5 Units
- Speaker
- Dial Lamp

Specifications:
Coverage ...... See Comments
Selectivity ...... 3 kHz -6dB.
Sensitivity 4µV
Modes .... AM/CW
Antenna Input . Terminals 50 ohm

Circuit Complement:
Dual gate MOSFET mixer stage, ceramic filter, zener diode regulated oscillator power supply for good stability.

Comments:
Ranges: .55-1.55, 1.5-4, 5-12.5 and 10-30 MHz.

Mfg. In:	United States 1972–1981
Voltages:	120/240 VAC 50/60 Hz 8W
Readout:	Analog
Physical:	14.5x5.34x8" 10 Lbs. 370x140x203mm 4.5 kg
Status:	Inactive Manufacturer Discontinued Model
Rarity:	Common

	New	Used
Price:	$70-139 Kit	$50-90
Rating:	⑨	⑨

Heathkit

SW-7800

General Coverage Communications Receiver Kit
Double Conversion Superheterodyne. Solid State.

Features:
- ¼" Head. Jack
- S-Meter
- Telescopic Whip
- AGC SLOW/FAST
- Mute Line
- Speaker
- RF Attenuator
- Record Jack
- Dial Lamp

Specifications:
Coverage ...... 150 - 30000 kHz
Selectivity ...... 5/2.5 kHz -6dB.
Sensitivity ...... <.35µV SSB 10 dB S/N
Audio Out ...... 1 W 8 ohms
Modes .... AM/LSB/USB/CW
Image Rej .... >55 dB
Antenna Input . Terms. & SO230 50 ohm

Accessories:
.4 kHz CW Filter   2.1 kHz SSB Filter   2.1 kHz Deluxe SSB Filter

Comments:
Quite drifty. This receiver was not warmly received by radio reviewers or the listening community.

Mfg. In:	United States 1985-1990
Voltages:	120 VAC or 13.8 VDC .75 Amps
Readout:	00001. Digital LED
Physical:	11.5x4.875x11.5" 9 Lbs. 292x123x292mm 4 kg
Status:	Inactive Manufacturer Discontinued Model
Rarity:	Common
Reviews:	Passport 1985/86, 89-91 WRTH 1986 QST April 1985

	New	Used
Price:	$300-350 Kit	$100-150
Rating:	⑨	⑨

41 Howard Radio Company

The **Howard Radio Company** was a significant receiver manufacturer prior to World War II. Their advertising included the slogan, *America's Oldest Radio Manufacturer.* Howard was a significant force in the amateur market prior to World War II. They had an impressive selection of communications receiver.

An interesting marketing program called the *Howard Progressive Series* allowed the customer to initially purchase an inexpensive, simple receiver. They could then send the radio back to the manufacturer at a later date to add factory-installed optional features. For example, the model 435 could be returned to the factory and upgraded to the model 436. For an additional charge it could even be upgraded to a 437.

Howard Radio Company
1731-35 Belmont Ave.
Chicago, Illinois 1938-1948

HOWARD

430

General Coverage Communications Receiver
Single Conversion Superheterodyne. 6 Tubes.

Features:
- ¼" Head. Jack
- S-Meter
- Speaker 6"
- Logging Scale
- Standby
- Mute Line
- AVC ON/OFF
- Bandspread 0-100
- Dial Lamp

Specifications:
Coverage ...... 550 - 42000 kHz
IFs .... 465 kHz
Audio .... 2 W
Modes .... AM/CW
Antenna Input . Terminals

Circuit Complement:
6K8G Converter, 6K7 IF, 6Q7G Detector/1st Audio/AVC, 6C5 BFO, 41 Audio Output and 5W4 Rectifier.

Accessories:
1-820 External Speaker 600 S-Meter 610 6 VDC Power Supply

Comments:
Ranges: .55-1.7, 1.7-5.5, 5.5-18 and 16-42 MHz. The headphone jack is on the rear panel. The Howard Type 610 "B" power pack shown right converted 6 volts DC to 300 volts DC for operation of several Howard receiver models. A four prong plug from the power pack plugged into a socket on the back of the receiver. Automobiles of the day featured 6 volt batteries.

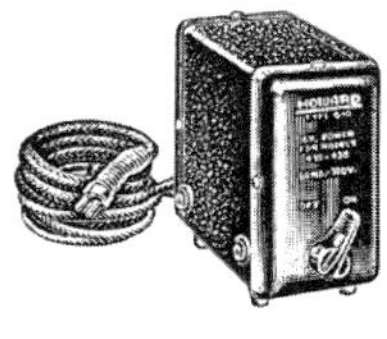

Made In: United States 1938-1940

Voltages: 105-125 VAC 60 Hz
50 W

Readout: Analog

Physical:

Status: Inactive Manufacturer
Discontinued Model

Rarity: Very Scarce

	New	Used
Price:	$30-43	⑤
Rating:	⑥	⑥

H·O·W·A·R·D

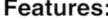

435-A

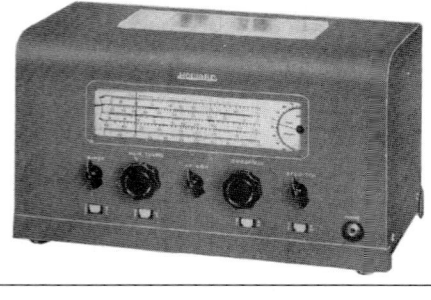

General Coverage Communications Receiver
Single Conversion Superheterodyne. 7 Tubes.

Features:
- ¼" Head. Jack
- Standby
- Dial Lamp
- BFO
- Mute Line
- AVC
- Speaker 6.5"
- Flywheel Tuning
- Logging Scale
- Bandspread 0-100

Specifications:
Coverage ...... 540 - 43000 kHz
IFs .... 465 kHz
Modes .... AM/CW
Antenna Input . Terminals

Circuit Complement:
6SD7GT, 6SA7, 6SK7, 6SQ7, 6K6G, 6J5 and 5Y3G.

Accessories:
3-820 External Speaker 610 6VDC Power Pack 650 Preselector
S-Meter

Comments:
This receiver has four bands. The headphone jack is on the rear panel. The steel cabinet has a gray wrinkle finish. Shown to the right, is the model 435-A with the S-Meter retrofitted.

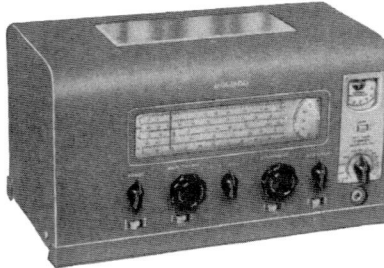

Variants:
Model **435**, earlier production, has 6 tubes. Model **445** was an AC-DC version of the 435A.

Made In: United States 1941-1942

Voltages: 105-125 VAC 60 Hz
50 W

Readout: Analog

Physical:

Status: Inactive Manufacturer
Discontinued Model

Rarity: Very Scarce

	New	Used
Price:	$30-37	⑤
Rating:	⑥	⑥

H·O·W·A·R·D

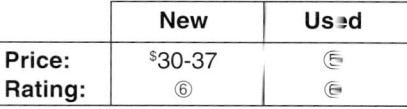

436

General Coverage Communications Receiver
Single Conversion Superheterodyne. 7 Tubes.

Features:
- ¼" Head. Jack
- Standby
- Dial Lamp
- S-Meter
- Mute Line
- AVC
- Speaker 6.5"
- Flywheel Tuning
- Logging Scale
- Bandspread 0-100

Specifications:
Coverage ...... 540 - 43000 kHz
IFs .... 465 kHz
Audio .... 2.5 W
Modes .... AM/CW
Antenna Input . Terminals

Circuit Complement:
6K8G Mixer, 6SK7 IF Amp, 6SQ7 Detector/1st IF, 6K6G Output, 6C5 BFO, 6H6G NL and 80 Rectifier.

Accessories:
3-820 External Speaker 610 6VDC Power Pack 650 Preselector
S-Meter

Comments:
Ranges: .54-1.7, 1.7-5.5, 5.6-18 and 17-43 MHz. The headphone jack is on the rear panel.

Variants:
Model **436-A**, later production, has 8 tubes, ANL and optional S-Meter. As part of the "Progressive Series", the 436-A could be upgraded to the 437-A by the factory.

Made In: United States 1939-1940

Voltages: 105-125 VAC 60 Hz
50 W

Readout: Analog

Physical:

Status: Inactive Manufacturer
Discontinued Model

Rarity: Very Scarce

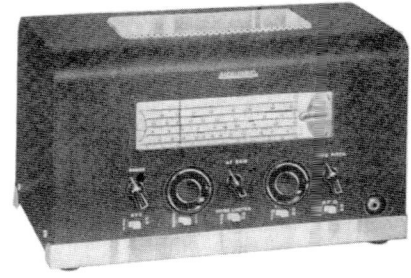

	New	Used
Price:	$40	⑤
Rating:	⑥	⑥

HOWARD

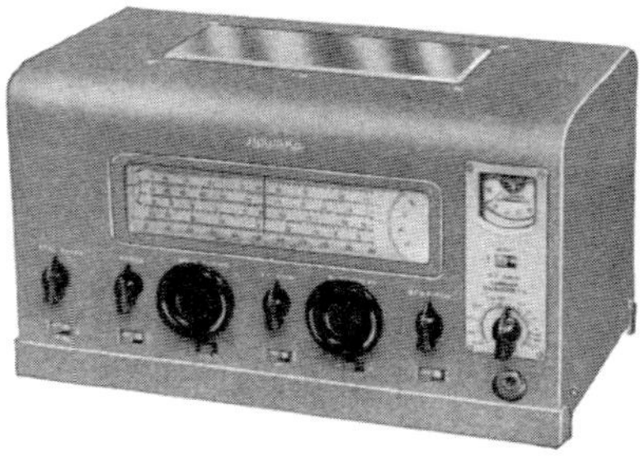

437-A

General Coverage Communications Receiver
Single Conversion Superheterodyne. 9 Tubes.

Features:
- ¼" Head. Jack
- Standby
- Dial Lamp
- S-Meter
- Mute Line
- AVC
- Speaker
- Flywheel Tuning
- Logging Scale
- Bandspread 0-100

Specifications:
Coverage ...... 540 - 43000 kHz
IFs .... 465 kHz
Modes .... AM/CW
Antenna Input . Terminals

Circuit Complement:
6SK7 RF Amp, 6K8G Mixer Osc., 6SK7 IF Amp, 6SK7 IF Amp, 6SQ7 Detector/1st AF, 6K6G Output, 6C5 BFO, 6H6G NL and 80 Rectifier.

Accessories:
3-820 External Speaker
S-Meter (shown)
Crystal Filter
610 6 VDC Power Pack
650 Preselector

Comments:
Ranges: .54-1.7, 1.7-5.5, 5.6-18 and 17-43 MHz. This model is similar to the 435-A and 436-A, but with an additional IF stage and a Crystal Phasing Control added.

Variants:
Model **437** was earlier production.

Made In: United States 1939-1941

Voltages: 105-125 VAC 60 Hz
60 W

Readout: Analog

Physical:

Status: Inactive Manufacturer
Discontinued Model

Rarity: Very Scarce

	New	Used
Price:	$55-62	⑤
Rating:	⑥	⑥

HOWARD

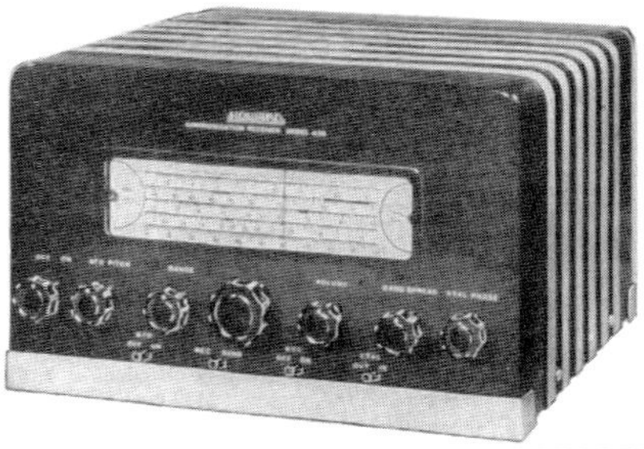

438

General Coverage Communications Receiver
Single Conversion Superheterodyne. 8 Tubes.

Features:
- ¼" Head. Jack
- Standby
- Dial Lamp
- Ext. Speaker Jack
- S-Meter
- Mute Line
- AVC ON/OFF
- Speaker 6"
- Flywheel Tuning
- BFO
- Logging Scale
- Bandspread 0-100
- RF Gain

Specifications:
Coverage ...... 550 - 43000 kHz
IFs .... 465 kHz
Audio 2W
Modes .... AM/CW
Antenna Input . Terminals

Circuit Complement:
6K7 RF, 6K8G Converter, 6K7 IF, 6K7 IF, 6Q7G Detector/AVC/1st Audio, 41 AF Output and 80 Rectifier.

Accessories:
External Speaker
610 6 VDC Power Pack
60 S-Meter
Crystal

Comments:
The headphone jack is on the rear panel.

Made In: United States 1939-1940

Voltages: 105-125 VAC 60 Hz

Readout: Analog

Physical:

Status: Inactive Manufacturer
Discontinued Model

Rarity: Very Scarce

Reviews: *Electric Radio* June 1992

	New	Used
Price:	$50	⑤
Rating:	⑥	⑥

HOWARD

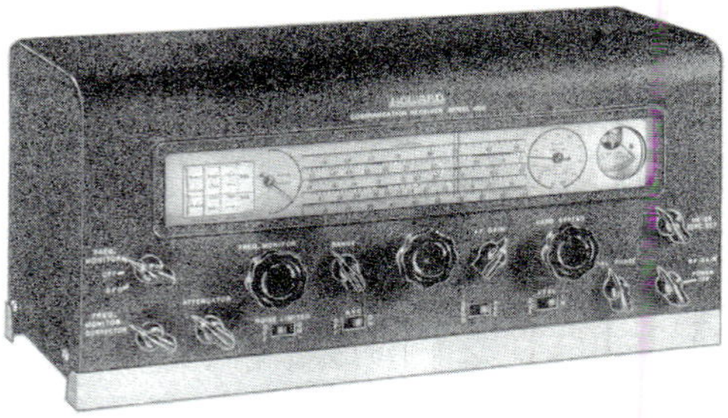

460

General Coverage Communications Receiver
Single Conversion Superheterodyne. 10 Tubes.

Features:
- ¼" Head. Jack
- Standby
- Dial Lamp
- RF Gain
- S-Meter
- Mute Line
- AVC
- Standby
- Logging Scale
- Flywheel Tuning
- BFO
- NL
- Attenuator
- Bandspread
- Frequency Monitor

Specifications:
Coverage ......550 - 42000 kHz
Audio ....5 ohm 4W
Antenna Input Terminals
Modes .... AM/CW
IFs .... 465 kHz

Circuit Complement:
Crystal filter (crystal optional).

Comments:
Ranges: .55-1.6, 1.7-5.5, 5.5-18 and 18-42 MHz. The model 460 had two BFO systems and a special extended electrical bandspread that can be calibrated against the frequency monitor. This allowed better than average frequency resolution. The 460 was sold with the crystal filter for $90 or without for $80.

Made In: United States 1939-1940

Voltages: 105-125 VAC 60 Hz

Readout: Analog

Physical:

Status: Inactive Manufacturer
Discontinued Model

Rarity: Very Scarce

	New	Used
Price:	$90	⑤
Rating:	⑥	⑥

HOWARD

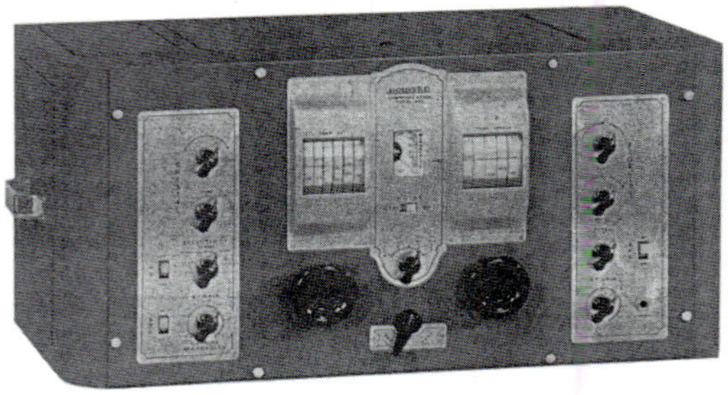

490

General Coverage Communications Receiver
Single Conversion Superheterodyne. 14 Tubes.

Features:
- ¼" Head. Jack
- Standby
- Dial Lamp
- RF Gain
- S-Meter
- Mute Line
- AVC
- Standby
- Ext. Speaker 10"
- Flywheel Tuning
- BFO
- NL
- Logging Scale
- Bandspread
- Hinged Top Cover
- 500 Ohm Output Jack

Specifications:
Coverage ......540 - 44000 kHz
Selectivity ......5 Position
Audio ....5 ohm 8W
Modes .... AM/CW
IFs .... 465 kHz
Antenna Input . Terminals

Circuit Complement:
6AB7 RF, 6AB7 RF, 6SA7 Mixer, 6SA7 H.F. Osc., 6SK7 IF, 6SK7 IF, 6H6 Detector/NL/AVC, 6SF5 1st Audio, 6J5 Phase Inverter, 6J5 BFO, 7E7 Carrier Level Meter, 6K6 AF Output, 6K6 AF Output and 5Y3G Rectifier.

Accessories:
10" Howard-Jensen Speaker

Comments:
This model was initially supplied with a 10" external speaker in cabinet. The bandspread is for the 80, 40, 20 and 10 meter amateur bands. The cabinet is finished in a blue gray wrinkle finish. This model has a 19" wide front panel and may be rack mounted. A highly interesting and collectable model.

Made In: United States 1940-1941

Voltages: 105-125 VAC 60 Hz

Readout: Analog

Physical: 21.675x11x13.5" 57 Lbs.
550x280x343mm 25.8 kg

Status: Inactive Manufacturer
Discontinued Model

Rarity: Very Scarce

	New	Used
Price:	$149-158	⑤
Rating:	⑥	⑥

Born in 1931 in Kyoto Prefecture, Japan, Mr. Tokuzo Inoue founded the Inoue Electric Factory in 1954. In 1964 the Inoue Electric Factory Company Limited was established.

In 1965 the all transistorized FDAM-1 50 MHz transceiver was introduced ushering the move towards all solid state transceivers.

In 1967 the Inoue's first H.F. amateur band transceiver was introduced consisting of the IC-700R receiver and the IC-700T transmitter. In 1970 the Inoue IC-20 12 channel VHF transceiver was made. The 24 channel IC-21 followed in 1974.

The Company name was changed to **ICOM** in 1978. Icom America, headquartered in Bellevue, Washington was established in October of 1979. By the late 1970's the amateur market became acquainted with Icom's high quality standards and state-of-the-art communications products. The shortwave community came to know Icom in 1982 with the introduction of the successful R-70 general coverage receiver. The even more highly acclaimed R-71 followed. The R-7000 and R-7100 VHF receivers offered sophistication never before available to the hobby market. For many, the R-9000 wideband model defines the ultimate receiver.

Mr. Tokuzo Inoue, the Founder, President and Chairman of Icom overseas a company with ¥24 billion in annual sales.

ICOM Inc.
6-9-16, Kamihigashi
Hirano-ku
Osaka 547, Japan

ICOM America Inc.
2380 116th Ave. N.E.
Bellevue, WA 98004

INOUE

IC-700R

Amateur Band Communications Receiver
Single Conversion Superheterodyne. Solid State.

Features:
- ¼" Head. Jack
- S-Meter
- 500 Hz CW Audio Filter
- RF Gain
- NB
- BFO
- Mute
- Fixed Position

Specifications:

Coverage ......Ham. See Comments.	Modes .... AM/SSB/CW
Selectivity ......2.4 kHz -6dB.	Stability .... ±100 Hz
Sensitivity ......<1µV	IF Rejection .... >60 dB
Image Rej......>60 dB	

Circuit Complement:
This all transistor receiver uses FETs in the RF and oscillator stages and a 9 MHz crystal filter.

Comments:
Ranges: 3.5-4, 7-7.5, 14-14.5, 21-21.5, 28-28.5, 28.5-29, 29-29.5 and 10-10.5 MHz. Requires a speaker. This was Inoue's first receiver. The IC-700R matches the IC-700T transmitter.

Made In:	Japan	1967-1968
Voltages:	110/240 VAC or 12 VDC	
Readout:	(analog scale) Analog Lin.	
Physical:	11x6.5x9.5" 12 Lbs	
	280x165x241mm 5.44 kg	
Status:	Active Manufacturer	
	Discontinued Model	
Rarity:	Extremely Scarce	

	New	Used
Price:	①	②
Rating:	⑥	⑥

○ ICOM

PCR1000

Wideband Communications Computer Receiver
Triple Conversion Superheterodyne. Solid State.

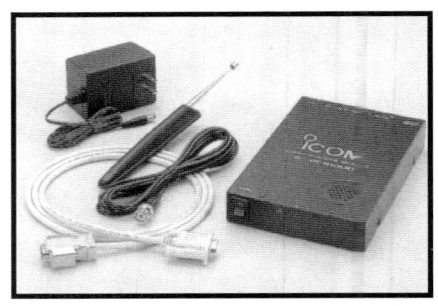

Features:
- S-Meter
- AGC
- Clock
- Packet Jack
- Squelch
- Scan
- NB
- Speaker Jack
- Voice Scan
- Sweep
- Speaker
- Band Scope ±200kHz
- IF Shift ±1.2 kHz
- Unlimited Memories
- Attenuator -20 dB

Specifications:
Coverage ...... 100 kHz - 1.3 GHz
Selectivity ...... 15/6/2.8 kHz [50/230]
Sensitivity ...... <.28μV 1.8-28 MHz
Audio Out ...... 0.2 W 4-8 ohms
I/O DB-9F

Modes .... AM/SSB/CW/FM/NFM
Stability .... ±3 PPM
IFs .... 266.7, 10.7 MHz, 450 kHz
Environment ... 0° to +50°C. (32°-122°F)
Antenna Input . BNC

Accessories:
OPC-131 DC Power Cable   UT-106 DSP Option

Comments:
You can choose from three interface screes: Communications Receiver Screen, Component Receiver Screen or Radio Screen. System requirements: PC with 486x4, Pentium 100 or higher CPU, Windows 3.1 or higher, including Windows 95, 16 Mb RAM, 3.5" FD and 10 Mb HD space. The American version comes with the BC123A AC adapter, RS232 cable and telescopic antenna.

Variants:
The American version **PCR1000-02** has the cellular frequencies blocked. The American unblocked, full coverage government version is called the **PCR1000-07**.

Made In:	Japan	1997 1998
Voltages:	12.8 VDC 0.7A	
Readout:	 1 Hz.	PC
Physical:	5x1.2x7.9" 2.3 Lbs.	
	126x30x200mm 1 kg	
Status:	Active Manufacturer	
	Active Model	
Reviews:	*S.W. Magazine* October 1997	
	Popular Comm. March 1998	
Rarity:	Too new.	

	New	Used
Price:	$500-600	$300-340
Rating:	⑧	Ⓔ

⊙ ICOM

R-70

General Coverage Communications Receiver
Quadruple Conversion Superheterodyne. Solid State.

Features:
- ¼" Head. Jack
- RF Gain
- Mute
- Line Out Jack
- Two VFOs
- Dial Lock
- S-Meter
- Squelch
- Tone
- Ext. Spkr Jack
- Preamp
- Monitor
- NB
- Notch
- Dimmer
- RIT ±800 Hz
- Speaker
- Scope Jack
- PBT
- Three Tuning Rates
- AGC OFF/SLO/FST
- Converter In Jack
- Attenuator

Specifications:
Coverage ...... 100 - 30000 kHz
Selectivity ...... 6/2.3/_ kHz -6dB.
Sensitivity ...... <.16μV SSB 10 db S/N^
Image Rej ...... >60 dB
Audio Out ...... 3 W 8 ohms 10% dist.
Dyn.Range .... 100 dB @500 Hz bw.

Modes .... AM/SSB/CW/RTTY
Stability .... ±250 Hz (1-60 mins.)
IF Rejection .... >60 dB
IF .... 70.4515, 9.0115, .455, 9.0115
Environment ... -10° to +60° C.
Antenna Input . SO-239 and Terminals

Circuit Complement:
43 Integrated Circuits, 77 transistors, 14 FETs and 180 diodes.

Accessories:
FL-63 CW Filter   FL-44 SSB Filter   EX-257 FM Mode Option
DC Kit

Comments:
Utilizes 10 Hz step digital VFO. ^Sensitivity is 1μV below 1.6 MHz. Interfaces to IC-720A amateur transceiver. The R-70 is a good receiver, but relatively complex to operate.

Made In:	Japan	1982-1984
Voltages:	100/117/220-240 VAC	
	50/60 Hz 25 W	
Readout:	00000.1 Digital Flor.	
Physical:	11.3x4.4x10.9" 16.3 Lbs.	
	286x111x276mm 7.4 kg	
Status:	Active Manufacturer	
	Discontinued Model	
Rarity:	Common	
Reviews:	*QST* June 1983	
	Popular Comm. Dec. 1983	
	Mon. Times January 1983	
	WRTH 1983, 1984	
	73 February 1982	
	Proceedings 1989	

	New	Used
Price:	$600-700	$390-410
Rating:	★★★★	★★★★

 **ICOM**

ICOM R-71A

General Coverage Communications Receiver
Quadruple Conversion Superheterodyne. Solid State.

Features:
- ¼" Head. Jack
- RF Gain
- Mute
- Line Out Jack
- Memory Scan
- Dial Lock
- S-Meter
- 32 Memories
- Tone
- Ext. Spkr Jack
- Sweep
- Speaker
- PBT[1]
- Notch
- Dimmer
- Squelch
- 10 Hz VFO
- Keypad
- Two VFOs
- Three Tuning Rates
- AGC OFF/SLO/FST
- Preamp
- Dial Tension Adjust
- Attenuator

Specifications:
Coverage ...... 100 - 30000 kHz
Selectivity ...... 6/2.3/_/_ kHz -6dB.
Sensitivity ...... <.15µV SSB 10 db S/N^
IF Rejection ... >60 dB
Audio Out ...... 2 W 8 ohms 10% dist.

Modes .... AM/SSB/CW/RTTY
Stability .... ±250 Hz (1-60 mins.)
IF .... 70.4515, 9.0115, .455, 9.0115 MHz
Antenna Input . SO239 and Terminals
Environment ... -10° to +60° C.

Accessories:
RC-11 Remote	EX-257 FM Mode Option	CR-64 High Stability
EX-310 Voice Synthz.	EX-309 Intfc. Option	CK-70 DC Option
FL-32A 500Hz CW Filter	FL-63A 250Hz CW Filter	FL-44A 2.4 kHz SSB Filter
CT-17 Level Converter	UX-14 CI-IV/CI-V Conv.	MB-12 Mobile Bracket

Comments: [1]The Passband Tuning was deleted temporarily from production (04/89) due to a patent issue. Used units without PBT sell for $50 less. Memories store freq. and mode. Volatile firmware programming. An excellent SSB receiver.
Variants: Model **R-71E** is the European version (**R-71D** is the German version).

Made In:	Japan	1984-1996
Voltages:	100/117/220-240 VAC 50/60 Hz 25 W	
Readout:	`00000.1`	Digital Flor.
Physical:	11.75x4.4x10.9" 17 Lbs. 298x112x277mm 7.7 kg	
Status:	Active Manufacturer Discontinued Model	
Rarity:	Abundant	
Reviews:	IBS-RDI Whitepaper	
	Passport 1985/86-1997	
	WRTH 1985	
	Mon. Times May 1984, 1995	
	CQ August 1984	
	S.W. Magazine January 1992	
	Buyer's Guide to Amateur R.	

	New	Used
Price:	$689-1280	$560-670
Rating:	★★★★★	★★★★★

ICOM R-72

General Coverage Communications Receiver
Double Conversion Superheterodyne. Solid State.

Features:
- ¼" Head. Jack
- Dial Lock
- Record Jack
- Preamp
- Scan/Sweep
- S-Meter
- Atten. -10/20 dB
- Keypad
- Mute Line
- Tilt Bar
- AGC
- 99 Mems (F/M)
- NB
- Dimmer
- Fast/Slow Scan
- 24 Hour Clock-Timer
- 2 Band Edge Mem
- Three Tuning Rates
- Dial Adj. Brake
- Recorder Activation

Specifications:
Coverage ...... 100 - 30000 kHz
Selectivity ...... 6/2.3/_ kHz -6dB.
Sensitivity ...... <.16µV SSB 1.8-30 MHz
Image Rej ...... >70 dB
Audio Out ...... 2 W 8 ohms 10% dist.

Modes .... AM/SSB/CW
Stability .... ±200 Hz (1-60 mins.)
IF .... 70.4515, 9.0115, .455[FM] MHz
Dyn. Range .... 100 dB
Antenna Input . SO-239 and terminals

Accessories:
SP-3 Speaker	UI-8 FM Mode Option	UR-1 Receiver Protector
OPC-131 DC Kit	FL-100 500 Hz CW Filter	FL-101 250 Hz CW Filter
CR-64 High Stability	UT-36 Voice Synthesizer	CT-17 Level Converter
MB-23 Carry Handle	MB-5 Mobile Bracket	

Comments: The R-72 never gained wide popularity in the North American market. Some reviewers felt that the supplied bandwidths were inappropriate for DX'ing. Triple conversion in optional NB-FM mode. The Australian version covers 250 - 29900 kHz. The German version covers 150 - 26100 kHz. Units made after June 1996 may be for operation from 12 VDC only.

Made In:	Japan	1990-1998
Voltages:	100/117/220-240 VAC 50/60 Hz 25 W	
Readout:	`00000.01`	Digital LCD
Physical:	9.5x3.7x9" 10.6 Lbs. 241x94x229 mm 4.8 kg	
Status:	Active Manufacturer Active Model	
Rarity:	Common	
Reviews:	*Passport* 1991-1997	
	WRTH 1991	
	WRTH Buyer's Guide 1993	
	Mon. Times July 1990	
	SW Magazine May 1993	

	New	Used
Price:	$696-1099	$400-470
Rating:	★★	★★

○ICOM

R-100

Wideband Broadcast Receiver
Double Conversion Superheterodyne. Solid State.

Features:
- S-Indicator
- ANL
- Speaker Jack
- Backlit LCD
- AFC
- 100 Memories
- Keypad
- Speaker
- 8 Tune Steps
- Attenuator -20 dB
- Dimmer
- Ant. Sel. Jack
- Preamp 50-905 MHz
- 20 Band-Edge Mem.
- 24 Hour Clock-Timer

Specifications:
Coverage ......500 kHz - 1856 MHz^
Selectivity ......6/15/180 kHz -6dB.
Sensitivity ......<1.6 µV AM 1.6-50 MHz
Audio Out ......2.5 W 8 ohms 10% dist.

Modes .... AM/FM-N/FM-W
Stability .... ±3.5 PPM @1800 MHz.
Antenna Input . SO-239 / N / N

Accessories: SP-7 Speaker   SP-10 Speaker
SP-12 Speaker   CP-11 Cigar Lighter Cord   AH-7000 Antenna

Comments: Designed primarily for mobile use. ^Specifications guaranteed from .5-1800 MHz. Tuning steps: 1, 5, 8, 9, 10, 12.5, 20 or 25 kHz. Memories store: frequency, mode, step, and preamp/attenuator settings. Nine different scanning modes are supported. The French version covers .1-87.5 and 108.5-1856 MHz. Units manufactured for America after April 1995 lack coverage from 800 to 900 MHz. This was to comply with the Federal law prohibiting the reception of cellular frequencies. The American blocked versions are referred to as model **R-100-11**. The unblocked, or government version is referred to as **R-100-03**. This receiver is popular for tuning satellite transponders due to its wide spectrum coverage. There are three antenna inputs: .5-50 (SO-239), 50-905 (N) and 905-1800 (N) MHz.

Made In:	Japan	1990 1998

Voltages: 13.8 VDC ±15% 1.1 Amps

Readout: `00000.01` Digital LCD

Physical: 5.9x2x7.1" 3.1 Lbs.
150x50x181mm 1.4 kg

Status: Active Manufacturer
Active Model

Rarity: Scarce

Reviews: *WRTH* 1991
Mon. Times October 1990
73 June 1997

	New	Used
Price:	$599-760	$300-380
Rating:	★★★★	★★★

○ICOM

R-8500

Wideband Communications Receiver
Triple Conversion Superheterodyne. Solid State.

Features:
- ¼" Head. Jack
- RF Gain
- Clock-Timer
- Line Out Jack
- Scan
- FM AFC
- S/Tune Meter
- S-Meter Squelch
- Squelch
- Ext. Spkr Jack
- Sweep
- Dimmer
- IF Shift ±1.2 kHz
- 1000 Memories
- RS-232 Port
- 10.7 MHz IF Out
- Keypad
- NB
- Audio Peak Filter
- Three Tuning Rates
- Voice Scan Control
- Record Activation
- Atten. -10/20/30 dB
- Tune Drag Adj.

Specifications:
Coverage ....1 - 1999.99 MHz ^
Selectivity ...... 150/12/5.5/2.2 kHz -6dB.
Sensitivity ......-14 dBµ 2-30 MHz
Image Rej ......>60 dB 1.5-30 MHz
Audio Out ......2 W 4-8 ohms
Dyn.Range ....107 dB

Modes .... AM/SSB-CW/FM-W/FM-N
Stability .... ±100 Hz (<30 MHz)
IF Rejection .... >70 dB 1.6-30 MHz
IF .... 48.8, 10.7, .455 MHz. <30 MHz
Environment ... -10° to +50° C.
Antenna Input . SO-239 / RCA / N

Accessories:
TV-R7100 NTSC TV   FL-52A 500Hz CW Filter   CR-293 High Stability
UT-102 Voice Synth.   MB-23 Carry Handle   MB-12 Mobile Bracket

Comments: 800 standard memories (20 banks of 40 channels), 100 Skip Scans and 100 Auto Write Memory Scans. Five scan methods (10 to 20 ch/sec). Double conversion on FM wide.

Variants: In the U.S.A., model **R-8500-02** has the cellular frequencies of 824-849 and 869-894 MHz blocked. Model **R-8500-03** is an unblocked government version.

Made In:	Japan	1996-1998

Voltages: 117 VAC 60 Hz (AD-55A)
or 12 VDC 2 Amps

Readout: `.00000.01` Digital LCD

Physical: 11.3x4.4x12.16 17.7 Lbs
287x112x309mm 8 kg

Status: Active Manufacturer
Active Model

Rarity: Scarce

Reviews: *Passport* 1998
Popular Comm Feb. 1997
Mon. Times January 1997
Mon. Times July 1997
QST April 1997

	New	Used
Price:	$1749-1950	$1200-1300
Rating:	★★★★★	★★★★★

ICOM

R-9000

Wide Band Communications Receiver
Quadruple Conversion Superheterodyne. Solid State.

Features:
- ¼" Head. Jack
- S-Meter
- IF Shift
- Spectrum Display
- RF Gain
- 1000 Memories
- Notch
- Three Tuning Rates
- Bass
- Treble
- Dimmer
- AGC SLO/FST/OFF
- Line Out Jack
- Ext. Spkr Jack
- Squelch
- Preamp
- Mem. Scanning
- Sweep
- Keypad
- Attenuator -10/20 dB
- Rack Handles
- NB
- Video Out
- Dual Clock-Timer

Specifications:
Coverage ...... 100 kHz - 1999.8 MHz
Selectivity 150/15/6/2.4/.5 kHz-6dB
Sensitivity ...... <.16μV SSB 1.6-30 MHz
Stability .... ±25 Hz (.1-30 MHz)
Audio Out ...... 2.5W 4-8 ohms 10% dist.

Modes .... AM/SSB/CW/FM/RTTY
IF .... 48.8 nom., 10.7, .45, 10.7 MHz .1-30
Environment ... -10° to +50° C.
Antenna Input . SO239 / N / N

Accessories: CT-16 Satellite Intf.   CK-70 DC Option
UT-36 Voice Synthz.   CT-17 Level Converter   AH-7000 Antenna

Comments: FM selectivity: 6/15/30/150 kHz. Tuning steps: .01, .1, 1, 5, 9, 10, 12.5, 20, 25 and 100 kHz. The spectrum display shows: ±25, ±50 or ±100 kHz. CRT can also display TV video on some versions. Eight scanning methods are supported. Scans at 13 ch./sec. Memories store: frequency, mode and bandwidth. Sold primarily to the military and govt. markets. Sale to American hobbyist prohibited after 1995 because of cellular coverage. Versions for France, Germany and Australia have different coverage. A fabulous wideband receiver that is excellent on H.F.

Made In: Japan   1989-1998

Voltages: 100-120 VAC 60 Hz
(Non-USA: 220-240 VAC)

Readout: `00000.01` CRT
Physical: 16.7x5.9x14.4" 44.1 Lbs.
424x150x365mm 20 kg

Status: Active Manufacturer
Active Model

Rarity: Scarce

Reviews: 📖 *Passport* 1990-1998
WRTH 1990
WRTH Buyer's Guide 1993
Mon. Times August 1989
CQ February 1990
SW Magazine April 1989

	New	Used
Price:	$4699-6600	$3500-4000
Rating:	★★★★★	★★★★★

43 Japan Radio Company

The **Japan Radio Company** was established in 1915 and is one of the oldest and largest electronics manufacturing companies in the world. J.R.C. is a respected leader in marine radio equipment, mobile and satellite communications, computerized dam and river management systems, computer graphics, fiber-optics, radar, navigation equipment and avionics systems.

J.R.C. employes over 3,500 employees worldwide. J.R.C. sales in 1990 exceeded ¥108 billion.

J.R.C. is best known in North America for their respected line of NRD-5x5 receivers. This series started with the NRD-505 and continues to the current NRD-545. A line of quality HF amateur transceivers is also available.

Occasionally one may be lucky enough to find one of J.R.C.'s commercial receivers available in the hobbyist market such as the NRD-92 or NRD-93.

Please note that J.R.C. "commercial" sensitivity specifications are very conservative compared with "consumer" specifications used by other manufacturers.

Japan Radio Company Ltd.
Akasaka 2-chome,
Minato-ku, Tokyo
Japan

Japan Radio Company Ltd
New York Branch
430 Park Av.
New York, NY 10022

JRC *Japan Radio Co., Ltd*

NRD-1EL

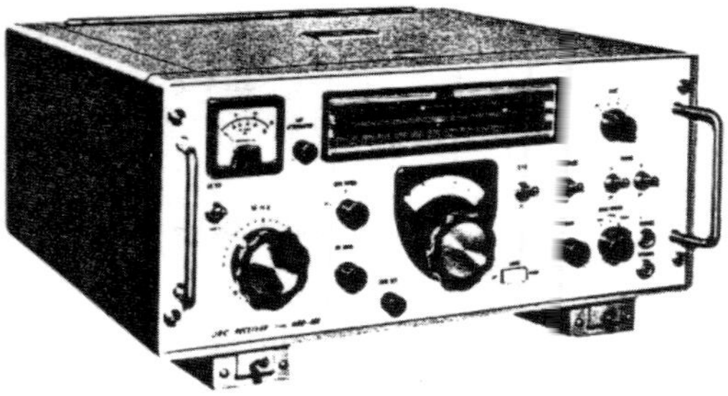

General Coverage Communications Receiver
Triple Conversion Superheterodyne. 20 Tubes plus Semiconductors.

Features:

• ¼" Head. Jack	• S/AF-Meter	• BFO	• AGC 4 POSITION
• RF Gain	• Mute	• Speaker Jack	• Calibrator 100 kHz
• Dial Lock	• Atten. Variable	• Rack Handles	• Modular Construction
• Zero Set	• IF Output	• Pretuner I/O	• Line Output

Specifications:
Coverage ...... 90 - 30000 kHz
Sensitivity ...... <2μV CW 2-30 MHz
Selectivity ...... 6/3/1 kHz -6dB.
Audio Out ...... 1 W 600 ohms
Modes .... AM/CW/MCW/LSB/USB
Image Reject .. >70 dB 2-13 MHz
Image Reject .. >50 dB 14-29 MHz

Accessories:
Cabinet (shown)

Comments:
The selection of any one MegaHertz band is accomplished via the large knob on the left. This selection is indicated on the top linear scale marked 0 to 29. A 0 to 10 "x100 kHz" scale, with ten division is below the MHz scale. The main tuning knob is scaled 0 to 99. Available in heavy cabinet (shown) or for rack mounting. Only double conversion above 7 MHz. This receiver requires, and was supplied with, an external SP-101 speaker. The NRD-1EL was designed for marine and commercial applications.

Variants:
Model **NRD-1EH** additionally includes a 0.5 kHz bandwidth.

Made In: Japan 965

Voltages: 90-120 VAC
50/60 Hz 120 VA

Readout: Analog Lin.

Physical: Cabinet Version:
19.25x9.5x15.7" 51 Lbs.
489x240x400mm 23 kg
Rack Version:
19x7.8x14.5" [illegible]1 Lbs.
489x199x370mm 19 kg

Status: Active Manufacturer
Discontinued Model

Rarity: Typically Unavailable

	New	Used
Price:	①	⑤
Rating:	⑥	⑥

JRC *Japan Radio Co., Ltd*

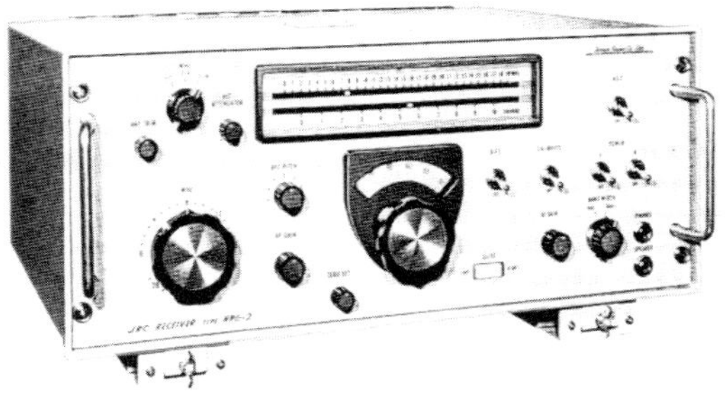

NRD-2

General Coverage Communications Receiver
Triple Conversion Superheterodyne. 15 Tubes plus Semiconductors.

Features:
• ¼" Head. Jack • BFO • Antenna Trimmer • AGC ON/OFF
• RF Gain • Mute • Speaker Jack • Calibrator 100 kHz
• Line Output • Atten. Variable • Rack Handles • Modular Construction
• Zero Set • IF Output • Pretuner I/O

Specifications:
Coverage ...... 90 - 30000 kHz
Sensitivity ...... <2μV CW 2-30 MHz
Selectivity ...... 6/1 kHz -6dB.
Audio Out ...... 1 W 600 ohms
Modes .... AM/CW/MCW/LSB/USB
Image Reject .. >70 dB 2-13 MHz
Image Reject .. >50 dB 14-29 MHz

Circuit Complement:
6BZ6 RF, 6BA6 Calibrator, 6U8A, 6U8A, 6AU6, 6BA6 1st VFO, 6BA6 2nd VFO, 6BA6 1st IF, 6BA6 2nd IF, 6U8A 3rd IF/AGC, 12AX7A AF, 6AQ5A AF, 6BA6 BFO, 12AU7A, VR150 plus diodes and transistors.

Accessories: Cabinet (shown) NXA-1532 Pretuner (90-2000 kHz).

Comments: The selection of any one MegaHertz band is accomplished via the large knob on the left. This selection is indicated on the top linear scale marked 0 to 29. A 0 to 10 "x100 kHz" scale, with ten division is below the MHz scale. The main tuning knob is scaled 0 to 99. Available in heavy cabinet (shown) or for rack mounting. Only double conversion above 7 MHz. This receiver requires, and was supplied with an external SP-101 speaker. The NRD-2 is a simpler version of the NRD-1 with a few less features.

Made In: Japan 1967

Voltages: 90-120 VAC
50/60 Hz 120 VA

Readout: Analog Lin.

Physical: Cabinet Version:
19.25x9.5x15.7" 64 Lbs.
489x240x400mm 29 kg
Rack Version:
19x7.8x14.5" 39.7 Lbs.
489x199x370mm 18 kg

Status: Active Manufacturer
Discontinued Model

Rarity: Typically Unavailable

	New	Used
Price:	①	⑤
Rating:	⑥	⑥

JRC *Japan Radio Co., Ltd*

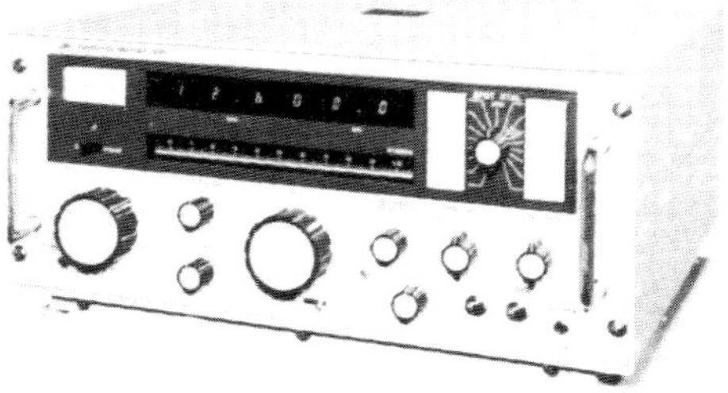

NRD-10

General Coverage Communications Receiver
Triple Conversion Superheterodyne. Solid State.

Features:
• ¼" Head. Jack • S-Meter • BFO ±150 Hz • AGC OFF/ON
• RF Gain • Mute • Speaker Jack • 16 Presets
• Dial Lock • Atten. -20 dB • Rack Handles • Modular Construction
• Analog Sub-Dial

Specifications:
Coverage ...... 100 - 30000 kHz
Image Reject . >70 dB .1-14 MHz
Image Reject . >50 dB 14-30 MHz
Modes .... AM/CW/MCW/LSB/USB
Selectivity .... 6/3/.5 kHz -6dB.

Circuit Complement:
28 Integrated Circuits, 37 Transistors and 57 Diodes.

Accessories:
Cabinet (shown) RTTY Demod Option

Comments:
Available in heavy cabinet (shown) or for rack mounting. Only double conversion on some frequencies. [¥180,000-250,000].

Made In: Japan 1974-1978

Voltages: 100/110/115/220/230 VAC
50/60 Hz 50W

Readout: 00000.1 Digital LED

Physical: Cabinet Version:
19.25x9.5x15.7" 44 Lbs.
489x240x400mm 20 kg
Rack Version:
19x7.8x14.5" 31 Lbs.
489x199x370mm 14 kg

Status: Active Manufacturer
Discontinued Model

Rarity: Typically Unavailable

	New	Used
Price:	①	$800
Rating:	⑥	⑥

JRC *Japan Radio Co., Ltd*

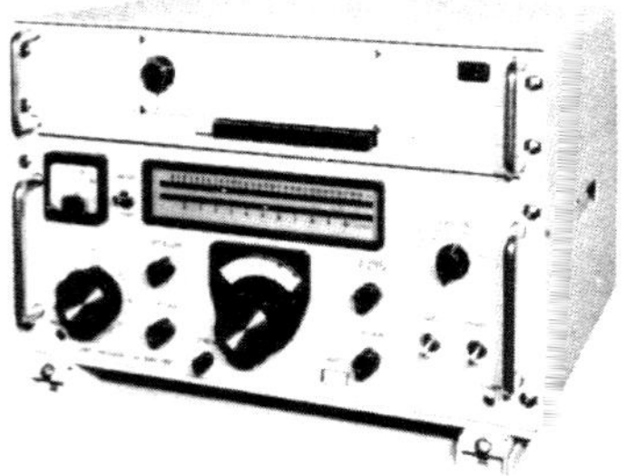

NRD-15K

General Coverage Communications Receiver
Triple Conversion Superheterodyne. Solid State.

Features:
- ¼" Head. Jack
- RF Gain
- Dial Lock
- BCB Reject
- S-Meter
- Mute
- Atten. -20 dB
- BFO
- Speaker Jack
- Rack Handles
- AGC OFF/ON
- 64 Memories
- Modular Construction

Specifications:
Coverage ...... 100 - 30000 kHz
Selectivity ...... 6/3/1/.5 kHz -6dB.
Modes .... AM/CW/MCW/LSB/USB
Environment ... -10° to +50°C 95% Humid

Circuit Complement:
7 FETs, 31 Transistors and 87 Diodes.

Accessories:
Cabinet (shown)
RTTY Demod Option

Comments:
Available in heavy cabinet (shown) or for rack mounting. Only double conversion on some frequencies. [¥250,000]

Variants:
Model **NRD-15J.**

Made In: Japan 197[illegible]-1977

Voltages: 90-120 VAC
50/60 Hz 115 VA

Readout: Analog Lin.

Physical: Cabinet Version:
19.25x13.4x16.4" 66.1 Lbs.
489x340x416mm 30 kg
Rack Version:
19x11.8x14.5" 44.1 Lbs.
480x300x370mm 20 kg

Status: Active Manufacturer
Discontinued Model

Rarity: Typically Unavailable

	New	Used
Price:	①	⑤
Rating:	⑥	⑥

JRC *Japan Radio Co., Ltd*

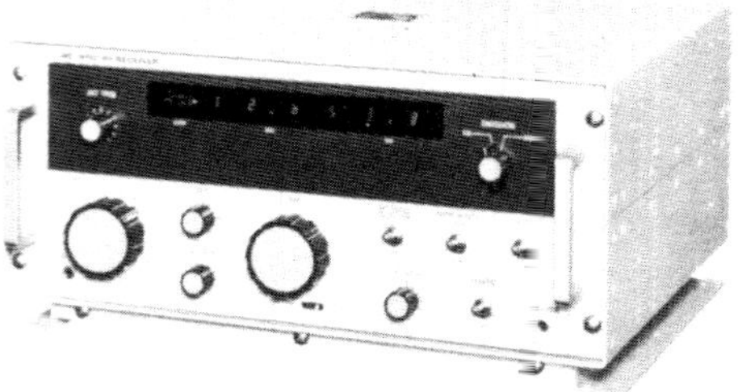

NRD-20

General Coverage Communications Receiver
Triple Conversion Superheterodyne. Solid State.

Features:
- ¼" Head. Jack
- RF Gain
- Dial Lock
- BFO
- Mute Line
- Atten. -20 dB
- AGC OFF/ON
- Speaker Jack
- Rack Handles
- Antenna Trimmer
- Modular Construction

Specifications:
Coverage ...... 100 - 30000 kHz
Modes .... AM/CW/MCW
Other specifications unknown.

Accessories:
Cabinet (shown)

Comments:
Available in heavy cabinet (shown) or for rack mounting. Only double conversion on some frequencies. [¥110,000].

Made In: Japan 1978

Voltages: 100/110/115/220/230 VAC
50/60 Hz 50W

Readout: 00000.1 Digital LED

Physical: Cabinet Version:
19.25x9.5x15.7" 37.5 Lbs.
489x240x400mm 17 kg
Rack Version:
19x7.8x14.5" 24.2 Lbs.
489x199x370mm 11 kg

Status: Active Manufacturer
Discontinued Model

Rarity: Typically Unavailable

	New	Used
Price:	①	⑤
Rating:	⑥	⑥

JRC *Japan Radio Co., Ltd*

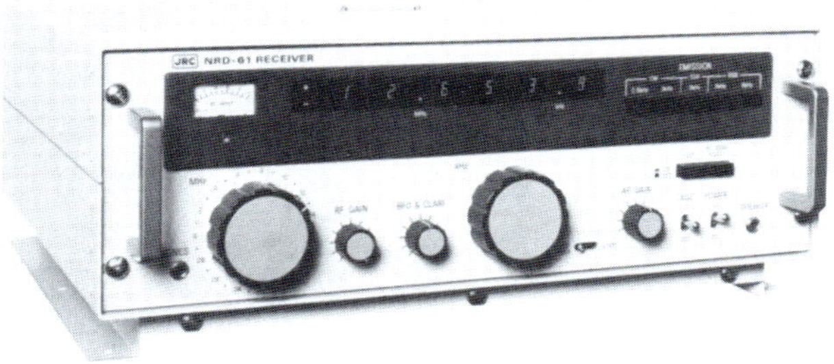

NRD-61

General Coverage Communications Receiver
Double Conversion Superheterodyne. Solid State.

Features:
- ¼" Head. Jack
- RF Gain
- Dial Lock
- Atten. -20 dB
- S-Meter
- Mute
- IF Out Jack
- Rack Handles
- BFO ±2.5 kHz
- Speaker Jack
- BCB Reject
- AGC OFF/ON
- Modular Construction
- Line Out 600 ohms

Specifications:
Coverage ...... 100 - 30000 kHz
Selectivity ...... 6/3/.5 kHz -6dB.
Sensitivity ...... <2μV CW .2-30 MHz
IF Reject .... >60 dB
Audio Out ...... 1 W 600 ohms
IF .... 70.455 MHz, 455 kHz
Modes .... AM/MCW/CW/SSB
Stability .... ±300 Hz/hr after warmup
Image Reject .. >60 dB
Environment ... -10° to +50°C 95% Humid
BFO Range .... ±2.5 kHz.
Antenna Input . SO-239

Accessories:
NDH-73 64 Ch. Memory   Cabinet (shown)

Comments:
Available in heavy cabinet (shown) or for rack mounting. Supplied with the external SP-101 speaker.

Variants:
Model **NRD-61A** 1981 has <2μV CW 1.6-30 MHz

Made In: Japan 1977-1979

Voltages: 100/110/220/230 VAC
50/60 Hz or 24 VDC

Readout: 00000.1 Digital LED

Physical: Cabinet Version:
19.25x7.5x15.7" 33 Lbs.
489x190x400mm 15 kg
Rack Version:
19x5.9x14.5" 20 Lbs.
480x149x370mm 9 kg

Status: Active Manufacturer
Discontinued Model

Rarity: Typically Unavailable

	New	Used
Price:	①	⑤
Rating:	⑥	⑥

JRC *Japan Radio Co., Ltd*

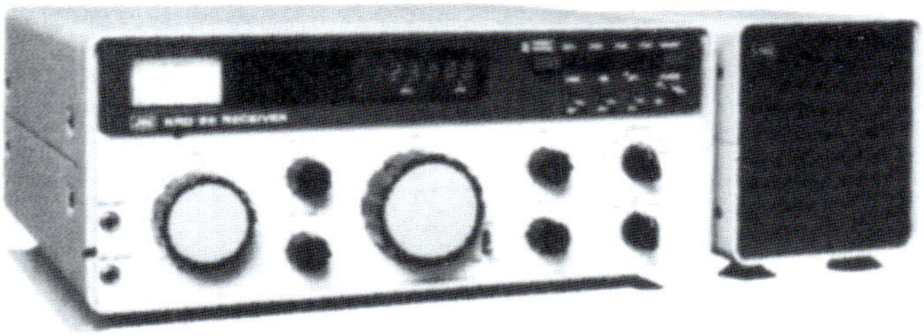

NRD-66

General Coverage Communications Receiver
Double Conversion Superheterodyne. Solid State.

Features:
- ¼" Head. Jack
- RF Gain
- Dial Lock
- Line Out Jack
- S-Meter
- Mute
- Atten. -20 dB
- BFO
- Speaker Jack
- 4 Channel Preset
- AGC
- Modular Construction

Specifications:
Coverage ...... 100 - 30000 kHz
Image Rej ...... >70 dB
Audio Out ...... 1 W 4 ohms
Other specifications unavailable.
Modes .... AM/LSB/USB/CW/RTTY
IF .... 70.455 MHz, 455 kHz.
Antenna Input . SO-239

Comments:
The NRD-66 seems to be the evolutionary missing-link between the NRD-505 and the NRD-515. It was sold strictly to the marine market, and was not marketed in North America. Supplied with an external speaker (shown).

Made In: Japan 1981-1982

Voltages: 100/110/115/220/230 VAC
50/60 Hz 60W or 24 VDC

Readout: 00000.1 Digital LED

Physical: 13.4x6.3x11.8" 24.2 Lbs.
340x160x300mm 11 kg

Status: Active Manufacturer
Discontinued Model

Rarity: Typically Unavailable

	New	Used
Price:	①	⑤
Rating:	⑥	⑥

JRC *Japan Radio Co., Ltd*

NRD-71

General Coverage Communications Receiver
Double Conversion Superheterodyne. Solid State.

Features:
- ¼" Head. Jack
- RF Gain
- Dial Lock
- S-Meter
- Mute
- Atten. 20 dB
- AGC OFF/ON
- Speaker Jack
- Rack Handles
- Antenna Trimmer
- BFO ±2.5 kHz
- Modular Construction

Specifications:
Coverage ...... 100 - 30000 kHz
Sensitivity ...... <2µV CW 1.6-30 MHz
Selectivity ...... 6/3/1/.5 kHz -6dB.
IF .... 70.455 MHz, 455 kHz

Modes .... AM/CW/MCW
Environment ... -10° to +50°C 95% Humd.
Image Rej. ...... >80 dB

Accessories:
Cabinet (shown)   NDH-71 Preset Option (shown)   RTTY Demod Option

Comments:
Available in heavy cabinet (shown) or for rack mounting. The optional NDH-71 Memory Option (shown) stores 64 channels. [¥220,000-300,000].

Variants:
The model NRD-70 (1972) has a similar appearance and the same frequency coverage.

Made In:	Japan	1975-1978
Voltages:	100/110/115/220/230 VAC 50/60 Hz 85W	
Readout:	00000.1 Digital LED	
Physical:	Cabinet Version: 19.25x9.5x15.7" 46.3 Lbs. 489x240x400mm 21 kg Rack Version: 19x7.8x14.5" 33 Lbs. 489x199x370mm 15 kg	
Status:	Active Manufacturer Discontinued Model	
Rarity:	Typically Unavailable	

	New	Used
Price:	①	$200
Rating:	⑥	⑥

JRC *Japan Radio Co., Ltd*

NRD-72

General Coverage Communications Receiver
Double Conversion Superheterodyne. Solid State.

Features:
- ¼" Head. Jack
- RF Gain
- Dial Lock
- Atten. 20 dB
- S-Meter
- Mute
- IF Out Jack
- AGC OFF/ON
- BFO ±2.5 kHz
- Speaker Jack
- UP/DN Tuning
- Rack Handles
- Line Out 600 ohm
- 2182 kHz Preset
- Two Tuning Speeds
- Modular Construction

Specifications:
Coverage ...... 100 - 30000 kHz
Selectivity ...... 6/3/.5 kHz -6dB.
Sensitivity ...... <2µV CW 1.6-30 MHz
IF Reject .... >60 dB
Audio Out ...... 1 W 600 ohms
IF .... 70.455 MHz, 455 kHz

Modes .... AM/MCW/CW/USB/LSB
Stability .... ±5 Hz for any 15 mins.
Image Reject .. >70 dB
Environment ... -10° to +50°C 95% Humd.
BFO Range .... ±2.5 kHz.
Ant. Jack .... 75 ohm

Accessories:
NDH-73 64 Ch. Memory   Cabinet (shown)

Comments:
Available in heavy cabinet (shown) or for rack mounting. Supplied with the external SP-101 speaker. [¥300,000-350,000]

Made In:	Japan	1978-1981
Voltages:	100/110/115/220/230 VAC 50/60 Hz or 24 VDC	
Readout:	00000.1 Digital LED	
Physical:	Cabinet Version: 19.25x7.5x15.7" 37.5 Lbs. 489x191x400mm 17 kg Rack Version: 19x5.9x14.5" 23.1 Lbs. 480x149x370mm 10.5 kg	
Status:	Active Manufacturer Discontinued Model	
Rarity:	Typically Unavailable	

	New	Used
Price:	①	$2600
Rating:	⑥	⑥

JRC **Japan Radio Co., Ltd**

NRD-73

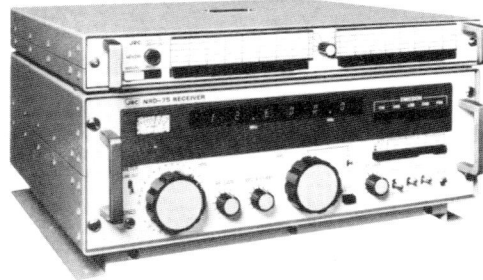

General Coverage Communications Receiver
Double Conversion Superheterodyne. Solid State.

Features:
- ¼" Head. Jack
- RF Gain
- Dial Lock
- Atten. -20 dB
- S-Meter
- Mute
- IF Out Jack
- AGC OFF/ON
- BFO ±2.5 kHz
- Speaker Jack
- UP/DN Tuning
- Rack Handles
- Line Out 600 ohms
- 64 Memories
- 2182 kHz Preset
- Modular Construction

Specifications:
Coverage ...... 100 - 30000 kHz
Selectivity ...... 6/3/.5 kHz -6dB.
Sensitivity ...... <2µV CW 1.6-30 MHz
IF Reject ....>60 dB
Audio Out ...... 1 W 600 ohms
IF ....70.455 MHz, 455 kHz

Modes .... AM/MCW/CW/USB/LSB
Stability .... ±5 Hz for any 15 mins.
Image Reject .. >70 dB
BFO Range .... ±2.5 kHz.
Environment ... -10° to +50°C 95% Humd.
Antenna Input . SO-239 75 ohm

Accessories:
NMB-101 FSK Demod
Cabinet (shown)
NDH-73 64 Ch. Memory (shown)
NDH-76 80 Ch. Scan. Memory

Comments:
Available in heavy cabinet (shown) or for rack mounting. Shown above with optional NDH-73 Memory Unit. Supplied with external SP-101 speaker. [¥350,000-400,000].

Made In:	Japan	1978-1982

Voltages: 100/110/115/220/230 VAC
50/60 Hz or 24 VDC

Readout: `00000.1` Digital LED

Physical: Cabinet Version:
19.25x7.5x15.7" 37.5 Lbs.
489x191x400mm 17 kg
Rack Version:
19x5.9x14.5" 24.2 Lbs.
489x149x370mm 11 kg

Status: Active Manufacturer
Discontinued Model
Rarity: Typically Unavailable

	New	Used
Price:	①	$3200
Rating:	⑥	⑥

JRC **Japan Radio Co., Ltd**

NRD-75

General Coverage Communications Receiver
Double Conversion Superheterodyne. Solid State.

Features:
- ¼" Head. Jack
- RF Gain
- Dial Lock
- Atten. -20 dB
- S/AF-Meter
- Mute
- IF Out Jack
- AGC OFF/ON
- BFO ±2.5 kHz
- Speaker Jack
- UP/DN Tuning
- Rack Handles
- Modular Construction
- 64 Memories
- Line Out 600 ohms

Specifications:
Coverage ...... 100 - 30000 kHz
Selectivity ...... 6/3/1/.3/_ kHz -6dB.
Sensitivity ...... <2µV CW 1.6-30 MHz
Audio Cut ...... 1 W 600 ohms
IF Reject ....>60 dB
IF ....70.455 MHz, 455 kHz

Modes .... AM/MCW/CW/USB/LSB
Stability .... ±5 Hz for any 15 mins.
Environment ... -10° to +50°C 95% Humd.
BFO Range .... ±2.5 kHz.
Image Reject .. >70 dB
Antenna Input . 75 ohm

Accessories:
NDH-73 64 Ch. Memory (shown)
Cabinet (shown)
NDH-76 80 Ch. Scan. Memory
NMB-101 FSK Demod

Comments:
Available in heavy cabinet (shown) or for rack mounting. Shown above with optional NDH-73 Memory Unit. Supplied with external SP-101 speaker. The NRD-75 is similar to the NRD-73, but includes four filter positions instead of three and the meter reads both Signal Strength or AF Output. [¥450,000 with NDH-76].

Made In:	Japan	1978-1981

Voltages: 100/110/115/220/230 VAC
50/60 Hz or 24 VDC

Readout: `00000.1` Digital LED

Physical: Cabinet Version:
19.25x7.5x15.7" 40 Lbs.
489x191x400mm 18 kg
Rack Version:
19x5.9x14.5" 26.5 Lbs.
483x149x370mm 12 kg

Status: Active Manufacturer
Discontinued Model
Rarity: Typically Unavailabl

	New	Used
Price:	①	$4100
Rating:	⑥	⑥

JRC Japan Radio Co., Ltd

NRD-91

General Coverage Communications Receiver
Double Conversion Superheterodyne. Solid State.

Features:
- ¼" Head. Jack
- RF Gain
- Mute
- Dial Lock
- Atten. -20 dB
- S/AF-Meter
- BCB Reject
- Dimmer (var)
- IF Out Jack
- Clarifier
- NB
- BFO
- Rack Handles
- UP/DN Tuning
- AGC OFF/SLO/FST
- Speaker Jack
- Line Out 600 ohms
- Modular Construction

Specifications:
Coverage ......90 - 30000 kHz
Selectivity ......6/3/.5 kHz -6dB.
Sensitivity ......<2µV CW 1.6-30 MHz
Image Rej ......>60 dB
Audio Out ......1 W 4 ohms
IF ....70.455 MHz, 455 kHz

Modes ....AM/SSB/CW/[AUX]
Stability ....±10x10⁻⁶ per hour
IF Rejection .... >60 dB
Environment ... -10° to +50°C 95% Humd.
BFO Range .... ±2 kHz.
Antenna Input . SO-239 50-75 ohm

Accessories:
Cabinet (shown)   FSK/FAX Crystal Osc.

Comments:
Available in heavy cabinet (shown) or for rack mounting. Supplied with external NVA-92 speaker. Extremely smooth 10 Hz step main tuning (10 kHz/rev.). [¥650,000-700,000].

Made In: Japan   1983-1995

Voltages: 100/110/200/220 VAC
50/60 Hz or 24 VDC

Readout: **00000.1** Digital LED

Physical: Cabinet Version:
19.25x7.5x12" 25 Lbs.
489x191x305mm 11.5 kg
Rack Version:
19x5.9x11.6" 15.5 Lbs.
480x149x294mm 7 kg

Status: Active Manufacturer
Discontinued Model
Rarity: Extremely Scarce

	New	Used
Price:	①	$2400
Rating:	⑥	⑥

JRC Japan Radio Co., Ltd

NRD-92

General Coverage Communications Receiver
Double Conversion Superheterodyne. Solid State.

Features:
- ¼" Head. Jack
- RF Gain
- Mute
- Dial Lock
- Atten. -20 dB
- S/AF-Meter
- AF Filter
- Dimmer (Var.)
- IF Out Jack
- Speaker Jack
- NB
- BFO
- Rack Handles
- UP/DN Tuning
- Clarifier ±120 Hz
- AGC OFF/SLO/FST
- Line Out 600 ohm
- Preset 500, 2182 kHz
- Modular Construction

Specifications:
Coverage ......90 - 30000 kHz
Selectivity ......6/3/.5/_ kHz -6dB.
Sensitivity ......<2µV CW 1.6-30 MHz
Image Rej ......>60 dB
Audio Out ......1 W 4 ohms
IF ....70.455 MHz, 455 kHz

Modes ....AM/SSB/CW/FAX/RTTY
Stability ....±5 Hz for any 15 mins.
IF Rejection .... >70 dB
Environment ... -10° to +50°C 95% Humd.
BFO Range .... ±2 kHz.
Antenna Input . SO-239 50-75 ohm

Accessories:
Cabinet (shown)   NDH-93 Scanning Mem.   NDH-95 Memory Timer
YF-455CB 200Hz Filter   YF-455FM 300 Hz Filter   YF-455DE 1 kHz Filter

Comments: Available in heavy cabinet (shown) or for rack mounting. Supplied with external NVA-92 speaker. Extremely smooth 10 Hz step main tuning (10 kHz/rev.). The Clarifier tunes in 1 Hz steps in all modes except CW and AM. [¥600,000 with NDH-93].

Variants: Model **NRD-92M** was a special type built to comply with the U.K. Home Office and FTZ German specifications (includes LSB/USB rather than just SSB).

Made In: Japan   1983-1995

Voltages: 100/110/115/220/230 VAC
50/60 Hz or 24 VDC

Readout: **00000.01** Digital LED

Physical: Cabinet Version:
19.25x7.5x13.4 33 Lbs.
489x190x340mm 15 kg
Rack Version:
19x5.86x13" 23 Lbs.
480x149x332mm 10.5 kg

Status: Active Manufacturer
Discontinued Model
Rarity: Very Scarce

	New	Used
Price:	$3600-3995	$2500-2700
Rating:	★★★★★	★★★★★

JRC *Japan Radio Co., Ltd*

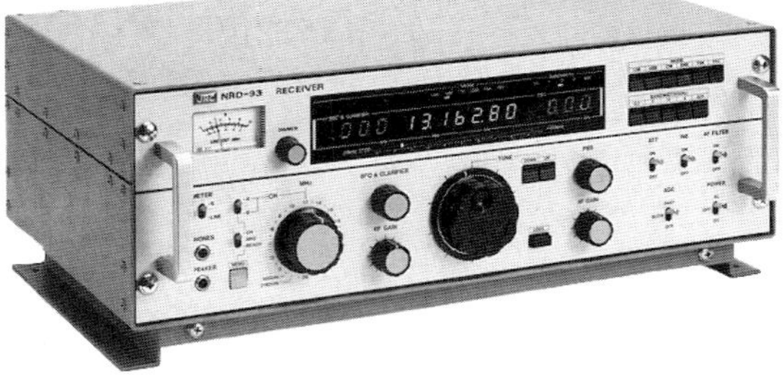

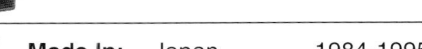

NRD-93

General Coverage Communications Receiver
Double Conversion Superheterodyne. Solid State.

Features:
- ¼" Head. Jack
- RF Gain
- Mute
- Dial Lock
- Atten. 20 dB
- Speaker Jack
- S/AF-Meter
- AF Filter
- Dimmer (Var.)
- IF Out Jack
- 60 Memories
- Analog Pointer Dial
- NB
- BFO ±2 kHz.
- PBS
- UP/DN Tuning
- Rack Handles
- AGC OFF/SLO/FST
- Line Out 600 ohm
- Preset 500, 2182 kHz
- Clarifier ±120 Hz
- Modular Construction

Specifications:
Coverage ...... 90 - 30000 kHz
Selectivity ...... 6/3/1/.3/_ kHz -6dB.
Sensitivity ...... <2μV CW 1.6-30 MHz
Image Rej ...... >70 dB
Audio Out ...... 1 W 4 ohms
IF .... 70.455 MHz, 455 kHz
Modes .... AM/SSB/CW/FAX/RTTY
Stability .... ±2 Hz for any 15 mins.
IF Rejection .... >80 dB
Environment ... -10° to +50°C 95% Humd.
Antenna Input . SO239 50-75 ohm

Accessories:
Cabinet (shown)
NDH-95 Memory Timer
NDH-93 Scanning Memory
YF-455CB 200Hz Filter

Comments:
Available in heavy cabinet (shown) or for rack mounting. Supplied with external NVA-92 speaker. Similar to the NRD-92 but with: one more bandwidth, 60 internal memories, PBT and slightly better rejection and stability specifications.

Made In: Japan 1984-1995

Voltages: 100/110/115/220/230 VAC 50/60 Hz or 24 VDC

Readout: 00000.01 Digital LED

Physical: 19.25x5.8x12" 33 Lbs. 489x149x305mm 15 kg

Status: Active Manufacturer Discontinued Model

Rarity: Very Scarce
Reviews: 📖 *Passport* 1985/86-1995
Mon. Times July 1995
Proceedings 1994-95

	New	Used
Price:	$5900-8000	$3500-5100
Rating:	★★★★★	★★★★★

JRC *Japan Radio Co., Ltd*

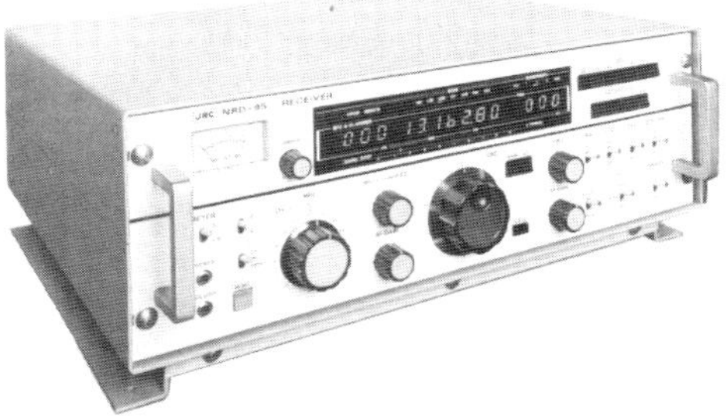

NRD-95

General Coverage Communications Receiver
Double Conversion Superheterodyne. Solid State.

Features:
- ¼" Head. Jack
- RF Gain
- Mute
- Dial Lock
- Atten. 20 dB
- Speaker Jack
- S/AF-Meter
- AF Filter
- Dimmer (Var.)
- IF Out Jack
- 60 Memories
- Analog Pointer Dial
- NB
- BFO ±2 kHz.
- PBS
- UP/DN Tuning
- Rack Handles
- AGC OFF/SLO/FST
- Line Out 600 ohm
- Preset 500, 2182 kHz
- Clarifier ±120 Hz
- Modular Construction

Specifications:
Coverage ...... 90 - 30000 kHz
Selectivity ...... 6/3/1/.3/_ kHz -6dB.
Sensitivity ...... <2μV CW 1.6-30 MHz
Image Rej ...... >70 dB
Audio Out ...... 1 W 4 ohms
IF .... 70.455 MHz, 455 kHz
Modes .... AM/SSB/CW/FAX/RTTY
Stability .... ±2 Hz for any 15 mins.
IF Rejection .... >80 dB
Environment ... -10° to +50°C 95% Humd.
Antenna Input . SO-239 50-75 ohm

Accessories: NDH-95 Memory Timer
ISB Option NCG-95 Telecontroller
Cabinet (shown) NDH-93 Scan. Mem.

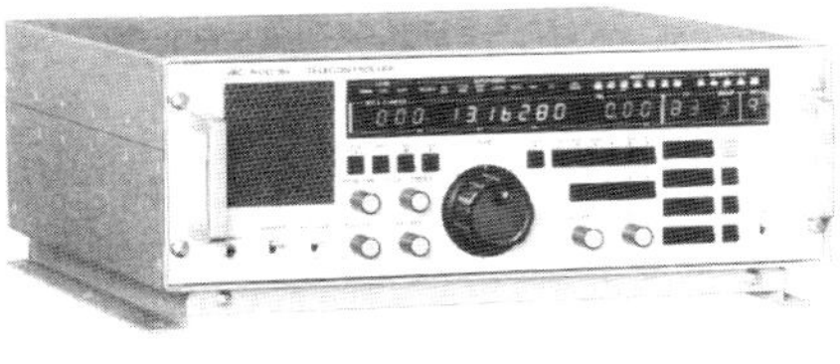

Comments:
Available with optional cabinet (shown) Supplied with external NVA-92 speaker. Similar to the NRD-93 but with option for ISB and for optional telecontrol via the NCG-95 Telecontroller (shown above).

Made In: Japan 1983-1990

Voltages: 100/110/115/220/230 VAC 50/60 Hz or 24 VDC

Readout: 00000.01 Digital LED

Physical: Cabinet Version:
19.25x7.5x15.7" 40 Lbs.
489x191x400mm 18 kg
Rack Version:
19x5.9x14.5" 26.5 Lbs.
489x149x370mm 12 kg

Status: Active Manufacturer Discontinued Model

Rarity: Typically Unavailable

	New	Used
Price:	[1]	[5]
Rating:	[6]	[6]

JRC *Japan Radio Co., Ltd*

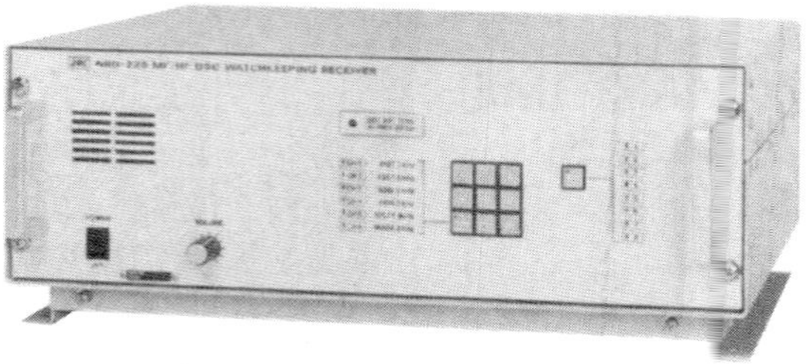

NRD-220

Fixed Frequency Communications Receiver
Triple Conversion Superheterodyne. Solid State.

Features:
- ¼" Head. Jack
- Speaker
- BFO
- Speaker Jack
- AGC OFF/ON
- Rack Handles
- B.I.T.E.

Specifications:
Coverage ...... Fixed. See Comments
Stability .... ±10 Hz
Audio Out ...... 1 W 4 ohms
Other specifications unknown.
Modes .... GMDSS RTTY

Comments:
This is a fixed channel receiver designed for the marine market. It receives the teletype DSC GMDSS distress/calling frequencies: 2187.5, 4207.5, 6312, 8414.5, 12577 and 16804.5 kHz.

Variants:
JRC also offers the smaller model **NRC-210** that covers only 2187.5 kHz.

Made In: Japan 1993 1996

Voltages: 100/110/1115/220/230 VAC 50/60 Hz 100W

Readout: None

Physical: Cabinet Version:
19.25x7.51x12" 30.3 Lbs.
489x191x305mm 14 kg
Rack Version:
19x5.86x12" 20.9 Lbs.
480x149x290mm 9.5 kg

Status: Active Manufacturer
Active Model

Rarity: Typically Unavailable

	New	Used
Price:	①	⑤
Rating:	⑥	⑥

JRC *Japan Radio Co., Ltd*

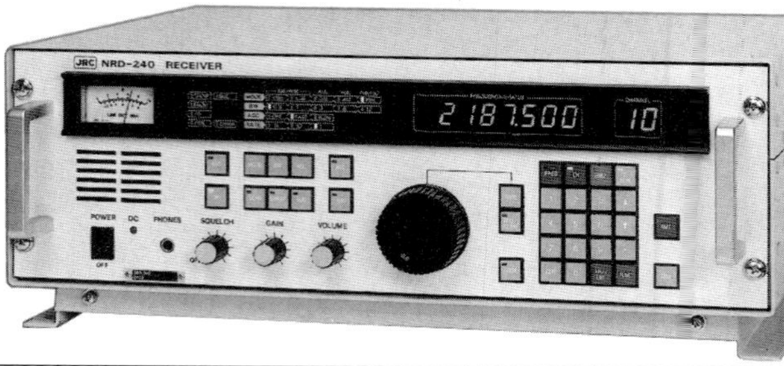

NRD-240

General Coverage Communications Receiver
Double Conversion Superheterodyne. Solid State.

Features:
- ¼" Head. Jack
- RF Gain
- Mute
- Dial Lock
- Atten. -20 dB
- Speaker
- 3 Tuning Rates
- S/AF-Meter
- BFO ±10 kHz
- Dimmer
- IF Out Jack
- Rack Handles
- B.I.T.E.
- Dial Lock
- NB
- Scan/Sweep
- PBS
- UP/DN Tuning
- 100 Memories
- RS-423A Control
- Squelch
- AGC OFF/SLO/FST
- Line Out 600 ohms
- Preset 2182 kHz
- Clarifier ±200 Hz
- Modular Construction
- GMDSS Certif.

Specifications:
Coverage ...... 90 - 30000 kHz
Selectivity ...... 6/3/1/.3/_ kHz -6dB.
Sensitivity ...... <2μV CW 1.6-30 MHz
Image Rej ...... >70 dB
Audio Out ...... 1 W 8 ohms
IF .... 70.455 MHz, 455 kHz
Modes .... AM/LSB/USB/CW/FAX/RTTY
Stability .... 0.3 PPM
IF Rejection .... >80 dB
BFO Range .... ±2 kHz.
Environment ... -15° to +55°C 93% Humd.
Antenna Input . SO-239 50 ohm

Accessories:
Cabinet (shown) NVA-92L External Speaker NDH-95 Memory Timer

Comments:
Available in heavy cabinet (shown) or for rack mounting. Operates on AC or DC supply by automatic switching. Memories store: frequency, mode, bandwidth, attenuator and AGC settings. The NRD-240 is designed in compliance with GMDSS carriage requirements.

Made In: Japan 1990-1997

Voltages: 100/110/115/220/230 VAC 50/60 Hz or 24 VDC 50W

Readout: 00000.001 Digital LED

Physical: Cabinet Version:
19.25x7.5x12" 34.1 Lbs.
489x191x305mm 15.5 kg
Rack Version:
19x5.8x11.4" 24.2 Lbs
480x149x290mm 11 kg

Status: Active Manufacturer
Discontinued Model

Rarity: Extremely Scarce

Reviews: 📖 *Passport* 1991

	New	Used
Price:	$8000	$4000-4800
Rating:	★★★★★	★★★★★

JRC *Japan Radio Co., Ltd*

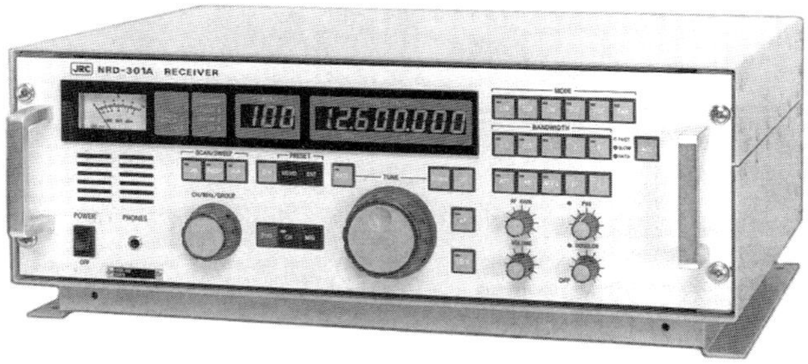

NRD-301A

General Coverage Communications Receiver
Double Conversion Superheterodyne. Solid State.

Features:
- ¼" Head. Jack
- RF Gain
- Mute
- Dial Lock
- Atten. -20 dB
- Speaker
- 8 Tuning Rates
- S/AF-Meter
- BFO ±10 kHz
- Dimmer
- IF Out Jack
- Rack Handles
- B.I.T.E.
- Dial Lock
- NB
- Scan/Sweep
- PBS
- UP/DN Tuning
- 300 Memories
- RS-422/423A
- Squelch
- AGC DATA/SLO/FST
- Line Out 600 ohms
- Preset 2182 kHz
- Clarifier ±200 Hz
- Modular Construction

Specifications:
Coverage ...... 90 - 30000 kHz
Selectivity ...... 6/3/.5/_/_ kHz -6dB.
Sensitivity ...... <2μV CW 1.6-30 MHz
Image Rej ...... >70 dB
Audio Out ...... 1 W 8 ohms
IF .... 70.455 MHz, 455 kHz
Modes .... AM/LSB/USB/CW/MCW/FAX/FSK
Stability .... ±20 Hz <13 MHz
Stability .... ±50 Hz >13 MHz
IF Reject .... >80 dB
Environment ... -10° to +50°C 93% Humd.
Antenna Input . SO239 50 ohm

Accessories:
Cabinet (shown) NVA-92L External Speaker NDH-95 Memory Timer
0.3 kHz Filter 1.0 kHz Filter

Comments:
Available in heavy cabinet (shown) or for rack mounting. Operates on AC or DC supply by automatic switching. Memories store: frequency, mode, bandwidth, attenuator and AGC settings.

Made In: Japan 1996-1998

Voltages: 100/110/115/220/230 VAC
50/60 Hz or 24 VDC 50W

Readout: 00000.001 Digital LED

Physical: Cabinet Version:
19.25x7.5x12" 37 Lbs.
489x191x305mm 17 kg
Rack Version:
19x5.8x11.4" 16.5 Lbs
480x149x290mm 7.5 kg

Status: Active Manufacturer
Active Model
Rarity: Extremely Scarce

	New	Used
Price:	$8600	[5]
Rating:	★★★★★	★★★★★

JRC *Japan Radio Co., Ltd*

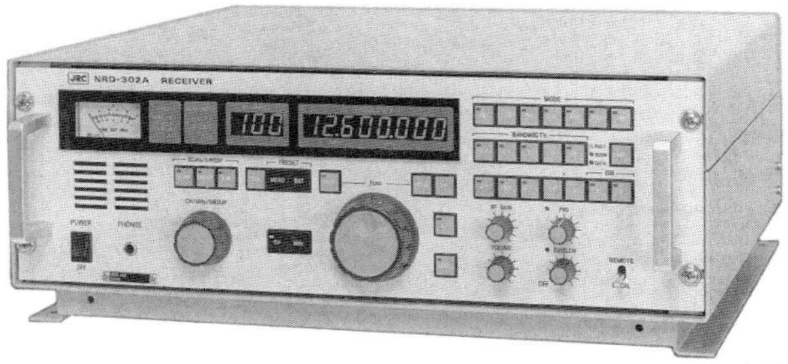

NRD-302A

General Coverage Communications Receiver
Double Conversion Superheterodyne. Solid State.

Features:
- ¼" Head. Jack
- RF Gain
- Mute
- Dial Lock
- Atten. -20 dB
- Speaker
- 8 Tuning Rates
- S/AF-Meter
- BFO ±10 kHz
- Dimmer
- IF Out Jack
- Rack Handles
- B.I.T.E.
- Dial Lock
- NB
- Scan/Sweep
- PBS
- UP/DN Tuning
- 300 Memories
- RS-422/423A
- Squelch
- AGC DATA/SLO/FST
- Line Out 600 ohms
- AF Filter
- Clarifier ±200 Hz
- Modular Construction
- GMDSS Certified
- Remote/Local Switch

Specifications:
Coverage ...... 90 - 30000 kHz
Selectivity ...... 6/3/1/.3/_ kHz -6dB.
Sensitivity ...... <2μV CW 1.6-30 MHz
Image Rej ...... >70 dB
Audio Out ...... 1 W 8 ohms
IF .... 70.455 MHz, 455 kHz
Modes .... AM/LSB/USB/CW/MCW/FAX/FSK
Stability .... ±20 Hz <13 MHz
Stability .... ±50 Hz >13 MHz
IF Reject .... >80 dB
Environment ... -15° to +55°C 93% Humd.
Antenna Input . SO239 50 ohm

Accessories:
Cabinet (shown) NVA-92L External Speaker NDH-95 Memory Timer
0.5 kHz Filter CND-88 ISB Two Channel CND-69 ISB Four Channel

Comments:
Available in heavy cabinet (shown) or for rack mounting. Operates on AC or DC supply by automatic switching. Memories store: frequency, mode, bandwidth, attenuator and AGC settings. The flagship of the J.R.C. receiver line.

Made In: Japan 1996-1998

Voltages: 100/110/115/220/230 VAC
50/60 Hz or 24 VDC 50W

Readout: 00000.001 Digital LED

Physical: Cabinet Version:
19.25x7.5x12" 37 Lbs.
489x191x305mm 17 kg
Rack Version:
19x5.8x11.4" 16.5 Lbs
480x149x290mm 7.5 kg

Status: Active Manufacturer
Active Model
Rarity: Typically Unavailable

	New	Used
Price:	$10750	[5]
Rating:	★★★★★	★★★★★

JRC *Japan Radio Co., Ltd*

NRD-345

General Coverage Communications Receiver
Double Conversion Superheterodyne. Solid State.

Features:
- ¼" Head. Jack
- 100 Memories
- Record Jack
- UP/DN Tune
- Scan & Sweep
- S-Meter
- BFO
- Ext. Spkr Jack
- Keypad
- Dial Lock
- Speaker
- Tone
- FAX Jack[1]
- RS-232 Jack
- NB
- Synchronous Det.
- 24 Hr. Clock/Timer
- AGC OFF/FST/SLO
- Attenuator -20 dB
- Dual VFOs

Specifications:

Coverage ...... 100 - 30000 kHz	Modes .... AM/LSB/USB/CW/FAX
Selectivity ...... 4/2/_ kHz -6dB.	IFs .... 44.85 MHz, 455 kHz
Sensitivity ...... <.3µV 1.8-30 MHz SSB	IF Rejection .... >70 dB 1.6-30 MHz
Image Rej ...... >70 dB 1.6-30 MHz	Dyn. Range .... >100 dB
Audio Out ...... 1 W 8 ohms	Antenna Input . SO239 and Terminals

Accessories:
CFL-231 300Hz Filter   CFL-232 500Hz Filter   CFL-233 1000Hz Filter
CFL-251 2400 Filter   CFL-218A 1800Hz Filter   CFQ-8673 Aux. Filter Board
6ZCJD00350 RS232 Cable

Comments: The VFO tuning resolution is 5 Hz. The memories store: frequency, mode, bandwidth, AGC, VFO, NB and ATT settings. The Lock button disables the manual tuning knob only. The built-in RS-232 jack is 4800 baud, 1 start bit, 8 data bits, 0 parity bits and 1 stop bit. The Pass button is used for memory channel lockouts. During memory scanning the pause duration may be set for 2 or 5 seconds. The RS-232 port is a DB-25. [1]The FAX Jack is simply a line out jack.

Made In:	Japan	1997-1998
Voltages:	12 VDC ±10% .8 Amp	
	Supplied in US with AC PS	
Readout:	`00000.01` LCD	
Physical:	9.9x4x9.4" 7.8 Lbs	
	250x100x238mm 3.5 kg	
Status:	Active Manufacturer	
	Active Model	
Rarity:	Too new.	
Reviews:	📖 *Passport* 1998	
	Shortwave Mag. May 1997	
	Monitoring Times Nov. 1997	
	WRTH 1998	

	New	Used
Price:	$799-899	$500-550
Rating:	★★★★★	★★★★★

JRC *Japan Radio Co., Ltd*

NRD-505

General Coverage Communications Receiver
Double Conversion Superheterodyne. Solid State.

Features:
- ¼" Head. Jack
- RF Gain
- Mute Jack
- Sidetone Jack
- Modular Design
- S-Meter
- BFO ±2.5 kHz
- NB
- Speaker Jack
- Two Tuning Speeds
- Analog Subdial
- RIT ±2.5 kHz
- IF Out Jack 455
- Attenuator -20 dB
- AGC OFF/FST/SLO
- Line Out 600 ohms
- VFO Out Jack
- Analog sub-dial

Specifications:

Coverage ...... 100 - 30000 kHz	Modes .... AM/LSB/USB/CW/RTTY
Selectivity ...... 4.4/2/_ kHz -6dB.	Stability .... <100Hz/hr. after warm-up
Sensitivity ...... <.5µV 1.6-30 MHz SSB	IF Rejection .... >70 dB
Image Rej ...... >70 dB	IF .... 70.455 MHz, 455 kHz.
Audio Out ...... 1 W 4 ohms	Antenna Input . SO239 50-75 ohm

Circuit Complement:
66 Integrated circuits, 18 FETs, 54 transistors and 103 diodes.

Accessories:
NVA-505 Speaker   CD4-8 4 Channel Preset   CW Filter 600 Hz.
CGA-26 Transceive VFO Converter

Comments:
Built to the highest construction standards. Matches the NSD-500 Transmitter. The optional CD48 four channel memory option mounts inside the NRD-505. This four channel memory stores frequency only. The memory channels are selected by push switches. Requires a speaker. Reportedly less than 1000 made, very collectable.

Made In:	Japan	1977-1979
Voltages:	100/115/200/230 VAC	
	50/60 Hz 50W	
Readout:	`00000.1` Digital LED	
Physical:	13.4x5.5x11.8" 22 Lbs.	
	340x140x300mm 10 kg	
Status:	Active Manufacturer	
	Discontinued Model	
Rarity:	Very Scarce	
Reviews:	*WRTH* 1980	

	New	Used
Price:	$2250-2275	$800-880
Rating:	★★★★★	★★★★★

 **Japan Radio Co., Ltd**

NRD-515

General Coverage Communications Receiver
Double Conversion Superheterodyne. Solid State.

Features:
- ¼" Head. Jack
- RF Gain
- Mute Jack
- Ext. Spkr Jack
- S-Meter
- BFO ±2 kHz
- RIT ±3 kHz
- Monitor
- PBT ±2 kHz
- NB
- Line Out Jack
- Dial Lock
- AGC OFF/FST/SLO
- UP/DN Tuning
- BC (MW) Tune
- Attenuator -10/20 dB

Specifications:
Coverage ...... 100 - 34000 kHz
Selectivity ...... 6/2.4/_/_ kHz -6dB.
Sensitivity ...... <.5µV 1.6-30 MHz SSB
Image Rej ...... >70 dB
Audio Out ...... 1 W 4 ohms

Modes .... AM/LSB/USB/CW/RTTY
Stability .... <50Hz/hour after warmup
IF Rejection .... >70 dB
IF .... 70.455 MHz, 455 kHz.
Antenna Input . SO239

Accessories:
CFL-230 300Hz Filter   CFL-260 600Hz Filter   NVA-515 Speaker
NCM-515 Keypad   NDH-515 Memory 24 Ch.   NDH-518 Memory 96 Ch.

Comments: Robustly built and straight-forward to operate, this receiver remains popular with utility DXers, tropical band enthusiast and general shortwave listeners. The '515' has a nearly cult-like following. It was the first JRC priced to enter the consumer market and enjoyed wide popularity. Matches the NSD-515 transmitter. The BC Tune function (on the BFO knob) serves as a manual preselector on the MW band. Requires a speaker. The optional external memory units store only frequency. Clean lines, easy operation, a solid feel and outstanding performance come together in this radio. Excellent for radioteletype, fax and Morse code. See *Chapter 107*.

Made In:	Japan	1979-1986
Voltages:	100/117/220/240 VAC 50/60 Hz 50W	
Readout:	**00000.1** Digital LED	
Physical:	13.4x5.5x11.8" 16.5 Lbs. 340x140x300mm 7.5 kg	
Status:	Active Manufacturer Discontinued Model	
Rarity:	Common	
Reviews:	📖*Passport* 1985/86, 1987 *WRTH* 1982 *QST* November 1981 *Radio Elec. World* May 1982 *Proceedings* 1988	

	New	Used
Price:	$900-1490	$650-680
Rating:	★★★★★	★★★★★

 **Japan Radio Co., Ltd**

NRD-525

General Coverage Communications Receiver
Double Conversion Superheterodyne. Solid State.

Features:
- ¼" Head. Jack
- RF Gain
- Mute Jack
- Line Out Jack
- Scan & Sweep
- UP/DN Tune
- S-Indicator
- BFO ±2 kHz
- AGC (3 Pos.)
- Ext. Spkr Jack
- Dial Lock
- Keypad
- PBS ±1 kHz
- Speaker
- 200 Memories
- Timer Terminals
- Dimmer (4 Pos.)
- Side Tone Jack
- NB
- Squelch
- RIT ±5 kHz
- 24 Hr. Clock/Timer
- Notch-30 dB
- Modular Design
- Tone
- ANL

Specifications:
Coverage ...... 90 - 34000 kHz
Selectivity ...... 4/2/_/_ kHz -6dB.[12FM]
Sensitivity ...... <.5µV 1.6-30 MHz SSB
Image Rej ...... >70 dB 1.6-30 MHz
Audio Out ...... 1 W 4 ohms

Modes .... AM/SSB/CW/FAX/FM/RTTY
Stability .... ±3 PPM
IF Rejection .... >70 dB 1.6-30 MHz
Antenna Input . SO239 and Terminals
Dyn. Range .... >100 dB

Accessories:
CFL-231 300Hz Filter   CFL-232 500Hz Filter   CFL-233 1000Hz Filter
CFL-218A 1800Hz Filter CMH-530 RTTY Unit   CMH-532 RS232 Intfc.
NVA-88 Speaker   CMK-165 VHF/UHF Conv.
6ZCJD00139 Printer Cable for CMH530

Comments: The memories store: frequency, mode, bandwidth, AGC and ATT settings. The optional CMK165 Converter adds: 34-60, 114-174 and 423-456 MHz in all modes. CMH530 internal RTTY demod unit decodes Baudot at 45 and 50 baud to the parallel print output port on back or into the optional CMH532. [¥90,000].

Made In:	Japan	1986-1992
Voltages:	100/120/220/240 VAC 50/60 Hz and 12-16 VDC	
Readout:	**00000.01** Digital Flur.	
Physical:	13x5.2x11.25" 19 Lbs. 330x132x285mm 8.5 kg	
Status:	Active Manufacturer Discontinued Model	
Rarity:	Abundant	
Reviews:	📖IBS-RDI Whitepaper 📖*Passport* 1987-1992 *WRTH* 1987 *QST* July 1988 *Proceedings* 1989	

	New	Used
Price:	$999-1249	$530-600
Rating:	★★★★★	★★★★★

JRC Japan Radio Co., Ltd

NRD-535

General Coverage Communications Receiver
Triple Conversion Superheterodyne. Solid State.

Features:
- ¼" Head. Jack
- RF Gain
- Mute
- Line Out Jack
- Scan & Sweep
- BFO ±2 kHz
- S-Meter
- ANL
- AGC (3 Pos.)
- Ext. Spkr Jack
- Dial Lock
- Modular Design
- PBS ±1 kHz
- Notch -40 dB
- 200 Memories
- Timer Terminals
- Keypad
- NB-1 & NB-2
- Tone
- RS-232 Port
- 24 Hr. Clock/Timer
- Dimmer (4 Pos.)

Specifications:
Coverage ...... 100 - 30000 kHz
Selectivity ...... 12/4/2/_ kHz -6dB.
Sensitivity ...... <.1µV 1.6-30 MHz SSB
Image Rej ......>70 dB 1.6-30 MHz
Audio Out ...... 1 W 4 ohms

Modes .... AM/SSB/CW/FAX/FM/RTTY
Stability .... ±10 PPM (±2PPM >1 hour)
IF Rejection .... >70 dB 1.6-30 MHz
Antenna Input . SO239 and Terminals

Accessories: NVA-319 Filter/Speaker
CFL-231 300Hz Filter   CFL-232 500Hz Filter   CFL-233 1000Hz Filter
CGD-135 Hi-Stability   CFL-218 1800Hz Filter   CFL-251 2400Hz Filter
CFL-243 BWC   CMF-78 ECSS Unit   CMH-530 RTTY Unit

Comments: 1 Hz tuning resolution and 10 Hz display resolution. An outstanding general coverage receiver. [¥150,000-180,000]. Also reviewed *SW Mag.* Nov. 1991.
Variants: Model **NRD-535V** includes the CMF-78. Model **NRD-535D** includes: CFL243, CMF78 and CFL233. Units after S/N 56005 had BWC operation in SSB <u>and</u> AM modes and on the Inter. <u>and</u> Wide filters. Also reviewed *Proceedings* 1992-93.

Made In:	Japan	1991-1998
Voltages:	100/120/220/240 VAC 50/60 Hz and 12 DC	
Readout:	00000.01	Digital Flur.
Physical:	13x5.2x11.25" 20 lbs. 330x132x285mm 9 kg	
Status:	Active Manufacturer Active Model	
Rarity:	Common	
Reviews:	📖*Passport* 1992- 998 📄IBS-RDI Whitepaper *WRTH* 1992 *WRTH Buyer's Guide* 1993 *Mon. Times* October 1991 *Pop Comm* December 1991 *ODXA* July 1991 *QST* May 1997	

	New	Used
Price:	$1200	$750-780
Rating:	★★★★★	★★★★★

JRC Japan Radio Co., Ltd

NRD-545

General Coverage Communications Receiver
Triple Conversion Superheterodyne. Solid State.

Features:
- ¼" Head. Jack
- RF Gain
- Mute
- Line Out Jack
- Scan & Sweep
- NB-1 & NB-2
- Tone
- S-Meter
- ANL
- AGC (3 Pos.)
- Ext. Spkr Jack
- Dial Lock
- ECSS
- Speaker
- PBS ±2.3 kHz
- Notch -40 dB
- 1000 Memories
- Timer Terminals
- Keypad
- Squelch
- Variable Torque
- All Mode DSP
- RTTY Demod
- RS-232 Port
- 24 Hr. Clock/Timer
- Dimmer (4 Pos.)
- Synchronous Det.

Specifications:
Coverage ...... 100 - 30000 kHz
Selectivity ...... See Comments.
Sensitivity ...... <.32µV 1.6-30 MHz SSB
Image Rej ......>70 dB
Notch Atten. ..>40 dB
Audio Out ...... 1 W 4 ohms
Dyn. Range ... 106 dB

Modes .... AM/SSB/CW/FAX/FM/RTTY
Stability .... ±10 PPM (±2PPM >1 hour)
IF Rejection .... >70 dB
IFs .... 70.455 MHz, 455, 20.22 kHz
RS-232 Intfc ... 4800/1/8/N/1
Environment ... 0° to +50°C
Antenna Input . SO239 and Terminals

Accessories:
CGD-197 Hi-Stability   CHE-199 VHF-UHF Converter   NVA-319 Speaker
Comments: Bandwidth may be adjusted continuously from 40 Hz to 10 kHz in 10 Hz steps. Tuning steps: 1, 10, 100 Hz, 1, 5, 6.25, 9, 10, 12.5, 20, 25, 30 and 100 kHz. DSP is used for NB, Notch, PBS, AGC, BFO, RF Gain, Squelch, Tone, S-Meter and Bandwidth. The RTTY demod (37-75 baud) output is to the RS232 port. Awesome.

Made In:	Japan	1998
Voltages:	100/120/220/240 VAC 50/60 Hz and 12VDC	
Readout:	00000.01	Digital
Physical:	13x5.2x11.25" 16.5 Lbs. 330x132x285mm 7.5 kg	
Status:	Active Manufacturer Active Model	
Rarity:	Too new.	

	New	Used
Price:	①	€
Rating:	⑧	€

JRC | Japan Radio Co., Ltd

NRD-1000

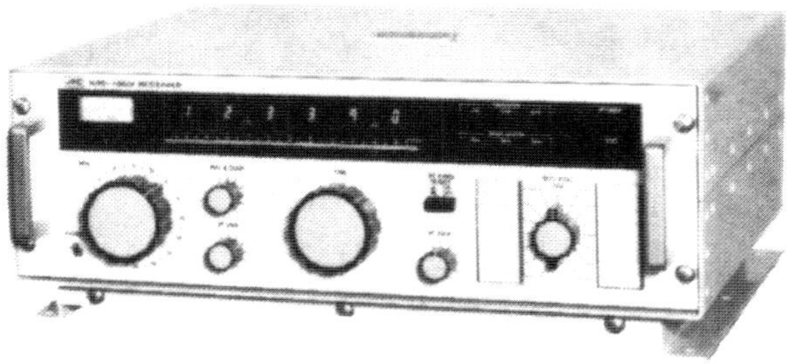

General Coverage Communications Receiver
Triple Conversion Superheterodyne. Solid State.

Features:
- ¼" Head. Jack
- S-Meter
- BFO
- AGC OFF/ON
- RF Gain
- Mute
- 14 Presets
- Line Out 600 ohms
- Dial Lock
- IF Out Jack
- Rack Handles
- Modular Construction
- Atten. -20 dB

Specifications:
Coverage ...... 100 - 30000 kHz   Modes .... AM/SSB/CW/MCW
Other specifications unknown.

Accessories:
Cabinet (shown)

Comments:
Double conversion on some frequencies.

Made In: Japan   1981-1984

Voltages: 100/110/115/220/230 VAC
50/60 Hz

Readout: `00000.1` Digital LED

Physical: Cabinet Version:
19.25x7.5x15.7" 35.2 Lbs.
489x191x400mm 16 kg
Rack Version:
19x5.8x12.6" 22 Lbs.
480x149x320mm 10 kg

Status: Active Manufacturer
Discontinued Model
Rarity: Typically Unavailable

	New	Used
Price:	①	⑤
Rating:	⑥	⑥

JRC | Japan Radio Co., Ltd

NRD-1003A

General Coverage Communications Receiver
Double Conversion Superheterodyne. Solid State.

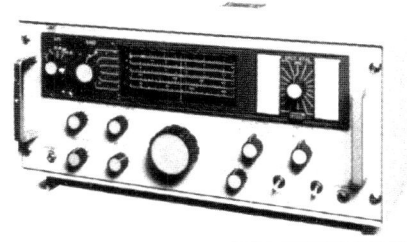

Features:
- ¼" Head. Jack
- BFO
- AGC OFF/ON
- Modular Construction
- RF Gain
- Mute
- Speaker Jack
- 17 Presets
- Atten. -20 dB
- Rack Handles

Specifications:
Coverage ...... 100 - 28000 kHz   Modes .... AM/CW/MCW/LSB/USB
Other specifications unknown.

Accessories:
Cabinet (shown)

Comments:
Available in heavy cabinet (shown) or for rack mounting.

Made In: Japan   1978

Voltages: 100/110/117/220 VAC
50/60 Hz or 24 VDC 2A

Readout: Analog

Physical: Cabinet Version:
19.25x9.5x15.7" 40 Lbs.
489x240x400mm 18 kg
Rack Version:
19x7.83x14.5" 26.5 Lbs.
480x199x370mm 12 kg

Status: Active Manufacturer
Discontinued Model
Rarity: Typically Unavailable

	New	Used
Price:	①	⑤
Rating:	⑥	⑥

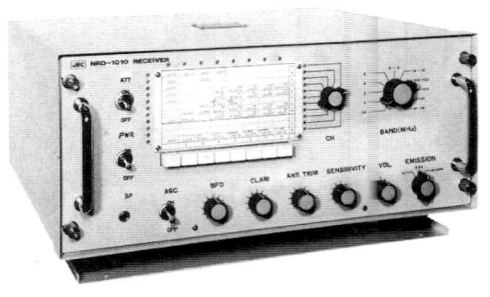

JRC   Japan Radio Co., Ltd

NRD-1010

Fixed Frequency Communications Receiver
Triple Conversion Superheterodyne. Solid State.

Features:
- ¼" Head. Jack
- BFO
- AGC OFF/ON
- Antenna Trimmer
- RF Gain
- Mute
- Speaker Jack
- Atten. -20 dB
- Rack Handles
- Fine Tuning

Specifications:
Coverage ......88 Ch. (1.6 - 28 MHz)
Selectivity ......6/2.4 kHz -6dB.
Sensitivity ......<2µV CW
Audio Out ......1 W 4 ohms

Modes .... AM/CW/SSB
Image Ratio .... >40 dB
IF .... 14.5-15.5, 3-4, .455 MHz

Comments:
This is a fixed channel receiver designed for the marine market. The 88 crystal positions are selected by pressing one of the 8 push buttons along the horizontal access and then rotating the 11 position switch on the vertical axis. The band range switch must be manually selected for preselection purposes, then the Antenna Trimmer must be adjusted. The associated frequency or channel number can be written on the paper chart mounted on the front of the receiver. The NRD-1010 may also be rack mounted.

Made In:	Japan	1974-1976

Voltages: 100/220 VAC 50/60 Hz 100W

Readout: None

Physical: 19x7.8x16.1" 39.7 Lbs. 480x199x410mm 18 kg

Status: Active Manufacturer Discontinued Model

Rarity: Typically Unavailable

	New	Used
Price:	①	⑤
Rating:	⑥	⑥

JRC   Japan Radio Co., Ltd

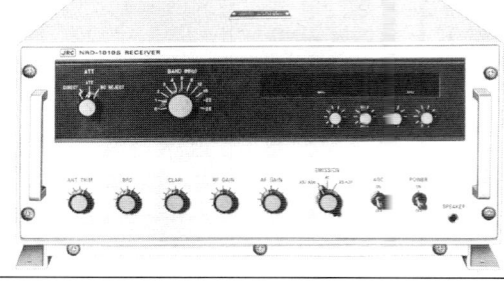

NRD-1010S

General Coverage Communications Receiver
Triple Conversion Superheterodyne. Solid State.

Features:
- ¼" Head. Jack
- BFO
- AGC OFF/ON
- Antenna Trimmer
- RF Gain
- Mute
- Speaker Jack
- BCB Reject
- Atten. -20 dB
- Rack Handles
- Fine Tuning

Specifications:
Coverage ......100 - 26000 kHz
Selectivity ......6/2.4 kHz -6dB.
Sensitivity ......<2µV CW 1-26 MHz
Audio Out ......1 W 4 ohms

Modes .... AM/CW/SSB
Image Ratio .... >40 dB
IF .... 14.5-15.5, 3-4, .455 MHz

Accessories:
Cabinet (shown)

Comments:
The JRC NRD-1010S is the synthesized version of the NRD-1010. Instead of selecting fixed crystal controlled frequencies, a synthesizer has been added. The frequency is selected by rotary knobs located below the digital display. Frequency selection may be made down to 100 Hz. Range selection must still be made via the 11 position band switch. This model is double conversion on some frequencies.

Made In:	Japan	1977

Voltages: 100/220 VAC 50/60 Hz 100W

Readout: 00000.1 Digital LED

Physical: 19x7.8x14.5" 39.7 Lbs. 480x199x370mm 18 kg

Status: Active Manufacturer Discontinued Model

Rarity: Typically Unavailable

	New	Used
Price:	①	⑤
Rating:	⑥	⑥

Kenwood's forerunner company, Kasuga Musen Denki Shokai Limited, was founded in 1946. In 1952 the first communications grade receiver was produced. This was the 6 tube Trio model 6R-4. In 1955 the company began the mass production of audio, communications and test equipment. The Trio model 9R4J was a nine tube receiver kit that featured an S-Meter, BFO and covered from 550 to 30000 kHz. It also sold in Germany as the Jennen 9R-4J and is dead ringer for the Lafayette HE-10.

In 1960 the Trio Company changed its name to Trio Limited. The assembled model 9R-59 General Coverage Receiver was introduced in 1961. Trio successfully manufactured this and other models for other private label firms such as Lafayette. In 1963 the overseas brandname of Kenwood is established for hi-fi equipment in Los Angeles.

In 1975 Trio-Kenwood Communications Inc. was established in Gardena, California to market amateur radio equipment in the United States.

In 1986 the company changed its name to Kenwood Corporation.

Kenwood hit three home runs in a row with the R-1000, R-2000 and R-5000 receivers.

Trio Corporation
Chofu-Chidoricho 74
Ohta-ku, Tokyo,
Japan

Trio-Kenwood Communications
116 East Alondra
Gardena, CA 90248   1976-1980

Trio-Kenwood Communications
1111 West Walnut
Compton, CA 90220   1981-1986

Kenwood Communications
2201 E. Dominguez St.
Long Beach, CA 90801   1987-1998

9R-59

General Coverage Communications Receiver
Single Conversion Superheterodyne. 9 Tubes.

Features:
- ¼" Head. Jack
- S-Meter
- Mute Line
- Bandspread
- Dial Light
- RF Gain
- Standby
- Antenna Trimmer
- ANL
- BFO
- AVC/MVC
- Q-Multiplier

Specifications:
Coverage ......550 - 30000 kHz.   Modes .... AM/CW-SSB
Sensitivity ......10 µV 20 dB S/N @10 MHz Antenna Input . Terminals
Audio ....4 or 8 ohms 1.5W

Circuit Complement:
6BA6 RF Amp, 6BE6 Osc., 6AV6 Q-Multiplier, 6BE6 Mixer, 6BA6 IF Amp, 6BA6 IF Amp, 6AV6 Detector/ANL/AF Amp, 6AQ5 Power Amp and 5Y3 (or 5CG4) Rectifier. Ore RF and IF Stages.

Comments:
Ranges: .54-1.605, 1.6-4.8, 4.8-14.5 and 10.5-30 MHz. Bandspread is provided for 80, 40, 20 ,15 and 10 meters. The Trio 9R-59 appears to possibly be the same as Lafayette model HA-230 and HE-30. Although the 9R-59 is very scarce in the United States, it is quite common in Europe. This radio requires a speaker. It is difficult to tune SSB on this receiver.

Variants:
This model was also sold as the **Jennen-Trio JR-101**. Please also see the Lafayette HE-30.

Made In:	Japan	1961-1962

Voltages: 110/220 VAC 50/60 Hz
50VA

Readout: [analog dial] Analog

Physical: 15x7x10" 20.5 Lbs
381x178x254mm 9.3 kg

Status: Active Manufacturer
Discontinued Model

Rarity: Very Scarce

	New	**Used**
Price:	$160	$80-100
Rating:	⑦	⑦

9R-59DE

General Coverage Communications Receiver
Single Conversion Superheterodyne. 8 Tubes plus Semiconductors.

Features:
- ¼" Head. Jack
- Dial Lamp
- ANL
- S-Meter
- RF Gain
- BFO
- Mute Line
- Standby
- AVC
- Bandspread
- Antenna Trimmer

Specifications:
Coverage ......550 - 30000 kHz.
Sensitivity ......2 µV 10 dB S/N
Selectivity ......4 kHz -6dB.
Audio Out ......1.5W 8 ohms 10% dist.

Modes .... AM/CW-SSB
IF .... 455 kHz
Antenna Input . Terminals

Circuit Complement:
6BA6 RF Amp, 6BA6 RF Amp, 6BE6 Mixer, (2) 6AQ8 Oscillator/BFO/AF Amp, 6AQ5 AF Output plus 7 diodes.

Accessories:
SP-5D Speaker

Comments:
Ranges: .55-1.6, 1.6-4.8, 4.8-14.5 and 10.5-30 MHz. Bandspread on 80, 40, 20, 15 and 10 meters. Dark gray case.

Variants:
Model **9R-59DS**.

Made In:	Japan	1968-1969
Voltages:	115/230 VAC 50/60 Hz 45W	
Readout:	Analog	
Physical:	15x7x10" 17.6 Lbs. 380x180x260 mm 8 Kg	
Status:	Active Manufacturer Discontinued Model	
Rarity:	Very Scarce	

	New	Used
Price:	$200-270	$150-170
Rating:	⑦	⑦

ER-202

General Coverage Communications Receiver
Single Conversion Superheterodyne. 14 Tubes.

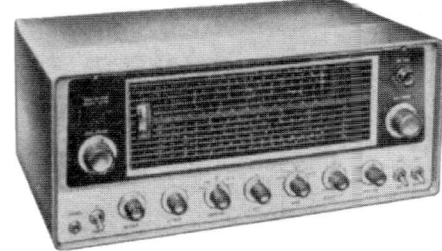

Features:
- ¼" Head. Jack
- RF Gain
- Record Jack
- S-Meter
- ANL
- MVC/AVC
- Bandspread
- BFO
- Crystal Cal. 100 kHz
- Q-Multiplier

Specifications:
Coverage ......550 - 30000 kHz +2M
Selectivity ......10 kHz -60dB.
Sensitivity ......<3 µV
Audio Out ......1.5 W 8 ohms

Modes .... AM/CW/SSB
IF .... 455 kHz
Antenna Input . Terminals

Circuit Complement:
6AQ8 2M RF Amp, 6AU6 2M Mixer, 6AQ8 6M Oscillator, 6BA6 RF Amp, 6BE6 Mixer, 6AQ8 Oscillator Buffer, (2) 6BA6 IF Amp, 6AL5 AM Detector/ANL, 6BE6 Product Detector, 6AQ8 AF Amp/BFO, 6AQ5 AF Amp, 6AQ8 Q-Multiplier/Calibrator, 6CA4 Rectifier, 0A2 Voltage Regulator and 1N60 Crystal Calibrator.

Accessories:
Crystal for Calibrator

Comments:
Ranges: .54-1.6, 1.6-4.8, 4.8-14.5, 10.5-30 and 142-148 MHz.

Variants:
This model was sold in the U.S.A. as the Lafayette HE-80. The Lafayette HE-80 version included 6 meters instead of 2 meters and ran from 110 VAC rather than 220 VAC.

Made In:	Japan	1963-1964
Voltages:	220 VAC 50/60 Hz 65W	
Readout:	Analog	
Physical:	17x7.5x10" 24 Lbs. 430x190x260mm 10.7kg	
Status:	Active Manufacturer Discontinued Model	
Rarity:	Extremely Scarce	

	New	Used
Price:	$140	$65-70
Rating:	★★	⑦

JR-310

Amateur Band Communications Receiver
Double Conversion Superheterodyne.

Features:
- ¼" Head. Jack
- S-Meter
- IF Tune
- RF Tune
- RF Gain
- ANL
- Mute Line
- AVC
- BFO
- RIT
- Standby

Specifications:

Coverage ......Ham Bands	Modes .... AM/CW-SSB
Selectivity ......3 kHz -6 dB	Stability .... ±2 kHz in 1-60 mins.
Sensitivity ......<1µV 10 dB S/N	Stability .... <100Hz/30 min after warmup
Image Rej. ......>50 dB	

Circuit Complement:
High stability VFO with 2 FETs and 2 transistors.

Comments:
Coverage is amateur bands from 80 to 10 meters in seven bands. WWV reception at 15 MHz is also supported. Dial accuracy is ±1 kHz. One dial rotation covers 25 kHz. This model was sold in Europe, but not in the United States.

Made In:	Japan	1969-1970
Voltages:		
Readout:	Analog Lin.	
Physical:	13x7.1x12.4" 16 Lbs. 330x180x315mm 7.3 kg	
Status:	Active Manufacturer Discontinued Model	
Rarity:	Extremely Scarce	

	New	Used
Price:	£65 UK	⑤
Rating:	★★★	⑦

JR-500SE

Amateur Band Communications Receiver
Double Conversion Superheterodyne. 7 Tubes plus Semiconductors.

Features:
- ¼" Head. Jack
- S-Meter
- Preselector
- Standby
- RF Gain
- ANL
- Mute Line
- AVC
- BFO
- Line Out 500 ohms

Specifications:

Coverage ......Ham Bands	Modes .... AM/CW-SSB
Selectivity ......3 kHz -6 dB	IF .... 8.5-9.1 MHz, 455 kHz
Sensitivity ......<1.5µV 14 MHz	Image Rej .... >40 dB
Audio Out ......1W 8/500 ohms	Antenna Input . Terminals. (hole for SO239)

Circuit Complement:
6BZ6 RF Amp, 6BL8 1st Mixer/Local Osc, 6BE6 2nd Mixer, 6BA6 1st IF, 6BA6 2nd IF, 1N60 AM Detector, SW05S ANL, 6AQ8 Product Detector/BFO, 6BM8 AF Amp/PA, 2SC185 Buffer and 2SC185 VFO.

Accessories:
SP-5D Speaker

Comments:
Coverage: 3.5-4, 7-7.3, 14-14.35, 21-21.45, 28-28.5, 28.5-29.1 and 29.1-29.7 MHz plus 9.6-10 MHz for WWV. Provides ±1 kHz dial accuracy. Tuning is at 50 kHz per revolution of main tuning knob. Dark gray metal cabinet with dark green panel. See Allied A2516.

Made In:	Japan	1969-1970
Voltages:	115/230 VAC 65 W	
Readout:	Analog	
Physical:	13x7x10" 16 Lbs. 362x163x322mm 7.3 Kg	
Status:	Active Manufacturer Discontinued Model	
Rarity:	Very Scarce	
Reviews:	*Radio Comm.* June 1969	

	New	Used
Price:	£65 UK	⑤
Rating:	★★★	⑦

 KENWOOD

QR-666

General Coverage Communications Receiver
Double Conversion Superheterodyne. Solid State.

Features:
- S-Meter
- Light
- ANL
- Speaker 4"
- Bandspread
- RF Gain
- BFO
- Mute Line
- ¼" Head. Jack
- Standby Switch
- Dial Light Switch
- Dial Light
- Telescopic Ant.
- Antenna Trimmer
- Calibrator 500 kHz

Specifications:
Coverage ...... 170-410 & 525-30000 kHz.
Selectivity ...... 5/2.5 kHz -6dB.
Sensitivity ...... 1-3 µV (2-30 MHz).
Image Ratio ... >70 dB (2-12 MHz).

Modes .... AM/CW-SSB
Stability .... See Note 1
Audio Out .... 1.5W 8 ohms 10% distort.
Antenna Input . SO-239 & Terminals.

Circuit Complement:
4 FETs, 16 transistors and 24 diodes.

Accessories:
FM/IF FM Band Option 87.5-108 MHz (internal board).
QR-6MK Calibrator 500 kHz

Comments:
Note 1. Stability: 2 MHz <10 kHz, 5 MHz <15 kHz, 12 MHz <20 kHz, 24 MHz <25 kHz.
Ranges: .17-.41, .525-1.25, 1.25-3, 3-7.5-18 and 18-30 MHz. Amateur bandspread:
3.5-4, 7-7.5, 14-14.6, 21-21.5 and 28-30 MHz. May be run from eight D cells. An
optional SWL bandspread drum was also offered.

Made In:	Japan	1974-1976
Voltages:	100/117/220/240 AC or 12~15 VDC or 3 D cells.	
Readout:	Analog	
Physical:	14.25x6.4x12.8" 7 Lbs. 362x163x322mm 7.3 Kg	
Status:	Active Manufacturer Discontinued Model	
Rarity:	Scarce	
Reviews:	*WRTH* 1976	

	New	Used
Price:	$129	$60-80
Rating:	★★★	★

 KENWOOD

R-300

General Coverage Communications Receiver
Double Conversion Superheterodyne. Solid State.

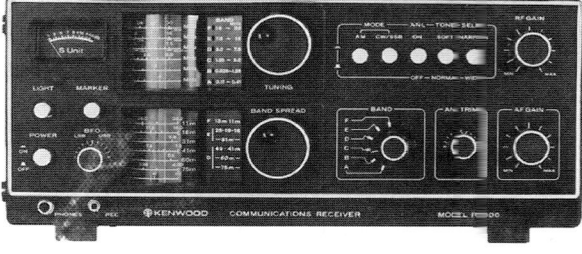

Features:
- ¼" Head. Jack
- Dial Lamp
- ANL
- Speaker 4"
- S-Meter
- RF Gain
- BFO
- Bandspread
- Record Jack
- Tone (2 Pos.)
- Antenna Trimmer
- Ext. Spkr Jack
- Calibrator 500 kHz

Specifications:
Coverage ...... 170-410 & 525-30000 kHz.
Selectivity ...... 5/2.5 kHz -6dB.
Sensitivity ...... 1-1.5 µV (2-30 MHz).
Image Ratio ... >40-55 dB (2-12 MHz).
Audio Out ...... 1.5W 10% distortion

Modes .... AM/CW-SSB
Stability .... *unspecified*
IF .... 4.034 MHz, 455 kHz
Antenna Input . Terminals. (hole for SO239)

Comments:
Ranges: .17-.41, .525-1.25, 3-7.5, 7.5-18 and 18-30 MHz. Two different bandspread
drums were developed. The amateur version was shipped to some areas and the
shortwave version to other areas. Amateur bandspread: 3.5-4, 7-7.5, 14-14.6, 21-
21.5 and 28-30 MHz. Shortwave bandspread: 3.82-4, 7-7.5, 9.4-9.8, 11.7-12, 15-
15.5, 17.6-18, 21.4-21.8 and 25.6-26.2 MHz. The R-300 may be run from eight D
cells. These cells are accessible through a trap door on the bottom of the receiver.

Made In:	Japan	1976-1979
Voltages:	117/220 AC 8W 12~16 VDC or 8 D cells.	
Readout	Analog	
Physical:	14.25x6.4x12.8" 7 Lbs. 362x162x325mm 7.7 kg	
Status:	Active Manufacturer Discontinued Model	
Rarity:	Scarce	
Reviews:	*73* June 1977	

	New	Used
Price:	$239	$80-95
Rating:	★★★	★

R-599

Amateur Band Communications Receiver
Double Conversion Superheterodyne. Solid State.

Features:
• ¼" Head. Jack • S-Meter • Standby • AGC SLO/MED/FST
• RF Gain • Preselector • R.I.T. ±2 kHz • Calibrator 100/25 kHz
• Squelch • NB

Specifications:
Coverage ...... 160-10M ham bands   Modes .... CW/LSB/USB/AM/FM[1]
Selectivity ...... 5/2.5/.5 kHz-6dB.[25 FM]   Stability .... ±100Hz after 15 mins.
Sensitivity ...... <.5µV (160-10M).   IF Rejection .... >50 dB
Audio Out ...... 1W 8Ω 10% dist.   Image Ratio .... >50 dB
Antenna Input SO-239

Circuit Complement:
2 Integrated circuits, 8 FETs, 30 transistors and 53 diodes.

Accessories:
S-599 Speaker   CC-29 2M Converter   CC-69 6M Converter
FM-599D[1] FM

Comments: Matches T-599A Transmitter. Dial accuracy ±1 kHz. Original ranges: 1.8-2.3, 3.5-4.0, 7-7.5, 14-14.5, 21-21.5 and 28-29.7 plus the CB band of 26.8-27.4. After March 1978 the CB band was removed and labeled AUX.

Variants: The model **R-599A** changed the dial skirt from 25 kHz/rev. to 100 kHz/rev., had VFO light and Noise Blanker. The crystal filters was changed from 4 to 8 pole. The later U.S. model **R-599D** 1976 $459-500 and European model **R-599S** were darker in color. The R-599D was reviewed *CQ* March 1978.

Made In:	Japan	1970-1975
Voltages:	100/120/220/240 AC 15W or 12~15 VDC	
Readout:	⊞⊞⊞⊞ Analog Lin.	
Physical:	10.6x5.5x12.4" 13 Lbs. 269x140x315mm 6 kg	
Status:	Active Manufacturer Discontinued Model	
Rarity:	Common	
Reviews:	QST October 1974 QST June 1971	

	New	Used
Price:	$298-459	$100-120
Rating:	★★★★	★★

R-600

General Coverage Communications Receiver
Triple Conversion Superheterodyne. Solid State.

Features:
• ¼" Head. Jack • S-Meter • Tone Control • Attenuator -20 dB
• Noise Blanker • Mute Terminal • Carrying Handle • Record Jack
• Speaker

Specifications:
Coverage ...... 150 - 30000 kHz.   Modes .... AM/USB/LSB-CW
Selectivity ...... 6/2.7 kHz -6dB.   Stability .... ±2 kHz
Sensitivity ...... <.5µV (2-30 MHz).   IF Rejection .... >60 dB
Audio Out ...... 1.5W 10% distortion   Image Ratio .... >60 dB
Antenna Input SO-239 & Terminals.

Accessories:
DCK-1 12VDC Kit

Comments:
A communications receiver for the casual listener. Operates with thirty 1 MHz bands. Lacks stability found in models R-1000 and up. Some European production units cover 150 - 26000 kHz.

Made In:	Japan	1982-1985
Voltages:	100/120/220/240 AC 50/60 Hz	
Readout:	**00001.** Digital LED	
Physical:	12.8x4.5x7.8" 10 Lbs. 325x114x198 4.5 kg	
Status:	Active Manufacturer Discontinued Model	
Rarity:	Common	
Reviews:	Passport 1987 WRTH 1983 Mon. Times May 1983	

	New	Used
Price:	$329-399	$180-250
Rating:	★★★	★★★

 KENWOOD

R-820

Amateur Band Communications Receiver
Double Conversion Superheterodyne. Solid State.

Features:
- ¼" Head. Jack
- RF Gain
- V.B.T.
- Noise Blanker
- R.I.T.
- I.F. Shift
- Notch Filter
- Preselector
- Tone Control
- Mute
- 4 Fixed Channels
- Atten.-10/20/30/40 dB
- S-Meter
- Monitor Level
- Calibrator 25 kHz
- Analog sub-dial

Specifications:
Coverage ...... 160-10M Ham Bands
Selectivity ...... 6/2.4/_/_ kHz -6dB.
Sensitivity ...... <.25μV
Ant. Jack SO-239
Notch Attn ..... 50 dB or greater
Antenna Input SO-239

Modes .... AM/CW/LSB/USB/RTTY
Stability .... ±1 kHz or better.
IF Rejection .... >80 dB[1]
IF Rejection .... >50 dB[2]
[1](160, 80, 40, 20, 15, 10 & 19M) [2](49, 31, 25 & 16M)

Circuit Complement:
40 Integrated circuits, 34 FETs, 89 transistors and 170 diodes.

Accessories:
| SP-820 Speaker | YG-88A 6kHz AM Filter | YG-455C 500Hz CW Filter |
| VFO-820 VFO | YG-88C 500Hz CW Filter | YG-455CN 250Hz CW Filter |

Comments:
Supplied amateur band coverage: 1.8-2, 3.5-4, 7-7.5, 14-14.5, 21-21.5 and 28-30 MHz. Supplied shortwave broadcast band coverage: 5.9-6.4, 9.4-9.9, 11.5-12, 15-15.5 and 17.7-18.2 MHz. Matches the TS-820S transceiver. An excellent receiver but inexpensive on the used market because it is not general coverage.

Made In:	Japan	1978-1979
Voltages:	100/120/220/240 VAC	
	50/60 Hz or 12-15 VDC	
Readout:	**00000.1**	Digital Phs.
Physical:	13.2x6x14.5" 26.- Lbs.	
	336x167x397mm 12 kg	
Status:	Active Manufacturer	
	Discontinued Model	
Rarity:	Scarce	
Reviews:	QST July 1979	

	New	Used
Price:	$1049	$250-320
Rating:	★★★★	★★★

 KENWOOD

R-1000

General Coverage Communications Receiver
Double Conversion Superheterodyne. Solid State.

Features:
- ¼" Head. Jack
- S-Meter
- Speaker 4"
- Atten. -20/40/60 dB
- Noise Blanker
- Tone Control
- Record Jack
- Recorder Activation
- Dimmer
- Tilt-Carry Handle
- 12 Hour Clock-Timer
- Analog Sub-dial

Specifications:
Coverage ...... 200 - 30000 kHz.
Selectivity ...... 12/6/2.7 kHz -6dB.
Sensitivity ...... <.5μV SSB 2-30 MHz.
Image Ratio ...>60 dB (1.8-30 MHz).
Audio .... 1.5W 8 ohms 10% dist.

Modes .... AM/USB/LSB/CW
Stability .... ±2 kHz, ±300 Hz/30 min.
IF Rejection .... >70 dB (1.8-30 MHz).
IF .... 48 MHz, 455 kHz
Antenna Input . SO-239 & Terminals.

Circuit Complement:
40 Integrated circuits, 11 FETs, 64 transistors and 71 diodes.

Accessories:
SP-100 Speaker (shown)   DCK-1 DC Kit

Comments:
A solid performer. Very stable after warmup and even suitable for quality RTTY reception. The AVC release time is a bit long. The somewhat ugly carry handle may be swivelled down to angle the receiver, or swung up over the top of the receiver. Shown above with optional SP-100 Speaker. A good receiver.

Made In:	Japan	1979-1985
Voltages:	100/120/220/240 VAC	
	50/60 HZ 20W	
Readout:	**00001.**	Digital Phs.
Physical:	12.8x4.5x8.6" 12 Lbs.	
	300x115x218mm 5.5 kg	
Status:	Active Manufacturer	
	Discontinued Model	
Rarity:	Abundant	
Reviews:	Passport 1937	
	WRTH 1980	
	QST December 1980	
	Commun. World 1981	

	New	Used
Price:	$430-499	$250-300
Rating:	★★★★	★★★★★

KENWOOD

R-2000

General Coverage Communications Receiver
Triple Conversion Superheterodyne. Solid State.

Features:

- ¼' Head. Jack
- 3 Tuning Rates
- Dial Lock
- Dimmer
- S-Meter
- 10 Memories
- 4" Speaker
- AGC FAST/SLOW
- Tone Control
- Tilt Bar
- Memory Scan
- Noise Blankers
- Record Jack
- Carrying Handle
- Squelch
- Sweep
- Atten. -10/20/30 dB
- Record Contacts
- Mute line
- Dual Clock-Timer 24 Hr.

Specifications:

Coverage ...... 150 - 30000 kHz.
Selectivity ...... 6/2.7/_ kHz -6dB [15 FM]
Sensitivity ...... <.4µV CW (2-30 MHz)
Stability .... ±50 Hz after 1 hour
IF .... 45.85, 9.85, .455 MHz
Audio .... 1.5W 8 ohms

Modes .... AM/FM/USB/LSB/CW
Accuracy .... ±10x10⁻⁶ or better.
IF Rejection .... >70 dB (1.8-30 MHz).
Image Ratio .... >70 dB (1.8-30 MHz).
Antenna Input . SO-239 & Terminals.

Accessories:

SP-430 Speaker   YG-455C 500Hz CW Filter   VC-10 VHF Converter
DCK-1 DC Kit   SP-100 Speaker

Comments: A good value. A popular modification was the removal of the tone control in favor of an R I.T. to decrease the VFO tuning increment from the supplied 50 Hz. This made the R-2000 very suitable for RTTY reception. No keypad entry. Memories store: frequency and mode. The optional VC-10 Internal Converter covers 118-174 MHz in all modes with full frequency readout.

Variants: Sold as **Trio R-2000** in the U.K.

Made In:	Japan	1983-1992

Voltages:	100/120/220/240 AC

Readout: `00000.1` Digital Phs.

Physical:	14.8x4.5x8.3" 12 Lbs. 376x114x211mm 5.4 kg
Status:	Active Manufacturer Discontinued Model
Rarity:	Abundant
Reviews:	📖 *Passport* 1987-1994 *WRTH* 1983 *WRTH Buyer's Guide* 1993 *Radio Elec.World* Feb. 1986 *Buyer's Guide to Amateur R.* *Proceedings* 1989

	New	Used
Price:	$499-699	$380-450
Rating:	★★★★	★★★★

KENWOOD

R-5000

General Coverage Communications Receiver
Triple Conversion Superheterodyne. Solid State.

Features:

- ¼' Head. Jack
- Dual VFOs
- Dial Lock
- Mute
- Speaker 4"
- Carrying Handle
- 100 Memories
- I.F. Shift
- AGC (2 Pos.)
- S-Meter
- Tone Control
- Memory Scan
- Notch Filter
- Keypad Entry
- Record-out Jack
- Squelch
- Band Scan
- Atten. -10/20/30 dB
- Dual Noise Blankers
- Record Activation
- Dual 24 Hr. Clock Timer

Specifications:

Coverage ...... 100 - 30000 kHz.
Selectivity ...... 6/2/_/_ [12 FM] kHz -6dB.
Sensitivity ...... <.25µV CW 1.8-30 MHz.
IF .... 58.115, 8.83, .455 MHz.
Audio .... 1.5 W 8 ohms

Modes .... AM/FM/USB/LSB/CW
Accuracy .... ±10x10⁻⁶ or better.
IF Rejection .... >70 dB (1.8-30 MHz).
Image Ratio .... >70 dB (1.8-30 MHz).
Antenna Input . SO-239 & Terminals.

Accessories:

SP-430 Speaker   YK-88A1 6kHz AM Filter   YK-88SN 1.8kHz SSB Filter
DCK-2 DC Kit   YK-88C 500Hz CW Filter   YK-88CN 270Hz CW Filter
VS-1 Voice Synthz.   IF-232/IC-10 Interface   VC-20 VHF Converter

Comments:
The 2 x 5 keypad format takes getting used to. Memories store: frequency, mode and antenna input. The supplied 6 kHz AM filter is only marginal. The optional VC-20 nternal converter covers 108-174 MHz in all modes with full readout. The non-U.S.A. version operates from 120/220/240 50/60 Hz.

Made In:	Japan	1987-1996

Voltages:	120 AC 60 Hz 40W USA

Readout: `00000.01` Digital Phs.

Physical:	10.6x3.8x10.6" 12 Lbs. 269x96x269mm 5.4 kg
Status:	Active Manufacturer Discontinued Model
Rarity:	Abundant
Reviews:	📄 IBS-RDI Whitepaper 📖 *Passport* 1988-1998 *WRTH* 1987, 1989 *WRTH Buyer's Guide* 1993 *QST* February 1988 *ODXA* July 1988 *SW Magazine* June 1987 *Proceedings* 1990

	New	Used
Price:	$799-1019	$600-650
Rating:	★★★★★	★★★★★

KENWOOD

RZ-1

Wideband Broadcast Receiver
Triple Conversion Superheterodyne. Solid State.

Features:
- Mini Head. Jack
- Squelch
- Speaker
- 100 Memories
- Sweep
- Beep
- Memory Scan
- Attenuator
- Illuminated Keys
- Ext. Speaker Jack
- Line Out (Stereo)
- Alpha Memories

Specifications:
Coverage ...... 500 kHz - 905 MHz.
Selectivity ...... 250/10/7 kHz -6dB.
Sensitivity ...... <5µV AM
IF .... 45.75, 10.7, .455 MHz.
Audio .... 2 W 8Ω

Modes .... AM/FM-N/FM-W/C3F(TV)
IF Rejection .... >40 dB (60-905 MHz).
Environment ... -10° to +60° C.
Video Output .. 1 V p-p 75 ohm NTSC
Antenna Input . SO-239 & Motorola

Accessories:
SP-41 Speaker   SP-50B Speaker

Comments:
This receiver is designed for mobile operation. An under-dash car bracket is supplied. Only double conversion in FM-W mode. FM reception is stereo to the line outputs. Scanning is manual, timer or carrier operated. Memories may be lock-out of the scan sequence. Tuning steps: 5, 12.5, 20 or 25 kHz. The NTSC television video output is 1 Vp-P. and is only on the USA version. A lithium battery backs-up the memories. As with most wideband receivers, the shortwave band is not exceptional. However, the RZ-1 has sufficient performance to monitor the major worldband broadcasters while on the road. It also provides VHF-UHF coverage and even TV audio. If a monitor is connected the TV video may be viewed.

| **Made In:** | Japan | 1988-1992 |

Voltages: 11-16 VDC 1 Amp

Readout: `00005.` Digital LCD

Physical: 7.09x1.97x6.22" 3.3 Lbs
180x50x158mm 1.5 kg

Status: Active Manufacturer
Inactive Model

Rarity: Scarce

Reviews: *Mon. Times* January 1989
SW Magazine April 1988

	New	**Used**
Price:	$500-560	$270-280
Rating:	★★★	★★★

45 Kingsley Radio

In 1938 with the war looming, the Australian Military realized that they d d not have a modern H.F. communications receiver. They also knew that if the war widened they may not be able to rely upon equipment being readily available from Britain and America. A program was put in place to explore the feasibility of building such equipment in Australia. When faced with the design and production of a modern receiver, the proprietor of **Kingsley Radio** (est. 1931), Howard Kingsley Love, looked at what was being produced on the world market and decided that the National HRO was an excellent design anc one that could be replicated, with some redesign in Australia. The result was the Kingsley K/CR/11. It was known in the Royal Australian Air Force, who was the main user, as the AR.7. The name stuck. The AR.7 is thus a near-copy of the HRO and attained a near 100% Australian manufactured content. It was a major achievement of Australian industry in the late 1930's and was considered one of the better wartime HF receivers in the Pacific Theater. After the war they were used by the Australian Department of Civil Aviation for air-ground communications, and eventually many found their way into Australia ham shacks. Unfortunately many were destroyed by well-intentioned amateurs in the 1960's who attempted to fit them with product detectors and more stable oscillators for single sideband reception. Original specimens, in working order, are difficult to find.

Please see the K/CR/12 in *Chapter 105 Briefly Mentioned*.

Kingsley Radio
225 Trafalgar Street
Petersham, NSW
Australia 1931-1948

Kingsley Radio
380 St. Kilda Road
Melbourne, VIC
Australia 1931-1948

K/CR/11

General Coverage Communications Receiver
Single Conversion Superheterodyne. 9 Tubes.

Features:
• ¼" Head. Jack (2) • S-Meter • Standby • AVC
• RF Gain •NL

Specifications:
Coverage ...... 140 - 25000 kHz — Modes .... AM/CW
Selectivity ...... Variable.
Sensitivity ...... 2 µV — IF .... 455 kHz
Audio Out ...... 2W 600 ohms — Antenna Input . Terminals

Circuit Complement:
6U7G 1st RF, 6U7G 2nd RF, 6K8G Converter/Mixer, 6U7G 1st IF, 6U7G 2nd IF, 6G8G Detector/AVC/1st Audio, 6V6G Audio O/P, 6CG8 BFO/Meter Driver and 5Y3 Rectifier. Separate power supply.

Comments:
This radio is a near copy of the famous National HRO. It utilizes plug-in coil racks. Supplied tuning coil ranges: A .14-.405, B .49-1.43, C 1.42-4.3, D 4.3-12.5 and E 12.5-25 MHz. Designed for 19 inch rack mounting.

Variants: This radio is actually better known at **AR.7** which is the Royal Australian Air Force name for this model. The variant model built for the Australian Army and Navy is called the **AUST. NO. 1**. This AUST. NO. 1 variant (which is shown in the photograph above) has a black front panel, while the more common AR.7 has a chrome front panel. A run was also produced for the Dutch Navy with the front panel labeled in Dutch.

Made In: Australia 1939-1945

Voltages: 240 VAC 50 Hz or 12 VDC (vibrator)

Readout: Micrometer Dial and Coil Chart.

Physical: 19x8.75x12" 483x222x305mm

Status: Inactive Manufacturer, Discontinued Model

Rarity: Typically Unavailable

	New	Used
Price:	①	⑤
Rating:	⑥	⑥

Knight Kit was a house label for kits offered by Allied Radio. They offered an incredible selection of low and medium priced shortwave kits. Many of the low-end regenerative models had colorful names such as *Ocean Hopper, Space Spanner* and *Span Master.* A representative sample is illustrated in this chapter.

The Knight Kit *Star Roamer* superheterodyne was produced in prodigious numbers in the late 1960's and 1970's. Many young radio enthusiasts (including the author) had one of these units geared towards the neophyte.

Please see *Chapter 105 Briefly Mentioned* for additional Knight Kit models.

In most cases, Knight Kit models were not produced in the same prodigious numbers as Heath Kit. Therefore they are more difficult to find and collect.

Allied Radio Corp.
100 N. Western Av.
Chicago, IL 60680

R-55

General Coverage Communications Receiver Kit
Single Conversion Superheterodyne. 6 Tubes.

Features:
- ¼" Head. Jack
- BFO
- Mute Line
- S-Meter
- MVC/AVC
- NL
- Speaker 3.5"
- Standby
- Dial Lamp
- Bandspread
- Antenna Trimmer
- Flywheel Tuning

Specifications:
Coverage ...... 530 - 33000 kHz +6M   Modes .... AM/SSB/CW
IF .... 1650 kHz   Antenna Input . Terminals 52 ohms

Circuit Complement:
6BE6 Mixer/Osc, 6BZ6 1st IF Amp, 6AW8 2nd IF Amp/Audio Driver, 6AW8 BFO/ Audio Output, 6AL5 Detector-AVC/NL and 6X4 Rectifier.

Accessories:
Crystal Calibrator 100 kHz

Comments:
Ranges: .53-1.9, 1.8-6.3, 6-14.5, 11.5-33 and 47-54 MHz. Bandspread: 3.5-4, 7-7.3, 14-14.4, 20.5-21.5, 27-30 and 47-54 MHz. Matches the T-60 Transmitter. Allied #83YU935J.

Made In: 1961-1963

Voltages: 117 VAC 60Hz 60 W

Readout: Analog

Physical: 14.25x8.625x11" 8 Lbs. 377x210x280mm 3.6 kg

Status: Inactive Manufacturer Discontinued Model

Rarity: Very Scarce

Reviews: *73* December 1961
QST September 1961
CQ August 1961

	New	Used
Price:	$60-68	$40-70
Rating:	⑨	⑨

R-55A

General Coverage Communications Receiver Kit
Single Conversion Superheterodyne. 6 Tubes.

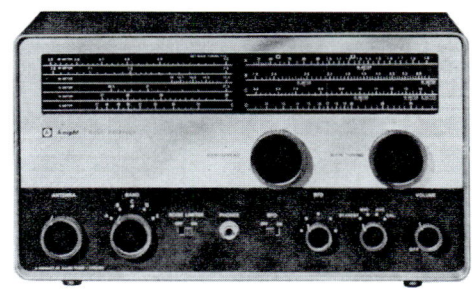

Features:
- ¼" Head. Jack
- S-Meter
- Speaker 3.5"
- Bandspread
- BFO
- MVC/AVC
- Standby
- Antenna Trimmer
- Mute Line
- NL
- Dial Lamp
- Flywheel Tuning

Specifications:
Coverage ......530 - 36000 kHz +6M   Modes .... AM/SSB/CW
IF ....1650 kHz   Antenna Input . Terminals 52 ohms

Circuit Complement:
6BE6 Mixer/Osc, 6BZ6 1st IF Amp, 6AW8 2nd IF Amp/Audio Driver, 6AW8 BFO/
Audio Output, 6AL5 Detector-AVC/NL and EZ90 Rectifier.

Accessories:
Crystal Calibrator

Comments:
Ranges: .53-1.9, 1.8-6.3, 6-14.5, 11.5-36 and 47-54 MHz.   Matches the T-60
Transmitter. The R-55A is similar to the R-55, but note the reversal of the colors on
the front panel. Allied #83YU417G. [¥51,000].

Made In:	1965-1968
Voltages:	110-130 VAC 60 Hz 60W
Readout:	⬜⬜⬛⬜ Analog
Physical:	14.75x8.375x10.25" 20 Lbs. 377x213x260mm 9.1 kg
Status:	Inactive Manufacturer Discontinued Model
Rarity:	Scarce

	New	**Used**
Price:	$60-65	$50-55
Rating:	⑨	⑨

R-100

General Coverage Communications Receiver Kit
Single Conversion Superheterodyne. 9 Tubes.

Features:
- ¼" Head. Jack
- Speaker
- RF Gain
- Bandspread
- BFO
- MVC/AVC
- Standby
- Antenna Trimmer
- Mute Line
- ANL
- Q-Multiplier

Specifications:
Coverage ......540 - 30000 kHz +6M   Modes .... AM/SSB/CW
Selectivity ......4.5 - .3 kHz -6dB.   IF .... 455 kHz
Sensitivity ......<1.5µV 10 dB S/N   Antenna Input . Terminals & Coaxial
Audio Out ......8 ohms .5 W

Circuit Complement:
6BZ6 RF Amp, 6BH8 Mixer/HFO, 6AZ8 1st IF Amp, 6AZ8 2nd IF/1st Audio, 6BC7
Detector AVC/ANL, 6AW8A BFO/Audio, 12AX7 Q-Multiplier, 6X4 Rectifier and 0B2
Voltage Regulator.

Accessories:
83Y728 Speaker   83Y727 X-10 Crystal Calibrator
83Y727 S-Meter (shown on radio)

Comments:
Ranges: .54-1.65, 1.6-4.6, 4.4-12.4 and 12-30 MHz. Bandspread: 3.5-4, 6.9-7.3,
14-14.4, 20.5-21.5 and 26.6-30 MHz. Uses printed circuit boards. Allied #83YU726.

Made In:	1959-1961
Voltages:	110-130 VAC 60 Hz 45W
Readout:	⬜⬜⬛⬜ Analog
Physical:	16x10x10.75" 25 Lbs. 280x254x273mm 11.3 kg
Status:	Inactive Manufacturer Discontinued Model
Rarity:	Scarce

	New	**Used**
Price:	$100-110	$50-120
Rating:	⑨	⑨

R-100A

General Coverage Communications Receiver Kit
Single Conversion Superheterodyne. 9 Tubes.

Features:
- ¼" Head. Jack
- Speaker
- RF Gain
- Bandspread
- BFO
- MVC/AVC
- Standby
- Antenna Trimmer
- Mute Line
- ANL
- Dial Lamp
- Q-Multiplier

Specifications:

Coverage ......540 - 30000 kHz +6M   Modes .... AM/SSB/CW
Selectivity ......4.5 - .3 kHz -6dB.   IF .... 455 kHz
Sensitivity ......<1.5µV 10 dB S/N   Antenna Input . Terminals & Coaxial
Audio Out ......8 ohms .5 W

Circuit Complement:
6BZ6 RF Amp, 6BH8 Mixer/HFO, 6AZ8 1st IF Amp, 6AZ8 2nd IF/1st Audio, 6BC7 Detector AVC/ANL, 6AW8 BFO/Audio, 12AX7 Q-Multiplier, 6X4 Rectifier and 0B2 Voltage Regulator.

Accessories:
S-8A Speaker   X-10 Crystal Calibrator
S-Meter (shown on radio)

Comments:
Ranges: .54-1.65, 1.6-4.6, 4.4-12.4 and 12-30 MHz. Bandspread: 3.5-4, 6.9-7.3, 14-14.4, 20.5-21.5 and 26.6-30 MHz. The R-100A matches the T-150 transmitter. Gray metal cabinet. Allied #83YU406DM.

Made In:	United States 1962-1968
Voltages:	110-130 VAC 45W 50/60 Hz
Readout:	Analog
Physical:	17.2x9.2x9.7" 25 Lbs. 437x234x246mm 11.3 kg
Status:	Inactive Manufacturer Discontinued Model
Rarity:	Scarce
Reviews:	73 July 1963

	New	Used
Price:	$100-110	$50 120
Rating:	⑨	⑨

R-195

General Coverage Communications Receiver Kit
Single Conversion Superheterodyne. Solid State.

Features:
- ¼" Head. Jack
- S-Meter
- Speaker 4"
- Antenna Trimmer
- Mute Line
- RF Gain
- Dial Lamp
- Bandspread
- AVC
- ANL
- Flywheel Tuning

Specifications:

Coverage ......200-420, 550-30000 kHz   Modes .... AM/SSB/CW
Selectivity ......4.5 kHz -6dB.   IF .... 455 kHz
Sensitivity ......<2µV 10 dB S/N   Antenna Input . Terminals
Audio .... 1 W

Circuit Complement:
FET Front-end. 13 Transistors.

Comments:
Ranges: .2-.42, .55-1.8, 1.8-4.8, 4.8-12 and 11-30 MHz. This is one of the more collectable Knight Kits.

Made In:	United States 1970-1971
Voltages:	117 VAC or 12 VDC
Readout:	Analog
Physical:	13.25x5.5x11" 10 Lbs. 336x140x279mm 4.5 kg
Status:	Inactive Manufacturer Discontinued Model
Rarity:	Very Scarce
Reviews:	QST October 1970

	New	Used
Price:	$90-100	$200 250
Rating:	⑨	⑨

Space Spanner

Shortwave Broadcast Receiver Kit
Regenerative. 3 Tubes.

Features:
- Head. Terminals • Speaker 4"
- Antenna Trimmer
- Regen. Tuning • Bandspread 0-100
- Speaker-Phones Switch

Specifications:
Coverage ...... See Comments
Modes .... AM
Audio Out ...... 0.5 W 8 ohms
Antenna Input Terminals

Circuit Complement:
12AT7 Detector/First Audio, 50C5 2nd Audio and 35W4 Second Audio Amp.

Comments:
Ranges: .54-1.7 and 6.5-17 MHz. The headphone tip jack and the speaker-phone slide switch are located on the back panel. Allied #83Y259.

Variants:
The later Space Spanner was Allied #83Y209 (1963) $19-20 new. Then the similar Allied 22A3209 (1967) $20 new followed.

Made In:	United States 1957-1962
Voltages:	105-125 VAC 50-60 Hz or 105-125 VDC
Readout:	Analog
Physical:	11x8x5.625" 6 Lbs 279x203x143mm 2.7 kg
Status:	Inactive Manufacturer Discontinued Model
Rarity:	Scarce

	New	Used
Price:	$16-19	$35-50
Rating:	⑨	⑨

Span Master
Y 258

General Coverage Communications Receiver Kit
Regenerative. 2 Tubes.

Features:
- Head. Terminals • Speaker 4"
- Phone-Speaker Switch
- Fine Regen. Tuning • Bandspread

Specifications:
Coverage ...... 540 - 30000 kHz
Modes .... AM/CW
Audio Out ...... 0.5 W 8 ohms
Antenna Input Terminals

Circuit Complement:
6BZ6 Regenerative Detector, 6AW8 Audio Driver/Audio Amp and selenium rectifier.

Comments:
Ranges: .54-1.6, 1.7-4.5, 4.8-14 and 13.5-30 MHz. The phone-speaker switch is on the rear panel. In gray-blue wood cabinet with aluminum grill. Allied #83YX258 and 83Y258.

Made In:	Japan 1957-1967
Voltages:	110-130 VAC 60 Hz
Readout:	Analog
Physical:	14x6.25x6.75" 7 Lbs. 355x158x171mm 3.1 kg
Status:	Inactive Manufacturer Discontinued Model
Rarity:	Common
Reviews:	*CQ* April 1959

	New	Used
Price:	$25-26	$35-50
Rating:	⑨	⑨

Span Master II

General Coverage Communications Receiver Kit
Single Conversion Superheterodyne. 5 Tubes.

Features:
• ¼" Head. Jack   • Speaker 3"   • BFO

Specifications:
Coverage ......550-30000 kHz
Modes ....AM/CW
Audio Out ......1 W 8 ohms
Antenna Input Terminals

Comments:
Ranges: .55-1.6, 1.5-4.5, 4-12 and 11.5-30 MHz.

Made In:	Japan	1969-1970
Voltages:	110-130 VAC 60 Hz	
Readout:	Analog	
Physical:	9.675x5.675x6' 6 Lbs. 246x144x152mm 2.7 kg	
Status:	Inactive Manufacturer Discontinued Model	
Rarity:	Scarce	

	New	**Used**
Price:	$30	$30-40
Rating:	⑨	⑨

Star Roamer

General Coverage Broadcast Receiver Kit
Single Conversion Superheterodyne. 4 Tubes

Features:
• ¼" Head. Jack   • S-Meter   • Speaker 4"   • Bandspread 0-100
• ANL   • Dial Lamp   • Antenna Trimmer   • AVC ON/OFF
• Sensitivity   • MW Ferrite Ant.   • CW Practice Jack

Specifications:
Coverage ......200-420, 550-30000 kHz Modes .... AM
Audio Out ......1 W 8 ohms   Antenna Input . Terminals

Circuit Complement:
6BE6 Converter, 6HR6 IF Amp, 12AX7 Audio Amp and 6AK6 Audio Output.

Comments:
Ranges: .2-.4, .55-1.8, 1.8-4.8, 4.8-12 and 12-30 MHz. This was a popular and affordable receiver kit that sold in large numbers, but was a very modest performer. Black metal cabinet. Allied #83YX102J.

Made In:	Japan and United States	1963-1972
Voltages:	110-130 VAC 60 Hz	
Readout:	Analog	
Physical:	12.25x5.5x8" 10 Lbs. 311x140x203mm 4.5 kg	
Status:	Inactive Manufacturer Discontinued Model	
Rarity:	Abundant	

	New	**Used**
Price:	$40-45	$20-50
Rating:	⑨	⑨

Star Roamer II

General Coverage Communications Receiver Kit
Single Conversion Superheterodyne. Solid State.

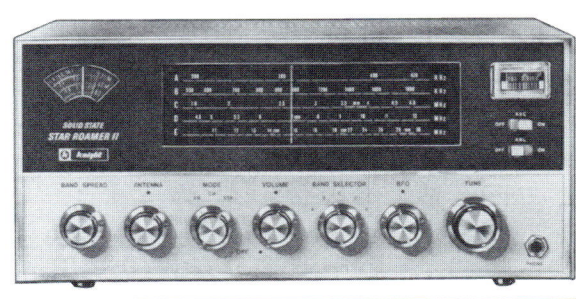

Features:
- ¼" Head. Jack
- ANL
- Bandspread
- S-Meter
- AVC ON/OFF
- Antenna Trimmer
- Speaker 4"
- Dial Lamp
- MW Ferrite Antenna
- CW Practice Jack

Specifications:
Coverage ......200-420, 550-30000 kHz
Selectivity ......5 kHz -6dB.
Audio Out ......1 W 8 ohms
Modes .... AM/CW/SSB
Sensitivity .... <1.5 µV
Ant. Jack .... Terminals

Circuit Complement:
FET Front-end.

Comments:
The solid state version of the famous Star Roamer.
Ranges: .2-.4, .55-1.8, 1.8-4.8, 4.8-12 and 11-30 MHz. Bandspread covers 80 - 10 meters.

Made In:	Japan	1971-1972

Voltages:	110-130 VAC

Readout:	Analog

Physical:	12.25x5.5x8" 10 Lbs.
	311x140x203mm 4.5 kg

Status:	Inactive Manufacturer
	Discontinued Model

Rarity:	Very Scarce

	New	**Used**
Price:	$70	$35-60
Rating:	⑨	⑨

Y-726

General Coverage Communications Receiver Kit
Single Conversion Superheterodyne. 9 Tubes.

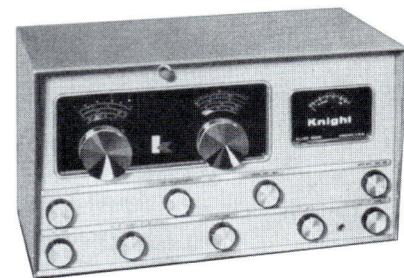

Features:
- ¼" Head. Jack
- ANL
- BFO
- S-Meter (option)
- Dial Lamp
- Q-Multiplier
- AVC/MVC
- Bandspread
- Mute

Specifications:
Coverage ......540 - 30000 kHz
Selectivity ......Var. .3-4.5 kHz -6dB.
Sensitivity ......<1.5µV CW
Audio Out ......1 W 8 ohms
Modes .... AM/CW
IF .... 455 kHz
Antenna Input Terminal and SO239

Circuit Complement:
6BZ6 RF Amp, 6BH8 Mixer/HF Osc, 12AX7 Q-Mult, 6AZ8 1st IF, 6AZ8 2nd IF/1st Audio, 6BC7 Detector/AVC/ANL, 6AW8A BFO/Audio Power Output, 6X4 Rectifier and 0B2 Voltage Regulator.

Accessories:
Y-728 Speaker   Y-727 S-Meter Kit (shown)
Y-256 Calibrator 100 kHz

Comments:
Ranges: .54-1.65, 1.6-4.6, 4.4-12.4 and 12-30 MHz. Bandspread: 3.5-4, 6.9-7.3, 14-14.4, 20.5-21.5 and 26.6-30 MHz. Has an improved metal vernier tuning dial with no strings.

Variants:
The Y-726 appears to be the same or similar to the R-100.

Made In:	United States 1958-1959

Voltages:	110-130 VAC 45 W

Readout:	Analog

Physical:	16x10.75x10" 26 Lbs.
	406x273x254mm 12 kg

Status:	Inactive Manufacturer
	Discontinued Model

Rarity:	Scarce

	New	**Used**
Price:	$105	$40-60
Rating:	⑨	⑨

47 KW Electronics

KW Electronics Ltd. produced an extensive and interesting series of amateur communications receivers and transceivers from the mid 1960's to the early 1970's. Although not well known in America, KW Electronics was a leading producer in the U.K. and Europe at large.

In the 1970's KW Electronics became an assembly plant for Granger Associates. They discontinued the manufacture of amateur radio products in favor of professional MF and HF communications equipment.

The company reportedly changed their name to KW Communications and then to KW Ten-Tec. The exact current status is unknown.

KW Electronics Ltd.
Vanguard Works
1 Heath St.
Dartford, Kent, England

KW ELECTRONICS

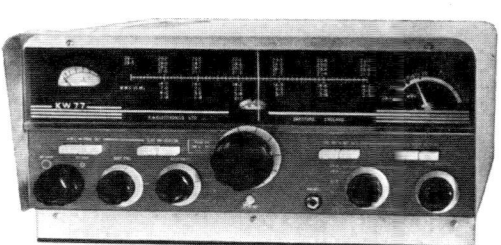

KW-77

Amateur Band Communications Receiver
Triple Conversion Superheterodyne. 16 Tubes plus Semiconductors.

Features:
- ¼" Head. Jack
- Flip Tilt Bar
- NL
- S-Meter
- Slot Filter
- Preselector
- Mute Line
- AVC
- Calibrator
- 50:1 Tune Reduction
- Audio Filter 3000 Hz

Specifications:

Coverage ...... Ham. See Comments.	Modes .... AM/USB/LSB/CW
Selectivity ...... 3.8/2.1/1/.5 kHz -6 dB	IF .... 5000-5600, 455 kHz
Sensitivity ...... 0.5 µV 10 dB S/N	Stability .... <100 Hz after warmup
Image Rej. ..... >65 dB	Antenna Input . 50/80 ohms
Audio .... 3 ohms	

Circuit Complement:
EF183 RF Amp, 6BE6 2nd Mixer/Osc., 6BA6 50 kHz IF Amp, 12AU7 S-Meter, ECL82 Audio Amp Output, EF91 Crystal Calibrator, ECF82 First Mixer & C.O., 6BE6 3rd Mixer/Osc., 6BE6 Product Detector, 6AT6 AM Detector/AVC Rectifier/AVC Amp, S35 Diode Noise Limiter and 0A2 HT Stabilizer.

Accessories:
Matching Speaker

Comments:
Ranges: 1.8-2, 3.5-4.1, 7-7.5, 14-14.5, 21-21.5 and 28-28.5, 28.5-29.1 and 29.1-29.7 MHz. The receiver is finished in grey Hammertone. A popular amateur receiver in England in the 1960's.

Made In: England 1964

Voltages: 110/220/240 VAC.

Readout: Analog Lin.

Physical: 16x6.75x12" 30 Lbs.
406x171x305mm 13.6 kg

Status: Inactive Manufacturer
Discontinued Model

Rarity: Typically Not Available

	New	Used
Price:	£120	☺
Rating:	⑥	☺

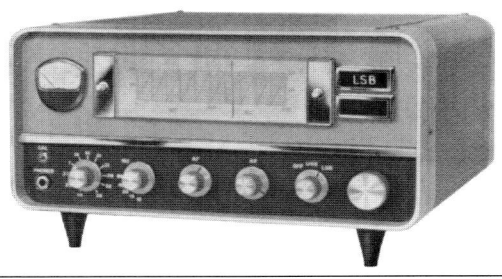

KW ELECTRONICS

KW-201

Amateur Band Communications Receiver
Double Conversion Superheterodyne. Tubes.

Features:
• ¼" Head. Jack   • S-Meter   • Mute Line

Specifications:
Coverage ......Ham. See Comments.   Modes .... USB/LSB/CW
Selectivity ......3.1 kHz -6 dB

Accessories:
Q-Multiplier

Comments:
Ranges: 160-10 Meters.
The optional Q-Multiplier provides 3.1-.2 kHz selectivity.

Variants:
Model **KW-202** 1972-1976 with Q-Multiplier and other features. $395 (shown right). The KW-202 matches the KW-204 transmitter.

Made In:	England	1967-1968

Voltages: 110/220/240 VAC.

Readout: [⎁⎁⎁⎁⎁▮▮] Analog Lin.

Physical: 13.75x6x12.5" 19 Lbs.
350x152x317mm 8.6 kg

Status: Inactive Manufacturer
Discontinued Model

Rarity: Typically Not Available

	New	**Used**
Price:	①	£150
Rating:	⑥	⑥

KW ELECTRONICS

KW-707

Amateur Band Communications Receiver
Double Conversion Superheterodyne. 16 Tubes plus Semiconductors.

Features:
• ¼" Head. Jack   • S-Meter   • Mute Line   • 50:1 Tune Reduction
• Flip Tilt Bar   • Slot Filter   • AVC   • NL
• Preselector   • BFO   • RF Gain   • Calibration Adjust

Specifications:
Coverage ......Ham. See Comments.   Modes .... AM/USB/LSB/CW
Selectivity ......5/2.1/.2 kHz -6 dB   IF .... 5000-5600, 455 kHz
Sensitivity ......0.5 μV 10 dB S/N   Stability .... <100 Hz after warmup
Image Rej. .....>65 dB   Antenna Input . 50/80 ohms
Audio ....3 ohms

Circuit Complement:
EF183 RF Amp, ECH82 First Mixer/C.O., 6BE6 2nd Mixer/Osc., 6BA6 IF Amp, 6BA6 IF Amp, (2) S35 Product Detector, EF91 Crystal Calibrator, ECL82 2nd Audio/Audio Output, (4) DD006 H.T. Rectifiers, 12AU7 S-Meter/1st Audio Amp, 12AU7 Carrier Insertion Oscillator, 6BA6 BFO, 6AL5 NL, ECF82 AVC Gate/Amp, S35 AM Detector, S35 AVC Rectifier and OA2 Stabilizer. The 2.1 kHz filter is mechanical.

Accessories:
Matching Speaker

Comments:
Ranges: 1.8-2, 3.5-4.1, 7-7.5, 14-14.5, 21-21.5, 28-28.5, 28.5-29.1 and 29.1-29.7 MHz. A kiloHertz counter is featured which enables a frequency check to the nearest 1 kHz. The receiver is finished in grey Hammertone.

Made In:	England	1964

Voltages: 110/220/240 VAC.

Readout: [⎁⎁⎁⎁▮⎁⎁] Analog Lin.

Physical: 16x6.75x12" 30 Lbs.
406x171x305mm 13.6 kg

Status: Inactive Manufacturer
Discontinued Model

Rarity: Typically Not Available

	New	**Used**
Price:	£165	②
Rating:	⑥	⑥

The **Lafayette Radio Company** was established in 1921.

Lafayette offered a number of different models during the late 50's and 60's. However, no model in the line stands out as a stellar performer. The HE-10 and HE-30 were extremely popular during the late fifties early sixties as first novice receivers and thousands were sold. Their last model, the BCR-101, was a technically interesting radio, but it suffered from severe mechanical instability, and tuning backlash. The use of an inner and outer knob made tuning cumbersome. Circuity problems were also evident.

The Lafayette chain of radio stores started faltering in the 1970's. The had invested heavily in 4 channel quadraphonic sound but conflicting non-compatible 4 channel formats drove the market so thin that few profits were ever realized. They also invested heavily in 23 channel Citizen Band transceivers and had accumulated a large inventory when the F.C.C. announced a ban on further sales of 23 channel CB's in favor of 40 channel CB's. Lafayette was stuck with millions of dollars of inventory and little time to sell it. Also a new breed of retailers emerged in the New York area during the 1970's which Lafayette either chose not to, or could not, respond to. Lafayette declared bankruptcy in the late 70's with 65 stores immediately closing. The Company eventually downsized to five stores which by the mid 1980's had disappeared.

Lafayette Radio Wire Te-evision
100 Sixth Avenue
New York 13, NY   1945-

Lafayette Radio
P.O. Box 54
Jamaica 31, NY   1958-1961

Lafayette Radio Electro■ics
111 Jericho Turnpike
P.O. Box 10
Syosset, NY 11791   1963-1981

Lafayette

BCR-101

General Coverage Communications Receiver
Double Conversion Superheterodyne. Solid State.

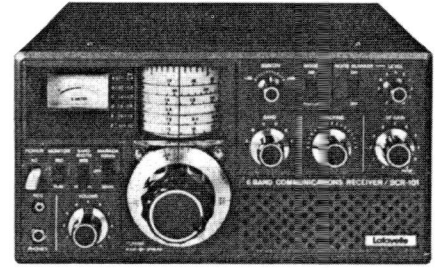

Features:
- ¼" Head. Jack
- RF Gain
- Mute
- IF Out Jack
- S-Meter
- NB
- Dial Lamp
- Tracking
- Record Jack
- BFO
- Speaker 5x2"
- Calibrator 50/500 kHz
- Ferrite MW Antenna
- Audio Input Jack

Specifications:
Coverage ...... 170 - 30000 kHz
Selectivity ..... 8/3 kHz -6dB.
Sensitivity ..... <2μV 10dB 3.5-30 MHz
Audio Out ...... 2 W 10% distortion
Antenna Input Terminals

Modes .... AM/LSB/USB/CW
IF .... 2.15 MHz, 455 kHz.
Image Rej. ...... >60 dB below 15 MHz
Image Rej. ...... >50 dB above 15 MHz

Accessories:
DC Power Cord #99-31106

Comments:
Front-facing oval speaker. Drum-type tuning dial. Ranges: .17-.4, .53-1.5, 1.4-4, 3.5-7.5. 7.5-15 and 15-30 MHz. Reception gap from 400-530 kHz. Single conversion below 4 MHz. The "Tracking" control is used to peak the RF and mixer tuned circuits when tuning frequencies above 3.5 MHz. This was required when changing frequency more than approximately 1 MHz. The "Monitor" switch allows input from an external audio source, such as a take recorder, to be played back through the receiver. This is technically very interesting radio. This radio was Lafayette #99-33805W.

Made In:	Japan 1378-1979
Voltages:	110 VAC 50/60 Hz 10W or 12 VDC @ 50C ma.
Readout:	Analog
Physical:	12x7x9.5" 13.5 Lbs. 301x241x178mm 6.1 kg
Status:	Inactive Manufacturer Discontinued Mcdel
Rarity:	Very Scarce
Reviews:	*Popular Elec.* ᵤul⁻ 1979

	New	Used
Price:	$250	$⁻5-85
Rating:	★★★	★★

Explor-Air Mark V

Shortwave Broadcast Receiver
Single Conversion Superheterodyne. Solid State.

Features:
• ¼" Head. Jack   • Speaker 4"   • Tone Control   • Ferrite MW Antenna
• Dial Lamp

Specifications:
Coverage ...... See Comments.
Sensitivity ...... <5μV
Selectivity ..... 1.5 kHz -30 dB.

Modes .... AM
IF .... 455 kHz
Antenna Input . Terminals

Comments:
Coverage: .55-1.6, 5.9-6.25, 9.45-9.8, 11.45-12 and 15.05-15.5 MHz. The Lafayette Explor-Air Mark V features a walnut grained metal cabinet. Lafayette #99E25850WX.

Made In:	Japan	1970-1972

Voltages: 105-125 VAC 50/60 Hz.

Readout: [⊔⊔⊔∎⊔] Analog

Physical: 14x6x8" 12 Lbs.
356x152x203mm 5.4 kg

Status: Inactive Manufacturer
Discontinued Model

Rarity: Scarce

Review: *Comm. Handbook* 1971

	New	Used
Price:	$50-55	$40-55
Rating:	★★	★

Explor-Air Mark VI

Shortwave Broadcast Receiver
Single Conversion Superheterodyne. Solid State.

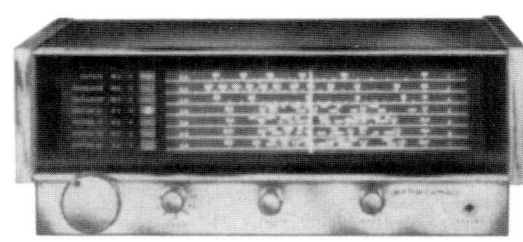

Features:
• ¼" Head. Jack   • 2 Speakers 4"   • Tone   • Ferrite MW Antenna
• Dial Lamp   • Speaker Jack   • Record Jack

Specifications:
Coverage ...... See Comments
Sensitivity ...... <2μV
Selectivity ..... 1.5 kHz -30 dB.
Audio .... 8 ohms

Modes .... AM
IF .... 455 kHz
Image Rej. ...... >45 dB
Antenna Input . Terminals

Circuit Complement:
17 Transistors, 6 diodes and 2 thermistors.

Comments:
Coverage: .15-.4, .55-1.6, 5.9-6.25, 9.45-9.85, 11.85-12.05 and 15.05-15.55 MHz plus 88-108 MHz FM broadcast band. The metal cabinet features brushed aluminum trim and control panel and oiled walnut wood sides. It is interesting and somewhat puzzling that the this model features 24 VDC input. Perhaps part of the run was produced for ship cabin use? Lafayette #99E26015WX.

Variants:
The announced **Explor-Air Mark VII** covers: .15-.4, .55-1.6, 5.9-6.25, 9.45-9.85, 11.85-12.05 and 15.05-15.55 MHz plus the FM broadcast band. $80 1972. It is unclear whether this model ever saw American distribution.

Made In:	Japan	1970-1972

Voltages: 105-125 VAC 50/60 Hz.
or 24 VDC

Readout: [⊔⊔⊔∎⊔] Analog

Physical: 14.75x6.125x9.9" 14 Lbs.

Status: Inactive Manufacturer
Discontinued Model

Rarity: Scarce

Review: *Comm. Handbook* 1971

	New	Used
Price:	$80	$40-55
Rating:	★★	★

HA-63

General Coverage Communications Receiver
Single Conversion Superheterodyne. 7 Tubes plus Semiconductors.

Features:
- ¼" Head. Jack
- S-Meter
- BFO
- Bandspread 0-100
- ANL
- AVC/MVC
- Mute Line
- Antenna Trimmer
- Dial Lamp
- Standby

Specifications:

Coverage ...... 550 - 31000 kHz	Modes .... AM/CW
Sensitivity ...... <1.5µV 10 dB S/N	IF .... 455 kHz
Selectivity ...... ±10 kHz -30 dB	Antenna Input . Terminals 50-70 ohms
Audio .... 1.5W 8 ohm	

Circuit Complement:
6BA6 RF Amp, 6BE6 Mixer, 6BE6 HF Osc., 6BA6 IF Amp, 6AV6 Detector/AVC/ANL/ 1st Audio, 6AV6 BFO and 6AR5 Audio Output and diode.

Accessories:
HE-48A Speaker

Comments: Ranges: .55-1.6, 1.6-4.8, 4.8-14.5 and 10.5-30 MHz. Requires a speaker. Lafayette #99A2534WX.

Variants:
Model **HA-63A** (shown right) 1966 $60 had minor circuit and cosmetic changes. Note different knobs.

Made In:	Japan	1963-1966
Voltages:	117 VAC 50/60 Hz 50W	
Readout:	Analog	
Physical:	13x8x10" 19 Lbs 330x203x254 8.3 kg	
Status:	Inactive Manufacturer Discontinued Model	
Rarity:	Scarce	

	New	Used
Price:	$50-65	$40-50
Rating:	★★★	⑥

HA-225

General Coverage Communications Receiver
Single Conversion Superheterodyne. 14 Tubes.

Features:
- ¼" Head. Jack
- S-Meter
- Dial Lamp
- Antenna Trimmer
- RF Gain
- ANL
- Bandspread
- Calibrator 100 kHz[1]
- Mute
- MVC/AVC
- Standby
- Speaker Terminals
- BFO
- Q-Multiplier
- Line Out 500 ohms

Specifications:

Coverage ...... 150 - 54000 kHz[1]	Modes .... AM/SSB-CW
Audio Out ...... 1.5 W 8 ohms	Antenna Input . SO239s and Terminals

Circuit Complement:
6AQ8 6M RF Amp, 6BL8 6M Mixer/6M Osc, 6BA6 RF Amp, 6BE6 Mixer, 6AQ8 Q-Mult/Calibrator, 6BA6 1st IF Amp, 6BA6 2nd IF Amp, 6AL5 AM Detector/ANL, 6BE6 Product Detector, 6AQA8 1st AF Amp/BFO, 6AQ5 AF Output, 0A2 Voltage Regulator and 6CA4 Rectifier. The HA-225 uses an AM diode detector and an SSB/ CW product detector.

Accessories:
HE-48A Speaker   Crystal for Calibrator

Comments:
Ranges: .15-.4, 1.6-4.8, 4.8-14.5, 10.5-30 and 48-54 MHz. [1] Note the gap of the MW band from .4-1.6 MHz and the gap from 30 to 48 MHz. Bandspread ranges: 3.5-3.7, 3.7-3.94, 7-7.17, 7.17-7.35, 14-14.4, 21-21.5 and 28-30 MHz. A separate SO-239 antenna jack is provided for 6 meters. Requires a speaker. Lafayette #99A2523WX. The crystal for the calibrator is optional.

Made In:	Japan	1964-1966
Voltages:	117 VAC 50/60 Hz 60W	
Readout:	Analog	
Physical:	17x7.5x10" 23 Lbs 432x191x254mm 10.4 kg	
Status:	Inactive Manufacturer Discontinued Model	
Rarity:	Scarce	
Reviews:	*QST* February 1970	

	New	Used
Price:	$120-130	$50-80
Rating:	★★	★★

HA-226

General Coverage Communications Receiver
Single Conversion Superheterodyne. 3 Tubes plus Semiconductors.

Features:
- ¼" Head. Jack
- S-Meter
- Speaker 4"
- MW Ferrite Antenna
- Logging Scale
- Dial Lamp

Specifications:
Coverage ......550 - 30000 kHz
IF . ...455 kHz
Modes .... AM/CW
Antenna Input . Terminals

Circuit Complement:
6BE6 Converter, 6BA6 IF Amp, 6BM8 1st/2nd Audio, a diode and silicon rectifier.

Comments:
Ranges: .55-1.6, 1.6-4.3, 4.3-12 and 11-30 MHz. Metal case with light blue finish. The tuning knob was smaller in some production. The HA-226 appears to be a repackaged HE-60 with an S-Meter added. Lafayette #99A2520WX. Please see Lafayette model HE-60 later in this chapter.

| **Made In:** | Japan | 1966-1967 |

Voltages: 105-125 VAC 50/60 Hz

Readout: Analog

Physical: 10.75x6x8" 12 Lbs
273x152203mm 5.5 kg

Status: Inactive Manufacturer
Discontinued Model

Rarity: Scarce

	New	**Used**
Price:	$40-50	$40-55
Rating:	★★	★★

HA-230

General Coverage Communications Receiver
Single Conversion Superheterodyne. 8 Tubes.

Features:
- ¼" Head. Jack
- S-Meter
- BFO
- Q-Multiplier
- RF Gain
- ANL
- Bandspread
- Antenna Trimmer
- Mute
- Dial Lamp
- Standby
- AVC ON/OFF

Specifications:
Coverage ......550 - 30000 kHz
Selectivity .....2.6 kHz -6dB.
Sensitivity .....<1.5μV 10 dB S/N
Audio Out ......1.3 W
Modes .... AM/SSB-CW
IF .... 455 kHz
Antenna Input . Terminals

Circuit Complement:
One RF and two IF stages.

Accessories:
HE-48A Speaker

Comments:
Ranges: .55-1.6, 1.6-4.8, 4.8-14.5 and 10.5-30 MHz. Bandspread for 80 through 10 meters. Requires a speaker. Always-on filament voltages on mixer and oscillator improve stability. Lafayette #99-2522WX.

Variants:
Model **KT-340** kit form $75. Lafayette #99-2521WX.

Made In: Japan   1964-1966

Voltages: 110-220VAC 50/60 Hz
45W

Readout: Analog

Physical: 15x7.5x10.2" 21 Lbs
380x190x260mm 9.5 kg

Status: Inactive Manufacturer
Discontinued Model

Rarity: Scarce

	New	**Used**
Price:	$70-90	$50-85
Rating:	★★★	★★

HA-350

Amateur Band Communications Receiver
Double Conversion Superheterodyne. 12 Tubes plus Semiconductors.

Features:
- ¼" Head. Jack
- S-Meter
- BFO
- Calibrator 100 kHz[1]
- RF Gain
- ANL
- Preselector
- Calibration Reset
- Mute
- Standby
- Dial Lamp
- Line Out 500 ohms

Specifications:
Coverage ...... Ham bands 80-10M
Selectivity ..... 2 kHz -6dB.
Sensitivity ..... <1µV 10 dB S/N
Audio Out ...... 1 W 8/500 ohm 10% dis.
Modes .... AM/LSB/USB/CW
IF .... 3500-4000, 455 kHz
Antenna Input . SO-239

Circuit Complement:
6BZ6 RF Amp, 6BL8 1st Mixer, 6BE6 2nd Mixer, 6BA6 VFO, 6BA6 IF Amp, 6BA6 IF Amp, 6AL5 AVC/ANL, 6AQ5 Product Detector/Calibrator, 6AVQ 1st Audio Amp, 6AQ5 Audio Output, 6BA6 BFO and OB2 Regulator. Crystal controlled first oscillator. Diode AM detector and CW/SSB product detector. Single conversion below 4 MHz.

Accessories:
HA-94 Speaker   HE-48 Speaker   Crystal for Calibrator

Comments:
Supplied coverage: 3.5-4, 7-7.5, 14-14.5, 21-21.5, 28.5-29.1 MHz and WWV at 15 MHz. Has mechanical filter. Lafayette #99-2524WX. [1]Crystal for calibrator not supplied.

Made In:	Japan	1964-1968

Voltages: 115 VAC 50/60 Hz

Readout: Analog Lin.

Physical: 15x7.5x10" 19 lbs
380x190x255mm 8.6 kg

Status: Inactive Manufacturer
Discontinued Model

Rarity: Scarce

Reviews: *CQ* February 1965
QST December 1964

	New	Used
Price:	$130-190	$5-95
Rating:	★★★	★

HA-500

Amateur Band Communications Receiver
Double Conversion Superheterodyne. 10 Tubes.

Features:
- ¼" Head. Jack
- S-Meter
- BFO
- Antenna Trimmer
- RF Gain
- ANL
- Line Out 500 ohm
- Calibrator 100 kHz
- Mute
- Dial Lamp
- Standby
- Logging Scale

Specifications:
Coverage ...... Ham. 80-6 Meters.
Sensitivity ...... <1.5µV 10 dB S/N
Selectivity ...... 4 kHz
IF Reject .... >40 dB
Audio .... 8/500 ohms 1W
Modes .... AM/CW-SSB
IF .... 2608, 455 kHz
Image Rej. ...... >40 dB
Antenna Input . Terminals

Circuit Complement:
6BZ6 RF Amp, 6AU6 1st Mixer, 6AQ8 Local Osc., 6BE6 Crystal Controlled 2nd Mixer, 6BA6 IF Amp, 6BA6 IF Amp, 6AQ8 Product Detector/BFO, 6AQ8 AM NL/Crystal Calibrator, 6BM8 Audio Amp and 0B2 Regulator. Two mechanical filters.

Accessories:
HE-48A Speaker

Comments:
Ranges: 3.5-4, 7-7.3, 14-14.35, 21-21.45, 28-29.7 and 50-54 MHz. Requires a speaker. Lafayette #99-2574WX.

Made In:	Japan	1967-1968

Voltages: 110-120 VAC 60 Hz.
65W

Readout: Analog

Physical: 15x7.5x10' 23 Lbs
380x190x255mm 10.4 kg

Status: Inactive Manufacturer
Discontinued Model

Rarity: Scarce

	New	Used
Price:	$130-150	$50-65
Rating:	★★★	★

HA-600A

General Coverage Communications Receiver
Single Conversion Superheterodyne. Solid State.

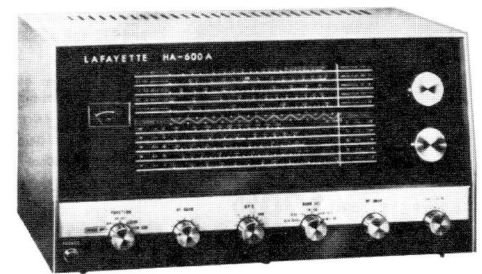

Features:
- ¼" Head. Jack
- S-Meter
- BFO ±2.5 kHz.
- Antenna Trimmer
- RF Gain
- ANL
- Bandspread
- Dial Lamp
- Mute
- Standby
- Record Jack

Specifications:

Coverage ...... 150 - 30000 kHz
Selectivity ...... ±2 kHz -6 dB.
Sensitivity ...... <1µV 10 dB S/N
Audio .... 3W 4/8/500 ohms

Modes .... AM/SSB-CW
IF .... 455 kHz
Antenna Input . 50-400 ohm

Circuit Complement:
10 Transistors, 2 FETs, 7 Diodes and 1 Zener Diode. Product detector and mechanical filter.

Accessories:
HE-48C Speaker

Comments:
Sleek angled front panel. Has rather poor image rejection. Requires a speaker. Ranges: .15-.4, .55-1.6, 1.6-5, 4.8-14.6, 10.5-30 MHz. Bandspread: 80/75, 40, 20, 15 and 10 meters. Lafayette #99E25991W.

Variants:
Model **HA-600** 1968 identical appearance, differences unknown. Model **HA-600T** 1968-1970 $100 Lafayette #99E25959WX. Reviewed in *Electronics Illustrated* July 1969.

Made In:	Japan	1970-1975

Voltages: 105-120 VAC 50/60 Hz
17 W or 12 VDC 500 ma

Readout: ⊔⊔⊔▮⊔ Analog

Physical: 15x8.25x9.75" 17 Lbs.
380x210x250mm 7.7 kg

Status: Inactive Manufacturer
Discontinued Model

Rarity: Scarce

Reviews: *WRTH* 1976
Popular Elec. Nov. 1972

	New	Used
Price:	$100-170	$75-95
Rating:	★★★	★★

HA-700

General Coverage Communications Receiver
Single Conversion Superheterodyne. 6 Tubes plus Semiconductors.

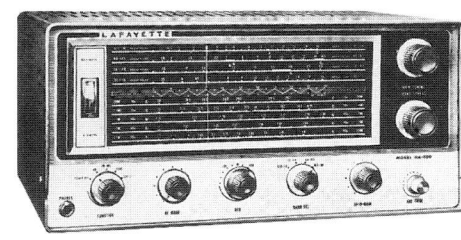

Features:
- ¼" Head. Jack
- S-Meter
- AVC
- Antenna Trimmer
- RF Gain
- ANL
- BFO
- Bandspread
- Mute
- Standby
- Dial Lamp

Specifications:

Coverage ...... 150 - 30000 kHz
Selectivity ...... 2.6 kHz -6dB.
Sensitivity ...... <1µV 6 dB
Audio Out ...... 1.3 W 4/8 ohm

Modes .... AM/CW-SSB
IF .... 455 kHz
Antenna Input . Terminals

Circuit Complement:
Tubes: (3) 6BA6 RF Amp/IF Amp, 6BL8 Mixer/Local Oscillator, 6AQ8 Product Detector/BFO, 6BM8 AF Amp. Diodes: (4) 1N60 AVC/AM Detector/S-Meter, SW-05S ANL and (2) FR-1K Rectifier.

Accessories:
HE48A Speaker

Comments:
Ranges: .15-.4, .55-1.6, 1.6-4.8, 4.8-14.6 and 10.5-30 MHz. Reception gap from .4 to .55 MHz. Requires a speaker.

Mfg. In:	Japan	1966-1968

Voltages: 110-120 VAC 50/60 Hz
45W

Readout: ⊔⊔⊔▮⊔ Analog

Physical: 15x7.675x10" 20.9 Lbs.
380x190x260mm 9.5 kg

Status: Inactive Manufacturer
Discontinued Model

Rarity: Common

	New	Used
Price:	$90	$60-95
Rating:	★★★	★

HA-800

Amateur Band Communications Receiver
Double Conversion Superheterodyne. Solid State.

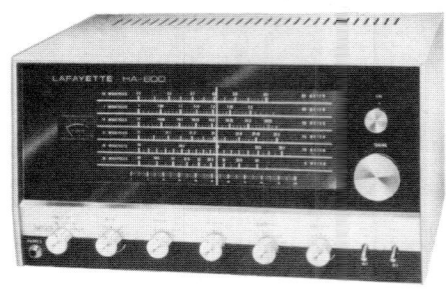

Features:
- ¼" Head. Jack
- S-Meter
- BFO ±2.5 kHz
- Antenna Trimmer
- RF Gain
- ANL
- Record Jack
- Crystal Calibrator[1]
- Mute
- Standby
- Dial Lamp
- Dial Lamp Switch

Specifications:

Coverage ......Ham. 80 - 6 Meters	Modes .... AM/SSB-CW
Selectivity .....4 kHz -6dB.	IF .... 2.608 MHz, 455 kHz
Sensitivity .....See Comments	Image Rej. .... >40 dB
Audio Out ......1 W 8/500 ohms	Antenna Input Terminals

Circuit Complement:
14 Transistors, 3 FETs and 7 diodes. Product detector for SSB/CW. Two mechanical filters.

Accessories:
HE-48B Speaker   Crystal for Calibrator

Comments:
Ranges: 3.5-4, 7-7.3, 14-14.35, 21-21.5, 28-29.7 and 50-54 MHz. Sensitivity: <.5 µV on 15/10M, <1µV on 80/40/20M and 2.5 µV on 6M. Requires a speaker. [1]The crystal calibrator requires optional 100 kHz crystal to operate. Lafayette #99E25942WX.

Made In:	Japan	1970-1971
Voltages:	117 VAC 50/60 Hz 8W or 12 VDC 7W	
Readout:	Analog	
Physical:	15x8.25x9.75" 13 Lbs. 380x210x250mm 7.3 kg	
Status:	Inactive Manufacturer Discontinued Model	
Rarity:	Scarce	
Reviews:	QST February 1970	

	New	Used
Price:	$130-150	$45-65
Rating:	★★★	★★

HA-800B

Amateur Band Communications Receiver
Double Conversion Superheterodyne. Solid State.

Features:
- ¼" Head. Jack
- S-Meter
- Record Jack
- Antenna Trimmer
- RF Gain
- ANL
- BFO
- Crystal Calibrator[1]
- Mute
- Dial Light
- Speaker Out Jack

Specifications:

Coverage ......Ham bands 80-6M	Modes .... AM/SSB/CW
Selectivity ......±6 kHz -60dB.	Antenna Input . Terminals
Sensitivity ......See Comments.	

Circuit Complement:
15 Transistors, 4 FETs, 1 Varactor and 13 diodes. Product detector and two mechanical filters.

Accessories:
HE-48C Speaker   Crystal for Calibrator

Comments:
Ranges: 3.5-4, 7-7.3, 14-14.35, 21-21.5, 28-29.7 and 50-54 MHz.
Sensitivity: <.5 µV on 15/10M, < 1µV on 80/40/20M and 2.5 µV on 6M. Lafayette #99R26155WX. Requires a speaker. [1]Calibrator requires optional 100 kHz filter to operate.

Made In:	Japan	1971-1974
Voltages:	117 VAC 50/60 Hz or 12 VDC	
Readout:	Analog	
Physical:	15x8.25x9.75" 16 Lbs. 380x210x250mm 7.3 kg	
Status:	Inactive Manufacturer Discontinued Model	
Rarity:	Scarce	

	New	Used
Price:	$130	$50-75
Rating:	★★★	★★

HE-10

General Coverage Communications Receiver
Single Conversion Superheterodyne. 9 Tubes

Features:
- ¼" Head. Jack
- S-Meter
- AVC/MVC
- Bandspread 0-100
- IF Gain
- ANL
- BFO
- Hinged Cover
- Mute
- Standby
- Dial Lamp

Specifications:

Coverage ...... 455 - 31000 kHz

Selectivity ...... 10 kHz -60 dB.

Sensitivity ...... <1.25µV 10 dB S/N

Audio Out ...... 1.5 W 4 ohm 10% dist

Modes .... AM/CW

IF .... 455 kHz

Image Rej. .... -40 dB @30 MHz.

Antenna Input . Terminals

Circuit Complement:
6BD6 RF Amp, 6BE6 Mixer, 6BE6 HF Osc, 6BD6 1st IF Amp, 6BD6 2nd IF Amp, 6AV6 Detector/AVC/AF Amp, 6AV6 BFO/ANL, 6AR5 Audio Output and 5Y3GT Rectifier. One RF and two IF stages.

Accessories:
HE48 Speaker

Comments:
Ranges: .55-1.6, 1.-4.8, 4.8-14.5 and 10.5-31 MHz. Gray metal cabinet. The HE-10 was fashioned after the Hallicrafters S-38 series, but is a superior radio.

Variants:
Model **KT-200WX** same in kit form $65. In Europe this model also apparently sold as the **Jennen 9R-4J**.

Made In:	Japan	1959-1963
Voltages:	105-125 VAC 50/60 Hz 50W	
Readout:	Analog	
Physical:	15.5x8.25x12" 20 Lbs. 394x210x305mm 9 kg	
Status:	Inactive Manufacturer Discontinued Model	
Rarity:	Scarce	
Reviews:	*73* May 1961	

	New	**Used**
Price:	$80	$45-65
Rating:	★★★	⑦

HE-30

General Coverage Communications Receiver
Single Conversion Superheterodyne. 9 Tubes.

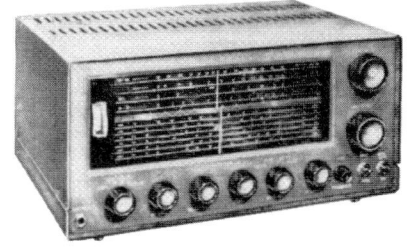

Features:
- ¼" Head. Jack
- S-Meter
- BFO
- Q-Multiplier
- Bandspread
- ANL
- AVC/MVC
- Antenna Trimmer
- Dial Lamp
- Standby

Specifications:

Coverage ...... 550 - 30000 kHz

Selectivity ...... 4 kHz (to 1.6 Q-Mult.)

Sensitivity ...... <1.6µV 10 dB S/N

Audio .... 4/8 ohms 1.5 W

Modes .... AM/CW/SSB

IF .... 455 kHz

Antenna Input . Terminals 50-400 ohms

Circuit Complement:
6BA6 RF Amp, 6BE6 Mixer, 6BE6 Osc, 6AV6 Q-Multiplier/BFO, (2) 6BA6 IF Amp, 6AV6 Detector/AF Amp/ANL, 6AQ5 Audio Output and 5Y3 Rectifier. One RF and two IF stages.

Accessories:
HE-48 Speaker   HE-48 Speaker

Comments:
Ranges: .55-1.6, 1.6-4.8, 4.8-14.5 and 10.5-30 MHz. Requires a speaker. Gray crackle finish metal cabinet with white lettering. Lafayette #99G2544WX. This was a very popular model with thousands sold.

Variants:
Model **KT-320** $75-80 same in kit form. Lafayette #99G2545WX. Also see the extremely similar Trio 9R59 in the Kenwood Chapter of this book.

Made In:	Japan	1961-1964
Voltages:	105-120 VAC 60 Hz 50W	
Readout:	Analog	
Physical:	15x7x10" 21 Lbs. 380x180x255mm 9.5 kg	
Status:	Inactive Manufacturer Discontinued Model	
Rarity:	Common	
Reviews:	*QST* November 1961 *Radio Bygones* #48	

	New	**Used**
Price:	$80-100	$55-75
Rating:	★★	★

HE-40

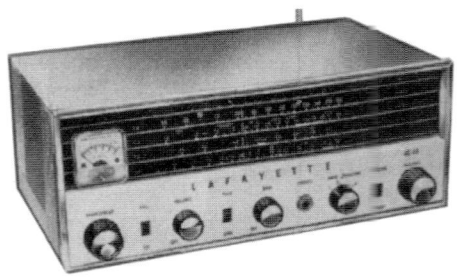

General Coverage Communications Receiver
Single Conversion Superheterodyne. 4 Tubes plus Semiconductors.

Features:
- ¼" Head. Jack
- S-Meter
- AVC
- Bandspread
- Speaker 5"
- ANL
- BFO
- Sensitivity
- Tone
- Standby
- Dial Lamp
- MW Ferrite Antenna
- Telescopic Whip 58"

Specifications:
Coverage ......550 - 30000 kHz   Modes .... AM/CW
IF ....455 kHz   Antenna Input . Terminals

Circuit Complement:
12BE6, 12BA6, 12AV6, 50C5 and silicon rectifier.

Comments:
Ranges: .55-1.6, 1.6-4.4, 4.5-11 and 11-30 MHz.

Variants:
This model was sold in England as the **HAM-I**. Model **KT-222** $40 was the same as the HE-40, but in kit form.

Made In:	Japan	1961-1963

Voltages:	105-125 VAC 50/60 Hz or 105-125 VDC

Readout:	Analog

Physical:	17x7.5x10" 12 Lbs. 431x190x254mm 5.5 kg

Status:	Inactive Manufacturer Discontinued Model

Rarity:	Scarce

Reviews:	73 December 1961

	New	**Used**
Price:	$50-55	$30-35
Rating:	★★	★

HE-60

General Coverage Communications Receiver
Single Conversion Superheterodyne. 3 Tubes plus Semiconductors.

Features:
- ¼" Head. Jack
- Dial Lamp
- Speaker
- MW Ferrite Antenna

Specifications:
Coverage ......550 - 30000 kHz   Modes .... AM/CW
IF ....455 kHz   Antenna Input . Terminals

Circuit Complement:
6BE6, 6BA6, 6BM8, a diode and silicon rectifier.

Comments:
Ranges: .55-1.6, 1.6-4.4, 4.5-12 and 11-30 MHz. Gray crackle finish. This receiver was announced in 1962 and appeared in the 1963 Lafayette Catalog. It was not advertised after that. The author questions whether this model was ever actually imported to the American market. It appears that is was repackaged with the addition of an S-Meter and sold in the U.S. as the HA-226.

Made In:	Japan	1962

Voltages:	115-125 VAC 50/60 Hz

Readout:	Analog

Physical:	10.75x5.7x8.2" 9 Lbs. 273x145x208mm 4.1 kg

Status:	Inactive Manufacturer Discontinued Model

Rarity:	Typically Unavailable.

	New	**Used**
Price:	$40	②
Rating:	⑥	⑥

HE-80

General Coverage Communications Receiver
Single Conversion Superheterodyne. 14 Tubes.

Features:
- ¼" Head. Jack
- S-Meter
- Bandspread
- Calibrator 100 kHz[1]
- RF Gain
- ANL
- BFO
- Q-Multiplier
- Record Jack
- MVC/AVC
- Dial Lamp
- Mute Line

Specifications:
Coverage ...... 550 - 30000 kHz +6M   Modes .... AM/CW/SSB/FM
Selectivity ...... 2.5 kHz -6dB & Var.   IF .... 455 kHz
Sensitivity ...... <1µV 10 dB S/N   Antenna Input . Terminals
Audio Out ...... 1.5 W 8/500 ohms

Circuit Complement:
6AQ8 6M RF Amp, 6AU6 6M Mixer, 6AQ8 6M Oscillator, 6BA6 RF Amp, 6BE6 Mixer, 6AQ8 Oscillator Buffer, (2) 6BA6 IF Amp, 6AL5 AM Detector/ANL, 6BE6 Product Detector, 6AQ8 AF Amp/BFO, 6AQ5 AF Amp, 6AQ8 Q-Multiplier/Calibrator, 6CA4 Rectifier, 0A2 Voltage Regulator and 1N60 Crystal Calibrator. Dual conversion on 6 meters. Product detector for SSB.

Accessories: Crystal for Calibrator   Speaker HE-48

Comments: This model was reportedly manufactured for Lafayette by Trio (later to become Kenwood) in Japan. Ranges: .54-1.6, 1.6-4.8, 4.8-14.5, 10.5-30 and 48-54 MHz. Variable selectivity is to ±700 Hz at -6 dB. Lafayette #99G2538WX. [1]Calibrator requires optional 100 kHz crystal to operate. Gray. Requires a speaker.

Variants: The similar model Trio ER-202 sold in Europe features 2 meter coverage instead of 6 meters. Please see the Kenwood chapter of this book.

Made In:	Japan	1963-1964

Voltages: 115 VAC 50/60 Hz
60W

Readout: |␣␣␣▮␣| Analog

Physical: 17x7.5x10" 23.5 Lbs.
430x190x260mm 10.7kg

Status: Inactive Manufacturer
Discontinued Model

Rarity: Common

	New	**Used**
Price:	$130-150	$60-90
Rating:	★★★	★★

KT-135
Explor-Air

General Coverage Communications Receiver Kit
Regenerative. 3 Tubes.

Features:
- ¼" Head. Jack
- Speaker 4"
- Antenna Tuner
- Bandspread
- Regeneration

Specifications:
Coverage ...... 550 - 30000 kHz   Modes .... AM/CW
Antenna Input Terminals

Circuit Complement:
12AT7 Regenerative Detector, 50C5 Audio Output and 35W4 Rectifier.

Accessories:
ML-150 Leatherette covered wooden cabinet 19E09068 (shown).

Comments:
Ranges: .55-1.6, 1.65-5.5, 5.5-16 and 16-30 MHz. Bandspread: 3.5-4, 7-7.3 and 12.5-14.4 MHz. The headphone jack is on the rear panel. Lafayette #19E09050 Earlier production featured a monotone, lighter, front panel. Different knobs were a so featured over the long production life of this model.

Made In: United States 1958-1970

Voltages: 105-125 VAC 50/60 Hz.
25 W

Readout: |␣␣␣▮␣| Analog

Physical: 10x7x5" 4 Lbs.
356x152x203mm 5.4 kg

Status: Inactive Manufacturer
Discontinued Model

Rarity: Scarce

	New	**Used**
Price:	$23	$25-35
Rating:	⑨	⑨

The **Lowe Electronics Limited** company was started in 1964, specializing in the supply of equipment for the radio amateur hobby market in the United Kingdom. Expansion led to the establishment in 1979 of a company headquarters and service center in Matlock, Derbyshire. Several Company owned retail branch outlets can be found throughout England.

For many years Lowe was the leading importer of Japanese receivers into the U.K. It was decided that a domestically produced receiver could be built with better performance and at a lower cost. The first receiver was introduced in 1987 and was sold predominantly within the United Kingdom. Additional models were produced with the unique HF-150 being perhaps the best known and most successful. This and other Lowe models have earned high marks for exceptional performance, solid, straight forward construction and excellent durability. This reputation has transformed Lowe from a domestic producer to a respected worldwide exporter with 85% of its (1993) production being sold outside the U.K.

Lowe Electronics Ltd.
Chesterfield Rd.
Matlock, Derbyshire DE4 5L≣
England   ₁979-1998

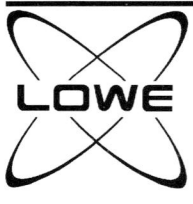

HF-125

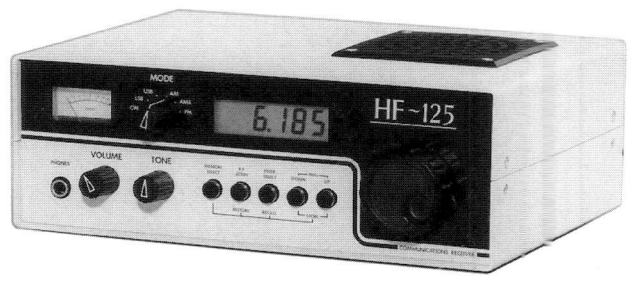

General Coverage Communications Receiver
Double Conversion Superheterodyne. Solid State.

Features:
- ¼" Head. Jack
- S-Meter
- Tone
- 30 Memories
- Speaker Jack
- Record Jack
- Tilt Stand
- Attenuator -20dB
- 400 Hz audio CW filter

Specifications:
Coverage ...... 30-30000 kHz
Selectivity ...... 10/7/4/2.5 kHz -6dB.
Sensitivity ...... <.3µV (SSB .5-30 MHz).
Audio Out ...... 0.75W 8 ohm 5% THD
Dyn.Range .... >90 dB at 50 kHz

Modes .... CW/LSB/USB/AM
Image Rej. ...... > 80 dB
IFs .... 45 MHz & 455 kHz
VFO Incr. .... 15.6 Hz
Antenna Input . SO-239 & Terminals

Accessories:
D125 Sync Det & FM K-125 Wired Keypad   P-125 Portable Option

Comments:
A receiver respected for its clean, solid design, good performance and ample filter selection. The P-125 Portable option consists of: an antenna preamp on a printed circuit board, telescopic whip antenna, eight NiCad cells and two battery holders. Also reviewed in *Practical Wireless* March & April 1987, *Electronics & Wireless World* April 1987.

Made In:	England	198₇-1988
Voltages:	12 VDC .25 Amp or internal batteries	
Readout:	`00001.`	Digital LCD
Physical:	10x4x8.3" 4.2 Lbs. 255x100x200	
Status:	Active Manufacture Discontinued Model	
Rarity:	Scarce	
Reviews:	IBS-RDI Whitepaper *Passport* 198₈-1989 *WRTH* 1987, 198₈ *Monitoring Times* April 1987 *WRTH Buyer's Guide* 1993	

	New	Used
Price:	$599	$3₅0
Rating:	★★★★	★★★

HF-150

General Coverage Communications Receiver
Double Conversion Superheterodyne. Solid State.

Features:
- ¼" Head. Jack
- Whip Preamp
- 2 Tuning Rates
- Speaker
- Metal chassis
- Whip Preamp
- 60 Memories
- Speaker Jack
- Atten. -20 dB
- Record Jack
- Sync Detect.

Specifications:

Coverage ......30 - 30000 kHz	Modes .... AM/LSB/USB
Selectivity ......7/2.5 kHz -6dB.	Stability .... <±30 Hz in 1 hour
Sensitivity ......<.5µV[1] (SSB .5-30 MHz).	IFs .... 45 MHz & 455 kHz
Audio Out ......1.6W 8 ohm 5% THD	Antenna Input . SO-239 & Terminals
Distortion ....<1% THD all modes	VFO Incr. .... 8 Hz & 60 Hz
Spur. Res. .....>65 dB	

Accessories:

AP-150 Audio Filter	MB-150 Mobile Bracket	C-150 Nylon Case
RK-150 Rack Shelf	IF-150 PC Interface	KPAD-1 Wired Keypad
BK-150 Backlit Kit	MPW Whip Aerial	KPAD-2 (for HF-150 Europa)
AK-150 Accs. Kit (Whip, Hand & Shoulder Straps, Nicads)		

Comments: [1] Sensitivity is .3µV with preamp on. An excellent, robustly built receiver with good filtering. The standard LCD is not backlit. Memories store frequency, mode and bandwidth. Sturdy metal alloy case.

Variants: Model **HF-150M** is a marine version (main circuit board is treated to be moisture resistant and the case is white). The **HF-150 Europa** 1998 replaces the HF-150 and includes a backlit LCD, an all black case and other refinements.

Made In:	England 1992-1997
Voltages:	12 VDC or internal batteries (8 x AA cells)
Readout:	`00001.` Digital LCD
Physical:	7.2x3.2x6.5" 3 Lbs. 185x80x160mm 1.3 kg
Status:	Active Manufacturer Discontinued Model
Rarity:	Common
Reviews:	IBS-RDI Whitepaper *Passport* 1993-1998 *WRTH* 1992, 1993 *WRTH Buyer's Guide* 1993 *Mon. Times* August 1992 *ODXA* February 1993 *SW Magazine* Jan. 1993 *Proceedings* 1992-93

	New	Used
Price:	$499-699	$340-390
Rating:	★★★★★	★★★★

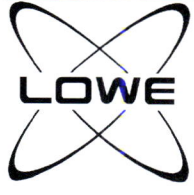

HF-225

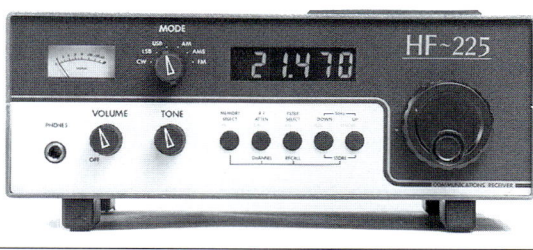

General Coverage Communications Receiver
Double Conversion Superheterodyne. Solid State.

Features:
- ¼" Head. Jack
- Front Tilt Feet
- VFO A/B
- Backlit LCD
- S-Meter
- Speaker
- 30 Memories
- Tone
- Speaker Jack
- Record Jack
- Attenuator -20 dB
- 200 Hz CW Audio Filter
- Speed-Sensitive Tune

Specifications:

Coverage ......30 - 30000 kHz	Modes .... CW/LSB/USB/AM
Selectivity ......10/7/4/2.2 kHz -6dB.	Stability .... <±30 Hz in 1 hour
Sensitivity ......<.5µV (SSB 2-30 MHz).	IFs .... 45 MHz & 455 kHz
Audio Out ......1.6W 8 ohm 5% THD	Antenna Input . SO-239 50 ohm & jacks
Distortion ....<1% THD all modes	VFO Incr. .... 8, 50, 125 Hz

Circuit Complement:

A dual conversion super het, using up-conversion to a HF first IF of 45 MHz and a second IF of 455 kHz for the selective filters. This provides good IF image rejection of all tuned frequencies in the HF band coupled with good filter shape factors in the 455 IF. Six band selecting filters are provided before the first mixer.

Accessories:

B-225 NiCad Pack	D-225 Sync Det & FM Mode	W-225 Tels. Whip & Amp
C-225 Carry Case	KPAD-1 Wired Keypad	S-225 Speaker

Comments: When in the memory channel mode, the main tuning knob allows rapid scrolling of the 30 memories. Memories 0-9 can be recalled through the keypad. Memories are maintained by a lithium cell. LCD is backlit. Also see model HF-225E.

Made In:	England 1990-1997
Voltages:	12 VDC 300ma Supplied with AC Power Supply
Readout:	`00001.` Digital LED
Physical:	10x4x8.3" 4.2 Lbs. 254x102x211mm 1.9 kg
Status:	Active Manufacturer Active Model
Rarity:	Common
Reviews:	IBS-RDI Whitepaper *Passport* 1990-1995 *WRTH* 1990 *WRTH Buyer's Guide* 1993 *ODXA* March 1993 *SW Magazine* June 1989 *Proceedings* 1992-93

	New	Used
Price:	$749-799	$490
Rating:	★★★★	★★★★

HF-225
Europa

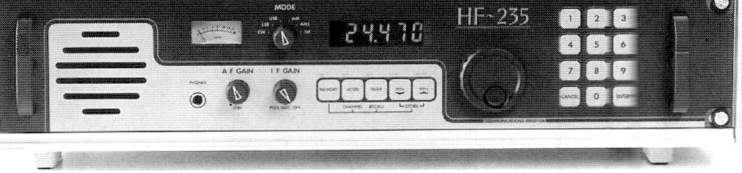

General Coverage Communications Receiver
Double Conversion Superheterodyne. Solid State.

Features:
- ¼" Head. Jack
- S-Meter
- Tone
- Attenuator -20 dB
- Front Tilt Feet
- Speaker
- Speaker Jack
- 200 Hz CW Audio Filter
- VFO A/B
- 30 Memories
- Record Jack
- Speed-Sensitive Tune
- Whip Preamp

Specifications:
Coverage ...... 30 - 30000 kHz
Selectivity ...... 7/4.5/3.5/2.2 kHz -6dB.
Sensitivity ...... <.5µV (SSB 2-30 MHz).
Audio Out ...... 1.6W 8 ohm 5% THD
Distortion .... <1% THD all modes

Modes .... CW/LSB/USB/AM
Stability .... <±30 Hz in 1 hour
IFs .... 45 MHz & 455 kHz
Antenna Input . SO-239 & jacks
VFO Incr. .... 8, 50, 125 Hz

Accessories:
B-225 Nicad Pack   C-225 Carry Case   W-225 Tels. Whip & Amp

Comments:
An optimized version of the HF-225, geared towards the serious DXer. Enhancements include: improved and tighter I.F. filters, improved filter chokes, filter selection diodes replaced by low capacitance switching diodes, filter select decoupling capacitors are bypassed by new chip capacitors. Additionally the KPAD-1 keypad and D-225 Sync AM and FM detector are featured as standard. This model is sometimes referred to as the HF-225E.

Made In:	England	1993-1997
Voltages:	12 VDC 300ma Supplied with AC Power Supply.	
Readout:	00001.	Digital LED
Physical:	10x4x8.3" 4.2 Lbs. 254x102x211mm 1.9 kg	
Status:	Active Manufacturer Discontinued Model	
Rarity:	Scarce	
Reviews:	Passport 1997 SW Magazine Sept. 1994 Proceedings 1994-95	

	New	Used
Price:	$969-1100	$590-620
Rating:	★★★★★	★★★

HF-235

General Coverage Communications Receiver
Double Conversion Superheterodyne. Solid State.

Features:
- ¼" Head. Jack
- S-Meter
- Tone
- 30 Memories
- Rack Handles
- Keypad
- Speaker 2"
- 200 Hz CW Audio Filter
- AGC ON/OFF
- VFO A/B
- Attenuator 20 dB
- Speed-sensitive Tune
- Mute Line
- Backlit LCD
- Line Out 600 ohm

Specifications:
Coverage ...... 30 - 30000 kHz
Selectivity ..... 10/7/4/2.2 kHz -6dB.
Sensitivity ...... <.5µV (SSB 2-30 MHz).
Audio Out ...... 3.5W 4 ohm 5% THD
Distortion .... <1% THD all modes

Modes .... CW/LSB/USB/AM
Stability .... <±30 Hz/hr. @20°C
IFs .... 45 MHz & 455 kHz
VFO Increment 8, 50, 125 Hz
Antenna Input . BNC 50 ohm

Accessories:
S-225 Speaker   D-225 NBFM/AM Sync.   HF-235/HS High Stability
DC-235 Cabinet   HF-235/RC RS232 Port.

Comments: This model was designed for the military-commercial market. When in the memory channel mode, the main tuning knob allows rapid scrolling of the 30 memories. Memories 0-9 can be recalled through the keypad. Memories are maintained by a lithium cell. A DB25M jack on the back provides the line out, mute, speaker out, DC power and AGC. Keypad entry to 1 kHz. The LCD display is backlit. Optional control by computer at 300/8/N/1 or 1200/8/N/1.
Variants: Model **HF-235/R** includes the RS-232 port. Model **HF-235/F** includes FAX mode. Model **HF-235/H** includes the high stability option.

Made In:	England	1992-1996
Voltages:	110-120 or 220-240 VAC or 20-40 VDC	
Readout:	00001.	Digital LCD
Physical:	19x3.5x12.6" 12 Lbs. 483x88x320mm 5.5 kg	
Status:	Active Manufacturer Discontinued Model	
Rarity:	Very Scarce	
Reviews:	Passport 1992-1996 IBS-RDI Whitepaper Mon. Times February 1992	

	New	Used
Price:	$1995	$600-950
Rating:	★★★★	★★★

LOWE

HF-250

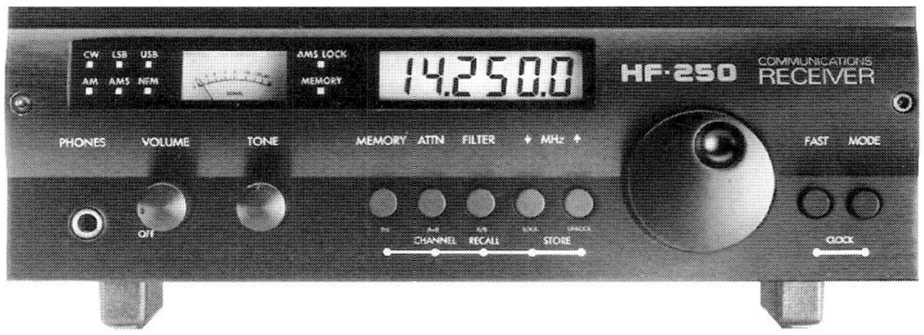

General Coverage Communications Receiver
Double Conversion Superheterodyne. Solid State.

Features:
- ¼" Head. Jack
- 255 Memories
- Speaker
- Mute Line
- S-Meter
- Front Tilt Feet
- Record Jack
- Speaker Jack
- Tone
- Backlit Display
- RS-232C Port
- Tilt Legs
- 200 Hz CW Audio Filter
- Attenuator -20 dB
- Dual Clock Timer
- FM Squelch

Specifications:
Coverage ......30 - 30000 kHz
Sensitivity ......<.2μV (SSB 2-30 MHz).
Selectivity ......10/7/4/2.2 kHz -6dB.
Audio Out ......1.6W 8 ohm 5% THD
Modes .... CW/LSB/USB/AM/FM
VFO Incr. .... 8 Hz
Stability .... <±10 Hz in 1 hour
Antenna Input . 50 ohm SO-239 & Terms.

Accessories:
DU250 FM/AM Sync. RC250 Infrared Remote WA250 Tels. Whip & Amp

Comments:
The antenna spring terminals are 600 ohms. The LCD is backlit. Memories store frequency and mode. This radios tunes in 8 Hz steps.

Variants:
The **HF-250E Europa** model was introduced in early 1998 and is optimized with improved filters and filter chokes.

Made In: England 1995-1997

Voltages: 10-15 VDC. Supplied with AC Power Supply

Readout: 00000.1 Digital LCD

Physical: 11x4.13x5.2" 6 Lbs. 280x105x205mm 2.7 kg
Status: Active Manufacturer, Active Model
Rarity: Common
Reviews: *Passport* 1996-1998
WRTH 1996
ODXA December 1995
NASWA May 1996
SW Magazine Sept. 1995

	New	Used
Price:	$1200	$700-720
Rating:	★★★	★★★

LOWE ELECTRONICS LTD.

SRX-30

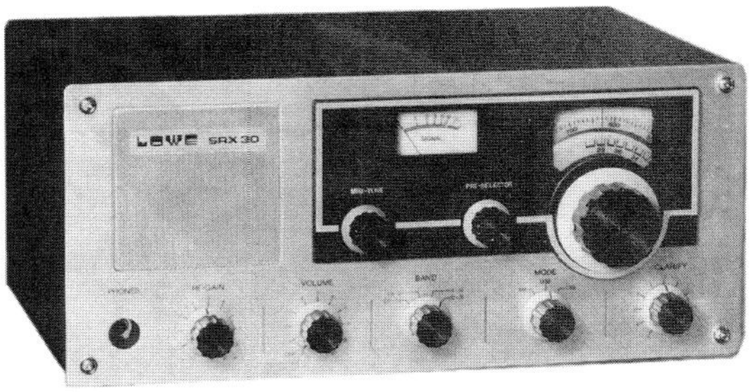

General Coverage Communications Receiver
Triple Conversion Superheterodyne. Solid State.

Features:
- ¼" Head. Jack
- Speaker
- BFO
- S-Meter
- Telescopic Whip
- Clarifier
- RF Gain
- Preselector
- Attenuator- 30 dB
- Ext. Speaker Jack

Specifications:
Coverage ......500 - 30000 kHz
Sensitivity ......<.3μV (SSB 2-30 MHz).
Selectivity ......5.5/3 kHz -6dB.
Audio ....8 ohms 2W 5% Dist.
Clarifier..........±2 to ±5 kHz
Modes .... AM/USB/LSB
IF Reject. .... >50 dB <20 MHz
IF Reject. .... >40 dB >20 MHz
Antenna Input . Terminals 75 ohm

Circuit Complement: Utilizes the Wadley drift cancelling system.

Comments: Ranges: .5-1.5, 1.5-5, 5-12 and 12-30 MHz. Tuning is accomplished with four knobs. First the Band knob is set for the proper range. The smaller Preselector knob and MHz Tune knobs are located under the S-Meter. The main tuning, or kHz knob is the large knob on the right. The concentric MHz and kHz scales are visible through the window. The kHz scale is calibrated from 000 to 1000 in steps of 10. The kHz knob has a 6:1 tuning ratio. This radio has a tuning accuracy of approximately ±5 kHz. The telescopic whip is mounted to the back panel. It may be swivelled down, but may not be removed. Audio via the internal speaker is only marginal. A sharp, accurate, compact, easy to use radio. It has been reported that this model came from the same Japanese manufacturer as the Drake SSR-1. The striking similarity of these two models supports this theory. See Drake Chapter.

Made In: Japan 1979-1980

Voltages: 100/117/220/240 VAC 50-60 Hz or 12 VDC

Readout: Analog Lin.

Physical: 12.8x5.5x9" 11 Lbs. 325x140x230mm 5 kg

Status: Active Manufacturer, Discontinued Model

Rarity: Extremely Scarce

Reviews: *Practical Wireless* Aug. 1979

	New	Used
Price:	£178	⑤
Rating:	⑥	⑥

LOWE ELECTRONICS LTD.

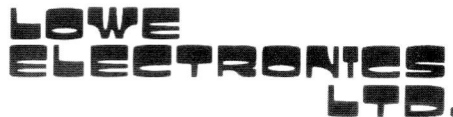

SRX-30D

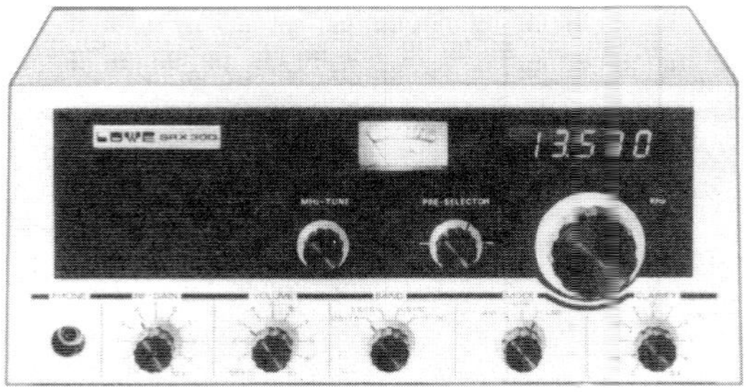

General Coverage Communications Receiver
Triple Conversion Superheterodyne. Solid State.

Features:
- ¼" Head. Jack
- S-Meter
- RF Gain
- Preselector
- Speaker
- Telescopic Whip
- Attenuator
- Ext. Speaker Jack
- BFO
- Clarifier ±2-5 kHz

Specifications:
Coverage ...... 200 - 30000 kHz
Sensitivity ...... <.3µV (SSB 2-30 MHz).
Selectivity ...... 5.5/3 kHz -6dB.
Audio .... 8 ohms 2W 5% Dist.
Clarifier .... ±2 to ±5 kHz
Modes .... AM/USB/LSB
IF Reject. .... >50 dB <20 MHz
IF Reject. .... >40 dB >20 MHz
Antenna Input . Terminals 75 ohm

Circuit Complement:
Utilizes the Wadley-Loop drift cancelling system. Frequency synthesis via Plessey SL6 1641 double balanced modular integrated circuit.

Comments:
Ranges: .2-1.5, 1.5-5, 5-12 and 12-30 MHz. The SRX-30D has green LEDs. There is one knob to select MHz and the large knob under the frequency display tunes kHz. This model was obviously the successor to the SRX-30. Note the addition of longwave coverage. The audio output and filtering was also improved.

Made In: Japan 1980-1982

Voltages: 100/117/220/240 VAC
50-60 Hz or 12 VDC

Readout: 00001. Digital

Physical: 12.8x5.5x9" 11 Lbs.
325x140x230mm 5 kg

Status: Active Manufacturer
Discontinued Model

Rarity: Very Scarce

	New	Used
Price:	£195	[5]
Rating:	★★★	★★★

The United States Air Force contracted with **LTV Temco** (Ling-Temco-Vought, Inc. of Dallas Texas) to repackage the famous Collins 51S-1 for airborne use. The receiver takes on a different look at first glance, but retains the same mechanical attributes as the 51S-1. Heavy shielding was added to withstand a nuclear electromagnetic pulse. Spectrum output was also added. If one refers to the photograph of the Collins 51S-1 shown earlier in this book, the knob orientation is virtually the same. The knob themselves were changed to be more unique. This was reportedly to facilitate operation if the radio man was blinded by a nuclear blast. Another less dramatic theory is that the different knobs simply allowed operation by someone with gloves on.

This radio does occasionally show up at hamfests and has been in the Fair Radio Catalog.

There was also a matching VHF/UHF receiver.

LTV Temco is now part of E-Systems Inc.

LTV Temco
Aerosystems Division
Greenville, TX   1964

LTV Temco

G133F

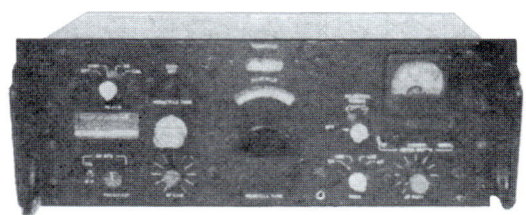

General Coverage Communications Receiver
Triple Conversion Superheterodyne. 18 Tubes plus Semiconductors.

Features:
- ¼" Head. Jack
- S-Meter
- BFO
- Calibrator 100 kHz
- RF Gain
- ANL
- Rejection Tuning
- Zero Set
- IF Out 500 kHz
- Standby
- Atten. -10 dB
- Line Out 150/600
- Mute Line
- Dial Lamp

Specifications:
Coverage ...... 2000 - 30000 kHz[1]
Selectivity ...... 5/2.4/.3 kHz -6dB.
Sensitivity ...... <.6µV SSB
Image Resp. ... >50 dB
Audio Out ...... 4 / 600 ohms

Modes .... AM/SSB/CW
Stability .... ±400-885 Hz
IF Rejection .... >80
Antenna Input . 50 ohms
Environment ... -10° to +80° C.

Accessories: G-186 Panoramic Display
Circuit Complement: 6DC5 RF Amp, 6EA8 1st Mixer/HF Osc., 6EA8 2nd Mixer/17.5 MHz Osc, 6EA8 3rd Mixer/Remote Gain Gate, (3) 6BA6 IF Amp, 12AX7 Q-Mult., 6BA6 AGC Amp, 6EA8 LF Mixer/LF Crystal Osc, 5670 IF Cathode Follow/AGC Cathode Follow, 6BF6 2nd Local AF Amp, 6AK5 2nd Line AF Amp, 12AX7 1st Line AF Amp/1st Local AF Amp, 6136 VFO, 6EA8 LF Mixer/Calib. Osc, 6BA6 BFO, 6136 VFO Amp plus semiconductors. Double conversion above 7 MHz.
Comments: This is a Collins 51S-1F repackaged and shielded by LTV Temco for military airborne use. [1]A low frequency tuner has also been added which will allow tuning the .2-1 and 1-2 MHz bands using the 28-29 and 29-30 MHz bands. (The 2.4 filter, which is mechanical, is measured at -3dB). [¥288,400].

Made In:	United States 1964-1978	
Voltages:	115/230 VAC 50-400 Hz 125W or 24 VDC 4.7W	
Readout:	001 ╷╻┃╻╷ Digital Mech. to 100 kHz then Analog Lin.	
Physical:	18.32x6.25x12.875" 26 Lbs 465x159x327mm 11.8 kg	
Status:	Active Manufacturer Discontinued Model	
Rarity:	Very Scarce	

	New	Used
Price:	①	$600-800
Rating:	⑥	⑥

The **Mackay Radio Company** has had a long and interesting history. They were owned for over ten years by International Telephone and Telegraph conglomerate. They then went private for a period of time and divided into two divisions. In 1995 the huge French company Thomson-CSF became the majority owner of Mackay Radio Systems. Thomson also acquired controlling interest in the British firm of Redifon, at around the same time.

Reportedly the Hammarlund SP-600-JX-32 variant was produced for Mackay Radio with the same specifications as the SP-600-JX-21. The author seeks more information on this point.

Mackay Radio Systems today provides communications equipment and integrated systems to military, paramilitary, maritime and civil customers for manpack, mobile, shipboard, tactical and strategic applications. The company offers technologically advanced equipment, systems planning, integration and turnkey installations. As part of Thomson-CSF, Mackay supports its customers with advanced HF and VHF technologies and features with MIL Spec/FED Standard compliant equipment and systems.

Mackay 3000 series receivers do appear on the surplus market from time to time and present excellent buying opportunities. Someday, no doubt, the 5050 series will also become available.

Mackay Radio & Telegraph Co.
Marine Division
345 Hudson St.
New York, NY   1954-1968

ITT Mackay Marine
133 Terminal Av.
Clark, NJ 07066   1968-1970

ITT Mackay Marine
2912 Wake Forest Rd.
Raleigh, NC 27611   1973-1978

Mackay Communications Inc.
Radio Systems Division
5301 Departure Dr.
Raleigh, NC 27658   1980-1995

Mackay Radio Systems Inc.
2721 Discovery Dr.
Raleigh, NC 27616   1995-1998

Mackay Radio

3007A

General Coverage Communications Receiver
Single Conversion Superheterodyne. 12 Tubes.

Features:
- ¼" Head. Jack
- RF Gain
- Dial Lamp
- Speaker
- BFO
- AVC ON/OFF
- Mute
- Bandspread
- Speaker Switch

Specifications:
Coverage ...... 85-550,1900-24000 kHz   Modes .... AM/CW
Selectivity ...... 3 Position   Antenna Input . SO239 75 ohms
Audio Out ...... 1 W 4/600 ohms

Circuit Complement:
(2) 6BA6, 6BA7, 6J7, 6SA7, (3) 6SG7, 6SQ7, 25L6 and 25Z6.

Comments:
Built for marine applications.

Made In: United States 1954

Voltages: 115VAC 50-60 Hz or
115 VDC

Readout: [⌶⌶⌶⌶⌶] Analog

Physical: 17.312x9.75x18
439x246x457mm

Status: Active Manufacturer
Discontinued Model

Rarity: Very Scarce

	New	Used
Price:	①	$200-250
Rating:	⑥	⑦

Mackay Marine Division **ITT**

3010B

General Coverage Communications Receiver
Triple Conversion Superheterodyne. 17 Tubes.

Features:
- ¼" Head. Jack
- S/AF Meter
- AGC ON/OFF
- Calibrator 100 kHz
- RF Gain
- BFO
- Mute
- Rack Handles
- Line Out Jack
- ANL
- Ext. Spkr Jack
- Hinged Top Cover
- Dial Lamp
- Attenuator
- Speaker Switch
- 455 IF Out Jack

Specifications:

Coverage ...... 10 - 30000 kHz	Modes .... AM/SSB/CW
Selectivity ...... 6/3.1/_ kHz -6dB.	Image Rej .... >80 dB
Sensitivity ...... <1µV 4-30 MHz CW	IF Rejection .... >80 dB
Audio Out ...... 1 W 4/600 ohms	Antenna Input . SO-239 75 ohms

Circuit Complement: See model 3010C
Accessories:
Mechanical Filters
Comments:
Ranges: .07-.21, .21-.7, .7-2, 2-4, 4-8, 8-16 and 16-30 MHz. A robustly built receiver designed for the marine market. An incredible 7½ foot fiber glass tape scale is provided for each 2 MHz band with calibration marks every 2 kHz. The MHz value is displayed digitally and the kHz positions are indicated on the linear analog tape display. With 15 bands, the effective dial length is over 100 feet. An outstanding general coverage receiver. Reduced sensitivity in the range of 10-70 kHz.
Variants:
Earlier model **3010** 1962 had a different colored dial tape (red and green).

Made In:	United States 1963-1969
Voltages:	115VAC 50-60 Hz 95 W
Readout:	Analog Lin.
Physical:	17.07x9.6x16" 45 Lbs. 433x244x406mm 20.4 kg
Status:	Active Manufacturer Discontinued Model
Rarity:	Very Scarce
Reviews:	*QST* April 1967

	New	Used
Price:	$1500-1600	$500-1000
Rating:	★★★★★	★★★★

Mackay Marine Division **ITT**

3010C

General Coverage Communications Receiver
Triple Conversion Superheterodyne. 17 Tubes.

Features:
- ¼" Head. Jack
- S/AF Meter
- AVC ON/OFF
- Calibrator 100 kHz
- RF Gain
- Speaker Switch
- Rack Handles
- Hinged Top Cover
- Line Out Jack
- 455 IF Out Jack
- Ext. Spkr Jack
- BFO
- Mute
- Standby
- Antenna Filter
- Dial Lamp.
- Attenuator
- ANL

Specifications:

Coverage ...... 10 - 30000 kHz	Modes .... AM/SSB/CW
Selectivity ...... 6/3.1/_ kHz -6dB.	Image Rej .... >80 dB
Sensitivity ...... <1µV 4-30 MHz CW	IF Rejection .... >80 dB
Audio Out ...... 1 W 4/600 ohms	Antenna Input . SO-239 75 ohms

Circuit Complement: 7788 RF Amp, 6C4 Sig. Cathode Follower, 6688 IF Amp, 6BL8 Sig. Cath. Follower/2nd Mixer, 6BE6 3rd Mixer, 6BA6 455 IF Amp, 6BA6 455 IF Amp, 6AV6 1st Audio Amp AVC Rectifier/2nd Detector, 12AT7 Product Detector/455 IF Out Cath Follower, 6BF5 Audio Output, 6BL8 Osc. Cath. Follower, 6U8 VFO Mixer/Crystal Osc., 6EW6 XVFO IF Amp, 6BA6 VFO, 6BA6 Crystal Calibrator, 6AU6 BFO and 0B2 Voltage Regulator.
Accessories: Mechanical Filters
Comments:
The 3010C which followed the 3010B added compensating capacitors for improved stability under temperature changes. It also features a crystal USB and LSB VFO and has Zener diode front-end protection. This 455 kHz IF receiver will accept Collins mechanical filters.

Made In:	United States 1969-1972
Voltages:	115VAC 50-60 Hz 95 W
Readout:	Analog Lin.
Physical:	17.07x9.6x16" 45 Lbs. 433x244x406mm 20.4 kg
Status:	Active Manufacturer Discontinued Model
Rarity:	Very Scarce
Reviews:	*Electric Radio* February 1997

	New	Used
Price:	$1800	$500-1100
Rating:	★★★★★	★★★★★

3020A

General Coverage Communications Receiver
Double Conversion Superheterodyne. Solid State.

Features:
- ¼" Head. Jack
- RF Gain
- Speaker
- Speaker Switch
- S/AF Meter
- Speech Clarifier
- Ext. Spkr Jack
- Attenuator -20 dB
- Preselector
- Mute
- Lock
- Line Out 600 ohm
- AGC (2 Pos.)
- Modular Construction

Specifications:
Coverage ...... 15 - 30000 kHz	Modes .... AM/LSB/USB/CW
Selectivity ...... 8/2/1/.4 kHz -6dB.	Stability .... <1Hz per MHz tuned
Sensitivity ...... <.4µV .1-30 MHz SSB	IF Rejection .... >70 dB
Image Rej ...... >70 dB	Environment ... 0° to +50°C 95% Humid.
Audio Out ...... 1 W 3.2 ohms (5%)	IF .... 92, 8 MHz
Line Out .... 600 ohm	Antenna Input . 50 ohm

Accessories:
Cabinet (shown)

Comments:
Tuning is via decadic switches with one switch per digit of frequency out to 100 Hz. Designed primarily for the commercial and marine market. [¥100000-¥185000].

Made In:	United States 1973-1977
Voltages:	115/230 ±15% VAC 50/60 Hz 90 W
Readout:	[00000.1] Digital Switch
Physical:	19x5.25x17" 36 Lbs. 482x133x425 mm 16.32 Kg.
Status:	Active Manufacturer Discontinued Model
Rarity:	Very Scarce
Reviews:	*QST* January 1974

	New	Used
Price:	$3700	$400-600
Rating:	★★★★	★★★

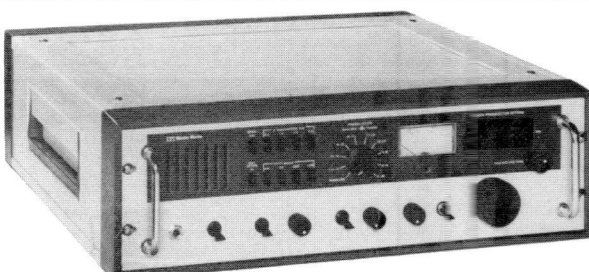

3021A

General Coverage Communications Receiver
Double Conversion Superheterodyne. Solid State.

Features:
- ¼" Head. Jack
- RF Gain
- Speaker
- Lock
- S/AF Meter
- Speech Clarifier
- Ext. Spkr Jack
- Speaker Switch
- Preselector
- Mute
- Rack Handles
- Attenuator -20 dB
- Line Out 600 ohm
- AGC FAST/SLOW
- Modular Construction
- 2 Tuning Rates

Specifications:
Coverage ...... 15 - 30000 kHz	Modes .... AM/LSB/USB/CW
Selectivity ...... 8/2/1/.4 kHz -6dB.	Stability .... <1Hz per MHz tuned
Sensitivity ...... <.4µV .1-30 MHz SSB	IF Rejection .... >70 dB
Image Rej ...... >70 dB	Environment ... 0° to +50°C 95% Humid.
IF .... 92, 8 MHz	Antenna Input . 50 ohm
Audio Out ...... 1 W 3.2 ohms (5%)	

Accessories:
Cabinet (shown)

Comments: The tuning is controlled by a three position toggle switch. In the uppermost position ("kHz"), the tuning rate is 10 kHz per rotation. In the center position ("MHz"), the rate is 2.5 MHz per rotation. In the bottom position ("Lock"), the tuning knob is inoperative. The Preselector is most noticeably effective below 4 MHz. It dramatically improves reception below 1 MHz making the 3021A a strong LW performer. Shown with optional cabinet. Designed primarily for the market.
Variants: Model **3021N** features a 6 rather than 8 kHz AM filter. Model **3021** differences unknown.

Made In:	United States 1973-1979
Voltages:	115/230 VAC ±15% 50/60 Hz 90 W
Readout:	[00000.1] Digital LED
Physical:	19x5.25x17" 30 Lbs. 482x133x425mm 13.6 kg.
Status:	Active Manufacturer Discontinued Model
Rarity:	Very Scarce

	New	Used
Price:	$4200-6000	$500-700
Rating:	★★★★	★★★

ITT Mackay

3040A

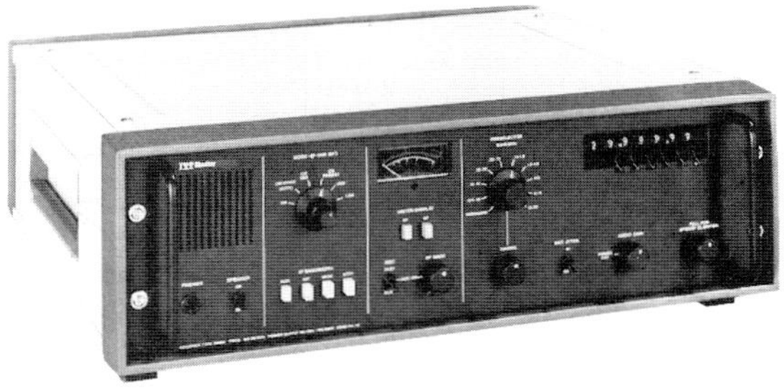

General Coverage Communications Receiver
Double Conversion Superheterodyne. Solid State.

Features:
- ¼" Head. Jack • S/AFMeter • Preselector • Speaker
- RF Gain • Speech Clarifier • Mute • AGC FST/SLO
- Line Out Jack • Ext. Spkr Jack • Rack Handles • Speaker Switch
- Modular Construction

Specifications:

Coverage	15 - 30000 kHz	Modes	AM/LSB/USB/CW/RTTY
Selectivity	8/2.3/2/.4 kHz -6dB.	Stability	<1 PPM (±10% PS)
Sensitivity	<.4µV .1-30 MHz SSB	IF Rejection	>80 dB
Image Rej	>80 dB	Environment	-15° to +55°C
Audio Out	2.5 W 3.2 ohms (10%)	BFO Range	± 2 kHz.
Line Out	600 ohm Adj. to +10 dBm	Spe. Clar. Rg.	± 50 Hz.

Accessories:
Cabinet (shown) Filter RTTY 500 Hz -3dB DC Power Supplies

Comments:
Tuning is via thumb-wheel switches. Shown with optional cabinet. Designed primarily for the commercial and marine market.

Variants:
Model **3030A** no LSB position.

Made In: United States 1985-1988

Voltages: 90-130/195-260 VAC
47-63 Hz

Readout: 00000.01 Digital Switch

Physical: 19x5.75x17" 19 Lbs.
493x133x432mm 8.6 Kg.

Status: Active Manufacturer
Discontinued Model

Rarity: Very Scarce

	New	Used
Price:	①	⑤
Rating:	⑥	⑥

ITT Mackay

3041A

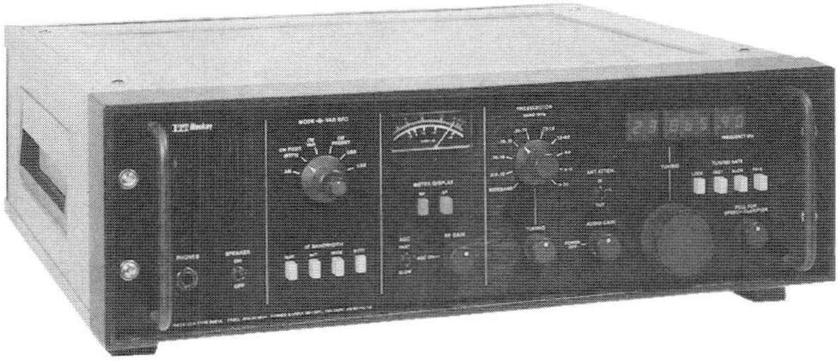

General Coverage Communications Receiver
Double Conversion Superheterodyne. Solid State.

Features:
- ¼" Head. Jack • S/AFMeter • Preselector • Speaker
- RF Gain • Speech Clarifier • Mute • AGC FST/SLO
- Line Out Jack • Ext. Spkr Jack • Rack Handles • 3 Tuning Rates
- Dial Lock • Attenuator • Speaker Switch • Modular Construction

Specifications:

Coverage	15 - 30000 kHz	Modes	AM/LSB/USB/CW/RTTY
Selectivity	8/2.3/2/.4 kHz -6dB.	Stability	<1 PPM (±10% PS)
Sensitivity	<.4µV .1-30 MHz SSB	IF Rejection	>80 dB
Image Rej	>80 dB	Environment	-15° to +55°C
Audio Out	3.5 W 3.2 ohms (10%)	BFO Range	±2 kHz.
Line Out	Adj. to +10 dBm	Spe. Clar. Rg.	± 50 Hz.

Accessories:
Cabinet (shown) Filter RTTY 500 Hz -3dB DC Power Supplies

Comments:
Shown with optional cabinet. Designed primarily for the commercial and marine market. [¥500,000].

Variants:
Model **3031A** has no LSB position.

Made In: United States 1985-1989

Voltages: 90-130/195-260 VAC
47-63 Hz 62W

Readout: 00000.01 Digital LED

Physical: 19x5.75x17" 19 Lbs.
493x133x432mm 8.6 Kg.

Status: Active Manufacturer
Discontinued Model

Rarity: Very Scarce

	New	Used
Price:	①	⑤
Rating:	⑥	⑥

MSR5050

General Coverage Communications Receiver
Double Conversion Superheterodyne. Solid State.

Features:
- ¼" Head. Jack
- RF Gain
- Line Out Jack
- Speaker Switch
- Mute
- S/AF Meter
- Keypad
- Ext. Spkr Jack
- BITE
- BFO
- Dimmer
- Rack Handles
- 99 Memories
- Speaker
- AGC FST/MED/SLO
- 3 Tuning Rates
- Modular Construction

Specifications:
Coverage ......10 - 30000 kHz	Modes .... AM/LSB/USB/CW/FSK
Selectivity .....6/2.7/_/_/_ kHz -6dB.	Stability .... ±1 part in 10^6
Sensitivity .....<.5µV 2-30 MHz SSB	IF Rejection .... >80 dB
Image Rej .....>80 dB	Environment ... -10° to +50°C 95% Humid

Accessories:
Cabinet (shown)   High Stability (±1 part in 10^8)
Filters   Remote FSK Modem
MSR6420 Remote Controller

Comments:
Designed for the commercial, military and diplomatic markets.

Made In:	United States 1979-1988
Voltages:	115/230 VAC ±15% 47-400 Hz 60W
Readout:	**00000.01** Digital LED
Physical:	19x5.25x17.5" 483x133x443mm
Status:	Active Manufacturer Discontinued Model
Rarity:	Very Scarce

	New	Used
Price:	$13000	②
Rating:	⑥	⑤

MSR5050A

General Coverage Communications Receiver
Double Conversion Superheterodyne. Solid State.

Features:
- ¼" Head. Jack
- RF Gain
- Line Out Jack
- RS-232/422/423
- Mute
- S/AF Meter
- Keypad
- Ext. Spkr Jack
- BITE
- Dial Lock
- BFO ±7.99 kHz
- Dimmer
- Rack Handles
- 100 Memories
- Speaker Switch
- Speaker
- AGC FST/MD/SLO/OFF
- 3 Tuning Rates
- Modular Construction

Specifications:
Coverage ......10 - 30000 kHz	Modes .... AM/LSB/USB/CW/FSK
Selectivity ......6/2.7/_/_/_ kHz -6dB.	Stability .... ±1 part in 10^6
Sensitivity ......<.5µV 2-30 MHz SSB	IF Rejection .... >80 dB
Image Rej ......>80 dB	Environment ... -10° to +55°C 95% Humid
Audio ....3W 5% Dist.	Noise Figure ... <16 dB

Accessories:
Cabinet   TADIL-A/LINK-11 Data   High Stability (±1 part in 10^8)
Filters   ISB Option
Remote FSK Modem
MSR6420 Remote Controller

Comments:
Designed for the commercial, military and diplomatic markets. The MSR6420 Remote Control Unit (shown right), can fully remote control up to 99 receivers.

Made In:	United States 1988-1998
Voltages:	115/230 VAC ±15% 47-400 Hz
Readout:	**00000.01** Digital LED
Physical:	19x5.2x19" 30 Lbs. 483x132x483mm 3.6 Kg.
Status:	Active Manufacturer Active Model
Rarity:	Very Scarce

	New	Used
Price:	①	②
Rating:	⑥	⑤

MSR5090

General Coverage Communications Receiver
Double ? Conversion Superheterodyne. Solid State.

Features:
- S/AF Indicator
- Mute
- RS-232/485
- BFO ± 3 kHz
- Squelch
- BITE
- Line Out Jack (4)
- Rack Handles
- 1000 Memories
- Speaker
- AGC FST/MED/SLO
- Modular
- Squelch
- DSP

Specifications:
Coverage ...... 10 - 30000 kHz
Selectivity ...... 6/2.75/.3/_¹ kHz
Sensitivity ...... <.5µV .5-30 MHz SSB
Image Rej ......>80 dB

Modes .... AM/LSB/USB/CW/FSK
Audio Out .... 600 ohms
IF Rejection .... >80 dB
Environment ... -10° to +55°C 95% Humid

Accessories:
MSR6490 Controller (shown)
MSR6420 Remote Controller

Four Channel ISB
Internal Preselector

Comments: ¹Other bandwidths are programmable. Uses DSP technology starting at the IF stage. Shown above with optional MSR6490 controller serving as front panel (which may be detached). Designed primarily for the commercial, military and diplomatic markets.

Variants
Model **MSR5091** is a dual receiver version of the MSR5090 configured in the same chassis with the addition of internal circuit cards. The MSR-5091 screen is shown to the right.

Made In: United States 1997-1998

Voltages: 90-270 VAC 40-400 Hz
20-32 VDC

Readout: `00000.001` Digital with optional MSR6490 shown.

Physical: 19x7x15" 30 Lbs.
483x180x360mm 13.6 Kg.

Status: Active Manufacturer
Active Model

Rarity: (Too new)

	New	Used
Price:	$5500A	②
Rating:	⑥	⑥

A company called the Wireless Telegraph and Signal company was formed in England in 1897. In 1890 this company would change its name to Marconi's Wireless Telegraph Company. This name would remain until 1963 when it was shorten to simply the **Marconi Company**.

In 1939 the Marconi Company designed a new receiver for the Royal Navy to compete with the popular National HRO. In 1940 the first 100 CR100s were produced.

Marconi International Marine Communications Company (MIMCo).

Please see *Chapter 105 Briefly Mentioned* for the Marconi M3000 Oceanic. For the Marconi model HR 101 please see Eddystone 910.

Marconi International Marine Co. Ltd.
Elettra House, Westway
Chelmsford, Essex CM1 33H
England   1964-1980

Marconi Communications Sys. Ltd.
Marconi House
Chelmsford, Essex CM1 1PL
England   1965-1980

GEC-Marconi Communications Ltd.
Marconi House, New Street
Chelmsford, Essex CM1 1PL
England   1992-1998

1017
Mercury

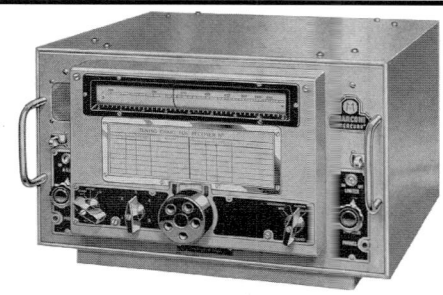

Marine L.F.-M.F. Communications Receiver
Double Conversion Superheterodyne. 12 Tubes.

Features:
- ¼ Head Jack (2)
- Speaker
- Dial Lamp
- 80:1 Tune Ratio
- RF Gain
- Rack Handles
- Mute Line
- Flywheel Tuning
- AGC ON/OFF
- Standby
- Speaker Output
- Logging Scale
- NL

Specifications:
Coverage ...... 15-40, 100-4000 kHz
Selectivity ...... 8/3/1 kHz -6dB
Audio Out ...... 1W 3-5 ohms

Modes .... AM/CW/MCW
IF .... 4500, 85 kHz
Antenna Input . 75 ohm

Circuit Complement:
KTW61 1st SF Amp, KTW61 2nd SF Amp, X61M 1st Frequency Changer, X61M Second Frequency Changer, KTW61 1st IF Amp, 2nd IF Amp, D63M Final Detector, X61M BFO, D63M Noise Limiter, DH63 1st LF Amp, L63 LF Output Amp and VR-150 Voltage Stabilizer.

Comments: This Marconi Mercury has five tuning ranges. Each is displayed on 10.25" turret type drum. The provided power supply (Marconi type 889A or 966A) is separate and weights 21 Lbs. A paper tuning chart is available on the front of the receiver to record stations from the logging scale. The logging scale consists of a 0-40 scale engraved on the calibration escutcheon and a calibrated disk mounted on the main tuning knob. Please note the limited frequency of this receiver which is designed strictly for marine applications. It was a commonly used receiver at British coastal stations in the late 1950's and early 1960's.

Made In: England   1953-1965

Voltages: 24/110/220 VDC or
230 VAC 65 W

Readout: Analog

Physical: 17.9x11.25x18.25" 55 Lbs.
454x285x463mm 25 kg
The power supply is
separate and is 21 Lbs.

Status: Active Manufacturer
Discontinued Model

Rarity: Extremely Scarce

	New	Used
Price:	①	②
Rating:	⑥	⑥

1018
Electra

Generat Coverage Communications Receiver
Single Conversion Superheterodyne. 13 Tubes.

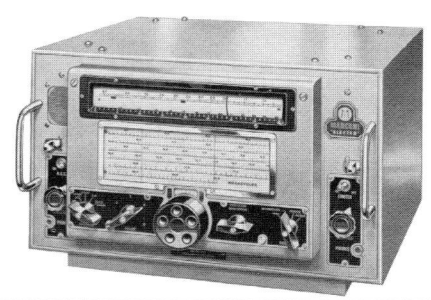

Features:
- ¼ Head Jack (2)
- RF Gain
- AGC ON/OFF
- Bandspread
- Speaker
- Rack Handles
- Standby
- NL
- Dial Lamp
- Mute Line
- Speaker Output
- 80:1 Tune Ratio
- Flywheel Tuning
- Logging Scale

Specifications:
Coverage ...... 250-520, 1500-25000 kHz
Selectivity ...... 8/3/1 kHz -6dB
Audio Out ...... 1W 3-5 ohms
Modes .... AM/CW/MCW
IF .... 690 kHz
Antenna Input . 75 ohm

Circuit Complement:
KTW61 1st SF Amp, KTW61 2nd SF Amp, X61M 1st Frequency Changer, L63 1st Frequency Changer Oscillator, KTW61 1st IF Amp, 2nd IF Amp, D63M Final Detector, X61M BFO, D63M NL, DH63 1st LF Amp, L63 LF Output Amp, KTW61 Calibrating Oscillator and VR-150 Voltage Stabilizer.

Comments:
The Marconi Electra has five tuning ranges. Each is displayed on 10.25" turret type drum. Above the main tuning knob is a bandspread of the six major H.F. marine bands. The bandspread ranges are: 4.4-4.45, 6.15-6.55, 8.15-8.9, 12.3-13.1, 16.3-18 and 21.7-23 MHz. The provided power supply (Marconi type 889A or 966A) for this receiver is separate and weighs 21 Lbs.

Made In:	England	1953-1965
Voltages:	24/110/220 VDC or 230 VAC 65 W	
Readout:	Analog	
Physical:	17.9x11.25x18.25" 55 Lbs. 454x285x463mm 25 kg The power supply is separate and is 21 Lbs.	
Status:	Active Manufacturer Discontinued Model	
Rarity:	Extremely Scarce	

	New	Used
Price:	①	②
Rating:	⑥	⑥

2207C
Atalanta

General Coverage Communications Receiver
Double Conversion Superheterodyne. 13 Tubes.

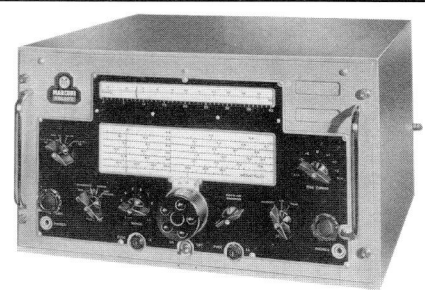

Features:
- ¼ Head Jack (2)
- RF Gain
- AGC
- Speaker Grill
- Rack Handles
- Mute Line
- Bandspread
- Flywheel Tuning
- Dial Lamp
- Fine Tuning ±3 kHz.
- Calibrator 700 kHz
- NL
- BFO

Specifications:
Coverage ...... 15 - 28000 kHz
Selectivity ...... 8/3/1/.1 kHz -6dB
Sensitivity ...... 2-3 µV 20 dB
IF .... 700, 85 kHz
Audio Out ...... 1W 3-5 ohms
Modes .... AM/SSB/CW/MCW
Image Rej .... >70 dB 3-7.5 MHz
Image Rej .... >55 dB 7.5-15 MHz
Image Rej .... >30 dB 15-28 MHz
Antenna Input . 75 ohm

Circuit Complement: EF85 1st RF Amp, EF85 2nd RF Amp, ECH81 1st Freq. Change, EF85 1st Freq. Change Osc., ECH81 2nd Freq. Change & Osc., W77 1st IF Amp, Z77 2nd IF Amp, D77 Final Det., ECH81 BFO, B309 AF Amp, N37 Output Amp, QS75/20 Voltage Stab. and Z77 Calib. Osc. (Optional ECH81 Muting).

Accessories: 1362A Meter   2432A Switching Unit
2361 Duplex Rejector   2203A 24 VDC PS   2202A 115/220-250 VAC PS

Comments: The main ranges are on a rotating turret: .015-.025, .025-.1, .1-.2, .2-.4, .4-.8, .8-1.7, 1.7-3.6, 3.6-7.5, 7.5-15 and 15-28 MHz. The bandspread is on the large flat scale over the main tuning knob and covers six HF marine bands: 4.1-4.4, 6.1-6.7, 8.2-8.8, 12.2-13.3, 16.4-17.6 and 21.8-23 MHz. Designed for the marine market. A speaker grill is located at the top right hand corner, although the speaker was not normally fitted. This receiver may be rack mounted.

Made In:	England	1958-1967
Voltages:	110/220 VDC 45 W or optional power supplies.	
Readout:	Analog	
Physical:	Cabinet Version: 19.5x12.6x19.8" 78 Lbs 495x319x503mm 35.4 kg	
Status:	Active Manufacturer Discontinued Model	
Rarity:	Extremely Scarce	

	New	Used
Price:	①	②
Rating:	⑥	⑥

Apollo

General Coverage Communications Receiver
Double Conversion Superheterodyne. Solid State.

Features:
- ¼ Head Jack (2)
- RF Gain
- Speaker
- Speaker
- Rack Handles
- NL Variable
- BFO
- Dial Lock
- Mute Line
- Atten. -10/20/30 dB
- AGC OFF/LNG/SHT
- Line Out 600 Ohm

Specifications:
Coverage ...... 15 - 28000 kHz
Selectivity ...... 5 Position
Sensitivity ...... 1µV 10 dB S/N 3-28 MHz
IF Rejection ... >90 dB .15-.16 MHz
IF Rejection ... >60 dB >.16 MHz
Audio Out 2W
Antenna Input 75 ohm

Modes .... AM/SSB/CW/RTTY
Stability .... 1×10^8 per day
IF .... 1100, 100 kHz
Image Reject .. >60 dB .015-15 MHz
Image Reject .. >45 dB 15-28 MHz
Environment ... -10° to +55° C.

Accessories:
High Stability Option   24 Fixed Channel Option

Comments:
Ranges: .015-.03, .03-.065, .065-.14, .14-.3, .3-.675, .675-.14, 1.4-3, 3-6, 6-15 and 15-28 MHz. Single conversion below 1.4 MHz. The widest filter is 10 kHz, and the most narrow is 300 Hz. The Noise Limiter is variable. Designed for the marine market. Like many marine receivers, the fuses are accessible from the front panel.

Made In:	England	197■-1982
Voltages:	110/115/120/220/2■0/240 VAC 45-65 Hz 9■W	
Readout:	**00000.1** Digi■al	
Physical:	Cabinet Version: 19.6x12.6x18.25' 7⁵ Lbs 497x318x463mm ■4 kg Rack Version: 19x10.5x18" 55 Lb■. 483x265x457mm 2■ kg	
Status:	Active Manufacture■ Discontinued Mo■e	
Rarity:	Typically Unavailab e	

	New	Us■d
Price:	①	⑥
Rating:	⑥	⑥

Marconi

CR100/7

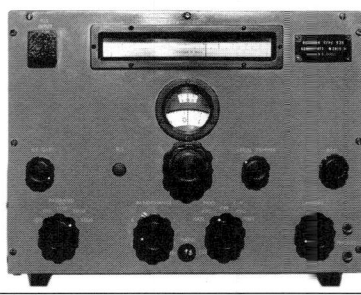

General Coverage Communications Receiver
Single Conversion Superheterodyne. 11 Tubes.

Features:
- ¼ Head Jack (2)
- IF Gain
- BFO
- Standby
- Dial Lamp
- AVC ON/OFF
- Antenna Trimmer
- Line Out 600 ohm.

Specifications:
Coverage ...... 60 - 32000 kHz
Selectivity ...... 6/3/1.2/.3/.1 kHz -6dB
Sensitivity ...... 2 µV typical
Audio Out ...... 1000 ohm 2W

Modes .... AM/CW/MCW
IF .... 465 kHz
Antenna Input . 100 ohm

Circuit Complement:
KTW62 1st RF, KTW62 2nd RF, KTW62 HF Osc., KTW62 1st IF, KTW62 2nd IF, KTW62 3rd IF, DH63 Detector/AVC/1st Audio, KT63 Audio Output, KTW62 BFO and U50 Rectifier.

Comments: Ranges: .06-.16, .16-.42, .5-1.4, 1.4-4, 4-11 and 11-30 MHz. The RIS INPUT (upper left) stands for radar interference suppression. It was connected to the ship's local radar. The RIS knob could be adjusted to reduce radar interference. This radio was made for the British Royal Navy and was used extensively during World War II.

Variants: The CR100/7 is also known as the Navy **B28** or Admiralty Pattern B28. Eight variants, starting in 1939 were produced, most with minor differences. Only the model **CR100/2** had sidetone facilities so a Morse operator could hear his own signal. Models **CR100/4**, **CR100/5**, **CR100/6** and **CR100/8** also had 1000 ohm speaker output. Earlier models had 3 ohm output.

Made In:	England	1■4■-1945
Voltages:	200-250 VAC 50 H■ 85W or 6 VDC 4A, 25■ V■C .1A	
Readout:	Ana■g	
Physical:	16x16.5x12.5" 8■ Lbs. 410x420x320mm ■■ kg	
Status:	Active Manufac■ure Discontinued Mode■	
Rarity:	Extremely Scarce	

	New	Us■d
Price:	①	②
Rating:	⑥	⑥

Marconi

CR150/3

General Coverage Communications Receiver
Double Conversion Superheterodyne. 12 Tubes.

Features:
- ¼ Head Jack (2)
- S-Meter
- BFO
- Antenna Trimmer
- IF Gain
- Standby
- AGC ON/OFF
- Line Out 600 ohm
- Rack Handles
- Dial Lamp
- Hinged Top Cover

Specifications:

Coverage ......2000 - 60000 kHz	Modes .... AM/CW/MCW
Selectivity ...... 13/8/3/1 kHz -6dB	IF .... 1600, 465 kHz
Audio Out ......3 ohm 0.2W	Antenna Input . 75-100 ohm

Circuit Complement:
EF50, EF50, EF50, EF50, X66, 6K7G, 6K7G, DH63, 6KG, DH63, L63, STV280/40 and U52.

Comments:
Handbook reference is T2148/1. Photo courtesy of *Radio Bygones* and GEC-Marconi Ltd.

Variants:
Model **CR150/6** (1950) used 7 pin tubes instead of octals and featured six crystal positions for the reception of six fixed frequencies.

Made In:	England 1943-1950
Voltages:	300 VDC 65 ma and 6.3 VDC 3.7 Amps
Readout:	Analog
Physical:	20.5x17x14" 62 Lbs. 521x431x355mm 28 kg
Status:	Active Manufacturer Discontinued Model
Rarity:	Extremely Scarce

	New	Used
Price:	①	②
Rating:	⑥	⑥

CR300/1

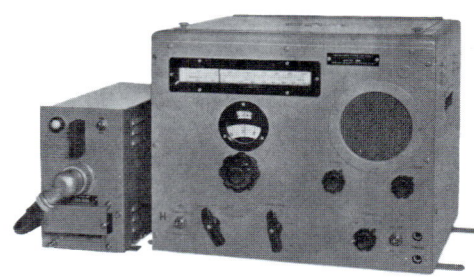

General Coverage Communications Receiver
Single Conversion Superheterodyne. 8 Tubes.

Features:
- ¼ Head Jack (2)
- Speaker
- Standby
- Calibrator 500 kHz
- RF Gain
- Carry Handles
- BFO
- Hinged Cover
- AGC ON/OFF
- Mute Line
- Dial Lamp
- NL

Specifications:

Coverage ......15 - 25000 kHz	Modes .... AM/CW/MCW
Selectivity ......5/4/2 kHz -6dB	IF .... 570, 98 kHz[1]
Sensitivity ......2-5 µV 20 dB	Antenna Input . 100 ohm
Audio Out ......3-5 ohms 2W	

Circuit Complement: ARTH2 Signal Frequency Amp, X66 (or 6K8) Frequency Changer, KTW61 IF Amp, KTW61 IF Amp, DH63 Second Detector/AGC Rectifier/L.F. Amp, KTW61 BFO, 6V6G Output, KTW61 Calibration Oscillator (and OZ4 HT Fullwave Rectifier in 889 Power Supply). [1] The 98 kHz IF is used in the 1st two ranges, and the 570 kHz IF is used if the other six ranges.

Comments: Ranges: .015-.085, .21-.55, .375-1, 1-2.6, 2.6-6.8, 6.8-17 and 15-25 MHz. The separate model 889 power supply measures 6.375x8.675x13.25" 21 Lbs. (162x220x336mm 9.5 kg). Finished in "Marconi gray". The CR 300/1 was built primarily for the Royal Navy.

Variants: Model **CR300/2** substitutes the ARTH2 with a KTW61. It was designed for merchant ships. Its calibrator emits harmonics on 690 kHz intervals. Please note that the earlier and different **CR200** (Navy B29) was a five tube longwave receiver covering only 15 to 550 kHz.

Made In:	England 1943-1946
Voltages:	
Readout:	Analog
Physical:	18.75x13.675x15.5" 55 Lbs 476x348x394mm 25 kg (see Comments for P.S.)
Status:	Active Manufacturer Discontinued Model
Rarity:	Extremely Scarce

	New	Used
Price:	①	②
Rating:	⑥	⑥

Canadian Marconi

CSR-5

General Coverage Communications Receiver
Single Conversion Superheterodyne. 11 Tubes.

Features:
- ¼ Head Jack
- RF Gain
- Dial Lamp
- Standby
- BFO
- Rack Handles
- NL
- Tone
- Speaker Jack
- Panoramic Jack
- Fixed Crystal Position
- AVC ON/OFF
- Line Out 500 ohms

Specifications:
Coverage ...... See Comments.
Selectivity ...... 4 Position
IF .... 575 kHz
Modes .... AM/CW/MCW
Antenna Input . Terminals 70/500 Ohms

Circuit Complement: 6SK7 1st RF Amp, 6SK7 2nd RF Amp, 6K8 Mixer and Crystal Controlled Osc., 6SK7 1st IF Amp, 6SK7 2nd IF Amp, 6B8 Diode Detector, 6H6 NL Rectifier/AVC Rectifier, 9002 HF Conversion Osc., 6SK7 BFO, VR150-30 Voltage Regulator and 6F6 Pentode Power Output. Temperature compensation and voltage stabilization provide a high degree of stability.

Accessories:
110-540 VP3 Pwr.Supl. WE-11 Pwr. Supl 106-704 Shock Mounts
110-823 Speaker 110-386 Rack Speaker Cabinet (shown)

Comments:
Ranges: .079-.207, .195-.518, 1.5-3.5, 3.55-7.65, 6-16.1 and 14.9-30.3 MHz. A single crystal socket is available on the front of the receiver (lower, center) for fixed operation. The CSR-5 was designed for the Royal Canadian Navy. [¥123,600].
Variants: Model **CSR-5A** (shown). Differences unknown.

Made In:	Canada	1942
Voltages:	Requires 12 VAC or VDC @2.3A and 250V @115ma	
Readout:	Analog	
Physical:	Cabinet Version: 20.25x10.5x13.675" 68 Lbs. 514x259x347mm 30.8 kg Rack Version: 19x8.75x13.25" 58 Lbs 483x222x336mm 26.3 kg	
Status:	Active Manufacturer Discontinued Model	
Rarity:	Very Scarce	

	New	Used
Price:	①	②
Rating:	⑥	⑦

H2540

General Coverage Communications Receiver
Double Conversion Superheterodyne. Solid State.

Features:
- ¼ Head Jack
- AGC 4 Pos.
- Remote Port
- Speaker
- Rack Handles
- BFO
- Speaker Switch
- Atten. -10/20/30 dB
- AF/S/FSK Indicator

Specifications:
Coverage ...... 15 - 30000 kHz
Selectivity ...... 6/3/1.2/.5/.3 kHz
IF Rejection ... >80 dB
Image Rej ...... >100 dB
Accuracy .... ±.5 Hz.
Modes .... AM/SSB/CW/ISB/RTTY
Stability .... 1 x 10⁸ per day
Environment ... -10° to +55°C 95% Humid.
Antenna Input . 50 ohms

Accessories:
High Stability Option
H6800 Remote Controller
Computer Port

Comments:
Reduced performance 15 - 1500 kHz. 1 Hz tuning and display resolution. Tuning via knob or decade switches.

Variants:
Later model **H2541** adds a keypad and deletes the decade switches. This receiver is shown to the right.

Made In:	England	1982-1987
Voltages:	100-250 VAC 45-65 Hz 135W	
Readout:	00000.001 Digital	
Physical:	19x8.75x21" 55 Lbs. 483x222x533mm 25 kg	
Status:	Active Manufacturer Discontinued Model	
Rarity:	Typically Unavailable	

	New	Used
Price:	①	②
Rating:	⑥	⑥

H2542

General Coverage Communications Receiver
Double Conversion Superheterodyne. Solid State.

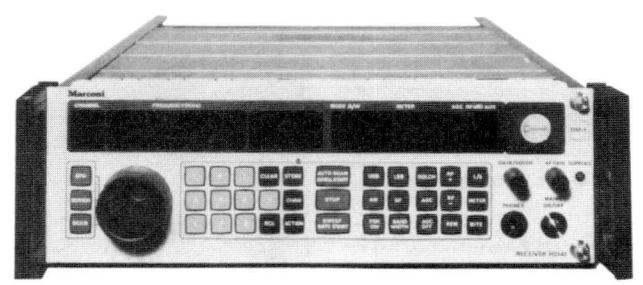

Features:
- ¼ Head Jack
- AGC.
- Squelch
- Keypad
- B.I.T.E.
- Speaker
- Rack Handles
- Dimmer
- Scan
- BFO
- Speaker Switch
- 1 MHz Out Jack
- Sweep
- Atten. -10/20/30 dB
- AF/S/FSK Indicator
- Memories
- Modular Construction

Specifications:
Coverage ...... 15 - 30000 kHz
Stability .... 1 x 10⁸ per day
Image Rej ...... >100 dB
Environment .. -10° to +55° C.

Modes .... AM/SSB/CW/ISB/RTTY
IF Rejection .... >80 dB
Antenna Input . 50 ohms

Comments:
Reduced performance 15 - 1500 kHz. 1 Hz tuning and display resolution.

Made In:	England	1985-1994
Voltages:	100-125 200-250 VAC 50-60 Hz 100W	
Readout:	`00000.001` Digital	
Physical:	19x5.25x23.1" 33 Lbs. 483x133x588mm 15 kg	
Status:	Active Manufacturer Discontinued Model	
Rarity:	Typically Unavailable	

	New	Used
Price:	①	②
Rating:	⑥	⑥

H2550

General Coverage Communications Receiver
Double Conversion Superheterodyne. Solid State.

Features:
- ¼ Head Jack
- RF Gain
- Mute
- Speaker
- Keypad
- 256 Memories
- B.I.T.E.
- DSP
- Rack Handles
- Modular Construction
- Atten. -6/12/18 dB
- RS232/422/485

Specifications:
Coverage ...... 10 - 30000 kHz
Stability .... 3 parts 10⁹ per day
Sensitivity ...... <.5µV 13 dB SINAD
Image Rej ...... >100 dB

Modes .... AM/SSB/CW/ISB/RTTY
IF Rejection .... >80 dB
Environment ... -10° to +55° C.

Accessories:
Internal Preselector   ISB (4 Channel)   Various Front Panels
MIL-STD-1 553B Bus   ALE & Multimode Modems

Comments:
1 Hz tuning and display resolution.

Made In:	England	1995-1998
Voltages:	115/230 VAC 50-60 Hz	
Readout:	`00000.001` Digital Phs.	
Physical:	19x5.2x25" 44 Lbs. 483x133x635mm 20 kg	
Status:	Active Manufacturer Active Model	
Rarity:	Typically Unavailable	

	New	Used
Price:	①	②
Rating:	⑥	⑥

Monitor

Marine Communications Receiver
Single Conversion Superheterodyne. 10 Tubes.

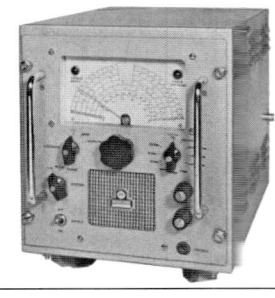

Features:
- ¼ Head Jack
- RF Gain
- Dial Lamp
- BFO
- Speaker
- Standby
- Rack Handles
- Speaker Jack
- 500 kHz Preset
- AVC ON/OFF

Specifications:

Coverage ...... See Comments.
Selectivity ...... 4 Position
Audio .... 0.5W 15 ohms

Modes .... AM/CW/MCW
IF .... 2700, 700 kHz
Antenna Input . Terminals 70/500 Ohms

Circuit Complement:
EF97 RF Amp, EF98 1st Oscillator, ECH83 1st Frequency Changer, ECH83 2nd Frequency Changer/2nd Oscillator, ECH83 1st IF Amp/AF Amp, EBF83 2nd IF Amp/AGC Detector/Detector, D77 AGC Detector, ECH83 BFO/AF Amp and (2) PL84 Output. The preceding tube functions are for 4-22.7 MHz double conversion operation. This radio is single conversion below 3.8 MHz. and operates as a T.R.F. on the 500 kHz preset.

Accessories: 115/230 VAC Transformer

Comments: Ranges: .165-.235, .405-.535, 1.6-2.5, 2.5-3.8, 4.063-4.438, 6.2005-6.525, 8.195-8.815, 12.33-13.2, 16.47-17.36 and 22-22.72 MHz. Red lamp activates during 500 kHz watch.

Variants: Auxiliaries rack version (shown right).

Made In:	England 1956-1969
Voltages:	220 VDC or 24 VDC
Readout:	⊞⊞⊞⊞ Analog
Physical:	10.375x12.75x16.4" 35 Lbs 270x324x416mm 15.9 kg
Status:	Active Manufacturer Discontinued Model
Rarity:	Extremely Scarce

	New	Used
Price:	①	②
Rating:	⑥	⑦

Nebula

General Coverage Communications Receiver
Triple Conversion Superheterodyne. Solid State.

Features:
- ¼" Head. Jack
- RF Gain
- BFO ± 5kHz
- Speaker
- S/AF/FSK Meter
- Dial Lamp
- Rack Handles
- Line Out 600 ohms
- Line Out Jack
- Speaker Jack
- Mute Line
- Attenuator -10/20 dB
- AGC ON/OFF
- IF Out Jack

Specifications:

Coverage ...... 10 - 30000 kHz
Selectivity ...... 8/3/2.65/1.3/.4kHz -6dB.
Sensitivity ...... <1µV 10 dB S/N CW
Audio Out ...... 1 W 3 ohm 5% Dist.
Environment .. 0 to +50° C

Modes .... AM/SSB/CW
IF .... 1235-1335, 250, 100 kHz.
Image Rej. .... >70 dB over 18 MHz
Stability .... 20 Hz after warm-up
Antenna Input . BNC and Terminals

Accessories:
DC/AC Converter

Comments:
Ranges: 20-30, 10-20, 4-10, 1.6-4, .68-1.65, .29-.68, .125-.295, .53-.126, .023-.054 and .010-.023 MHz. The careful reader will note that this is a variant of the Eddystone EC-958 badged under the Marconi Marine label.

Made In:	England 1963-1974
Voltages:	100/125 or 200/250 VAC 40-60 Hz
Readout:	⊞⊞⊞⊞ Analog Lin.
Physical:	Cabinet Version: 19.75x6.5x18" 50 Lbs. 502x165x457mm 22.7 kg Rack Version: 19x5.25x16.2" 43.5 Lbs. 483x133x411mm 19.6 kg
Status:	Active Manufacturer Discontinued Model
Rarity:	Typically Unavailable

	New	Used
Price:	①	②
Rating:	⑥	⑥

53 McKay Dymek

The **McKay Dymek Company** produced an interesting and somewhat unusual line of receivers in the 1970's and 1980's. Radically different in appearance than traditional communications receivers, McKay Dymek models more closely resembled home stereo receivers. They produced superb audio. Tuning was also different. A separate knob was provided for each digit. This approach provides fast, accurate tuning, but is not efficient for simply "tuning the band". Many receivers found their way into government installations including the F.A.A. where fixed operation was called for.

McKay Dymek even sold an analog longwave, medium wave, FM, and five shortwave band portable called the DR-11 (which was manufactured by the Selena Company of Latvia).

The DR300 series was the last line of H.F. receivers produced by McKay Dymek. The DR333 and DR360 were gaining sales momentum when production was abruptly halted in October of 1994. The unexpected discontinuance of a very vital and very specialized integrated circuit from Plessey of England prevented further production.

The Company continues to supply high quality active antennas and other communications products to the commercial, broadcast, military and hobby markets.

McKay Dymek Co.
675 N. Park Av.
Pamona, CA 91766   1977-1978

McKay Dymek Co.
111 South College Av.
Claremont, CA 91711   1979-1980

Stoner Communications, Inc.
McKay Dymek Div.
9119 Milliken Av.
Rancho Cucamonga, CA 91730
1981-1992

Stoner Communications, Inc.
300 N. Eighth St.
Lakeside, OR 97449   1992-1998

McKay Dymek

DR22C-6

General Coverage Communications Receiver
Triple Conversion Superheterodyne. Solid State.

Features:
- ¼" Head. Jack   • S-Meter   • NL   • Fixed Notch 5000 Hz.
- 4" Speaker   • Mute Line   • Speaker Jack   • IF Out Jack 455 kHz
- Fine Tuning ±5 kHz

Specifications:
Coverage ...... 50 - 29700 kHz
Selectivity ...... 8/4 kHz -6dB.
Sensitivity ...... <.5µV SSB .4-20 MHz
Audio Out ...... 4 ohms 2 W 10% dist.
IFs .... 30, 10.7 MHz, 455 kHz

Modes .... AM/LSB/USB
Stability .... ±40 Hz. after ½ hour
Image Reject .. 70 dB
Antenna Input . RCA & Terminals

Circuit Complement:
43 Integrated circuits, 16 FETs, 18 transistors and 54 diodes.

Accessories:
DS111 Speaker   19RM22 Rack Mount Kit   DP40 RF Preselector

Comments:
One knob is featured for each digit of display except last digit which displays either 0 or 5 kHz. The Fine Tuning knob on this digit is for tuning stations not on an even 5 kHz channel. This interesting approach may be cumbersome for those who like to tune up and down the bands. This series of receivers has a distinct "home stereo" look to it and is a departure from the traditional communications receiver design. **Variants:** It is believed that the prior model **DR22** was very similar, but with no NL switch above the headphone jack $895-$995.

Made In:	United States 1977-1980	
Voltages:	110-120/220-240 VAC 50/60 Hz 30 W	
Readout:	`00005.` Digital LED	
Physical:	17.5x5.1x15" 20 Lbs. 430x130x370mm 9 kg	
Status:	Active Manufacturer Discontinued Model	
Rarity:	Scarce	
Reviews:	*WRTH* 1978	

	New	Used
Price:	$995-1250	$200-300
Rating:	★★★	★★★

MK McKay Dymek

DR33C-6

General Coverage Communications Receiver
Triple Conversion Superheterodyne. Solid State.

Features:
- ¼" Head. Jack
- 4" Speaker
- 4 Tuning Steps
- S-Meter
- Mute Line
- Speaker Jack
- ANL
- IF Out Jack
- Attenuator
- Fixed Notch 5000 Hz.
- Line Out 600 ohms
- Fine Tuning ±5 kHz

Specifications:

Coverage ......50-29700 kHz	Modes .... AM/LSB/USB/CW
Selectivity ......8/4/_ kHz -6dB.	Stability .... ±40 Hz. after ½ hour
Sensitivity ......<.5µV SSB/CW .4-20 MHz	Image Reject .. 70 dB
Audio Out ......2W 4 ohms 10% dist.	Antenna Input . RCA or BNC & Terminals

Circuit Complement:
56 Integrated circuits, 31 FETs, 20 transistors and 66 diodes. Collins mechanical filters.

Accessories:

DS111 Speaker	19RM22 Rack Mount Kit	DP40 RF Preselector
375 Hz Filter	1200 Hz Filter	

Comments: Features a large 0.5" (13 mm) LEDs. The 5 position "Band" control selects: ".05-29.7 MHz Preamp" (broadband preamp), ".05-29.7 MHz" (full coverage no preamp), "Local" (engages 30 dB attenuator), "2.5-29.7 MHz" (rejects signals below 2.5 MHz), 2.5-29.7 Preamp" (preamp and rejection below 2.5 MHz). The advertising literature and the owner's manual both indicate that a BNC jack is used as the antenna input. However, the same literature clearly shows an RCA phono jack (and terminals) for this use. Sold for $800 in 1987 as a closeout.

Made In:	United States 1975-1987
Voltages:	110-120/220-240 VAC 50/60 Hz 30 W
Readout:	**00000.1** Digital LED
Physical:	17.5x5.1x15" 16 Lbs. 430x130x370mm 7.3 kg
Status:	Active Manufacturer Discontinued Model
Rarity:	Scarce
Reviews:	📖Passport 1937 WRTH 1980 Commun. World 1980 Popular Elec. October 1979 Radex 1979 QST September 1979

	New	Used
Price:	$1000-1995	$400-650
Rating:	★★★	★★★

MK McKay Dymek

DR44

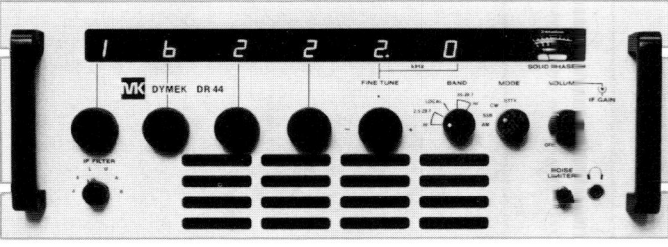

General Coverage Communications Receiver
Triple Conversion Superheterodyne. Solid State.

Features:
- ¼" Head. Jack
- 4" Speaker
- IF Gain
- S-Meter
- Mute Line
- Speaker Jack
- ANL
- IF Out Jack
- Fixed Notch 5000 Hz.
- Fine Tuning ±5 kHz

Specifications:

Coverage ......50 - 29700 kHz	Modes .... AM/SSB/CW/RTTY
Selectivity .....8/4/2.5/_ kHz -6dB.	Stability .... ±40 Hz. after ½ hour
Sensitivity .....<.5µV CW .4-20 MHz	Image Reject.. 70 dB
Audio Out ......2 W 4 ohms 10% dist.	Antenna Input RCA
Environment .0 to +50°C 95% Humid	

Circuit Complement:
56 Integrated circuits, 31 FETs, 20 transistors and 66 diodes.

Accessories:

DS111 Speaker	DP4044 RF Preselector
375 Hz Filter	1200 Hz Filter

Comments:
The frequency is selected by rotating individual knobs per digit. The first knob selects 0,10 or 20 MHz. The next knob selects the single MHz position, the third knob the 100 kHz position, the fourth knob the 10 kHz position and the fifth knob (Fine Tune) selects the 1 kHz and 100 Hz position. This model was used by the Federal Aviation Administration.

Made In:	United States 1975-1987
Voltages:	110-120/220-240 VAC 50/60 Hz 30 W
Readout:	**00000.1** Digital LED
Physical:	19x7x15" 16 Lbs. 480x180x370 mm 7.3 Kg
Status:	Active Manufacturer Discontinued Model
Rarity:	Very Scarce

	New	Used
Price:	$1600-1900	$390-480
Rating:	★★★	★★★

☒ McKay Dymek

DR44-6

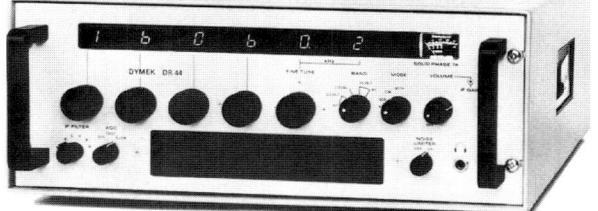

General Coverage Communications Receiver
Triple Conversion Superheterodyne. Solid State.

Features:
- ¼" Head. Jack
- 4" Speaker
- IF Gain
- Rack Handles
- S-Meter
- Mute Line
- Speaker Jack
- ANL
- IF Out Jack
- 4 Tuning Steps
- Fixed Notch 5000 Hz.
- AGC OFF/FST/SLO
- Fine Tuning ±5 kHz

Specifications:
Coverage ......50 - 29700 kHz
Selectivity ......8/4/_/_ kHz -6dB.
Sensitivity ......<.5µV SSB/CW .4-20 MHz
Audio Out ......2 W 10% distortion

Modes .... AM/LSB/USB
Stability .... ±40 Hz. after ½ hour
Image Reject .. 70 dB
Antenna Input . RCA

Circuit Complement:
56 Integrated circuits, 31 FETs, 20 transistors and 66 diodes.

Accessories:
DS111 Speaker   DC719 Cabinet (shown)   DP4044 RF Preselector
375 Hz Filter   1200 Hz Filter

Comments:
The frequency is selected by rotating individual knobs per digit. The first knob selects 0,10 or 20 MHz. The next knob selects the single MHz position, the third knob the 100 kHz position, the forth knob the 10 kHz position and the fifth knob (Fine Tune) selects the 1 kHz and 100 Hz position. Note the addition of the AGC knob on the left, not found in the DR44. Shown in optional cabinet. This model was actively marketed to the marine and military markets. The DR44-6 was also used by the F.A.A.

Made In:	United States 1975-1987	
Voltages:	110-120/220-240 VAC 50/60 Hz 30 W	
Readout:	`00000.1` Digital LED	
Physical:	17.5x5.1x15" 16 Lbs. 444x129x381mm 7.2 kg	
Status:	Active Manufacturer Discontinued Model	
Rarity:	Very Scarce	
Reviews:	*Radio Electronics*	

	New	Used
Price:	$1900-2200	$460-530
Rating:	★★★	★★★

☒ McKay Dymek

DR55-6

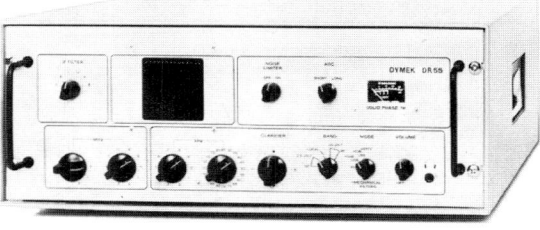

General Coverage Communications Receiver
Triple Conversion Superheterodyne. Solid State.

Features:
- ¼" Head. Jack
- 4" Speaker
- AGC (2 Pos).
- S-Meter
- Mute Line
- Rack Handles
- ANL
- IF Out Jack
- Fine Tuning ±5 kHz
- Fixed Notch 5000 Hz.
- Speaker Jack

Specifications:
Coverage ......50-29700 kHz
Selectivity ......8/4/_/_ kHz -6dB.
Sensitivity ......<.5µV CW .4-20 MHz
Audio Out ......2 W 4 ohms 10% dist.
Environment ..0 to +50°C 95% Humid

Modes .... AM/LSB/USB
Stability .... ±40 Hz. after ½ hour
Image Reject .. 70 dB
Antenna Input . RCA 50 ohm

Circuit Complement:
43 Integrated circuits, 16 FETs, 18 transistors and 54 diodes.

Accessories:
DS111 Speaker   DC719 Cabinet (shown)   DP4044 RF Preselector
375 Hz Filter   1200 Hz Filter

Comments: Tuning is accomplished with a knob for: 10s of MHz., MHz, 100 Hz and 5 kHz. A Fine Tuning knob is featured (±5 kHz) for frequencies off 5 kHz increments. Closed-out at $650 in 1987. Shown in optional cabinet. This model was actively marketed to the marine and military markets.

Variants: A version of this receiver was manufactured in Brazil under license from McKay Dymek Company. Forty units were made for use in the country's coastal communications stations. Labeling was in Portuguese. It was model **RC-02-10C**.

Made In:	United States 1974-1988	
Voltages:	110-120/220-240 VAC 50/60 Hz 30 W	
Readout:	`●●●●●.` Knobs	
Physical:	19x7x15" 16 Lbs. 483x178x381mm 7.2 kg	
Status:	Active Manufacturer Discontinued Model	
Rarity:	Very Scarce	

	New	Used
Price:	$800-1295	$300-380
Rating:	★★★	★★★

MK McKay Dymek

DR101-6

General Coverage Communications Receiver
Triple Conversion Superheterodyne. Solid State.

Features:
- ¼" Head. Jack
- 4" Speaker
- IF Gain
- S-Meter
- Mute Line
- 3 Tuning Rates
- ANL
- IF Out Jack
- VRIT
- Fixed Notch 5000 Hz.
- Speaker Jack
- AGC OFF/FST/SLO

Specifications:
Coverage ...... 50 - 29700 kHz
Selectivity ...... 8/4 kHz -6dB.
Sensitivity ...... <.5µV SSB/CW .4-20 MHz
Audio Out ...... 2 W 4 ohms 10% dist.
Environment .. 0 to +50°C.

Modes .... AM/LSB/USB/RTTY
Stability .... ±40 Hz. after ½ hour
Image Reject .. 70 dB
Antenna Input . RCA & Terminals

Accessories:
DS111 Speaker   Rack Mounting Kit   DP40 RF Preselector
375 Hz Filter   1200 Hz Filter

Comments:
The DR101 features Variable Rate Incremental Tuning that provides electronic scan tuning in 100 Hz steps. The speed of the tuning is adjustable by how far to the right or the left you set the VRIT knob. One of the earliest slewing receivers. Consequently this model may be used for band scanning unlike the DR22, DR33, etc. The 4 kHz filter is Collins.

Made In:	United States 1979-1983	
Voltages:	110-120/220-240 VAC 50/60 Hz 30 W	
Readout:	**00001.** Digital LED	
Physical:	17.5x5.1x15" 16 Lbs. 444x130x381mm 7.3 kg	
Status:	Active Manufacturer Discontinued Model	
Rarity:	Very Scarce	
Reviews:	📖 Passport 1987 WRTH 1982	

	New	Used
Price:	$1150-1850	$400-620
Rating:	★★★★	★★★★

MK McKay Dymek

DR333

General Coverage Communications Computer Receiver
Double Conversion Superheterodyne. Solid State.

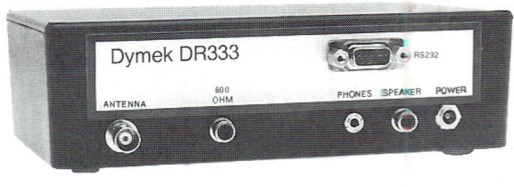

Features:
- Head. Jack
- Squelch
- Clock 24 Hr.
- PBT ±600 Hz.
- PC S-Indication
- Speaker Jack
- RF Gain
- Operation via mouse and/or keyboard
- AGC
- 9999 Memories
- 999 Band Scans with auto logging
- Attenuator
- Spectrum Display

Specifications:
Coverage ...... 10 - 30000 kHz
Selectivity ...... 6/2.7/_/_ kHz -6dB.
Sensitivity ...... <.5µV SSB
Audio Out .... 6 W 10% dist.
IFs .... 455 kHz, 40.04 MHz

Modes .... AM/LSB/USB/CW/FM
Third Order ..... +35 dBm
IF Reject .... >70 dB
Image Reject .. 90 dB
Antenna Input . BNC 50 ohm

Accessories:
.4 kHz Filter   1.2 kHz Filter

Comments:
This receiver lacks any knobs or controls and is operated exclusively through a PC. System requirements: DOS 3.1 or higher, 256K memory. Software supplied on two 5¼" diskettes. Connection to the PC is via the serial port (9600 baud, DB-9). The supplied 2.7 kHz filter is a Collins. The spectrum analyzer features 400 stops and is near real-time. A 15 character label can be stored with each memory. Please see **Displays** at the bottom of the following page.
Variants: The model **DR300** was the receiver on printed circuit board with no enclosure (OEM). Only the 6 kHz filter was supplied as standard.

Made In:	United States 1992-1994	
Voltages:	11 - 16 VDC 1A	
Readout:	**1 Hz.** Digital on PC	
Physical:	7.4x2.25x4.75' 2 Lbs. 188x57x121mm .9 kg	
Status:	Active Manufacturer Discontinued Model	
Rarity:	Very Scarce	
Reviews:	Audio September 1992 Funk February 1993 NASWA September 1992 Proceedings 1991	

	New	Used
Price:	$1199	$390-500
Rating:	★★★★	★★★

McKay Dymek

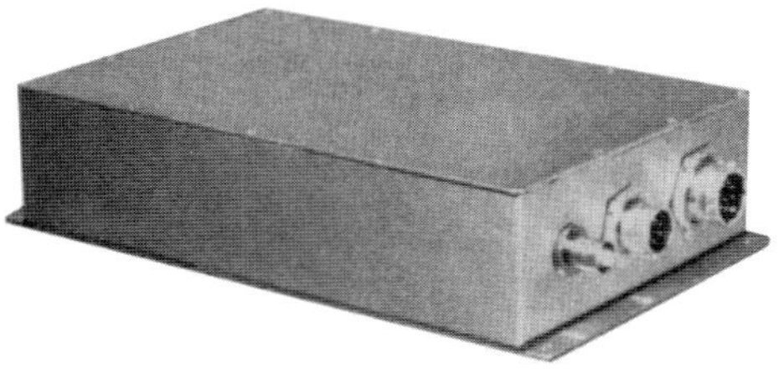

DR360

General Coverage Communications Computer Receiver
Double Conversion Superheterodyne. Solid State.

Features:
- IF Out Jack
- Squelch
- Clock 24 Hr.
- PBT ±600 Hz.
- PC S-Indication
- Speaker Jack
- RF Gain
- Operation via mouse and/or keyboard
- AGC
- 9999 Memories
- 999 Band Scans with auto logging
- Attenuator
- Spectrum Display

Specifications:

Coverage	10 - 30000 kHz	Modes	AM/LSB/USB/CW/FM
Selectivity	6/2.7/.5/_ kHz -6dB.	Third Order	+35 dBm
Sensitivity	<.5µV SSB	IF Reject	>70 dB
Audio Out	.6 W 10% distortion	Image Reject	90 dB
IFs	455 kHz, 40.04 MHz	Antenna Input	BNC 50 ohm

Comments:
The DR360 is the military version of the DR333. All inputs and outputs are furnished via three weather and dust resistant connectors. The heavy duty enclosure is RF and weather gasketed. Standard outputs are: 455 kHz IF, 600 ohms unbalanced, .6 watt audio. Supplied filters include: 6, 2.7 and .5 kHz -6dB. The 2.7 filter is a Collins. One optional filter may be added. A 1 PPM TCXO is also featured on this model. The antenna jack is a BNC. All other specifications and capabilities are the same as the model DR-333.

Made In: United States 1992-1994

Voltages: 11 - 16 VDC 1A

Readout: 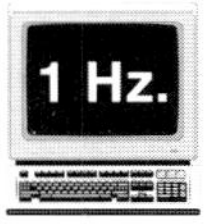

Digital on PC

Physical: 9x2.25x5" 3 Lbs.
229x57x127mm 1.4 kg

Status: Active Manufacturer
Discontinued Model

Rarity: Typically Unavailable

	New	Used
Price:	①	⑤
Rating:	⑥	⑥

◆ DR333 and DR360 Displays

The DR333 and DR360 feature several modes of operation:

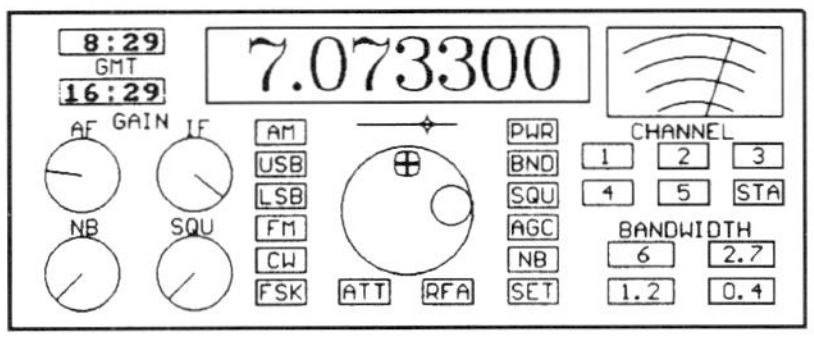

Standard Operating display with S-meter, clock and 10 Hz freq. resolution.

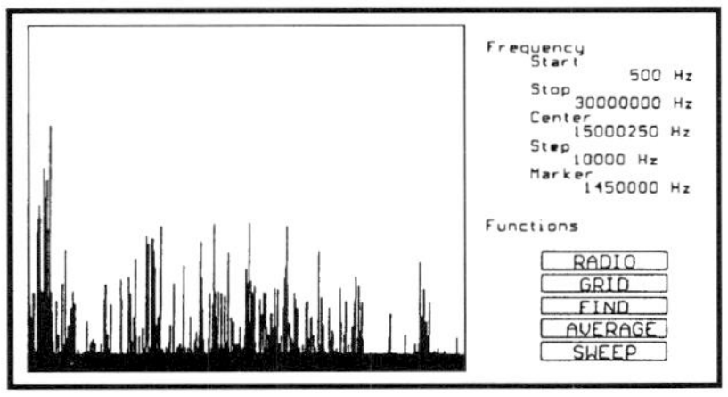

Spectrum Analyzer display with up to 400 stops.

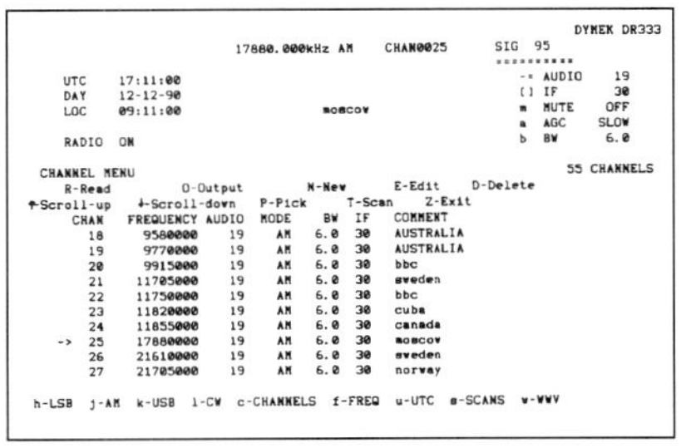

Memory Scan Menu display showing channel, freq., mode, bandwidth, etc.

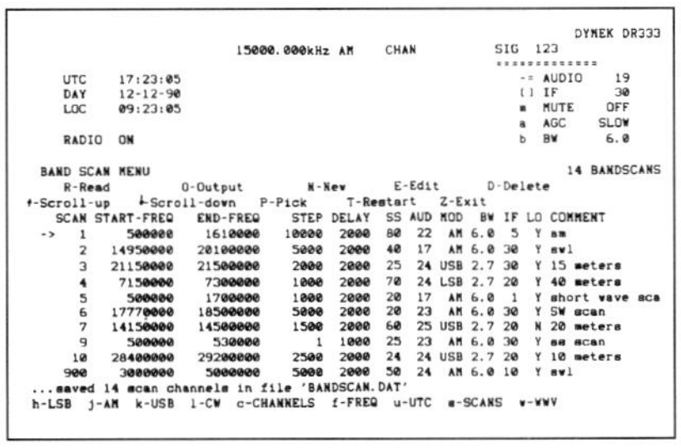

Band Scan Menu display showing start, stop frequency, step, delay, etc.

The **Midland Company** has for many years produced a broad line of electronic parts, accessory items, test devices, CBs and CB accessories. In the late 1960's and early 1970's they produced a fairly nondescript line of shortwave receivers. Midland manufactured perhaps the most consistently poor offering of shortwave receivers ever produced. Many featured cardboard back panels *and* bottom panels. Their performance was below average, even for their price class.

Midland no longer manufactures shortwave receivers. Today Midland's other radio products exhibit a much higher quality of construction and are respected in the industry.

Midland International Corp.
1909 Vernon St.
N. Kansas City, MO 64113

Midland
111 South College Av.
Claremont, CA 91711

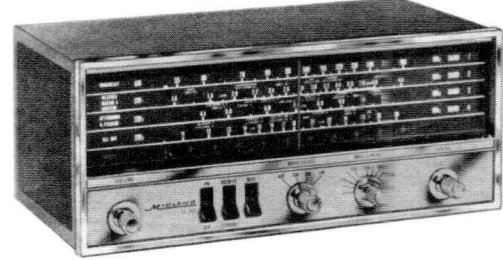

11-500

General Coverage Communications Receiver
Single Conversion Superheterodyne. Solid State.

Features:
- ¼" Head. Jack
- Dial Lamp
- BFO
- MW Ferrite Antenna
- Ext. Spkr Jack
- Standby
- AVC
- Bandspread 0-100
- Speaker 4"

Specifications:
Coverage ...... 550 - 30000 kHz   Modes .... AM/CW
Antenna Input Terminals

Circuit Complement:
9 Transistors and 3 diodes. Has three IF stages.

Comments:
Ranges: .55-1.6, 1.6-4.4, 4.5-11 and 11-30 MHz. Black metal cabinet.

Made In:	Japan	1969-1970
Voltages:	117 VAC	
Readout:		Analog
Physical:	12x5.25x4.75' 5 Lbs.	
	305x133x120mm 2.3 kg	
Status:	Active Manufacturer	
	Discontinued Model	
Rarity:	Scarce	
Reviews:	*CQ* February 1969	

	New	Used
Price:	$45	$20-30
Rating:	★★	★

MIDLAND

11-520

General Coverage Communications Receiver
Single Conversion Superheterodyne. Solid State.

Features:
- ¼" Head. Jack
- S-Meter
- BFO
- MW Ferrite Antenna
- Ext. Spkr Jack
- Fine Tuning
- AVC
- Dial Lamp
- AFC [FM]
- Speaker
- Phono Input
- Tuner Output
- Tone Control

Specifications:
Coverage ...... 550 - 30000 kHz & FM   Modes .... AM/CW
Antenna Input Terminals

Circuit Complement:
11 Transistors and 6 diodes.

Comments:
Ranges: .55-1.6, 2-5, 4.8-12 and 11-30 MHz plus FM 88-108 MHz. Features a black vinyl-covered cabinet.

Made In:	Japan	1969-1970

Voltages: 117 VAC

Readout: Analog

Physical: 14x7x8" 5 Lbs.
355x177x203mm 2.3 kg

Status: Active Manufacturer
Discontinued Model

Rarity: Scarce

	New	Used
Price:	$80-90	$35-45
Rating:	★★	★

MIDLAND

11-530

General Coverage Broadcast Receiver
Single Conversion Superheterodyne. Solid State.

Features:
- ¼" Head. Jack
- S-Meter
- Tone Control
- Speaker 5"
- Ext. Spkr Jack
- Bandspread
- AVC
- Dial Lamp
- AFC [FM]
- Fine Tuning
- Band Indicator Lamps

Specifications:
Coverage ...... See Comments.   Modes .... AM
Antenna Input Terminals

Circuit Complement:
15 Transistors, 6 diodes, 2 varistors, 1 thermistor, 2 silicon rectifiers and 3 auxiliary transistors.

Comments:
Seven bands are featured. Ranges: .55-1.6, 1.6-4, 4-12, 30-50, 88-108 (FM), 108-135 (VHF Air) and 144-174 MHz (VHF High band). Please note that the coverage of the shortwave band ends at 12 MHz. Features a black vinyl-covered metal cabinet with chrome trim.

Made In:	Japan	1969-1970

Voltages: 117 VAC

Readout: Analog

Physical: 13x5x8" 11 Lbs.
330x127x203mm 5 kg

Status: Active Manufacturer
Discontinued Model

Rarity: Scarce

	New	Used
Price:	$90-100	$35-45
Rating:	★★	★

13-900

General Coverage Communications Receiver
Single Conversion Superheterodyne. Solid State.

Features:
- ¼" Head. Jack
- Ext. Spkr Jack
- Speaker
- Dial Lamp
- Standby
- BFO
- AVC
- MW Ferrite Antenna
- Bandspread 0-100

Specifications:
Coverage ......540 - 30000 kHz   Modes .... AM/CW
Antenna Input Terminals

Comments:
Reception gap from 1.6-2 MHz. The 13-900 appears to be an updated version of the model 11-500.

Made In:	Japan	1972-1973

Voltages: 117 VAC

Readout: Analog

Physical: 12x5.25x4.75" 5 Lbs.
305x133x120mm 2.3 kg

Status: Active Manufacturer
Discontinued Model

Rarity: Scarce

	New	**Used**
Price:	$63	$25-35
Rating:	★★	★

13-910

General Coverage Communications Receiver
Single Conversion Superheterodyne. Solid State.

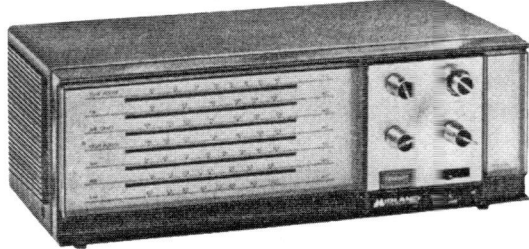

Features:
- ¼" Head. Jack
- Ext. Spkr Jack
- S-Meter
- Speaker
- Dial Lamp
- AFC [FM]
- Fine Tuning

Specifications:
Coverage ......See Comments   Modes .... AM
Selectivity ......8 kHz -6dB.   Antenna Input . Terminals
Sensitivity ......<2 - 3 µV

Comments:
Ranges: .54-1.6, 1.6-4, 4-12, 30-45 (VHF Low Band), 88-108 [FM], 108-135 (VHF Air) and 145-175 MHz (VHF High Band). Note that the shortwave coverage of this receiver is very limited, stopping at 12 MHz. The Midland 13-910 is not a serious shortwave radio, but may be of interest for those wanting to sample the VHF air and public service bands.

Made In:	Japan	1972-1973

Voltages: 117 VAC

Readout: Analog

Physical: 13x5x8" 11 Lbs.
330x127x203mm 5 kg

Status: Active Manufacturer
Discontinued Model

Rarity: Very Scarce

Reviews: *Popular Elec.* December 1972

	New	**Used**
Price:	$105	$30-40
Rating:	★	★

---

55 Morrow Radio

The **Morrow Radio Company** was founded by Ray Morrow W7AWE in the early 1950's. The first product was an amateur band radio converter rather than a receiver. The Morrow Radio Manufacturing Company is best remembered for its line of mobile amateur band receivers, converters and transmitters.

The Company also produced the Morrow FTR Fixed Tuned Receiver. This compact mobile fixed tuned superheterodyne receiver has an input frequency of 1525 kHz. It is designed to work with the Morrow 5BRF converter to form a complete mobile receiver. The Morrow FTR also works with the 5BR, 5BRLN or any of the 2 and 3BR series converters. In the late 1950's the company produced products for the C.B. market including an early handheld C.B. transceiver. The company closed down in 1962.

In the early 1970's Ray's two sons started Morrow Electronics in Salem. According to reliable sources, the Morrow family lost ownership in the company and the company named was changed to Arnav Systems. Meanwhile the Morrow family formed another new company called "II Morrow" (pronounced "tomorrow") is still in business today selling aircraft navigation and communications products.

The November 1992 issue of *Electric Radio* features an interesting history of Ray Morrow and the Morrow Radio Company.

Morrow Radio Manufacturing Co.
2794 Market St.
Salem, Oregon 1956-1962

MORROW
RADIO MANUFACTURING CO.
SALEM, ORE.

Falcon

Amateur Band Communications Receiver
Double Conversion Superheterodyne. Tubes.

Features:
- BFO
- RF Gain
- Mute
- Dial Lamp
- AVC ON/OFF
- Squelsh
- Standby
- Antenna Trimmer
- NL

Specifications:
Coverage ...... Ham. See Comments.
Modes .... AM/CW
Selectivity ...... 4 kHz -6dB.
Sensitivity ...... <1μV CW 14 dB S/N

Circuit Complement:
Crystal controlled second mixer.

Accessories:
PW-115 115 VAC PS. BCT MW Tuner

Comments: Ranges: 80 to 10 meters. Matches the model MB-560A transmitter. The Falcon appears to be the Morrow MBR-5 sans S-Meter and calibrator. Requires an external power supply. Designed for mobile use. Requires a speaker.

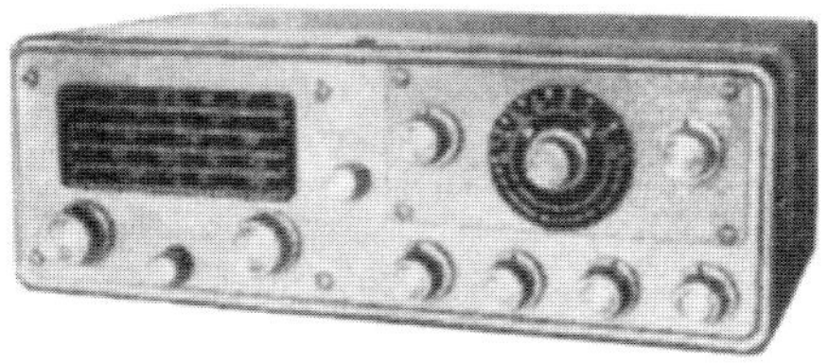

Variants: The **Falcon BCT** included the Conelrad monitor and broadcast tuner accessory which is installed to the right of amateur dial (shown) $189.

Made In: United States 1957-1959

Voltages: Requires Power Supply

Readout: Analog

Physical: 11.75x4x7.125" 6.5 Lbs.
298x102x181mm 2.95 kg

Status: Inactive Manufacturer
Discontinued Model

Rarity: Very Scarce

	New	Used
Price:	$169	⑤
Rating:	⑦	⑦

MB-6

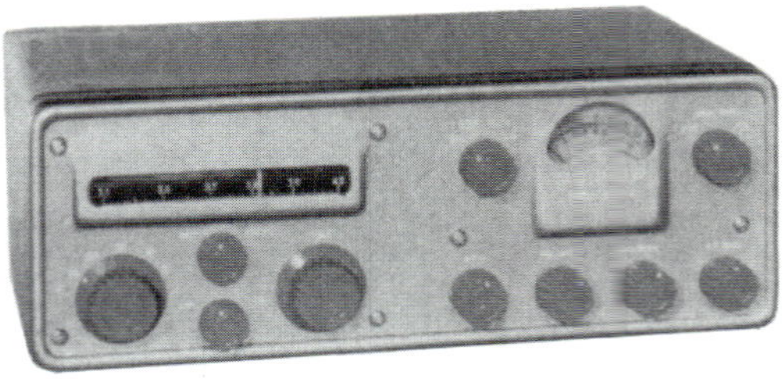

Amateur Band Communications Receiver
Double Conversion Superheterodyne. 13 Tubes.

Features:
- S-Meter
- BFO
- Dial Lamp
- Calibrator 100 kHz.
- RF Gain
- AVC
- Squelch
- Antenna Trimmer
- Mute
- Standby
- NL

Specifications:
Coverage ...... Ham. See Comments.   Modes .... AM/CW
Selectivity ...... 4 kHz -6dB.
Sensitivity ...... <1µV CW

Circuit Complement:
Crystal controlled second mixer.

Accessories:
RAP-250S Power Supply & Speaker

Comments:
Ranges: 80 to 10 meters. Turret-type band display. Matches MB-565 Transmitter. Designed for mobile use. Requires an external power supply. The Morrow RAP-250S 115 VAC power supply also serves as an elevated stand for the radio and features two built-in speakers. This accessory is shown above.

Made In: United States 1958-1959

Voltages: Requires Power Supply

Readout: Analog

Physical: 11.5x4.125x7.25" 11 Lbs.
292x105x184mm 5 kg

Status: Inactive Manufacturer
Discontinued Model

Rarity: Very Scarce

	New	Used
Price:	$185-239	[5]
Rating:	[7]	[7]

MBR-5

Amateur Band Communications Receiver
Double Conversion Superheterodyne. 13 Tubes.

Features:
- ¼" Head. Jack[1]
- S-Meter
- BFO
- Calibrator 100 kHz.
- RF Gain
- AVC ON/OFF
- Squelch
- NL
- Mute
- Standby

Specifications:
Coverage ...... Ham. See Comments   Modes .... AM/CW
Selectivity ...... 4 kHz -6dB.   Sensitivity .... <.5µV CW
IF .... 1525, 200 kHz

Circuit Complement:
6BZ6 RF, 12AT7 Mixer/Osc, 6BJ6 IF, 6BE6 Mixer/Crystal Osc, 6BJ6 IF, 6T8 Detector/BFO, 6AL5 Noise Rectifier, 6AL5 Noise Limiter, 12AX7 Audio Amp/Squelch, 6C4 Audio Amp, 6AQ5 Audio Output, 6BJ6 Crystal Cal. and 12AT7 Noise Amp/S-Meter.

Accessories:
PW-115 115 VAC PS.

Comments:
Ranges: 3.5-4, 7-7.3, 14-14.35, 21-21.45 and 28-29.7 MHz. Matches MB-560 Transmitter. Designed for mobile use. [1]The ¼" headphone jack is on the power supply - speaker.

Made In: United States 1955-1957

Voltages: Includes P.S. for 6/12 VDC.

Readout: Analog

Physical: 11.75x4x6.5" 11 Lbs.
299x102x165mm 5 kg

Status: Inactive Manufacturer
Discontinued Model

Rarity: Very Scarce

Reviews: *QST* May 1956

	New	Used
Price:	$225	[5]
Rating:	[7]	[7]

The **Mosley** company has been a respected manufacturer of amateur antennas for over fifty years.

In 1961 they briefly entered the amateur equipment market with their first and last offering, the Mosley CM-1 amateur receiver. This innovative design used five tubes and four diodes. The unique aspect of the circuit was that all five tubes were 6AW8A's. Early production of the this model featured an all light-gray front panel, unlike the photo shown below which is dark on the upper third.

Mosley continues to produce quality antennas for the amateur radio market.

Mosley Electronics Inc.
4610 N. Lindbergh Blvd.
Bridgeton, MO   1960-1974

Mosley Electronics Inc.
1344 Baur Blvd.
St. Louis, MO 63132   1990-1998

CM-1

Amateur Band Communications Receiver
Double Conversion Superheterodyne. 5 Tubes plus Semiconductors

Features:
- ¼" Head. Jack
- S-Meter
- Preselector
- BFO
- RF Gain
- ANL
- Dial Lamp
- Standby
- Mute Line
- VOX Terminals
- Speaker Terminals

Specifications:
Coverage ...... Ham Bands 80-10M.
Selectivity ...... 2.5 kHz -6dB.
Sensitivity ...... <.5µV 10 dB S/N
Image Rej ...... >35 dB
Audio Out ...... 0.5W 6% Distortion

Modes .... AM/SSB/CW
Stability .... <500 Hz after 1 min.
IF Rejection .... >35 dB
Antenna Input . Terminals

Circuit Complement:
Amazingly, all five tubes are type 6AW8A. Crystal controlled first oscillator, diode detector for AM and product detector for CW/SSB. 5 Tubes and 4 semiconductors.
Accessories: CMS-1 Speaker   CV-160 160M Converter (shown below)
Comments:
Ranges: 3.49-4.14, 6.86-7.51, 13.86-14.51, 20.86-21.51, 28.49-29.14, 29.09-29.74 and 27.99-28.64 MHz and WWV at 15 MHz. Frequency coverage from 14.49-15.14 MHz is achieved by placing the preselector knob to 15 MHz and the band switch to 40 meters. Frequency is read from the A scale of the dial. Requires a speaker.
Variants: The export version operates from 230 VAC 50-60 Hz.

Mfg. In:	United States 1961-1962
Voltages:	115 VAC 50/60 Hz 33W
Readout:	Analog Lin.
Physical:	10.5x7x8" 267x190x203mm
Status:	Active Manufacturer Discontinued Model
Rarity:	Very Scarce
Reviews:	*Electric Radio* Sept. 1995

	New	Used
Price:	$170-183	$120-200
Rating:	★★★	⑦

The **Multi-Products Company**, founded in 1947, is best known for its line of compact mobile amateur receivers. First produced in the early fifties, production continued until the early sixties.

The Elmac A54 transmitter was the contemporary of the PMR-6A and can still be found on the used market, often with the receiver. The AF-68 transmitter is often found with the PMR-8.

The Elmac company also marine radios, manufactured garage door and gate door controls and other special electronic products. The company was acquired in 1976 by the Stanley Door Company.

The July 1992 issue of *Electric Radio* contains an interesting article on the Multi-Elmac Company.

Multi-Products Company
559 East Ten Mile Road
Hazel Park, Michigan   1953-1956

Multi-Products Company
21470 Coolidge Highway
Oak Park 37, Michigan   1957-1962

PMR-6A

Amateur Band Communications Receiver
Double Conversion Superheterodyne. 10 Tubes.

Features:
- NL
- BFO
- Antenna Trimmer • Speaker Terminals
- AVC ON/OFF
- Dial Lamp

Specifications:
Coverage ...... Ham 160-10M & MW
Selectivity ...... 3 kHz -6dB.
Sensitivity ...... <1µV
Audio Out ...... 3.5W 3-6 ohms

Modes .... AM/CW-SSB
IF .... 1600, 455 kHz
Antenna Input . Motorola 50-72 ohms

Circuit Complement:
6BJ6 RF Amp, 6BE6 1st Osc, 6C4 HF Osc, 6BE6 2nd Mixer Osc, 6BA6 1st IF Amp, 6BA6 2nd IF Amp, 6AL5 Detector/NL, 12AT7 1st Audio Amp/BFO, 6BK5 Audio Output and OB2 Voltage Regulator. Has ten tuned circuits.

Accessories:
PSR-116 115 VAC PS   PSR-116S 115 VAC PS & S-Meter
PSR-6 6 VDC PS   PSR-12 12 VDC PS   ESS-2 Ext. S-Meter

Comments:
Designed for mobile use. Ranges: .59-2, 3.5-4, 6.9-7.4, 13.95-14.45, 20.95-21.65 and 28-29.7 MHz. Requires a speaker. Gray metal cabinet.

Variants:
Model **PMR-12A** is the 12 VDC version.

Made In:	United States 1953-1954	
Voltages:	Requires Power Supply See Accessories.	
Readout:	⨅⨅⨅⨅ Analog	
Physical:	6x4.5x8.5" 6.5 Lbs 152x114x216mm 2.9 kg	
Status:	Inactive Manufacturer Discontinued Model	
Rarity:	Scarce	

	New	Used
Price:	$135	$70-110
Rating:	⑥	◌

PMR-7

Amateur Band Communications Receiver
Double Conversion Superheterodyne. 10 Tubes.

Features:
- ¼ Head. Jack
- Squelch
- Dial Lamp
- Speaker Terminals
- RF Gain
- ANL
- BFO
- Antenna Trimmer
- AVC ON/OFF

Specifications:

Coverage ......Ham 160-10M & MW		Modes .... AM/CW-SSB	
Selectivity ......3 kHz -6dB.		IF .... 2238, 262 kHz	
Sensitivity ......<.5µV CW		Antenna Input . Motorola 50-72 ohms	
Audio Out ......3 - 6 ohms			

Circuit Complement:
6BZ6 RF Amp, 6BE6 1st Mixer, 6C4 HF Osc, 6BJ6 1st IF Amp, 6X8 2nd Mixer, 6BA6 2nd IF Amp, 6BJ6 Detector/AVC/ANL, 6AN8 1st Audio/Squelch, 12AU7 BFO/Amp and 6AQ5 Audio Output.

Accessories:
PSR-117 117 VAC PS   PSR-612 6/12 VDC PS   ESS-3 Ext. S-Meter

Comments:
Ranges: .54-1.6, 1.8-2, 3.5-4, 7-7.3, 14-14.4, 21-21.45 and 28-29.7 MHz. Headphone jack is on the rear panel. Designed for mobile use.

Variants:
Model **PMR-7A** 1958 $159.

Made In:	United States 1955-1957
Voltages:	Requires Power Supply See Accessories.
Readout:	Analog
Physical:	7x4.675x11" 8.5 Lbs. 178x119x279mm 3.8 kg
Status:	Inactive Manufacturer Discontinued Model
Rarity:	Scarce
Reviews:	*QST* July 1956 *CQ* January 1957

	New	Used
Price:	$159	$75-110
Rating:	⑥	⑦

PMR-8

Amateur Band Communications Receiver
Double Conversion Superheterodyne. 9 Tubes.

Features:
- ¼ Head. Jack
- ANL
- BFO
- Antenna Trimmer
- RF Gain
- Dial Lamp

Specifications:

Coverage ......Ham 80-6M +MW		Modes .... AM/CW-SSB	
Selectivity ......±3 kHz -3 dB.		IF .... 2238, 262 kHz	
Sensitivity ......<.5µV SSB		Antenna Input . Motorola 50-72 ohms	
Audio Out ......1 W 10% dist. 3.2 ohms			

Circuit Complement:
6DK6 RF Amp, 6BE6 1st Mixer, 6C4 HF Osc, 6BE6 2nd Mixer/Crystal Osc, 6BA6 2nd IF Amp, 6BJ7 Detector/AVC/ANL, 6AW8 1st Audio/BFO, 6AQ5 Audio Out and 3TF4 Filament Regulator. The circuit consists of one stage of RF amplification, a first mixer and stabilized oscillator.

Accessories:
AS-1 Speaker 4"   ESS-3 Ext. S-Meter   M1010 PS 6/12VDC 115VAC

Comments:
Designed for mobile use. Ranges: .54-1.6, 3.5-4, 7-7.3, 14-14.4, 21-21.45, 28-29.7 and 50-52 MHz. Headphone jack is on the rear panel. Dark gray metal cabinet.

Made In:	United States 1960-1961
Voltages:	Requires PS See Accessories.
Readout:	Analog
Physical:	7x4.675x11" 8.5 Lbs. 178x119x279mm 3.8 kg
Status:	Inactive Manufacturer Discontinued Model
Rarity:	Scarce
Reviews:	*73* January 1963

	New	Used
Price:	$190	$80-130
Rating:	★★★★	⑦

58 Murphy

The striking **Murphy** B40 series was designed and built for the British Royal Navy, in a fashion somewhat similar to the battleships that is was to serve in. Regarded by some as "ugly duckling", while others referred to them as the "lighthouse receiver" or "towering infernos". The innovative vertical drum tuning display differentiates this receiver from most others. The B40 series was used in both shore and shipboard installations and were also sold to other Commonwealth navies. Many served well until replaced, in some instances, by the Racal RA-17.

Murphy
London, England

MURPHY

B40D

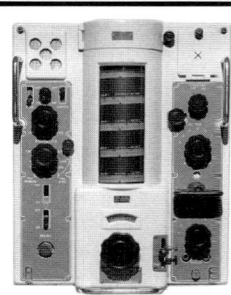

General Coverage Communications Receiver
Single Conversion Superheterodyne. 15 Tubes plus Semiconductors.

Features:
- ¼" Head. Jack (2)
- RF Gain
- Standby
- AGC ON/OFF
- Standby
- BFO
- Dial Lamp
- ANL ON/OFF/ADJ
- Speaker
- Dial Lock
- Calibrator
- Logging Scale 0-100
- Osc. Trimmer
- Rack Handles
- Fiduciary Adjust
- Fixed Crystal Position
- Speaker Switch
- Line Out 600 Ohms

Specifications:

Coverage ......640 - 30500 kHz	Modes .... AM/CW/FSK	
Selectivity ......8/3/1 kHz -6 dB	Sensitivity .... 2 µV CW	
Audio Out ......2.5 W 600 ohms	IF .... 500 kHz	

Circuit Complement: CV4014/6AM6 1st RF, CV454/6BA6 2nd RF, CV2128/6AJ8 Frequency Converter, CV4014/6AM6 Local Osc., CV131/6CQ6 1st IF, CV131/6CQ6 2nd IF, CV131/6CQ6 3rd IF, CV140/6AL5 AGC/Detector, CV140/6AL5 Noise Limiter, CV131/6CQ6 BFO, CV454/6BA6 1st AF, CV2136/6BW6 Audio O/P, 0A2 Voltage Reg. and (2) CV493/6X4 Rectifiers. Has two RF and three IF stages.

Comments: Ranges: .64-1.65, 1.57-4.1, 3.9-10, 9.5-18.5 and 17.6-30.5 MHz. The switchable BFO offsets which can be set up for 1.5 kHz above and below the IF frequency were designed for FSK use. However this feature would later make the receiver quite suitable for USB or LSB operation.

Variants: Models **B40A**, **B40B** and **B40C** were earlier production with no Osc. Trimmer Control or FSK position. Model **B41** is a longwave version covering 15 to 700 kHz. Model **62B** was also a longwave version. Both were used in submarines.

Made In:	England 1950-1954
Voltages:	240 VAC 50 Hz
Readout:	Analog
Physical:	13.5x19.5x16" 114 Lbs. 343x495x406mm 51.7 kg
Status:	Inactive Manufacturer Discontinued Model
Rarity:	Typically Unavailable

	New	Used
Price:	£500	⑤
Rating:	⑥	⑤

The **National Toy Company** was established in 1914 producing parts and toys. In 1916 the National Company Inc. was established to better reflect its broad product line.

National's first amateur receivers included the SW-5 Thrillbox and the SW-3, both designed by the famous James Millen. The SW-3 started many of today's "old timers" in amateur radio. In the 1930's National introduced many innovative mechanical and electrical designs that came together in the first HRO model. In 1939 the company went public.

Like Hallicrafters and Hammarlund, National produced large quantities of equipment for the armed forces during World War II. Immediately after the conclusion of the war, National briefly ventured into consumer electronics. In the 1950's and 1960's the company became a significant contractor for military electronics. Its grip on the amateur market was slipping however, as hams chose the smaller, and in most cases, better offerings from Drake and Collins. The HRO-500 and HRO-600 were revolutionary for their time, but were not enough to save the company.

The value of nearly all National receivers continues to climb. The HRO series is particularly sought after by collectors. Prices for the HRO-500 and HRO-600 in good condition are rising steadily with no end in sight.

National Co. Inc.
61 Sherman St.
Malden, Mass.   1947-1957

National Radio Co.
National Company Inc.
37 Washington St.
Melrose 76, Mass.   1961-1971

NATIONAL  COMPANY

FB-7-A

Amateur Band Communications Receiver
Single Conversion Superheterodyne. 7 Tubes.

Features:
• Standby   • Hinged Top Cover
Specifications:
Coverage ...... See Comments   Modes .... AM/CW
Antenna Input Terminals
Circuit Complement:
For the 2.5-volt type: 57 Mixer, 24 High Freq. Osc., 58 IF, 58 IF, 56 Detector, 59 AF Output and 24 BFO. For the 6.3-volt type: 77 Mixer, 37 Detector, 36 HFO, 78 IF, 78 IF, 89 AF Output and 36 BFO.
Accessories:
5897 AB 115 VAC Power Supply

Comments:
This model was sold without tubes, plug-in coils, speaker or power supply. General coverage coils: FBAA 18-34 MHz, FBA 11.4-19.5, FBB 7-11.7, FBC 4-7.3 MHz, FBD 2.4-4.2 MHz, FBE 1.5-2.5 and FBF .9-1.5 MHz. Bandspread coils: AB-10 10 meters, AB-20 20 meters, AB-40 40 meters, AB-80 80 meters and AB-160 160 meters.

Variants:
Earlier model **FB-7** was introduced in 1933 ($55 new). The models **FBA** and **FBX-A** are the same as above, but with single channel crystal filter unit. Both were also sold without tubes, coils, speaker or power supply.

Made In: United States 1935-1936

Voltages: See Comments.

Readout: ⊞⊞⊞ Coil Chart

Physical:

Status: Inactive Manufacturer
Discontinued Model

Rarity: Very Scarce

	New	Used
Price:	$62	$250-380
Rating:	⑥	⑦

FRR-24

General Coverage Diversity Communications Receiver
Double Conversion Superheterodyne. Tubes.

Features:
• ¼" Head. Jacks • S/Audio Meters
• AGC • Diversity
• IF Output

Specifications:
Coverage ...... 2000 - 32000 kHz
Modes .... CW/MCW/SSB/FAX
Environment .. 0° to +50°C 95% humid.

Comments:
The National AN/FRR-24 is triple diversity receiver contained in four rack cabinets. The complete receiving equipment system consists of a group of three radio receivers together with switching-combining circuits to receive signals in single channel, space or frequency dual or triple diversity operation. The frequency display is a tape projection to fogged glass arrangement. Each rack weighs approximately 300 Lbs. This receiver configuration was designed for point-to-point applications.

Made In:	United States 1950-1952	
Voltages:	110 VAC	
Readout:	`00002.±`	Projection of tape to glass
Physical:	{Four 19" Rack Cabinets}	

Status:	Inactive Manufacturer Discontinued Model	
Rarity:	Extremely Scarce	

	New	Used
Price:	①	$1500-2000
Rating:	⑥	⑦

FRR-59A

General Coverage Communications Receiver
Triple Conversion Superheterodyne. 64 Tubes plus Semiconductors.

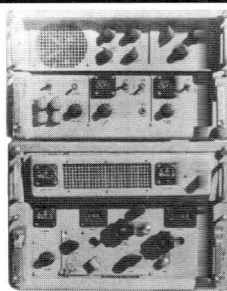

Features:
• ¼" Head. Jack • Standby • BFO • Antenna Trimmer
• AGC • RF Gain • NL • Calibrator 1000 kHz
• Rack Handles

Specifications:
Coverage ...... 2000 - 34000 kHz
Selectivity ...... 12/3/1/.35 kHz -6dB.
Image Rej. ..... 8-35 dB
Audio Out ...... 15 mw 600 ohm

Modes .... AM/LSB/USB/CW/FSK
IF .... 1625-1725, 220, 80 kHz
Stability .... 1×10^{-7} per day
Antenna Input . 50 ohms

Circuit Complement:
64 Tubes and 24 semiconductors.

Comments:
Ranges: 2-4, 4-8, 8-16 and 16-32 MHz. Dial accuracy is better than 100 Hz. This behemoth was envisioned to replace the R-390A. It was not as reliable as the venerable R-390A because of the heat it generated and its complexity. This model experienced considerable downtime in field use. It was designed primarily for shore operation. Reportedly a very good performer, but certainly a maintenance challenge (not to mention a lifting challenge). Asking prices have been seen over at over $2000, but the usual trading range is much lower.

Variants:
Model **WRR-2A** is similar, but is designed for shipboard use and is mounted in a cradle.

Made In:	United States 1950-1965	
Voltages:	105-125 VAC 50-60 Hz 250W	
Readout:	`00000.1`	Mechanical
Physical:	25x22x24" 250 Lbs. 635x558x610mm 113 kg	
Status:	Inactive Manufacturer Discontinued Model	
Rarity:	Scarce	

	New	Used
Price:	$10,000	$350-700
Rating:	⑥	⑦

NATIONAL  COMPANY

HRO

General Coverage Communications Receiver
Single Conversion Superheterodyne. 9 Tubes.

Features:
- ¼" Head. Jack
- S-Meter
- AVC
- RF Gain
- BFO
- S-Meter Switch

Specifications:
Coverage ......1700 - 30000 kHz
IF ....456 kHz
Audio Out ......7000 ohms 1.5 W

Modes .... AM/CW/MCW
Antenna Input . Terminals

Accessories: MCS Table Speaker   697 PS 115/230 VAC

Comments: Utilizes plug-in coil racks. Supplied tuning coil ranges: A 14-30, B 7-14.4, C 3.5-7.3, D 1.7-4 MHz. Other coils: E 900-2050, F 480-960, G 180-430, H 100-200, and J 50-100 kHz. Requires external 697 power supply. Requires a speaker. Black metal cabinet. The original HRO series spanned over ten years with several variations in design including switches and S-Meter design. This model was used in World War II and the design was copied by several countries including Japan, Australia, New Zealand and Germany. It is reported that HRO stands for "helluva rush order".

Variants: The model **HRO-C** configuration features a 697 power supply, SPC coil storage unit and speaker built into an MCS tabletop rack. Model **HRO Junior** (1936-1942) $165-198 new, was a simplified, less expensive version of the HRO. It did not include the crystal filter, signal strength meter and bandspread coils. This table model included only one set of 14 to 30 MHz coils. Other coils for the HRO Jr. include: JA 14-30, JB 7-14.4, JC 3.5-7.3, JD 1.7-4 MHz.

Made In: United States 1935-1946

Voltages: 115 VAC 50/60 Hz 75W using 697 PS.

Readout: Micrometer Dial and Coil Chart.

Physical: Cabinet Version:
17.25x9x12" 51 Lbs
438x228x305mm 23.1 kg
Rack Version:
19x8.75x12" 53 Lbs.
483x222x305mm 24 kg

Status: Inactive Manufacturer Discontinued Model

Rarity: Scarce

Reviews; *Radio Bygones* #2

	New	Used
Price:	$168-329	$250-300
Rating:	⑥	⑦

NATIONAL  COMPANY

HRO-5A1

General Coverage Communications Receiver
Single Conversion Superheterodyne. 12 Tubes.

Features:
- ¼" Head. Jack
- S-Meter
- AVC
- RF Gain
- NL
- Hinged Cover
- Speaker Out

Specifications:
Coverage ......1700 - 30000 kHz
Selectivity ......3/.2 kHz -6dB.
Audio Out ......3 W

Modes .... AM/CW/MCW
Sensitivity .... <1µV SSB/CW
Antenna Input . Terminals 500 ohms

Accessories:
MCS Table Speaker   RFSH Rack Speaker   697 PS 115/230 VAC
SFU-697 Rack PS   MRR Table Rack   SPU-686S 6VDC Vibrapack

Comments:
Utilizes plug-in coil racks. Supplied tuning coil ranges: A 14-30, B 7-14.4, C 3.5-7.3, D 1.7-4 MHz. Other coils: E 900-2050, F 480-960, G 180-430, H 100-200, and J 50-100 kHz. Also: AA 27.5-30, AB 25-35 MHz and AC 21-21.5 MHz. Requires external 697 PS. Requires a speaker. Black metal cabinet. A very quiet receiver. [¥83,000]

Variants:
Model **HRO-5A** was prior model (1945). Model **HRO-5RA-1** was rack version. The model **HRO-5C** configuration features a 697 power supply and SPU697 speaker built into an MRR table-top rack. The HRO-5C variant is shown in *Chapter 107.*

Made In: United States 1946-1947

Voltages: 115 VAC 50/60 Hz 75W using 697 PS.

Readout: Micrometer Dial and Coil Chart.

Physical: Cabinet Version:
17.25x9x12" 51 Lbs
438x228x305mm 23.1 kg
Rack Version:
19x8.75x12" 53 Lbs.
483x222x305mm 24 kg

Status: Inactive Manufacturer Discontinued Model

Rarity: Scarce

	New	Used
Price:	$275-307	$150-300
Rating:	⑥	⑦

HRO-7

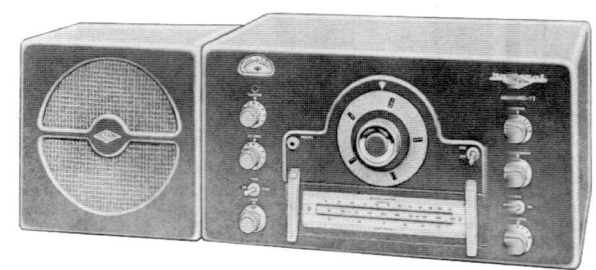

General Coverage Communications Receiver
Single Conversion Superheterodyne. 12 Tubes.

Features:
- ¼" Head. Jack   • S-Meter   • RF Gain   • Hinged Top Cover
- ANL • AVC   • Tone (2 Pos.) • Phono Input   • Phono/Radio Switch

Specifications:

Coverage ...... 1700 - 30000 kHz   Modes .... AM/CW/MCW
Selectivity ...... 6 Pos. .2-6.5 kHz   IF .... 456 kHz
Sensitivity ...... <1µV   Image Rej .... >30 dB @ 30 MHz.
Audio Out ...... 5000 Ohm   Antenna Input . Terminals

Circuit Complement: 6K7 1st RF Amp, 6K7 2nd RF Amp, 6J7 1st Detector, 6C4
High Frequency Oscillator, 6K7 1st IF Amp, 6K7 2nd IF Amp, 6H6 Diode Detector/
AVC, 6H6 Noise Limiter, 6SJ7 1st AF Amp, 6J7 BFO and 0A2 Voltage Regulator.
Two tuned RF stages and two IF stages.

Accessories: 697 PS 115/230 VAC   SPU697 PS Rack 115/230VAC
686S PS 6VDC   MCR Speaker (shown)   Coils - see Comments
RFSH Rack Speaker   MRR Rack Mounting   NFM-07 NBFM Adapter

Comments: This model is often referred to as the HRO-7T to indicate table version
as opposed to the rack version. Supplied tuning coil ranges: A 14-30, B 7-14.4, C
3.5-7.3, D 1.7-4 MHz. Other coils: E 900-2050, F 480-960, G 180-430, H 100-200,
and J 50-100 kHz. Requires a speaker (MCR shown). Power supply is separate.
The 500 degree micrometer dial has an effective length of 90 inches. [¥130,000].
Variants: Rack version is model **HRO-7R**. Model **HRO-7C** includes desk rack, 10"
rack speaker and coil compartment $358 (20.75x29x14.75").

Made In:	United States 1947-1949
Voltages:	115/230 VAC 50/60 Hz using 697 PS.
Readout:	Micrometer Dial and Coil Chart.
Physical:	65 Lbs. 29.5 KG
Status:	Inactive Manufacturer Discontinued Model
Rarity:	Scarce

	New	Used
Price:	$279-313	$200-300
Rating:	⑥	⑦

HRO-50

General Coverage Communications Receiver
Single Conversion Superheterodyne. 14 Tubes.

Features:
- ¼" Head. Jack   • S-Meter   • Antenna Trimmer • Hinged Top Cover
- Dial Lamp   • Bandspread   • Phono Input Tips • Dimmer   • BFO
                                    • AVC ON/OFF   • Standby   • Tone

Specifications:

Coverage ...... 1700 - 30000 kHz   Modes .... AM/CW
Sensitivity ...... <1µV 6 dB S/N   IF .... 455 kHz
Selectivity ...... 15-.4 kHz 5 Pos.   Antenna Input . Terminals
Audio Out ...... 8 W 8/500 ohm

Circuit Complement: 6BA6 1st RF Amp, 6BA6 2nd RF, 6BE6 Mixer, 6C4 HF Osc,
0B2 Regulator, 6K7 1st IF, 6K7 2nd IF, 6H6 Detector/AVC, 6H6 BFO, 6J7 NL, 6H6
1st Audio, 6SJ7 Phase Inv/S-Meter, 6SN7 Audio, (2) 6V6 Rectifier, 5V4G Assc
Crystal Cal, 6AQ5 NFM IF amp, 6SK7 Radio Detector and 6H6 Ratio Detector.

Accessories:
NFM-50 NBFM   XCU Calibrator 100/1000   SOJ-3 Select-O-Ject (shown)
50TS Speaker (shown) 50RS Rack Speaker   MRR2 Rack   650S 6VPS

Comments: Utilizes plug-in coil racks. Supplied coil ranges: A 14-
30, B 7-14.4, C 3.5-7.3, D 1.7-4 MHz. Other coils: E 900-2050, F 480-
960, G 180-430, H 100-200 and J 50-100 kHz. Also: AA 27.5-30,
AB 25-35 and AC 21-21.5 MHz. Built in power supply. Gray.
Variants: The **HRO-50C** configuration features a speaker and ten
coil compartment built into a tabletop rack. Model **HRO-50R** is rack
version. Model **HRO-50-T1** 1951 $384 (also rack version **HRO-50-R1**).

Made In:	United States 1949-1950
Voltages:	115/230 VAC 50/60 Hz
Readout:	Analog and Micrometer
Physical:	Cabinet Version: 19.75x10.125x16.5" 95 Lbs. 501x257x420mm 43 kg Rack Version: 19x10.5x17.5" 483x267x444mm
Status:	Inactive Manufacturer Discontinued Model
Rarity:	Scarce

	New	Used
Price:	$335-359	$200-400
Rating:	⑥	⑦

HRO-60

General Coverage Communications Receiver
Double Conversion Superheterodyne. 18 Tubes.

Features:
- ¼" Head. Jack
- RF Gain
- Tone
- S-Meter
- Dial Lamp
- BFO
- Antenna Trimmer
- Dimmer
- Bandspread
- Hinged Top Cover
- Phono Input Jack
- AVC ON/OFF

Specifications:
Coverage ...... 50-35000 kHz 50-54 MHz   Modes .... AM/CW
Selectivity ...... 6 Position Crystal Fil.   IF .... 2010, 455 kHz
Sensitivity ...... <1.7µV 6 dB 2-30 MHz.   Image Rej .... >80dB 1.7-14.4 MHz
Audio Out ...... 8 W 8/500 ohms   Antenna Input . Terminals 50-300 ohms

Circuit Complement:
6BA6 1st RF Amp, 6BA6 2nd RF Amp, 6BE6 1st Freq Converter, 6C4 HF Osc, 6BE6 2nd Freq Converter, 6SG7 1st IF Amp, 6SG7 2nd IF Amp, 6SG7 3rd IF Amp, 6H6 Det/AVC, 6H6 NL, 6SN7GT S-Meter/Phase Inverter, 6SJ7 1st AF Amp, (2) 6V6GT Audio Out, 6SJ7 BFO, 0B2 Voltage Reg, 4H4C Current Reg, 5V4G Rectifier. Two RF stages and twelve tuned circuits at 455 kHz. Single conversion below 7 MHz.

Accessories: HRO-60TS Spkr.(shown)   HRO-60RS Rack Speaker
NFM-83-50 NBFM   SOJ-3 Select-O-Ject   XCU-60-2 Calibrator 100/1000
MRR-2 Rack   650S 6VDC Power Supply

Comments: Uses plug-in coil racks. Supplied coil ranges: A 14-30, B 7-14.4, C 3.5-7.3, D 1.7-4 MHz. Other coils: E 900-2050, F 480-960, G 180-430, H 100-200 and J 50-100 kHz. Also: AA 27.5-30, AB 25-35, AC 21-21.5, AD 50-54 MHz. [¥80,000].

Variants: Model **HRO-60R** is rack version. **HRO-60C** includes MRR-2 & HRO-60RS.

Mfg. In: United States 1952-1964

Voltages: 115/230 VAC 50-60 Hz

Readout: Analog and Micrometer

Physical: Cabinet Version:
19.75x10.125x16.5" 85 Lbs.
502x257x419mm 38.5 kg
Rack Version:
19x10.5x17.6"
483x267x447mm

Status: Inactive Manufacturer
Discontinued Model

Rarity: Scarce

	New	Used
Price:	$483-745	$300-400
Rating:	⑥	⑦

HRO-500

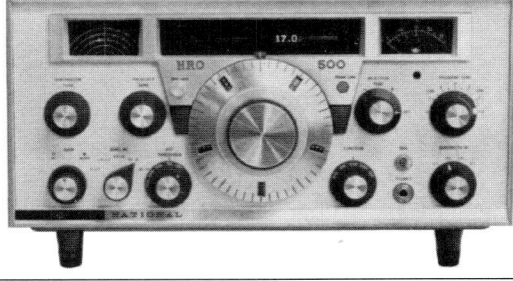

General Coverage Communications Receiver
Double Conversion Superheterodyne. Solid State.

Features:
- ¼" Head. Jack
- RF Gain
- Mute
- IF Out Jack
- HFO Out Jack
- S-Meter
- Dial Lamp
- Notch
- Ext. Spkr. Term.
- VFO Out Jack
- PBT (.5/2.5 kHz)
- Preselector
- 2 Tune Speeds
- Line Out 600 ohm
- IF Out Jack
- Dial Lock
- Calibrator 50 kHz
- AGC OFF/ON
- Atten. -10/20/30
- BFO Output Jack

Specifications:
Coverage ...... 5 - 30000 kHz   Modes .... AM/LSB/USB/CW/FSK
Selectivity ...... 8/5/2.5/.5 kHz -6dB.   Stability .... <100 Hz per day
Sensitivity ...... <1µV SSB/CW   IF .... 2.75-3.25 MHz, 230 kHz.
Image Rej ...... 50-80 dB   Antenna Input . SO239 and RCA
Audio Out ...... 1 W 4.4% dist. 3.2 ohm   Environment ... -20° to +45° ° 90% Humid

Circuit Complement: 37 Transistors and 21 diodes. Triple conversion below 4 MHz.

Accessories:
RMK-5 Rack Kit   LF-10 LF Preselector   HRO-500TS Speaker
WPC-5 Carry Case   BP-8 NiCad Battery

Comments: Digital readout to 500 kHz. Analog to 1 kHz marks (24 ft/MHz). Tuning at 10 or 50 kHz per main knob revolution. Operations is via sixty 500 kHz bands. Requires speaker. PBT on .5 and 2.5 kHz filters only. An outstanding receiver. VLF sensitivity improves from 25-50 µV to 1 µV with addition of LF-10. [¥134,000] Also reviewed in *CQ* November 1965.

Variants: **HRO-500P** Portable version with built-in speaker, BP-8 and WPC-5.

Made In: United States 1964-1973

Voltages: 115/230 VAC 50/60Hz 22W or 12.6 VDC 200 ma.

Readout: 500. Digital Mech. & Micrometer

Physical: 16.5x7.675x12.75" 32 Lbs.
419x195x324mm 14.5 kg

Status: Inactive Manufacturer
Discontinued Model

Rarity: Very Scarce

Reviews: *Monitor. Times* March 1985
Popular Elec. August 1965
WRTH 1976
Electric Radio August 1996
Electric Radio March 1997
73 August 1965

	New	Used
Price:	$1300-1995	$600-1500
Rating:	★★★★★	★★★★★

NATIONAL RADIO COMPANY, INC.
NRCI

HRO-600

General Coverage Communications Receiver
Double Conversion Superheterodyne. Solid State.

Features:
- ¼" Head. Jack
- S/AF-Meter
- Dial Lock
- AGC SLO/MED/FST
- RF Gain
- BFO ±3 kHz
- Preselector
- Rack Ears • Atten.
- Mute
- IF Out Jack
- Ext. Spkr Jack
- Line Out 600 ohm

Specifications:
Coverage ...... 16 - 30000 kHz
Selectivity ...... 8/2.4/2/1/.35 kHz -6dB.
Sensitivity .... 375µV SSB/CW
Image Rej ...... >90 dB
Audio Out ...... 1 W 10% dist.

Modes .... AM/LSB/USB/CW/FSK
Stability .... ± 20 Hz after 30 min.
IF Rejection .... >90 dB
Antenna Input . SO239
Environment ... -20° to +55°C 95% Humid

Accessories:
Rack Kit	650 FSK Converter	610 DC PS
620 Noise Blanker	640 Preselector	660 Diversity
630 IBS Converter	670 Remote	

Comments: Had motorized preselector tuning with bypass. Manufactured with three different VFO assemblies. Type 601 (shown above) featured a large conventional tuning knob and Nixie tube display for the 100 kHz, 10 kHz, 1 kHz and 100 Hz positions. Type 602 featured thumbwheel digits for the same positions (shown bottom right). Type 603 with four channel fixed frequency module is very scarce. The National HRO-600 a rare and highly collectable model. Caution: replacement Nixie tubes are becoming hard to find.

Made In: United States 1971-1981

Voltages: 117/234 VAC 50W or 12.6 VDC 200 ma.

Readout: `00 0000.1` Digital Mech
or `00 000.1` Digital Nixie

Physical: 17x5.25x15.5" 39 Lbs. 432x133x394mm 17.7 kg

Status: Inactive Manufacturer Discontinued Model

Rarity: Very Scarce

Reviews: *Comm. World* 1977

	New	Used
Price:	$2900-5900	$1800-3400
Rating:	★★★★★	★★★★

NBS-2

General Coverage Communications Receiver
Single Conversion Superheterodyne. 17 Tubes.

Features:
- ¼" Head. Jack
- S-Meter
- HFO INT/EXT
- Antenna Trimmer
- RF Gain
- Dial Lamp
- CWO INT/EXT
- AVC FST/SLO
- Mute
- Standby
- NL
- Bandspread 0-100
- Tone

Specifications:
Coverage ......200-400, 540-30000 kHz Modes .... AM/CW
Antenna Input SO-239

Comments:
Ranges: .2-.4, .54-1.6, 1.4-4.3, 4.3-12 and 12-30 MHz. The bandspread dial only contains a 0 to 1000 scale. Requires a speaker. The NBS-1 has a dramatic brushed silver front panel. Designed for rack mounting. Reportedly produced for the National Bureau of Standards and/or the C.I.A. The dials glow orange.

Variants:
Model **NBS-1** is similar, but with a dark front panel and normal bandspread dial. The tuning ranges are: .54-1.6, 1.6-4.3, 4.3-12, 12 -30 MHz and 48-56 MHz. Note the coverage of the six meter amateur band. The 10, 20, 40 and 80 meter amateur bands are also on the bandspread. The NBS-1 has a phono input jack. Tubes used in the NBS-1 include: 6J5, 6SA7, (5) 6SG7, 6H6, 6AC7, 6H6, 6SJ7, 6SJ7, 6J5, (2) 6V6GT, 5U4 and OD3.

Made In: United States 1950's

Voltages:

Readout: Analog

Physical: 19" wide

Status: Inactive Manufacturer Discontinued Model

Rarity: Extremely Scarce

	New	Used
Price:	①	$400-700
Rating:	⑥	⑦

NATIONAL COMPANY

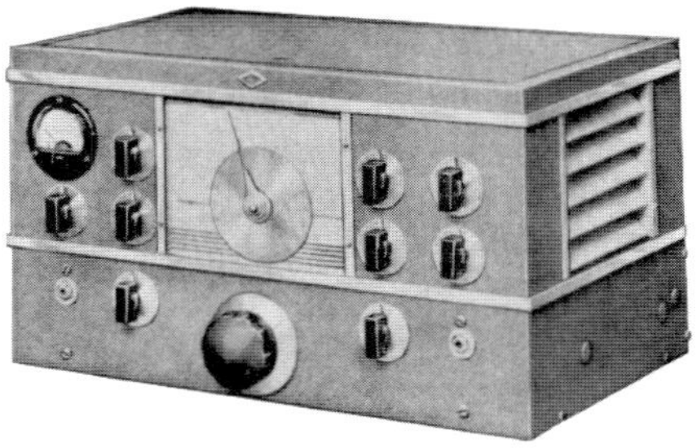

NC-2-40C

General Coverage Communications Receiver
Single Conversion Superheterodyne. 12 Tubes.

Features:
- ¼" Head. Jack
- S-Meter
- Standby
- Bandspread
- RF Gain
- Tone
- BFO
- Hinged Top Cover
- AVC
- Dial Lamp
- Speaker Jack

Specifications:
Coverage ...... 490 - 30000 kHz
Selectivity ...... 6 Pos. 2 - 7.5 kHz -6dB.
Sensitivity ...... 1µV for 1W output
Audio Out ...... 8 W
Modes .... AM/CW
IF .... 455 kHz
Antenna Input . Terminals

Circuit Complement:
6SK7 RF Amp, 6K8 Mixer, 6J5 H.F. Osc, 6K7 1st IF Amp, 6SK7 2nd IF Amp, 6SL7GT 2nd Detector/Limiter, 6SJ7 AVC, 6SJ7 BFO, 6SN7GT Amp/Phase Inverter, (2) 6V6GT Power Output and 5Y3GT Rectifier.

Accessories:
NC-2TS Table Speaker NC-2RS Rack Speaker

Comments:
Ranges: .49-1, 1-2, 1.7-4, 3.5-7.3, 7-14.4 and 14-30 MHz. Requires a speaker. This model is sometimes referred to as the NC-2-40CT to indicate table version as opposed to the rack version.

Variants: Model **NC-2-40CR** is rack version. Model **NC-2-40CS** is identical but covers from 200 to 400 and 1000 to 30000 kHz. Model **NC-2-40CSR** is rack version of NC-2-40CS.

Made In: United States 1946-1947

Voltages: 110-120 or 220-240 AC
50-60 Hz or external batteries

Readout: Analog

Physical: 85 Lbs
38.5 kg

Status: Inactive Manufacturer
Discontinued Model

Rarity: Scarce

	New	Used
Price:	$225-250	⑤
Rating:	⑥	⑦

NATIONAL COMPANY

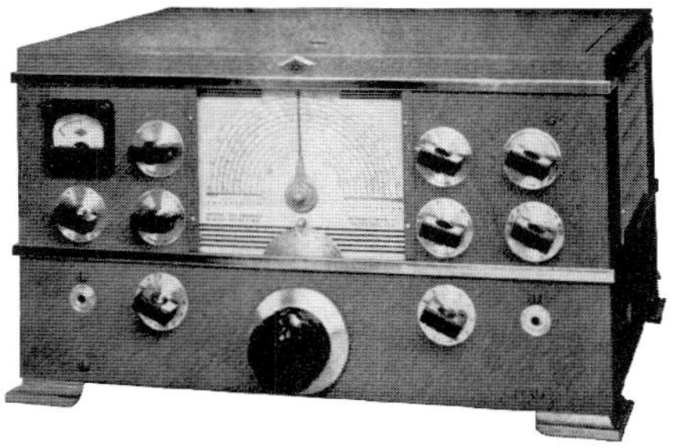

NC-2-40D

General Coverage Communications Receiver
Single Conversion Superheterodyne. 12 Tubes.

Features:
- ¼" Head. Jack
- S-Meter
- Standby
- Bandspread
- RF Gain
- Tone
- NL
- Hinged Top Cover
- Dial Lamp
- 60:1 Tune Ratio
- Phono Input Jack

Specifications:
Coverage ...... 490 - 30000 kHz
Selectivity ...... 6/4/2/1/.2 kHz -6dB.
Sensitivity ...... 1-2.6µV 6 dB
Audio Out ...... 8 W 10000 ohm
Modes .... AM/CW
IF .... 455 kHz
Antenna Input . Terminals

Circuit Complement:
6SK7 RF Amp, 6K8 Mixer, 6J5 Osc, 6K7 1st IF Amp, 6SK7 2nd IF Amp, 6SL7GT 2nd Detector/Limiter, 6SJ7 CW-Osc, 6V6GT AVC, 6SN7GT AF-Phase Inverter, (2) 6V6 Power Output and 5Y3GT Rectifier.

Accessories:
NC-2TS Speaker NC-2RS Rack Speaker

Comments:
Ranges: .48-1.04, .92-2.1, 1.68-4.05, 3.4-4.05, 3.4-7.4, 6.9-7.35, 6.65-14.6, 13.8-14.46, 13.9-31 and 26.9-30.05 MHz. Requires a speaker. This model is sometimes referred to as the NC-2-40DT to indicate table version as opposed to the rack version. Two-tone gray metal cabinet.

Variants:
Model **NC-2-40DR** is rack mount version.

Made In: United States 1947-1948

Voltages: 110-120/ 220-240VAC 70W
50-60 Hz or external batteries

Readout: Analog

Physical: 19.25x10.675x15.5" 65 Lbs.
489x271x393mm 29.5 kg

Status: Inactive Manufacturer
Discontinued Model

Rarity: Scarce

	New	Used
Price:	$225-242	⑤
Rating:	⑥	⑦

National

NC-33

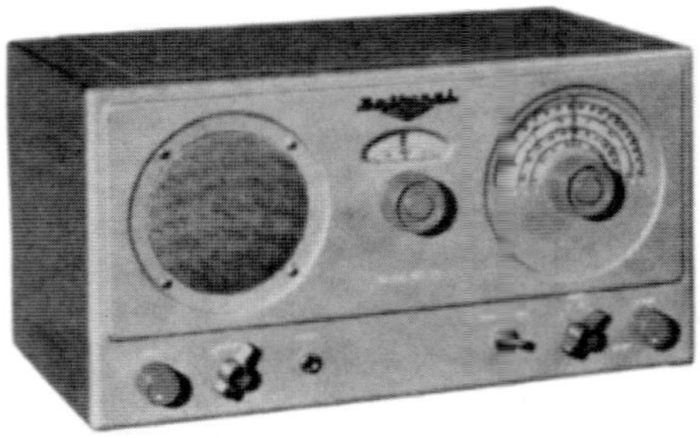

General Coverage Communications Receiver
Single Conversion Superheterodyne. 6 Tubes.

Features:
- ¼" Head. Jack
- Speaker 5"
- Standby
- BFO
- Dial Lamp
- ANL
- Bandspread 0-100

Specifications:
Coverage ...... 500 - 35000 kHz
IF .... 456 kHz
Audio Out ...... 1.5 W 3.2 ohm
Modes .... AM/CW
Antenna Input . Terminals

Circuit Complement:
12SA7 Converter, 12SG7 IF Amp, 12H6 2nd Detector/AVC/ANL, 12SL7 1st Audio/BFO, 35L6GT Audio Output and 35Z5GT Rectifier.

Comments:
Ranges: .5-1.42, 1.42-4.2, 4-12 and 12-35 MHz. Gray metal cabinet.

Made In: United States 1948-1950

Voltages: 105-125 V 50-60 Hz
AC or DC.

Readout: Analog

Physical: 16.675x8.75x8.5" 21 Lbs.
423x222x216mm 9.5 kg

Status: Inactive Manufacturer
Discontinued Model

Rarity: Scarce

	New	Used
Price:	$58-66	$40-60
Rating:	⑥	⑦

NATIONAL COMPANY

NC-45

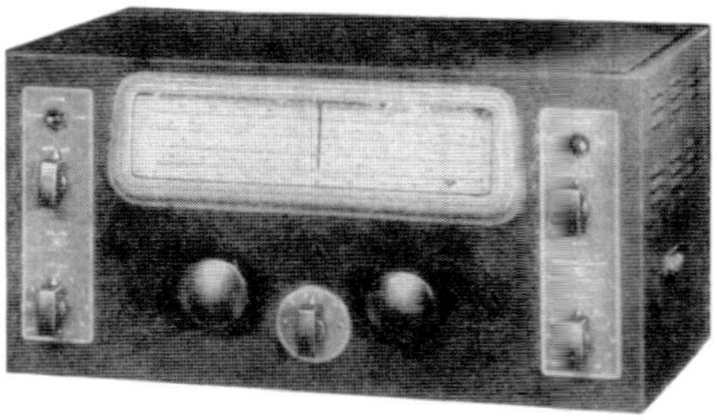

General Coverage Communications Receiver
Single Conversion Superheterodyne. 8 Tubes.

Features:
- ¼" Head. Jack
- RF Gain
- Standby
- Dial Lamp
- AVC
- NL
- Bandspread
- Hinged Top Cover

Specifications:
Coverage ...... 550 - 30000 kHz
IF .... 456 kHz
Modes .... AM/CW
Antenna Input . Terminals

Circuit Complement:
6K8 Converter, 6L7 IF Amp, 6L7 IF, 6SQ7 Detector/AVC/Audio, 6H6 NL, 6J7 BFO, 25L6 AF Output and 25Z5 Rectifier.

Accessories:
NC-44TS Speaker
RRA Rack Adapter

Comments:
This radio has four ranges.

Variants:
Model **NC-45A** only operates from 105-130 VAC (no DC). Model **NC-45B** is a battery version. Prior models **NC-44, NC-44A** (shown right) and **NC-44B** (1938-1941) were very similar in function and appearance but did not have the NL.

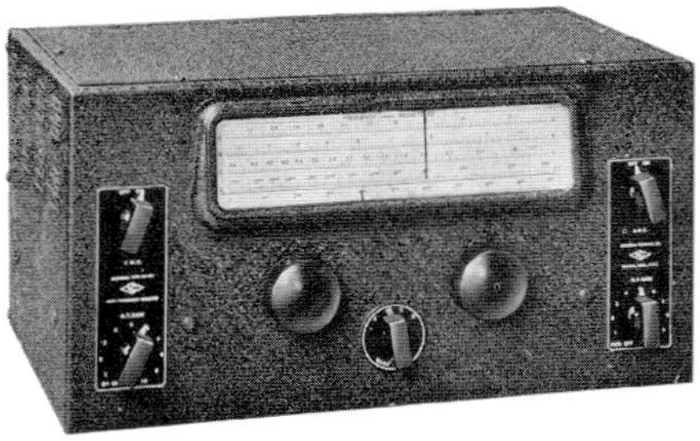

Made In: United States 1942-1945

Voltages: 105-130 VAC or
105-130 VDC

Readout: Analog

Physical:

Status: Inactive Manufacturer
Discontinued Model

Rarity: Scarce

	New	Used
Price:	$50-84	⑤
Rating:	⑥	⑦

NATIONAL COMPANY

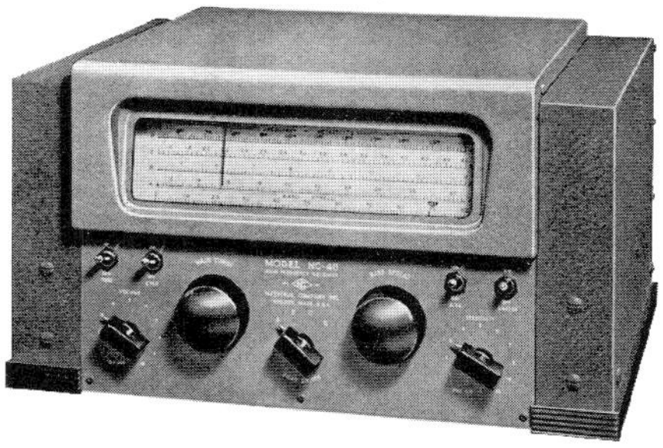

NC-46

General Coverage Communications Receiver
Single Conversion Superheterodyne. 10 Tubes.

Features:
- ¼" Head. Jack
- Sensitivity
- AVC ON/OFF
- Dial Lamp
- Mute Line
- Tone HI/LO
- NL
- Bandspread 0-100
- Hinged Top Cover

Specifications:
Coverage ...... 550 - 30000 kHz.
Selectivity ...... 4 kHz -6dB
Sensitivity ...... <5 µV
Audio Out ...... 3 W 8 ohm
Modes .... AM/CW
IF .... 455 kHz
Antenna Input . Terminals

Circuit Complement:
6K8 Converter, 6SG7 1st IF Amp, 6H6 2nd Detector, 6SF7 AVC, 6SJ7 BFO, 6SC7 Amp Phase Inverter, (2) 25L6GT Power Output and 25Z5 Rectifier.

Accessories:
NC-46TS Speaker

Comments:
Ranges: .54-1.6, 1.55-4.6, 4.4-12 and 11.5-30 MHz. Bandspread: 3.5-4, 7-7.3, 14-14.4 and 28-30 MHz. Requires a speaker.

Made In: United States 1946-1948

Voltages: 110-130 VAC 50/60Hz 55W

Readout: Analog

Physical: 17.375x9.5x12.375" 32 Lbs. 441x241x314mm 14.5 kg

Status: Inactive Manufacturer Discontinued Model

Rarity: Scarce

	New	Used
Price:	$98-107	⑤
Rating:	⑥	⑦

National NC

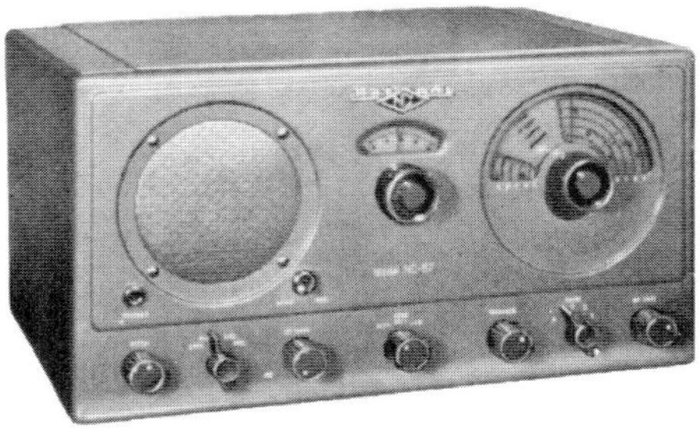

NC-57

General Coverage Communications Receiver
Single Conversion Superheterodyne. 9 Tubes.

Features:
- ¼" Head. Jack
- Speaker 5"
- Standby
- Mute Line
- S-Meter Jack
- BFO
- ANL
- MVC/AVC
- RF Gain
- Tone LO/MED/HI
- Dial Lamp
- Antenna Trimmer
- Bandspread 0-100
- Hinged Top Cover

Specifications:
Coverage ...... 540 - 54000 kHz
Selectivity ...... 4.2 kHz -6dB
Audio Out ...... 1.5 W 3.2 ohm
Modes .... AM/CW
IF .... 455 kHz
Antenna Input . Terminals

Circuit Complement:
6SG7 RF Amp, 6SB7Y Converter, (2) 6SG7 IF Amp, 6H6 Detector/AVC/ANL, 6SN7 Audio Amp/BFO, 6V6GT Audio Amp, 5Y3GT Rectifier and 0D3/VR150 Voltage Regulator. One RF and two IF stages.

Accessories:
SM-57 Ext. S-Meter NC-686S 6VDC PS Select-O-Ject

Comments:
Ranges: .54-1.55, 1.55-4.4, 4.4-12, 12-35 and 35-54 MHz. Gray metal cabinet.

Variants:
Model **NC-57B** has different tubes 1951 $99. Model **NC-57C** covers 190-410 and 540-35000 kHz 1951 $99. Model **NC-57M** marine version covers 200-400 and 540-35000 kHz and is AC/DC 1950-1951 $99. [¥40,000].

Made In: United States 1947-1950

Voltages: 105-130 VAC 50-60 Hz 84W or 110-130 VDC

Readout: Analog

Physical: 16.5x11.75x8.75" 31 Lbs. 419x298x222mm 14 kg

Status: Inactive Manufacturer Discontinued Model

Rarity: Scarce

	New	Used
Price:	$90-100	$50-100
Rating:	⑥	⑦

National

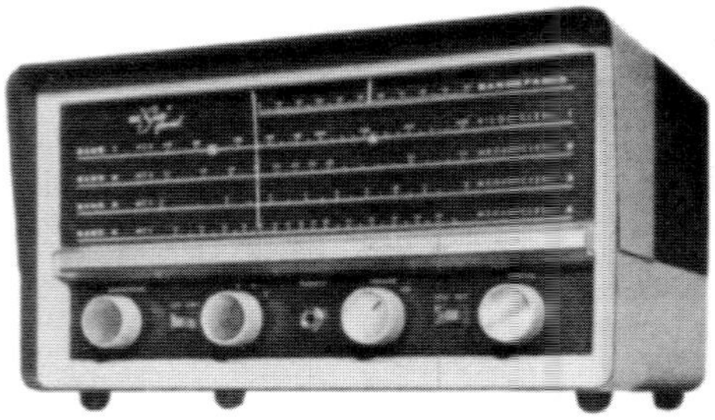

NC-60
Sixty Special

General Coverage Communications Receiver
Single Conversion Superheterodyne. 5 Tubes.

Features:
- ¼" Head. Jack
- Standby
- Speaker 5"
- Dial Lamp
- Bandspread 0-100

Specifications:
Coverage ...... 540 - 31000 kHz
IF .... 455 kHz
Modes .... AM/CW
Antenna Input . Terminals 50-300 ohms

Circuit Complement:
12BE6 Osc/Converter, 12BA6 IF Amp/CW Osc, 12AV6 Detector/AVC/1st Audio Amp, 50C5 Power Amp and 35W4 Rectifier.

Comments:
Ranges: .55-1.6, 1.6-4.5, 4-12, and 10.5-31 MHz. The 0-100 bandspread is the top scale on the radio with the small pointed. This receiver was produced in both blue and gray cabinets.

Variants:
Model **NC-60 Special A** and **NC-60 Special B** differences are unknown.

Made In: United States 1959-1964

Voltages: 110-120 VAC or
110-120 VDC

Readout: Analog

Physical: 13.5x7.675x8.675 12 Lbs.
343x195x220mm 5.4 kg

Status: Inactive Manufacturer
Discontinued Model

Rarity: Scarce

	New	Used
Price:	$60	$40-70
Rating:	★★★	★★

NATIONAL RADIO COMPANY, INC.
NRCI

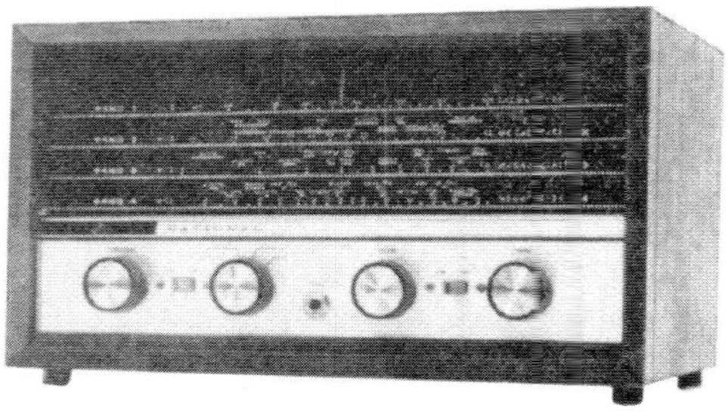

NC-77X

General Coverage Communications Receiver
Single Conversion Superheterodyne. 5 Tubes.

Features:
- ¼" Head. Jack
- Speaker 5"
- Dial Lamp
- Standby
- BFO Switch
- Bandspread 0-100

Specifications:
Coverage ...... 540 - 31000 kHz
IF .... 455 kHz
Audio Out ...... 1 W
Modes .... AM/CW
Antenna Input . Terminals

Circuit Complement:
12BE6 Converter, 12BA6 IF Amp, 12AV6 Detector Amp, 50C5 Audio Output and 35W4 Rectifier.

Comments:
The bandspread is on a 0 to 100 scale located at the top of the radio. It has a small pointer. The owner's manual includes a dial scale calibration cross-reference chart for the 80, 40, 20, 15 and 10 meter ham bands and the 16, 19, 25, 31 and 49 meter bands. This type of chart allows the listener to extrapolate the 0-100 bandspread dial to the actual received frequency.

Variants:
Model **NC-77XW** (shown) features a walnut case $90.

Made In: United States 1963-1967

Voltages: 117 VAC 60 Hz

Readout: Analog

Physical: 13.5x7.675x9" 18 Lbs
343x195x229mm 8.1 kg

Status: Inactive Manufacturer
Discontinued Model

Rarity: Very Scarce

	New	Used
Price:	$70	⑤
Rating:	⑥	⑥

NATIONAL COMPANY

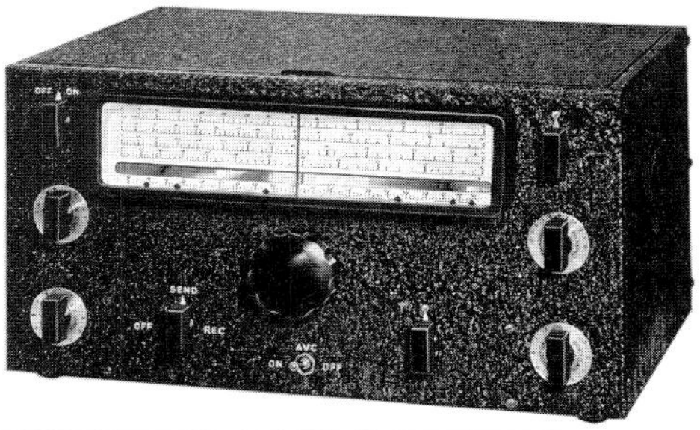

NC-80X

General Coverage Communications Receiver
Single Conversion Superheterodyne. 10 Tubes.

Features:
- ¼' Head. Jack
- RF Gain
- Freq. Markers
- Standby
- Dial Lamp
- BFO ±20 kHz
- AVC ON/OFF
- NL
- Dual Tuning Ratio
- Logging Scale 0-100
- Hinged Top Cover

Specifications:
Coverage ...... 550 - 30000 kHz
Selectivity ...... Variable
Audio 2.5W
Modes .... AM/CW
IF .... 1560 kHz
Antenna Input . Terminals

Circuit Complement:
6L7 (or 6J7) 1st Detector, 6J7 HF Osc., 6K7 IF, 6K7 1st IF, 6K7 2nd IF, 6K7 3rd IF, 6C5 2nd Detector, 6B8 AVC, 6J7 BFO, 25L6 AF Output and 25Z5 Rectifier. Crystal filter.

Accessories:
DCS-8 External Speaker Cabinet

Comments:
Ranges: .55-1.5, 1.7-4.6, 4.5-12 and 12-30 MHz. Supplied with an external, unmounted 8" speaker. The headphone jack is on the rear panel.

Variants:
Model **NC-80XB** is a battery version (6V heater, 135 V B+). It has no rectifier tube and employs a 6V6G output tube instead of the 25L6.

Made In: United States 1937-1939

Voltages: 110-120 VAC or 110-120 VDC

Readout: Analog

Physical:

Status: Inactive Manufacturer
Discontinued Model

Rarity: Very Scarce

	New	Used
Price:	$88-147	⑤
Rating:	⑥	⑦

NATIONAL COMPANY

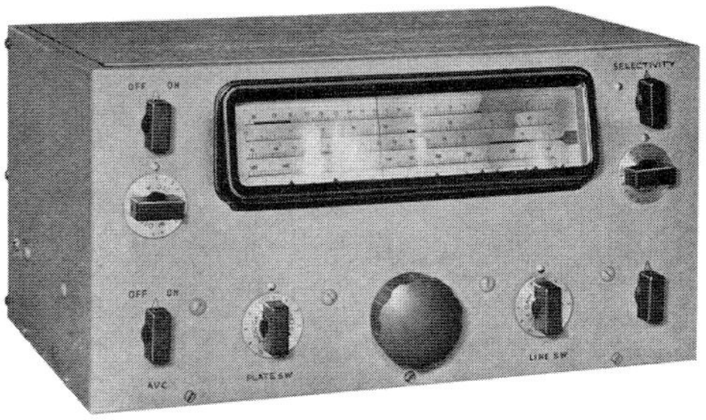

NC-81X

Amateur Band Communications Receiver
Single Conversion Superheterodyne. 10 Tubes.

Features:
- ¼" Head. Jack
- RF Gain
- Freq. Markers
- Standby
- Dial Lamp
- BFO ±20 kHz
- AVC ON/OFF
- NL
- Dual Tuning Ratio
- Logging Scale 0-100
- Hinged Top Cover

Specifications:
Coverage ...... 160-10M Ham Bands
Selectivity ...... 7-0.2 kHz Variable
Audio 2.5W
Modes .... AM/CW
IF .... 1560 kHz
Antenna Input . Terminals

Circuit Complement:
6L7 (or 6J7) 1st Detector, 6J7 HF Osc., 6K7 IF, 6K7 1st IF, 6K7 2nd IF, 6K7 3rd IF, 6C5 2nd Detector, 6B8 AVC, 6J7 BFO, 25L6 AF Output and 25Z5 Rectifier. Crystal filter.

Accessories:
DCS-8 External Speaker Cabinet

Comments: Ranges: 1.7-2, 3.5-4, 7-7.3, 14-14.4 and 28-30 MHz. Supplied with an external, unmounted 8" speaker. The headphone jack is on the rear panel. The NC-81X employs the same circuit as the NC-80X, but is fitted with a special tuning condenser and special coils to provide full bandspread on the 10, 20, 40, 80 and 160 meter amateur bands. Frequencies between these bands are not covered.

Variants:
Model **NC-81XB** is a battery version (6V heater, 135 V B+). It has no rectifier tube and employs a 6V6G output tube instead of the 25L6.

Made In: United States 1937-1939

Voltages: 110-120 VAC or 110-120 VDC

Readout: Analog

Physical:

Status: Inactive Manufacturer
Discontinued Model

Rarity: Very Scarce

	New	Used
Price:	$88-147	⑤
Rating:	⑥	⑦

NC-88
World Master

General Coverage Communications Receiver
Single Conversion Superheterodyne. 9 Tubes.

Features:
- ¼" Head. Jack
- Sensitivity
- Phono Input Jack
- Antenna Trimmer
- Speaker 5"
- Pitch
- Tone HI/LO
- Bandspread
- Standby
- Dial Lamp
- NL

Specifications:
Coverage ......540 - 40000 kHz
Selectivity ......5.2 kHz -6dB.
Sensitivity ......<5µV 10 dB S/N
Audio Out ......1.5W 4-8 ohms
Modes .... AM/CW
IF .... 455 kHz
Image Rej .... 18-67 dB
Antenna Input . Terminals

Circuit Complement:
6BA6 RF Amp, 6BE6 Mixer, 6C4 Osc, 6BD6 1st IF Amp, 6BD6 2nd IF Amp, 6AL5 2nd Detector/AGC/ANL, 12AX7 1st Audio/CW Osc, 6AQ5 Audio Output and 5Y3GT Rectifier. One RF and two IF stages.

Comments:
Ranges: .54-1.6, 1.6-4.7, 4.7-14 and 14-40 MHz. Bandspread: 3.5-4, 6.9-7.3, 14-14.35, 20.4-21.5 and 27-30 MHz. Model **NC-88SW** is an SWL version with bandspread calibrated for the 17, 19, 25, 31 and 49 shortwave meter bands. Gray metal cabinet. Similar to the NC-98 sans S-Meter and crystal filter.

Made In:	United States 1953-1956
Voltages:	105/120 VAC 50-60 Hz 62W
Readout:	Analog
Physical:	16.5x8.75x10.5 30 Lbs. 419x222x269mm 13.6 kg
Status:	Inactive Manufacturer Discontinued Model
Rarity:	Scarce

	New	Used
Price:	$100-130	$70-110
Rating:	⑥	⑦

NC-98

General Coverage Communications Receiver
Single Conversion Superheterodyne. 9 Tubes.

Features:
- ¼" Head. Jack
- S-Meter
- Phono Input Jack
- Antenna Trimmer
- RF Gain
- Pitch
- Tone HI/LO
- Bandspread
- Standby
- Dial Lamp
- Sensitivity
- Hinged Top Cover

Specifications:
Coverage ......540 - 40000 kHz
Selectivity ......3 position 5.2-.2 kHz
Sensitivity ......<5µV 10 dB S/N
Audio Out ......3.2 ohm
Modes .... AM/CW
Image Rej .... 18-67 dB
IF .... 455 kHz
Antenna Input . Terminals

Circuit Complement:
6BA6 RF Amp, 6BE6 Mixer, 6C4 Osc, 6BD6 1st IF Amp, 6BD6 2nd IF Amp, 6AL5 2nd Detector/AGC/ANL, 12AX7 1st Audio/CW Osc., 6AQ5 Audio Output, 5Y3GT Rectifier. One RF and two IF Stages.

Accessories:
NC-98TS Speaker   NFM-83-50 FM Adapter

Comments:
Ranges: .54-1.6, 1.6-4.7, 4.7-14, and 14-40 MHz. Bandspread: 3.5-4, 6.9-7.3, 14-14.35, 20.4-21.5 and 27-30 MHz. The NC-98 is similar to the NC-88 but with S-Meter and Crystal Filter.

Variants:
Model **NC-98SW** is an SWL version with bandspread calibrated for the 17, 19, 25, 31 and 49 shortwave meter bands.

Made In:	United States 1954-1956
Voltages:	105-130 VAC 50-60 Hz 62W
Readout:	Analog
Physical:	16.5x8.75x10.5' 30 Lbs. 419x222x269mm 13.6 kg
Status:	Inactive Manufacturer Discontinued Model
Rarity:	Scarce
Reviews:	*QST* August 1954

	New	Used
Price:	$150	$90-120
Rating:	⑥	⑦

NATIONAL COMPANY

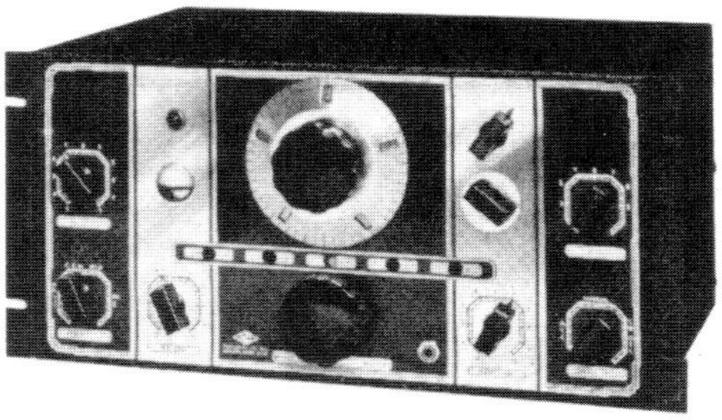

NC-100X

General Coverage Communications Receiver
Single Conversion Superheterodyne. 12 Tubes.

Features:
• ¼" Head. Jack • Tuning Eye • Standby • External Speaker
• RF Gain • Tone • AVC

Specifications:
Coverage ...... 540 - 30000 kHz
Selectivity ...... variable
Audio Out ...... 10 W
Modes .... AM/CW
Antenna Input . Terminals 500 ohms

Circuit Complement:
6K7 RF Preselection, 6J7 1st Detector, 6K7 1st IF Amp, 6K7 2nd IF Amp, 6C5 2nd Detector, 6J7 I.F. Oscillator, 6E5 Tuning Indicator, 6J7 BFO, 6J7 AVC, (2) 6F6 Power Output and 80 Rectifier. Crystal filter.

Accessories:
RRA Rack Mounting Brackets (shown).

Comments:
Ranges: .54-1.3, 1.3-2.8, 2.7-6.4, 5.3-14.4 and 13.5-30 MHz.

Variants:
Model **NC-100XS** is the same, but with a 12" Rola speaker and crystal filter (new $260). Model **NC-100** 1936-1938 (new $105-200) does not have a crystal filter. Model **NC-100S** (1932) is the same as the NC-100, but with 12" Rola speaker (new $222). The NC-100 was also produced for Russia (Lend Lease Act).

Made In: United States 1937-1938

Voltages: 115 VAC 60 Hz

Readout: Micrometer Dial

Physical:

Status: Inactive Manufacturer
Discontinued Model

Rarity: Very Scarce

	New	Used
Price:	$127-237	⑤
Rating:	⑥	⑦

NATIONAL COMPANY

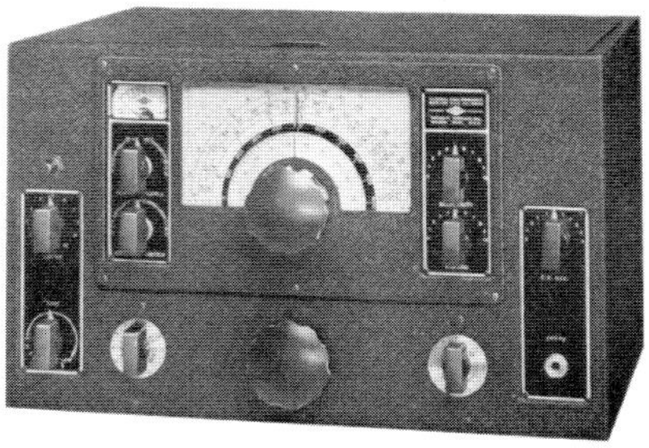

NC-100XA

General Coverage Communications Receiver
Single Conversion Superheterodyne. 11 Tubes.

Features:
• ¼" Head. Jack • S-Meter • Standby • Hinged Top Cover
• RF Gain • Dial Lamp • Bandspread • Tone • NL

Specifications:
Coverage ...... 540 - 30000 kHz
IF .... 455 kHz
Audio Out ...... 500 ohm 7 W
Modes .... AM/CW
Antenna Input . Terminals 500 ohms

Circuit Complement: 6K7 RF Amp, 6J7 1st Detector, 6J7 HF Oscillator, 6K7 1st IF Amp, 6SK7 2nd IF Amp, 6C8G 2nd Detector/Limiter, 6F8G AVC 1st Audio, 6J7 BFO, (2) 6F6G Power Output and 80 Rectifier.

Accessories: RRA Rack Mounting Kit

Comments: Ranges: .54-1.3, 1.3-2.8, 2.7-6.4, 6-14 and 13.5-30 MHz. Supplied with a matching 10" external speaker.

Variants: Model **NC-100SA** features 12" Rola G-12 external speaker. The **NC-100A** (right) is the same, but without a crystal filter. Model **NC-100XSA** has crystal filter and 12" speaker. The **NC-100AB** is a battery version with 8" speaker and only one 6F6G power output tube. The **NC-100XAB** is the same as the NC-100AB, but with a crystal filter.

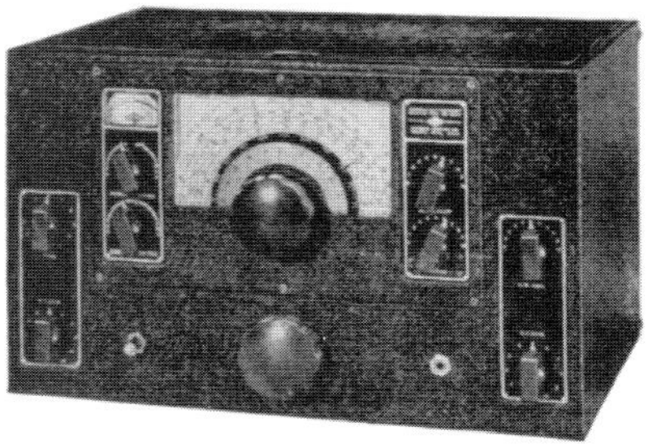

Made In: United States 1940-1942

Voltages: 115 VAC 60 Hz

Readout: Analog

Physical:

Status: Inactive Manufacturer
Discontinued Model

Rarity: Very Scarce

	New	Used
Price:	$220-237	⑤
Rating:	⑥	⑦

NATIONAL  COMPANY

NC-101X

Amateur Band Communications Receiver
Single Conversion Superheterodyne. 12 Tubes.

Features:
- ¼" Head. Jack
- RF Gain
- Dial Lamp
- S-Meter
- Tone
- Mute Line
- Standby
- NL
- Speaker 10"
- Bandspread
- Hinged Top Cover

Specifications:
Coverage ...... Ham See Comments.
IF .... 455 kHz
Audio Out ...... 500 ohm 7 W
Modes .... AM/CW
Antenna Input . Terminals 500 ohms

Circuit Complement:
6K7 RF Amp, 6J7 Mixer, 6K7 HF Oscillator, 6K7 1st IF Amp, 6K7 2nd IF Amp, 6C5 Detector, 6J7 AVC, 6J7 BFO, (2) 6F6 Power Output and 80 Rectifier. Crystal filter.

Accessories:
RRA Relay Rack Mount

Comments:
Ranges: 1.7-2.05, 3.5-4, 7-7.3, 14-14.5 and 28-30 MHz.

Variants:
Some earlier production units include an 6E5 tuning eye rather than an S-Meter. The model **NC-101XA** (new $236) has the same features and amateur band coverage as the NC-101X, but with a direct reading dial and cabinet similar to the NC-100XA. Model **NC-101XB** and **NC-101XAB** are battery versions (9 volt heat and 180 volt B required, with 8" speaker).

Made In:	United States 1935-1942
Voltages:	115 VAC 60 Hz
Readout:	Micrometer Dial
Physical:	
Status:	Inactive Manufacturer Discontinued Model
Rarity:	Very Scarce

	New	Used
Price:	$125-236	$220-300
Rating:	⑥	◯

NC-105

General Coverage Communications Receiver
Single Conversion Superheterodyne. 6 Tubes.

Features:
- ¼" Head. Jack
- Speaker 5"
- Pitch
- S-Meter
- Dial Lamp
- Standby
- RF Gain
- Line Out Jack
- Bandspread 0-100
- ANL

Specifications:
Coverage ...... 550 - 30000 kHz
Selectivity ...... 0.5-5 kHz -6dB.
Audio Out ...... 3.2 ohm
Modes .... AM/CW/SSB
IF .... 455 kHz
Antenna Input . Terminals 50-300 ohms

Circuit Complement:
6BE6 Converter, 6BA6 IF Amp, 6BA6 IF Amp/Q-Multiplier, 6T8 AM Detector/ANL/Audio Amp, 6AW8 Audio Amp/Detector and 6X4 Rectifier. Product detector.

Comments:
Ranges: .54-1.6, 1.6-4.5, 4-12 and 11-30 MHz. Blue-gray metal cabinet.

Variants:
Model **NC-105W** (shown) features a walnut enclosure $140.

Made In:	United States 1961-1964
Voltages:	105-125 VAC 50-60 Hz 49W
Readout:	Analog
Physical:	13.5x7.675x8.675" 25 Lbs. 343x195x220mm 1.3 kg
Status:	Inactive Manufacturer Discontinued Model
Rarity:	Scarce
Reviews:	QST April 1962 CQ September 1962

	New	Used
Price:	$90-120	$45-65
Rating:	⑥	◯

NC-109

General Coverage Communications Receiver
Single Conversion Superheterodyne. 11 Tubes.

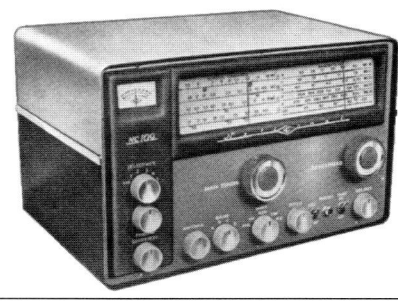

Features:
- ¼" Head. Jack
- S-Meter
- Antenna Trimmer
- Bandspread
- Sensitivity
- Dial Lamp
- Tone HI/LO
- Accessory Socket
- Mute
- Standby
- ANL
- Flywheel Tuning
- BFO

Specifications:
Coverage ......540 - 40000 kHz   Modes .... AM/CW/SSB
Selectivity ......6 Position 5.2-.2 kHz   Stability .... <.01%
Sensitivity ......<1-2 μV 10 dB S/N   IF .... 455 kHz
Audio Out ......1.5 W 3.2 ohm 10% dist. Antenna Input . Terminals 50-300 ohms

Circuit Complement:
6BA6 RF Amp, 6BE6 Mixer, 6BA6 1st IF Amp, 6BA6 2nd IF Amp, 6AL5 Detector/AGC/NL, 6BE6 Detector/BFO, 12AT7 AF/S-Meter, 6AQ5 Output, 6C4 HF Oscillator, 5Y3GT Rectifier and 0B2 Voltage Regulator. Separate product detector for SSB.

Accessories:
NTS-1 Speaker   XCU-109 Calibrator 100 kHz NFM-83-50 FM Adapter

Comments:
Ranges: .54-1.6, 1.6-4.7, 4.7-15, and 14-40 MHz. Bandspread: 3.5-4, 6.9-7.3, 14-14.35, 20.4-21.5 and 27-30 MHz. Requires a speaker. Promoted as "America's lowest priced SSB receiver". See model NC-188. [¥92,700].

Made In:	United States	1957-1960
Voltages:	105-130 VAC 50-60 Hz	
Readout:	Analog	
Physical:	16.875x10x10.875" 32 Lbs. 428x254x276mm 14.5 kg	
Status:	Inactive Manufacturer Discontinued Model	
Rarity:	Common	
Reviews:	*QST* January 1958	

	New	Used
Price:	$170-206	$100-140
Rating:	⑥	⑦

NATIONAL ⬥NC⬥ COMPANY

NC-120

General Coverage Communications Receiver
Single Conversion Superheterodyne. 11 Tubes.

Features:
- ¼" Head. Jack
- S-Meter
- Standby
- BFO
- RF Gain
- Dial Lamp
- Bandspread
- Rack Handles
- NL
- Speaker Output
- Tone Control
- Hinged Top Cover

Specifications:
Coverage ......540 - 30000 kHz   Modes .... AM/CW/MCW
Selectivity ......5 Position   IF .... 455 kHz
Audio Out ......600 ohm 3 W   Antenna Input . Terminals 500 ohms

Circuit Complement:
6K7 RF Amp, 6K7 2nd RF Amp, 6J7 1st Detector, 6K7 1st IF Amp, 6K7 2nd IF Amp, 6C8G 2nd Detector Limited, 6F8G 1st Audio Amp/AVC, 6K6GT/G 2nd Audio Amp, 6J5 High Frequency Osc., 6J7 BFO and 5Z3 Rectifier.

Comments: Ranges: .54-1.3, 1.3-2.8, 2.8-6.4, 6.4-14 and 14-30 MHz. This model is more well known, and more plentiful under its military nomenclature. See variants below. May also be run from an external power source supplying 6.3 VDC at 3.45 amps and 180 VDC at 30 ma.

Variants:
Military models are: **RAO-1, RAO-2, RAO-3, RAO-4, RAO-5, RAO-6, RAO-7, RAO-8** and **RAO-9**. Model RAO-7 is shown. The RAO-7 is similar to the RAO-2/6/9 except for minor electrical changes in the grid circuits of the first and second RF amp tubes. The RAO-2/6 differ in the cabinet, panel, type of coil carriage, connector and they contain an S-meter, but do not have the panoramic adapter circuit.

Made In:	United States	1940-1944
Voltages:	115 VAC 50-60 Hz 60W	
Readout:	Analog	
Physical:	19x12.5x17.75" 75 Lbs 483x317x450mm 34 kg	
Status:	Inactive Manufacturer Discontinued Model	
Rarity:	Very Scarce	

	New	Used
Price:	①	$150-250
Rating:	⑥	⑦

NATIONAL RADIO COMPANY, INC.
NRCI

NC-121

General Coverage Communications Receiver
Double Conversion Superheterodyne.

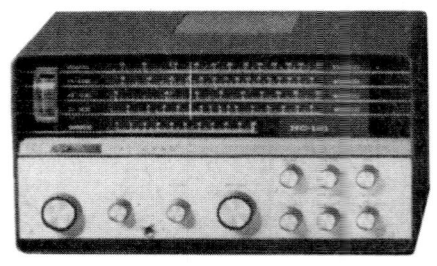

Features:
- ¼" Head. Jack
- S-Meter
- NL
- Q-Multiplier
- RF Gain
- Speaker 5"
- Dial Lamp
- Bandspread
- ANL
- Standby
- AGC
- Tuner Output
- BFO

Specifications:
Coverage ...... 540 - 30000 kHz   Modes .... AM/CW-SSB
Antenna Input Terminals

Comments:
Ranges: .55-1.6, 1.6-4.5, 4-12, and 11-30 MHz. Requires speaker.

Variants:
Model **NC-121W** features a walnut case. $160 new.

Made In:	United States 1963-1964	
Voltages:	117 VAC	
Readout:	Analog	
Physical:	13.5x7.675x9" 28 Lbs. 342x195x229mm 2.7 kg	
Status:	Inactive Manufacturer Discontinued Model	
Rarity:	Scarce	

	New	**Used**
Price:	$130	$70-80
Rating:	⑥	○

NC-125

General Coverage Communications Receiver
Single Conversion Superheterodyne. 8 Tubes

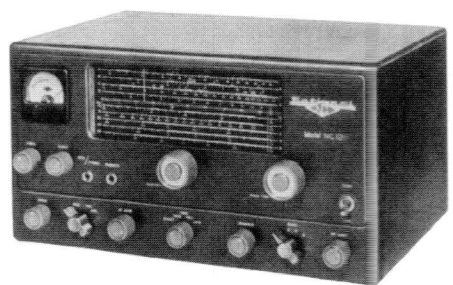

Features:
- ¼" Head. Jack
- S-Meter Jack
- RF Gain
- Antenna Trimmer
- Pitch
- Tone 4 pos.
- Bandspread
- Phono Input Jack
- Standby
- ANL
- AVC
- Hinged Cover
- Dial Lamp

Specifications:
Coverage ...... 550 - 36000 kHz   Modes .... AM/CW
Selectivity ...... 4.1 kHz   Sensitivity .... <4 µV 10 dB S/N
Audio Out ...... 1.5 W 3.2 ohm   Antenna Input . Terminals 50-300 ohms

Circuit Complement:
6SG7 RF Amp, 6SB7Y Osc/Mixer, 6SG7 2nd IF, 6H6 2nd Detector/AVC/ANL, 6SL7GT 1st Audio/CWO, 6V6GT Audio Output, 0D3 Voltage Regulator and 5Y3GT Rectifier.

Accessories:
NC-125TS Speaker   686S 6VDC PS   NFM73 NBFM

Comments:
The National Select-O-Ject is built into this model. Ranges: .55-1.6, 1.5-4.4, 4.4-12 and 12-36 MHz. Bandspread: 3.5-4, 6.9-7.3, 13.9-14.4, 19.9-21.5 and 26-30 MHz. Gray steel cabinet.

Made In:	United States 1950-1956	
Voltages:	105-125 VAC 50-60 Hz	
Readout:	Analog	
Physical:	16.5x11.75x8.25" 31 Lbs. 419x298x209mm 14 kg	
Status:	Inactive Manufacturer Discontinued Model	
Rarity:	Scarce	

	New	**Used**
Price:	$150-200	$75-150
Rating:	⑥	⑥

NC-140

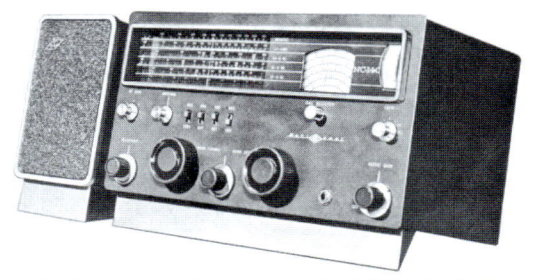

General Coverage Communications Receiver
Double Conversion Superheterodyne. 8 Tubes.

Features:
- ¼" Head. Jack
- BFO
- Standby
- S-Meter
- AGC
- ANL
- RF Gain
- Bandspread
- Dial Lamp
- Antenna Trimmer
- Tilt Bar

Specifications:
Coverage ......540 - 30000 kHz
Selectivity ......8 & Var. .5-5 kHz
IFs ....2215, 230 kHz
Audio Out ......3.2 ohm 3W

Modes .... AM/CW
Sensitivity .... <1 µV
Antenna Input . Terminals 50 ohms

Circuit Complement:
6BZ6 RF Amp, 12BE6 1st Mixer HF Osc, 12BE6 2nd Mixer/2nd Conversion Osc., 12BA6 IF Amp, 12AV6 AM Detector/Audio Amp, 6BN8 ANL/BFO, 6CW5 Output Amp and 5Y3GT Rectifier.

Accessories:
NTS-3B Speaker   XCU-109 Calibrator 1000 kHz

Comments:
Shown with optional NTS-3B speaker. Ranges: .54-1.6, 1.6-4, 4-10, 10-20 and 20-30 MHz. Amateur bandspread: 3.5-4, 6.9-7.5, 13.8-14.5, 21-21.5 and 26.6-30 MHz. Shortwave bandspread: 5.9-6.3, 8.6-10, 11.7-12, 14.6-15.4, 16.4-18 and 21.5-22.1 MHz.

| **Made In:** | United States 1963 |

| **Voltages:** | 105-120 VAC 50-60 Hz 75W |

| **Readout:** | Analog |

| **Physical:** | 15.75x8.75x9" 28 Lbs. 400x225x229mm 12.7 kg |

| **Status:** | Inactive Manufacturer Discontinued Model |

| **Rarity:** | Scarce |

	New	**Used**
Price:	$190	$45-70
Rating:	⑥	⑦

NC-155

Amateur Band Communications Receiver
Double Conversion Superheterodyne. 10 Tubes.

Features:
- ¼" Head. Jack
- BFO
- Dial Lamp
- S-Meter Jack
- Tilt Bar
- Flip Foot
- RF Gain
- ANL
- AGC
- Antenna Trimmer
- Standby
- 60:1 Tuning Ratio

Specifications:
Coverage ......Ham. See Comments.
Selectivity ......5/3/.6 kHz.
Sensitivity ......<1µV 10 dB S/N
Audio Out ......1.5 W 3.2 ohm 10% dis.

Modes .... AM/CW/SSB
Stability .... <1 kHz per 5 hours
IFs .... 2215, 230 kHz
Antenna Input . Terminals

Circuit Complement:
6BZ6 RF Amp, 6BE6 1st Converter, 6BE6 2nd Converter, 6BA6 IF Amp/S-Meter, 6BA6 IF Amp, 6T8 AM Detector/ANL/Audio Amp, 12AX7 SSB Detector/BFO, 6CW5 Audio Amp, 5Y3GT Rectifier and 0B2 Regulator.

Accessories:
NTS-3B Speaker   XCU-109 Crystal Calibrator 1000 kHz

Comments:
Ranges: 3.5-4, 7-7.3, 14-14.4, 21-21.5, 28-29.7 and 50-54 MHz. Blue gray metal cabinet. Requires a speaker.

| **Made In:** | United States 1961-1964 |

| **Voltages:** | 105-120 VAC 50-60 Hz 75W |

| **Readout:** | Analog |

| **Physical:** | 15.5x8.675x9" 25 Lbs. 394x220x228mm 11.3 kg |

| **Status:** | Inactive Manufacturer Discontinued Model |

| **Rarity:** | Scarce |

| **Reviews:** | *QST* July 1962 *CQ* January 1963 *73* June 1962 |

	New	**Used**
Price:	$150-200	$80-165
Rating:	⑥	⑦

NC-173T

General Coverage Communications Receiver
Single Conversion Superheterodyne. 13 Tubes.

Features:
- ¼" Head. Jack
- S-Meter
- Antenna Trimmer
- Bandspread
- RF Gain
- Dial Lamp
- Tone
- Phono Input Jack
- Mute
- AVC ON/OFF
- AF Filter
- Hinged Top Cover
- Standby

Specifications:

Coverage ....54-31 & 48-56 MHz
Selectivity ......6 position
Sensitivity ......<3.5 μV
Audio Out ......3.5W 8 / 500 ohms
Modes .... AM/CW
IF .... 455 kHz
Antenna Input . Terminals

Circuit Complement:
6SG7 Tuned RF, 6SA7 1st Detector, 6J5 Osc, (2) 6SG7 IF, 6H6 2nd Detector/AVC, 6AC7 AVC, 6SJ7 BFO, 6H6 NL, 6SJ7 Audio, 6V6 Output, VR150 Regulator and 5Y3GT/G Rectifier.

Accessories:
NC-173TS Speaker   NC-173RS Rack Speaker NFM-73 FM Adaptor

Comments:
Ranges: .54-1.6, 1.6-4.3, 4.3-12, 12-31 MHz and 48-56 MHz. Bandspread: 80, 40, 20, 10 and 6 meters. Requires a speaker. Has the functions of the Select-O-Ject audio filter built in. Similar to the NC-183, but with one stage RF and single ended audio output. Built-in power supply. Gray metal cabinet.

Variants: Model **NC-173R** is rack version.

Made In:	United States 19-7-1950
Voltages:	115/230 VAC 50-60 Hz 83W
Readout:	[analog dial icon] Analog
Physical:	19.75x10.125x12.5" 42 Lbs. 501x257x317mm 19 kg
Status:	Inactive Manufacturer Discontinued Model
Rarity:	Common
Reviews:	*QST* July 1947

	New	**Used**
Price:	$190-200	$100-140
Rating:	⑥	⑦

NC-183

General Coverage Communications Receiver
Single Conversion Superheterodyne. 16 Tubes.

Features:
- ¼" Head. Jack
- S-Meter
- Antenna Trimmer
- Hinged Top Cover
- RF Gain
- Dial Lamp
- Tone
- Phono Input Jack
- Mute
- Standby
- Bandspread
- AVC
- NL

Specifications:

Coverage ....54-31 & 48-56 MHz
Selectivity ......6 Position 3.5-.1 kHz
Sensitivity ......<4μV 10 dB
Audio Out ......8 W 8/500 ohm
Modes .... AM/CW
IF .... 455 kHz
Image Rej. ...... >40dB at 28 MHz
Antenna Input . Terminals

Circuit Complement: 6SG7 1st RF, 6SG7 2nd RF, 6SA7 1st Detector, 6J5 Osc, (2) 6SG7 IF, 6H6 2nd Detector, 6SJ7 BFO, 6AC7 AVC, 6H6 NL, 6SJ7 AF, 6J5 Phase Inverter, (2) 6V6GT Audio Out, VR150 Regulator and 5U4G Rectifier. Two RF stages and push-pull audio.

Accessories:
NC-183TS Speaker   NC-183DRS Rack Speaker NFM-83 FM Adaptor Select-O-Ject

Comments: Ranges: .54-1.6, 1.6-4.3, 4.3-12, 12-31 and 48-56 MHz. Bandspread: 80, 40, 20, 11-10 and 6 meters. Requires a speaker. The Select-O-Ject was an external audio notch filter. Gray metal cabinet. The table top model (shown) is also referred to as the NC-183T.

Variants: Model **NC-183R** rack mount version.

Made In:	United States 19-7-1952
Voltages:	110/120/220/240 VAC 50-60 Hz
Readout:	[analog dial icon] Analog
Physical:	19.75x10.25x16.75" 65 Lbs. 501x260x425mm 29.4 kg
Status:	Inactive Manufacturer Discontinued Model
Rarity:	Scarce

	New	**Used**
Price:	$268-279	$150-290
Rating:	⑥	○

NC-183D

General Coverage Communications Receiver
Single Conversion Superheterodyne. 17 Tubes.

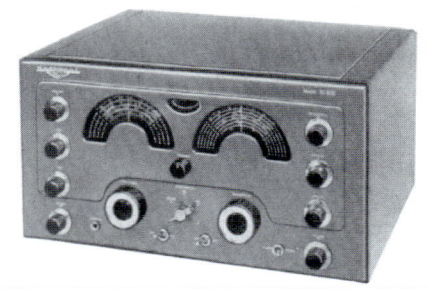

Features:
- ¼" Head. Jack
- S-Meter
- Antenna Trimmer
- Hinged Top Cover
- RF Gain
- Dial Lamp
- Tone
- Phono Input Jack
- Mute
- Standby
- Bandspread
- AVC
- NL

Specifications:

Coverage ....54-31 & 48-56 MHz	Modes .... AM/CW
Selectivity ......6 Position 3.5-.1 kHz	IF .... 1720, 455 kHz
Sensitivity ......<3.5µV 10 dB	Image Rej. ...... >40dB at 28 MHz
Audio Out ......8 W 8/500 ohm	Antenna Input . Terminals 50-300 ohms

Circuit Complement: 6BA6 1st RF Amp, 6BA6 2nd RF Amp, 6BE6 1st Converter, 6BE6 2nd Converter, 6BA6 1st IF Amp, 6BA6 2nd IF Amp, 6BA6 3rd IF Amp, 6AL5 2nd Detector/AVC, 6AH6 AVC Amp, 6SJ7 BFO, 6AL5 NL, 6SJ7 1st Audio, 6SN7 Phase Inverter/S-Meter Amp, (2) 6V6GT Audio Out, VR150 Regulator and 5U4G Rectifier. Two RF stages and push-pull audio. Single conversion below 4.4 MHz.

Accessories:
NC-183TS Speaker   NC-183DRS Rack Speaker   NFM-83-50 FM Adaptor
Select-O-Ject

Comments: Ranges: .54-1.6, 1.6-4.3, 4.3-12, 12-31 and 48-56 MHz. Bandspread: 3.5-4, 6.9-7.3, 14-14.4, 20-21.5, 26.5-30 MHz and 48-56 MHz. Requires a speaker. Gray metal cabinet. This model is a later, dual conversion version of the NC-183.
Variants: Model **NC-183DR** is the rack version. Model **NC-183MR** is the military rack version.

Made In:	United States 1952-1957
Voltages:	110/120/220/240 VAC 50-60 Hz 120 W
Readout:	⊥⊥⊥⊥ ▮ ⊥⊥ Analog
Physical:	19.75x10.25x16.75" 65 Lbs. 501x260x425mm 29.4 kg
Status:	Inactive Manufacturer Discontinued Model
Rarity:	Scarce
Reviews:	*Electric Radio* August 1992

	New	**Used**
Price:	$370	$150-290
Rating:	⑥	⑦

NC-188

General Coverage Communications Receiver
Single Conversion Superheterodyne. 9 Tubes.

Features:
- ¼" Head. Jack
- S-Meter
- Antenna Trimmer
- Bandspread
- Sensitivity
- Dial Lamp
- Tone HI/LO
- BFO
- Standby
- ANL

Specifications:

Coverage ......540 - 40000 kHz	Modes .... AM/CW
Selectivity ......5.2 kHz - 6dB	IF .... 455 kHz
Sensitivity ......<2.5 µV 10 dB S/N	Antenna Input . Terminals
Audio Out ......1.5 W	

Circuit Complement:
6BA6 RF Amp, 6BE6 Mixer, 6BA6 1st IF Amp, 6BA6 2nd IF Amp, 6AL5 Detector/ AGC/NL, 6C4 RF Osc. 5Y3 Rectifier, 12AT7 CW Osc/S-Meter/1st Audio Amp, 6AQ5 Audio Out Amp. Has an RF amp stage, two IF stages and two audio stages.

Accessories:
NTS-1 Speaker

Comments:
Ranges: .54-1.6, 1.6-4.7, 4.6-15 and 14- 40 MHz. Bandspread: 3.5-4, 6.9-7.3, 14-14.35, 20.4-21.5 and 27-30 MHz. Requires a speaker. Similar to NC-109 but without the crystal filter and product detector. See model NC-109.

Made In:	United States 1957-1960
Voltages:	105-130 VAC 50/60 Hz
Readout:	⊥⊥⊥⊥ ▮ ⊥⊥ Analog
Physical:	16.82x10x10.875" 427x254x276mm
Status:	Inactive Manufacturer Discontinued Model
Rarity:	Scarce

	New	**Used**
Price:	$140-160	$80-100
Rating:	⑥	⑦

NC-190

General Coverage Communications Receiver
Double Conversion Superheterodyne. 10 Tubes.

Features:
- ¼" Head. Jack
- S-Meter
- Antenna Trimmer
- Bandspread
- RF Gain
- Dial Lamp
- BFO
- Front Tilt Legs
- Mute
- Standby
- ANL
- AGC ON/OFF

Specifications:

Coverage ......540 - 30000 kHz
Selectivity ......5/3/.6 kHz
Sensitivity ......<1 µV 10 dB S/N
Audio Out ......1W 3.2 ohm

Modes .... AM/CW/SSB
IF .... 2215, 230 kHz
Antenna Input . RCA & Terminals

Circuit Complement:
6BZ6 RF Amp, 6BE6 HF Converter, 6BE6 2nd Converter, 6BA6 IF Amp, 6BA6 IF Amp, 12AX7A Product Detector/BFO, 6T8 AM Detector/ANL/AGC/Audio Amp, 6CW5 Audio Out, 5Y3GT Rectifier and OB2 Reg. Single conversion below 4 MHz.

Accessories:
NTS-3B Speaker   XCU-109 Crystal Calibrator 1000 kHz

Comments: Ranges: .54-1.6, 1.6-4, 4-10, 10-20 and 20-30 MHz. The Dial Selector knobs selects either amateur or shortwave bandspread. The bandspread is on the rotary dial to the right of the main tuning dial. Amateur bands are shown in red: 3.5-4, 6.9-7.5, 13.8-14.5, 21-21.5 and 26.6-30 MHz. The shortwave broadcast bands are shown in blue: 5.9-6.3, 8.6-10, 11.7-12, 14.6-15.4, 16.4-18 and 21.5-22.1 MHz. Requires a speaker. The NC-190 has National's "cosmic" blue metal cabinet. 60 to 1 Bandspread tuning ratio.

Made In:	United States 1961-1965
Voltages:	105-125 VAC 50-60 Hz 75W
Readout:	Analog
Physical:	15.75x8.75x9" 26 Lbs. 400x222x228mm 11.8 kg
Status:	Inactive Manufacturer Discontinued Model
Rarity:	Common
Reviews:	73 June 1962 QST October 1961

	New	**Used**
Price:	$200-220	$100-170
Rating:	⑥	⑦

NATIONAL  COMPANY

NC-200

General Coverage Communications Receiver
Single Conversion Superheterodyne. 12 Tubes.

Features:
- ¼" Head. Jack
- S-Meter
- Standby
- Bandspread
- RF Gain
- Tone
- NL
- Hinged Top Cover
- Dial Lamp
- 30:1 Tune Ratio

Specifications:

Coverage ......490 - 30000 kHz
Selectivity ......Variable kHz -6dB.
Sensitivity ......1-2.6 µV 6 dB
Audio Out ......10000 ohm 8 W

Modes .... AM/CW
IF .... 455 kHz
Antenna Input . Terminals

Circuit Complement:
6SK7 RF Amp, 6K8 1st Detector, 6J5 HF Oscillator, 6K7 1st IF Amp, 6SK7 2nd IF Amp, 6C8G 2nd Detector/Limiter, 6SJ7 AVC, 6SJ7 BFO, 6F8G Amp/Phase Inverter, (2) 6V6 Power Output and 5Y3 Rectifier. Crystal filter.

Accessories: NC-2TS Speaker 10"   NC-2RS Rack Speaker 10"

Comments: Ranges: .48-1.04, .92-2.1, 1.68-4.05, 3.4-4.05, 3.4-7.4, 6.9-7.35, 6.65-14.6, 13.8-14.46, 13.9-31 and 26.9-30.05 MHz. The ranges of 3.4-7.4, 6.9-7.35, 13.8-14.46 and 26.9-30.05 are bandspreaded amateur bands and are printed on the scale in red. The large center knob is used for tuning and band changing. To change bands, you pull this knob out by ¼ inch, and select the desired range. Then you press the knob back in for normal tuning. The NC-200 requires a speaker. Two-tone gray metal cabinet. Silver anniversary model.

Variants: Model **NC-200 RG** is rack mount version.

Made In:	United States 1940-1942
Voltages:	110-120 VAC 50-60Hz or external batteries
Readout:	Analog
Physical:	65 Lbs. 29.5 kg
Status:	Inactive Manufacturer Discontinued Model
Rarity:	Very Scarce

	New	**Used**
Price:	$147-265	$250
Rating:	⑥	⑦

NC-270

Amateur Band Communications Receiver
Double Conversion Superheterodyne. 10 Tubes.

Features:
- ¼" Head. Jack
- S-Meter
- Notch
- Antenna Trimmer
- RF Gain
- Dial Lamp
- BFO
- Calibrator 100 kHz
- Mute
- Standby
- Tilt Foot
- AGC
- ANL

Specifications:
Coverage ...... Ham. See Comments. Modes .... AM/SSB/CW
Selectivity ...... 5/3/2.5/.6 kHz -6dB. IFs .... 2215, 230 kHz.
Sensitivity ...... <1.5 µV 10 dB S/N Antenna Input . Terminals and RCA Jack
Audio Out ...... 1.5W 3.2 ohm 10% dist.

Circuit Complement:
6BZ6 RF Amp, 6BE6 Converter, 6BE6 Converter, 6BA6 IF, 6BA6 Product Detector, 6T8 Detector/AVC/ANL/Audio, 6CW5 AF Output, 12AU7 Calibrator/BFO, 5Y3GT Rectifier and 0B2 Voltage Regulator. Utilizes product and diode detectors.

Accessories:
NTS-3 Speaker

Comments:
Ranges: 3.5-4, 7-7.3, 14-14.4, 21-21.5, 28-29.7 and 50-54 MHz. This was the first model to feature National's dramatic "cosmic" blue metal cabinet.

Made In:	United States 1960-1964
Voltages:	105-125 VAC 50-60 Hz 75W
Readout:	Analog
Physical:	15.75x8.75x9" 28 Lbs. 400x222x228mm 12.7 kg
Status:	Inactive Manufacturer Discontinued Model
Rarity:	Scarce
Reviews:	QST January 1961 CQ May 1961

	New	Used
Price:	$180-280	$90-140
Rating:	⑥	⑦

NC-300

Amateur Band Communications Receiver
Double Conversion Superheterodyne. 13 Tubes.

Features:
- ¼" Head. Jack
- S-Meter
- Phono Jack
- Antenna Trimmer
- RF Gain
- Dial Lamp
- Tone
- Cal Reset
- Mute
- 40:1 Tune Ratio
- Flywheel Tuning

Specifications:
Coverage ...... Ham. See Comments. Modes .... AM/SSB/CW
Selectivity ...... 8/3.5/.5 kHz -6dB. Image Rej .... 50-60 dB
Sensitivity ...... <1.5 µV IF .... 2215, 80 kHz.
Audio Out ...... 1W 8 ohms Antenna Input . Terminals 50-300 ohms

Circuit Complement:
6BZ6 RF Amp, 6AH6 HF Osc., 6BA7 1st Mixer, 6BE6 2nd Converter , 6BJ6 1st IF Amp, 6BJ6 2nd IF Amp, 6AL5 Detector/NL, 6BE6 CW Osc/Het Detector, 12AT7 S-Meter/AF Amp, 6AQ5 Audio Amp, 5Y3GT Rectifier, 0B2 Voltage Regulator and 4H4C Current Stabilizer.

Accessories:
NC-300C2 2M Conv. NC-300C6 6M Conv. NC-300C1 1.25M Conv.
NTS-300TS Speaker XCU-300 Calibrator NC-300CC Conv. Cabinet

Comments:
Ranges: 1.8-2, 3.5-4, 7-7.3, 14-14.4, 21-21.5, 26.5-27.5 and 28-29.7 MHz. The radio dial is also pre-calibrated for the use of optional converters: 49.5-54.5, 143.5-148.5 and 220-225 MHz. Turret-type band display. Two-tone gray enameled cabinet. Requires a speaker.

Made In:	United States 1955-1958
Voltages:	105-130 VAC 50-60 Hz 60W
Readout:	Analog
Physical:	19.5x11.25x15" 57 Lbs. 495x286x381mm 25.9 kg
Status:	Inactive Manufacturer Discontinued Model
Rarity:	Common
Reviews:	QST January 1956 CQ October 1955 Electric Radio October 1990

	New	Used
Price:	$350-400	$150-350
Rating:	⑥	⑦

NC-303

Amateur Band Communications Receiver
Double Conversion Superheterodyne. 15 Tubes.

Features:
- ¼" Head. Jack
- S-Meter
- ANL
- Antenna Trimmer
- RF Gain
- Dial Lamp
- Manual NL
- Tone Hi/Nor/Lo/Peak
- Mute
- Cal Reset
- Q-Multiplier
- Vernier Fine Tuning
- Standby

Specifications:
Coverage ...... Ham. See Comments.
Selectivity ...... 8/3.5/2/.4 kHz -6dB.
Sensitivity ...... 1.5µV 10 dB S/N
Audio Out ...... 1 W 10% dist.

Modes .... AM/SSB/CW
IF .... 2175, 455 kHz
Image Rej .... 50-80 dB
Antenna Input . Terminals 50-70 ohms

Circuit Complement:
6BZ6 RF Amp, 6AH6 HF Osc, 6BA7 1st Mixer, 6BE6 2nd Converter, 12AX7 Q-Multiplier, 6BJ6 1st IF Amp, 6BJ6 2nd IF Amp, 6AL5 Det/NL/AGC, 6BE6 Product Detector, 12AT7 S-Meter/AF Amp, 6AQ5 Audio Amp, 5Y3GT High Voltage Rectifier, OB2 Voltage Reg., 4H4C Current Stabilizer and 1N1692 AGC Clamp Diode.

Accessories:
NTS-2S Speaker
NC-300C2 2M Conv.

XCU-300 Calibrator
NC-300C6A 6M Conv.

XCU-303 Calibrator
NC-300C1 1.25M Conv.

Comments: Ranges: 1.8-2, 3.5-4, 7-7.3, 14-14.4, 21-21.5, 26.5-27.5 and 28-29.7 MHz. Turret-type band display. The radio dial is also pre-calibrated for the use of converters: 49.5-54.5, 143.5-148.5 and 220-225 MHz. The XCU-303 also provides WWV reception at 10 MHz.

Made In:	United States 1958-1962
Voltages:	105-130 VAC 50-60 Hz 70W
Readout:	Analog
Physical:	19.75x11.25x15" 60 Lbs. 502x286x381mm 27 kg
Status:	Inactive Manufacturer Discontinued Model
Rarity:	Common
Reviews:	QST April 1959 CQ December 1958

	New	Used
Price:	$449	$150-190
Rating:	⑥	⑦

NC-400

General Coverage Communications Receiver
Double Conversion Superheterodyne. 18 Tubes.

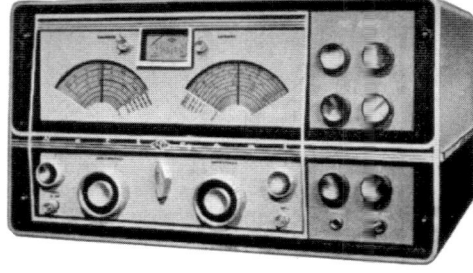

Features:
- ¼" Head. Jack
- S-Meter
- Manual NL
- Antenna Trimmer
- RF Gain
- Dial Lamp
- IF Out Jack
- Mute
- Tone
- Bandspread
- Cal Reset
- HFO In Jack
- 5 Fixed Channels

Specifications:
Coverage ...... 540 - 31000 kHz
Selectivity ...... 16/8/4/3.5 kHz -6dB.
Sensitivity ...... 1µV 10 dB S/N
IF .... 1720, 455 kHz
Audio Out ...... 1 W 3.2 ohms 10% dist.

Modes .... AM/SSB/CW
Stability .... .002% after warm-up
Image Rej .... 80-100 dB
Antenna Input . SO-239 & Terminals

Circuit Complement:
6BZ6 1st RF Amp, 6BZ6 2nd RF Amp, 6BE6 1st Mixer 455 kHz, 6BE6 1st Mixer 1730 kHz, 6BE6 2nd Converter/Soc, 6BZ7 HF Osc, 6BA6 1st IF Amp, 6BA6 2nd IF Amp, 3rd IF Amp, 4H4C Fil. Reg. 6AL5 AM Detector/ANL, 6BE6 Het Det. 6AL5 Manual NL, 6U8 BFO, 12AT7 S-Meter/1st Audio, 6AQ5 Audio Amp, OB2 Voltage Regulator and 5U4GB Rectifier. Two tuned RF stages with three tuned circuits on all bands. Single conversion below 7 MHz.

Accessories:
NTS-2 Speaker
MX-400 50 Fixed Freq.

XCU-400 Calibrator 100/1000 kHz
NC-400 DMK Diversity Modification Kit

Comments: Ranges: .54-1.1, 1.1-2.1, 2.1-4.1, 4.1-7, 6.9-12.2, 11.8-20.4 and 19.6-31 MHz. Bandspread: 3.5-4.1, 7-7.3, 13.7-14.5, 20.8-21.6 and 28-30 MHz. Has 5 crystal positions. Gray and black metal cabinet. Were used in F.B.I. field offices.

Made In:	United States 1959-1963
Voltages:	115/230 VAC 50 60 Hz 138 W
Readout:	Analog
Physical:	19x11x16" 482x279x406mm
Status:	Inactive Manufacturer Discontinued Model
Rarity:	Very Scarce
Reviews:	QST February 1960 CQ April 1960 Electric Radio August 1995

	New	Used
Price:	$895-995	$500-900
Rating:	⑥	⑦

R-1230/FLR

General Coverage Countermeasures Receiver
Double Conversion Superheterodyne. 32 Tubes.

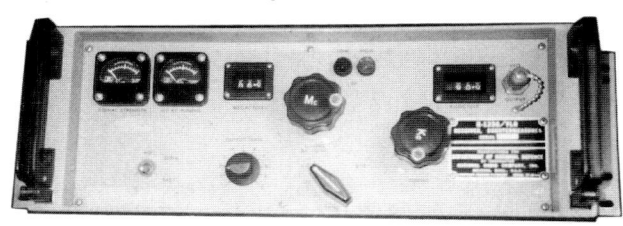

Features:
- ¼" Head. Jack   • S-Meter   • AGC SLOW/FAST • 100 kHz Tuning Meter
- DF Lo/Hi Lamps • Rack Handles   • Dial Spinners   • VFO Out Jack

Specifications:
Coverage ...... 2000 - 32000 kHz
Modes .... AM   Antenna Input . SO239

Circuit Complement:
16 Miniature tubes and 16 nuvistors. Barlow-Wadley upconversion.

Comments: Ranges: 2-4, 4-8, 8-16 and 16-32 MHz. The large center knob tunes the 10 MHz, 1 MHz and 100 kHz positions which are displayed on an odometer type display to the right of the two meters. The right meter is labeled 100 KC TUNING. This shows when tuning lock is achieved. Then the 10 kHz, 1 kHz and 100 Hz positions are tuned by the second large knob and is displayed on the second odometer display. This is *not* a typical or complete receiver. It provides wideband output for further processing such as for direction finding. A separate power supply can support up to six receivers when used in a direction finding configuration. It is believed that the R-1230 was the last tube-type military receiver produced by National. This extremely interesting receiver has a plate on the front that states: R-1230/FLR RECEIVER, COUNTERMEASURES. MANUFACTURED FOR DE-PARTMENT OF NATIONAL DEFENCE BY CONTRACTOR NATIONAL RADIO COMPANY. It may have been produced for the British or Canadian government. The tuning capacitor is reportedly gold plated.

Made In:	United States
Voltages:	(Separate power supply)

Readout: | 00000.1 | Dig. Mech.

Physical: 19"x5.75x?"
483x146x?mm

Status: Inactive Manufacturer
Discontinued Model

Rarity: Extremely Scarce

	New	**Used**
Price:	①	⑤
Rating:	⑥	⑦

R-1490/GRR-17

General Coverage Communications Receiver
Double Conversion Superheterodyne. Solid State.

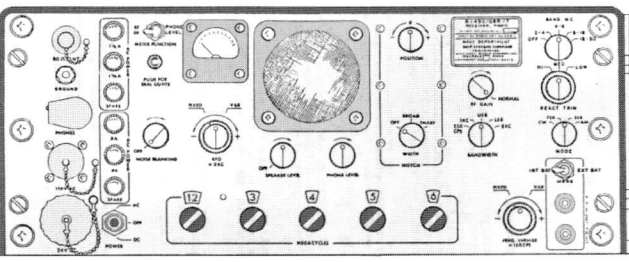

Features:
- ¼" Head. Jack   • S/AF-Meter   • RF Gain   • Modular Construction
- Speaker   • Notch Filter   • BFO ±3 kHz   • Vernier ±150 Hz
- Dial Lamp   • NB   • Antenna Trimmer   • Dial Lamp Switch
- Calibrator

Specifications:
Coverage ...... 2000 - 30000 kHz   Modes .... AM/CW/MCW/LSB/USB/FSK
Selectivity ...... 8/3/1/.35 kHz -6dB.   Stability .... ±1 x 10⁻⁷ per day
Sensitivity ...... 1.5µV 10 dB S/N CW   Environment ... -40° to +65°C 95% Humid
IF .... 112, 5 MHz.   Antenna Input . N 50 ohms
Audio Out ...... 1 W

Comments:
This receiver is tuned via decadic (ten position) rotary knobs with one knob for each digit down to 100 Hz. A separate vernier knob permits fine tuning. The R-1490 has two hinged flanges which swing out and forward from each side. These allow for mounting in a 19" rack. Separate headphone jack and speaker volume controls are provided. This receiver may be found with its Transit Case (D43611G1). A built-in dry loop (switching function only) is provided to drive a radioteletype printer. This receiver was produced in fairly small numbers for the U.S. Marine Corps. An upconversion design with very high IF frequencies.

Made In:	United States 1968-1970
Voltages:	115 VAC 50-400 Hz or 24 VDC

Readout: | ●●●●●.❶ | Dig. Knobs

Physical: 17x8x12.125" 43.6 Lbs.
431x203x308mm 19.8 kg

Status: Inactive Manufacturer
Discontinued Model

Rarity: Very Scarce

	New	**Used**
Price:	$23,000	$450-950
Rating:	⑥	⑦

NATIONAL COMPANY

SW-3U

Communications Receiver
Regenerative. 3 Tubes.

Features:
• ¼" Head. Jack • Hinged Top Cover

Specifications:
Coverage ...... See Comments   Antenna Input . Terminals

Circuit Complement:
The SW-3U receiver employs one RF stage transformer coupled to a regenerative detector and one stage of impedance coupled audio.

Comments:
The SW-3U was sold without coils, tubes or power supply. The circuit of the SW-3U is arranged for either battery or AC operation without coil substitution. For AC operation the 5886-AB power supply is required. For DC operation, external batteries are used. Audio is only available to the 20,000 ohm headphone output. Please see *Chapter 107 Additional Information* for details on the appropriate required coils for the SW-3U and variants. This receiver is also referred to as the Universal SW-3.

Variants:
Earlier model **SW-3** was introduced in 1933. The model **ACSW-3** was designed for AC operation only using two 58 and one 27 tube. It requires the 5880-AB AC power supply. The model **6DC SW-3** was for 6VDC operation and the model **2DC SW-3** as for 2 VDC operation.

Made In: United States 1940-1941

Voltages: See Comments.

Readout: External Coil Chart

Physical: 15 Lbs.
6.8 KG

Status: Inactive Manufacturer
Discontinued Model

Rarity: Very Scarce

	New	Used
Price:	$32-38	⑤
Rating:	⑥	⑦

National

SW-54

General Coverage Communications Receiver
Single Conversion Superheterodyne. 5 Tubes.

Features:
• Phone Tip Jack • Standby • Bandspread 0-100
• Speaker • Dial Lamp • Spkr/Phones Switch

Specifications:
Coverage ...... 540 - 30000 kHz   Modes .... AM/CW
Selectivity ...... 5 kHz -6dB.   IF .... 455 kHz
Sensitivity ...... 15-50µV for 50 mw out   Image Rej. ...... 8-35 dB
Audio Out ...... 1.8 W   Antenna Input . Terminals 50-300 ohms

Circuit Complement:
12BE6 Converter, 12BA6 CW Oscillator/IF Amp, 12AV6 2nd Detector/1st Audio/AVC, 50C5 Audio Out and 35Z5 Rectifier.

Comments:
Ranges: .54-1.6, 1.6-4.7, 4.6-14.5 and 12-30 MHz. The SW-54 has a red dial pointer, and red slide switches. A popular model fondly remembered by baby-boomers. [¥30,000].

Made In: United States 1950-1957

Voltages: 105/130 VAC 50-60Hz 25W or 105/130 VDC

Readout: Analog

Physical: 11x7x7" 13 Lbs.
279x178x178mm 5.9 kg

Status: Inactive Manufacturer
Discontinued Model

Rarity: Common

	New	Used
Price:	$50-60	$40-65
Rating:	⑥	⑦

The Nera Company of Norway has a been a manufacturer of marine electronic equipment since 1919. The company has also produced television and broadcast transmitters, microwave radio links and mobile radio communications equipment.

By the late 1960's Nera had stopped producing their own MF/HF communications receivers for ships. This may be, in part, to Norway's early embracement of MARISAT technology.

Nera Telecommunications today continues to be a leading manufacture of marine satellite communications systems.

Nera Aksjeselskapet
Pilestredet 75 C
P.O.B. 7033 H
Oslo 3, Norway

Nera Telecommunications
Bergerveien 12
N-13061 Billingstad, Norway

AKSJESELSKAPET NERA

M-200A

General Coverage Communications Receiver
Single Conversion Superheterodyne. 7 Tubes plus Semiconductors.

Features:
- ¼" Head. Jack
- Ext. Spkr Jack
- Line Out
- S-Meter
- Standby
- Rack Handles
- AVC
- Dial Lamp
- Bandspread ±
- Spinner Knob

Specifications:
Coverage ...... 100 - 25000 kHz
Selectivity ...... ±2 kHz -7 dB
IF .... 335 kHz
Audio Out ...... 4 ohm .5W
Modes .... AM/CW
Sensitivity .... <5 µV 20 dB S/N
Antenna Input . 75 ohm unbalanced

Circuit Complement:
(2) 12BA6, 12BA7, 35C5, 12AL5, (2) 12AT7 and 1N34A semiconductor.

Comments:
Ranges: .1-.27, .39-1.2, 2.7-4.2, 4.2-10 and 10-25 MHz. Note the reception gap from 270-390 kHz (see IF). A switch on the front panel of this receiver can select power from either 115 VAC or operation from a 24 volt battery. This model was supplied with its own metal cabinet or may be mounted in a standard 19 inch rack. The radio is dark cobalt grey, hard enamel finish with white letters on the front panel. The cabinet is gray hammertone. This radio is classified as a marine reserve receiver.

Made In: Norway 1966

Voltages: 110 VAC 50/60 Hz or 110 VDC or 24 VDC

Readout: Analog

Physical: Cabinet Version:
21.5x14x13.25" 45 Lbs
545x355x335mm 20.5 kg
Rack Version
19x10.7x13.25" 40 Lbs
483x265x335mm 18 kg

Status: Active Manufacturer
Discontinued Model

Rarity: Typically Unavailable

	New	Used
Price:	①	⑤
Rating:	⑥	⑥

61 Norlin Communications

Norlin Communications was a Washington belt-way defense contractor that manufactured a range of advanced communications equipment for the surveillance market. The company was previously known as Astro Communications Labs. It has been reported, but not confirmed, that both companies were fronts for Watkins-Johnson, producing equipment almost exclusively for the N.S.A.

The author does not know the exact fate of Norlin, only that many employees found their way across town to Racal or Watkins-Johnson.

Norlin Communications
9125 Gaither Rd.
Gaithersburg, MD 20760   1986-1991

SR-2070

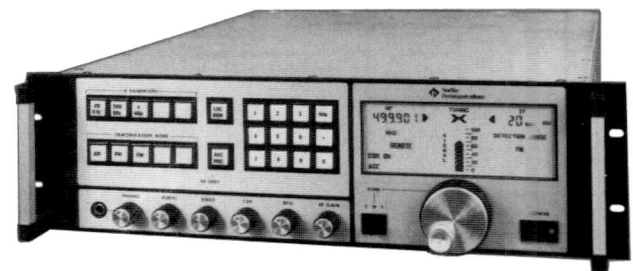

General Coverage Communications Receiver
Double Conversion Superheterodyne. Solid State.

Features:
- ¼" Head. Jack
- S-Meter
- BFO ±2 kHz
- Three Tune Rates
- RF Gain
- ANL
- Keypad
- Atten. -20/40 dB
- Mute
- Rack Handles
- Remote Port
- Modular Construction
- IF Out Jack
- Line Out 600 ohm

Specifications:
Coverage ...... 500 - 30000 kHz
Selectivity ...... 16/12/8/4/2/1 kHz -6dB[1]
Image Rej ...... >80 dB
Audio Out ...... 1 W 8 ohms
Environment .. 0° to +50° C.

Modes .... AM/LSB/USB/CW/FM
Stability .... 1×10^{-6} per day
IF Rejection .... >90 dB
Antenna Input . 50 ohm

Accessories:
DU2090 Display Memory Unit

Comments:
The LCD display is backlit. Designed for military and surveillance applications. Meets MIL-STD-461, Class 1A. [1]Only 3 IF filters are supplied as standard.

Variants:
Model SR-2093 (shown) is a VHF/UHF version with an identical appearance to the SR-2070. Model SR-2090 covers 20-500 MHz with different IF bandwidths and 1 kHz display resolution.

Made In:	United States 1979-1988	
Voltages:	115/230/240 VAC 47-400 Hz 100 W	
Readout:	`00000.01` Digital LCD	
Physical:	19x5.25x21" 55 Lbs. 483x133x533mm 24.8 kg	
Status:	Inactive Manufacturer Discontinued Model	
Rarity:	Extremely Scarce	

	New	Used
Price:	①	②
Rating:	⑥	⑥

Panasonic is one of the most recognized names in consumer electronics in the world. In shortwave circles, Panasonic is best known for their portable models. However, from the mid 1970's to the late 1980's they did offer several tabletop digital communications receivers.

The RF-4800 and RF-4900 may be best described as analog radios with integrated digital counters. These substantial radios afforded pleasant audio, but lacked stability.

Panasonic's current involvement in the worldband market is limited to one digital and one analog portable.

Panasonic
Matsushita Electric Trading Co. Ltd.
C.P.O. 288
Osaka 530-91, Japan

Panasonic Co.
One Panasonic Way
Secaucus, NJ 07094

Panasonic

RF-3100

General Coverage Communications Receiver
Double Conversion Superheterodyne. Solid State.

Features:
- ¼" Head. Jack
- RF Gain
- Light
- ANL
- S-Meter
- BFO
- Bass & Treble
- Tilt Bar
- Meter Lamp
- Speaker 3.5"
- FM 88-108 MHz.
- Light Switch
- Telescopic Antenna
- Record Jack
- Speaker Jack
- Shoulder Strap

Specifications:
Coverage ...... 525 - 30000 kHz. +FM
Selectivity ...... 7/3 kHz -6dB. [200 FM]
Sensitivity 1-1.4µV
IF .... 10.695 MHz, 455 kHz

Modes .... AM/SSB
Image Ratio .... 30-60 dB
Antenna Input . SO-239 & Terminals

Comments:
There is an antenna switch on the back that permits the selection from the built-in medium wave antenna to an external medium wave antenna or from a low to a high impedance antenna on shortwave. The filters are somewhat sloppy and stability is marginal, making this model unsuitable for single sideband reception. The audio is quite good yielding pleasant reception of amplitude modulated signals. It is single conversion on medium wave. This model has pleasant audio. The RF-3100 may be more properly classified as a portable than a communications receiver.

Variants:
Sold outside the U.S.A. as model **DR-31** or **RFB-30**.

Made In:	Japan 1982-1985
Voltages:	120 AC 60 Hz or 8 D cells
Readout:	00001. Digital Phs.
Physical:	14.6x4.8x9.5" 8 Lbs. 371x122x241mm 3.6 kg
Status:	Active Manufacturer Discontinued Model
Rarity:	Abundant
Reviews:	📖 Passport 1987 WRTH 1983

	New	Used
Price:	$289-319	$150-160
Rating:	★★★	★★★

Panasonic

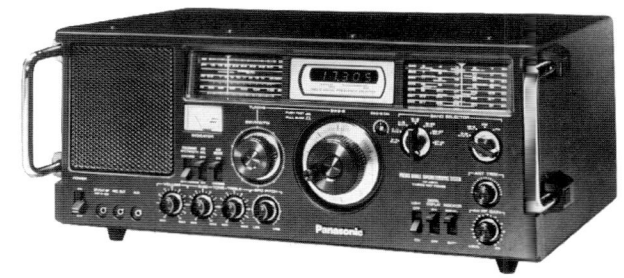

RF-4800

General Coverage Communications Receiver
Double Conversion Superheterodyne. Solid State.

Features:
- Mini Head. Jack
- ANL
- RF Gain
- Dial Lamp
- Spinner Knob
- S/Batt-Meter
- Bass & Treble
- BFO
- Handles
- Antenna Trimmer
- Analog Sub-dial
- Speaker 4"
- FM 88-108 MHz.
- Two Tuning Speeds
- Aux Audio-In Jack
- Record Jack
- MW Ferrite Antenna

Specifications:
Coverage ...... 525 - 31000 kHz. +FM
Selectivity ...... 5/3.4 kHz -6dB. [200 FM]
Sensitivity ...... 1-1.4μV 3-30 MHz.
Image Ratio ... 30-60 dB
I.F. .... 2 MHz, 455 kHz
Modes .... AM/SSB/CW
Stability .... <±500 Hz after 30 min.
Antenna Input . Terminals
Audio Out .... 2 W 10% distortion

Circuit Complement:
5 ICs, 3 FETs and 34 transistors. Single conversion below 3 MHz.

Comments:
Bands: .525-1.6, 1.6-3, 3-7, 7-11, 11-15, 15-19, 19-23, 23-27 and 27-31 MHz. and 88-108 MHz FM. The RF-4800 may be properly called an analog receiver with a digital counter display. The RF-4800 was not very stable. The band selector switches are the first components to wear on this model. The eight internal D cells may be assessed through two trap doors on the bottom of the receiver.
Variants: Sold outside the U.S.A. as model **DR-48**. The Japanese domestic version is model **RJX-4800**.

Made In:	Japan	1976-1978
Voltages:	120 AC 60 Hz 10W or 12 VDC or 8 D cells	
Readout:	`00001.`	LED
Physical:	19x7.875x14" 20 lbs. 482x200x354mm 8 kg	
Status:	Active Manufactuer Discontinued Model	
Rarity:	Abundant	
Reviews:	*WRTH* 1978	

	New	Used
Price:	$450	$150-170
Rating:	★★★	★

Panasonic

RF-4900

General Coverage Communications Receiver
Double Conversion Superheterodyne. Solid State.

Features:
- Mini Head. Jack
- ANL
- RF Gain
- Dial Lamp
- Spinner Knob
- S/Batt-Meter
- Bass & Treble
- BFO
- FM 88-108 MHz.
- Antenna Trimmer
- Analog Sub-dial
- Speaker 4"
- Handles
- Two Tune Speeds
- Aux Audio-In Jack
- Record Jack
- MW Ferrite Antenna

Specifications:
Coverage ...... 525 - 30000 kHz. +FM
Selectivity ...... 5/3.4 kHz -6dB.
Sensitivity .... 8-1.4μV 3-30 MHz.
Image Ratio ... 30-60 dB
Modes .... AM/SSB/CW
Stability .... <±500 Hz after 30 min.
Antenna Input . SO-239 & Terminals
Audio Out .... 2 W 10% distortion

Circuit Complement:
4 Integrated circuits, 3 FETs and 39 transistors. Single conversion below 3 MHz.

Comments:
Bands: .525-1.6, 1.6-3, 3-7, 7-11, 11-15, 15-19, 19-23, 22-26 and 26-30 MHz and 88-108 MHz FM. This radio is best described as an analog radio with a digital counter built-in. The readout is green in color. Like its predecessor model, the RF-4800, stability is not good. Black metal cabinet.
Variants: Also sold outside the U.S.A. as model **DR-49**. Some DR-49 rear panels features a second ferrite antenna for the European longwave broadcast band. The "LBS" DR-49 variant is 110-125/220-240 VAC and has 455 kHz second IF. The "LBE" DR-49 variant is strictly 240 VAC and has a 462 kHz second IF.

Made In:	Japan	1979-1983
Voltages:	120 AC 60 Hz 10W or 12 VDC or 8 D cells	
Readout:	`00001.`	Digital Phs.
Physical:	19x7.875x14" 17.7 Lbs. 482x200x354mm 8 kg	
Status:	Active Manufactuer Discontinued Model	
Rarity:	Abundant	
Reviews:	*Passport* 1987 *WRTH* 1980	

	New	Used
Price:	$399-479	$180-190
Rating:	★★★	★★

Panasonic

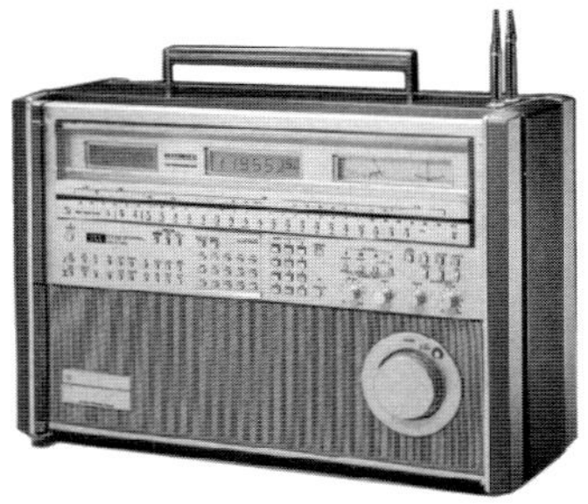

RF-9000

General Coverage Communications Receiver
Double Conversion Superheterodyne. Solid State.

Features:

• Mini Head. Jack	• S-Meter	• FM Tune Meter	• Two Tune Speeds
• ANL	• Bass & Treble	• 15 Memories	• Aux Audio-In Jack
• RF Gain	• BFO	• FM AFC	• Speaker 7x5" & 2.6"
• Dial Lamp	• FM 88-108 MHz.	• Carry Handle	• MW Ferrite Antenna
• Keypad Entry	• 4 Event Timer	• 12/24 Hr. Clock	• Telescopic Antennas
• Dial Lock	• Loudness	• Record Jack	• Scanning

Specifications:

Coverage ...... 150 - 30000 kHz. +FM
Selectivity ...... 4.8/3.2/2.4 kHz -6 dB
Sensitivity 1μV
Image Ratio ... 30-60 dB
Modes .... AM/LSB/USB/CW
Audio Out .... 2 W 10% distortion
Antenna Input . SO-239 & Terminals

Circuit Complement:
Single conversion below 2.9009 MHz.

Comments:
Ranges: .15-.42, .52-1.611, 1.611-2.9009 and 2.901-30 MHz plus 87.5-108 MHz FM. Two AA cells required for clock. The clock-timer supports four events in seven days. This technology showpiece model was not produced in large numbers. Its price was commensurate with its physical size, and number of buttons, but not with its performance. Very collectable.

Variants:
Sold as model **DR-90** is some parts of the world.

Made In: Japan 1982-1985

Voltages: 100~110/115~127/200~220 230~250 VAC or 8 D cells or 12 VDC

Readout: 00001. Digital LCD

Physical: 20.6x14.3x8.125" 45 Lbs. 523x363x206mm 20.4 kg

Status: Active Manufacturer Discontinued Model

Rarity: Very Scarce

Reviews: 📖 *Passport* 1987, *WRTH* 1982

	New	Used
Price:	$2800-3800	⑤
Rating:	★★★	★★

Panasonic

RF-B600

General Coverage Communications Receiver
Double Conversion Superheterodyne. Solid State.

Features:

• S/Batt. Meter	• ¼" Head. Jack	• 9 Memories	• Telescopic Antenna
• ANL	• Bass & Treble	• Dial Lock	• Two Tune Speeds
• RF Gain	• BFO	• Speaker 3.5"	• Ferrite MW Antenna
• Light	• Record Jack	• Keypad	• FM 87.5-108 MHz.
• Tilt Stand	• Carry Handle	• Memory Scan	• Zone Scan ± 150 kHz
• Mute Jack	• DC Input Jack	• Display Switch	• Ext. Speaker Jack

Specifications:

Coverage ...... 150 - 29999 kHz. +FM
Selectivity ...... 7/3 kHz -6dB.
Sensitivity ...... 1.2 μV S/N 6 dB @6 MHz
Image Ratio ... 50 dB at 6 MHz
I.F. .... 39.9-40 MHz, 450 kHz
Modes .... AM/LSB/USB
Stability .... <±50 Hz after warm-up
Antenna Input . SO-239 & Terminals
Audio Out .... 2 W 10% distortion

Circuit Complement: 15 Integrated circuits, 9 FET's and 96 transistors. Single conversion below 1.611 MHz.

Accessories: RP-952 Car PS Adapter

Comments: Memories store frequency only. Built-in ferrite antenna for the medium wave band. Frequency gap from 420 to 520 kHz. The digital display may be turned off to conserve battery life. There is a switch on the back of the radio that permits the selection of the telescopic whip antenna or an external shortwave antenna. Note that the 12VDC input jack is center negative. List price $699.

Variants: Also sold outside the U.S.A. as model **DR-B600.**

Made In: Japan 1984-1989

Voltages: 110-115/115-127/200-220/ 230-250 VAC 50/60 Hz 16W 12 VDC 3xAA cells 8xD cells

Readout: 00001. Digital Flour.

Physical: 14.81x4.81x11.5" 10.2 Lbs 376x122x292mm 4.6 kg

Status: Active Manufacturer Discontinued Model

Rarity: Abundant

Reviews: *Pop. Comm.* March 1984, *Mon. Times* September 1988

	New	Used
Price:	$440-650	$250-280
Rating:	★★★	★★★

Phase Track Ltd. was established in the early 1980's.

The Liniplex receiver series is different from most receivers. A patented synchronous demodulation method for distortion free reception of double sideband signals provides audio that is unusually clear and stable.

The Liniplex F1 and F2 are fixed receivers that cannot be tuned. Rather, a crystal is required for each frequency desired. Crystal controlled reception affords several advantages. It affords very high resistance to vibration and audio feedback effects, very high local oscillator purity resulting in a minimum of receiver noise and spurious responses, very cost effective, nonvolatile frequency retention, minimized power and space requirements and reduced circuit complexity.

Phase Track products are used extensively by the British Forces Broadcasting Service.

Phase Track Ltd.
Unit B3,
Robert Cort Ind. Est., Reading,
Berkshire,
England

Phase Track Ltd.
16 Britten Road, Reading,
Berkshire, RG2 0AU
England

Phase Track Ltd.
132 Queens Road, Reading,
Berkshire, RG1 4DG
England

LINIPLEX F1

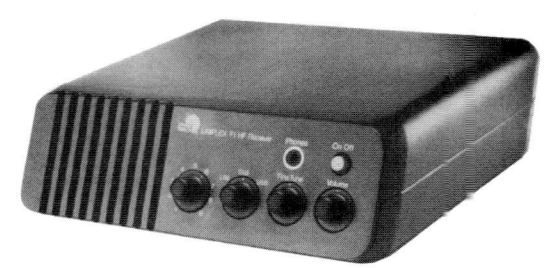

Fixed High Frequency Broadcast Receiver
Double Conversion Superheterodyne. Solid State.

Features:
- ¼" Head. Jack
- Fine Tuning
- Speaker
- 9 Fixed Channels
- Speaker Jack
- Record Jack
- Synchronous Demodulation

Specifications:
Coverage ...... Fixed within 2-26.5 MHz
Selectivity ...... 4 kHz -6dB.
Sensitivity ...... <1.5 µV -10 dB S/N SSB
Audio Out .... .25 W 1% dist..
Environment .. -10° to +35° C
Modes .... AM/(LSB/USB ECSS)
AM Passband . 50-3400 Hz, -3 dB
Spurious Res.. <-60 dB
IF .... 35.4 MHz, 455 kHz
Antenna Input . 300/1500 ohm

Accessories:
Line Output Option

Comments:
This receiver covers 9 frequencies in the range of 2 to 26.5 MHz with one crystal required for each frequency. Supplied with 9 crystals for B.B.C. frequencies. A successful implementation of synchronous tuning. LSB/DSB/USB mode is self-tuned, or may be hard-selected. Has sturdy cast aluminum case.

Variants:
Model **Liniplex F1-2** is a 19" rack mounted configuration consisting of two receivers. Shown right.

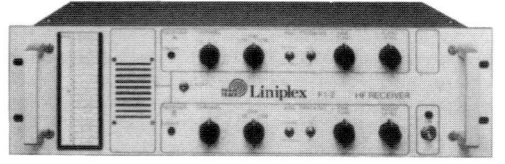

Made In:	England	1983-1988
Voltages:	120/240 VAC 50-60 Hz or NiCad battery	
Readout:	[Channel] Ch. Display	
Physical:	8x2.5x7.5" 4.2 Lbs 193x63x203mm 1.35 Kg	
Status:	Active Manufacturer Discontinued Model	
Rarity:	Very Scarce	
Reviews:	📖 *Passport* 1985/6-1988 *WRTH* 1985 *Pop. Comm.* August 1986	

	New	Used
Price:	$700-928	$200-350
Rating:	⑩	③

PHASE TRACK

LINIPLEX F2

Fixed High Frequency Receiver
Double Conversion Superheterodyne. Solid State.

Features:
- ¼" Head. Jack
- S-Meter
- AGC FST/SLO
- Line Out Jack
- Fine Tuning
- Speaker
- Synchronous Demodulation
- Ant. Imp. Switch
- 8 Fixed Channels

Specifications:
Coverage ...... Fixed within .15-26.5 MHz
Selectivity ...... 4 kHz -6dB.
Sensitivity ...... <1.5 µV -10 dB S/N SSB
Audio Out .... 5 W 6 ohms
Environment .. -10° to +45° C

Modes .... AM/(LSB/USB ECSS)
AM Passband . 50-3400 Hz, -3 dB
Spurious Res .. <-60 dB
IF .... 35.4 MHz, 455 kHz
Antenna Input . SO239s 50/200 ohm

Accessories:
OSC-1 Synthesizer

Comments:
Housed in a high quality blue aluminum cabinet. The first rotary switch selects one of the eight channels or the "EXT OSC" position for use with the optional OSC-1 Synthesizer. The OSC-1, which is housed in the same cabinet size, permits continuous coverage reception from 2 to 22 MHz. The frequency is selected by five decade rotary switches on the front of the OSC-1.

Made In: England   1989-1991

Voltages: 90-265 VAC 50/60 Hz
or NiCad battery

Readout: ⌜Channel⌟ Ch. Display

Physical: 11.3x2.8x11" 5.3 Lbs.
288x72x280mm 2.4 Kg

Status: Active Manufacturer
Discontinued Model

Rarity: Very Scarce

Reviews: *SW Magazine* Aug. 1989

	New	Used
Price:	$1600	②
Rating:	⑩	⑩

The **Philips** company was founded in the Netherlands by Gerard Philips in 1891, to produce carbon-filament incandescent lamps. By 1900, the factory was one of the largest producers in Europe. Diversification came early, and electron tube production was added in the 1910's. The company went multinational early, and had offices and factories in more than twenty countries by 1930.

Radio parts and complete radio equipment for both broadcasting and communications became a major division in the Philips line of products in the 1920's and 1930's. By World War II the Philips company had more than 45,000 employees.

Today Philips has 270,000 employees in 150 countries worldwide and produces a wide variety of electrical and electronics products.

Philips Telecommunicatie Industrie
P.O. Box 32
Hilversum, Holland

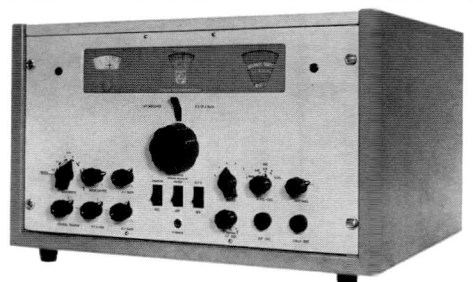

8RO 501/00

General Coverage Communications Receiver
Double Conversion Superheterodyne. 13 Tubes plus Semiconductors.

Features:
- ¼" Head. Jack
- RF Gain
- NL
- Dial Lock
- S-Meter
- Mute
- Standby
- Dial Lamp
- BFO
- Antenna Trimmer
- Line Out 600 ohm
- 455 IF Output
- Hinged Top Cover
- Bandspread 0-100
- AGC SHT/MED/LNG
- Calibrator 500 kHz.

Specifications:

Coverage ...... 200 - 31200 kHz	Modes .... AM/CW/MCW
Selectivity ...... 10/6/3/2/.9/.2 kHz -6dB.	Stability .... ±300Hz/MHz after warm
Sensitivity ...... <.5 µV 20 dB S/N CW	Image Rej .... >80 dB <27 MHz
Audio Out ...... 5 ohms 2.5 W	Antenna Input . 75 ohms

Circuit Complement: (2) 6BY7 RF Amp, 6BY7 IF Amp, 6BY7 Calibrator, (2) 6AJ8 Mixer, 6AJ8 BFO/Detector, (2) 6BJ6 IF Amp, 6BJ6 AF Amp, 6AL5 AVC/Detector, 6BQ5 AF Power Amp and 0A2 Stabilizer plus 8 semiconductors.
Accessories: 8RY 504 FSK Option 8RY 507 ISB Option 8RY 505 Twinplex Opt.
Comments: Ranges: .200-.54, .47-1.128, 1.1-2.64, 2.6-6.24, 6-14.4 and 13-31.2 MHz. Only the band-in-use is displayed. This version includes a cabinet. This model is the successor to the BX 925A.
Variants: Model **8RO 501/50** is rack mount version. Model **8RO 501/01** is the same as the 8RO 501/00 but with coverage from 225-31200 kHz. Model **8RO 501/51** is rack mount version of the 8RO 501/01. Model **8RO 501/02** has one different capacitor allowing the AF output to be fully reducible to zero. Model **8RO 501/52** has magnetic latches.

Made In:	Holland 1963-1968
Voltages:	90/110/125/145/220/245 40-60 Hz VAC 65 W
Readout:	[|||||] Analog
Physical:	Cabinet Version: 21x14x18" 70 Lbs. 527x327x448mm 32 kg Rack Version: 19x10.5x15.4 483x266x390mm
Status:	Active Manufacture Discontinued Model
Rarity:	Extremely Scarce

	New	Used
Price:	$1600	€
Rating:	⑥	€

BX 925A/09

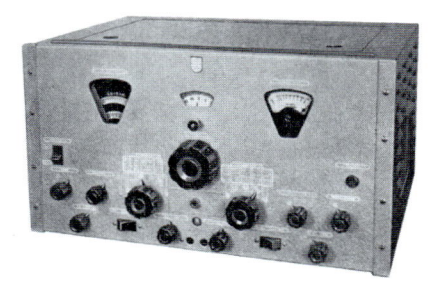

General Coverage Communications Receiver
Single Conversion Superheterodyne. 16 Tubes.

Features:
- ¼" Head. Jack
- S-Meter
- Hinged Top Cover
- Radio Phono Switch
- RF Gain
- Mute
- Antenna Trimmer
- Bandspread 0-100
- NL
- Standby
- Line Out 600 ohm
- AGC SHT/MED/LNG
- Dial Lock
- Dial Lamp
- Motorized Tuning
- Calibrator 500 kHz.
- Phono Input
- BFO
- Fixed Crystal Position

Specifications:

Coverage ......210-540, 1450-32000 kHz		Modes .... AM/CW/MCW	
Selectivity ......13/8.4/6/2.6/.9 kHz -6dB.		Stability .... <0.1 %	
Sensitivity ......<5 µV 10 dB S/N		Image Rej .... >75 dB 3.5-9.1 MHz	
IF ....75 kHz		Image Rej .... >66 dB 9.1-13.7 MHz	
Audio Out ......5 ohms 1.3 W		Antenna Input . 75 ohms	

Circuit Complement:
6BA6 1st RF Amp, 6BA6 2nd RF Amp, 6BE6 Mixer, 6BA6 RF Osc., 6BA6 1st IF Amp, 6BA6 2nd IF Amp, 6AL5 Detector/AVC, 6AL5 AVC Diode, 6BA6 AF Amp, 6AQ5 AF Output Tube, 6AL5 NL, 5Y3GT Rectifier, 0D3 Voltage Stabilizer, 6BA6 BFO, 6BA6 Calibrator and 6AQ5 Voltage Stabilizer.

Comments:
Ranges: .210-.54, .1.45-3.6, 3.5-9.1, 9.1-13.7, 13.7-20.7 and 20.7-32 MHz. There is provision for one crystal for fixed frequency operation. Only the band-in-use is displayed. The Radio-Phonograph switch is on the back panel. Shipped with NE53653 external loudspeaker. Not produced in large numbers.

Made In: Holland   1955

Voltages: 100/125/145/200/220 40-60 Hz VAC 110 W or 280 V and 6 V batteries

Readout: [|||| ▮ |]   Analog

Physical: Cabinet Version:
19.25x10.8x16.25" 70 Lbs.
490x275x415mm 32 kg
Rack Version:
19x10.5x14.75
483x267x375mm

Status: Active Manufacturer Discontinued Model

Rarity: Extremely Scarce

	New	Used
Price:	①	⑤
Rating:	⑥	⑥

RO 150

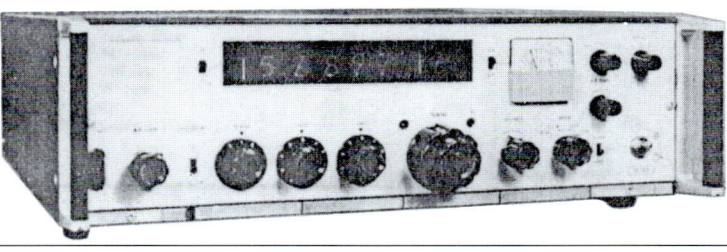

General Coverage Communications Receiver
Double Conversion Superheterodyne. Solid State.

Features:
- ¼" Head. Jack
- S/AF-Meter
- BFO ±3 kHz
- Fine Tuning ±70 Hz
- RF Gain
- Mute
- Speaker
- Attenuator
- NL
- Standby
- Dial Lock
- Line Out 600 ohm
- AGC
- IF Out 1.6 MHz
- Line Output
- Modular Construction
- Rack Handles
- Antenna Switch
- Speaker Switch
- 5 MHz Ref. Input

Specifications:

Coverage ......200 - 30000 kHz		Modes .... AM/CW/MCW/USB/LSB	
Selectivity ......6/3.1/.6 kHz -6dB.		Stability .... <50 Hz/8 Hours	
Sensitivity ......<0.5 µV 10 dB S/N CW		Image Rej .... >90 dB (100 typical)	
IF ....71.6 MHz, 1.6 MHz		IF Rejection .... >90 dB	
Audio Out ......0.5 W		Antenna Input . 50-75 ohms	

Accessories: Filters

Comments: Tuning is by means of decade switches for the 10 MHz, 1 MHz and 100 kHz steps combined with free tuning facilities over a 100 kHz band. The frequency display shows the received antenna frequency with an accuracy of 1 Hz. A 10 Hz display may be switch selected from the front panel if preferred. The antenna frequency is presented in 7 or 8 digits, 3 of which are supplied by the first three decade switches. The frequency of the free tuning oscillator is counted and presented as the last of 4 or 5 digits. A frequency locking system holds the received frequency at the last setting. Environment 0° to +50° C.

Variants: Model **RO 250** is similar, but with an LED frequency display.

Made In: Holland   1975

Voltages: 110/127/200/220/240 47-63 Hz VAC 85 W

Readout: [00000.001]   Digital Nixie

Physical: 19x5.2621.6" 53 Lbs
484x134x550mm 24 kg

Status: Active Manufacturer Discontinued Model

Rarity: Typically Unavailable

	New	Used
Price:	①	⑤
Rating:	⑥	⑥

Philmore is one of the oldest electronics parts and accessories manufacturer in the world. The company was established in 1921.

In 1959 Philmore introduced the CR-5AC shortwave receiver kit. This would be their only serious offering in this area. In 1970 Philmore did offer a simple two band, regenerative kit called the model 7001CR. Please see *Chapter 105 Briefly Mentioned*.

The company is currently located in Rockford, Illinois and is a division of the LKG Industries Company.

Philmore Manufacturing Co. Inc.
130-01 Jamaica Avenue
Richmond Hill 18, NY   1959

Philmore

CR-5AC

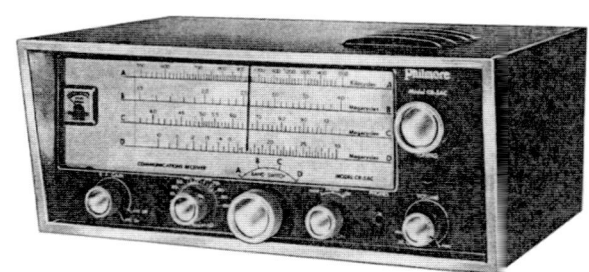

General Coverage Communications Receiver Kit
Single Conversion Superheterodyne. 5 Tubes.

Features:
- ¼" Head. Jack
- S-Meter
- BFO
- Bandspread 0-100
- RF Gain
- NL
- Standby
- Aux. Power Jack
- Dial Lamp
- Speaker
- AVC ON/OFF
- Q-Multiplier Jack
- Flywheel Tuning

Specifications:
Coverage ...... 550 - 30000 kHz
IF .... 455 kHz
Audio Out ...... 8 ohms

Modes .... AM/CW
Antenna Input . Terminals

Circuit Complement:
6BE6 Converter, 6AZ8 IF Amp/S-Meter, 6AV6 Demodulator/AVC/Audio Amp/BFO/NL, 6AQ5 Audio Output and 6X4 Rectifier.

Comments:
Ranges: .55-1.5, 1.5-4, 4-10 and 10-30 MHz. The Noise Limiter is activated by a slide switch on the rear panel. The 7½ inch dial has the amateur bands shaded and the Conelrad indicators at 640 and 1240 kHz.

Made In: United States 1959-1961

Voltages: 105-125 VAC 50-60 Hz
45 W

Readout: Analog

Physical: 15.1875x6.25x10"
385x157x254mm

Status: Active Manufacturer
Discontinued Model

Rarity: Very Scarce

	New	Used
Price:	$52	$40-60
Rating:	⑨	⑨

Pierson Electronic was established in 1946. It was founded by Karl E. Pierson W6BGH.

The company only produced one receiver, the KP-81. Pierson went on to form Pierson Holt Electronics with Eskil Holt in 1955. He remained the Chief Engineer when Automation Electronics acquired Pierson Holt.

Please also see the *Pierson Holt* chapter.

Pierson Electronic Corp.
533 East Fifth Street
Los Angeles 13, California 1946-1948

PIERSON ELECTRONIC CORP.

KP-81

General Coverage Communications Receiver
Double Conversion Superheterodyne. 20 Tubes.

Features:
- ¼" Head. Jack
- RF Gain
- Mute
- Standby
- Tilt Stand
- S-Meter
- Noise Silencer
- Dial Locks
- BFO
- AVC ON/OFF
- High Audio Filter
- Lo Audio Filter
- Bandspread
- Squelch
- 10" Speaker (in P.S.)
- Calibrator 500 kHz
- Phono Input Jack
- Comm/Normal Switch

Specifications:
Coverage ...... 540 - 40000 kHz.
Selectivity ...... 3-.5 kHz -6dB.
Audio Out ...... 12 W

Modes .... AM/CW
IFs .... 465 kHz.
Antenna Input . Terminals 75 ohm

Circuit Complement:
7A7 R Meter, 7C7 Squelch, 7B4 1st Audio Amp, 7C7 Crystal Calibrator Osc., 7A7 1st RF Amp, 7A7 2nd RF Amp, 7B4 Mixer, 7B5 HF Osc., 7A7 1st IF Amp, 6L7 2nd IF Amp, 7A6 Noise Rectifier, 7H7 Noise Amp, 7A7 Noise Control, 7H7 3rd IF Amp, 7A4 Detector, 7C7 BFO, 7Y4 AVC Detector Rectifier, 5U4G Power Supply Rectifier and (2) 7C5 Audio Output Amp. Crystal filter. 2 RF and 3 IF stages.

Comments: Ranges: .54-1.7, 1.7-5.5, 5.5-12, 12-20 and 20-40 MHz. The Communications-Normal switch attenuates all audio frequencies above 1500 Hz. This robustly built commercial grade receiver featured a separate speaker-power supply. Only a few hundred were produced and they are very scarce. This receiver may be mounted in a standard 19" rack. Some production had a slightly different appearance with the dial locks below the tuning knobs, upside down S-Meter, etc.

Made In: United States 1946-1947

Voltages: 115/230 VAC 50-60 Hz. 125 VA

Readout: Analog

Physical: 20x11.5x15.675" 120 Lbs. 508x292x398mm 54.5 kg

Status: Inactive Manufacturer Discontinued Model

Rarity: Very Scarce

	New	Used
Price:	$320	$650-750
Rating:	⑥	⑥

67 Pierson-Holt

Pierson-Holt Electronics
2308 West Washington Blvd.
Venice, California 1955-1956

Automation Electronics Inc.
1500 West Verdugo Ave.
Burbank, California 1957-1961

Pierson-Holt Electronics was established in 1954. This was the third radio company Karl Pierson was involved with. Pierson-Delane was established in 1937 and continued to 1943. After the war Pierson Electronic Corporation produced one receiver, the KP-81. Please see the preceding *Pierson Electronic Chapter*.

Karl E. Pierson W6BGH met Eskil Holt in 1955 and formed Pierson Holt Electronics.

Pierson-Holt Electronics was sold to Automation Electronics in 1957. Automation Electronics literature at this time refers to a "New improved Model A". It is unclear exactly what improvements were made. Perhaps it is a reference to the addition of the 7HTF Ballast tube.

The Automation Electronics literature also mentions . . .

> *Special Models* and a *Continuous coverage model tuning from above 40 M.C. through the broadcast band. This model has an additional bandspread dial arrangement. Other variations include two and six meter bands.*

It is not certain whether these special models were ever produced. The regular KE-93 was produced until 1960.

PIERSON-HOLT

KE-93

Amateur Band Communications Receiver
Double Conversion Superheterodyne. 12 or 13 Tubes.

Features:
- Fine Tuning
- Calibrator
- BFO ±3 kHz
- Antenna Trimmer
- RF Gain
- Squelch (AM)
- Ext. S-Meter Jack
- ANL
- Mute

Specifications:

Coverage ...... Ham. See Comments.	Modes .... AM/SSB/CW
Selectivity ...... 3 kHz -6dB.	Sensitivity .... <1µV
IFs .... 2.2 MHz, 265 kHz.	Image Rej. ...... >80 dB
Audio Out ...... 3 W 10% distortion	Antenna Input . 50 ohm

Circuit Complement: 6BZ6 RF, 6BE6 Mixer, 3CB6 Osc., 6BE6 Converter, 6BA6 IF Amp, 6BE6 IF Amp, 6AL5 2nd Detector/AVC, 12AX7 AF BFO, 6AQ5 Output, 6AU6 Noise Amp, 6AL5 Noise Rectifier Damper and 6BA8A Squelch. A 7HTF Ballast tube was added to later production. Even later, the 3CB6 replaced the 3BA6.

Accessories:
6-12 VDC P.S. 12 VDC PS Ext. S-Meter KE-110 AC PS/Speaker

Comments: Designed for mobile use. Seven band revolving drum turret tuning dial. Coverage: 1650-3500, 3500-4030, 6990-7310, 13970-14360, 20990-21450 and 27950-30000 kHz plus MW 550-1650 kHz. The KE-93 initially included a power supply and external speaker. Later the power supply and speaker were sold as options. The later production units, labelled Pierson KE-93 had a different finish and are more plentiful than the original Pierson-Holt units. This model no longer frequently appears on the used market. As a result, an average used price is difficult to establish.

Made In: United States 1955-1961

Voltages: See Comments.

Readout: Analog Lin.

Physical: 6.125x5.125x9" 10.5 Lbs.
156x130x229mm 7.7 kg

Status: Inactive Manufacturer
Discontinued Model

Rarity: Very Scarce

Reviews: *QST* May 1958
CQ June 1960
Electric Radio August 1997

	New	Used
Price:	$199	⑤
Rating:	★★★★	⑦

68 Plessey

Plessey is a recognized name in the British defense industry. They have many divisions including Plessey Military Communications, Plessey Radar and Plessey marine.

Plessey is now part of the giant Siemens multinational corporation headquartered Germany.

Plessey also manufactured products for S.A.I.T.

Plessey Company Ltd.
Radio Systems Division
Ilford, Essex
England   1965-1970

Plessey Electronic Systems
Avionics & Communications
Vicarage Ln., Ilford,
England   1979-1990

Siemens Plessey Electronic Sys.
Grange Road
Christchurch, Dorset
England BH23 4JE   1992-1998

PLESSEY Electronics

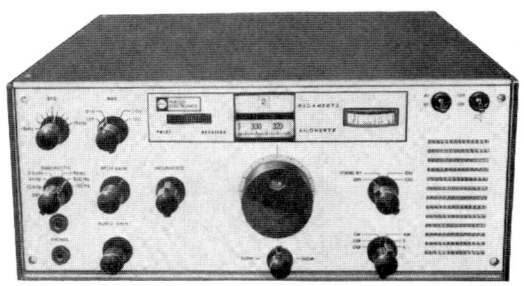

PR-155

General Coverage Communications Receiver
Triple Conversion Superheterodyne. Solid State.

Features:
- ¼" Head. Jack
- RF Gain
- IF Out 100 kHz
- Standby
- S/AF-Meter
- Mute
- Ext. 1 MHz Jack
- Speaker Switch
- BFO ±8 kHz
- Speaker
- 2 Tuning Speeds
- Line Out 600 ohm
- Line Out 150 ohm
- AGC SHT/MED/LNG

Specifications:

Coverage ...... 15 - 30000 kHz	Modes .... AM/LSB/USB/CW
Selectivity ...... 12/6/3.5/1.4/.3/.15 kHz -6dB.	Stability .... ±30 Hz (after 4 hours)
Sensitivity ...... <.5µV 20 dB S/N CW	IF Rejection .... >75 dB
Image Rej ......>100 dB <20 MHz	Antenna Input . SO-239 75 ohm
IF ....37.3, 10.7 MHz, 100 kHz	Environment ... -20° to +50° C. 95% Humid
Audio Out ......0.4 W 15 ohms	

Accessories:

RTTY Mode	FAX Mode	ISB Mode

Comments:
The MHz is selected by a rotary knob to the left of the main tuning knob. The kiloHertz film scale provides an amazing 70 inches of dial per MHz yielding an impressive 350 Hz tuning resolution. The main knob tuning can be set for 6 or 60 kHz per revolution. Brackets are supplied for mounting in a standard 19 inch rack. Designed for fixed commercial use.

Variants:
Also sold as the **S.A.I.T.** model **MR 1402.**

Made In: England   1967-1970

Voltages: 100-125 200-250 VAC 40W
               48-420 Hz or 24 VDC

Readout: [⌊⌊⌊⌊⌊▮⌊⌊]   Analog Lin.

Physical: 16.75x7x17" 38 Lbs.
               425x178x432mm 17.2 kg

Status: Active Manufacturer
               Discontinued Model

Rarity: Extremely Scarce

	New	Used
Price:	$5000A	$1000A
Rating:	⑥	⑥

PLESSEY Electronics

PR-1550

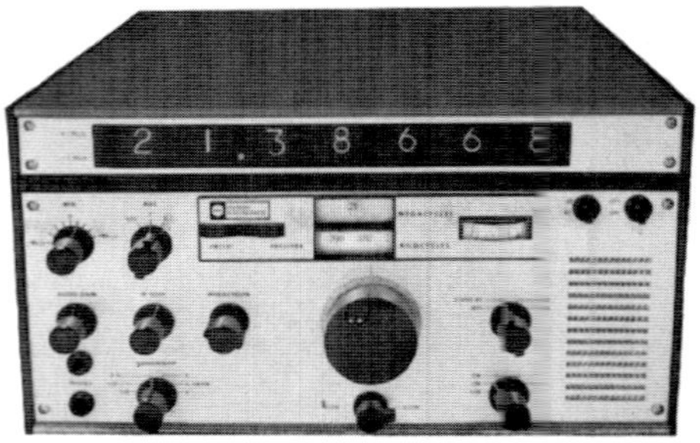

General Coverage Communications Receiver
Double Conversion Superheterodyne. Solid State.

Features:
- ¼" Head. Jack
- RF Gain
- IF Out 100 kHz
- S/AF-Meter
- Mute
- 2 Tuning Speeds
- BFO ±8 kHz
- Speaker
- Line Out 600 ohm
- AGC SLO/FST

Specifications:
Coverage ...... 15 - 30100 kHz
Selectivity ...... 6/3/.3/_/_/_ kHz -6dB.
Sensitivity ...... <.5μV 20 dB S/N CW
Image Rej ...... >100 dB <20 MHz
IF .... 37.3 MHz, 100 kHz
Audio Out ...... 0.4 W 15 ohms
Modes .... AM/LSB/USB/CW
Stability .... ±30 Hz (after 4 hours)
IF Rejection .... >75 dB
Antenna Input . 75 ohm
Environment ... -20° to +50° C. 95% Humid

Accessories:
RTTY Mode
FAX Mode

Comments:
The MHz is selected by a rotary knob to the left of the main tuning knob. The position of the Selectivity knob and the Audio Gain have been reversed from the PR155. Fewer filters are provided as standard equipment. Performance specifications are lower below 60 kHz. Designed for fixed commercial use.

Made In: England 1968-1971

Voltages: 100-125 200-250 VAC 65W
48-420 Hz or 24 VDC

Readout: Analog Lin.
00000.01 & Digital Nixie

Physical: 16.75x8.75x17" 50 Lbs.
425x222x432mm 23 kg

Status: Active Manufacturer
Discontinued Model

Rarity: Extremely Scarce

	New	Used
Price:	①	⑤
Rating:	⑥	⑥

PLESSEY Electronics

PR-1553

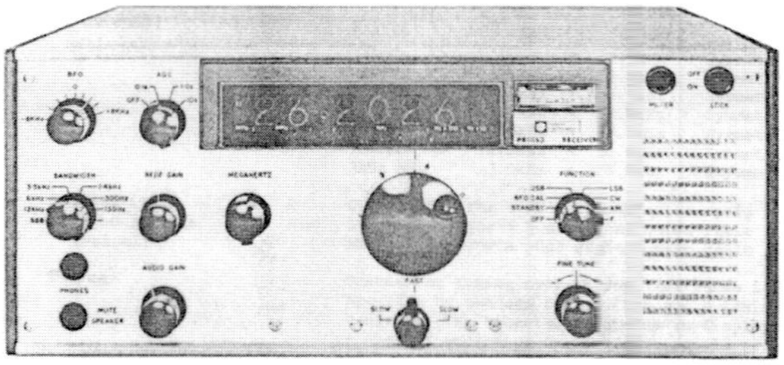

General Coverage Communications Receiver
Double Conversion Superheterodyne. Solid State.

Features:
- ¼" Head. Jack
- RF Gain
- IF Out 100 kHz
- ANL
- S/AF-Meter
- Mute
- Speaker Switch
- BFO ±8 kHz
- Speaker
- 2 Tuning Speeds
- AGC SHT/MED/LNG
- Line Out 150 ohm
- Line Out 600 ohm

Specifications:
Coverage ...... 15 - 30100 kHz
Selectivity ...... 12/6/3.5/1.4/.3/.15kHz -6dB.
Sensitivity ...... <.5μV 20 dB S/N CW
Image Rej ...... >80 dB <15 MHz
IF .... 37.3 MHz, 100 kHz
Audio Out ...... 0.4 W 15 ohms
Modes .... AM/LSB/USB/CW
Stability .... <30 Hz/hr. (after 5 hours)
IF Rejection .... >75 dB
Environment ... -20° to +50° C. 95% Humid
Antenna Input . 75 ohm

Comments:
The MHz is selected by a rotary knob to the left of the main tuning knob. Designed for fixed commercial use.

Made In: England

Voltages: 100-125 200-250 VAC 64W
48-420 Hz

Readout: 00000.01 Digital Nixie

Physical: 16.75x7x17" 38 Lbs.
425x178x431mm 17.3 kg

Status: Active Manufacturer
Discontinued Model

Rarity: Extremely Scarce

	New	Used
Price:	①	⑤
Rating:	⑥	⑥

PLESSEY Electronics

PR-2250B

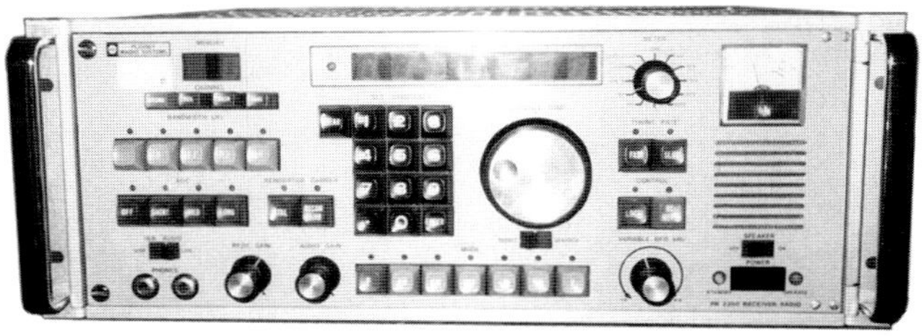

General Coverage Communications Receiver
Double Conversion Superheterodyne. Solid State.

Features:
- ¼" Head. Jack (2)
- RF Gain
- Keypad
- Rack Handles
- Standby
- S/AF/Test Meter
- Mute
- 16 Memories
- B.I.T.E.
- Speaker Switch
- BFO ±8 kHz
- Speaker
- Scanning
- IF Output
- Line Outputs (2)
- AGC SHT/MED/LNG/OFF
- Local/Remote
- Modular Construction
- Fast/Slow Tuning

Specifications:
Coverage ...... 10 - 30000 kHz
Selectivity ...... 8/6/3/1.2/.3/.1 kHz -3dB.
Sensitivity ...... <1μV 15 dB S/N SSB
Audio .... 4/8 ohm 2W
Modes .... AM/LSB/USB/CW/ISB/FSK
IF .... 65, 1.4 MHz
Environment ... -20° to +55° C. 95% Humid
Antenna Input . 50 ohms

Comments:
Memories store: frequency, mode, bandwidth and AGC settings.

Variants:
Model **PR-2250A** is not remotely controllable. Model **PR-2250C** is not remotely controllable and has just 3 bandwidths. Model **PR-2250D** is remotely controllable, does not have selectable sub-octave filters and has 3 bandwidths. Model **PR-2250E** is not remotely controllable, does not have selectable sub-octave filters and has 5 bandwidths. Model **PR-2250F** is remotely controllable, does not have selectable sub-octave filters and has 5 bandwidths. Model **PR-2251A** is a slave receiver with selectable sub-octave filters and 5 bandwidths. The **PR-2252** is for use only by remote control.

Made In: England 1978-1985

Voltages: 105-130 or 200-250 VAC
45-450 Hz 65 VA

Readout: 00000.01 Digital LED

Physical: 17.375x7x18.125" 44 Lbs.
440x178x460mm 20 kg

Status: Active Manufacturer
Discontinued Model

Rarity: Typically Unavailable

	New	Used
Price:	①	⑤
Rating:	⑥	⑥

PLESSEY Electronics

PRS-2282

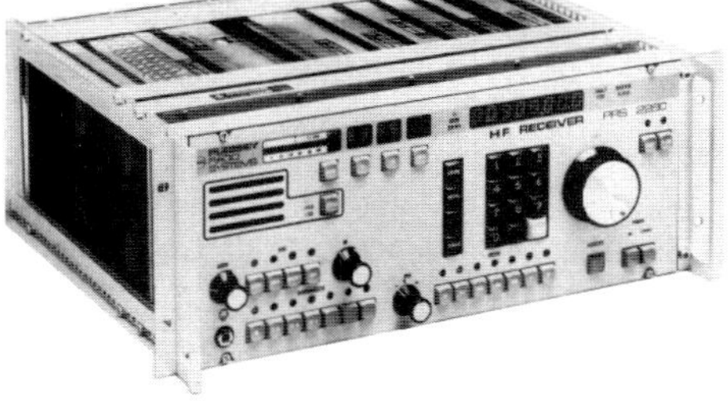

General Coverage Communications Receiver
Double Conversion Superheterodyne. Solid State.

Features:
- ¼" Head. Jack
- RF Gain
- Keypad
- Rack Handles
- S-Indicator
- Mute
- 100 Memories
- B.I.T.E.
- BFO
- Speaker
- Scanning
- Fast/Slow Tuning
- AGC SHT/MED/LNG/OFF
- Local/Remote
- Modular Construction

Specifications:
Coverage ...... 10 - 30000 kHz
Selectivity ...... 8/2.7/1/.5/.3/.1 kHz -6dB.
Stability .... <30 Hz/hr. (after 5 hours)
Modes .... AM/LSB/USB/CW

Accessories:
ISB Option

Comments:
The PRS-2280 series replaces the PR-2250.

Variants:
Model **RS-2281** is a remote controlled version. Model **PRS-2283** is a slave receiver.

Made In: England 1987

Voltages:

Readout: 00000.001 Digital LED

Status: Active Manufacturer
Discontinued Model

Rarity: Typically Unavailable

	New	Used
Price:	①	⑤
Rating:	⑥	⑥

69 Quality U.S. Technologies

The initial production of this receiver was exclusively marketed and sold on the conservative shortwave program *For the People*, on WWCR hosted by Chuck Harder. The main marketing attribute of this receiver was that it was American made. According to the manufacturer, 75% of the parts (by dollar value) were of American origin. Over 1,000 units of the American Electrola DX-100 were sold in this manner.

The manufacturer then elected to offer this model through traditional radio dealers, under the **Quality U.S.** label. Sales did not reach the expected levels. Lacking single sideband capability and suffering from excessive microprocessor noise, the 8A never found a strong following among mainstream listeners or DX'ers.

The announced computer interface was never developed. The company apparently folded sometime in 1994.

Quality U.S. Technologies, Inc.
206 Center Ave.
Pittsburgh, PA 15202   1992-1994

Quality U.S. Technologies, Inc.
8033 Bennett St.
Pittsburgh, PA 152221   1992-1994

8A
World Access

General Coverage Broadcast Receiver
Double Conversion Superheterodyne. Solid State.

Features:
- 1/8" Head. Jack
- Speaker 4"
- 12 Event Timer
- Clock
- Clock-Timer
- 60 Memories
- MW Ferrite Ant.
- Sweep
- Line Out
- Keypad
- Record Activation
- Telescopic Whip
- Wood Cabinet
- Sleep Function

Specifications:
Coverage ...... 150 - 30000 kHz +FM   Modes .... AM
Selectivity ...... 7 kHz -6dB. [20 kHz FM]   IF .... 45, .455 MHz [10.7 FM]
Sensitivity ...... <2 µV 20 dB SINAD   Antenna Input . Terminals
Audio Out .... .75 W <2% THD

Circuit Complement:
Single conversion on FM.

Comments:
Includes FM broadcast band 88-108 MHz. The claim was that it was made of 75% (by dollar value) American parts. An interesting effort, but the radio had shortcomings. Perhaps the most serious problem was very high microprocessor noise. The membrane keypad lacked tactile feedback. A rear-mounted DIN jack provided line out and remote recorder activation. The clock may be set for 12 or 24 hour format. Uses an outboard AC adapter.

Variants: The initial production was purchased by *For the People* and sold as **American Electrola DXC-100.** The Quality U.S. World Access model 8A (also referred to as model R-8X) was sold through radio dealers.

Made In: United States 1993-1994

Voltages: 9 VAC (Supplied with PS for 110 VAC operation).

Readout: `00001.` Digital LED

Physical: 12x7x5" 4 Lbs.
305x178x127mm 1.8 kg

Status: Inactive Manufacturer
Discontinued Model

Rarity: Scarce

Reviews: *Passport* 1994
Mon. Times September 1993

	New	Used
Price:	$260-315	$60-70
Rating:	★★	★

The **Racal Radio Company** is an established name in British communications. The name Racal was derived from the first name of the two people who designed the first receiver; Sir Raymond Brown and George Calder Cunningham. These former Plessey employees started the company in 1950 as Racal Ltd.

Racal was the first commercial receiver manufacturer to fully embrace the famous Wadley loop circuit that stabilizes the received frequency and eliminates drift. This ingenious circuit was developed by Dr. Trevor Wadley of South Africa. See the Racal RA17.

Racal Communications has several assembly plants throughout the world. Racal has a tendency to slightly rename model numbers, or add a suffix based on country of manufacture. The American made RA71 is similar to British made RA17. Likewise with the RA117 and the RA6117A. The tube line-up and other circuit differences were typically adjusted for the location where the receiver would be used. Therefore one should carefully review the **Variants** portion of each receiver entry in this chapter.

Racal continues to manufacture modems, receivers, transmitters and other advanced products for the military and commercial sector.

Racal Engineering Ltd.
Western Rd.
Bracknell, Berkshire
England   1951-1979

Racal Communications Ltd.
Western Rd.
Bracknell, Berkshire
England   1979-1980

Racal Radio Ltd.
472 Basingstoke Rd.
Reading, Berkshire
England RG2 0QF   1992-1997

Racal Communications Inc.
8440 Second Ave.
Silver Spring, MD   1965-1971

Racal Communications Inc.
5 Research Place
Rockville, MD 20850   1972-1997

RA17

General Coverage Communications Receiver
Triple Conversion Superheterodyne. 23 Tubes.

Features:
- ¼" Head.Jack (2)
- S/AF Meter
- RF Gain
- Calibrator 100 kHz.
- Rack Handles
- IF Output
- BFO ±8 kHz
- Atten. -10/20/30/40dB
- AVC SLOW/FAST
- Dial Lock
- Rack Handles
- IF Output
- Mute
- Speaker 2.5"
- Speaker Switch
- Antenna Trimmer
- Standby
- NL

Specifications:
Coverage ......980 - 30000 kHz
Selectivity 8/3/1.2/.75/.3/.1 kHz-6dB
Sensitivity ......<1µV CW
Audio ....1W USA (50mw others)

Modes .... AM/CW
Stability .... <150 Hz after 3 hours
Image Rej .... >60 dB
Antenna Input . 75 ohms

Accessories:
RA37 LF Converter   RA63 SSB Adapter   RA66 Panoramic Display
RA98 ISB Adapter   RA137 LF Conv. (late)

Comments: Ranges: 1-2, 2-4, 4-8, 8-16 and 16-30 MHz. The left tuning knob is for kHz and the right tuning knob is for MHz. The kHz dial is a 6 foot 35mm film scale yielding very high accuracy +55° kHz). Stability improved in later production. First communications receiver to use the famous Wadley loop drift cancelling system. 4 Filters are L-C, 2 are crystal. Heavy duty cast aluminium chassis. Light battleship gray. Over 10,000 RA17s were reportedly made. The RA137 covers 10-980 kHz.

Variants: American versions **RA17C**, **RA17C-2** have slightly different tubes and antenna connections. **RA17C-3** AN/URR-501, **RA17C-12** AN/URR-501A U.S. version. Model **RA17L** later production with bandwidths of: 13/6.5/3/1.2/.3/.1 kHz.

Made In:	England	1957-1962
Voltages:	100-125/200-250 VAC	
	45-65 Hz 85W	
Readout:	Analog Lin.	
Physical:	Cabinet Version:	
	20.5x12x21.875" 96 Lbs.	
	520x305x556mm 44 kg	
	Rack Version:	
	19x10.5x20.125" 67 Lbs	
	483x267x510mm 30.5 kg	
Status:	Active Manufacturer	
	Discontinued Model	
Rarity:	Scarce	
Reviews:	QST March 1958 & Oct. 1965	
	Radio Bygones #25	

	New	**Used**
Price:	$1200-2400	$450-600
Rating:	⑦	⑦

RA71

General Coverage Communications Receiver
Triple Conversion Superheterodyne. Tubes.

Features:
- ¼" Head. Jack
- S-Meter
- Notch Filter
- AGC Short/Long
- Rack Handles

Specifications:

Coverage ...... 500 - 30000 kHz	Modes .... AM/LSB/USB/CW
Selectivity ...... 8/2.5/.5 kHz	Antenna Input . SO239
Stability .... ±150 Hz	

Circuit Complement:
Partial list: 6AU6 Osc, 6AU6 Harmonic Generator, 6EH7 RF Amp, 6AS6 Mixer, 6AU6 1st VFO, 6AU6 Amp, 6688 1st IF Mixer, 6AU6 IF Amp, 6688 2nd IF Mixer, 6AU6 IF Amp, 6BE6 2nd IF Mixer, 6AU6 2nd VFO, 6BA6 3rd IF Amp, EK90 BFO/Product Detector, 12AU7 Notch Filter, 6AL5 Detector/NL, 6BA6 IF Amp, 12AU7 Audio Amp, 6AL5 AVC Ret. and 6AK6 Audio Output.

Accessories:
See RA17

Comments:
The American made RA71 is a direct adaptation of the British made RA17. The left tuning knob is for kHz and the right tuning knob is for MHz. The 60 inch kHz dial is a film scale yielding very high accuracy with a calibration mark every kHz.

Made In:	United States 1965-1966
Voltages:	100-125/200-250 VAC 45-65 Hz 85W
Readout:	Analog Lin.
Physical:	20.5x12x21.875" 90 Lbs. 521x305x556mm 41 kg
Status:	Active Manufacturer Discontinued Model
Rarity:	Very Scarce
Reviews:	*QST* October 1965

	New	Used
Price:	$1200	$300-420
Rating:	★★★★	⑦

RA117

General Coverage Communications Receiver
Triple Conversion Superheterodyne. 23 Tubes.

Features:
- ¼" Head.Jack (2)
- S & AF Meter
- RF Gain
- Calibrator 100 kHz.
- Rack Handles
- Mute Line
- BFO ±8 kHz
- AVC Short/Long
- RF 2-3 MHz In
- Attenuator
- Rack Handles
- Dial Lock
- NL
- Speaker 2.5"
- IF Out 100 kHz
- 1 MHz I/O
- Line Out 600 ohm

Specifications:

Coverage ...... 1000 - 30000 kHz	Modes .... AM/CW
Selectivity 13/6.5/3/1./.3/.1 kHz-6dB	Stability .... <50 Hz after warmup
Sensitivity ...... <1µV 18 dB S/N CW	Intermod Rej ... >100 dB
Image Rej ...... >60 dB	Antenna Input . 75 ohms
Audio Out ...... 1 W 3 ohm 5% Dist.	Line Outs .... 3 mW & 10 mW 600 ohms

Circuit Complement: 6AK5W Crystal Osc., 6AK5W Harmonic Gen., 6ES8 RF Amp, 6AS6 Harmonic Mix, 6BA6 1st VFO, 6BA6 37.5 MHz Amp, 6688 1st Mix, 6BA6 37.5 MHz Amp, 6688 2nd Mix, 6BA6 37.5 MHz Amp, 6AK5W 2nd VFO Amp, 12AT7 2nd VFO, 6BE6W Calibrator, 6BA6 1st IF Amp, 6BA6 Calibrator, 6BA6 2nd IF Amp, 6BA6 IF Output, 6AL5 AVC, 6AK5W BFO, 6AL5 Noise Det, 6AQ5 Audio Out, 12AT7 AF Output, 6688 3rd Mixer, 6BE6W 4th Mixer and 6AK5W 1.7 MHz Crystal Osc.

Accessories: RA237 LF Converter (10-980 kHz). RA298 SSB/ISB Option

Comments: Ranges: 1-2, 2-4, 4-8, 8-16 and 16-30 MHz. The left tuning knob is for MHz and the right tuning knob is for kHz. The kHz dial is a film scale yielding very high accuracy. Light battleship gray. This model was designed for the U.S. market.

Variants: Model **RA6117A** is the American version manufactured in the U.S.A.

Made In:	England
Voltages:	110-125/200-250 VAC 45-65 Hz 100W
Readout:	Analog Lin.
Physical:	Cabinet Version: 20.5x12x21.875" 92 Lbs. 520x305x556mm 42 kg Rack Version: 19x10.5x21.125" 62 Lbs. 483x267x510mm 28 kg
Status:	Active Manufacturer Discontinued Model
Rarity:	Very Scarce

	New	Used
Price:	①	$350-550
Rating:	★★★★	⑦

RA153A

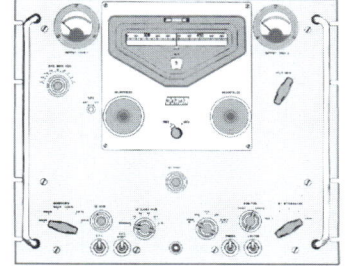

General Coverage Communications Receiver
Double Conversion Superheterodyne. Tubes.

Features:
- ¼" Head. Jack
- Rack Handles
- NL
- 1 MHz I/O
- AF Meters
- Mute Line
- Attenuator
- Dial Lock
- RF Gain
- BFO ±8 kHz
- Rack Handles
- Line Out 600 ohm
- VFO In
- AVC Short/Long
- Calibrator 100 kHz
- IF Out 100 kHz (2)

Specifications:
Coverage ...... 1000 - 30000 kHz
Selectivity 8/3/1.2/.75/.3/.1kHz-6dB
Sensitivity ...... <1µV 17 dB S/N CW
Image Rej ...... >60 dB
Audio Out ...... 3 and 600 ohm
Modes .... AM/CW/MCW
Intermod Rej ... >100 dB
Stability .... ±50 Hz.
Line Out .... 3/600 ohms
Antenna Inputs (2) SO-239 75 ohms

Circuit Complement: Main chassis: (2) 6EH7 RF Amp, (2) 6688 1st VFO Buffer, (2) 6688 1st Mixer, (2) 6AK5W 40 MHz Buffer Amp, (2) 6688 37.5 MHz Buffer, (2) 6688 2nd Mixer, 6AL5W NL and 12AT7WA Audio Output.

Accessories: MA126 Phase & Amp. MA190 Indicator   MA173 DF Indicator
Comments: Ranges: 1-3, 3-4, 4-8, 8-16 and 16-30 MHz. The RA153A is a twin channel receiver having two entirely separate signal paths. This is useful for direction finding applications. The left tuning knob is for MHz and the right tuning knob is for kHz. The kHz dial is a film scale yielding very high accuracy. The power supply unit is separate.
Variants: The **RA153B** has separate non-switched AF stages. Model **RA253A** is similar, but for diversity applications.

Made In:	England 1967-1969
Voltages:	110-125/200-250 VAC 45-65 Hz 160W
Readout:	Analog Lin.
Physical:	Cabinet Version: 20.5x24x22.5" 180 Lbs. 520x610x570mm 82 kg Rack (including PS): 19x26.25x21.125" 130 Lbs. 483x667x536mm 62 kg
Status:	Active Manufacturer Discontinued Model
Rarity:	Extremely Scarce

	New	Used
Price:	①	⑤
Rating:	⑥	⑥

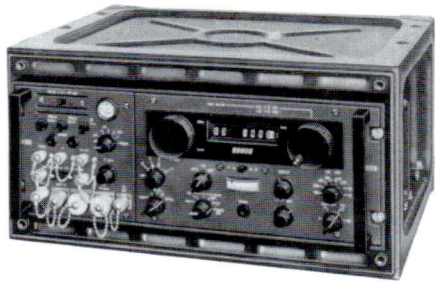

COMMUNICATIONS, INC.

RA329B

General Coverage Communications Receiver
Double Conversion Superheterodyne. Solid State.

Features:
- ¼" Head. Jack
- Mute
- IF Out 100 kHz
- Cal. Fine Tune
- Standby
- S/AF Meter
- RF Tuning
- 1 MHz Input
- Spinner Knob
- Speaker
- RF Gain
- BFO ±3 kHz
- 2nd VFO Input
- Dial Lamp
- Speaker Switch
- Calibrator 100 kHz
- AGC SHT/MED/LNG
- 2nd VFO Output
- Line Out 600 ohm
- Attenuator

Specifications:
Coverage ...... 1000 - 30000 kHz
Selectivity ...... 13/3/1/.2 kHz -3dB.
Sensitivity ...... <1µV 15 dB S+N/N
Audio Out ...... 600 ohms 10 mW
Modes .... AM/USB/LSB/FSK/FM
Stability .... ±50Hz/8 hr. after warm.
Environment ... -5 to +55°C
Antenna Input . BNC 50-70 ohms

Accessories:
RA337 LF Converter   RA298 ISB/SSB Adapter   RA366 Panoramic Adapter
MA197 Preselector   MA350 Freq. Generator

Comments:
The RA329B is a military version of the RA 217D receiver combined with the MA 323 FSK-Loudspeaker Terminating unit, all mounted in a rugged field case. The MA323 component includes an FSK converter supporting shift of between 85 and 850 Hz and speeds of up to 150 baud. Note that the connections for power supplies, antenna jacks and external adapters are all brought to the front. The left tuning knob selects MHz and the right tuning knob (with spinner) selects the kHz. There are separate amps and controls for the line and headphone outputs.

Made In:	England 1969-1972
Voltages:	110-125/200-250 VAC 45-400 Hz or 12/24 VDC
Readout:	00001. Digital
Physical:	Cabinet Version: 20x10.5x19.5" 90 Lbs. 500x270x495mm 40 kg Rack Version: 19x7x15" 50 Lbs. 483x175x380mm 22 kg
Status:	Active Manufacturer Discontinued Model
Rarity:	Typically Unavailable

	New	Used
Price:	①	⑤
Rating:	⑥	⑥

RACAL
COMMUNICATIONS, INC.

RA1218

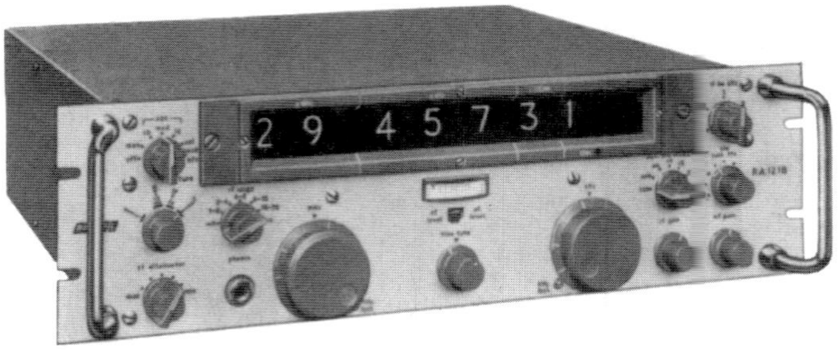

General Coverage Communications Receiver
Triple Conversion Superheterodyne. Solid State.

Features:

• ¼" Head. Jack	• S/AF Meter	• RF Gain	• Dial Locks
• Rack Handles	• Mute	• BFO ±8 kHz	• AGC[1]
• Attenuator	• Rack Handles	• IF Out Jack	• Fine Tuning ±200 Hz
• VFO Out Jack	• 1 MHz Out Jack	• 1 MHz Input Jack	• 1.7 MHz I/O Jack
• Line Out Jack	• Pan Out Jack	• Mute Line	

Specifications:

Coverage ...... 1000 - 30000 kHz
Selectivity ...... 8/3/.2/_/_ kHz. -3dB
Sensitivity ...... <1µV 15 dB S/N CW
Audio .... 600 ohm 1 & 10 mw
Modes .... AM/CW/LSB/USB/RTTY
Stability .... ±50 Hz after 2 hours on
Environment ... -5 to +55°C 95% Humid

Accessories:

13 kHz Filter	MA358 Crystal Osc.	RA316 FSK Converter
6.0 kHz Filter	RA337 LF Converter	RA326 Dual Div. FSK Conv.
1.2 kHz Filter	RA298 ISB/SSB Adapter	RA366 Panoramic Adapter
0.5 kHz Filter	MA397 Pre-Sel. Protect	

Comments:
The left tuning knob is for MHz and the right tuning knob is for kHz. [1]AGC positions are: Off, Manual, Short, Medium and Long. Audio output at 1 watt at 15 ohms (for a loudspeaker) available as an option. The optional RA337 LF converter permits reception as low as 3 kHz.

Made In: England 1968-1970

Voltages: 100-125/200-250 VAC
45-400 Hz 60W

Readout: 00000.01 Digital Nixie

Physical: 19x5.25x19" 50 Lbs.
483x133x483mm 23 kg

Status: Active Manufacturer
Discontinued Model

Rarity: Extremely Scarce

Reviews: *Radio Comm.* January 1968

	New	Used
Price:	①	⑤
Rating:	⑥	⑥

RACAL
COMMUNICATIONS, INC.

RA1219

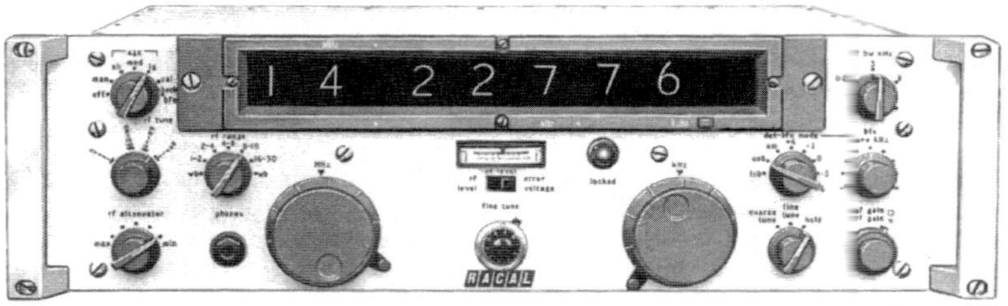

General Coverage Communications Receiver
Triple Conversion Superheterodyne. Solid State.

Features:

• ¼" Head. Jack	• S/AF Meter	• RF Gain	• Dial Locks
• Rack Handles	• Mute	• BFO ±3 kHz	• AGC SHT/MED/LNG
• Attenuator	• Rack Handles	• IF Out Jack	• Fine Tuning ±500 Hz
• VFO Out Jack	• 1 MHz Out Jack	• 1 MHz Input Jack	• 1.7 MHz I/O Jack
• Line Out Jack	• Pan Out Jack	• Mute Line	

Specifications:

Coverage ...... 1000 - 30000 kHz
Selectivity ...... 8/3/.2/_/_ kHz. -3dB
Sensitivity ...... <1µV 15 dB S/N CW
Audio .... 600 ohm 1 & 10 mw
Modes .... AM/CW/LSB/USB/RTTY
Stability .... ±3 Hz after 4 hours
Environment ... -5 to +55°C 95% Humid

Accessories:

13 kHz Filter	RA298 ISB/SSB Adapter	RA316 FSK Converter
6.0 kHz Filter	RA337 LF Converter	RA326 Dual Div. FSK Conv.
1.2 kHz Filter	RA366 Panoramic Adapter	
0.5 kHz Filter		

Comments:
The model RA1219 differs from the RA1218 in that the second VFO can be controlled by a frequency stabilizer which is integral within the receiver. The receiver can be commanded to display the frequency to 1 Hz using the Fine Tune with the Fine Display modes. The **RA337** is an LF version covering 9 - 980 kHz. Model **RA1220** has the same appearance as the RA1219. Differences are unknown.

Made In: England 1969-1970

Voltages: 100-125/200-250 VAC
45-400 Hz 60W

Readout: 00000.001 Digital Nixie

Physical: 19x5.25x19" 50 Lbs.
483x133x483mm 23 kg

Status: Active Manufacturer
Discontinued Model

Rarity: Extremely Scarce

	New	Used
Price:	①	⑤
Rating:	⑥	⑥

RA1771

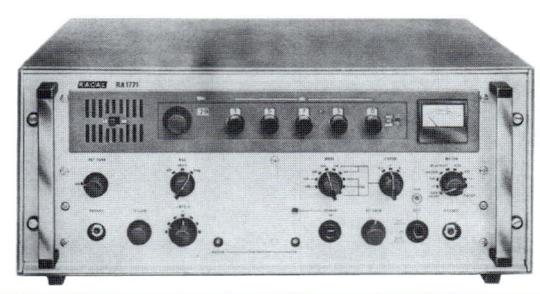

General Coverage Communications Receiver
Double Conversion Superheterodyne. Solid State.

Features:
- ¼" Head. Jack
- S & AF Meter
- RF Gain
- AGC OFF/SHRT/LNG
- Rack Handles
- Mute Line
- BFO ±3 kHz
- Speaker Out
- 1.4 MHz IF Out
- Line Out Jack
- Speaker
- Speaker Switch
- 34 MHz I/O Jack
- 1 MHz I/O Jack
- Local Osc. I/O

Specifications:
Coverage ......15 - 30000 kHz
Selectivity ......8/3/.3 kHz -3dB.
Sensitivity ......<.25 V .5-30 MHz SSB
Image Rej ......>80 dB
Audio Out ......1 W 8 ohms

Modes .... AM/SSB/CW/MCW
Stability .... ±1.5 x 10^6
IF Rejection .... >80 dB
Antenna Input . BNC 50-75 ohms
Environment ... -10 to 55°C 95% Humid

Accessories:
High Stability TCXO   AFC Option   FSK Demodulator
ISB Option   9400 Standard   9420 Standard

Comments:
A separate tuning knob is provided for MHz and then each position to 10 Hz has its own ten position rotary knob.

Made In: England   1978-1981

Voltages: 110-125/200-250 VAC
45-65 Hz. 60-90 W

Readout: ●●●●●●.●⊕ Knobs

Physical: Rack Version:
19x7x16.14" 45 Lbs.
483x178x410mm 21 kg
Cabinet Version:
19.5x8.65x17.5" 61.5 Lbs
495x220x445mm 28 kg

Status: Active Manufacturer
Discontinued Model
Rarity: Extremely Scarce

	New	Used
Price:	①	⑤
Rating:	⑥	⑥

RA1772

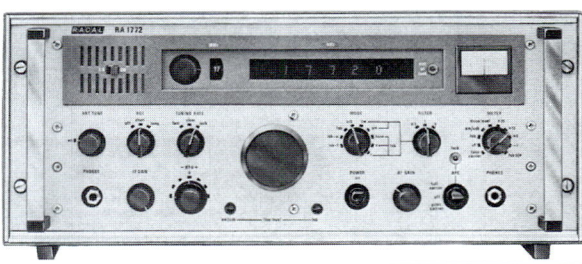

General Coverage Communications Receiver
Double Conversion Superheterodyne. Solid State.

Features:
- ¼" Head. Jack
- S & AF Meter
- RF Gain
- AGC (OFF/SHRT/LNG)
- Rack Handles
- Mute Line
- BFO ±3 kHz
- Two Tuning Speeds
- 1.4 MHz IF Out
- Line Out Jack
- Speaker
- Dial Lock
- 34 MHz I/O Jack
- 1 MHz I/O Jack
- Local Osc. I/O
- Speaker Out
- Speaker Switch

Specifications:
Coverage ......15 - 30000 kHz
Selectivity ......8/3/.3 kHz -6dB.
Sensitivity ......<.1µV .5-30 MHz
Image Rej ......>80 dB
Audio Out ......1 W 8 ohms

Modes .... AM/SSB/CW/MCW
Stability .... ±1.5 x 10^6
IF Rejection .... >80 dB
Antenna Input . BNC 50-75 ohms
Environment ... -10 to 55°C 95% Humid

Accessories:
High Stability TCXO   AFC   FSK Demodulator
ISB Option   9400 Standard   9420 Standard

Comments:
A separate tuning knob is provided for MHz which displays in a window to the left of the digital display. The main tuning knob selects the remaining digits. [¥188,000]
Variants: Model **RA1778** features 12 memories selected by a rotary switch from the front panel. Perhaps the same knob as used for MHz selection on the RA-1772? Model **RA1779** is very similar to the RA1772, but appears to be later production [¥325,000]. Also please see models RA6772 and RA8772.

Made In: England   1978-1981

Voltages: 110-125/200-250 VAC
45-65 Hz. 60-90 W

Readout: 00 000.01 Digital
Mechanical & Digital LED

Physical: Rack Version:
19x7x16.14" 45 Lbs.
483x178x410mm 21 kg
Cabinet Version:
19.5x8.65x17.5" 61.5 Lbs
495x220x445mm 28 kg

Status: Active Manufacturer
Discontinued Model
Rarity: Extremely Scarce

	New	Used
Price:	①	$750-900
Rating:	⑥	⑥

COMMUNICATIONS, INC.

RA1784

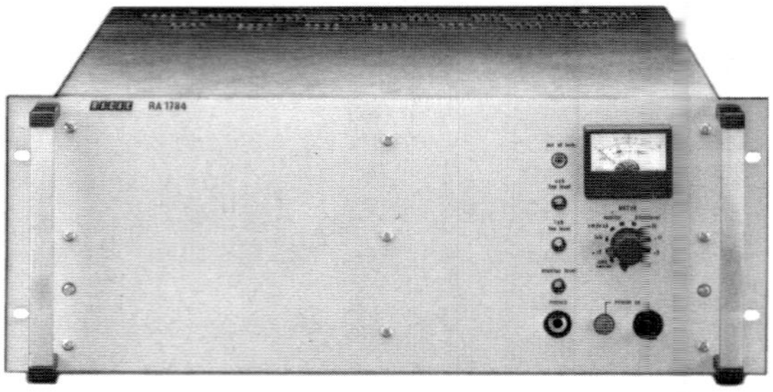

General Coverage Communications Receiver
Double Conversion Superheterodyne. Solid State.

Features:
- ¼" Head. Jack
- Rack Handles
- S/AF Meter
- Mute
- BFO ±4 kHz
- Line Out
- IF Out
- AGC LNG/SHT

Specifications:
Coverage ...... 15 - 30000 kHz
Selectivity ...... 3/_/_/_/_ kHz -3dB.
Sensitivity ...... <1µV 10 dB S+N/N
Environment .. -10° to 55°C
Audio .... 8 ohms 1W
Modes .... AM/LSB/USB/CW
Stability .... ±1x10^6
BFO Resol. ..... 10 Hz
Antenna Input . BNC 50-70 ohms

Accessories:
ISB Option
Filters
MA 1072 Controller (shown below)
DA77060 Cabinet

Comments:
The RA1784 must be manually controlled via the MA1072 receiver controller or via the MA1072 and a computer. The control protocol used is SCORE (Serial Control of Racal Equip.).
Variants: The **RA1781** is conceptually the same and similar in appearance to the RA1784.

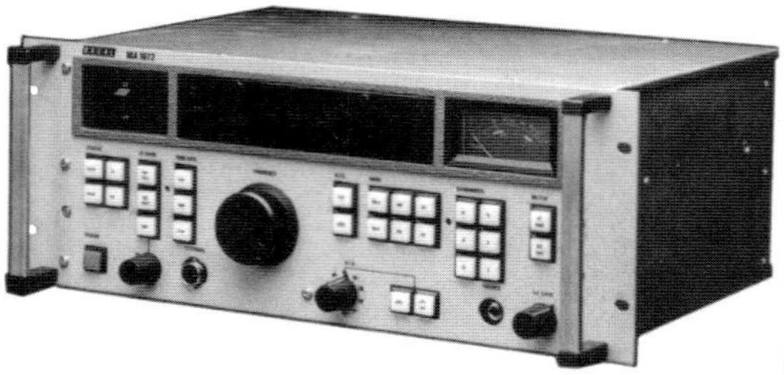

Made In: England 1977

Voltages: 10-125 or 200-250 VAC 45-65 Hz 60-90 VA

Readout: The RA1784 must be controlled by the MA 1072 which has 10 Hz readout.

Physical: 19x7x18.25" 55 Lbs 483x178x464mm 25 kg

Status: Active Manufacturer Discontinued Model

Rarity: Typically Unavailable.

	New	Used
Price:	①	⑤
Rating:	⑥	⑥

COMMUNICATIONS, INC.

RA1792

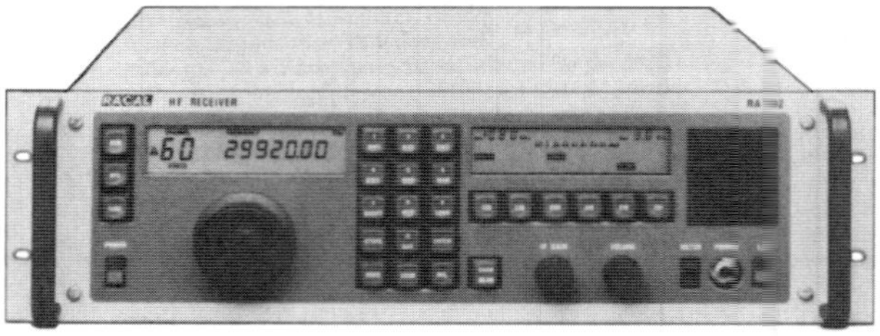

General Coverage Communications Receiver
Double Conversion Superheterodyne. Solid State.

Features:
- ¼" Head. Jack
- Memory Scan
- Rack Handles
- Scanning
- B.I.T.E.
- S/AF Indicator
- Keypad
- Mute
- IF Out 455 kHz
- IF Gain
- BFO ±8 kHz
- Speaker
- Flywheel Tuning
- 100 Memories
- AGC
- TCXO Freq. Standard
- VRIT Tuning

Specifications:
Coverage ...... 100 - 30000 kHz
Selectivity ...... 6/3/1.2/.3/_/_ kHz
Sensitivity ...... <.35µV .5-30 MHz SSB
Stability .... ±1.5x10^6
Modes .... AM/LSB/USB/CW/MCW/FM
Environment ... -10 to +55°C 95% Humid
Antenna Input . BNC 50-70 ohms

Accessories:
Cabinet
ISB Option
FSK Option
9442 Oven Crystal
LA1519 Extended Control
MA1075 Remote Controller

Comments:
Memories are organized in ten banks of ten channels each. Memory scanning must be within a bank. Memories store frequency and mode. Adjustable dwell time on channel scanning. Remote operation is available via LA1519/LA1520 using the SCORE (Serial Control Of Racal Equipment) system. [¥279,00-288,000].

Made In: United States 1979-1990

Voltages: 110-125/200-250VAC 60VA or 18-32 VDC 40W

Readout: 00000.01 Digital LCD

Physical: 19x5.25x18" 31 Lbs 483x133x458mm 14kg

Status: Active Manufacturer Discontinued Model

Rarity: Very Scarce

	New	Used
Price:	£8000	£1000-1200
Rating:	⑥	⑥

RACAL

RA3701

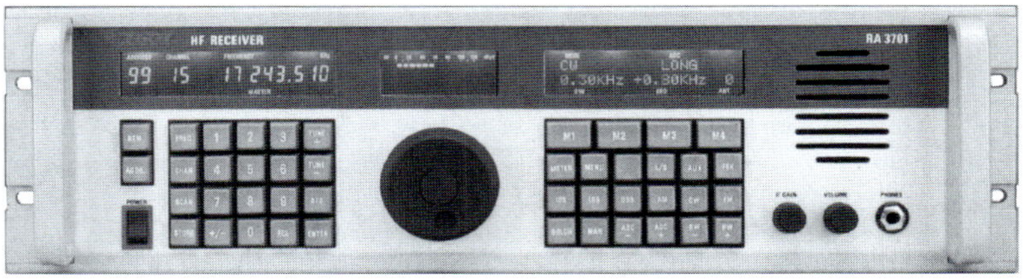

General Coverage Communications Receiver
Double Conversion Superheterodyne. Solid State.

Features:
- ¼" Head. Jack
- Memory Scan
- Rack Handles
- Scan & Sweep
- Speaker Jack
- S/AF Indicator
- Keypad
- Mute
- Speaker Switch
- IF Out Jack
- RF Gain
- BFO ±9.99 kHz
- Speaker
- B.I.T.E.
- Master/Slave
- Modular Construction
- AGC SHT/MED/LNG
- 100 Memories
- Line Out 600 ohms
- IEEE 488/Serial Port

Specifications:
Coverage ...... 15 - 30000 kHz
Selectivity ...... 12/6/2.4/1/.3 kHz
Sensitivity ...... 1µV for S+N/N 16 dB
Image Rej ...... >90 dB
Audio Out ...... 0.2 W 8 ohms
Modes .... AM/LSB/USB/CW/MCW/FM
Stability .... ±$1.5x10^6$
IF Rejection .... >90 dB
Environment ... -10 to +55°C 95% Humid
Antenna Input . BNC 50 ohms

Accessories:
ISB Module
IF Filter Module
FSK Module
MA2232 Tune Aid
9420 High Stability
MA3700 Receiver Controller
9442 High Stability

Comments:
Each memory stores: frequency, mode, bandwidth, AGC and BFO settings. The display above the right keypad alphanumerically displays: mode, bandwidth, AGC, BFO, dwell length and attenuator settings. It is also used for BITE readout.
Variants: Model **RA3702** is a dual receiver with front panel controls. Model **RA3703** is a remotely controlled single receiver. Model **RA3704** is a remotely controlled, dual receiver.

Made In: England 1987-1998

Voltages: 100/120/220/240 VAC
45-65 Hz. 60W

Readout: 00000.001 Digital

Physical: 19x5.25x17.7" 31 Lbs.
483x133x450mm 14 kg

Status: Active Manufacturer
Active Model

Rarity: Very Scarce

	New	Used
Price:	①	⑤
Rating:	⑥	⑥

RACAL
COMMUNICATIONS, INC.

RA3791

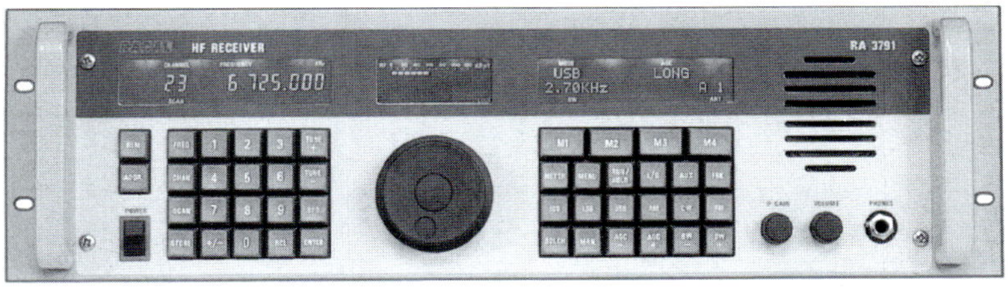

General Coverage Communications Receiver
Double Conversion Superheterodyne. Solid State.

Features:
- ¼" Head. Jack
- Memory Scan
- Rack Handles
- Notch
- RS423/232 Port
- IF Out Jack
- RF Gain
- Keypad
- Mute
- DSP
- RF Gain
- 100 Memories
- PBT
- BFO ±8 kHz
- Speaker
- Scan & Sweep
- Speaker Jack
- Squelch
- S/AF/FSK Indicator
- AGC SHT/MED/LNG
- B.I.T.E.
- Modular Construction
- Line Out 60 ohms

Specifications:
Coverage ...... 10 - 30000 kHz
Selectivity ...... See Comments.
Sensitivity ...... 1µV for S+N/N 16 dB
Image Rej ...... >90 dB
Audio Out ...... 1W 8 ohms
Modes .... AM/LSB/USB/CW/MCW/FM
Stability .... ±$1.5x10^6$
IF Rejection .... >90 dB
Environment ... -10 to +55°C 95% Humid
Antenna Input . 50 ohm

Accessories:
TCXO ±$7x10^7$
ISB Option
Oven TCXO ±$1x10^7$
Digital IF Out
Hi-Stab. Oven TCXO ±$3x10^3$
MA3790 Receiver Controller

Comments: The receiver has digital filters with the following bandwidths: 12/6/3/2.7/1/.3 kHz. Up to 100 different filters my be user configured from the front panel. The display above the right keypad alphanumerically displays: mode, bandwidth, AGC, BFO, dwell length and attenuator settings (and used for BITE readout). Blue.
Variants: Model **RA3792** is a dual receiver with front panel controls. The **RA3793** is a remotely controlled single receiver. **RA3794** is a remotely controlled, dual recv.

Made In: England 1995-1998

Voltages: 90-132 and 175-264 VAC
47-63 Hz 50W

Readout: 00000.001 Digital

Physical: 19x5.25x17.7"
483x133x450mm

Status: Active Manufacturer
Active Model

Rarity: Very Scarce

	New	Used
Price:	$8700	$4000+
Rating:	⑥	⑥

RA6217

General Coverage Communications Receiver
Double Conversion Superheterodyne. Solid State.

Features:
- ¼" Head. Jack
- S/AF Meter
- RF Gain
- Atten.-10/20/30/40 dB
- Mute
- Dimmer
- BFO ±3.5 kHz
- AGC SHT/MED/LNG
- IF Out 455 kHz
- 1.6 MHz Out
- 1.145 MHz I/O
- Ext. VFO
- Cal. Fine Tune
- Spinner Knob
- Dial Lamp
- Line Out 600 ohm

Specifications:

Coverage ...... 980 - 30000 kHz	Modes .... AM/SSB/CW/MCW/FM
Selectivity ...... 13/6/3/1/.2 kHz -3dB.	Stability .... ±50Hz, ±5Hz after warm.
Sensitivity ...... <.5µV 15 dB S+N/N	IF Rejection .... >80 dB
Image Rej ...... >80 dB	Environment ... 0 to +55°C
Audio Out ...... 600 ohms 10 mW	Antenna Input . BNC 50-70 ohms

Accessories:
RA6337 LF Adapter   MA350B Synthesizer   RA6366 Panoramic Display
SA77T Comparator   MA168 Diversity Switch   RA6367 Dual Panoramic D.

Comments: RA6337 LF adapter provides 3 - 980 kHz coverage. Designed for rack mounting. Widely used by the F.B.I., F.C.C. and other government agencies.

Variants: This model sold initially as the British made **RA217** and **RA1217**. Model **RA6217B** tuning via manual or synthesizer. Model **RA6217D** is designed for remote control operation. Model **RA6217F** is a slave receiver. Model **RA6217G** has simplified mode and bandwidth operation. Model **RA6217Q** has a wide IF output with a 100 kHz bandwidth and no detectors or AF circuits. Model **RA6218** use an electronic digital display. Model **RA6225** is a slave receiver.

Made In:	United States 1965-1976
Voltages:	110-125/200-250 VAC 48-420 Hz
Readout:	00001. Digital
Physical:	19x3.5x17" 25 Lbs. 483x89x432mm 11.3 kg
Status:	Active Manufacturer Discontinued Model
Rarity:	Very Scarce

	New	Used
Price:	$5795-6295	$500-300
Rating:	★★★★	★★★★

RA6772

General Coverage Communications Receiver
Double Conversion Superheterodyne. Solid State.

Features:
- ¼" Head.Jack (2)
- S & AF Meter
- IF Gain
- Two Tuning Rates
- Rack Handles
- Mute Line
- BFO ±3 kHz
- TCXO Freq. Standard
- Speaker Switch
- Line Out 600 ohm
- 1 MHz In/Out
- AGC OFF/SHT/LNG
- 1st IF Out
- 2nd IF Out
- RF Tune
- Speaker   • Dial Lock

Specifications:

Coverage ...... 15 - 30000 kHz	Modes .... AM/LSB/USB/CW
Selectivity ...... 13/8/3/1/.3/_ kHz	Stability .... $\pm1.5\times10^6$
Sensitivity ...... <.35µV .5-30 MHz SSB	IF Rejection .... >80 dB
Image Rej ...... >80 dB	Antenna Input . BNC 50-70 ohms
Audio Out ...... 1 W 8 ohms	

Accessories:
9400 Freq. Standard   9420 Freq. Standard   FSK Board

Comments: The RF Tune knob is required only when the receiver is operated in close proximity to strong interfering signals. It is switched out of the circuit when set to WB (wideband). Tuning drag may be adjusted. The small MHz dial is just to the left of the frequency display.

Variants: Model **RA6772A-1** with ISB installed, **RA6772A-2** with FSK Demod installed, **RA6772A-3** with ISB & FSK. **RA6772A-4** with AFC option installed, **RA6772-5** with ISB & AFC, **RA6772A-6** with AFC & FSK installed, **RA6772A-7** with AFC, ISB and FSK options installed. A model **RA6772-P** is the parent receiver to drive two model **RA6230** slave receivers for direction finding.

Made In:	United States 1975-1979
Voltages:	110-125/200-250 VAC 45-420 Hz 60-90W
Readout:	00000.01 Digital LED
Physical:	19x7x16.3" 45 Lbs. 483x178x410mm 20.4 kg
Status:	Active Manufacture Discontinued Mode
Rarity:	Very Scarce
Reviews:	*Ham Radio* October 1978

	New	Used
Price:	$5995	$700-390
Rating:	★★★★	★★★★

COMMUNICATIONS, INC.

RA6772E

(R-2130/GRR)

General Coverage Communications Receiver
Double Conversion Superheterodyne. Solid State.

Features:
- ¼" Head. Jack
- S & AF Meter
- IF Gain
- Line Out 600 ohm
- Rack Handles
- Mute
- BFO ±8 kHz
- AGC OFF/SHT/MED/LNG
- 5 MHz Ref In
- IF Out 455 kHz
- Two Tuning Rates

Specifications:
Coverage ...... 1000 - 30000 kHz.
Selectivity ...... 16/8/3.2/1/.3 kHz -3dB.
Sensitivity ...... <.5μV 10 dB S+N/N
Image Rej ...... >80 dB
BFO Range ... ± 8 kHz
BFO Resol. ... 10 Hz
3rd Order Int .. >=35 dBm

Modes .... AM/LSB/USB/CW/FM
Stability .... ±1x10^6
IF Rejection .... >80 dB
Dyn. Range .... >120 dB
Environment ... 0 to 50°C 95% Humid.
Antenna Input . BNC 50-70 ohms

Accessories:
RS232C/MIL188 Port   ISB Option   MA6003 Remote Unit
High Stability

Comments:
The highly shielded RA6772E is Tempest qualified. The BFO has a separate three digit LED display. The RA6772E computer controllable via the optional MA6003 Remote unit. The RA6772E is more similar to the RA6778 than it is the RA6772.

Made In:	United States
Voltages:	120/240 VAC 48-420 Hz. 75W
Readout:	`00000.01` Digital LED
Physical:	19x8.75x19.9" 45 Lbs. 483x222x506mm 20.3 kg
Status:	Active Manufacturer Discontinued Model
Rarity:	Very Scarce

	New	Used
Price:	$5500	$490-850
Rating:	★★★★	★★★★

COMMUNICATIONS, INC.

RA6778A

General Coverage Communications Receiver
Double Conversion Superheterodyne. Solid State.

Features:
- ¼" Head. Jack
- S & AF Meter
- IF Gain
- Two Tuning Rates
- Rack Handles
- Mute
- BFO ±8 kHz
- 1.4 MHz IF Out
- Line Out
- 50 Memories[1]
- Speaker
- Speaker Switch
- 5 MHz Ref Jack

Specifications:
Coverage ...... 15 - 30000 kHz
Selectivity ...... 16/8/3.2/1/.3 kHz -3dB.
Sensitivity ...... <.5μV 10 dB S+N/N
Image Rej ...... >80 dB
BFO Range ... ± 8 kHz
Environment .. 0° to 50°C 95% Humid.

Modes .... AM/LSB/USB/CW/FM
Stability .... ±1x10^6
IF Rejection .... >80 dB
Dyn. Range .... >120 dB
BFO Resol. ...... 10 Hz
Antenna Input . BNC 50-70 ohms

Accessories:
ISB Option   Filters   Preamp   Computer Control Option

Comments: May be tuned in 10 or 100 Hz increments. [1]Some units have no memories, some 32 memories and some have 50 memories. The memories store: frequency, mode, bandwidth, AGC and BFO settings. The RA6778 (unlike the RA6790!) is Tempest qualified (extremely low emissions from the receiver). It is one of the most heavily shielded receivers ever produced.

Variants: Model **RA6778B** has one less RF amplifier. Model **RA6778C**, later production, 10-30000 kHz with different front panel. Caution: Model **RA6778Q** is a considerably different receiver designed primarily for remote operation (1980).

Made In:	United States 1977-1982
Voltages:	120/240 VAC 48-420 Hz. 75W
Readout:	`00000.01` Digital LED
Physical:	19x8.75x19.9" 45 Lbs. 483x222x506mm 20.3 kg
Status:	Active Manufacturer Discontinued Model
Rarity:	Very Scarce

	New	Used
Price:	$7000	$1400-1900
Rating:	⑥	⑥

RACAL
COMMUNICATIONS, INC.

RA6790/GM

General Coverage Communications Receiver
Double Conversion Superheterodyne. Solid State.

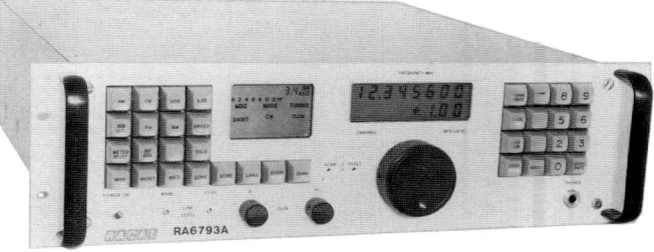
RACAL RA6790/GM

Features:
- ¼" Head. Jack
- S & AF Meter
- Keypad
- RF Gain
- Rack Handles
- Mute
- AGC (3 Pos)
- 3 Tuning Rates
- Line Out Jack (2)
- Ext. Spkr Jack
- I.F. Out Jack
- Lock
- BFO ±8 kHz
- B.I.T.E.

Specifications:
Coverage ...... 500 - 30000 kHz
Selectivity ...... 20/6/3.2/1/.3 kHz -3dB.
Sensitivity ...... <.5µV .5-30 MHz SSB
Image Rej ...... >80 dB
BFO Range ... ± 8 kHz
Audio Out ...... 1 W 8 ohms
Dyn. Range ... RF >180 dB/Hz

Modes .... AM/LSB/USB/CW/FM
Stability .... ±.5 x 10^8 per 10°C.
IF Rejection .... >80 dB
BFO Resol. ..... 10 Hz.
Line Outs .... 1 mW 600 ohms
Environment ... 0° to 50°C 95% Humid.
Antenna Input . N

Accessories:
MA6004 Control Unit.

Preselector	ISB 2 Channel	VLF Extension (1 - 500 kHz)
I.F. Converter	ISB 4 Channel	LF Extension (10 - 500 kHz)
Ext. Ref. Jack	Very NBFM	High Stability Option

Remote Control via ASCII, MIL-STD-188C, RS-232/422/423 or IEEE 488.

Comments: Often sold on the surplus market without *any* I.F. filters. [¥150,000].
Variants: Earlier model **RA6790** similar, but without LED on the front panel for Fault. Model **R-2174(P)** is a military version. **R-2174A(P)** is later production with 18 bandwidths. Later model **R-2174B(P)** covers 1-30000 kHz.

Mfg. In:	United States 1979-1988
Voltages:	110/120/200/240 VAC 48-420 Hz 40 W
Readout:	`00000.001` Digital LCD.
Physical:	19x5.25x18.5" 32 Lbs. 483x133x470mm 14.5 kg
Status:	Active Manufacturer Discontinued Model
Rarity:	Common
Reviews:	*WRTH* 1982

	New	Used
Price:	$5000-6000	$760-1600
Rating:	★★★★★	★★★★★

RACAL
COMMUNICATIONS, INC.

RA6793A

General Coverage Communications Receiver
Double Conversion Superheterodyne. Solid State.

RACAL RA6793A

Features:
- ¼" Head. Jack
- S & AF Meter
- Keypad
- RF Gain
- Rack Handles
- Mute
- AGC (3 Pos.)
- 3 Tuning Rates
- Line Out Jack (2)
- Ext. Spkr Jack
- I.F. Out Jack
- Lock
- B.I.T.E.
- BFO ±8 kHz
- 100 Memories
- Memory Scan
- Sweep

Specifications:
Coverage ...... 500 - 30000 kHz
Selectivity ...... 20/6/3.2/1/.3 kHz -3dB.
Sensitivity ...... <.5µV .5-30 MHz SSB
Image Rej ...... >80 dB
Audio Out ...... 1 W 8 ohms
Dyn. Range ... RF >180 dB/Hz
3rd Ord. Int. ... +30 dBm

Modes .... AM/LSB/USB/CW/FM
Stability .... ±.5 x 10^8 per 10°C.
IF Rejection .... >80 dB
Environment ... 0° to 50°C 95% Humid.
Line Outs .... 1 mW 600 ohms
Antenna Input . 50 ohms N

Accessories:
I.F. Converter	ISB 2 Channel	VLF Extension (1 - 500 kHz)
RS-232C/422/432	ISB 4 Channel	LF Extension (10 - 500 kHz)
Preselector	Very NBFM	High Stability Option
	MA6004 Command Control Unit (shown below)	

Comments: Uses Hall-effect keypad (not membrane like RA6790).
Variants:
The **R-2320** is a militarized version of the RA6793A with 2nd head. jack and C.O.R.

RACAL MA6004

Mfg. In:	United States 1991-1998
Voltages:	115/230 VAC 48-420 Hz 40 W
Readout:	`00000.001` Digital LCD.
Physical:	19x5.25x18.5" 32 Lbs. 483x133x470mm 14.5 kg
Status:	Active Manufacturer Active Model
Rarity:	Scarce
Reviews:	*WRTH* 1982

	New	Used
Price:	①	$2200+
Rating:	★★★★★	★★★★★

RACAL
COMMUNICATIONS, INC.

RA6830

General Coverage Communications Receiver
Double Conversion Superheterodyne. Solid State.

Features:

- ¼" Head. Jack
- Rack Handles
- Speaker Jack
- IF Gain
- Spinner Knobs
- S Indicator
- Mute
- B.I.T.E.
- Tuning Lock
- Keypad
- BFO ±8 kHz
- IF Out 455 kHz
- Ref. Input Jack
- Modular
- 2 Tuning Rates
- AGC SHT/MED/LNG
- Line Out 600 ohm (2)

Specifications:

Coverage ...... 100 - 30000 kHz
Selectivity ...... 20/6/3.2/1/.3 kHz -3dB.
Sensitivity ...... <.5µV 10 dB S+N/N
Image Rej ...... >80 dB
BFO Range ... ± 8 kHz
Audio Out ...... 1 W 8 ohms
Line Out (2) ... 600 ohms 1 mW
Modes .... AM/LSB/USB/CW/FM
Stability .... $\pm5\times10^8$ per 10° C.
IF Rejection .... >80 dB
Dyn. Range .... RF >180 dB/Hz
BFO Resol. ..... 10 Hz
Environment ... 0 to 50°C 95% Humid.
Antenna Input . BNC 50-70 ohms

Accessories:

High Stability RS232C/MIL188C Muting
ISB Option

Comments:

The RA6830 tunes in 100, 30 or 1 Hz increments. This "half rack" size permits two receivers to be mounted side-by-side in a 19" rack. Also has been produced with a single horizontal rack handle.

Made In: United States 1986-1998

Voltages: 115/230 VAC 48-420 Hz. 40W

Readout: 00000.001 Digital LED

Physical: 9.5x5.25x22" 32 Lbs. 244x133x560mm 14.5 kg

Status: Active Manufacturer
Active Model

Rarity: Extremely Scarce

	New	Used
Price:	①	⑤
Rating:	⑥	⑥

RACAL
COMMUNICATIONS, INC.

RA8772CB

General Coverage Communications Receiver
Double? Conversion Superheterodyne. Solid State.

Features:

- ¼" Head. Jack
- Rack Handles
- 34 MHz I/O
- IF Gain
- Speaker
- S/AF Indicator
- Mute
- 1 MHz I/O
- Tuning Lock
- Speaker Switch
- AFC
- BFO
- IF Out Jack
- Ref. Input Jack
- RF Tune
- Modular
- 3 Tuning Rates
- AGC OFF/SHT/LNG
- Line Out
- Dial Lock

Specifications:

Coverage ...... 15 - 30000 kHz
Selectivity ...... 8/3/.3/.1/_ kHz -3dB.
Image Rej ...... >80 dB
Audio Out ...... 1 W 8 ohms
Line Out (2) ... 600 ohms 1 mW
Modes .... AM/LSB/USB/CW/ISB
IF Rejection .... >80 dB
Environment ... -10 to 55°C 95% Humid
Antenna Input . BNC 50-70 ohms

Comments:

The MHz position is selected by a rotary knob to the left of the digital display. The MHz position shows through a very small window to the right of this knob. The remaining six digits are tuned from the large center knob and displayed on the red LEDs. This rare model was made in Canada. It is similar to model RA1772, but with one more tuning speed and two more filter positions.

Made In: Canada 1979-1980

Voltages: 100/105/110/115/120/200 210/220/230/240/250 VAC

Readout: 00000.001 Digital Mechanical & Digital LED

Physical: 19x7x16.14" 45 Lbs. 483x178x410mm 21 kg

Status: Active Manufacturer
Discontinued Model

Rarity: Extremely Scarce

	New	Used
Price:	①	⑤
Rating:	⑥	⑥

Prior to World War II, R.C.A. was an active manufacturer of radios for the hobby market. The AR77 was a very popular communications receiver, regularly advertised in *QST*.

Perhaps R.C.A.'s most famous receiver is the model AR88. Most AR88 production was sent to England and Russia during World War II as part of the Lend Lease Act. The AR88 in RAF service was called the R1556. The AR88D was called the R1556A and the AR88LF was referred to as the R1556B.

After the war, the AR88 became a staple of British DXer's in the 1950's and 1960's. Many were sold new-in-the-box to the English hobbyist market. In fact, the AR88 is far more common in the United Kingdom than it is in America.

After World War II, R.C.A. focused more on television, components and other areas. They produced a few receivers for the RCA Radio Marine division, but many were relabeled radios made by other manufacturers such as Drake.

Please see *Chapter 105 Briefly Mentioned* for additional R.C.A. receivers.

Radio Corporation of America
RCA Manufacturing Co. Inc.
Front & Cooper Streets
Camden, NJ 08102

Radio Corporation of America
Radiomarine Products
75 Varick St.
New York 13, NY

AR77

General Coverage Communications Receiver
Single Conversion Superheterodyne. 10 Tubes.

Features:
- ¼" Head Jack (2)
- S-Meter
- AVC/MVC
- Antenna Trimmer
- BFO
- RF Gain
- NL Variable
- Standby
- Mute Line
- Flywheel Tuning
- Dial Lamp
- Bandspread
- Speaker Output
- Negative Feedback Control

Specifications:
Coverage ...... 540 - 31000 kHz   Modes .... AM/CW
Selectivity ...... 6 Positions   Audio Out .... 3 W 2.5 ohms
Sensitivity ...... <2µV for .05W output   Antenna Input . Terminals
Circuit Complement: 6H6 2nd Detector, 6SJ7 BFO, 6SQ7 AF Amp/AVC, 6F6G Output, 6K7 1st IF Amp, 6K8 Detector Oscillator, 6SK7 2nd IF Amp, 6SK7 RF Amp, VR150 Voltage Regulator and 5Y3Z Rectifier. Crystal filter.

Accessories:
MI-8303 External Speaker   MI-8303A Rack Speaker
MI-8304 Rack Mounting Kit   MI-8314A External Speaker

Comments:
Ranges: .54-1.34, 1.34-3.3, 3.3-5.8, 5.8-10.2, 10.2-18 and 18-31 MHz. Bandspread ranges: 80, 40, 20, and 10 meter amateur bands. The headphone jack is on the back panel.
Variants: Model **AR77E** is the export version operating from 105-130/140-160/195-250 VAC 50/60 Hz. The steel blue color model **GR-10** was produced for the Royal Canadian Air Force.

Made In: United States 1940-1942

Voltages: 105-125 VAC 50/60 Hz
70 W

Readout: Analog

Physical: Cabinet Version:
20.13x10.5x11.75 49 Lbs.
511x26.7x298mm 22.2 kg
Rack Version:
19x10.5x11.25
482x267x285mm

Status: Active Manufacturer
Discontinued Model

Rarity: Very Scarce

Reviews: *Electric Radio* April 1996

	New	Used
Price:	$139	②
Rating:	⑥	⑦

AR88

General Coverage Communications Receiver
Single Conversion Superheterodyne. 14 Tubes.

Features:
- ¼" Head. Jack
- AVC/MVC
- Tone Control
- Antenna Trimmer
- BFO
- RF Gain
- NL Variable
- Standby
- Mute Line
- Flywheel Tuning
- Dial Lamp
- Line Out 600 Ohms

Specifications:
Coverage ...... 535 - 32000 kHz   Modes .... AM/CW
Selectivity ...... 13.5/7/2.5/1/.5 kHz -6dB.   IF .... 455 kHz
Sensitivity ...... <2.2µV 6dB S/N   Antenna Input . Terminals
Audio Out ...... 2.5 W 2.5 ohms

Circuit Complement: 6SG7 1st RF Amp, 6SG7 2nd RF Amp, 6SG7 1st IF Amp, 6SG7 2nd IF Amp, 6SG7 3rd IF Amp, 6J5 Osc., 6J5 BFO, 6SA7 1st Det., 6H6 2nd Det./ AVC, 6H6 NL, 6SJ7 1st Audio, 6K6GT Output, 5Y3GT Rect. and VR150 Voltage Reg.
Comments: Ranges: .535-1.6, 1.57-4.55, 4.45-12.15, 11.9-16.6, 16.1-22.7 and 22-32 MHz. May also be run from external batteries: 250-300 volts and 6 volt. This model was produced in large numbers for the war effort. Most were sent to England and Russia so they are hard to find in the United States. A heavy receiver!
Variants: Model **AR88D** is the cabinet version. Model **CR88** has crystal phasing (a few with S-Meters) 1946. **CR88A** later production and S-Meter. Model **AR88LF** covers .073-.55 and 1.48-30.5 MHz with 735 kHz IF. Later models **CR91** and **CR91A** have the same coverage as the AR88LF. Model **SC-88** (R-320/FRC) like CR-88 but shows only band in use. **CR88B** last version 1951 has a 50 kHz calibrator. Model **DR89** is a triple diversity configuration.

Made In: United States 1941-1945

Voltages: 100-117/117-135/135-165/ 190-230/200-260 VAC 50/60 Hz 100 W. Or batteries. See Comments.

Readout: [diagram] Analog

Physical: 19.3x11x19.3" 110 Lbs. 490x280x490mm 50 kg

Status: Active Manufacturer Discontinued Model

Rarity: Very Scarce

Reviews: *Buyer's Guide to Amateur R. Practical Wireless* Oct. 1995 *Electric Radio* July 1990

	New	Used
Price:	①	$200-550
Rating:	⑥	⑦

AR-8516

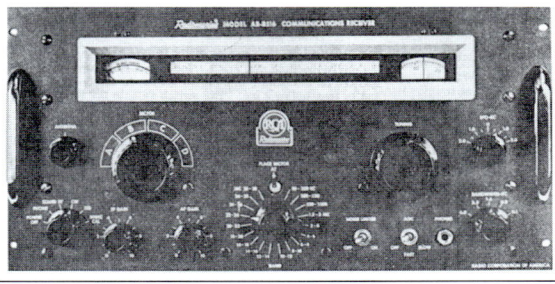

General Coverage Communications Receiver
Triple Conversion Superheterodyne. 18 Tubes.

Features:
- ¼" Head. Jack
- S-Meter
- Tone Control
- Antenna Trimmer
- BFO ±2 kHz
- NL
- Standby
- Crystal Cal. 500 kHz
- Mute Line
- IF Out Jack
- Dial Lamp
- Line Out 600 Ohms
- Preselector
- Rack Handles
- AGC OFF/FST/SLO

Specifications:
Coverage ...... 80 - 30000 kHz   Modes .... AM/CW/SSB
Selectivity ...... 6/3/1.5/.8/.1 kHz -6dB.   IF .... See Comments
Sensitivity ...... <1µV 4-30 MHz -6 dB   Stability .... <500 Hz typical
Audio Out ...... 3.2/600 ohms 1W   Antenna Input . Terminals

Circuit Complement:
3BZ6 RF Amp, 3CB6 1st Mixer, 3CB6 HF Crystal Osc., 3BZ6 RF-IF Amp, 3CB6 2nd Mixer, 7AU7 LF Osc. 44 kHz BFO, 3CB6 VFO, 3BZ6 1st 455 IF Amp, 3BZ6 2nd 455 IF Amp, 3BZ6 3rd 455 IF Amp, 2AL5 Detector/AGC Rectifier, 7AU7 AF-500 kHz Crystal Osc., 5U8 455 kHz BFO, 3BE6 3rd Mixer, 5U8 45 kHz IF Amp, 3AL5 AGC Rectifier-Noise Limiter, 7AU7 Audio Amp and 12CU5 Audio Output.

Accessories:
RM-290 Speaker   RM-289 115/230VAC   RM-288 Resistor Unit 230VDC
Comments: Ranges: .08-.2, .2-.52, .52-1.3, 1.2-3, 2-4, 4-6, 6-8, 8-10, 10-12, 12-14, 14-16, 16-18, 18-20, 20-22, 22-24, 24-26, 26-28 and 28-30 MHz. IF's: 1.09-3.09 or 2-4 MHz, 455 kHz and 45 kHz. The 3 kHz position is a 3.1 kHz Collins mechanical filter (F-455-H-31). Linear frequency display above 1.2 MHz. For marine use.

Made In: United States 1964-1965

Voltages: 115 VAC 45-60 Hz or 115 VDC

Readout: [diagram] Analog

Physical: Cabinet Version: 22x11.375x17.5" 91 Lbs 558x289x445mm 41 kg Rack Version: 19x9.5x16" 61 Lbs 482x241x406mm 28 kg

Status: Active Manufacturer Discontinued Model

Rarity: Very Scarce

	New	Used
Price:	①	②
Rating:	⑥	⑦

CRM-R6A

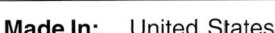

General Coverage Communications Receiver
Double Conversion Superheterodyne. 16 Tubes

Features:
- ¼" Head. Jack
- BFO ±2 kHz
- Mute Line
- Preselector
- S-Meter
- NL
- IF Out Jack
- Rack Handles
- RF Gain
- Standby
- AF Filter 100 Hz
- 42:1 Tune Ratio
- Antenna Trimmer
- Calibrator 100 kHz
- Line Out 600 Ohms
- AGC OFF/FST/SLO

Specifications:
Coverage ......80 - 30000 kHz	Modes .... AM/CW/SSB
Selectivity ......6/3/_/_ kHz -6dB.	IF .... 825, 455 kHz
Sensitivity ......<1-2µV S/N -6 dB CW	Stability .... <500 Hz typical
Audio Out ......3.2/600 ohms 1W	Antenna Input . Terminals

Accessories:
MI-555820A 1.5 kHz Filter   MI555820B 0.5 kHz Filter
MI-555151 External Speaker

Comments:
Ranges: .08-.2, .2-.52, .52-1.3, 1.1-3, 2-4, 4-6, 6-8, 8-10, 10-12, 12-14, 14-16, 16-18, 18-20, 20-22, 22-24, 24-26, 26-28 and 28-30 MHz. Linear frequency display above 1.2 MHz. Designed for shipboard or shore marine use. This receiver requires a speaker. Single conversion from .52-4 MHz. This model replaced the AR-8516. In fact later production models of the AR-8516 has the lighter gray color of the CRM-R6A.

Made In: United States

Voltages: 115/230 VAC 50/60 Hz 70W

Readout: ‖‖‖‖‖ Analog

Physical: Cabinet Version:
22x11.375x17.5" 92 Lbs
558x289x445mm 42 kg
Rack Version:
19x9.5x16" 61 Lbs
482x241x406mm 28 kg

Status: Active Manufacturer
Discontinued Model

Rarity: Very Scarce

	New	Used
Price:	$1795	②
Rating:	⑥	⑦

R-203A/SR

General Coverage Communications Receiver
Single Conversion Superheterodyne. 10 Tubes.

Features:
- ¼" Head Jack (2)
- BFO
- Speaker Output
- AVC/MVC
- RF Gain
- Hinged Top Cover
- BFO
- Speaker
- Bandspread ±25
- Speaker Switch

Specifications:
Coverage ......See Comments	Modes .... AM/CW/MCW
IF ....1700 kHz	Antenna Input . Terminals
Audio Out ......2 W	

Circuit Complement:
6SQ7, (5) 6SG7, (2) 6J5, 25L6GT and 25ZG.

Accessories:
F-91/U Noise Suppressor (shown)   MX-1024 Fixed Resistor

Comments:
Ranges: .085-.22, .21-.55, 1.9-5.4, 5.2-12 and 11.5-25 MHz. The receiver cabinet is mounted on four rubber shock-mounts which are attached to a metal base. Operation from 230 VAC or 230 VDC may be accomplished with the MX-1024 fixed resistor.

Variants:
Model **R-203/SR** has the Power On-Off switch on the RF Gain knob. Its optional Noise Suppressor is RM-8. R.C.A. model **AR-8506-B** is the same as the R-203/SR.

Made In: United States

Voltages: 115 VAC or 115 VDC

Readout: ‖‖‖‖‖ Analog

Physical: 21x11.75x13.125 77 Lbs.
534x298x333mm 34.5 kg

Status: Active Manufacturer
Discontinued Model

Rarity: Very Scarce

	New	Used
Price:	①	②
Rating:	⑥	⑦

SRR-13

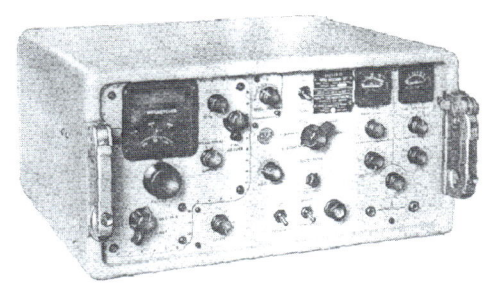

General Coverage Communications Receiver
Double Conversion Superheterodyne. 29 Tubes plus Semiconductors.

Features:

• ¼" Head. Jack (2)	• S/AF Meter	• RF Gain	• Antenna Trimmer
• Dial Lamp	• NL	• Standby	• Logging Scale 1000
• Mute Line	• IF Out Jack	• Dimmer	• Line Out 600 Ohms
• Squelch	• Rack Handles	• Dial Lamp Switch	• Modular Construction
• Calibrator	• BFO (Vernier)	• Cal. Adjust	• Attenuator

Specifications:

Coverage ......2000 - 32000 kHz   Modes .... AM/CW/FSK
Selectivity ......8/3.2/1 kHz -6dB.   IF .... 200, 1600 kHz
Audio Out ......600 ohms 0.2W

Circuit Complement: 5899 Preamp, 5899 RF Amp, 5636 Mix., 5719 Mix. Cath. Fol., 5840 HFO/Local Osc., 5636 1st IF Amp/2nd Mix., (3) 5899 2nd IF Amp, 5636 BFO/Mix., 5719 BFO, 5647 Diode Det., 5647 AGC Delay Diode, 5718 2nd IF Cath. Fol., 5647 NL, 5718 1st Audio , 5647 Squelch, 5719 DC Amp, 5718 2nd Audio, (2) 5647 Series Diode Lim., 5719 Audio Driver, 5902 Audio Output, (2) 5718 Crystal Cal., (2) 6X4 Rect., 5644 Reg. and 1N458 Tuning Meter Diode. Tubes are subminiature.
Comments: Ranges: 2-4, 4-8, 8-16, 16-24 and 24-32 MHz. Designed for Navy.
Variants: Model **SRR-13A** was later production. Models **SRR-11** and **SRR-11A** are longwave versions covering only 14-600 kHz, have IFs of 60 and 200 kHz and have only 28 tubes. Model **SRR-12** and **SRR-12A** cover 250 - 8000 kHz. Equivalent fixed version of the SR-11/12/13 are: **FRR-21**, **FRR-22** and **FRR-23**. Mobile version are: **MRR-1**, **MRR-2** and **MRR-3.** Tubes are soldered in place in all models.

Made In: United States 1950-1961

Voltages: 105-125 VAC
50-60 or 400 Hz.

Readout: | 00001. | Projection of tape to glass

Physical: 18.5x8.5x17" 77 Lbs.
470x216x432mm 35 kg.

Status: Active Manufacturer
Discontinued Model

Rarity: Very Scarce

Reviews: *Electric Radio* July 1991

	New	Used
Price:	①	②
Rating:	⑥	⑦

SSB-R3

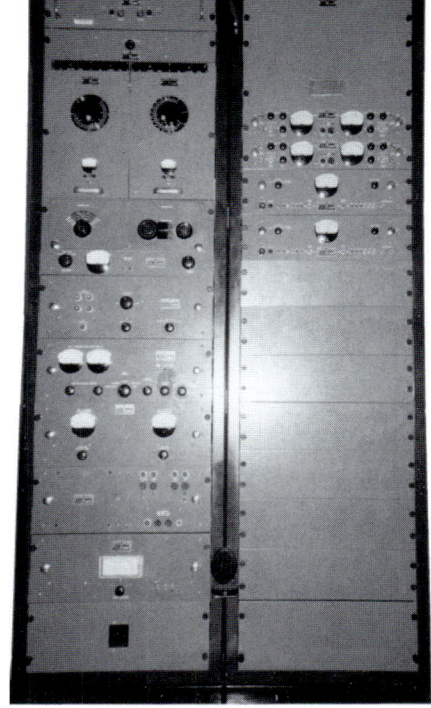

General Coverage Diversity Communications Receiver
Double Conversion Superheterodyne. 79 Tubes and Semiconductors.

Features:
• ¼" Head. Jacks   • S/Audio Meters
• AGC   • Diversity

Specifications:
Coverage ......2800 - 28000 kHz
Modes ....CW/MCW/SSB/ISB/FAX
Selectivity ......12/9/6.2/4 kHz -3 dB.
Image Ratio ...90 dB @ 28 MHz.
Sensitivity ......<.2µV 10 dB S/N
Environment ..0° to +50°C 95% humid.

Comments:
Ranges: 2.8-5, 5-9, 9-16 and 16-28 MHz. The SSB-R3 is a commercial receiver designed for diversity and independent sideband applications. The standard receiver configuration requires two full height 19" rack cabinets to house.

This receiver is expressly designed for feeder and point-to-point applications. In early 1997 over 20 of these receivers were decommissioned from the Voice of America relay station at Kavala, Greece. Several have been sold to the hobby surplus market. The SSB-R3 may qualify as the largest modern shortwave receiver ever.
Variants: The SSB-R3 may be configured for FSK and other applications

Made In: United States

Voltages: 115-130 or 210 - 260 VAC
50-60 Hz 650 W

Readout: |⌊⌊⌊⌊▮⌊⌋| Analog

Physical: 44x84x22" 967 Lbs.
1118x2133x558mm 438 kg.

Status: Active Manufacturer
Discontinued Model

Rarity: Extremely Scarce

	New	Used
Price:	①	$1500-2000
Rating:	⑥	⑦

72 Raytheon

In the 1950's and 1960's **Raytheon** was the international export agent for Hallicrafters radios.

Raytheon today is a leading defense contractor with 1995 revenues of $11.7 billion. They produce sophisticated communications and electronic equipment for government, military and commercial markets.

Raytheon did produce the model 1230 H.F. communications receiver which was advertised heavily in *QST* in the early 1970's. It is doubtful many reached the hobbyist market because of the high price. The author would enjoy hearing from anyone who is lucky enough to possess a Raytheon 1230.

Raytheon Company
6380 Hollister Ave.
Goleta, CA 93017   1972-1974

1230

General Coverage Communications Receiver
Double Conversion Superheterodyne. Solid State.

Features:
- ¼" Head. Jack
- S/AF-Meter
- Rack Handles
- Main & Fine Knobs
- AGC FST/SLO
- RF Gain
- Dial Locks
- Dual Tuning Rates
- Mute
- Tilt Bar
- 1 MHz Output
- AFSK Output Jack

Specifications:

Coverage ...... 2000 - 30000 kHz.	Modes .... AM/SSB/CW/FSK/FM
Selectivity ...... 16/6/3.2/.5 kHz -6dB.	Dial Accuracy . 1.0 Hz.
Sensitivity ...... 0.6 µV SSB	Stability .... ±1 Hz per day
Audio Out 2W	Image Reject .. 50 dB @30 MHz
IF Rejection ... >80 dB	

Accessories:

ISB Option	FSK Adapter	Rack Mount

Comments:
Rear panel jacks for AF outputs, 1 MHz clock output, AFSK output. Designed for the commercial market. Appeared in full page ads in *QST* from 1972 through 1974.

Variants:
Model **1230A.** Differences unknown.

Mfg. In:	United States 1972-1975
Voltages:	110/220 AC 50-400 Hz

Readout:	00000.001 Digital LED
Physical:	3.5x17x18.75" 28 Lbs. 88x432x476mm 12.7 kg
Status:	Active Manufacturer Discontinued Model
Rarity:	Extremely Scarce

	New	**Used**
Price:	$3930-4475	⑤
Rating:	⑥	⑥

Radio Shack was established in 1921 on Brattle Street in Boston. By the late 1940's it had become a leading distributor of electronic parts to hobbyists and industry. In 1947 it branched into the new area of hi-fi. In 1954 it began marketing under the **Realistic** private label. Radio Shack began importing from Japan in the late 50's and opened an engineering office in Tokyo in 1961. By the end of 1962 there were nine stores located in the North East, but the Company was in debt and losing money. In 1963 Radio Shack was purchased by Tandy Corporation. Under the guidance of Charles Tandy, Radio Shack would expand into the largest chain of electronics stores in history. For a detailed account of this amazing story read, *Tandy's Money Machine - How Charles Tandy Built Radio Shack Into the World's Largest Electronics Chain* by Irvin Farman.

Although Realistic receivers have never been the darlings of the radio press, or advanced DXers, the Company's ubiquitous distribution has served to provide shortwave radios to virtually every corner of the country. Reported problems with the DX-300 and DX-302 were not necessarily of poor design. In many cases the factory did not take the necessary time to properly align these sets. Owners who have these sets realigned often report dramatic performance improvement. The Realistic AX-190 and SX-190 are listed in the Allied chapter of this book. Realistic is a trademark of Tandy Corporation.

Radio Shack
167 Washington St.
Boston 8, Mass   1958

Radio Shack
730 Commonwealth Av.
Boston, MA 02215   1965-1971

Radio Shack Div. of Tandy Corp.
One Tandy Center
Ft. Worth, TX 76102   1972-1998

REALISTIC

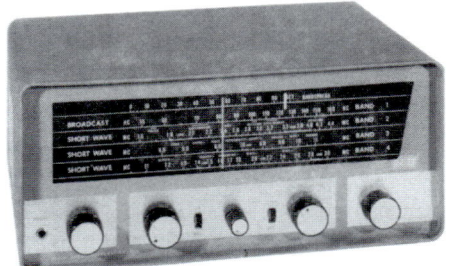

DX-75

General Coverage Communications Receiver
Single Conversion Superheterodyne. 4 Tubes plus Semiconductors.

Features:
- ¼" Head. Jack
- Speaker
- ANL
- Bandspread 0-100
- BFO
- Standby
- Dial Lamp

Specifications:
Coverage ...... 550 - 30000 kHz
Selectivity ...... ±4.5 kHz -6dB.
Audio 1.5W
IF .... 455 kHz
Modes .... AM/CW
Sensitivity .... 3.5 µV 6 dB @ 20 MHz
Antenna Input . Terminals

Circuit Complement:
6CL8A, 12BA6, 12AV6, 6AQ5A plus diodes.

Comments:
Ranges: .55-1.6, 1.6-4.4, 4.5-11 and 11-30 MHz. Vinyl clad steel cabinet.
Variants: The **Regency WT-4** (shown right) appears to be identical except for the cabinet color. The "WT" stands for "World Traveler." The Regency WT-4 features a blue vinyl clad steel. It was featured in the 1966 Lafayette Catalog. [Photo by Joe Veras N4QB].

Made In:	Japan	1965-1966

Voltages: 105-120 VAC 60 Hz
35W

Readout: Analog

Physical: 19.53x6.31x14.25" 15 Lbs
496x160x362mm 6.8 kg

Status: Active Manufacturer
Discontinued Model

Rarity: Very Scarce

	New	**Used**
Price:	$70	$50-60
Rating:	⑥	⑥

REALISTIC

DX-100

General Coverage Communications Receiver
Single Conversion Superheterodyne. Solid State.

Features:
- ¼" Head. Jack
- S-Meter
- Speaker
- ANL
- BFO
- Fine Tuning
- Standby
- Dial Lamp

Specifications:
Coverage ......520 - 30000 kHz   Modes .... AM/SSB-CW
Antenna Input Terminals

Comments:
Main tuning bands: 55-1.62, 1.55-4.5, 4.5-13 and 13-30 MHz. Ceramic IF filter.
Radio Shack #20-206.

Made In:	Taiwan	1981-1984

Voltages:	117 VAC or 12 VDC

Readout:  Analog

Physical:	12x5.75x8"
	305x146x203mm

Status:	Active Manufacturer
	Discontinued Model

Rarity:	Common

	New	Used
Price:	$100	$30-35
Rating:	★★★	★

REALISTIC

DX-120
Star Patrol

General Coverage Communications Receiver
Single Conversion Superheterodyne. Solid State.

Features:
- ¼" Head. Jack
- S-Meter
- Fine Tuning
- Speaker 3x5"
- ANL
- AVC (2 Pcs.)
- RF Gain
- Bandspread 0-100
- Dial Lamp
- Ext. Spkr. Jack
- BFO

Specifications:
Coverage ......535 - 30000 kHz   Modes .... AM/SSB-CW
IF ....455 kHz   Antenna Input . Terminals

Circuit Complement:
Utilizes an FET front-end.

Accessories:
20-1500 SP-150 Speaker 20-1501 DC Pack (8 x D)

Comments:
Radio Shack #20-120.

Made In:	Japan	1970-1971

Voltages:	117 VAC or
	12 VDC

Readout:  Analog

Physical:	12.5x5.25x7.75" 10 Lbs
	317x133x167mm 4.5 kg

Status:	Active Manufacturer
	Discontinued Model

Rarity:	Common

	New	Used
Price:	$70	$30-40
Rating:	★★★	★★

DX-150

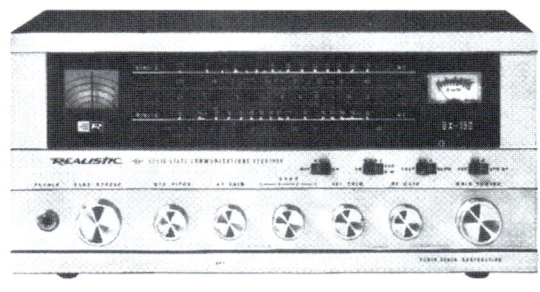

General Coverage Communications Receiver
Single Conversion Superheterodyne. Solid State.

Features:
- ¼" Head. Jack
- ANL
- Mute Line
- Dial Lamp
- S-Meter
- AVC FST/SLO
- Ext. Spkr. Jack
- Standby
- Bandspread
- RF Gain
- BFO
- Speaker 3x5"
- Antenna Trimmer
- DC Input Jack

Specifications:
Coverage ......535 - 30000 kHz
Selectivity ......4.5 kHz -6dB.
Sensitivity ......0.5 µV @ 30 MHz

Modes .... AM/SSB-CW
IF .... 455 kHz
Antenna Input . Terminals

Circuit Complement:
30 Semiconductors.

Accessories:
20-1500 SP-150 Speaker   20-1501 DC Pack (8 x D)

Comments:
Ranges: .535-1.6, 1.55-4.5, 4.5-13 and 13-30 MHz. Bandspread bands: 3.3-4.1, 4.85-5.1, 5.75-6.25, 6.6-7.4, 8.4-10, 9.5-12.5, 13.7-14.4, 14.7-15.5, 16.8-18.1, 19.8-22 and 25.5-30 MHz. Radio Shack #20-150

Made In:	Japan	1967-1969
Voltages:	120 VAC 50/60 Hz 10W or 12 VDC	
Readout:	Analog	
Physical:	14.2x6.5x9.25" 14 Lbs 360x165x235mm 6.4 kg	
Status:	Active Manufacturer Discontinued Model	
Rarity:	Common	
Reviews:	*Pop. Elec.* Dec. 1967 *QST* March 1968	

	New	Used
Price:	$120	$45-55
Rating:	★★	★

DX-150A

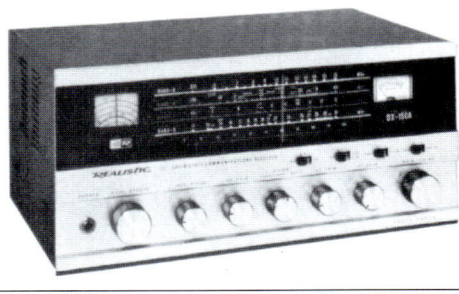

General Coverage Communications Receiver
Single Conversion Superheterodyne. Solid State.

Features:
- ¼" Head. Jack
- ANL
- Mute Line
- DC Input Jack
- S-Meter
- AVC FST/SLO
- Ext. Spkr. Jack
- Standby
- Bandspread
- RF Gain
- BFO
- Speaker 3x5"
- Antenna Trimmer
- Dial Lamp

Specifications:
Coverage ......535 - 30000 kHz
Selectivity ......4.5 kHz -6dB.
Sensitivity ......0.5 µV @ 30 MHz

Modes .... AM/SSB-CW

Antenna Input . Terminals

Circuit Complement:
Has FET front-end, product detector for SSB/CW. 30 Semiconductors.

Accessories:
20-1500 SP-150 Speaker   20-1501 DC Pack (8 x D)

Comments:
Main tuning bands: 535-1.6, 1.55-4.5, 4.5-13 and 13-30 MHz. Bandspread bands: 3.5-4, 7-7.3, 14-14.35, 21-21.4, 28-29.5 MHz and CB channels 1-23. Similar to the DX-150, but with a FET front-end. Radio Shack #20-150.

Made In:	Japan	1969-1972
Voltages:	120 VAC 50/60 Hz 10W or 12 VDC	
Readout:	Analog	
Physical:	14.2x6.5x9.25" 15 Lbs 360x165x235mm 6.8 kg	
Status:	Active Manufacturer Discontinued Model	
Rarity:	Common	
Reviews:	*Pop. Elec.* August 1970 *QST* September 1970	

	New	Used
Price:	$120	$55-65
Rating:	★★	★

REALISTIC

DX-150B

General Coverage Communications Receiver
Single Conversion Superheterodyne. Solid State.

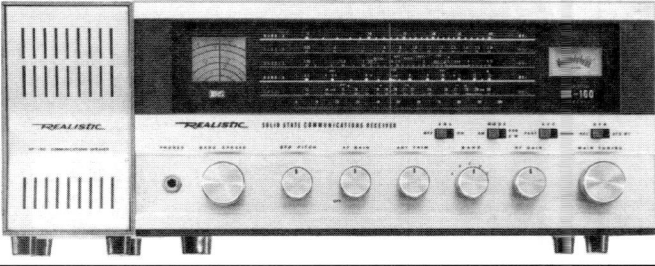

Features:
- ¼" Head. Jack
- ANL
- Mute Terminal
- Dial Lamp
- S-Meter
- AVC FST/SLO
- Ext. Spkr. Jack
- Bandspread
- RF Gain
- BFO
- Speaker
- Antenna Trimmer
- Standby

Specifications:
Coverage ...... 535 - 30000 kHz
Selectivity ...... 4.5 kHz -6dB.
Sensitivity ...... 0.5 µV @ 30 MHz

Modes .... AM/SSB-CW
Antenna Input . Terminals

Circuit Complement:
Utilizes five FETs and has a mechanical filter.

Accessories:
20-1501 DC Pack (8 x D)

Comments:
Ranges: .535-1.6, 1.55-4.5, 4.5-13 and 13-30 MHz. Bandspread bands: 3.3-4.1, 4.85-5.1, 5.75-6.25, 6.6-7.4, 8.4-10, 9.5-12.5, 13.7-14.4, 14.7-15.5, 16.8-18.1, 19.8-22 and 25.5-30 MHz. Supplied with matching external speaker. Radio Shack #20-151.

Made In:	Japan	1972-1974
Voltages:	120 VAC 50/60 Hz or 12 VDC	
Readout:	Analog	
Physical:	14.2x6.5x9.25" 15 Lbs 360x165x235mm 6.8 kg	
Status:	Active Manufacturer Discontinued Model	
Rarity:	Abundant	
Reviews:	*Pop. Elec.* Nov. 1972	

	New	Used
Price:	$120-140	$60-65
Rating:	★★	★★

REALISTIC

DX-160

General Coverage Communications Receiver
Single Conversion Superheterodyne. Solid State.

Features:
- ¼" Head. Jack
- ANL
- Mute Terminal
- Standby
- S-Meter
- AVC FST/SLO
- Ext. Spkr. Jack
- Bandspread
- RF Gain
- BFO
- External Speaker
- Antenna Trimmer
- Dial Lamp

Specifications:
Coverage ...... 150 - 30000 kHz[1]
Selectivity ...... 4 kHz -6dB.
Sensitivity ...... 3-4µV 16-30 MHz
Audio .... 0.7W 8 ohms

Modes .... AM/SSB-CW
Image Rej .... -20 to -35 dB 4.5-30 MHz
Antenna Input . Terminals

Circuit Complement:
1 Integrated circuit, 5 FETs, 6 transistors and 15 diodes.

Accessories:
20-1501 DC Pack (8 x D)

Comments:
Ranges: .15-.4, .535-1.6, 1.55-4.5, 4.5-13 and 13-30 MHz. [1]Coverage gap from .4-.535 MHz. Bandspread bands: 3.5-4, 7-7.3, 14-14.35, 21-21.4 and 28-29.7 MHz plus CB channels 1-23. Supplied with matching external speaker. Radio Shack #20-152. The European version operates from 220-240 VAC.

Made In:	Japan	1975-1980
Voltages:	120 VAC 50/60 Hz 6W or 12 VDC 30-180 ma	
Readout:	Analog	
Physical:	14.5x6.5x9.25" 15 Lbs 368x165x235mm 6.8 kg	
Status:	Active Manufacturer Discontinued Model	
Rarity:	Common	

	New	Used
Price:	$160	$70-75
Rating:	★★	★★

DX-200

General Coverage Communications Receiver
Single Conversion Superheterodyne. Solid State.

Features:
- ¼" Head. Jack
- S-Meter
- Bandspread
- Calibrator 500 kHz
- ANL
- AGC FST/SLO
- RF Gain
- Antenna Trimmer
- Mute Terminal
- Ext. Spkr. Jack
- Dial Lamp
- Standby

Specifications:
Coverage ...... 150 - 30000 kHz[1]
Selectivity ...... 4 kHz -6dB.
Sensitivity ...... 1µV (.25-21MHz).
S/N Ratio .... 45 dB @ 7MHz
Audio Out ...... 1.5 W 10% distortion
BFO Range ... ±3 kHz.

Modes .... AM/SSB-CW
Image Ratio .... 30 dB @ 7 MHz
Image Ratio .... 15 dB @ 21 MHz
Antenna Input . Terminals
IF Freq .... 455 kHz
Hum & Noise .. 3 mV at speaker

Circuit Complement:
4 Integrated circuits, 5 FETs, 8 transistors, 16 diodes and 5 LEDs.

Comments:
Ranges: .15-.4, .52-4.5, 4.5-13 and 13-30 MHz. [1] Coverage gap from .4-.52 MHz. Bandspread bands: 3.3-4.1, 4.85-5.1, 5.75-6.25, 6.6-7.4, 8.4-10, 9.5-12.5, 13.7-14.4, 14.7-15.5, 16.8-18.1, 19.8-22 and 25.5-30 MHz. Radio Shack #20-205. The European/Australian version operates from 220/240 VAC 50 Hz.

Made In:	Japan	1981-1983
Voltages:	120 VAC 50/60 Hz	
	12 VDC	
Readout:	Analog	
Physical:	14.5x6.5x9.25" 15 Lbs	
	368x165x235mm 6.8 kg	
Status:	Active Manufacturer	
	Discontinued Model	
Rarity:	Common	
Reviews:	*73* April 1981	

	New	Used
Price:	$230	$80-90
Rating:	★★	★★

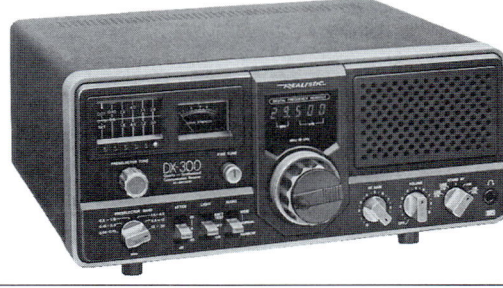

DX-300

General Coverage Communications Receiver
Triple Conversion Superheterodyne. Solid State.

Features:
- ¼" Head. Jack
- S-Meter
- Preselector
- Fine Tuning
- ANL
- RF Gain
- Dial Lamp
- Atten. -20/40 dB
- Mute Terminal
- Ext. Spkr. Jack
- Speaker
- Standby
- Record Jack
- Key Input Jack
- Light Switch
- Audio Filt. 3 Pos.

Specifications:
Coverage ...... 10 - 30000 kHz
Selectivity ...... 6 kHz -6dB.
Sensitivity .... 3µV (.9-28.1MHz).
Stability .... 1 kHz per hour
Audio Out ...... 1.5W 8 ohm 10% dist.
BFO Range ... ±3 kHz.

Modes .... AM/LSB/USB-CW
Image Ratio .... 70 dB @ 3.1-28.1 MHz
Image Ratio .... 80 dB @ .1-3.1 MHz
IF .... 54.5-55.5, 3-2 MHz, 455 kHz
Environment ... 0 to +43°C.
Antenna Input . SO-239 & Terminals

Circuit Complement:
4 Integrated circuits, 10 FETs, 27 transistors, 35 diodes and 1 thermistor.

Comments:
Can hold eight "C" cells for portable operation. This model lacked adequate stability and image rejection. Note that the switch for Wide/Normal/Narrow is merely an audio filter circuit and not an IF filter selection. Radio Shack #20-204. European version is 220/240 VAC 50 Hz.

Made In:	Japan	1979-1980
Voltages:	120 VAC 50/60 Hz or	
	12 VDC or 8 C cells.	
Readout:	00001. Digital LED	
Physical:	14.2x6.5x9.25" 13.5 Lbs	
	362x165x235mm 6 kg	
Status:	Active Manufacturer	
	Discontinued Model	
Rarity:	Abundant	
Reviews:	*WRTH* 1980	
	Pop. Elect. November 1979	
	73 September 1980	
	Comm. World 1980	

	New	Used
Price:	$380	$120-140
Rating:	★★	★

REALISTIC

DX-302

General Coverage Communications Receiver
Triple Conversion Superheterodyne. Solid State.

Features:
- ¼" Head. Jack
- ANL
- Mute Terminal
- Record Jack
- S-Meter
- RF Gain
- Ext. Spkr. Jack
- Standby
- Preselector
- Dial Lamp
- Speaker
- Light Switch
- Fine Tuning
- Atten. -20/40 dB
- Key Input Jack

Specifications:

Coverage ...... 10 - 30000 kHz
Selectivity ...... 7/5 kHz -6dB.
Sensitivity .... 3µV 3.1-28.1MHz.
Stability .... 1 kHz per hour
Audio Out ...... 1.5W 8 ohm 10% dist.
BFO Range ... ±1 kHz.

Modes .... AM/LSB/USB-CW
Image Ratio .... 60 dB @ 3.1-28.1 MHz
Image Ratio .... 70 dB @ .1-3.1 MHz
IF .... 54.5-55.5, 3-2 MHz, 455 kHz
Environment ... 0 to +43°C.
Antenna Input . SO-239 & Terminals

Circuit Complement:
4 Integrated circuits, 39 transistors and 35 diodes.

Comments:
Can hold eight "C" cells for portable operation. This model lacked adequate stability and image rejection. Note that, unlike the DX-300, the switch for Wide/Narrow on the DX-302 is for IF filter selection. Radio Shack #20-220. European version is 220/240 VAC 50 Hz.

Made In:	Japan	198 -1982
Voltages:	120 VAC 50/60 Hz or 12 VDC or 8 C cells.	
Readout:	00001.	Digital LED
Physical:	14.2x6.5x9.25" 13.5 Lbs 362x165x235mm 6 kg	
Status:	Active Manufacture Discontinued Model	
Rarity:	Abundant	
Reviews:	QST August 1981 QST September 1981 73 November 1980	

	New	Used
Price:	$400	$150-170
Rating:	★★★	★★

Radio Shack®

DX-394

General Coverage Communications Receiver
Double Conversion Superheterodyne. Solid State.

Features:
- Mini Head. Jack
- NB
- Speaker 3"
- Record Jack
- Lock
- S-Meter
- Keypad
- Speaker Jack
- Atten. -20 dB
- Sleep 30/60 min
- 160 Memories
- RF Gain
- Scan
- Standby
- VRIT Tuning
- Fine Tuning (10 Hz)
- Timer (5 Event)
- Sweep
- Clock Dual Time
- Telescopic Antenna

Specifications:

Coverage ...... 150 - 30000 kHz
Selectivity ...... 7.2/6/5.7 kHz
Sensitivity .... 1µV CW
IF .... 45 MHz, 455 kHz
Audio Out ...... 0.8W 10% THD

Modes .... AM/LSB/USB/CW
Image Reject .. >80 dB
IF Reject .... >80 dB
Antenna Input . SO-239 & RCA

Comments:
Tuning step rate: .1, 1, 5, and 10 (9) kHz. The LCD is backlit. The bandwidths are mode-dependent. 7 kHz is for AM and 5.7 or 6 kHz is for SSB. The AC supply is built-in. The DX-394 makes a nice bedside radio. Radio Shack #20-224.

Made In:	Japan	1995-1998
Voltages:	120 VAC 60 Hz 13W or 13.8 VDC 8W	
Readout:	00000.1	Digital LCD
Physical:	9.125x3.5x7.785" 4.6 Lbs 232x89x198mm 2 kg	
Status:	Active Manufacture Active Model	
Rarity:	Common	
Reviews:	Passport 1997-1998 WRTH 1996 Mon. Times June 1996 NASWA January 1996 QST October 1997	

	New	Used
Price:	$250-400	$140-150
Rating:	★★★	★★★

---

74 Redifon Ltd.

The **Redifon Company** has been a leader in British communications equipment for decades.

Several years ago, the Redifon Company was acquired by the French electronics giant Thomson-CSF. Equipment continues to be manufactured under the Redifon name.

The author would like to hear from any reader who has information on the R1000 series.

Redifon is perhaps now best known for their production of flight simulators and air traffic control equipment.

Redifon Ltd.
Broomhill Rd.
London S.W.18
England 1950-1980

Rediffusion Radio Systems Ltd.
Newton Rd.
Crawley,
West Sussex RH10 2PY
England 1985-1998

Redifon

R50M

General Coverage Communications Receiver
Single Conversion Superheterodyne. 15 Tubes.

Features:
- ¼" Head. Jack
- S-Meter
- Standby
- 80:1 Ratio Tuning
- RF Gain
- BFO
- Mute Line
- Hinged Top Cover
- Logging Scale

Specifications:

Coverage ...... 13 - 32000 kHz
Selectivity ...... 5 Positions
Sensitivity ...... <1 to 5µV 10 dB S/N
Audio Out ...... 2 W 3 or 12 ohms
Modes .... AM/CW
IF .... See Comments
Antenna Input . 75 ohm

Circuit Complement:
EF39 RF Amp, EF39 RF Amp, ECH35 Frequency Changer, L63 Oscillator, (3) EF39 IF Amp, EB34 Detector AVC, EB34 NL, EF37 BFO, EF37 AF Amp and 6V6G Output. The external AC power supply uses: S130 Stabilizer and 5Z4G Rectifier. The IF is 465 kHz on all ranges except .0135-.026 and .24-.6 MHz. On these two bands the IF is 110 kHz.

Accessories:
DC Power Supplies for 6, 12, 24, 110 and 220 VDC.

Comments:
Ranges: .0135-.026, .095-.25, .24-.6, .585-1.55, 1.5-4, 3.8-8, 7.7-16 and 15.5-32 MHz. The standard R50M is supplied with a gray crackle metal cabinet. The power supply is in a separate chassis measuring 17x6.5x7" 26 Lbs.

Made In: England 1955-1957

Voltages: 100-125 or 200-250 VAC
40-60 Hz 80 W

Readout: Analog

Physical: Cabinet Version:
21x14.5x21.5" 92 Lbs.
530x370x550mm 41.5 kg
Rack Version:
19x12.25x21" 55 Lbs
485x310x530mm 25 kg

Status: Active Manufacturer
Discontinued Model

Rarity: Typically Unavailable

	New	Used
Price:	①	⑤
Rating:	⑥	⑥

Redifon

R146

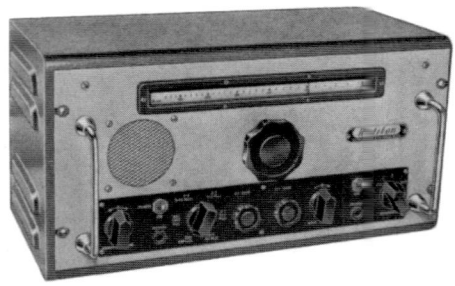

General Coverage Communications Receiver
Single Conversion Superheterodyne. 9 Tubes.

Features:
- ¼" Head. Jack
- S-Meter
- Standby
- Antenna Trimmer
- RF Gain
- BFO ±4 kHz
- Mute Line
- Rack Handles
- Speaker
- Speaker Output

Specifications:

Coverage ...... 250 - 24000 kHz	Modes .... AM/CW
Selectivity ...... 12/7/1 kHz -6 dB	IF .... 570 kHz
Sensitivity ...... <10-100µV 50 mw out	Image Rej. ...... 60 dB @ 4 MHz
Audio Out ...... 0.05 W 3 ohms	Image Rej. ...... 45 dB @ 13 MHz
IF Reject .... 40 dB	Image Rej. ...... 22 dB @ 20 MHz

Circuit Complement:
12AC6 RF Amp, 12AH8 Mixer, 12AU6 1st Oscillator, 12AC6 IF Amp, 12AC6 IF Amp, 12AU6 BFO, 12AU6 Audio Driver, 25L6GT Audio Output and 25L6GT Audio Output.

Comments:
Ranges: .25-.53, .625-1.52, 1.5-3.75, 3.7-9.2, 9.1-16.6 and 16.5-24 MHz. Note the reception gap from 530 to 625 kHz (note IF). Only the band-in-use is shown on the turret type display. The R146 was designed for marine reserve and general use. The power unit required for AC Mains or DC mains operation is external.

Made In: England 1958-1961

Voltages: 110/125 or 200/250 VDC or 110/125 or 200/250 VAC 50/60 Hz or 24 VDC.

Readout:  Analog

Physical: 17x8.75x10.75" 28 Lbs 432x222x273mm 12.7 kg

Status: Active Manufacture Discontinued Mode

Rarity: Typically Unavailable

	New	Used
Price:	①	€
Rating:	⑥	€

Redifon

R408

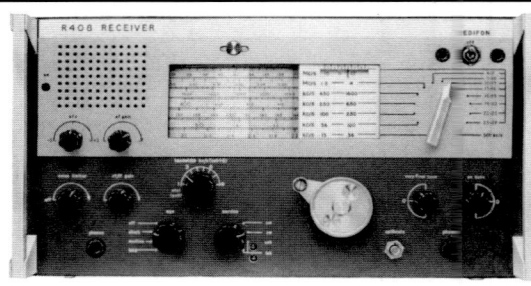

General Coverage Communications Receiver
Double Conversion Superheterodyne. Solid State.

Features:
- ¼" Head. Jack
- BFO ±3 kHz
- Speaker
- 500 kHz Preset
- RF Gain
- ANL (Variable)
- Antenna Trimmer
- Calibrator 100 kHz
- Mute
- Fine Tuning
- Semi-Modular
- AGC OFF/SHT/MED/LNG
- IF Out Jack
- AGC Out Jack
- Dial Lamp
- Line Out 600 ohm
- Rack Handles
- Fiduciary Adjust
- Ext. Speaker Jack

Specifications:

Coverage ...... 13 - 28000 kHz	Modes .... AM/CW/LSB/USB
Selectivity ...... 8/3/1 kHz -6dB.	Stability .... ±40 Hz after warm-up
Sensitivity ...... <1µV above 650 kHz	IF Rejection .... >90 dB 3-28 MHz
Image Rej ...... >70 dB 3-28 MHz	IF .... See Comments
Audio Out ...... 1.5 W 3 ohms	Environment ... -15° to +55° C. 95% Humid
BFO Range ... ±3 kHz.	Antenna Input . 75 ohm

Comments:
Ranges: .013-.036, .036-.1, .1-.25, .25-.65, .65-1.6, 1.5-4, 4-7, 7-10, 10-13, 13-16, 16-19, 19-22, 22-25 and 25-28 MHz. Single conversion below 650 kHz. A 90 inch film tuning scale, in conjunction with the 100 kHz crystal calibrator, provides very accurate frequency readout. A 24 VDC version of this receiver was also produced. The first IF is 470, 1500 or 4500 kHz depending on frequency. The second IF is 80 kHz. Selectivity is continuously variable from 800 Hz to 8 kHz for AM and 800 Hz to 4 kHz for SSB. This receiver is designed for the marine market.
Variants: This model was also sold as the Hagenuk E 408 R.

Made In: England

Voltages: 110/125 or 200/250 VAC 50/60 Hz 17W

Readout: Analog

Physical: 17.5x8.75x19.75" 55 Lbs. 445x223x502mm 25 kg

Status: Active Manufacture Discontinued Mode

Rarity: Typically Unavailable

	New	Used
Price:	①	€
Rating:	⑥	€

REDIFON SPT

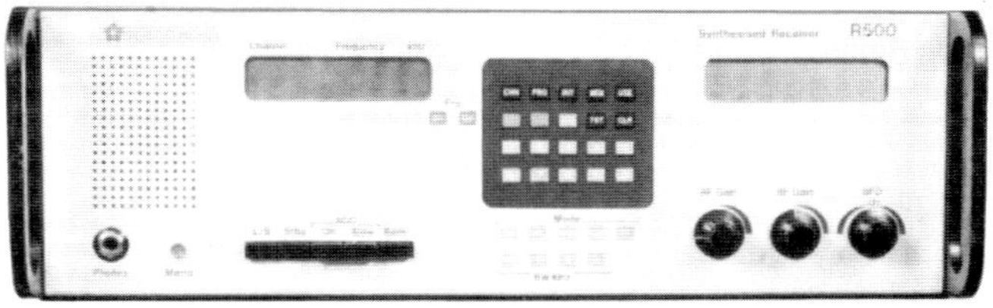

R500

General Coverage Communications Receiver
Single Conversion Superheterodyne. Solid State.

Features:
- ¼" Head. Jack
- RF Gain
- Mute
- Line Output
- Speaker
- 63 Memories
- Standby
- Up/Down Tuning
- BFO
- Rack Handles
- Keypad
- AGC ON/SLO
- Modular Construction
- Fine Tuning

Specifications:
Coverage ...... 10 - 30000 kHz
Selectivity ...... 6/3/1/.3 kHz -6dB.
Sensitivity ...... <1µV 10 dB S/N
IF Rej. .... 100 dB (2-20 MHz)
Audio .... 8 ohms
Modes .... AM/CW/LSB/USB/MCW
Stability .... 5 x 10^{-8} from -10 to +55° C
Image Rej. ...... 80 dB (2-20 MHz)
Environment ... -10° to +55° C. 95% Humid

Accessories:
FSK Option   Remote Operation Option   24 VDC Option

Comments:
This receiver performs with reduced specifications from 10 to 60 and 1000 to 1600 kHz. Tuning is via keypad or memories. Fine tuning (10 Hz) via Up and Down keys is also supported. The receiver mutes during tuning. The left LCD indicates channel and frequencies. The right LCD indicates: mode, bandwidth, mute status, AGC and other control parameters. RF tuning is accomplished by the microprocessor and a motor drive system.

Variants:
Model **R500N** supplied to New Zealand Navy.

Made In: England 1981-1998

Voltages: 100/105/110/115/120/125 or 200/210/220/230/240/250VAC 47-63 Hz 60W

Readout: 00000.01 Digital LCD

Physical: 19x5.25x17.5" 41.4 Lbs. 483x133x445mm 18.8 kg

Status: Active Manufacturer
Active Model

Rarity: Typically Unavailable

	New	Used
Price:	①	⑤
Rating:	⑥	⑥

REDIFFUSION
Radio Systems

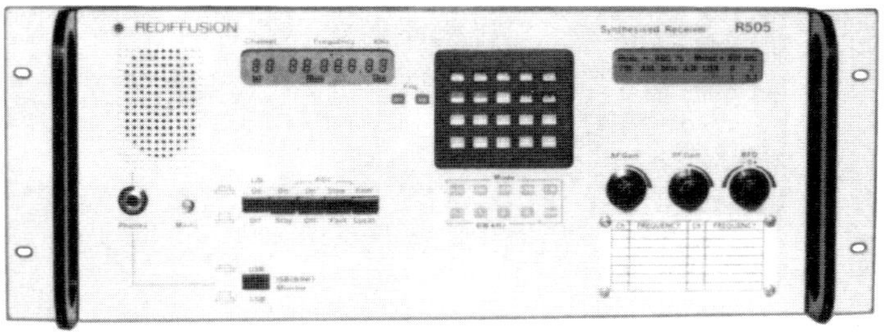

R505

General Coverage Communications Receiver
Single Conversion Superheterodyne. Solid State.

Features:
- ¼" Head. Jack
- RF Gain
- Mute
- Fine Tuning
- Speaker
- 63 Memories
- Standby
- BFO
- Rack Handles
- Keypad
- AGC ON/SLO
- Modular Construction
- LSB/USB Mon. Switch

Specifications:
Coverage ...... 60 - 30000 kHz
Selectivity ...... 6/3/1/.3 kHz -6dB.
Sensitivity ...... <1µV 10 dB S/N
Modes .... AM/CW/LSB/USB/MCW/ISB
Stability .... 5 x 10^{-8} from -10 to +55° C
Environment ... -10° to +45° C. 95% Humid

Accessories:
FSK Option   Remote Operation Option

Comments:
This receiver performs with reduced specifications from 1000 to 1600 kHz. Tuning is via keypad or memories. Fine tuning (10 Hz) via Up and Down keys is also supported. The receiver mutes during tuning. The left LCD indicates channel and frequencies. The right LCD indicates: mode, bandwidth, mute status, AGC and other control parameters. RF tuning is accomplished by the microprocessor and a motor drive system. Similar to the R500, but with ISB capability. The headphone jack can be selected to monitor either ISB channel.

Made In: England 1986-1998

Voltages: 100/105/110/115/120/125 or 200/210/220/230/240/250VAC 47-63 Hz 60W

Readout: 00000.01 Digital LCD

Physical: 19x7x17.5" 50 Lbs. 483x177x445mm 22.7 kg

Status: Active Manufacturer
Active Model

Rarity: Typically Unavailable

	New	Used
Price:	①	⑤
Rating:	⑥	⑥

REDIFON

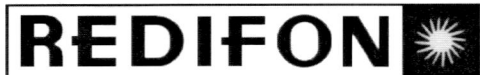

R550
Altair

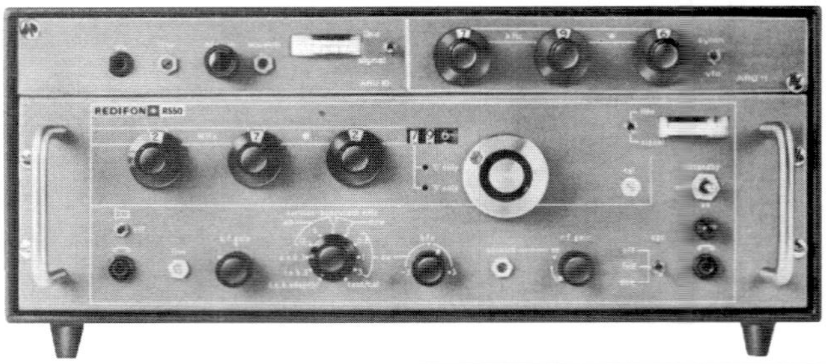

General Coverage Communications Receiver
Double Conversion Superheterodyne. Solid State.

Features:
• ¼" Head. Jack (2) • S/AF Meter • BFO ±3 kHz • AGC OFF/FST/SLO
• RF Gain • Squelch • Rack Handles • Modular Construction
• Mute • Standby • Sidetone • Line Out 600 ohm
• IF Out 1.4 MHz • Speaker • Ext Speaker Jack • Speaker Jack

Specifications:
Coverage ...... 200 - 30000 kHz
Selectivity ...... 12/6/3/1/.2 kHz -6dB.
Sensitivity ...... <1µV above 1 MHz
Image Rej ...... >80 dB
Audio Out ...... 1.5 W 3 ohms
BFO Range ... ±3 kHz.
Modes .... AM/CW/LSB/USB/MCW
Stability .... ±6 to ±40Hz after warm-up
IF .... 38, 1.4 MHz
Environment ... -15° to +45° C. 95% Humid
Antenna Input . 50 ohm

Accessories:
ARU.10 ISB Adapter   ARU.11 Synthesizer (shown in photo)

Comments:
Reception down to 80 kHz is possible but with reduced performance. Frequency selection is via decadic knobs for the 10 MHz, 1 MHz and 100 kHz positions. Mechanical digital readout via a VFO is provided for the 10 kHz, 1 kHz and 100 Hz positions. With the optional ARU.11 Frequency synthesizer (shown above on top of the R550 receiver) the 10 kHz, 1 kHz and 100 Hz positions may be directly selected by decadic knobs or the VFO. Designed for maritime and commercial applications.
Variants: Model **R551N** is a similar configuration covering 100 - 30000 kHz.

Made In: England

Voltages: 100/125 200/250 VAC
47-63 Hz 40W

Readout: ●●● 00.1 Digital

Physical: 19x5x15" 35 Lbs
483x127x380mm 17 kg

Status: Active Manufacturer
Discontinued Model

Rarity: Typically Unavailable

	New	Used
Price:	$3000-4000	⑤
Rating:	⑥	⑥

REDIFON

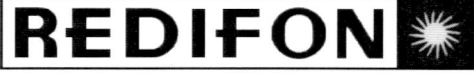

R1007

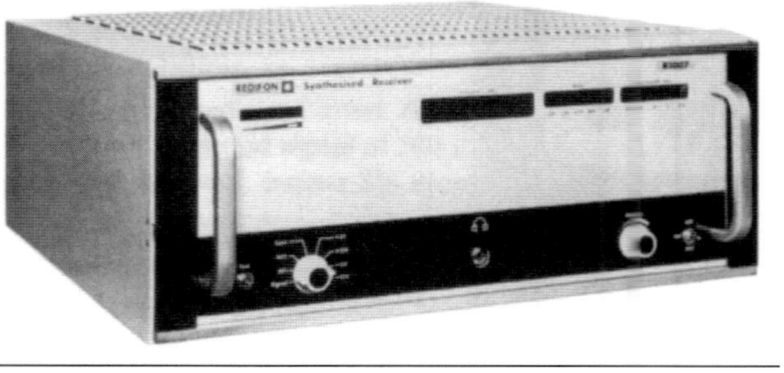

General Coverage Communications Receiver
Double Conversion Superheterodyne. Solid State.

Features:
• ¼" Head. Jack • S Indicator • 96 Memories • Rack Handles

Specifications:
Coverage ...... 15 - 30000 kHz   Modes .... AM/CW/LSB/USB/MCW

Comments:
The Redifon R1007 is a remote control receiver. The only local controls are a mode selection switch and an AF Gain. The R1007 must therefore be operated by the RC1000 Remote Unit. Up to 99 receivers may be controlled. Control of the R1007 can be effected over any distance using ordinary phone lines or radio links as the data transfer rate has been kept as low as 300 baud (300-9600 baud). The use of ASCII code for control information ensures that the R1007 may be interfaced with standard computer peripherals. The RC1000 Remote Control (shown below) allows complete remote capability. It features manual tuning, keypad entry, 10 Hz frequency readout, mode selection, RF gain, BFO, S Indicator, speaker, speaker switch etc. Please note that although the RC1000 looks like a receiver, it is merely a controller.

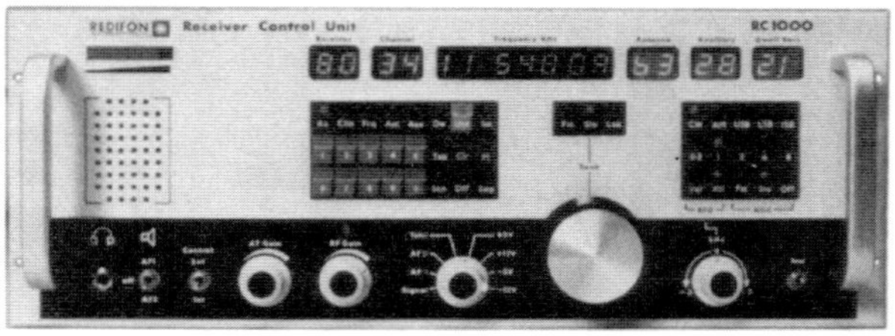

Made In: England 1981

Voltages: 100/125 200/250 VAC
47-63 Hz

Readout: 00000.01 Digital LED

Physical: 19x7x?"
483x180x?mm

Status: Active Manufacturer
Discontinued Model

Rarity: Typically Unavailable

	New	Used
Price:	①	⑤
Rating:	⑥	⑥

75 Rees Mace Marine

The **Rees Mace Marine Company** produced marine communications equipment for the British Royal Navy. The CAT receiver shown below was part of a system called 619/CAT with the 619 consisting of the MF/HF transmitters and the CAT being the receiver. The Murphy Company (please see *Chapter 58*) also built this configuration as the 618/CAS with the 618 again being the transmitter side, and the CAS being the receiver side.

The Pye Company acquired Rees Mace Marine in the early fifties. The CAT can also be found badged with the Pye label.

The Rees Mace Marine logo is particularly interesting with the pair of dolphins and Neptune's spear.

The author wishes to thank Geoffrey Arnold of *Radio Bygones* for his permission to use material from Ben Nock's article on Rees Mace that appeared in issue #49.

Rees Mace Marine
Oulton Broad Plant
Lowestoft, England

REES MACE MARINE

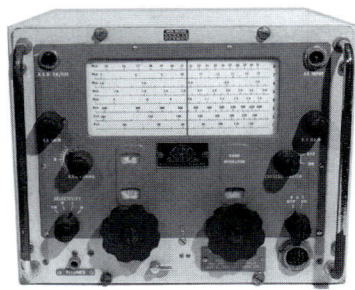

CAT

General Coverage Communications Receiver
Double Conversion Superheterodyne. 12 Tubes.

Features:
- ¼" Head. Jack
- RF Gain
- Ext. Spkr Jack
- Dial Lock
- ANL
- BFO ±5 kHz
- AGC ON/OFF
- Mute Line
- Line Output
- Rack Handles
- 48:1 Ratio Tuning
- Logging Scale

Specifications:
Coverage ...... 555 - 31000 kHz
Selectivity ...... 6/4/.95/.7 kHz -6dB
Audio Out ...... 500 ohms 2W
Modes .... AM/CW
IFs .... 1.4 MHz, 460 kHz
Antenna Input . Coaxial on Front Panel

Circuit Complement:
6BA6 RF Amp, ECH81 1st Mixer/Crystal Oscillator, ECH81 2nd Mixer/2nd Local Oscillator, EF92 IF Amp, EF92 IF Amp, EB91 Detector/AGC, EB91 Noise Limiter, EF92 AF Amp, 6CH6 AF Amp, QS150/45 HT Regulator and EF92 BFO.

Accessories:
Speaker

Comments:
Ranges: .055-.125, .1-.26, .26-.66, .65-1.55, 1.5-3.4, 3.4-7.2, 7-15 and 15-31 MHz. The handbook reference number is BR2169. The antenna input, headphone and audio outputs are all available via the front panel. Physical dimensions indicated are approximate.

Variants:
This model was also sold as the **Pye CAT**. The **Murphy CAS** was similar.

Made In:	England	1952-1957
Voltages:	Requires 6 VAC 3.5A and 245 VDC at 135 ma.	
Readout:	Analog	
Physical:	12x12x12" 44 Lbs 305x305x35mm 20 kg	
Status:	Inactive Manufacturer Discontinued Model	
Rarity:	Typically Not Available	
Reviews:	*Radio Bygones* #49	

	New	Used
Price:	①	⑤
Rating:	⑥	⑥

Since the inception of the company in 1961, **Harris RF Communications** has been providing breakthrough communications technology to the navies of the world.

Harris is the recognized world leader in the development of revolutionary H.F. waveforms, many of which have been adopted as international military standards. Harris pioneered the serial-tone modem that makes fast, error-free data transmission over H.F. radio. The company co-developed Automatic Link Establishment (ALE) which has simplified the operation of HF radio systems.

Harris Corporation has supplied a large number of models RF-590 and R-2368 to the United States Navy and other government agencies.

RF Communications, Inc.
1680 University Ave.
Rochester 10, NY   1964-1970

Harris Corporation
RF Communications Group
1680 University Ave.
Rochester, NY 14610   1985-1998

 RF COMMUNICATIONS, INC.

RF-501

Single Channel Single Sideband Receiver
Double Conversion Superheterodyne.

Features:
- ¼" Head. Jack
- RF Gain
- Rack Handles
- Line Out 600

Specifications:

Coverage ...... 1 Chan. 1.6 - 28 MHz.
Selectivity ...... 2.1 kHz -6dB.
Sensitivity ...... <1µV 10 db S/N
Audio Resp. .. 350 - 2450 Hz
Audio Out ...... 1 W 3.2 ohms

Modes .... SSB
Stability .... $\pm 3 \times 10^7$ per day.
IF Rejection .... 70 dB
Image Rej. ...... 65 dB
Antenna Input . 50 ohm

Comments:
The RF-501 receiver is capable of single channel, single sideband operation in the range of 1.6 to 28 MHz. The 2.1 kHz filter is mechanical. The RF-501 is designed for fixed commercial applications.

Made In: United States 1964-1966

Voltages: 115/230 VAC 50/60 Hz 65W

Readout:  No Display

Physical: 19x5.25x6" 15 Lbs. 483x133x152mm 6.8 kg

Status: Active Manufacturer Discontinued Model

Rarity: Very Scarce

	New	**Used**
Price:	①	$200
Rating:	⑥	⑥

 **RF COMMUNICATIONS, INC.**

RF-503

Six Channel Communications Receiver
Double Conversion Superheterodyne.

Features:
- ¼" Head. Jack
- S-Meter
- Rack Handles
- Line Out 600
- RF Gain
- Speaker 4x6"
- AGC FAST/SLOW

Specifications:

Coverage ......6 Chan. 1.6 - 25 MHz.	Modes .... AM/LSB/USB
Selectivity ......7/2.1 kHz -6dB.	Stability .... $\pm 2 \times 10^6$
Sensitivity ......<1µV 10 dB S/N SSB	IF Rejection .... 70 dB
Audio Resp. ..350 - 2450 Hz	Image Rej. ...... 65 dB
Audio Out ......1.5 W 3.2 ohms	Antenna Input . 50 ohm

Comments:
The RF-503 receiver is capable of six channel, LSB/USB/AM operation in the range of 1.6 to 25 MHz. The 2.1 kHz filter only is mechanical. The RF-503 is designed for fixed commercial applications.

Made In: United States 1964-

Voltages: 115/230 VAC 50/60 Hz 70W

Readout: ⌈ Channel ⌉ Channel

Physical: 19x7x8" 17 Lbs. 483x178x203mm 7.7 kg

Status: Active Manufacturer Discontinued Model

Rarity: Very Scarce

	New	Used
Price:	①	$250
Rating:	⑥	⑥

 RF COMMUNICATIONS, INC.

RF-505

General Coverage Communications Receiver
Double Conversion Superheterodyne. Solid State.

Features:
- ¼" Head. Jack
- S/AF Meter (2)
- Squelch
- NB
- RF Gain
- Preselector
- Rack Handles
- Speaker
- ISB
- Modular Construction

Specifications:

Coverage ......1600 - 30000 kHz	Modes .... AM/LSB/USB/CW/ISB
Dyn. Range ...125 dB	Stability .... ± 1 PPM
Sensitivity ......<.5µV 10 db S/N	IF Rejection .... -70 dB
Environment ..-18 to +55°C	Antenna Input . 50 ohm

Accessories:
Cabinet   RF-507 Manual Preselector

Comments:
The RF-505 is designed for fixed commercial applications, especially ISB. Frequency selection is via decadic knobs. Synthesizer tunes down to 100 Hz steps. [¥250,000].

Made In: United States 1969-1974

Voltages: 100-260 VAC 48-1000 Hz or 10-40 VDC

Readout: 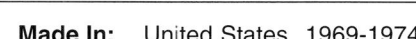 Mechanical

Physical: 19x5.25x13.825" 29.25 Lbs. 483x133x352 13.25 kg

Status: Active Manufacturer Discontinued Model

Rarity: Scarce

	New	Used
Price:	①	$250-420
Rating:	⑥	⑥

RF-505A

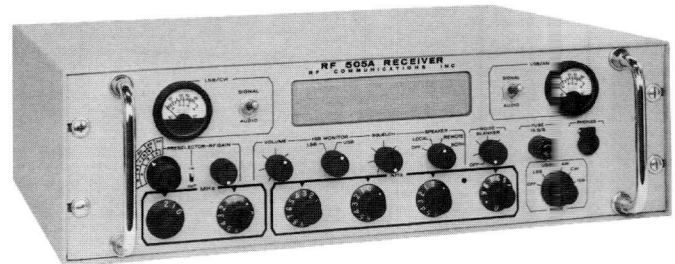

General Coverage Communications Receiver
Double Conversion Superheterodyne. Solid State.

Features:
- ¼" Head. Jack
- S/AF Meter (2)
- Squelch
- NB
- RF Gain
- Preselector
- Rack Handles
- Modular Construction
- Speaker
- ISB
- Ext. Freq. Standard

Specifications:

Coverage ...... 10 - 30000 kHz	Modes .... AM/LSB/USB/CW/ISB
Selectivity ...... 10/3.3/_ kHz -6dB.	Stability .... $\pm 1 \times 10^6$
Sensitivity ...... <.5µV	IF Rejection .... -70 dB
Audio Out ...... 3 W 3.2/600 ohm	Image Reject .. -70 dB
IF .... 156 MHz, 500 kHz	Environment ... -28 to +65°C 95% Humid.
Dyn. Range ... 125 dB	Antenna Input . 50 ohm

Accessories:

RF506 NB	RF507 Preselector	RF508 Hi Stability $\pm 1 \times 10^8$
RF509 CW Filter	RF511 Speaker	RF512 Desk Cabinet
RF513 Stack Kit	RF514 Rack Kit	RF520 Mute Module
RF522 RF Protector	RF523 RTTY Filter	RF524 RTTY Filter Narrow
RF525 SSB Filter	RF780 Remote Controller	

Comments:
Designed for fixed commercial applications, especially ISB. Frequency selection is via decadic knobs. Synthesizer tunes down to 100 Hz steps. The RF-505A is the companion to the RF-130 Transmitter. Some were produced in blue.

Made In: United States 1974-1981

Voltages: 100-260 VAC 48-1000 Hz
6W or 10-40 VDC

Readout: ●●●●●.① Mechanical

Physical: 19x5.25x13.825" 29.25 Lbs.
483x133x352 13.25 kg

Status: Active Manufacturer
Discontinued Model

Rarity: Common

	New	Used
Price:	$5320-8000	$500-600
Rating:	⑥	⑥

RF-550

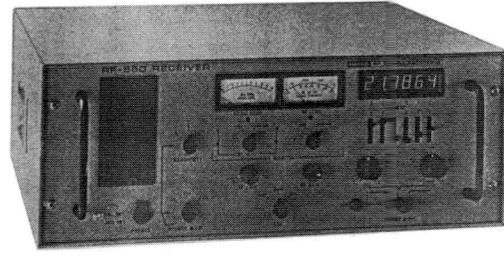

General Coverage Communications Receiver
Double Conversion Superheterodyne. Solid State.

Features:
- ¼" Head. Jack
- S-Meter
- AF Meter
- Line Out 600 ohm
- RF Gain
- Remote Port
- Rack Handles
- Modular Construction
- BFO ±1 kHz
- Speaker
- AGC Off/Fast/Slow/Coherent/External

Specifications:

Coverage ...... 100 - 30000 kHz	Modes .... AM/LSB/USB/CW/ISB
Selectivity ...... 20/6/3.2/.5/= kHz -3 dB	Stability .... $\pm 1 \times 10^6$ per day
Sensitivity ...... <.15 µV CW	Image Reject .. 100 dB
Audio Out ...... 2.5W 8 ohm 5% Dist.	Environment ... -10 to +55°C 95% Humid
IF .... 158.25, 1.75 MHz	Antenna Input . BNC 50 ohms

Circuit Complement:
7 plug-in modules and 11 plug-in printed wiring boards.

Accessories:

RF-551 Preselector	RF-557 Scan Control Unit	RF-560 1 MHz Standard
RF-562 ISB 4 Chan.	RF-794RE Remote Unit	.4 kHz Filter
2.8 kHz Filter	5.7 kHz Filter	1.2 kHz Filter

Comments:
This receiver employs a unique tuning method. Under each digit of the LED frequency display can be found a vertically oriented ten-position slide switch. Each digit is independently selected by sliding the switch to the correct position. The RF-550 matches the RF-130 transmitter.

Made In: United States 1977-1980

Voltages: 115/230 VAC 47-420 Hz
75 W

Readout: 00000.1 Digital LED

Physical: Cabinet Version
19.5x7.5x18.5" 48 Lbs
495x191x470mm 21.8 kg
Rack Version:
19x7x18.5" 45 Lbs
483x178x470mm 20.4 kg

Status: Active Manufacturer
Discontinued Model

Rarity: Typically Unavailable

	New	Used
Price:	$13,000	$650-790
Rating:	⑥	⑥

 **HARRIS**

RF-590

General Coverage Communications Receiver
Double Conversion Superheterodyne. Solid State.

Features:
- ¼" Head. Jack
- RF Gain
- Speaker
- B.I.T.E.
- AGC ON/OFF

- S/AF Meter
- Squelch
- Scan
- 1/5/10 MHz I/O
- Speaker Switch

- 100 Memories
- Rack Handles
- Sweep
- BFO ±9.99 kHz

- IF Out 455 kHz
- Modular Construction
- Speaker Out Jack
- Line Out 600 ohm

Specifications:
Coverage ...... 10 - 30000 kHz
Selectivity ...... 6/3.2/2.8/1/.3/_ kHz -3dB.
Sensitivity ...... <.15μV CW
Audio Out ...... 2W 8 ohm 5% Dist.
Image Reject . -100 dB

Modes .... AM/LSB/USB/CW
Stability .... ± 1 x 10^6
IF Rejection .... -100 dB
Environment ... -10 to +55°C 95% Humid.
Antenna .... BNC

Accessories:
RF-551A Preselector RF-575 Diversity Combiner RF-592 Remote Control Option
RF-593 High Stability RF-595-01 ISB Option RF-595-02 ISB Delay Option
RF-598 ISB-4 Option RF-597 Noise Blanker RF-596-01 Half-Oct. Filter

Comments:
Dwell time is adjustable. The RF-590 enjoys wide use in the American military, especially the Navy. The RF-590 also replaced the Collins 51S-1's in many American embassies worldwide. RF-590/A recently coming into the surplus market.
Variants: Model **RF-590A** is later production. Model **R-2368/URR** is a variant produced for the Navy. It includes two channel ISB (B8E) as standard.

Made In: United States 1982-1998

Voltages: 100-260 VAC 48-1000 Hz
6W or 10-40 VDC

Readout: `00000.001` Digital

Physical: 19x5.25x20.5" 40 Lbs.
483x133x520mm 18.5 kg

Status: Active Manufacturer
Active Model

Rarity: Scarce

	New	Used
Price:	$9700-17000	$1800-2200
Rating:	⑥	⑥

77 R.F.T. Köpenick

VEB Kombinat Nachrichtenelektronik
Köpenik
German Democratic Republic

R.F.T. Köpenick
Köpenik
German Democratic Republic

R.F.T. was a leading producer of military communications equipment in the former East Germany. The EKD series was the final grouping of receivers produced before the company ceased operation.

The EKV series preceded the EKD's and were properly classified as "boat anchors". Their size and weight dwarfed even the infamous Collins R-390s

Late in its life, the company also produced an interesting analog amateur band receiver called the AFE 12. This was sold either as a kit or fully assembled. It covered only the 80 and 160 meter bands in CW and LSB.

The EKD 700 was on the drawing board, but the Company was a casualty of the tumultuous political and economic changes that occurred in the former East Germany in the early 1990's. It has been rumored that another German communications manufacturer bought R.F.T. just to scrap it, and/or eliminate any patent conflict issues.

AFE 12

Amateur Band Communications Receiver
Single Conversion Superheterodyne. Solid State.

Features:
- ¼" Head. Jack
- S-Meter
- BFO ±200 Hz
- Speaker
- RF Gain
- Speaker Jack

Specifications:

Coverage	See Comments.	Modes	LSB/CW
Sensitivity	<1µV 10 dB S/N CW	IF	200 kHz
Selectivity	2.35 kHz	Antenna Input	75 ohm

Comments:
Coverage: 1.81-1.95, 3.5-3.8 MHz. Please note the very limited amateur band coverage and note that only the LSB and CW modes of reception are supported. This radio was sold assembled and in kit form.

Made In: East Germany

Voltages: 12.6 VDC

Readout: Analog Lin.

Physical: 13.7x4.3x9" 6.6 Lbs"
350x110x230mm 3 kg

Status: Inactive Manufacturer
Discontinued Model

Rarity: Typically Not Available

	New	Used
Price:	①	⑤
Rating:	⑥	⑥

EKD 100

General Coverage Communications Receiver
Double Conversion Superheterodyne. Solid State.

Features:
- Head. Jack
- RF Gain
- Speaker Jack
- Sync. Detection
- S-Meter
- Fine Tuning
- Line Out Jack
- Rack Handles
- BFO
- IF Out Jack
- Speaker
- Modular Construction
- FAX/RTTY Demod.

Specifications:

Coverage ......14 - 30000 kHz	Modes .... AM/SSB/CW/FAX/ISB
Selectivity ......3.4/2.7/6/3/1.4/.5/.1 kHz	Environment ... -25° to +55° C.

Comments:
Tuning is via ten position (decade) thumbwheels down to 10 Hz. Produced for the East German military, especially the Navy.

Made In:	East Germany 1982 A
Voltages:	115/125/220/240 VAC 50/60 Hz or 12/24 VDC
Readout:	00000.01 Digital
Physical:	Cabinet Version: 21.25x7.8x15.55" 540x198x395mm
Status:	Inactive Manufacturer Discontinued Model
Rarity:	Extremely Scarce

	New	Used
Price:	①	⑤
Rating:	⑥	⑥

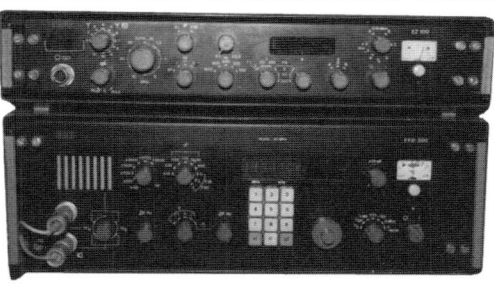

EKD 300

General Coverage Communications Receiver
Double Conversion Superheterodyne. Solid State.

Features:
- Head. Jack
- RF Gain
- Speaker
- IF Out Jack
- S-Meter
- Fine Tuning
- BFO
- Sync. Detection
- Rack Handles
- Keypad
- Line Out Jack
- Modular Construction
- Speaker Jack
- FAX/RTTY Demod.

Specifications:

Coverage ......14 - 30000 kHz	Modes .... AM/SSB/CW/FAX/ISB
Selectivity ......3.4/2.7/6/3/1.4/.5/.1 kHz	Stability .... 5×10^{-7} -10 to +50° C.
Sensitivity ......<.5µV CW	IF Rejection .... >80 dB
Image Rej ......>80 dB	Environment ... -25° to +55° C.
Audio Out ......0.5 W 8 ohms	

Accessories:
EZ100 Preselector/Diversity/FSK Mux. (shown)
EZ111 Preselector/Diversity/FSK Mux.

Comments:
Similar to the EKD100, but with a manual tuning knob and keypad entry.

Made In:	East Germany 1986 A
Voltages:	115/125/220/240 VAC 50/60 Hz or 12/24 VDC
Readout:	00000.01 Digital
Physical:	Cabinet Version: 21.25x7.8x15.55" 540x198x395mm
Status:	Inactive Manufacturer Discontinued Model
Rarity:	Extremely Scarce
Reviews:	*Funk* December 1991

	New	Used
Price:	①	⑤
Rating:	⑥	⑥

EKD 511

General Coverage Communications Receiver
Double Conversion Superheterodyne. Solid State.

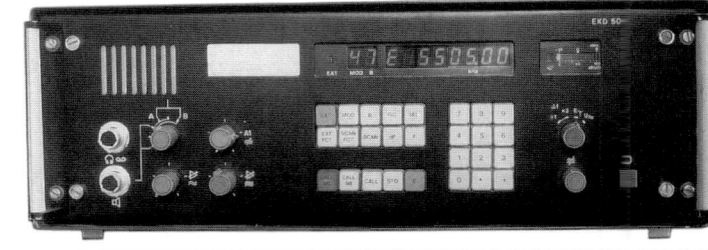

Features:

• Head. Jack	• S-Indicator	• Rack Handles	• Modular Construction
• RF Gain	• Fine Tuning	• Keypad	• Speaker Jack
• Speaker	• 99 Memories	• Serial Port	• Scan
• Sweep	• IF Out	• Sweep	• FAX/RTTY Demod.
• Clock	• Sync. Detection	• Line Out (2)	

Specifications:

Coverage ...... 14 - 30000 kHz	Modes .... AM/SSB/CW/FAX/ISB
Selectivity ...... See Comments.	Stability .... 5 x 10⁻⁷ -10 to +50° C.
Sensitivity ...... <.5µV CW	IF Rejection .... >80 dB
Image Rej ...... >80 dB	IF .... 70.2 MHz, 200 kHz
Audio Out ...... 0.5 W 8 ohms	Environment ... -25° to +55° C.

Stability: 5×10^{-7} -10 to +50° C.

Comments:
Bandwidths for the EKD 511: 6/3.1/1.75/.75/.4/.15 and +.25-3 and -.25-3 kHz. The EKD511 may be used as a master or slave receiver. It features built-in RTTY and FAX demodulators. The two line outputs are at 200 ohms. A built-in automatic preselector with 14 band-filters may be switched off in a bypass mode.

Variants:
Model **EKD 512** has slightly different bandwidths in position 7 and 8: 6/3.1/1.75/.75/.4/.15 and ±.25 - 3 kHz. The model **EKD 514** covers 10 - 30000 kHz with SSB bandwidths ±.25 - 3 kHz. Model **EKD 515** also covers 10 - 30000 kHz, but with SSB bandwidths ±.25 - 6 kHz.

Made In:	East Germany 1987 A
Voltages:	127/220 VAC 55W
	45/65 Hz or 12/24 VDC
Readout:	**00000.01** Digital
Physical:	Cabinet Version:
	21.25x7.8x15.55" 55 Lbs.
	540x198x395mm 25 kg
Status:	Inactive Manufacturer
	Discontinued Model
Rarity:	Extremely Scarce
Reviews:	*Funk* May 1992

	New	Used
Price:	3500 DM	$1500
Rating:	⑥	⑥

EKV 10

General Coverage Communications Receiver
Triple Conversion Superheterodyne. Tubes.

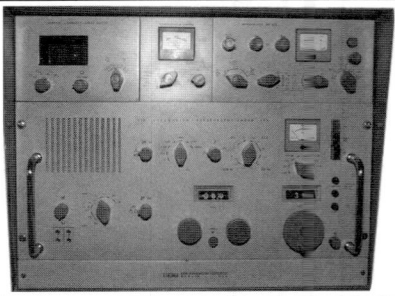

Features:

• Head. Jack (2)	• S-Meter	• Rack Handles	• Speaker
• RF Gain	• Squelch	• Line Out Jack	• IF Out Jack
• Speaker Jack			

Specifications:

Coverage ...... 1600 - 30000 kHz	Modes .... AM/SSB/CW/
Sensitivity ...... <.5µV CW	Stability .... <100 Hz after 2 hours
Selectivity ...... 6/2.7/3/1.4/.5/.15/_ kHz	IFs .... 38.3, 3.2, 0.2 MHz
Audio .... 12 ohms .5W	Antenna Input . 75 ohm

Accessories:

LZ01 LW Converter 14-535 kHz	DM031 External ISB Demod 250-6000 Hz
AAD02 Diversity Unit	DM032 External ISB Demod 300-3400 Hz
DM011 FSK Converter	DM023 Display Unit for FSK & Twinplex

Comments: Tuning is via three knobs. The first small round select the ten MHz and MHz digits. The second small round knobs selects the 100 kHz digit. The large round knob selects the 10 kHz, 1 kHz and 100 Hz digits. Preselector ranges: 1.6-3.3, 3.3-6.9, 6.9-14.4 and 14.4-30 MHz. Produced primarily for the East German military.

Variants: Model **EKV 11** includes LZ01, AAD, DM011 options. Model **EKV 12** includes AAD, DM023 and DM031. Model **EKV 13** (shown) includes: LZ01, AADI, DM023 and DM031. Model **EKV 14** includes AAD and DM023. Model **EKV 15** includes: AAD and DM011. Supplementary filters are different. Models **EKV 01, EKV 02** and **EKV 03** are configured in a small cabinets without the accessory units.

Made In:	East Germany
Voltages:	115/125/220/240 VAC
	50/60 Hz or 12/24 VDC
Readout:	**000 00.1** Digital Mech.
Physical:	20.8x16.3x16.1" 110 Lbs.
	530x415x410mm 51 kg
Status:	Inactive Manufacturer
	Discontinued Model
Rarity:	Typically Not Available

	New	Used
Price:	①	⑤
Rating:	⑥	⑥

The **Radio Manufacturing Engineers Company** was established in 1932. Their first receiver was the RME 9. The Company's best years were before World War II. Their most popular model was the RME 69. Please see *Chapter 105 Briefly Mentioned* for the RME 70.

The company merged with Electro-Voice in 1953. Electro-Voice is still in business.

The author is seeking further information on their last model, the RME 6902 allegedly produced in the early 1960's.

Radio Mfg. Engineers Inc.
304 First Ave.
Peoria 6, Illinois   1934-1949

Radio Mfg. Engineers
Div. of Electro-Voice Inc.
Washington, Illinois   1955-1956

RME Division
Electro-Voice Inc.
Buchanan, Michigan

RME

RME 43

General Coverage Communications Receiver
Single Conversion Superheterodyne. 9 Tubes.

Features:
- ¼" Head. Jack
- S-Meter
- AVC
- Bandspread
- RF Gain
- ANL
- Tone Switch
- Mute Line
- Standby
- Dial Lamp
- Ext. Speaker
- Two Tuning Speeds
- BFO
- Hinged Top Cover

Specifications:
Coverage ...... 550 - 33000 kHz
Selectivity ...... 5 Position
Audio .... 3 Watts
Modes .... AM/CW
IF .... 455 kHz.
Antenna Input . Terminals

Circuit Complement:
7B7 RF Amp, 7J7 Converter, 7B7 1st Amp, 7B7 2nd IF Amp, 7B6 Detector/BFO, 7C7 AF Amp, 7A6 NL, 7C5 Power Output and 80 Rectifier.

Accessories:
Speaker

Comments:
Ranges: .54-1.6, 1.6-2.9, 2.9-5.4, 5.4-9.8, 9.8-18 and 18-33 MHz. Requires a speaker. The bandspread is in the smaller inner white window.

Variants:
Earlier model **RME 41** was similar sans S-Meter and crystal filter.

Made In: United States 1945-1946

Voltages: 110-120 VAC 50/60 Hz

Readout: Analog

Physical: 22.19x12x11" 44.25 Lbs.
563x305x279mm 20 kg

Status: Inactive Manufacturer
Discontinued Model

Rarity: Scarce

	New	Used
Price:	$110	$120-220
Rating:	⑦	⑦

RME

RME 45

General Coverage Communications Receiver
Single Conversion Superheterodyne. 9 Tubes.

Features:
- ¼" Head. Jack
- S-Meter
- AVC
- Bandspread
- RF Gain
- ANL
- Tone Switch
- Mute Line
- Standby
- Dial Lamp
- Ext. Speaker
- Two Tuning Speeds
- BFO

Specifications:

Coverage ......550 - 33000 kHz		Modes .... AM/CW	
Selectivity ......5 Position		IF .... 455 kHz.	
Audio ....3 Watts		Antenna Input . Terminals	

Circuit Complement:
7B7 RF Amp, 7J7 Converter, 7B7 1st Amp, 7B7 2nd IF Amp, 7B6 Detector/BFO, 7C7 AF Amp, 7A6 NL, 7C5 Power Output and 80 Rectifier.

Accessories:
Speaker   DB-22A Preselector   NBF-4 NBFM Adapter
XC-2 Fixed Freq. Opt.

Comments:
Ranges: .54-1.6, 1.6-2.9, 2.9-5.4, 5.4-9.8, 9.8-18 and 18-33 MHz.   Requires a speaker. The bandspread is in the smaller inner white window.

Variants: A universal voltage variant was also produced which operated from 115/230 VAC 25/60 Hz. Model **RME 45B** had 10 tubes using a 5Y3G instead of an 80 plus adding VR-150 Regulator. It also featured two-speed tuning and calibrated bandspread 1947. Also see model RME 79.

Made In:	United States 19-6-1948
Voltages:	110-120 VAC 50/60 Hz 90W
Readout:	Analog
Physical:	22.19x12x11" 44.25 Lbs. 563x305x279mm 20 kg
Status:	Inactive Manufacturer Discontinued Model
Rarity:	Scarce
Reviews:	*Electric Radio* July 1996

	New	**Used**
Price:	$166-199	$100-220
Rating:	⑦	⑦

RME

RME 50

General Coverage Communications Receiver
Single Conversion Superheterodyne. 12 Tubes.

Features:
- ¼" Head. Jack
- S-Meter
- Speaker 5"
- Antenna Trimmer
- RF Gain
- ANL
- Tone Switch
- Mute Line
- Standby
- Bandspread
- Extl. Speaker
- Hinged Cover
- BFO
- Dial Lamp

Specifications:

Coverage ......550 - 33000 kHz		Modes .... AM/CW/NBFM	
Selectivity ......5 position		IF .... 455 kHz.	
Sensitivity ......< 2µV		Antenna Input . Terminals	
Audio Out ......1.5 W			

Circuit Complement:
6BJ6 RF Amp, 6U8A Osc/Mixer, 6BJ6 1st IF Amp, 6BJ6 2nd IF Amp, 6AU6 BFO, 6AU6 1st AF Amp, 6V6 AF Output, 6AL5 2nd Detector/ANL, 6AL5 Ratio Detector, 6BJ6 Limiter, VR-150 Voltage Regulator and 5Y3GT Rectifier.

Accessories:
DB22A Preselector in type "S" (small) or standard cabinet (both shown to the right).

Comments:
Gray metal cabinet. Sold with External Speaker 8".

Variants:
Model **RME 50A**.

Made In:	United States 1952-1954
Voltages:	115 VAC 50/60 Hz
Readout:	Analog
Physical:	22.19x12x11" 58 Lbs. 563x305x279mm 26.3 kg
Status:	Inactive Manufacturer Discontinued Model
Rarity:	Scarce

	New	**Used**
Price:	$188-197	$100-170
Rating:	⑦	⑦

RME 69

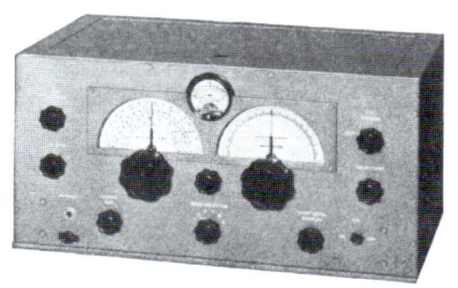

General Coverage Communications Receiver
Single Conversion Superheterodyne. 9 Tubes.

Features:
- ¼" Head. Jack
- S-Meter
- Speaker 5"
- Bandspread 0-100
- BFO
- Mute Line
- Standby
- Tone
- Hinged Cover

Specifications:
Coverage ......550 - 31500 kHz
IF ....465 kHz.
Audio Out ......2.5W 600/4000 ohms
Modes .... AM/CW
Antenna Input . Terminals

Circuit Complement:
6D6 RF Amp, 6C6 Mixer, 6BJ6 1st IF Amp, 6D6 IF, 6D6 IF, 6B7 Detector/AVC/Audio, 6D6 BFO, 42 AF Out and 80 Rectifier. Crystal filter.

Accessories: DB20 Preselector

Comments: Note that the intermediate frequency is 465 kHz, not 455 kHz. Ranges: .55-1.5, 1.43-3.1, 3.03-6.9, 6.1-13.2, 12.8-21.1 and 15.8-31.58 MHz.

Variants: There are four versions: Model **RME 69 AC** has no knob under the headphone jack. Model **RME 69 AC/Battery** (shown) has a small Battery Switch knob under the headphone jack and a BFO knob in the lower right. Model **RME 69 AC with Noise Silencer** has a BFO knob under the headphone jack and Noise Silencer knob in the lower right. The model **RME 69 AC/Battery** with **Noise Silencer** has a small battery switch knob under the headphone jack and a BFO knob with a concentric Silencer knob in the in the lower right. Most production was with a black cabinet with no lettering on the front panel to indicate the function of the knobs!

Made In:	United States 1936-1940
Voltages:	115 VAC 50/60 Hz
Readout:	Analog
Physical:	19x9.25x10.625" 32 Lbs. 483x235x270mm 14.5 kg
Status:	Inactive Manufacturer Discontinued Model
Rarity:	Scarce
Reviews:	*Electric Radio* May 1992

	New	Used
Price:	$135	$150-350
Rating:	⑦	⑦

RME 79

General Coverage Communications Receiver
Single Conversion Superheterodyne. 12 Tubes.

Features:
- ¼" Head. Jack
- S-Meter
- AVC
- Bandspread
- RF Gain
- ANL
- Tone Switch
- Mute Line
- Standby
- Dial Lamp
- Ext. Speaker
- Two Tuning Speeds
- BFO
- Hinged Cover

Specifications:
Coverage ......540 - 30000 kHz
Selectivity ......5 Position
Audio ....3 Watts
Modes .... AM/CW
Antenna Input . Terminals

Circuit Complement:
6BA6 RF Amp, 6BA6 RF Amp, 6BE6 Mixer, 6C4 HF Oscillator, 6BA6 IF, 6BA6 IF, 6AU6 BFO, 6AL5 2nd Detector/ANL, 6AU6 Audio, 6AQ5 Output, 5Y3GT Rectifier and VR150 Voltage Regulator.

Accessories:
RME-79S SSB Adapter

Comments:
Ranges: .54-1.6, 1.6-2.9, 2.9-5.4, 5.4-9.8, 9.8-18 and 18-30 MHz. This model is similar in appearance and has the identical knobs positions of the earlier RME 45 and RME 50, but features a geared "Dial-O-Matic" vernier tuning knob.

Made In:	United States 1953-1954
Voltages:	110-120 VAC 50/60 Hz
Readout:	Analog
Physical:	22x11x11" 58 Lbs 558x280x280mm 26.23 kg
Status:	Inactive Manufacturer Discontinued Model
Rarity:	Scarce

	New	Used
Price:	$287	$140-160
Rating:	⑦	⑦

RME 84

General Coverage Communications Receiver
Single Conversion Superheterodyne. 8 Tubes.

Features:
- ¼" Head. Jack
- Standby
- Speaker 5"
- Antenna Trimmer
- RF Gain
- ANL
- Tone Switch
- Mute Line
- BFO
- Hinged Top
- Dial Lamp
- Battery Input Jack

Specifications:

Coverage ......540 - 44000 kHz	Modes .... AM/CW
Selectivity ......2.8 kHz -6dB.	IF .... 455 kHz.
Sensitivity ......<2 µV	Antenna Input . Terminals
Audio Out ......1.1 W	

Circuit Complement:
7B7 RF Amp, 7S7 Converter, 7B7 1st IF Amp, 7B7 2nd IF Amp, 7K7 Detector/AVC/AF, 7K7 NL/BFO, 6G6G Audio Output and 5Y3GT Rectifier.

Accessories:
CM-1 External S-Meter   VP-5 Power Pack for 6 VDC
DB-22A Preselector   VHF-152 2/6/10 Meter Converter

Comments:
Ranges: .54-1.65, 1.65-5, 5-15 and 15-44 MHz

Variants:
Model **RME 84A** differences unknown.

Made In:	United States 1946-1949
Voltages:	110-120VAC 50/60 Hz 62W or A & B Batteries
Readout:	⊞ Analog
Physical:	18x9.5x9.75" 28 Lbs. 457x241x248mm 2.7 kg
Status:	Inactive Manufacturer Discontinued Model
Rarity:	Scarce

	New	Used
Price:	$98-110	$50-90
Rating:	⑦	○

RME 99

General Coverage Communications Receiver
Single Conversion Superheterodyne. 12 Tubes.

Features:
- ¼" Head. Jack
- S-Meter
- NL
- Bandspread
- BFO
- Mute Line
- Standby
- BS Logging Scale
- Dial Lamp
- Hinged Top Cover

Specifications:

Coverage ......550 - 33000 kHz	Modes .... AM/CW
Selectivity ......8.8/4/2.8/1/.4/.14 kHz	IF .... 465 kHz.
Audio Out ......2.5W 600/4000 ohms	Antenna Input . Terminals

Circuit Complement:
7A7 RF Amp, 7B8 1st Detector, 7A4 HF Oscillator, 7A7 1st IF Amp, 7A7 2nd IF Amp, 7A7 3rd IF Amp, 7F7 1st AF Amp/2nd Detector, 7C5 Output Amp, 7A4 BFO, 7A6 NL, VR150 Voltage Regulator and 80 Rectifier. Crystal filter.

Comments:
Ranges: .55-1.6, 1.6-2.95, 2.95-5.45, 5.45-9.8, 9.8-18.5 and 18.5-30 MHz. The large dial in the center of the radio is actually the bandspread dial covering the 80, 40, 20 and 10 meter amateur bands. The small Main Tuning knob and Main Tuning Dial are located to the right of the large center dial.

Variants: The **RME 99 Deluxe** has a sloping front panel.

Made In:	United States 1940-1941
Voltages:	115 VAC 50/60 Hz
Readout:	⊞ Analog
Physical:	19x9.25x10.625" 32 Lbs. 483x235x270mm 14.5 kg
Status:	Inactive Manufacturer Discontinued Model
Rarity:	Very Scarce
Reviews:	*Radio* June 1940

	New	Used
Price:	$138-139	○
Rating:	⑦	○

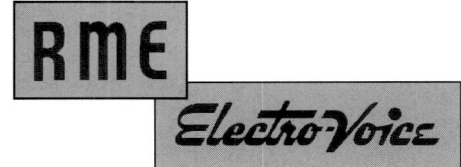

RME 4300

Amateur Band Communications Receiver
Single Conversion Superheterodyne. 8 Tubes.

Features:
- ¼" Head. Jack
- RF Gain
- Mute Terminals
- Standby
- S-Meter
- ANL
- Dial Lamp
- 2 Tuning Rates
- BFO
- Cal. Adjust
- Hinged Top
- Antenna Trimmer
- 75:1 Tune Ratio
- Ext. Spkr Terminals

Specifications:

Coverage ...... Ham. See Comments.	Modes .... AM/CW/MCW/SSB
Selectivity ...... 4 pos.	Image Ratio .... >40 dB <14.35 MHz
Sensitivity ...... <2µV 10 dB S/N	Image Ratio .... >25 dB >21 MHz
Calibration 0.02%	Drift .... .01%
Audio Out ...... 1.5 W 4 ohms	Antenna Input . Terminals 50-600 ohms
IF .... 455 kHz	

Circuit Complement:
6CB6 RF Amp, 6U8 Converter Osc., 6CB6 IF, 6U8 IF/BFO, 6T8 Detector/ANL/AF, 6AQ5 Output, 0A2 Regulator and 5Y3 Rectifier. One RF and two IF stages.

Accessories:
4302 Speaker   4301 Sideband Detector   8933 Calibrator 100 kHz

Comments:
Coverage: 1.8-2, 3.5-4, 7-7.3, 14-14.35, 21-21.5 and 27-29.7 MHz. SSB performance is enhanced with the 4301 Sideband Detector. Requires a speaker.

Made In: United States 1955-1957

Voltages: 117 VAC 50/60 Hz 65W

Readout: [dial] Analog

Physical: 16.5x10x10" 28 Lbs.
419x254x254mm 12.7 kg

Status: Inactive Manufacturer
Discontinued Model

Rarity: Scarce

Reviews: *QST* October 1956

	New	Used
Price:	$194	$100-160
Rating:	⑦	⑦

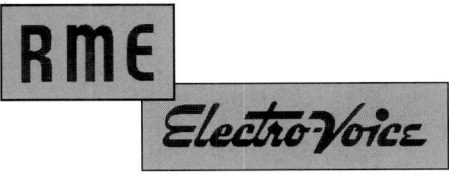

RME 4350

Amateur Band Communications Receiver
Double Conversion Superheterodyne. 9 Tubes.

Features:
- ¼" Head. Jack
- RF Gain
- Mute Terminals
- Standby
- S-Meter
- ANL
- 75:1 Tune Ratio
- BFO
- 2 Tuning Speeds
- Antenna Trimmer
- Hinged Top Cover
- Ext. Spkr Terminals
- 4301 I/O Jacks

Specifications:

Coverage ...... Ham. See Comments.	Modes .... AM/USB/LSB
Selectivity ...... 2.8 kHz -6dB.	Stability .... .01 per cent
Sensitivity ...... <2µV 10 dB S/N	Image Ratio .... >54 dB
Audio Out ...... 1.5 W 4	IF .... 2195, 455 kHz.
Calibration 0.02%	Antenna Input . Terminals

Circuit Complement:
6BZ6 RF Amp, 6U8A 1st Mixer, 6CB6 1st Amp, 6U8A 2nd Mixer, 6U8 2nd IF Amp/BFO, 6T8 2nd Detector/ANL/1st AF, 6AQ5 AF Output, 5Y3 Rectifier and 0A2 Voltage Regulator.

Accessories:
4302 Speaker   4301 Sideband Detector   8933 Calibrator 100 kHz

Comments:
Ranges: 1.8-2, 3.5-4, 7-7.3, 14-14.34, 21-21.5 and 27-29.7 MHz. Requires a speaker.

Made In: United States 1957-1958

Voltages: 117 VAC 50/60 Hz 65W

Readout: [dial] Analog

Physical: 16.5x10x10" 32 Lbs.
419x254x254mm 14.5 kg

Status: Inactive Manufacturer
Discontinued Model

Rarity: Scarce

Reviews: *CQ* September 1958

	New	Used
Price:	$229-249	$100-210
Rating:	⑦	⑦

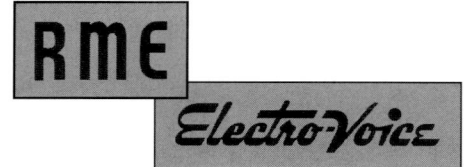

RME 4350A

Amateur Band Communications Receiver
Double Conversion Superheterodyne. 10 Tubes.

Features:
- ¼" Head. Jack
- S-Meter
- BFO
- Ext. Spkr Terminals
- RF Gain
- ANL
- 2 Tuning Speeds
- Calibrator 100 kHz.
- Mute Terminals
- 75:1 Tune Ratio
- Antenna Trimmer
- 4301 I/O Jacks
- Standby

Specifications:

Coverage ......Ham. See Comments.	Modes .... AM/CW[1]
Selectivity ......2.8 kHz -6dB.	Stability .... .01 per cent
Sensitivity ......<1 to 2µV 10 db S/N	Image Ratio .... >54 dB
Audio Out ......1.5 W 4 ohms	IF .... 2195, 455 kHz.
Calibration 0.02%	Antenna Input . Terminals

Circuit Complement:
6BZ6 RF Amp, 6U8A 1st Mixer, 6CB6 1st Amp, 6U8 2nd Mixer, 6U8 2nd IF Amp/BFO, 6T8 2nd Detector/ANL/1st AF, 6AQ5 AF Output, 5Y3 Rectifier, 0A2 Voltage Regulator and 6CB6 Crystal Calibrator.

Accessories:
4302 Speaker   4301 Sideband Detector   DB23 Preselector

Comments:
Ranges: 1.8-2, 3.5-4, 7-7.3, 14-14.34, 21-21.5 and 27-29.7 MHz. Note that the "A" model now includes the Calibrator, previously optional on the RME 4350. The toggle switch for this can be seen below the S-Meter. [1]The LSB/USB mode positions will not function without the optional 4301 Sideband Detector. Requires a speaker.

Made In:	United States 1958-1959
Voltages:	117 VAC 50/60 Hz 65W
Readout:	Analog
Physical:	16.5x10x10" 32 Lbs. 419x254x254mm 4.5 kg
Status:	Inactive Manufacturer Discontinued Model
Rarity:	Scarce
Reviews:	QST September 1958 CQ September 1958

	New	Used
Price:	$249	$100-240
Rating:	⑦	○

RME 6900

Amateur Band Communications Receiver
Double Conversion Superheterodyne. 12 Tubes plus Semiconductors.

Features:
- ¼" Head. Jack
- S-Meter
- T Notch
- Calibrator 100 kHz.
- RF Gain
- ANL
- Antenna Trimmer
- Two Tuning Speeds
- Mute Terminals
- Standby
- Line Out 500 ohm
- Ext. Spkr Terminals
- AGC
- Cal. Adjust
- 54:1 Tune Ratio
- Hinged Top Cover

Specifications:

Coverage ......Ham. See Comments.	Modes .... AM/USB/LSB
Selectivity ......3.6/2/.5 kHz -6dB.	Stability .... <.005% (after 15 mins.)
Sensitivity ......<1µV 10 dB S/N	Notch Rej. ...... >40 dB
Image Ratio ...56 dB	IF .... 2195, 57 kHz.
Audio Out ......1 W 4 ohms	Antenna Input . Terminals & SO-239

Circuit Complement:
6BA6 RF Amp, 6U8A 1st Mixer/Stabilizer. Osc, 6U8A 2nd Mixer/Crystal Osc, 6C4 1st IF & T Notch, 6BA6 2nd IF, 6BA6 3rd IF, 6AL5 NL, 6T8 2nd Detector/1st Audio, 6AQ5 AF Out, 6CB6 Crystal Calibrator Osc, 12AT7 Product Detector/BFO and 2 type 1N1763 diode power rectifiers.

Accessories:
6901 Speaker

Comments:
Coverage: 3.5-4, 7-7.3, 10-11, 14-14.4, 21-21.5 and 28-29.7 MHz. Features an 11½ inch drum type tuning dial. Requires a speaker. Gray and charcoal finish.

Mfg. In:	United States 1959-1962
Voltages:	117 VAC 50/60 Hz 55W
Readout:	Analog Lin.
Physical:	17x9.75x12.125" 35 Lbs. 432x248x308mm 5.9 kg
Status:	Inactive Manufacturer Discontinued Model
Rarity:	Very Scarce
Reviews:	QST February 1961

	New	Used
Price:	$350-369	$150-350
Rating:	⑦	○

The **Rohde & Schwarz Company** is an established and respected German manufacturer of communications equipment.

Rohde & Schwarz became best known to the listening community with the development (1953-1955) of the famous EK 07 general coverage receiver. This, the "Collins R-390" of Germany, was of the highest technology of the time and offered frequency accuracy and performance that was to be a standard for nearly twenty years.

Today Rohde & Schwarz is engaged in the development, manufacturing and marketing of professional radio communications products. They provide system solutions in the fields of: global communications, ATC and airport communications, avionics and naval communications.

Rohde & Schwarz GmbH & Co. KG
Postfach 80 14 69
D-81614 München
Germany   1979-1998

ROHDE & SCHWARZ

EK 07

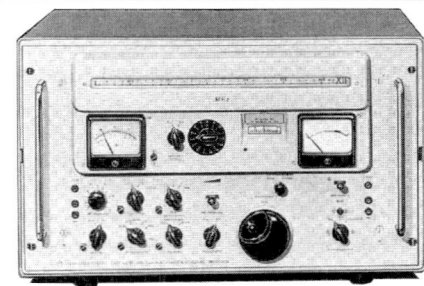

General Coverage Communications Receiver
Double Conversion Superheterodyne. 27 Tubes.

Features:
- ¼" Head. Jack (2)
- S&AF Meters
- NB
- Rack Handles
- RF Gain
- BFO ±3 kHz
- Flywheel Tuning
- 30:1 Tuning Ratio
- Calibrator
- Diversity Mode
- ANL
- Modular Construction
- Mute
- Osc. Out Jack
- Spinner Knob
- Line Out 600 ohm
- AGC (3 Pos.)
- IF Out Jack
- Tube Checker
- Flywheel

Specifications:
Coverage ...... 500 - 30100 kHz
Selectivity ...... 12/6/3/1.5/.6/.3 kHz
Sensitivity ...... <.2µV 3.1-30.1 MHz
Image Rej ...... >70 - 80 dB
Audio Output . 2W 15 ohm 10% THD

Modes .... AM/CW
IF .... 3.3 MHz and 300 kHz.
IF Rejection .... >75 dB
Antenna Input . Coaxial 50-75 ohm

Accessories: NZ-10 SSB Adapter

Comments: Ranges: 3.1-6.1, 6.1-9.1, 9.1-12.1, 12.1-15.1, 15.1-18.1, 18.1-21.1, 21.1-24.1, 24.1-27.1, 27.1-30.1 and .5-1.1, 1.1-2.1, 2.1-3.1 MHz. Turret-type band display. Dual concentric tuning knobs. Frequency accuracy is 1 kHz. The S Meter is calibrated in µV. Requires a speaker. Incredible craftsmanship is evident in this classic receiver. The meter on the left checks performance of each tube (an early version of B.I.T.E.). Sometimes called the R-390 of Germany, the workmanship is compared (better?) to Collins. One of the best tube radios ever made.

Variants: **EK 07/D** refers to the German version. Suffix numbers **/D1** and **/D2** refer to different connectors on the rear panel. The **EK 11** has a large ext. synthesizer.

Made In: Germany   1956-1966

Voltages: 115/125/220/235 VAC
47-63 Hz 130 W

Readout: Analog Lin.

Physical: 21.5x13x21.6" 146 Lbs.
540x330x550mm 66.3 kg

Status: Active Manufacturer
Discontinued Model

Rarity: Extremely Scarce

	New	Used
Price:	$6000	$2000-2500
Rating:	★★★★★	⑦

ROHDE & SCHWARZ

EK 047

General Coverage Communications Receiver
Double Conversion Superheterodyne. Solid State.

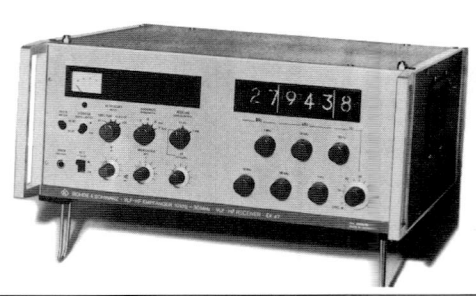

Features:
- ¼" Head. Jack
- S/AF Meter
- NB
- Rack Handles
- RF Gain
- BFO 1.2 kHz
- Tilt Bar
- Modular Construction
- Speaker
- Speaker Switch
- Semi-Modular
- Line Out 600 ohms

Specifications:
Coverage ...... 10 - 30000 kHz
Selectivity ...... _/_/_/_/_ kHz
Sensitivity ..... <.1µV CW
Image Rej ...... >80 dB
IFs .... 73.03 MHz, 30 kHz
Audio 2W

Modes .... ISB/AM/LSB/USB/CW
Stability .... 5x10⁻⁸ per month
Enviroment ..... -25 to +55° C. 95% Humid.
IF Rejection .... >90 dB
Antenna Input . 50 ohm

Accessories:
Preselector   Filters   NZ47 Telegraphy Demod.
Rack Mount

Comments:
Frequency tuning is achieved through separate knobs for: 10 MHz, 1 MHz, 100 kHz, 10 kHz, 1 kHz and 100 Hz. Available Siemens mechanical filter values were: 6, 3.1, 2.7, 1.5, 1, .6, .3 and .15 kHz.

Variants:
Also sold as **Siemens E 401.**

Made In:	Germany 1969-1975
Voltages:	100/120/220/240 VAC 45-60 Hz 84VA
Readout:	**00000.1** Digital Nixie
Physical:	17.5x7.7x18.1" 66 Lbs. 445x195x460 mm 30 kg
Status:	Active Manufacturer Discontinued Model
Rarity:	Extremely Scarce
Reviews:	*Funk* March 1989

	New	Used
Price:	20000 DM	3000 DM
Rating:	⑥	⑥

ROHDE & SCHWARZ

EK 049

General Coverage Communications Receiver
Double Conversion Superheterodyne. Solid State.

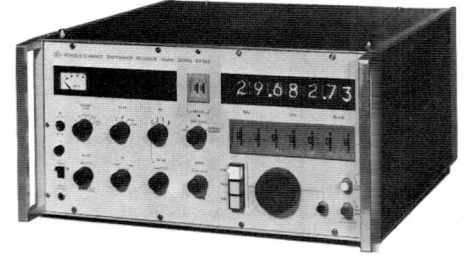

Features:
- ¼" Head. Jack
- S/AF Meter
- NB
- Rack Handles
- RF Gain
- BFO
- AGC (4 Pos.)
- Three Tuning Rates
- Remote Bus
- AFC
- Dial Lock

Specifications:
Coverage ...... 10 - 30000 kHz
Selectivity ...... 6 Position
Sensitivity ...... <.1µV CW
Image Rej ...... >80 dB
IF Rejection ... >80 dB

Modes .... ISB/AM/LSB/USB/CW
Stability .... 5x10⁻⁸ per month
Enviroment ..... -25 to +45° C. 95% Humid.
Antenna Input . 50 ohm

Accessories:
Preselector Unit   Filters   NZ Telegraphy Demodulator
Rack Mount   Notch Filter / AFC   N403 Remote Controller

Comments:
Frequency tuning is achieved through separate decadic thumbwheel knobs for: 10 MHz, 1 MHz, 100 kHz, 10 kHz, 1 kHz, 100 Hz and 10 Hz or via manual tuning knobs. Up to 30 receivers may be controlled remotely.

Variants:
See Siemens E403.

Made In:	Germany 1986-1988
Voltages:	100/120/220/240 VAC
Readout:	**00000.01** Digital Nixie
Physical:	
Status:	Active Manufacturer Discontinued Model
Rarity:	Extremely Scarce

	New	Used
Price:	①	⑤
Rating:	⑥	⑥

ROHDE & SCHWARZ

EK 056

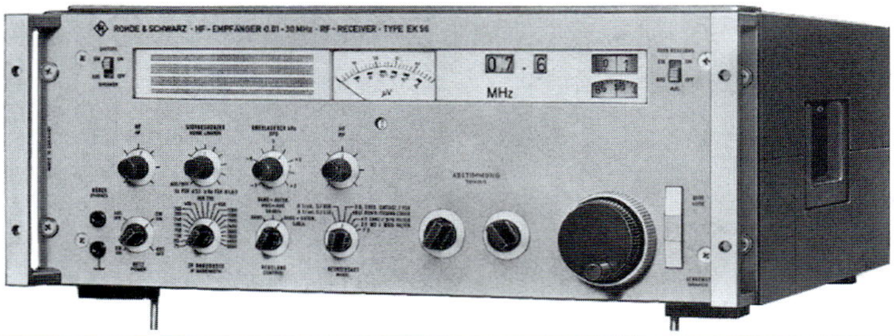

General Coverage Communications Receiver
Triple Conversion Superheterodyne. Solid State.

Features:
- ¼" Head. Jack
- S/AF Meter
- NB
- Rack Handles
- RF Gain
- BFO
- Tilt Bar
- Speaker
- Carry Handles
- AGC
- Speaker Switch

Specifications:
Coverage ......10 - 30000 kHz
Selectivity ......20 Pos. .075-12 kHz
Sensitivity ......<.8µV 20 dB S/N CW
Image Rej ......>80 dB

Modes .... AM/LSB/USB/CW
Stability .... 5 Hz/day
Dyn. Range .... >80 dB

Comments:
The frequency is displayed digitally through the 100 kHz position, then via an analog dial. The analog dial is very accurate providing better than 1 kHz resolution. This interesting model was featured in a full page ad in the June 1974 issue of *QST*.

Made In:	Germany	1972-1975

Voltages:

Readout: **00.1 MHz** ⊔⊔⊔⊔⊔▌⊔

Digital to 100 kHz, then Analog Linear.

Status: Active Manufacturer
Discontinued Model

Rarity: Extremely Scarce

	New	Used
Price:	$7000A	⑤
Rating:	⑥	⑥

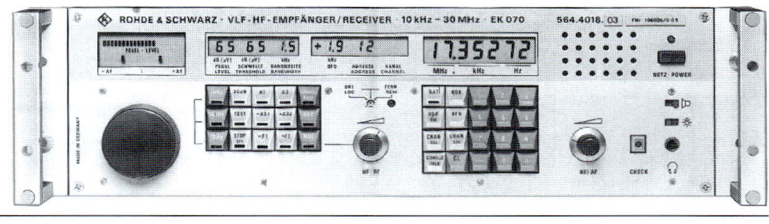

ROHDE & SCHWARZ

EK 070

General Coverage Communications Receiver
Double Conversion Superheterodyne. Solid State.

Features:
- ¼" Head. Jack
- S-Indicator
- Speaker
- BFO ±3.1 kHz
- RF Gain
- 30 Memories
- BITE
- IEEE 488 Bus
- I.F. Out Jack
- Squelch
- Atten. -20 dB
- Rack Handles
- AGC/MGC
- Speaker Switch
- Keypad
- 3 Tuning Rates

Specifications:
Coverage ......10 - 30000 kHz
Selectivity ......12/6/3/1.5/.6/.3/.15 kHz -3dB.
Sensitivity ......<.7µV CW
Image Rej ......>80 dB
Audio Output .1W 5 0hm

Modes .... AM/LSB/USB/CW/MCW
Stability .... <3x10⁻⁷ @25°C >10 min.
IF Rejection .... >80 dB
IF .... 81.4 MHz, 1.4 MHz
Environment ... -25° to +55° C. 95% Humid.

Accessories:
ISB Option   RTTY Option   FM Option
FDM Option

Comments:
Memories store frequency, mode and bandwidth. The alternate liquid crystal displays show bandwidth, BFO offset, memory, etc. Shown right is earlier production with LED frequency display, analog S-Meter and carry handle. Light gray. An outstanding receiver.

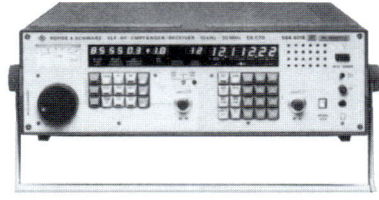

Made In:	Germany	1979-1998

Voltages: 100/120/220/240 VAC
47-440 Hz 65 VA

Readout: **00000.01** Digital LCD.

Physical: 19x5.9x20" 44 Lbs.
484x149x507mm 20 kg

Status: Active Manufacturer
Active Model

Rarity: Extremely Scarce

	New	Used
Price:	$10000-15000	$6500-8500
Rating:	⑥	⑥

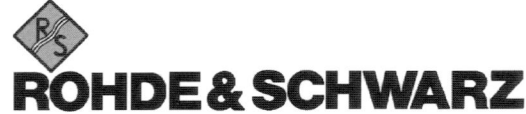

ROHDE & SCHWARZ

EK 071

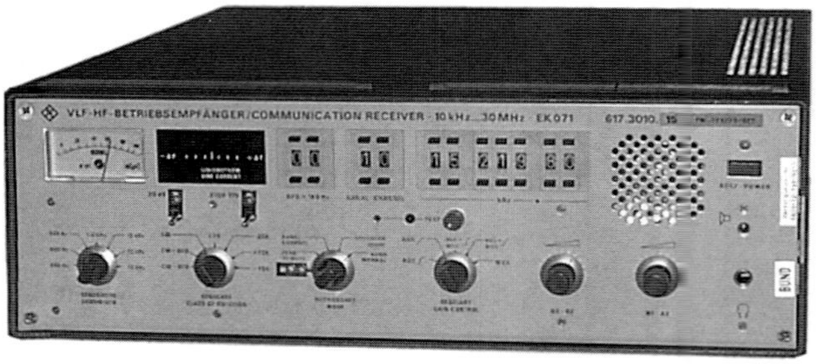

General Coverage Communications Receiver
Double Conversion Superheterodyne. Solid State.

Features:
- ¼" Head. Jack
- RF Gain
- AGC/MGC
- S-Meter
- Squelch
- Speaker Switch
- Speaker
- Atten. -20 dB
- BFO
- Carry Handle

Specifications:
Coverage ...... 10 - 30000 kHz
Modes .... AM/LSB/USB/CW/FSK

Comments:
The frequency is selected and display by seven ten position (BCD) decade switches yielding a display accuracy of 1 Hz. This receiver may be operated remotely. A two BCD switch sets the address. The BFO offset is also adjusted and displayed by two BCD switches.

This model is appearing on the used market in Europe and Japan, but is not seen in North America.

The author welcomes further technical information on this interesting receiver.

Made In: Germany

Voltages: 100/120/220/240 VAC
47-440 Hz

Readout: 00000.01 Digi. Switch

Physical: 19x5.9x17.1"
484x149x435mm

Status: Active Manufacturer
Discontinued Model

Rarity: Typically Unavailable

	New	Used
Price:	①	$2500-7400
Rating:	⑥	⑥

ROHDE & SCHWARZ

EK 085

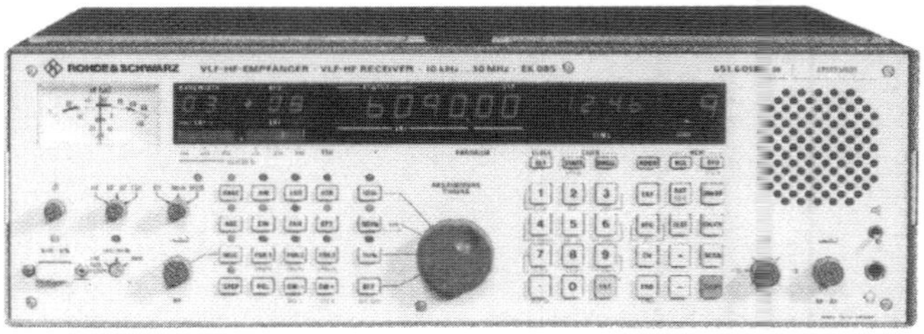

General Coverage Communications Receiver
Double Conversion Superheterodyne. Solid State.

Features:
- ¼" Head. Jack
- RF Gain
- B.I.T.E.
- Clock-Timer
- Speaker Switch
- S-Meter
- 100 Memories
- Carry Handle
- Dimmer
- B.I.T.E.
- Squelch
- Speaker
- Modular
- BFO ±1.5 kHz
- RS232 or IEEE488
- Keypad
- 600 Ohm Line Out
- I.F. Out 1.44 MHz
- 5 MHz Ref. Jack

Specifications:
Coverage ...... 10 - 30000 kHz
Selectivity ...... 4/2.4/.3-2.7/1.2/.15 kHz
Sensitivity ...... <.2µV CW with LPF
Audio .... 2W at 4 ohms
Environment .. -25 to +55° C. 95% Humid.
Modes .... AM/LSB/USB/CW/RTTY
Stability .... 3 x 10^{-7}
IF .... 80.64, 1.44 MHz
Antenna Input . BNC 50 ohm

Accessories:
GA001 Speaker
Filters
GM853C2 ISB Modem
Cabinet
GM085P1 FAX Demodulator

Comments:
The EK 085 may be tuned in 10, 100 or 1000 Hz increments. It has a built-in FSK demodulation supporting shifts of ±42.5 kHz, ±85 Hz and ±425 Hz.

Made In: Germany 1987-1998

Voltages: 100/120/220/240 VAC 50W
47-420 Hz or 19-31 VDC

Readout: 00000.01 Digital LCD.

Physical: Rack Version:
19x5.2x21.25" 39.7 Lbs.
483x132x540 18 kg
Cabinet Version:
18.5x6.5x19.9"
471x164x505mm

Status: Active Manufacturer
Active Model

Rarity: Typically Unavailable

	New	Used
Price:	①	⑤
Rating:	⑥	⑥

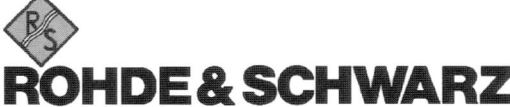

ROHDE & SCHWARZ

EK 851C2

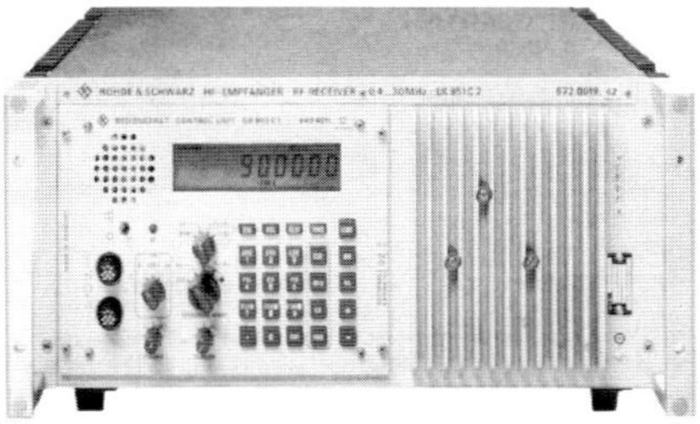

General Coverage Communications Receiver
Double Conversion Superheterodyne. Solid State.

Features:

- Head. Jack
- RF Gain
- I.F. Out Jack
- B.I.T.E.
- 100 Memories
- Rack Handles
- Squelch
- Speaker
- Preselector
- Keypad
- Speaker Switch

Specifications:

Coverage ...... 400 - 30000 kHz
Selectivity ...... 2.4/.3-2.7/.3 kHz -3dB.
Modes .... AM/LSB/USB/CW/RTTY
Sensitivity .... <.4µV CW

Accessories:

ISB Option
IN852C1 Power Supply 115/220 VAC

Comments:

This receiver matches the YK852 transmitter. It has a built-in FSK demodulation supporting shifts of ±42.5 kHz, ±85 Hz and ±425 Hz.

Made In: Germany 1993-1998

Voltages: 19 to 31 VDC
or AC with IN852C1 P.S.

Readout: 00000.01 Digital LCD.

Physical: 16.7x6.9x15.7" 62 Lbs
426x176x400mm 28 kg

Status: Active Manufacturer
Active Model

Rarity: Typically Unavailable

	New	Used
Price:	①	⑤
Rating:	⑥	⑥

ROHDE & SCHWARZ

EK 890

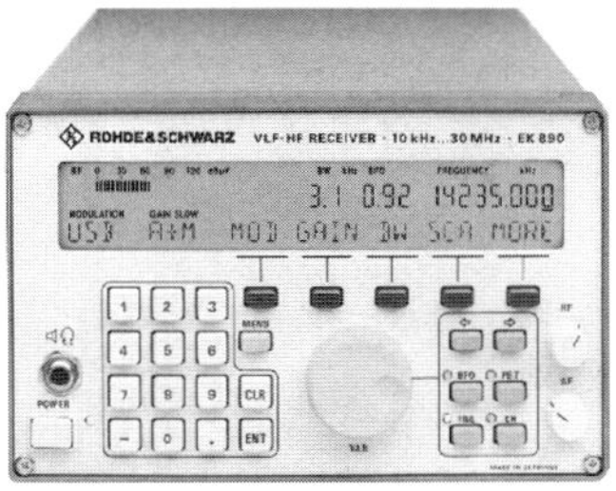

General Coverage Communications Receiver
Double Conversion Superheterodyne. Solid State.

Features:

- ¼" Head. Jack
- RF Gain
- I.F. Out Jack
- S-Indicator
- 1000 Memories
- Scan & Sweep
- PBT
- BFO
- Ext. Ref. Input 1/5/10 MHz
- Built in Self-Test
- RS232/485 Port

Specifications:

Coverage ...... 10 - 30000 kHz
Selectivity ...... 8/3.1/_/_/_/_ kHz -3dB.
Sensitivity ...... <.1µV SSB
Image Rej ...... >90 dB
Line Out .... 600 ohms
Modes .... AM/LSB/USB/CW/FAX
Stability .... ±5 PPM (±1 PPM optional)
IF Rejection .... >90 dB
IP3 .... >35 dBm
Antenna Input . BNC 50 ohm

Accessories: Various Filters

FK890 Preselector
GC890 BCD Interface
GM890 Signal Processor
GH890 TTY Loop Supply
GA890L1 Speaker
GB899 Remote Control

Comments: This receiver may also be rack mounted in a single or dual configuration. Memories store all receiver parameters. In master-slave operation, the EK 890 can control up to 99 slave receivers.

Variants: The EK 890 also available for computer control only without front control panel. Either configuration can be dual-mounted in a 19 inch rack . The EK 890 is shown right in a dual configuration.

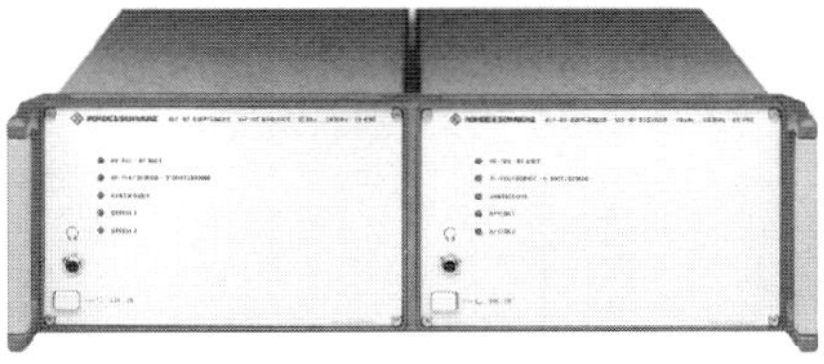

Made In: Germany 1991-1998

Voltages: 100/120/220/240 VAC
47 - 420 Hz 25 VA

Readout: 00000.001 Digital LCD.

Physical: 8.3x5.2x18.1 17.7 Lbs.
211x132x460mm 8 kg

Status: Active Manufacturer
Active Model

Rarity: Extremely Scarce

	New	Used
Price:	①	⑤
Rating:	⑥	⑥

ROHDE & SCHWARZ

EK 891

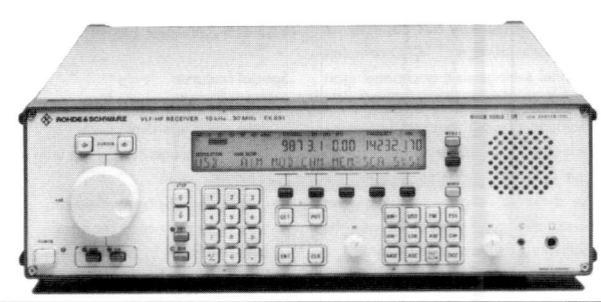

General Coverage Communications Receiver
Double Conversion Superheterodyne. Solid State.

Features:
- ¼" Head. Jack
- S-Indicator
- PBT
- Built in Self-Test
- RF Gain
- 1000 Memories
- Speaker
- RS232/485 Port
- I.F. Out Jack
- Scan & Sweep
- Ext. Ref. Input 1/5/10 MHz
- BFO
- Speaker Switch
- Keypad

Specifications:

Coverage ...... 10 - 30000 kHz	Modes .... AM/LSB/USB/CW/FAX
Selectivity ...... 6 Pos. .2-8 kHz -3dB.	Stability .... ±5 PPM (±1 PPM optional)
Sensitivity ...... <.3µV CW	IF Rejection .... >90 dB
Image Rej ...... >90 dB	Environment ... -25 to +55°C. 95% Humid.
Line Out .... 600 ohms	Antenna Input . BNC 50 ohms
BFO Resol. ... 10 Hz.	

Accessories:

FK890 Preselector	GM890 Signal Processor	GH890 TTY Loop Supply
GC890 BCD Interface	GB899 Remote Control	6 kHz Filter
1.5 kHz Filter	600 Hz Filter	200 Hz Filter

Comments:
Memories store all receiver parameters. May be rack mounted.

Variants:
The EK 891 is also available for controlling up to 99 slave receivers.

Made In:	Germany 1992-1998
Voltages:	100/120/220/240 VAC 47 - 420 Hz 25 VA
Readout:	**00000.001** Digital LCD.
Physical:	16.7x5.2x18.1" 44 Lbs 426x132x460mm 20 kg
Status:	Active Manufacturer Active Model
Rarity:	Extremely Scarce

	New	Used
Price:	①	⑤
Rating:	⑥	⑥

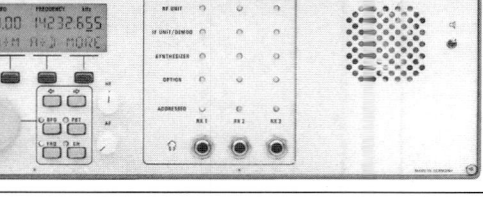

ROHDE & SCHWARZ

EK 893

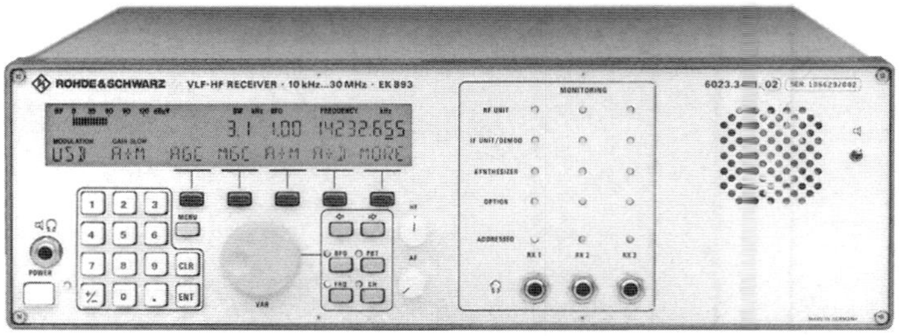

General Coverage Communications Receiver (Triple)
Triple Conversion Superheterodyne. Solid State.

Features:
- ¼" Head. Jack (4)
- S-Indicator
- PBT
- Built-in Self-Test
- RF Gain
- 1000 Memories
- Speaker
- RS232/485 Port
- I.F. Out Jack
- Scan & Sweep
- Ext. Ref. 1/5/10 MHz
- Keypad
- Speaker Switch
- Tilt Legs

Specifications:

Coverage ...... 10 - 30000 kHz	Modes .... AM/LSB/USB/CW/FAX
Selectivity ...... 8/3.1/_/_/_/_ kHz -3dB.	Stability .... ±5 PPM (±1 PPM optional)
Sensitivity ...... <.1µV SSB	IF Rejection .... >90 dB
Image Rej ...... >90 dB	Antenna Input . BNC
Line Out .... 600 ohms	Environment ... -25 to +55° C. 95% Humid.

Accessories:

	UX890 IF Converter	GM890 Signal Processor
Hi Stability Option	6 kHz Filter	1.5 kHz Filter
600 Hz Filter	200 Hz Filter	

Comments: The EK 893 is designed to mount in a 19" rack and provides three separate and independent receivers operated via a common processor and power supply. Simultaneous reception of three different signals, in three different modes, is possible. Memories store all receiver parameters. BFO resolution is 10 Hz.

Variants: Also available for computer control only without front control panel.

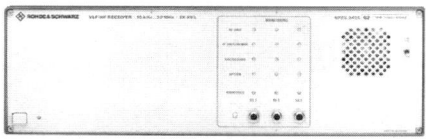

Made In:	Germany 1992-1998
Voltages:	100/120/220/240 VAC 47 - 420 Hz 25 VA
Readout:	**00000.001** Digital LCD.
Physical:	16.7x5.2x18.1" 44 Lbs 426x132x460mm 20 kg
Status:	Active Manufacturer Active Model
Rarity:	Typically Unavailable

	New	Used
Price:	①	⑤
Rating:	⑥	⑥

ROHDE & SCHWARZ

EK 895

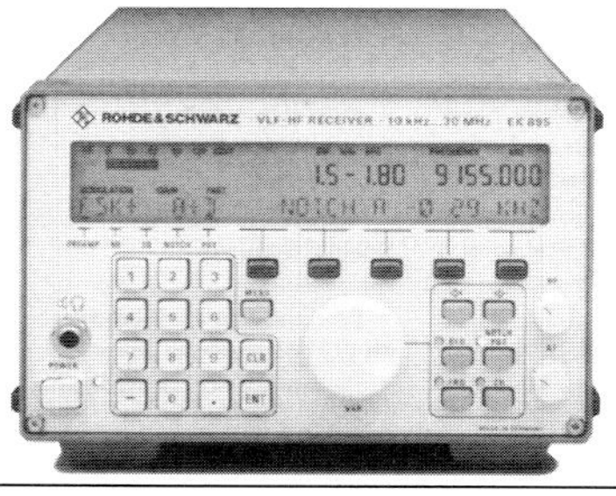

General Coverage Communications Receiver
Triple Conversion Superheterodyne. Solid State.

Features:

- ¼" Head. Jack
- RF Gain
- I.F. Out Jack
- NB
- Keypad
- S-Indicator
- 1000 Memories
- Scan & Sweep
- Notch
- Tilt Legs
- PBT
- BFO
- Preamp
- Ext. Ref. 1/5/10 MHz
- Built-in Self-Test
- RS232/485 Port
- Squelch

Specifications:

Coverage ...... 10 - 30000 kHz
Selectivity ...... 8/3.1/_/_/_/_ kHz -3dB.
Sensitivity ...... <.1µV SSB
Image Rej ...... >90 dB
Line Out .... 600 ohms
Modes .... AM/LSB/USB/CW/FAX
Stability .... ±5 PPM (±1 PPM optional)
IF Rejection .... >90 dB
BFO Resol. ..... 10 Hz.
Antenna Input . BNC 50 ohm

Accessories:

FK890 Preselector, GM890 Signal Processor, GH890 TTY Loop Supply
GC890 BCD Interface, GB899 Remote Control, GB890 Control Panel (shown)
1.5 kHz Filter, 600 Hz Filter, 200 Hz Filter
6 kHz Filter

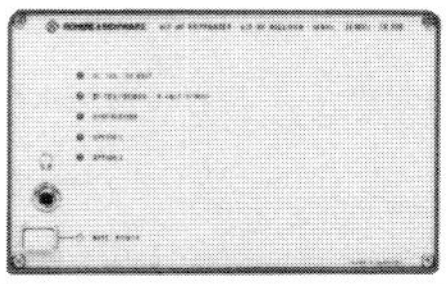

Comments: The EK895 is an enhanced version of the EK 890. Newly added features include: Preamp, Noise Blanker, Squelch, Notch Filter and PBT.
Variants: Also available for computer control only without the GB890 front control panel as shown to the right.

Made In: Germany 1993-1998

Voltages: 100/120/220/240 VAC
47 - 420 Hz

Readout: 00000.001 Digital LCD.

Physical: 8.3x5.2x18.1 17.7 Lbs.
211x132x460mm 8 kg

Status: Active Manufacturer
Active Model

Rarity: Typically Unavailable

	New	Used
Price:	$7800A	⑤
Rating:	⑥	⑥

ROHDE & SCHWARZ

EK 896

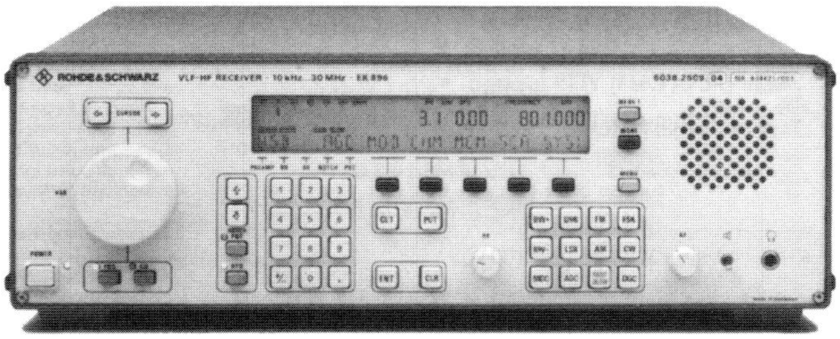

General Coverage Communications Receiver
Double Conversion Superheterodyne. Solid State.

Features:

- ¼" Head. Jack
- RF Gain
- I.F. Out Jack
- NB
- Speaker
- S-Indicator
- 1000 Memories
- Scan & Sweep
- Notch
- Speaker Switch
- PBT
- BFO
- Preamp
- Keypad
- Ext. Ref. 1/5/10 MHz
- Built-in Self-Test
- RS232/485 Port
- Squelch
- Tilt Legs

Specifications:

Coverage ...... 10 - 30000 kHz
Selectivity ...... 13 from 8-.15 kHz -3dB.
Sensitivity ...... <.2µV CW
Line Out .... 600 ohms
Enviroment .... -25 to +55° C. 95% Humid.
Modes .... AM/LSB/USB/CW/FAX/ISB
Stability .... ±5 PPM
BFO Resol. ..... 10 Hz.
Antenna Input . BNC 50 ohm

Accessories:

FK890 Preselector, GM890 Signal Processor, GH890 TTY Loop Supply
GC890 BCD Interface, GB899 Remote Control, 6 kHz Filter
1.5 kHz Filter, 600 Hz Filter, 200 Hz Filter

Comments:
The EK 896 can control up to 99 slave receivers. May be rack mounted. It has a very similar appearance to the EK 891. However the EK 896 uses DSP and has more bandwidths.

Made In: Germany 1994-1998

Voltages: 100/120/220/240 VAC
47 - 420 Hz 25 to 75W

Readout: 00000.001 Digital LCD.

Physical: 16.8x5.2x18.1" 24 Lbs
426x132x460mm 11 kg

Status: Active Manufacturer
Active Model

Rarity: Typically Unavailable

	New	Used
Price:	$12500A	⑤
Rating:	⑥	⑥

ROHDE & SCHWARZ

ESH 2

General Coverage Communications Test Receiver
Double Conversion Superheterodyne. Solid State.

Features:
- ¼" Head. Jack
- S-Meter
- Dial Lock
- Built-in Self-Test
- RF Gain
- Speaker
- Carry Handle
- Attenuator
- I.F. Out Jack
- Scan & Sweep
- 75 MHz Output

Specifications:
Coverage ...... 9 - 30000 kHz   Modes .... AM/LSB/USB/CW/FM
Selectivity ...... 10/2.4/.5/.2 kHz -3dB.

Accessories:

| Active Probe | Passive Probe | Clamp-on RF Current Probe |
| Rod Antenna | Loop Antenna | Inductive Probe |

Comments:
This is **not** a traditional shortwave receiver. It is a general purpose test receiver designed for the following applications: Field-strength measurements in conjunction with the design and manufacture of antennas and radio interference measurements. Remote frequency measurements and selective voltage measurements for laboratory and test departments.

Variants:
Sold in the U.S.A. as **Polarad ESH 2**. Sold by Polarad Electronics, Inc., 5 Delaware Dr., Lake Success, NY 11042.

Made In:	Germany	1987-1998
Voltages:	100/120/220/240 VAC	
	47 - 420 Hz or Batteries.	
Readout:	00000.1	Digital LCD.
Physical:		
Status:	Active Manufacturer	
	Active Model	
Rarity:	Extremely Scarce	

	New	**Used**
Price:	①	$6900-7400
Rating:	⑥	⑥

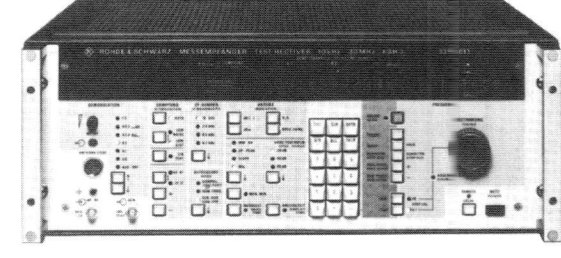

ROHDE & SCHWARZ

ESH 3

General Coverage Communications Test Receiver
Double Conversion Superheterodyne. Solid State.

Features:
- ¼" Head. Jack
- S-Meter
- IEEE488 Bus
- Built in Self-Test
- RF Gain
- Speaker
- Rack Handles
- Alpha-Numeric Data
- I.F. Out Jack
- Scan & Sweep
- 75 MHz Output
- Keypad

Specifications:
Coverage ...... 9 - 30000 kHz   Modes .... AM/LSB/USB/CW/FM
Selectivity ...... 10/2.4/.5/.2 kHz -3dB.

Accessories:

| Active Probe | Passive Probe | Clamp-on RF Current Probe |
| Rod Antenna | Loop Antenna | Inductive Probe |

Comments:
This is **not** a traditional shortwave receiver. It is a test receiver designed for the following applications: Field-strength measurements in conjunction with the design and manufacture of antennas. Radio interference measurements. Remote frequency measurements and selective voltage measurements for laboratory and test departments.

Variants:
Sold in the U.S.A. as **Polarad ESH 3**. Sold by Polarad Electronics, Inc., 5 Delaware Dr., Lake Success, NY 11042.

Made In:	Germany	1987-1998
Voltages:	100/120/220/240 VAC	
	47 - 420 Hz	
Readout:	00000.1	Digital LED.
Physical:	19.3x8x20.3" 56 Lbs	
	492x205x514mm 25 kg	
Status:	Active Manufacturer	
	Active Model	
Rarity:	Extremely Scarce	

	New	**Used**
Price:	$65,000	$35,000
Rating:	⑥	⑥

The **Rosetta Laboratories'** WinRadio is totally configured on a PC computer board that fits inside the computer.

The Windows user interface is friendly and the WinRadio offers excellent frequency coverage. Like most wide band receivers in this price range, its performance above 30 MHz is more impressive than below 30 MHz. The manufacturer continues to improve the product and add new capabilities. The optional WinRadio "Digital Suite" adds FAX, PACKET and ACARS demodulation, as well as CTCSS and DTMF decoding.

Rosetta Laboratories Pty. Ltd.
222 St. Kilda Rd.
St. Kilda, Victoria
3182, Australia   1995-1998

WiNRADiO
WR-1001i

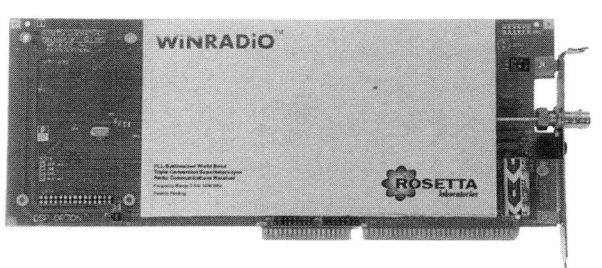

Wideband Communications Computer Receiver
Triple Conversion Superheterodyne. Solid State.

Features:
• Mini Head. Jack • Squelch   • 1000 Memories   • DX Switch
• BFO ±3 kHz   • Scan   • Sweep   • Signal Indicator
• Clock

Specifications:
Coverage ......500 kHz - 1.3 GHz   Modes .... AM/SSB/FM-W/FM-N
Selectivity ...... unspecified   Sensitivity .... <.5µV (CW 1.5-30 MHz)
Audio Out ......0.2 W 8 ohms   Ant. Jack .... BNC

Accessories: World Station Database Manager   Digital Suite

Comments: Tuning steps: 1 kHz to 1 MHz. The BFO can be tuned in 5 Hz steps. Memories store: frequency, mode, callsign, comment, group assignment and hotkey. System requirements: IBM PC with 386 or higher CPU, DOS 3.3 or higher or Windows 3.1 or higher, including Windows 95, 1 MB RAM (4 MB suggested) 16 bit slot and speakers or headphones. The American version has cellular frequencies blocked. The effective graphical user interface is shown below. Later production features 100 Hz readout and spectrum display capability.

Variants:
The professional model **WR-3001i** introduced in 1998, has DSP audio record-playback and covers from 150 kHz to 1.5 GHz with enhanced performance.

Made In:	Australia	1996-1998
Voltages:	N/A	
Readout:		PC
Physical:	12x5x.79" 1.25 Lbs.	
	300x125x20mm .56 kg	
Status:	Active Manufacturer	
	Active Model	
Rarity:	Scarce	
Reviews:	*SW Magazine* May 1996	
	WRTH 1998	

	New	Used
Price:	$500-700	$280-350
Rating:	★★★★	★★★★

The **S.A.I.T. Electronics** group of companies is headquartered in Brussels, Belgium and has offices worldwide. S.A.I.T. stands for Societe Anonyme International de Telegraphie. S.A.I.T. is in partnership with Radio Holland which does custom communications equipment installations for ocean going vessels. Radio Holland is one of the largest installer - systems integrators in the world.

S.A.I.T. did produce several receivers on their own. More recently, however, they have chosen to relabel models made by other manufacturers. Some examples include:

S.A.I.T. MR 1400	Eddystone EC958
S.A.I.T. MR 1402	Plessey PR155
S.A.I.T. MR 1414	Drake MSR-1
S.A.I.T. MR 1415	Drake MSR-2
S.A.I.T. MR 1431	Eddystone EC1830/1
S.A.I.T. MR 1541	Drake RR-1
S.A.I.T. MR 1542	Drake RR-2
S.A.I.T. MR 1543	Skanti R5000 (variant)
S.A.I.T. MR 14501	Skanti R5001
S.A.I.T. R8001	Skanti R8001
S.A.I.T. R8002	Skanti R8002

Some of these examples are shown in this Chapter.

S.A.I.T. Electronics
Société Anonyme
Chaussée De Ruisbroek 66
B-1190 Brussels,
Belgium   1980

S.A.I.T. Marine NV
Herentalsebaan 55
B-2100 Deurne, Antwerp,
Belgium

MR 1406A

General Coverage Communications Receiver
Double Conversion Superheterodyne. Solid State.

Features:
- ¼" Head. Jack
- S-Meter
- Fine Tuning
- AGC OFF/SLOW/MED/FST
- RF Gain
- ANL
- Scale Adjust
- Modular Construction
- Mute
- Speaker
- Speaker Switch
- Calibrator 1000 kHz
- IF Out 100 kHz
- Rack Handles
- Peak Tuning
- Spinner Knobs
- Line Out 600 Ohm

Specifications:
Coverage ...... 15 - 30000 kHz
Selectivity ...... 8/2/1/_ kHz -6dB
Sensitivity ...... <1µV 10 dB
Audio Out ...... 2 W 8 ohms

Modes .... AM/LSB/USB/CW
Stability .... >20 Hz
IF Rejection .... >90 dB 1.6-30 MHz
Antenna Input . 50 ohm

Accessories:
24 VDC Power Adapter   FSK Mode

Comments:
Analog tuning accuracy is quoted as 300 Hz.

Made In: Belgium

Voltages: 110/240 VAC or 24 VDC optional

Readout: `00000.01` Digital LED

Physical: 19.7x101x193" 77.1 Lbs
500x256x490mm 35 kg

Status: Active Manufacturer
Discontinued Model

Rarity: Typically Unavailable

	New	Used
Price:	①	②
Rating:	⑥	⑥

MR 1410

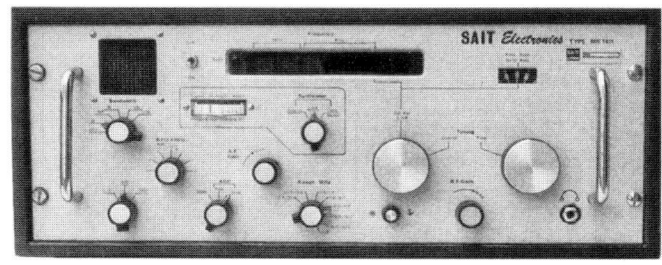

General Coverage Communications Receiver
Triple Conversion Superheterodyne. Solid State.

Features:
- ¼" Head. Jack
- S-Meter
- BFO ±2 kHz
- AGC OFF/SLO/FST
- RF Gain
- Dimmer
- IF Output
- Line Out 600 Ohm
- Mute
- NL
- Speaker Switch
- Fine Tuning
- Rack Handles
- Dial Lock
- Speaker Output

Specifications:
Coverage ...... 100 - 30000 kHz	Modes .... AM/LSB/USB/CW
Selectivity ...... 8.2/2.7/2.6/1.2/.3 kHz	IF Rejection .... >80 dB 1st IF
Sensitivity ...... <1µV 10 dB CW	IF Rejection .... >100 dB 2nd IF
Audio Out ...... 1 W 4 ohms	IF .... 10.1-14, 2.55-2.56, .455 MHz
Image Rej. ..... >60 dB	Environment ... -20° to +55°C.
Stability .... <20 Hz	Antenna Input . 50 ohm
Spurious Rej .>60 dB 1.6-30 MHz	

Accessories:
24 VDC PS   110-250 VDC PS

Comments:
This receiver employs double conversion above 4 MHz.

Variants:
This model was originally called the **Mentor M 1000M**. It was later renamed the MR 1410. The model **MR 1411** covered 10 - 30000 kHz. The MR1410 and MR1411 were originally designed by the Hatfield Company of Great Britain.

Made In:	Belgium 1976-1980
Voltages:	110-250 VAC 40-60 Hz or 24 VDC optional
Readout:	`00000.01` Digital LED
Physical:	Cabinet Version: 21x9.5x18" 50 Lbs. 534x242x457mm 25 kg Rack Version: 19x7x17.5" 472x178x444mm
Status:	Active Manufacturer Discontinued Model
Rarity:	Typically Unavailable

	New	Used
Price:	①	②
Rating:	⑥	⑥

MR 1543

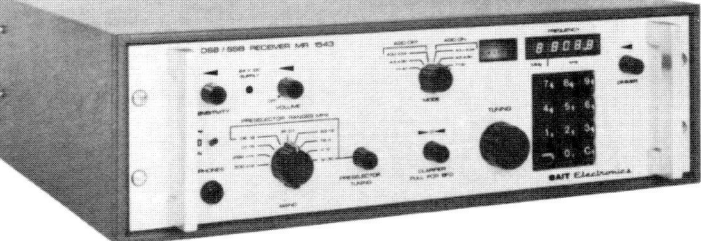

General Coverage Communications Receiver
Double Conversion Superheterodyne. Solid State.

Features:
- ¼" Head. Jack
- S-Meter
- BFO ±2 kHz
- AGC OFF/SLO/FST
- RF Gain
- Dimmer
- Keypad
- Modular Construction
- Mute
- Duplex Filters
- Speaker Switch
- 500/2182 kHz Presets
- Rack Handles
- Dial Lock
- Preselector
- Clarifier ±200 Hz
- Line Out 600 Ohm

Specifications:
Coverage ...... 10 - 30000 kHz	Modes .... AM/LSB/USB/CW
Selectivity ...... 5.4/2.4/1/.4 kHz -6dB	Stability .... 20 Hz
Sensitivity ...... <1.4µV 10 dB	Antenna Input . 50 ohm
Audio Out ...... 5 W 4 ohms	

Accessories:
IF Output

Comments:
Slightly reduced performance in the range of 10 to 60 kHz. If power from the AC mains fails an automatic change-over to 24 VDC occurs. This receiver is expressly designed for marine applications. A simplified, less expensive version of the Skanti R5000.

Made In:	Belgium 1980-
Voltages:	115/230 VAC 50/60 Hz or 24 VDC
Readout:	`00000.1` Digital LED
Physical:	Cabinet Version: 19.7x6x16.9" 53 Lbs. 502x152x430mm 24 kg Rack Version: 19x5.25x15.35" 33 Lbs. 483x133x390mm 15 kg
Status:	Active Manufacturer Discontinued Model
Rarity:	Typically Unavailable

	New	Used
Price:	①	②
Rating:	⑥	⑥

MR 1554

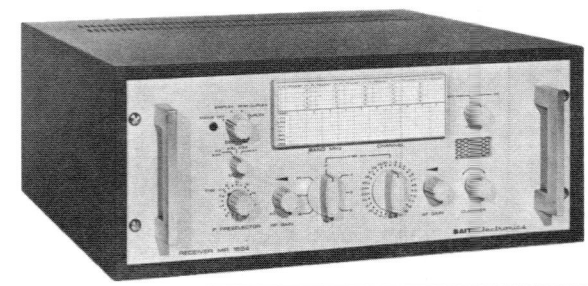

Fixed Frequency Communications Receiver
Double Conversion Superheterodyne. Solid State.

Features:
- Preselector
- RF Gain
- Speaker Output
- Clarifier ±10 Hz
- Mute
- Speaker
- Speaker Switch
- Preselector
- Rack Handles
- AGC

Specifications:

Coverage ...... See Comments		Modes .... AM/LSB/USB/CW	
Selectivity ...... 2.7 kHz -6dB		Stability .... 20 Hz/15 mins.	
Sensitivity ...... <3.5µV 20 dB		Antenna Input . 50 ohm	
Audio Out ...... 8 ohms 0.5 W			

Comments:
The MR 1554 is designed for shipboard use as a non-primary receiver. It covers 92 spot frequencies. The specific range is: 30 spot frequencies from 1605 to 4000 kHz. 10 Spots frequencies in each of the 4, 8, 12, 16 and 22 MHz HF marine bands and 6 spots each in the 6 and 25 MHz bands. Such a receiver is ideal quick, simple and accurate tuning to frequently used fixed channels. The associated frequency or channel number can be written on the paper chart mounted on the front of the receiver. The MR 1554 may also be rack mounted.

Made In: Belgium 1980-

Voltages: 12 or 24 VDC

Readout: None

Physical: 19.3x7.9x16.4" <1.9 Lbs
490x200x415mm 19 kg

Status: Active Manufacturer
Discontinued Model

Rarity: Typically Unavailable

	New	Used
Price:	①	②
Rating:	⑥	⑥

MR 14501

General Coverage Communications Receiver
Double Conversion Superheterodyne. Solid State.

Features:
- ¼" Head. Jack
- S-Meter
- BFO
- AGC OFF/SLO/FST
- RF Gain
- Dimmer
- Keypad
- Modular Construction
- Mute
- Duplex Filters
- Speaker Switch
- 500/2182 kHz Presets
- Rack Handles
- Dial Lock
- Preselector
- Line Out 600 Ohm

Specifications:

Coverage ...... 10 - 30000 kHz		Modes .... AM/LSB/USB/CW	
Selectivity ...... 5.4/2.4/1.4 kHz -6dB		Stability .... ±5 Hz	
Sensitivity ...... <1µV 10 dB		Antenna Input . 50 ohm	
Audio Out ...... 5 W 4 ohms			

Accessories:
IF Output   31 Memories

Comments:
Slightly reduced performance in the range of 10 to 60 kHz. If power from the AC mains fails an automatic change-over to 24 VDC occurs. This receiver is expressly designed for marine applications. See Skanti R-5001.

Made In: Belgium 1980-

Voltages: 115/230 VAC 50/60 Hz
or 24 VDC

Readout: 00000.1 Digital LED

Physical: Cabinet Version:
19.7x6x6.9" 55 Lbs.
502x152x430mm 25 kg
Rack Version:
19x5.25x15.35" 35 Lbs.
483x133x390mm 16 kg

Status: Active Manufacturer
Discontinued Model

Rarity: Typically Unavailable

	New	Used
Price:	①	②
Rating:	⑥	⑥

S 2000

General Coverage Communications Receiver
Double Conversion Superheterodyne. Solid State.

Features:
- ¼" Head. Jack
- S-Meter
- BFO ±2 kHz
- AGC OFF/SLO/FST
- RF Gain
- Dimmer
- Keypad
- Modular Construction
- Mute
- Preselector
- Speaker Switch
- Line Out 600 Ohm
- Rack Handles
- Dial Lock
- Clarifier ±200 Hz

Specifications:
Coverage ...... 10 - 30000 kHz   Modes .... AM/SSB/CW
Selectivity ...... 5.4/2.4/1/.4 kHz -6dB Stability .... ±5 Hz (15 min)
Sensitivity ...... <.5µV 1.6-30 MHz CW   Antenna Input . 50 ohm
Audio Out ...... 5 W 4 ohms

Accessories:
IF Output

Comments:
Slightly reduced performance in the range of 10 to 60 kHz. If power from the AC mains fails an automatic change-over to 24 VDC occurs. An 8 kHz bandwidth may be substituted for the 5.4 kHz filter in some units. The S 2000 appears to be a commercial, non-marine version of the MR 1543.

Made In: Belgium   1980-

Voltages: 115/230 VAC 50/60 Hz
or 24 VDC

Readout: `00000.1` Digital LED

Physical: Cabinet Version:
19.7x6x16.9" 53 Lbs.
502x152x430mm 24 kg
Rack Version:
19x5.25x15.35" 33 Lbs.
483x133x390mm 15 kg

Status: Active Manufacturer
Discontinued Model

Rarity: Typically Unavailable

	New	**Used**
Price:	①	②
Rating:	⑥	⑥

This company was formed as the Scott Transformer Company in 1924, but the named was changed to **E.H. Scott Radio Laboratories** Inc. in 1932. Prior to World War II, the Company produced high quality console radios.

The Company built at least one receiver model for the Navy during World War II. The model REE is shown below. This model was used in the crew's quarters rather than as a primary receiver in the "radio shack". After the War, a peace time version called the SLRM was sold. It had a maroon nameplate versus the black nameplate of the REE. These models are identical and the SLRM may simply be excess wartime production rebadged for civilian use. Both models are housed in an almost airtight metal enclosure. The heavy metal case was not designed to keep the elements out, but rather R.F. in. The Allies came to believe that German submarines were using the minute 455 kHz I.F. oscillator emissions from receivers to direction-find on ships and convoys. Marvin Hobbs, the Chief Engineer at E.H. Scott designed the circuit and cabinet with this in mind.

Hobbyists using these receivers for long continuous periods of time may wish to slide the chassis out a bit to facilitate air flow and prolong component life. Danger from direction-finding U-boats is now minimal.

E.H. Scott Radio Laboratories Inc.
4450 Ravenswood Ave.
Chicago 40, Illinois   1945-1947

E.H. Scott

REE

General Coverage Communications Receiver
Single Conversion Superheterodyne. 12 Tubes.

Features:
- ¼" Head. Jack
- RF Gain
- Tone
- Tuning Eye
- NL
- Dial Lamp
- Speaker
- BFO
- Speaker Switch
- Phono Input Terminals
- Mute Line
- Audio Terminals

Specifications:
Coverage ...... 540 - 18200 kHz
Selectivity ...... 5 Position
Audio .... 1.5W 60-600 ohms
Modes .... AM/CW
IF .... 455 kHz
Antenna Input . (Navy 49120)

Circuit Complement:
6K7 RF Amp, 12J5GT HF Osc., 12SA7 First Detector Mixer, 12SK7 1st IF Amp, 12SK7 2nd IF Amp, 12H6 2nd Detector/AVC/NL, 12SN7GT First Audio/CW Osc., 12SN7GT 2nd Audio/Phase Inverter, 25L6GT Output Audio Amp, 25L6GT Output Audio Amp, 1629 Tuning (eye) Indicator and 25Z6GT Rectifier.

Comments: The REE configuration consists of a type 6ZC-46270 radio receiver in a shock-mounted metal cabinet suitable for table top installation. Ranges: .54-1.6, 1.35-3.58, 3.4-8.8 and 8.5-186 MHz. A 0-1000 logging scale is provided. The radio is secured in the cabinet by eight knurled, thumb screws. The removal of these screws allows the receiver to be extracted from the front. Two fixed knobs are arranged on the front cabinet so the chassis may be removed without subjecting the operating controls to strain. Audio terminals provided 60/200/300/600 ohms output.

Variants: Model **SLRM** is the very similar post-War consumer version. In 1946 a deluxe consumer variant model **803** with horizontal pointer dial was offered.

Made In: United States 1944

Voltages: 115 VAC 60 Hz or 115 VDC 78W

Readout: ⊔⊔⊔⊔▪⊔ Analog

Physical: 18x12.5x16" 55 Lbs.
457x317x406mm 25 kg

Status: Inactive Manufacturer
Discontinued Model

Rarity: Very Scarce

	New	Used
Price:	①	④
Rating:	⑥	○

The **Sears Company** sold the venerable Yaesu FRG-7 under its own label via its mail-order division.

Sears
Sears Tower
Chicago, IL

Sears

412.36380700

General Coverage Communications Receiver
Triple Conversion Superheterodyne. Solid State.

Features:
- ¼" Head. Jack
- S-Meter
- Preselector
- Attenuator (3 Pos.)
- Fine Tuning
- ANL
- Tone (3 Pos.)
- Speaker
- Mute Line
- Dial Lamp
- Record Jack
- Ext. Spkr Jack
- Spinner Knob
- Carry Handle
- Fiduciary Adjust
- Dial Lamp Switch

Specifications:
Coverage ......500 - 29900 kHz
Selectivity ......6 kHz -6dB.
Sensitivity ......<.7µV SSB
Audio Out ......2 W

Modes .... AM/LSB/USB-CW
Stability .... <±500 Hz (any 30 mins.)
Ant. Jack .... SO-239 and terminals
Record Out ..... 50 mV fixed

Comments:
Ranges: .5-1.6, 1.6-4, 4-1 and 11-29.9 MHz. The 412.36380700 was manufactured for Sears by Yaesu and has a circuit virtually identical to the popular FRG-7. Unlike the FRG-7's gray cabinet, the Sear's version has a black cabinet. Tuning is a four step process. Select the appropriate band, set the preselector, set the MHz drum until the lock lamp goes off then tune the main tuning knob. The analog dial accuracy is 10 kHz or better. [1] Yaesu is an active manufacturer. May be difficult to sell on the used market simply because many people do not know what a Sears 412.36380700 is.

Made In:	Japan	1978-1980
Voltages:	117 VAC 50/60 Hz or 12 VDC or 8 D cells.	
Readout:	Analog Lin.	
Physical:	6.5x13.5x11.5" 16 Lbs. 340x153x285mm 7 kg	
Status:	Active Manufacturer[1] Discontinued Model	
Rarity:	Common	

	New	Used
Price:	$399	$180-200
Rating:	★★★★★	★★★★★

The huge German conglomerate **Siemens,** manufactures all types of consumer and commercial electronics. This company, which is over 100 years old is a world leader in automation, transportation, energy, telecommunications and information systems.

In recent years Siemens has been selling Sangean portables under their label.

For the Siemens E401, please see Rohde & Schwarz EK 047.

Siemens
Germany

SIEMENS

745 E 309a

General Coverage Communications Receiver
Single Conversion Superheterodyne.

Features:
- Head.Jack (2)
- S/AF-Meter
- Rack Handles
- Calibrator 100 kHz
- RF Gain
- Speaker
- Dial Lamp
- Fine Tuning
- Tone
- NL
- Speaker Switch
- Line Output 600 ohm
- IF Output Jack
- Dial Lock
- Flywheel Tuning

Specifications:
Coverage .... .255-.525, 1500-30300 kHz
Selectivity ...... Variable 4-.1 kHz -6dB.
Sensitivity ...... <.25µV 10 dB
Audio Out ...... 1.5W 5000 ohms
Modes .... AM/CW/MCW/FAX
IF .... 1326 kHz
IF Rejection .... >50 dB
Antenna Input . 60 ohm

Accessories:
A3A SSB Option   F1 FSK Option   F6 Twinplex Option
12/24 V Battery Input   FSE1300 Diversity

Comments:
Ranges: .25-.525, 1.5-3.2, 3.1-6.4, 6.1-10.3, 9.8-15.3, 14.8-20.3, 19.8-25.3 and 24.8-30.3 MHz. Cabinet version is shown. Headphone output is 100 ohms.

Made In:	Germany
Voltages:	110/125/220 VAC 40-60 Hz 75VA
Readout:	Analog
Physical:	Cabinet Version: 21.625x14x16" 83.3 Lbs 550x355x407mm 38 kg. Rack Version: 20.5x12x16" 520x304x407mm
Status:	Active Manufacture Discontinued Mode
Rarity:	Typically Unavailable

	New	Used
Price:	①	⑥
Rating:	⑥	⑥

SIEMENS

745 E 310a

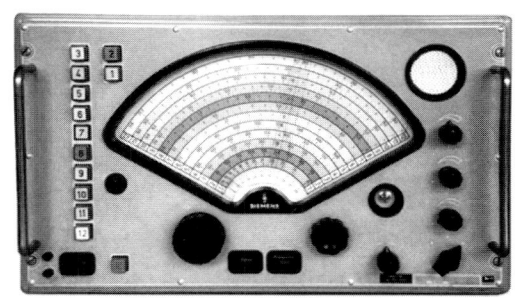

General Coverage Communications Receiver
Double Conversion Superheterodyne. 15 Tubes.

Features:
- ¼" Head.Jack (2)
- Tuning Eye
- Rack Handles
- Calibrator 100 kHz
- RF Gain
- Speaker
- Dial Lamp
- Fine Tuning
- Tone
- NL

Specifications:

Coverage ...... 14 - 30200 kHz
Selectivity ...... 6/2/1.6/.6 kHz -6dB.
Sensitivity ...... <.6µV 10 dB >1.5 MHz
IF Rejection ... >50 dB
Audio Out ...... 2W 5 ohms

Modes .... AM/CW
IF .... 1180, 100 kHz
Stability .... ±2x10⁻⁴ above 1.6 MHz
Image Rej. ...... >40 dB
Antenna Input . 75 ohm SO-239

Circuit Complement:
4x6BA6, 3x6BE6, 4x12AU7, 2x6AL5, 6AQ5 and 6CD7.

Comments:
Ranges: .014-.021, .085-.175, .17-.35, .34-.73, .72-1.54, 1.5-3.1, 3.1-6.3, 6-10.2, 9.7-15.2, 14.7-20.2, 19.7-25.2 and 24.7-30.2 MHz. This model was designed for the marine market. It was licensed as a class one receiver by Deutsche Bundespost. The twelve bands are selected by push-buttons on the left.

Variants:
Model **745 E 310b** is identical except that it features a C-type antenna jack rather than a SO-239 (U-type) antenna jack. Siemens model **E 566** appears to be a very similar model. It was referred to as the "rainbow" receiver among ham radio operators, referring to its colorful dial.

Made In:	Germany	1961-1965
Voltages:	110/125/220/250 VAC 40-60 Hz 100W	
Readout:	Analog	
Physical:	21x13x13" 70 Lbs 533x330x330mm 32 kg	
Status:	Active Manufacturer Discontinued Model	
Rarity:	Extremely Scarce	

	New	Used
Price:	①	⑤
Rating:	⑥	⑥

SIEMENS

E 311a

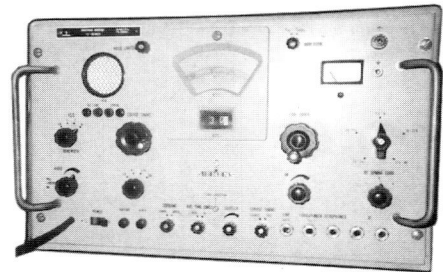

General Coverage Communications Receiver
Triple Conversion Superheterodyne. 16 Tubes.

Features:
- ¼" Head. Jack
- S-Meter
- Rack Handles
- Dial Lamp
- RF Gain
- Speaker
- Squelch
- Calibrator 100 kHz.
- Semi-Modular
- AVC/MVC
- Spinner Knobs
- Line Out 600 ohms
- NL
- IF Output Jack

Specifications:

Coverage ...... 1500 - 30100 kHz
Selectivity ...... 6/3/1/.3 kHz -6dB.
Sensitivity ...... <.3µV CW
Image Rej ...... >80dB
Audio Out ...... 1 W 5 ohms

Modes .... AM/SSB/CW
Stability .... ±20 Hz over 24 hours
IF .... 1.3 kHz-1.4 MHz, 370, 30 kHz
Antenna Input . SO-239 60 ohms

Circuit Complement:
(3) EF93, (5) ECH81 and (8) E88CC. It is interesting to note that only three types of tubes used in the E 311A.

Accessories:
FSE30 FSK Adapter   FAX Adapter   LF Converter 10-1500 kHz

Comments: Ranges: 1.5-3.4, 3.4-7.5, 7.5-15, 15-22.5 and 22.5-30.1 MHz. An incredibly accurate receiver that uses an interesting and unique mixture of analog and digital indicators. Readout to 100 Hz is supported. The tuning scale has an effective length of 935 feet (285 meters). The cord exiting from the lower left is the AC power cord. The E 311a has no rear panel connections.

Variants: Models **E 311e** and **E 311b** are very similar.

Made In:	Germany	1959-1961
Voltages:	110/125/220VAC 40-60 Hz. 70W	
Readout:	01. Analog Lin. for MHz and 100 kHz, then Digital Mech. for 10 kHz and 1 kHz, then Vertical Analog Lin. for 100 Hz.	
Physical:	18.75x11.7x15.375" 55 Lbs 475x295x390m 26 kg	
Status:	Active Manufacturer Discontinued Model	
Rarity:	Extremely Scarce	
Reviews:	*Radiowelt* October 1986	

	New	Used
Price:	15000DM	1100-2150DM
Rating:	⑥	⑥

SIEMENS

E 403

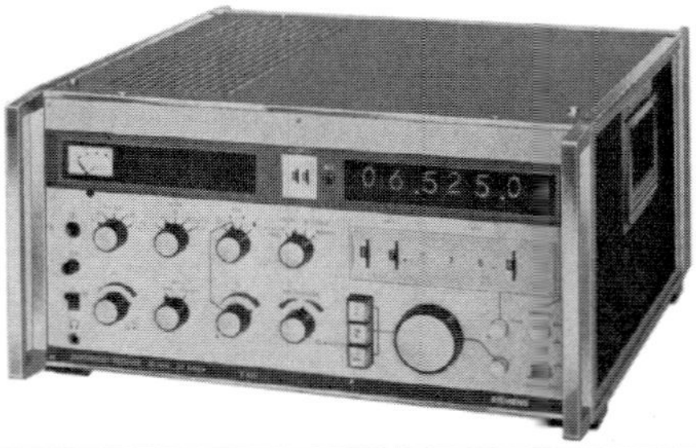

General Coverage Communications Receiver
Double Conversion Superheterodyne. Solid State.

Features:
- ¼" Head. Jack
- S-Meter
- NB
- Rack Handles
- RF Gain
- BFO
- Speaker
- Remote Bus
- AGC
- AFC
- Notch Filter
- Carry Handles
- IF Out Jack

Specifications:

Coverage ......10 - 30000 kHz	Modes .... AM/SSB/CW/FM/RTTY
Selectivity ......6 Position	Image Rej .... >80 dB
Sensitivity ......<.1µV CW 10 dB S/N	IF Rejection .... >80 dB
Freq. Drift ......<5x10^{-8}/month	Environment ... -25 to +45° C. 95% Humid.
Antenna Input 50 ohms	

Accessories:

Rack Mount   Notch Filter / AFC   N403 Remote Controller
ISB Module

Comments:
The E403 may be tuned in steps of 10, 100 Hz or 1 kHz by the rotary knob or via the decade switches. All functions are retained by core memory when the AC power source fails.

Made In:	Germany 1978-1988
Voltages:	100/125/220/240 VAC 45-100 Hz 140-160W
Readout:	**00000.01** Digital Nixie
Physical:	
Status:	Active Manufacturer Discontinued Model
Rarity:	Typically Unavailable

	New	Used
Price:	①	⑤
Rating:	⑥	⑥

SIEMENS

E 410

General Coverage Communications Receiver
Triple Conversion Superheterodyne. Solid State.

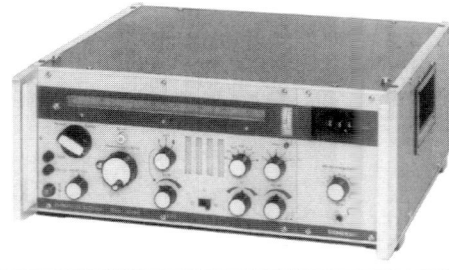

Features:
- ¼" Head. Jack
- S-Meter
- AGC (3 Pos.)
- RF Gain
- Speaker
- Rack Handles
- Spinner Knob

Specifications:

Coverage ......70 - 30000 kHz	Modes .... AM/SSB/CW
Selectivity ......6/3/1.5/1/.3 kHz -6dB.	IF .... 1.3-1.4 MHz, 370, 30 kHz
Sensitivity ......<.5µV CW 10 dB S/N	Environment ... -25 to +55° C. 95% Humid.
Image Rej ......>60 dB	

Accessories:
FSW401 Demod Option

Comments:
Ranges: .069-.155, .15-.335, .33-.74, .73-1.63, 1.6-3.5, 3.5-7.5, 7.5-15, 15-22.5 and 22.5-30 MHz. The Siemens E 410 is available with two different VFOs. One VFO selects the 10 kHz, 1 kHz and 100 Hz positions via ten position BCD switches and then provides a knob for 100 Hz tuning. The other VFO has a mechanical digital readout for the 10 kHz and 1 kHz positions and tick marks for the 100 Hz positions plus a tuning knob for fine tuning. Please note that tuning below 1.6 MHz is displayed on a nonlinear analog dial.

Made In:	Germany 1979-1981
Voltages:	110-240VAC 60W or 24 VDC 30W
Readout:	See Comments.
Physical:	17.5x7x17.3" 39.7 Lbs 445x177x440mm 18 kg
Status:	Active Manufacturer Discontinued Model
Rarity:	Typically Unavailable

	New	Used
Price:	5000 DM	1500 DM
Rating:	⑥	⑥

85 Skanti

The Danish company **Skandinavisk Teleindustri Skanti** was established in 1965. Skanti produces a broad line of maritime, commercial and military communications products. The company exports 95% of its production and is the largest manufacturer of HF-SSB marine radiotelephones in Europe. Due to reliable performance and cost-efficiency, Skanti HF equipment enjoys increasing popularity with military and paramilitary customers.

Skanti has also produced models sold under the Debeg and S.A.I.T. labels. Please see the Chapter on S.A.I.T. The S.A.I.T. company also assembled many series R5000 receivers in Belgium.

The Skanti R5000 and R8000 series of H.F. receivers have enjoyed widespread acceptance in the marine market.

Skandinavisk Teleindustri Skanti
Kirke Væløsevej 34
DK-3500 Værløse,
Denmark

SKANTI

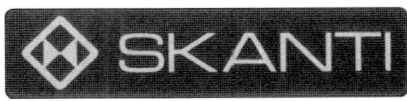

R5001

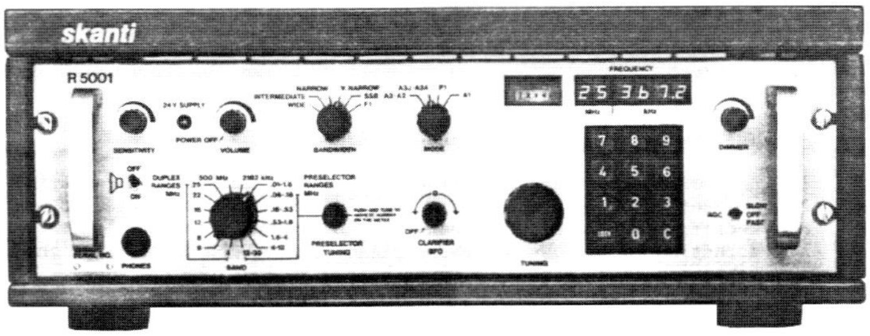

General Coverage Communications Receiver
Double Conversion Superheterodyne. Solid State.

Features:

- ¼" Head. Jack
- RF Gain
- Preselector
- Mute Line
- S-Meter
- BFO ±2 kHz
- Dimmer
- Sidetone Input
- Keypad
- Rack Handles
- Speaker Switch
- AGC OFF/SLO/FST
- 500/2182 kHz Presets
- Line Out 600 ohms

Specifications:

Coverage ...... 10 - 30000 kHz
Selectivity ...... 4/2.7/1.2/.5/.4/.1kHz -6dB.
Audio Out ...... 4 W 4 ohms
IF .... 38 MHz, 1.4 MHz.
Modes .... AM/USB/MCW/CW/RTTY
Sensitivity .... <.5μV 4-30 MHz SSB
Antenna Input . 50 ohm

Accessories: Cabinet (shown)

Comments: The R-5001 Version S-1 has an 11.2 MHz External master Osc. Input. Version S-2 does not. This receiver has seven duplex filters for the major H.F. maritime bands. These are used to filter-out duplex transmitter frequencies on maritime phone calls. The Duplex Ranges are selected off the other half of the rotary preselector knob and are labeled 4, 6, 8, 12, 16, 22 and 25 MHz. The associated frequencies are: 4355-4445, 6500-6596, 8710-8840, 13100-13350, 17230-17830, 22570-23430 and 25300-26300 kHz. This receiver is designed for maritime use. It is configured to be a main receiver.

Variants: Model **R5002** features LSB and USB. OEMs versions include: **DEBEG 7204** and S.A.I.T. MR 14501. Please see the S.A.I.T. Chapter. Model **R5000M** is for F1B telex only and has a blank front panel with operation by computer only.

Made In: Denmark 1973-1985

Voltages: 100/115/120 or 220/230/240 VAC 45W or 24 VDC

Readout: 00000.1 Digital LED

Physical: Rack Version:
19x5.2x15.3" 33 Lbs.
482x132x390mm 15.2 kg

Status: Active Manufacturer
Discontinued Model

Rarity: Very Scarce

	New	Used
Price:	①	⑤
Rating:	⑥	⑥

SKANTI

R8001

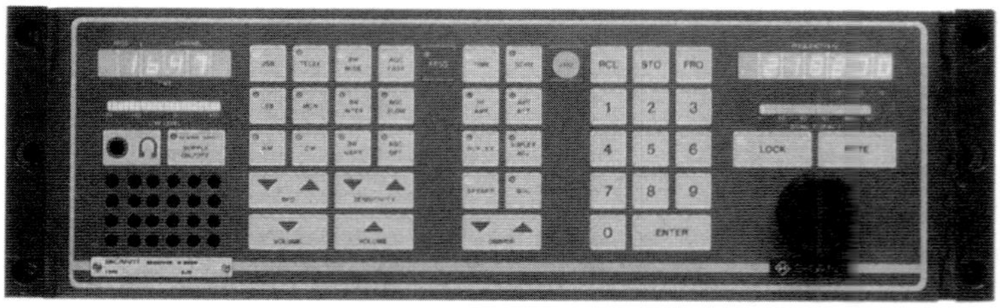

General Coverage Communications Receiver
Double Conversion Superheterodyne. Solid State.

Features:
- ¼" Head. Jack
- Remote Head
- Squelch
- 399 Memories
- Preamp
- RF Gain
- Keypad
- BFO ±3 kHz
- Flywheel Tuning
- B.I.T.E.
- Dimmer
- Attenuator -20 dB
- Rack Handles
- 2182 kHz Preset
- Clock-Timer
- Dial Lock
- S/AF Indicators
- 1.4 MHz IF Out Jack
- Scan
- Line Out 600 ohm
- AGC OFF/SLO/FST

Specifications:
Coverage ...... 10 - 30000 kHz
Selectivity ...... See Comments.
Sensitivity ...... <.45µV CW ±400 Hz
Audio Out ...... 0.5 W 4 ohms (4W ext.)
Environment .. 0 to +40°C.
Modes .... AM/USB/LSB/CW/RTTY
Stability .... 1.5 PPM 0 to +40°C
IF Reject .... >90 dB
Image Reject .. >80 dB
Antenna Input . 50 ohm

Accessories:
0.8 PPM 0.4 PPM

Comments: The R8001 has a removable and remotable control head. The control head may be up to 100 meters (330 feet) away from the radio itself. The keypad, and all switched functions are accessed via a membrane switches. Reduced performance from 10-100 kHz. The frequency display LEDs are yellow. Input protection is 30 V EMF for up to 15 minutes. Selectivity is 5.4/2.4/.5 kHz and .35-2.7 kHz. Manual tuning steps can be set for 10, 100 or 1000 Hz.
Variants: The **DEBEG 2081** is exactly the same except for having a beige color.

Made In: Denmark 1987-1989
Voltages: 100-120 or 220-240 VAC 50-60 Hz 50W or 24 VDC
Readout: 00000.01 Digital LED
Physical: 17x5.25x21.8" 34.2 Lbs. 432x132x553mm 15.5 kg
Status: Active Manufacturer Active Model
Rarity: Typically Unavailable

	New	Used
Price:	①	⑤
Rating:	⑥	⑥

SKANTI

R8003

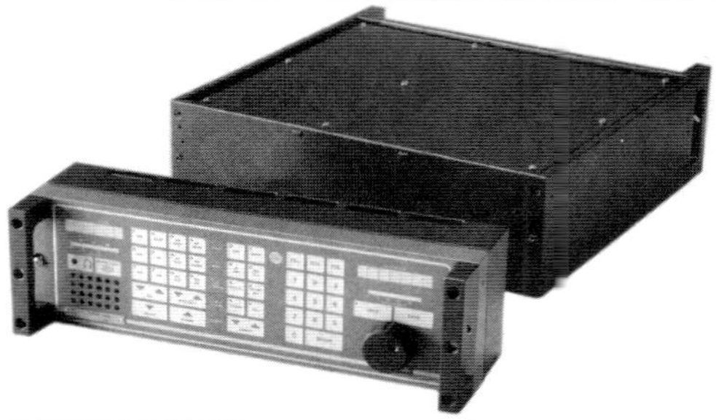

General Coverage Communications Receiver
Double Conversion Superheterodyne. Solid State.

Features:
- ¼" Head. Jack
- Remote Head
- Squelch
- 399 Memories
- Dial Lock
- Keypad
- BFO ±3 kHz
- Flywheel Tuning
- ISB Line Out
- Speaker Switch
- Attenuator -20 dB
- Rack Handles
- RS-232/RS-485
- B.I.T.E.
- Dimmer
- S/AF/FSK Indicators
- 1.4 MHz IF Out Jack
- Clock-Timer
- Modular
- Scan
- Notch
- Sweep

Specifications:
Coverage ...... 10 - 30000 kHz
Selectivity ...... See Comments.
Sensitivity ...... <.56µV CW ±400 Hz
Audio Out ...... 0.5 W 4 ohms (4W ext.)
Environment .. 0 to +40°C.
Modes .... AM/USB/LSB/ISB/CW/RTTY
Stability .... 1.5 PPM 0 to +40° C
IF Reject .... >90 dB
Image Reject .. >80 dB
Antenna Input . 50 ohm

Accessories:
PCB613 0.8 PPM PCB614 0.4 PPM PCB615 0.35 PPM PCB616 0.10 PPM

Comments: The R8003 also has a remotable head. Reduced performance from 10-100 kHz. The frequency display LEDs are red. Input protection is 30 V EMF for up to 15 minutes. Selectivity is .125/.4/1.2/3 kHz and .35-2.7 kHz. The Notch filter is 30 dB and adjustable in the range of 300-3000 Hz with digital readout. The R8003 has a built-in FSK demodulator with mark and space user selectable in the range of 1000-3000 Hz with data rates of 70-750 baud. Forty different scan programs may be devised with adjustable dwell, hold. Scan hold can be activated by squelch, signal threshold or external signal. ISB is standard.

Made In: Denmark 1994-1998
Voltages: 100-120 or 220-240 VAC 50-60 Hz 50W or 24 VDC
Readout: 00000.01 Digital LED
Physical: 17x5.25x21.8" 34.2 Lbs. 432x132x553mm 15.5 kg
Status: Active Manufacturer Active Model
Rarity: Typically Unavailable

	New	Used
Price:	①	⑤
Rating:	⑥	⑥

As one of the top three trademarks recognized worldwide, the Sony Corporation needs little introduction. Sony makes top drawer products in every aspect of consumer electronics. Sony enjoys annual sales of $45 billion.

Sony has been an established leader in the production of portable worldband radios for two decades. Although it has been ten years since Sony offered a full sized, tabletop shortwave radio, their offering of portables remains the most extensive and most respected in the world. It may be justifiably argued that every radio in this chapter is a portable. Each has a handle, may operate on batteries and has self contained antennas. However, because of their size and value, the typical use of these models is stationary or table top. Several readers of the preceding edition of this book felt they should be included.

Their last "table" model, the ICF-6800W produced over 15 years ago, still has a faithful following among those who enjoy music on shortwave. The last production version (the so-called "orange" variant) is seldom seen on the used market and usually sells quickly.

Sony Corporation
P.O. Box 10
Tokyo 149, Japan

Sony Consumer Products Co.
Sony Dr.
Park Ridge, NJ 07656

SONY®

CRF-320K

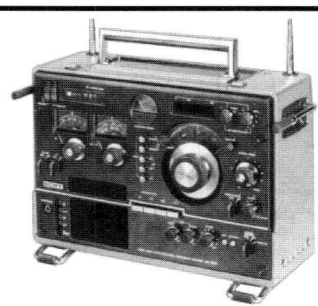

General Coverage Communications Receiver
Double Conversion Superheterodyne. Solid State.

Features:
- ¼" Head. Jack
- Speaker 4¾"
- RF Gain
- Dial Lamp
- Clock Timer
- S/Batt.-Meter
- Telescopic Ants.
- Muting
- Carry Handle
- Aux. Input Jack
- Bass
- NB
- AFC
- Fiduciary Adj.
- Record Jack
- Treble
- Ferrite MW Antenna
- Antenna Trimmer
- Timer Out Jack
- Local-DX Switch

Specifications:
Coverage ...... 150 - 30000 kHz[1] +FM
Selectivity ...... 8/6 kHz -6dB.
Sensitivity ...... <.7µV 6dB S/N 2-30MHz
Image Rej ...... >90 dB 1st @ 10 MHz
Modes .... AM/USB/LSB/CW
IFs .... 45.145 MHz, 455 kHz[2]
Antenna Input . Terminals

Accessories:
DCC-9 Car Battery Cord

Comments:
This radio has a separate analog tuner for the FM band, one for the LW/MW bands and a third tuner with analog and digital readout for shortwave. [1]There is a reception gap from 400 to 530 kHz. [2]10.7 MHz IF for FM. Features a black aluminum die-cast cabinet with chrome trim.

Made In:	Japan	1976-1980
Voltages:	100/120/220/240 VAC 10 W or 12 VDC. Requires 1 D cell for the clock and 8 D cells for portable use.	
Readout:	`00001.` Digital LED	
Physical:	17.75x12.12x8.125" 28 Lbs 451x308x207mm 13 kg	
Status:	Active Manufacturer Discontinued Model	
Rarity:	Very Scarce	

	New	Used
Price:	$1495	$500-650
Rating:	★★★	★★★

SONY®

CRF-330K

General Coverage Communications Receiver
Double Conversion Superheterodyne. Solid State.

Features:
- ¼" Head. Jack
- S/Batt.-Meter
- Bass
- Treble
- Speaker 4¾"
- Telescopic Ants.
- NB
- Ferrite MW Antenna
- RF Gain
- Muting
- AFC
- Antenna Trimmer
- Dial Lamp
- Carry Handle
- Fiduciary Adj.
- Cassette Recorder
- Clock Timer
- Aux. Input Jack
- Local-DX Switch
- Built-in Mic
- Timer Out Jack
- Record Jack

Specifications:
Coverage ...... 150 - 30000 kHz[1] +FM   Modes .... AM/USB/LSB/CW
Selectivity ...... 8/6 kHz -6dB.   IFs .... 45.145 MHz, 455 kHz[2]
Sensitivity ...... <.7µV 6dB S/N 2-30MHz   Antenna Input . Terminals
Image Rej ...... >90 dB 1st @ 10 MHz

Circuit Complement:
22 FETs, 21 Integrated Circuits and 93 Transistors.
Accessories: DCC-9 Car Battery Cord
Comments: This substantial receiver sports a pullout drawer at the base with an integrated cassette recorder. It can be controlled by the radio's timer for unattended recording. This radio has a separate analog tuner for the FM band, one for the LW/MW bands and a third tuner with analog and digital readout for shortwave. [1]There is a reception gap from 400 to 530 kHz. [2]10.7 MHz IF for FM. This unique radio enjoys strong collector demand. Features a black aluminum die-cast cabinet with chrome trim.

Made In:	Japan	1978-1981

Voltages: 100/120/220/240 VAC 12 W or 12 VDC. Requires 1 D cell for the clock and 8 D cells for portable use.

Readout: 00001. Digital LED

Physical: 17.75x13.5x8.125" 34 Lbs 459x343x206mm 15.4 kg

Status: Active Manufacturer Discontinued Model

Rarity: Very Scarce

	New	Used
Price:	$2495	$580-700
Rating:	★★★	★★★

SONY®

CRF-V21

General Coverage Communications Receiver
Double Conversion Superheterodyne. Solid State.

Features:
- Mini Ear Jack
- S-Indicator
- 350 Alpha Mems.
- RTTY/FAX Demod.
- Active Antenna
- Dial Lamp
- AFC (FM)
- Spectrum Display
- Speaker 4"
- Battery Check
- Carry Handle
- LCD Contrast Adj.
- Record Jack
- Scan/Sweep
- Keypad
- Clock/Timer 8 Event
- AF Filter
- Thermal Printer
- Speaker 2.75"
- Attenuator -30 dB
- Synchronous Det.

Specifications:
Coverage ...... 9 - 30000 kHz +FM   Modes .... AM/USB/LSB-CW/FM/RTTY/FAX
Selectivity ...... 6/3.5/2.7/14 kHz -6dB.   IF Reject .... > 80 dB
Sensitivity ...... <.17µV 6dB S/N 2-30 MHz   IFs .... 55.845 MHz, 455 kHz[2]
Image Rej ...... >70 dB   Antenna Input . TNC to Active Antenna

Accessories: AN-P1200 GOES Satellite Antenna
Comments: [2]10.7 MHz IF for FM/SAT. This technology show piece model is best known for its built-in spectrum display (selectable at 200 kHz or 5 MHz wide) and RTTY/FAX decoding. RTTY Baudot: 60/66/75/100 WPM. ASCII: 110/200/300/600 bps. FAX: 60/90/120/240 RPM, 576/288 IOC. Direct GOES satellite reception with optional AN-P1200 antenna. The thermal printer provides hardcopy of RTTY, FAX or GOES receptions. It can also print the LCD display contents including the spectrum display. The width of the spectrum display may be defined by the user. The receiver will then draw the display in near real-time. Officially classified as a portable but it is seldom seen at the beach.

Made In:	Japan	1989-1992

Voltages: 110/120/220/240 VAC or 6 VDC NiCad Pak Requires 2 AA cells for CPU

Readout: 00000.1 Digital LCD

Physical: 16.25x11.25x6.75 21 Lbs. 412x285x169mm 9.5 kg

Status: Active Manufacturer Discontinued Model

Rarity: Scarce

Reviews: 📖 *Passport* 1989-1993
WRTH 1988, 1990
SW Magazine October 1995

	New	Used
Price:	$3500-5100	$600-720
Rating:	★★★	★★★

SONY®

ICF-6700W

General Coverage Communications Receiver
Double Conversion Superheterodyne. Solid State.

Features:
- ¼" Head. Jack
- Mini Ear Jack
- Speaker 4"
- Record Jack
- Time Chart
- S/Batt.-Meter
- Dial Light
- Battery Check
- Timer Jack
- MPX Out Jack
- Bass
- AFC (FM)
- Carry Strap
- RF Gain
- Carry Handle
- Treble
- Preselector
- Telescopic Whip
- Ferrite MW Antenna

Specifications:

Coverage ...... 530 - 30000 kHz[1] +FM
Selectivity ...... 10/4 kHz -6dB.
Sensitivity ...... <.63µV 6 dB S/N
Audio .... 1W 8 ohms

Modes .... AM/USB/LSB-CW
IFs .... 10.700 MHz, 455 kHz
Antenna Input . SO-239 and Terminals

Circuit Complement:
5 FETs, 32 transistors and 19 diodes.

Accessories:
DCC-130 Car Battery Cord

Comments:
Ranges: .53-1.605, 1.6-10, 11.5-20, 20-30 MHz and 87.5-108 MHz FM. [1]There is a reception gap from 10400 to 11400 kHz (as a result of the 10.700 MHz 1st IF). The IF filters leave something to be desired.

Variants:
Model **ICF-6700L** covers longwave also.

Made In:	Japan	1978-1980
Voltages:	110/120/220/240 VAC 50/60 Hz or six D cells.	
Readout:	`00001.`	Digital LED
Physical:	18x7.5x9.25" 12 Lbs. 459x190x235mm 5.4 kg	
Status:	Active Manufacturer Discontinued Model	
Rarity:	Common	
Reviews:	📖 *Passport* 1987 *WRTH* 1980	

	New	Used
Price:	$365-440	$180-260
Rating:	★★★	★★★

SONY®

ICF-6800W

General Coverage Communications Receiver
Double Conversion Superheterodyne. Solid State.

Features:
- ¼" Head. Jack
- Mini Ear Jack
- Speaker 4"
- Record Jack
- Time Chart
- S-Meter
- Dial Light
- Battery Check
- Timer Jack
- Memo Light
- Bass
- AFC (FM)
- Carry Strap
- RF Gain
- Carry Handle
- Treble
- Preselector
- Telescopic Whip
- Ferrite MW Antenna

Specifications:

Coverage ...... 530 - 30000 kHz +FM
Selectivity ...... 9/4 kHz -6dB.
Sensitivity ...... <.63µV 6dB S/N
Image Rej ...... >50 to 60 dB SW

Modes .... AM/USB/LSB-CW
IFs .... 19.055 MHz, 455 kHz[2]
Antenna Input . SO-239 and Terminals.

Circuit Complement:
7 Integrated circuits, 57 transistors and 10 FETs.

Accessories: DCC-120 Car Battery Cord

Comments: This receiver has a dedicated following among shortwave content listeners who appreciate the radio's excellent fidelity. Not a fabulous DXing machine but one of best sounding communications receivers ever built. [2] 10.7 MHz IF for FM

Variants: The **ICF-6800W Orange** is later production (after serial number 30000) with improvements. Orange ink was used on the front panel for the model's name. This version boasts tighter IF bandwidths and adds an RF attenuator (on the back of the set). This model sold new for $700 in 1982 and is worth $400+ used. This variant is sometimes referred to as the **ICF-6800WA**.

Made In:	Japan	1980-1983
Voltages:	110/120/220/240 VAC 50/60 Hz or six D cells	
Readout:	`00001.`	Digital LED
Physical:	18x7.5x9.25" 12 Lbs. 460x190x235mm 5.4 kg	
Status:	Active Manufacturer Discontinued Model	
Rarity:	Scarce	
Reviews:	📖 *Passport* 1987 *WRTH* 1980-1983	

	New	Used
Price:	$539-650	$300-360
Rating:	★★★★★	★★★★

---

87 S.P. Radio

S.P. Radio A/S of Denmark is one of Europe's leading producer of marine communications equipment. S.P. Radio is a member of the S.A.I.T.-Radio Holland Group.

For more than half a century S.P. Radio A/S has been a market leader in maritime radio communications. S.P. Radio offers complete communications solutions in keeping with the latest technology.

S.P. Radio is known by ship board radio operators worldwide for its unique trademark colored green cabinets. These radios are fondly referred to as the *Green Wonders*.

S.P. Radio A/S
Porsvej 2
DK-9200 Aalborg SV, Denmark

Radio Holland
8943 Gulf Freeway
Houston, TX 77017-7004

S.P.RADIO A/S

R2122

General Coverage Communications Receiver
Double Conversion Superheterodyne. Solid State.

Features:
- ¼" Head. Jack
- RF Gain
- Mute
- Dimmer
- S-Indicator
- Keypad
- 100 Memories
- 2182 kHz Preset
- Speaker
- Clarifier
- Scan
- AGC/MGC
- Squelch
- Sweep

Specifications:
Coverage ...... 100 - 30000 kHz
Selectivity ...... 6.6/2.7-.3 kHz[1]
IF Rejection ... >70 dB
Image Rej ...... >90 dB
Audio Out ...... 8 ohms 5W
Modes .... AM/LSB/USB/CW
Stability .... ±1.3 PPM
Environment ... -15° to +55° C.
Antenna Input . 50 ohms

Comments:
In addition to the 100 user defined memories, the radio also has I.T.U. defined maritime telephony channels. Nylon coated green metal cabinet.

Made In: Denmark -1998

Voltages: 10.8 to 32 VDC
.7 Amps

Readout: 00000.1 Digital LED

Physical: 8.9x4.3x11.8" 9.3 Lbs
225x115x299mm 4.2 kg

Status: Active Manufacturer
Active Model

Rarity: Typically Unavailable

	New	Used
Price:	①	⑤
Rating:	⑥	⑥

S. P. RADIO A/S

RM2150

GMDSS Communications Receiver
Double Conversion Superheterodyne. Solid State.

Features:
- ¼" Head. Jack
- S-Indicator
- Speaker
- AGC/MGC
- RF Gain
- Keypad
- DSC Decoder
- Scan
- Dimmer
- 2187.5 kHz Preset

Specifications:
Coverage ......See Ccmments		Modes .... GMDSS DATA	
IF Rejection ...>70 dB		Environment ... -15° to +55° C.	
Image Rej......>90 dB		Antenna Input . 50 ohms	
Audio Out ......8 ohms 5W			

Comments:
This receiver covers only the GMDSS distress frequencies: 2187.5, 4207.5, 6312, 8414.5, 12577 and 16804.5 kHz. Nylon coated green metal cabinet.

Made In:	Denmark	-1998
Voltages:	24 VDC or 110/220 VAC 50/60 Hz	
Readout:	**00000.1** Digital LED	
Physical:	8.9x4.3x11.8" 9.3 Lbs 225x115x299mm 4.2 kg	
Status:	Active Manufacturer Active Model	
Rarity:	Typically Unavailable	

	New	**Used**
Price:	①	⑤
Rating:	⑥	⑥

S. P. RADIO A/S

RM2151

General Coverage Communications Receiver
Double Conversion Superheterodyne. Solid State.

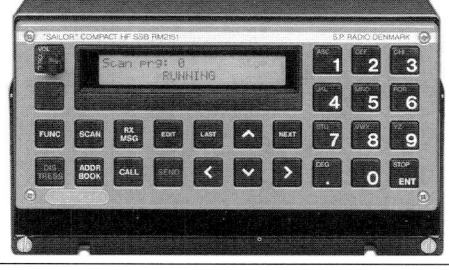

Features:
- ¼" Head. Jack
- Speaker
- AGC/MGC
- DSC Decoder
- RF Gain
- Keypad
- Clarifier
- Squelch
- Mute
- 100 Memories
- Scan
- Sweep
- Dimmer
- 2182 kHz Preset

Specifications:
Coverage ......100 - 30000 kHz		Modes .... AM/LSB/USB/CW	
Selectivity ......6.6/2.7-.3 kHz[1]		Stability .... ±1.3 PPM	
IF Rejection ...>70 dB		Environment ... -15° to +55° C.	
Image Rej......>90 dB		Antenna Input . 50 ohms	
Audio Out ......8 ohms 5W			

Comments:
Nylon coated green metal cabinet.

Made In:	Denmark	-1998
Voltages:	24 VDC or 110/220 VAC 50/60 Hz	
Readout:	**00000.1** Digital LED	
Physical:	8.9x4.3x11.8" 9.3 Lbs 225x115x299mm 4.2 kg	
Status:	Active Manufacturer Active Model	
Rarity:	Typically Unavailable	

	New	**Used**
Price:	①	⑤
Rating:	⑥	⑥

The **Squires-Sanders** company was a leading American manufacturer of quality communications equipment in the 1960's.

Clegg Laboratories of Mt. Tabor, New Jersey was a division of Squires & Sanders. They were a respected manufacturer of VHF amateur transmitters in the mid 1960's. In the late 1960's Squires-Sanders also produced high end CB radio equipment.

The SS-IBS is highly collectable and especially prized if found with the SS-1V panoramic display scope. The SS-1V has a strange sort of spectrum display that shows the entire 500 kHz band with an arrow moving as you change frequency.

Squires-Sanders, Inc.
475 Watchung Ave.
Watchung, NJ   1963-1964

Squires-Sanders, Inc.
Martinsville Road & Liberty Corners
Millington, NJ   1968

 **Squires-Sanders, Inc.**

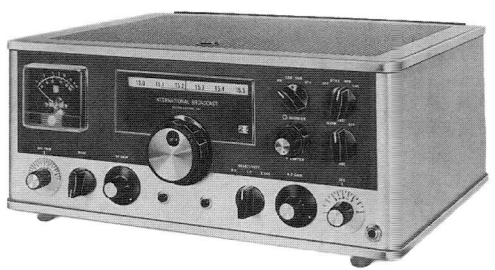

SS-IBS

Shortwave Broadcast Communications Receiver
Double Conversion Superheterodyne. 13 Tubes plus Semiconductors.

Features:
- ¼" Head. Jack   • S-Meter   • Antenna Trimmer   • AGC SLO/FST/OFF
- RF Gain   • BFO   • Calibrator 100 kHz
- Mute   • Noise Limiter   • Motor Driven Tuning

Specifications:
Coverage ...... SWBC. See Comments.   Modes .... AM/LSB/USB (CW)
Selectivity ...... 8/5/2.5 kHz   Stability .... <100 Hz within 1 hour
Sensitivity ...... <.5µV (1µV @7MHz).   IF Rejection .... 60 dB
Image Ratio ... 60 dB   Antenna Input . BNC
Audio Out ...... 1 W 10% distortion   Freq. Accuracy 1 kHz.

Accessories: SS-1RS Speaker   SS-1V Bandscope
Rack Mount Kit   Antenna Matcher   SS-1S Noise Silencer

Comments: Coverage: 3.5-4, 5.9-6.4, 7-7.5, 9.5-10, 11.5-12, 15-15.5, 17.5-18, 21.4-21.9 and 25.6-26.1 plus 5-5.5 and WWV 10 MHz. Frequency accuracy of 1 kHz is achieved with an odometer-type two digit display used in conjunction with the analog dial. Two buttons under the main knob provide motor assisted tuning. A highly advanced receiver designed for commercial and broadcast applications. Rarely seen in the used market. The optional SS-1V (shown right) is very scarce.

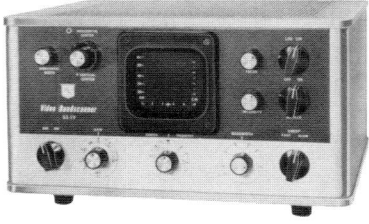

	Made In:	United States 1963-1967
	Voltages:	115/230 AC 50/60 Hz 55 W
	Readout:	[dial] 01. Analog Lin. & Mech. Dig.
	Physical:	16.25x7.75x13" 25 Lbs. 413x197x330mm 11.3 kg
	Status:	Inactive Manufacturer Discontinued Model
	Rarity:	Very Scarce

	New	Used
Price:	$1200	$700-850
Rating:	★★★★★	★★★★

Squires-Sanders, Inc.

SS-1R

Amateur Band Communications Receiver
Double Conversion Superheterodyne. 12 Tubes plus Semiconductors.

Features:

• ¼" Head. Jack	• S-Meter	• Antenna Trimmer	• AGC SLO/FST/OFF
• RF Gain	• BFO	• Standby	• Calibrator 100 kHz
• Mute	• Noise Limiter	• Dial Lamp	• Motor Driven Tuning

Specifications:

Coverage ......Ham. See Comments
Selectivity ......5/2.5/.5 kHz
Sensitivity ......<.5µV (1µV @7MHz).
Image Ratio ...60 dB
Audio Out ......1 W 10% distortion

Modes .... AM/LSB/USB/CW
Stability .... <100 Hz within 1 hour
IF Rejection .... 60 dB
Antenna Input . BNC 52 ohm
Freq. Accuracy 1 kHz.

Accessories:

SS-1RS Speaker   SS-1V Bandscope   SS-1S Noise Silencer
Rack Mount Kit

Comments:

Supplied coverage: 3.5-4,7-7.5, 14-14.5, 21-21.5 and 28.5-29 plus 5-5.5 and WWV 10 & WWV 15 MHz. Optional crystal positions provide for full 10 meter coverage plus two 500 kHz blocks. Frequency accuracy of 1 kHz is achieved through an odometer type two digit display used in conjunction with the linear analog dial. The SS-1R matches the SS-1T transmitter. Rarely seen in the used market. [¥800,000].

Variants:

The **SS-1R 701 Series** started in 1966 featured .25µV sensitivity, improved stability, and product detector plus extended 10 meter reception.

Made In: United States 1963-1967

Voltages: 115/230 AC 50/60 Hz
55 W

Readout: | Analog Lin.
& Mech. Dig.

Physical: 16.25x7.75x13" 25 Lbs.
413x197x330mm 11.3 kg

Status: Inactive Manufacturer
Discontinued Model

Rarity: Very Scarce

Reviews: *QST* May 1964
Electric Radio July 1994

	New	Used
Price:	$895-995	$550-750
Rating:	★★★★	★★★★

The **Standard Radio** is a leading supplier of amateur radio equipment in Japan. Standard amateur gear is now also distributed in the United States.

Some Standard amateur radio products are manufactured by Marantz Japan, Inc.

Standard made a number of analog portable receivers but only one communications receiver, the C-6500. The C-6500 was not aggressively promoted in the United States, although it was sold briefly by A.E.S.

Standard Radio Company
Tokyo,
Japan

 STANDARD

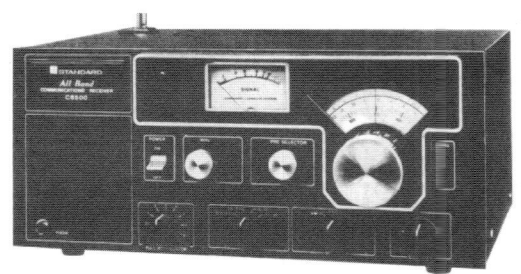

C-6500

General Coverage Communications Receiver
Double Conversion Superheterodyne. Solid State.

Features:
- ¼" Head. Jack
- Dial Lamp
- Line Out Jack
- S-Meter
- Speaker
- Tilt Legs
- Attenuator
- Preselector
- Mute Line
- Clarifier
- Telescopic Whip
- Dial Lamp Switch

Specifications:
Coverage ...... 500 - 30000 kHz
Selectivity ...... 7/4 kHz -6dB.
Sensitivity .... 5µV 1.5-30 MHz SSB
Audio Out ...... 1.5 W

Modes .... AM/USB/LSB-CW
Stability .... <500 Hz after warm-up
Antenna Input . 52-75 Ohms

Circuit Complement:
3 Integrated circuits, 16 transistors, 2 FETs and 30 diodes.

Comments:
An early example of the Wadley Loop system with an implementation similar to the FRG-7 (although the preselector is not calibrated). The C-6500 has 5 kHz tuning resolution. The Clarifier knob serves as a BFO for SSB and CW and as a fine tuning in the AM mode. The attenuator is activated by pulling out the volume knob. Ranges: .5-1.5, 1.5-5, 5-12 and 12-30 MHz. This radio was sold in Japan, Australia and Europe but was not actively marketed in the United States.

Made In: Japan 1977-1979

Voltages: 117 VAC or 12 VDC or 8 D cells

Readout: Analog Lin.

Physical: 13.4x11.4x6.1" 14 lbs.
340x290x156mm 6.4 kg

Status: Active Manufacturer
Discontinued Model

Rarity: Very Scarce

	New	Used
Price:	$250-400	$150-180
Rating:	⑥	○

The **Standard Radio & Telefon Company** was founded in 1938 at Bromma, a Stockholm suburb, as a subsidiary of the **Standard Telephone and Cables Ltd.**, for production of radio and telephone transmission equipment. Early in the history, the manufacture of radio equipment for maritime mobile use was adopted, and sometimes the equipment was marketed under different trademarks such as International Marine Radio Co. (I.M.R.C.).

S.R.T. was acquired by I.T.T. in the late 1940's and diversified its product lines into military radio equipment. SRT was the main HF radio supplier in later years to the Swedish Armed Forces and for several decades was also one of the main suppliers to the Swedish Telecommunications Adm. for shipboard and coast radio equipment. In 1990, the SRT operations was sold by I.T.T. to the French conglomerate Alcatel and shortly afterwards the radio business formed a new independent company Standard Radio Systems AB, which later was bought by Raytheon Marine. Currently, the maritime HF radio equipment is produced under the Raytheon Standard Radio Marine Label.

Author's Note: A very special thanks to Rolf Folkesson *SMØHP* and Karl-Arne Markström *SMØAOM* for suppling the information for this chapter. Stig Adolfsson, Olle Gerdes, Sixten Hansson and the Telecommunications Museum, Stockholm and the Museum of Military Vehicles, Malmköping also contributed.

Standard Radio & Telefon AB
Bromma, Sweden

Standard Radio & Telefon AB
Radio Systems
Siktgatan 11
Box 501
S-162 15 Vällingby, Sweden

Standard Radio & Telefon AB

CR90

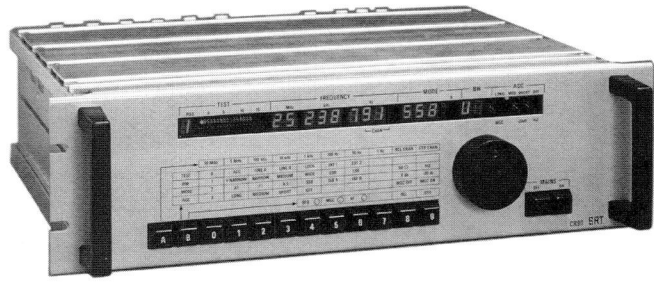

General Coverage Communications Receiver
Double Conversion Superheterodyne. Solid State.

Features:
- S-Indicator
- RF Gain
- Rack Handles
- 9 Memories
- Mute
- IF Output
- 5 MHz Ref I/O
- BFO
- Panoramic Output
- AGC SHT/MED/LNG/OFF
- Modular Construction
- Dual Line Outputs

Specifications:
Coverage ...... 10 - 30000 kHz
Selectivity ...... 6.86/1/.6/.3 kHz -3dB.
IF Reject .... >100 dB
Max Ant Input 30 V EMF
Audio 2W

Modes .... AM/LSB/USB/CW/FSK
Accuracy .... 1 x 10^{-7}
Image Reject .. >100 dB
Environment ... -40° to +55° C.
Antenna Input . BNC 50 ohms

Accessories:
ISB Option
100 Channel Memory
Diversity Adapter
FS90 FSK Demodulator
Serial Data Interface
Parallel Data Interface
FS91 FSK Demod & Squelch
CR90R Remote Control Unit

Comments: The CR90 is the successor the CR300. The CR90 became the core of "System 90" architecture from SRT, which included exciters, power amps, control devices and the early Automatic Link Establishment system (ARTRAC). The CR90 was an ideal system component in a computer controlled HF radio system, but it never became popular with radio operators due to its awkward front panel arrangement. The "A" button followed by 0-7 selects the digit to be tuned (10 MHz, 1 MHz, 100 kHz, etc.). The "B" button is pressed to select of one of four rows on a function matrix, and the "0-9" keys are used to select the column function.

Made In: Sweden   1979-1988

Voltages: 110/220 VAC 40-400 Hz
or 21-32 VDC

Readout: `00000.001` Digital LED

Physical: 19x5.25x12.6" 26 Lbs
482x133x320mm 12 kg

Status: Active Manufacturer
Discontinued Model

Rarity: Typically Unavailable

	New	**Used**
Price:	$9000	$1200-1900
Rating:	⑥	⑥

Standard Radio & Telefon AB

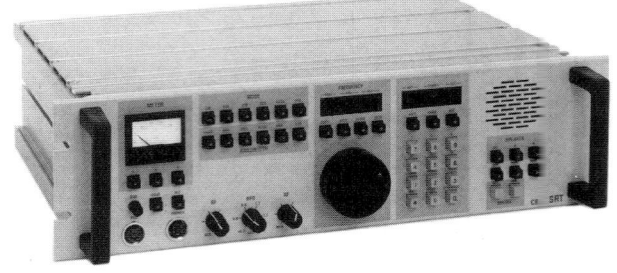

CR91

General Coverage Communications Receiver
Double Conversion Superheterodyne. Solid State.

Features:
- DIN Head. Jack
- S-Meter
- Dimmer
- AGC SHT/MED/LNG/OFF
- RF Gain
- Mute
- BFO
- Modular Construction
- Rack Handles
- IF Output
- Sweep
- Attenuator -20 dB
- Scan
- RS232 Port
- 100 Memories
- 3 Tune Rates

Specifications:
Coverage ...... 10 - 30000 kHz
Selectivity ...... 6.8/3/1.5/.6/.3 kHz -3dB.
IF Reject ....>100 dB
Max Ant Input 30 V EMF
Audio ....2W 4-8 ohms

Modes .... AM/LSB/USB/CW/FSK/ISB
Accuracy .... 1×10^{-7}
Image Reject .. >100 dB
Environment ... -30° to +55° C.
Antenna Input . BNC 50 ohms

Accessories:
SS91 Keypad Controller (shown)   CR91R Remote Unit   FS92 FSK Demod.

Comments:
The CR91 was a development from the CR90 to overcome the front-panel problems. If the CR90 was a nightmare for the radio operator, the CR91 was a dream. Its specifications came from signal intelligence requirements and it has a conventional front panel designed with the operator in mind. A few CR91s were exported for use in diplomatic networks and some found their way into coast stations.

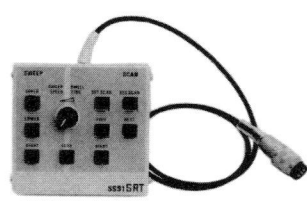

Made In:	Sweden	1982-1992

Voltages:	220 VAC 40-400 Hz
	60 VA

Readout: `00000.001` Digital LED

Physical:	19x5.25x12.6" 26 Lbs
	482x133x320mm 12 kg

Status:	Active Manufacturer
	Discontinued Model

Rarity:	Typically Unavailable

	New	Used
Price:	$12000	$1800-2500
Rating:	⑥	③

Standard Radio & Telefon AB

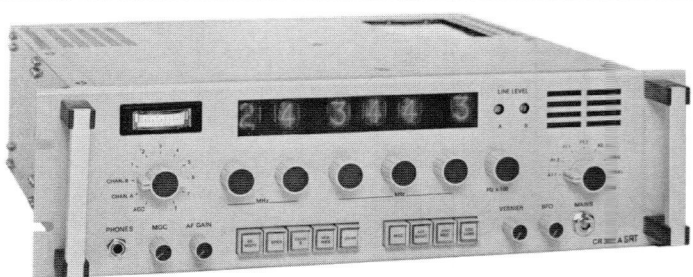

CR302

General Coverage Communications Receiver
Triple Conversion Superheterodyne. Solid State.

Features:
- ¼" Head. Jack
- S/AF/Test Meter
- Speaker
- AGC SHT/MED/LNG
- RF Gain
- Mute
- BFO ±2 kHz
- Modular Construction
- Rack Handles
- IF Output
- Speaker Switch
- Attenuator -20 dB
- Parallel Port
- Dual Line Out
- Remote Operation

Specifications:
Coverage ...... 10 - 30000 kHz
Selectivity ...... 6.8/3/1.5/.6/.3 kHz -3dB.
IF Reject ....>100 dB
Stability .... 10^{-7} after 4 mins.
Audio ....4-16 ohms 1 W

Modes .... AM/LSB/USB/CW/FSK/ISB
Image Reject .. >100 dB
Environment ... -30° to +55° C.
Max Ant Input . 20W
Antenna Input . 50 ohms

Comments: Tuning is provided via decadic knobs (one for each frequency digit to 100 Hz.). A vernier knob ±1 kHz is used for fine tuning. The CR300 receiver family was designed for the Swedish Air Force to replace older tube-type sets. It uses the same concept as the CR1000, but is considerably more integrated, packing the whole receiver, synthesizer, frequency standard and parallel remote control interface into a single 5¼" 19" rack space. The use of the CR300 spread outside the Swedish Armed Forces to merchant shipping, coast stations, point-to-point and diplomatic applications.

Variants:
Model **CR301** is the same sans ISB. Model **CR305** is a simplified version for marine applications.

Made In:	Sweden	1971-1987

Voltages:	105/115/127/220/230/240
	VAC 45-400 Hz 60 VA

Readout: `00000.1` Digital Nixie

Physical:	19x5.25x16.15" 40 Lbs
	482x133x410mm 18 kg

Status:	Active Manufacturer
	Discontinued Model

Rarity:	Typically Unavailable

	New	Used
Price:	$6000	⑤
Rating:	⑥	⑤

Standard Radio & Telefon AB

CR304

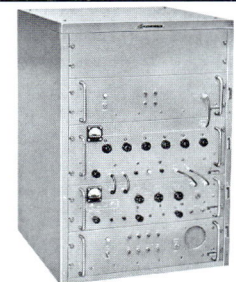

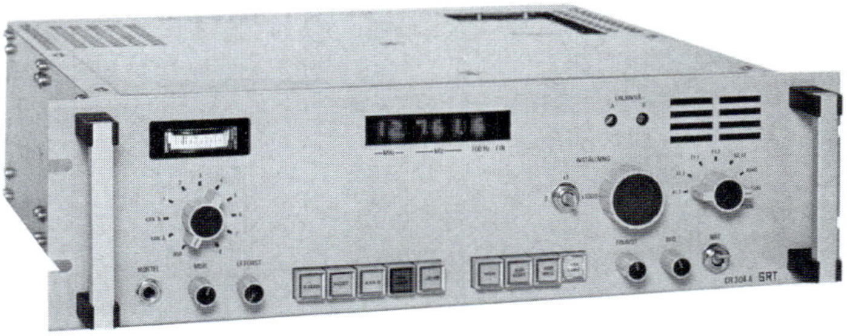

General Coverage Communications Receiver
Triple Conversion Superheterodyne. Solid State.

Features:
- ¼" Head. Jack
- RF Gain
- Rack Handles
- Parallel Port
- S/AF/Test Meter
- Mute
- IF Output
- 2 Tune Rates
- Speaker
- BFO ±2 kHz
- Speaker Switch
- Remote Operation
- AGC SHT/MED/LNG
- Modular Construction
- Attenuator -20 dB
- Dual Line Output

Specifications:
Coverage ...... 10 - 30000 kHz
Selectivity ...... 6.8/3/1.5/.6/.3 kHz -3dB.
IF Reject ....>100 dB
Stability ....10⁻⁷ after 4 mins.

Audio ....4-16 ohms 1 W

Modes .... AM/LSB/USB/CW/FSK/ISB
Image Reject .. >100 dB
Environment ... -30° to +55° C.
Max Ant Input . 20W
Antenna Input . 50 ohms

Comments:
The CR304 and CR303 are similar to the CR302 but utilize a single tuning knob rather than decadic switches.

Variants:
Model **CR303** is the same as the CR304 sans ISB. The **CR307** is similar to the CR303, but this version meets the requirements for marine radio receivers.

Made In:	Sweden	1971-1987
Voltages:	105/115/127/220/230/240 VAC 45-400 Hz 60 VA	
Readout:	`00000.1` Digital LED	
Physical:	19x5.25x16.15" 40 Lbs 482x133x410mm 18 kg	
Status:	Active Manufacturer Discontinued Model	
Rarity:	Typically Unavailable	

	New	Used
Price:	$7000	⑤
Rating:	⑥	⑥

Standard Radio & Telefon AB

CR1000

General Coverage Communications Receiver
Triple Conversion Superheterodyne. Solid State.

Features:
- ¼" Head. Jack (2)
- Ext. Spkr Jack
- BFO
- S-Meter
- Dial Lamp
- Speaker
- AGC 4 Position
- Standby
- Rack Handles
- Atten. -15/30/45 dB
- Speaker Switch
- Line Out 600 ohms

Specifications:
Coverage ...... 1000 - 30000 kHz
Selectivity ...... 7.4/7/1.5/.6 kHz
Sensitivity ......<2µV 10 dB S/N SSB
IF .... See Comments
Max Ant Input 10 V RMS
Audio Out ......6 ohms 1.5 W

Modes .... AM/CW/LSB/USB/FSK/ISB
Stability ....10⁻⁸ per 24 hours
IF Rejection ....>100 dB
Image Reject .. >100 dB
Environment ... -40° to +55° C.
Antenna Input . 50 ohms

Comments:
Intermediate frequencies are: first 130.7, 140.7 or 150.7 MHz, depending on the frequency band, second 10.7 MHz and third is 200 kHz. The CR1000 represented a complete new receiver concept, using a solid-state frequency synthesizer for generating all oscillators from a single frequency standard, having ISB and remote control capability as standard and using a VHF first IF. This concept was used for all future SRT receivers. Most CR1000's were installed in Swedish Army communications shelters, but a number were exported. The CTR1000 system was a combination of the CT1000 1 kW PA, the CTD1000 ISB exciter and the CR1000 receiver, sharing a common remote control.

Made In:	Sweden	1965-1975
Voltages:	220 VAC 50 Hz	
Readout:	`●●●●●.0` Digital Knob	
Physical:	Receiver Only: 19x5.1875x20" 482x133x530mm Oscillator: 19x7x20" 482x178x530mm Power Supply: 19x5.1875x20" 482x133x530mm	
Status:	Active Manufacturer Discontinued Model	
Rarity:	Typically Unavailable	

	New	Used
Price:	①	⑤
Rating:	⑥	⑥

Standard Radio & Telefon AB

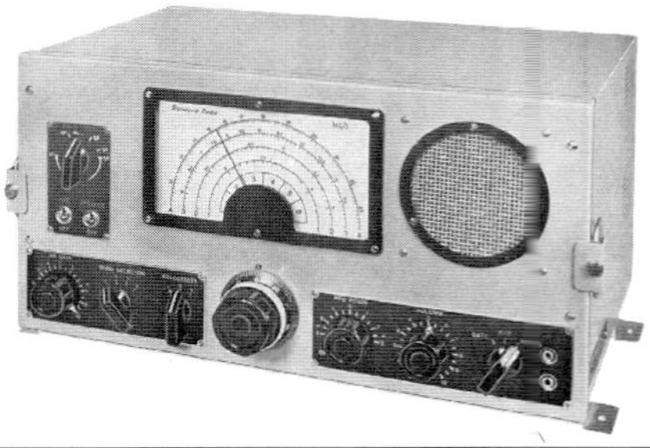

SR21B

General Coverage Communications Receiver
Single Conversion Superheterodyne. 6 Tubes.

Features:
- Head. Jack
- Ext. Spkr Jack
- BFO
- RF Gain
- Dial Lamp
- Speaker
- AVC
- Standby
- Mute Line
- Speaker Switch

Specifications:

Coverage	1600 - 23500 kHz	Modes	AM/CW
Selectivity	5/4/2.5 kHz	IF	790 kHz
Sensitivity	5μV	Antenna Input	Terminals
Audio Out	20/200/400 ohms		

Circuit Complement:
This receiver uses six ECH21 tubes!

Comments:
Ranges: 1.6-3.3, 3.2-6.2, 5.9-12 and 11.5-23.5 MHz. Maximum output to the 20/200/400 outputs is 0.3 watts. Headphone output is 4000 ohms. The receiver normally operates from the ships 110 VDC mains. A change over switch on the front panel allows the selection of power from a 65 volt emergency battery source. The SR21 was commonly used as the main receiver on Swedish flag ocean-going merchant ships well into the 1960's. The introduction of single sideband made the SR-21 series obsolete.

Variants:
Model **SR21** covers 1.6-17.2 MHz with ranges of 1.6-3, 2.8-5.3, 5-9.5 and 9.1-17.2 MHz. It has an 820 kHz IF. Model **SR21A** (1946) differences unknown.

Made In: Sweden 1946-1960

Voltages: 110/220 VDC or 65 VDC battery

Readout: Analog

Physical: 17.6x9.1x10.75" 35.2 Lbs. 448x230x273 mm 16 kg

Status: Active Manufacturer Discontinued Model

Rarity: Typically Unavailable

	New	Used
Price:	①	⑤
Rating:	⑥	⑥

Standard Radio & Telefon AB

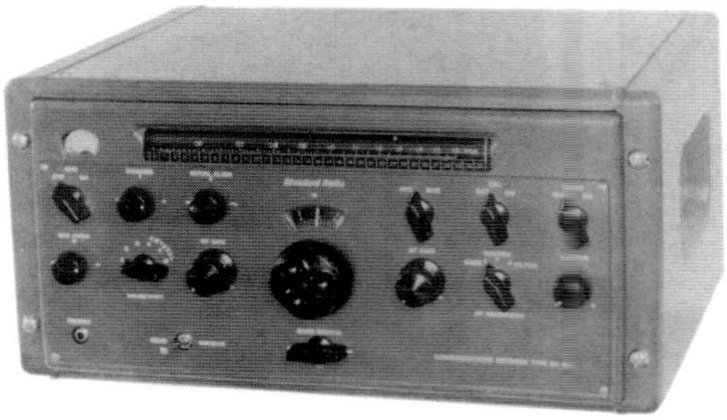

SR25

General Coverage Communications Receiver
Single Conversion Superheterodyne. 16 Tubes.

Features:
- ¼" Head. Jack
- RF Gain
- Ext. Spkr Jack
- Standby
- S-Meter
- ANL
- Dial Lamp
- BFO
- 64:1 Tune Ratio
- Mute Line
- Spinner Knob
- AGC/MGC
- Turret Type Display
- Calibrator 500 kHz.
- AF Bandwidth Filter
- Line Out 200 ohm

Specifications:

Coverage	520 - 30000 kHz	Modes	AM/CW
Selectivity	12.5/7.4/2.5/1.7/1/.5 kHz	Stability	<2 kHz/MHz after warmup
Sensitivity	5μV S/N 14 dB	Image Rej.	>100 dB @.52 MHz
IF	455 kHz	Image Rej.	>40 dB @30 MHz
Audio Out	2W 8 ohms, 200 ohms	Antenna Input	Terminals

Circuit Complement: 6BA6 1st RF Amp, 6BA6 2nd RF Amp, 6BE6 Mixer, 6C4 Osc., 6BA6 1st IF Amp, 6BA6 Second IF Amp, 6BA6 3rd IF Amp, 6AU6 BFO, 6AU6 AGC Amp, 6C4 Buffer Amp, 6AL5 AGC-MGC Delay Diodes, 6BA6 Calibrator Osc., 6AL5 NL, 6AU6 AF Amp, 6AQ5 Power Output, 0A2 Regulator, 0A51 Signal Detector diode and 0A51 AGC Detector diode.

Comments: Ranges: .52-1.6, 1.6-5, 5-10.8, 10.8-17, 17-23 and 23-30 MHz. The accuracy of the tuning scale is ±2 kHz below 10 MHz and ±5 kHz above 10 MHz. The SR25 was designed to military specifications to compete with he HRO and SuperPro receivers. Many SR25's were used in military FSK point-to-point systems and in some ship stations. An SR25 was lent to the historical Swedish Geophysical Expedition to Svalbard in 1958-59.

Made In: Sweden 1956-1965

Voltages: 110-127, 154-240 VAC 50/60 Hz

Readout: Analog

Physical: 19.5x9.8x15.125" 50.7 Lbs. 495x250x385mm 23 kg

Status: Active Manufacturer Discontinued Model

Rarity: Typically Unavailable

	New	Used
Price:	①	⑤
Rating:	⑥	⑥

91 Standard Telephones & Cables

Standard Telephones & Cables Pty. Ltd.
London, England

Standard Telephones and Cables Pty. Ltd. was a British company with a head office in London and branches in a number of countries throughout the British Empire.

The Australian branch had been making quite a lot of domestic, industrial and some military radio equipment from the 1920's onwards.

During the War years, they made not only communications receivers, but also mobile radio transceivers, telephones and quite a number of high-powered transmitters and antenna systems. During the 1950's and 1960's, S.T.C. Australia Branch was manufacturing broadcast transmitters from a few hundred watts, right up to 250 kW. They also produced a range of radio navigation beacons (NDB's).

S.T.C. was also active in the two-way radio market, and along with A.W.A., did quite a bit of pioneering work with the manufacture and use of VHF land-mobile communications equipment in Australia.

Standard Telephones & Cables Pty. Ltd.

A.679H

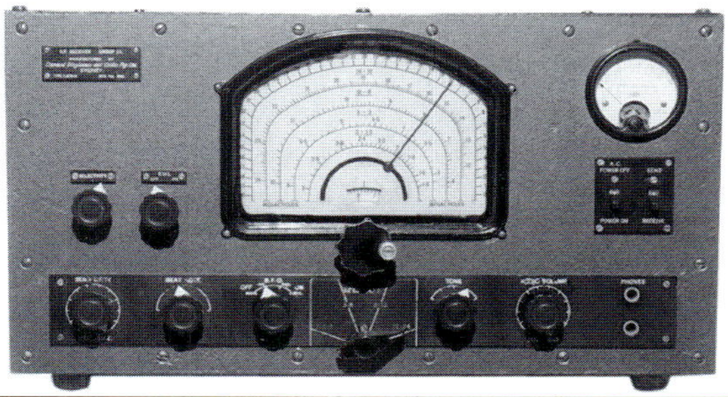

General Coverage Communications Receiver
Single Conversion Superheterodyne. 9 Tubes.

Features:
- ¼" Head. Jack (2)
- S-Meter
- AVC ON/OFF
- 80:1 Tune Ratio
- RF Gain
- BFO ±3 kHz
- Dial Lamp
- Line Out 600 ohms
- Logging Scale
- Tone 5 Position
- Standby
- Antenna Trimer
- Spinner Knob
- Dial Lamp Switch

Specifications:
Coverage ...... 1500 - 24000 kHz
Sensitivity ...... 2 μV
Audio Out ...... 2W 600 ohms
Modes .... AM/CW/MCW
IF .... 455 kHz
Antenna Input . Terminals 80 Ohms

Circuit Complement:
6U7G 1st RF, 6U7G 2nd RF, 6K8G Frequency Converter, 6U7G 1st IF, 6G8G 2nd IF /Detector, 6B8G Audio Amp, 6V6G Audio Output, 6J7G BFO and 5V4G Rectifier. Has a crystal filter.

Comments: Ranges: 1.5-3, 3-6, 6-12 and 12-24 MHz. This model is typically unavailable in North America and even scarce in Australia. The used value in Australia is in the range of $200-$300. Requires a speaker.

Variants: Model **A.679C** is earlier H.F. direction finding version. The model **A.679J** is a slightly updated version of the A.679H. It has small rack handles on it. The **AMR-300** (military designation) featured some minor upgrades. The AMR-300 was supplied to Australian and U.S. forces in the Pacific. After the War the AMR-300 was used as the main base receiver in several Royal Flying Doctor Service bases and were used through the fifties.

Made In: Australia 1939-1945

Voltages: 110/240 VAC 50-60 Hz

Readout: Analog

Physical:

Status: Inactive Manufacturer
Discontinued Model

Rarity: Typically Unavailable

	New	Used
Price:	①	⑤
Rating:	⑥	⑥

Sunair Electronics is a Florida Corporation organized in 1956. It is engaged in the design, manufacture and sale of high frequency single sideband communications equipment utilized for long range voice and data communications in fixed station, airborne, mobile and marine applications.

Sunair products are marketed both domestically and internationally and are primarily intended for strategic military and other governmental applications. Sales are executed through systems engineering companies, worldwide manufacturers and direct to the United States government for foreign military assistance.

Sunair products employ advanced solid state designs and computer controlled networking capabilities. Their principal products include: H.F. transceiver, H.F. receivers, H.F. exciters, automatic antenna couplers, digital modems, frequency management systems and H.F. airborne transceivers.

Sunair Electronics Inc.
3101 S.W. Third Av.
Fort Lauderdale, FL
33315   ⁻956-1998

GSR-920

General Coverage Communications Receiver
Double Conversion Superheterodyne. Solid State.

Features:
- ¼" Head. Jack
- RF Gain
- Speaker
- S-Meter
- Rack Handles
- Speaker Switch
- Dimmer
- AGC
- Line Out 600 ohm
- Modular Construction
- Ext. Speaker Jack

Specifications:
Coverage ...... 1600 - 29999 kHz
Selectivity ...... 8/3.3 kHz -6dB.
Sensitivity ..... <.5µV 10 dB S/N SSB
Image Rej ...... >80 dB
Audio Out ...... 2 W 8 ohms 10% dist.

Modes .... AM/USB/LSB
Stability .... 1 part 10^{-6}
IF Rejection .... >70 dB
Antenna Input . 50 ohm
Environment ... -30° to +65° C.

Accessories:
DC module (for 12/24 VDC operation)   ISB Option
TCXO 1 x 10^{-8}

Comments:
This receiver employes decadic knob tuning. It is designed for the military market. This model is still available from Sunair and is also seen on the used market from time to time.

Mfg. In:	United States 19⁻9-1998	
Voltages:	115/230 VAC 48-35 Hz 50W	
Readout:	●●●●●.● Knobs	
Physical:	18.25x6x18" 41 Lbs. 463x152x457mm 18.5 kg	
Status:	Active Manufactuer Active Model	
Rarity:	Scarce	

	New	Used
Price:	①	$450-700
Rating:	⑥	⑥

R-9200

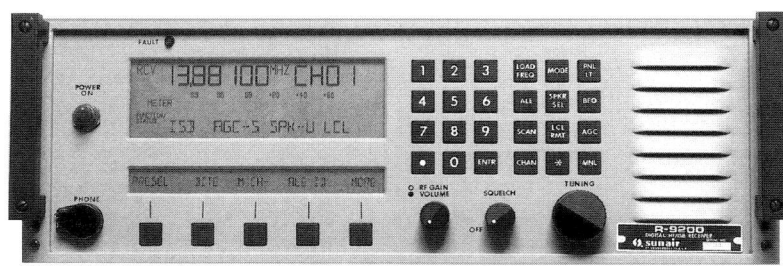

General Coverage Communications Receiver
Double Conversion Superheterodyne. Solid State.

Features:
- ¼" Head. Jack
- RF Gain
- Speaker
- Keypad
- Squelch
- Fault light
- Signal Indication
- Rack Handles
- 2nd IF Out Jack
- Scan
- Panel Lamp
- Panel Light Switch
- 128 Memories
- B.I.T.E
- BFO ±1.99 kHz
- Sweep
- ISB
- Modular Construction
- Ext. Speaker Jack
- AGC FST/MED/SLO
- Freq. Std. I/O 5MHz
- RS232/422/485 Jack

Specifications:
Coverage ...... 100 - 29999 kHz
Selectivity ...... 6/.3-3/.5 kHz -6dB.
Sensitivity ...... <.5µV SSB
Image Rej ...... >80 dB
Audio Out ...... 5W 8 ohms 5% dist.

Modes .... AM/USB/LSB/CW/ISB
Stability .... 1 part 10^{-6}
IF Rejection .... >80 dB
Antenna Input . 50 ohm
Environment ... -30° to +50°C.100% Hum.

Accessories:
High Stability   Filters
Remote Control   RCU-9310 Remote Controller Unit
Clock   F-9800 External Pre/Post Selector (shown below)

Comments: The built-in test feature provides fault isolation to the module level with descriptive readout on the front panel and individual module indication. A sophisticated and robust receiver designed for military, marine and commercial applications.

Mfg. In:	United States 1992-1998
Voltages:	115/230 VAC ±15% 26 VDC ±15%
Readout:	**00000.01** Digital LCD
Physical:	17.83x6x17.66" 39 Lbs. 454x152x449mm 39 kg
Status:	Active Manufacturer Active Model
Rarity:	Typically Unavailable

	New	**Used**
Price:	①	②
Rating:	⑥	⑥

Svenska Radio

Svenska Radio AB (S.R.A.) was founded in 1919 as the Marconi licensee in Sweden for radio communications technology as a joint venture between the Marconi company in England and the Swedish telecommunications company L.M. Ericsson which at the time, had no radio manufacturing capacity of its own. Svenska Radio made marine and military radio equipment from the beginning and HF equipment was added early to the product lines. Unfortunately, there is little information available about the pre-War H.F. equipment ... even in Sweden.

Svenska Radio was to take a key role in providing radio equipment for the Swedish military when Sweden was almost completely cutoff from import trade at the outbreak of *World War II*. Models MKL-940 (1941-1950) and MKL-941 (1944-1950) were essentially copies of the *Super-Pro,* produced from domestic materials. A model MK-46 was also produced from 1946-1960.

Production of H.F. equipment ended in he early 1970's when the company profile changed towards VHF and UHF land mobile and mobile telephony equipment. S.R.A. no longer exists as a separate company after merging in 1985 with the main Ericsson telecommunications corporation.

Author's Note: Thanks to Rolf Folkesson *SMØHP* and Karl-Arne Markström *SMØAOM* for supplying information for this chapter.

Svenska Radioaktiebolaget
Alströ mergatan 14
Stockholm 12, Sweden

SVENSKA RADIOAKTIEBOLAGET

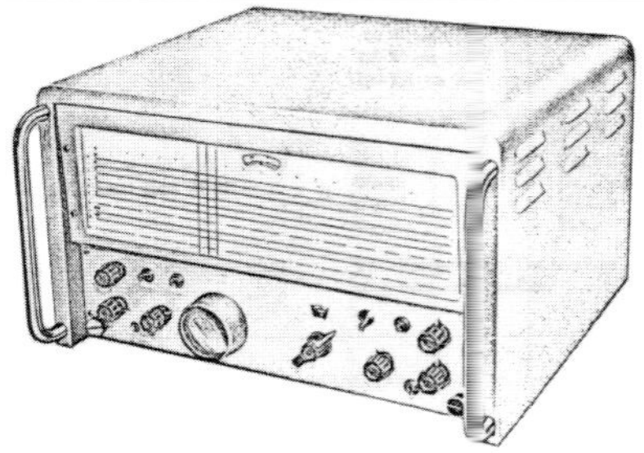

MR-201

General Coverage Communications Receiver
Double Conversion Superheterodyne. 12 Tubes.

Features:
- Head. Jack
- RF Gain
- Ext. Spkr Jack
- Standby
- Rack Handles
- ANL
- BFO
- 62:1 Tune Ratio
- Mute Line
- Antenna Trimmer
- Dial Lamp
- Bandspread

Specifications:

Coverage	85 - 28000 kHz	Modes	AM/CW/MCW
Selectivity	8/3/1.5/.5 kHz -6dB	Image Rej.	>70 dB 4-8 MHz
Sensitivity	5μV S/N .63-28 MHz	Image Rej.	>50 dB 8-12 MHz
IF	470, 70 kHz	Image Rej.	>40 dB 12-16 MHz
Audio Out	4 ohms 1.5W	Antenna Input	Terminals

Circuit Complement:
EF85 RF, ECH81 Converter, 6AU6 Osc., 6AU6 Osc., 6BA6 IF, 6BA6 IF Amp, 6AL5 Detector/Noise Limiter, 6AU6 BFO, 6AU6 LF and PL82 Output.

Comments:
Ranges: .085-.1, .1-.25, .25-.63, .63-1.6, 1.6-4, 4-8, 8-12, 12-16, 16-22 and 22-28 MHz. The MR-201 was designed to compete with the contemporary Standard Radio & Telefon models and with Eddystone. The receiver was not manufactured in large number because of the impact of upcoming introduction of single sideband. The receiver was designed to operate from either 110 VDC mains found on ships of that ear or 220 VAC or from an emergency battery pack. Finished in gray enamel.

Made In: Sweden 1959-1965

Voltages: 110/130/150/220 VAC 50/60 Hz or 110/220 VDC

Readout: Analog

Physical: 19x12.4x17.4" 55 Lbs
482x315x441mm 25 kg
Power Unit:
11.8x12.2x7.8" 25.3 Lbs
300x310x200mm 11.5 kg

Status: Inactive Manufacturer
Discontinued Model

Rarity: Typically Unavailable

	New	Used
Price:	①	⑤
Rating:	⑥	⑥

94 Swan

Swan Engineering, as it was first called, was founded by Herb Johnson W7GRA, in Arizona in late 1960. The first product was a single sideband transceiver called the SW-120. At that time the market was dominated by the expensive Collins KWM-2.

The Company quickly grew and moved to Oceanside, California and produced thousands of affordable single sideband transceivers for the growing amateur market.

Swan Electronics made only one receiver, the 600-R. The 600-R received only the amateur bands. It was designed to match the 600-T transmitter for those hams who still wished to operate "twins" or "separates".

The external 330 General Coverage Synthesizer was an accessory for the 600-R receiver. This gave the receiver continuous coverage from 3 to 30 MHz. Only a small number were made and they are very scarce.

Swan Electronics
A Subsidiary of Cubic Corporation
305 Airport Rd.
Oceanside, CA 92054   1968-

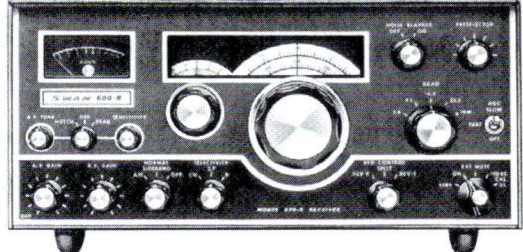

600-R

Amateur Band Communications Receiver
Single Conversion Superheterodyne. 7 Tube plus Semiconductors.

Features:
- ¼" Head. Jack
- S-Meter
- Preselector
- AGC FST/SLO/OFF
- RF Gain
- ANL
- Dial Lamp
- Calibrator 25/100 kHz
- Mute
- CW Sidetone
- Standby

Specifications:
Coverage ......Ham 80-10 Meters   Modes .... AM/SSB/CW/FSK
Selectivity ......2.7/_/_ kHz -6dB.   IF .... 5500 kHz
Sensitivity ......<.25µV 10 dB S/N   Image Rej. ...... >50 dB

Circuit Complement:
7 Tubes, 8 transistors and 12 diodes.

Accessories:
Speaker   NB-500 Noise Blanker   ICAF-500 Audio Notch Filter
AM Filter 4 kHz   330 GC Synthesizer (shown below)   CW Filter 500 Hz

Comments: Matches the Swan 600-T transmitter. Has built-in AC power supply. Features vernier tuning with virtually no backlash. Dial is calibrated every 2 kHz for 80 through 10 meters and every 5 kHz for 10 meters. The optional 330 GC Synthesizer covers: 3-5.4, 5.6-10, 10-16, 16-24 and 24-30 MHz.

Variants: Model **600-R Custom** includes the normally optional ICAF-500 Audio Notch Filter and NB-500 Noise Blanker $546 new.

Made In:	United States 1971-1974
Voltages:	120 VAC
	120 W
Readout:	Analog Lin.
Physical:	15x6.5x12" 23 Lbs
	380x165x305mm 10.4 kg
Status:	Active Manufacturer
	Discontinued Model
Rarity:	Very Scarce
Reviews:	QST January 1973

	New	**Used**
Price:	$395-440	$110-190
Rating:	★★★	★★

95 Taiyo Musen

The **Taiyo Musen Company** was established in 1947. The company manufactures equipment for: direction finding, LORAN, GPS, navigation, course plotting. They also make radio buoys, radar transponders and weather facsimile receivers. Their headoffice is in Tokyo with branches in Kobe, Kisyo and Fukuoka. The company employes 155.

Taiyo Musen is the world's leading manufacturer and supplier of radio direction finding equipment and they are commonly found on vessels of the Japanese fishing fleet. They are used for navigational purposes and for the location of transmitting fishing buoys. Over a dozen models are in their current product line. Many of these models operate exclusively above 25 MHz. These devices are also used in other countries to find the source of illegal transmissions.

Taiyo Musen Company Ltd.
20-7, 2-Chome, Ebisu-Nishi,
Shibuya-ku,
Tokyo 150, Japan

TAIYO MUSEN CO., LTD.

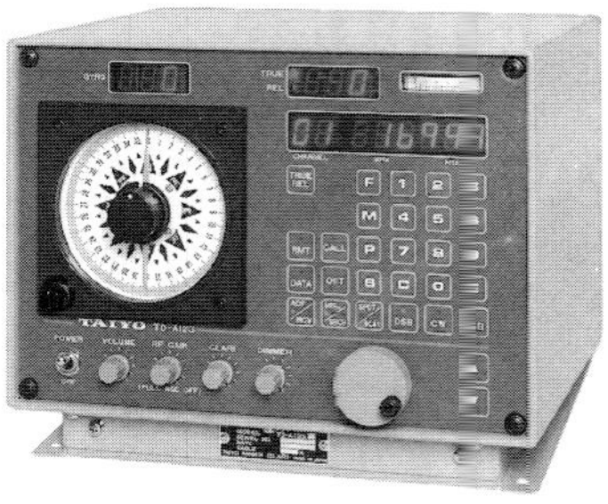

TD-A123

General Coverage Direction Finding Communications Receiver
Double Conversion Superheterodyne. Solid State.

Features:

• AGC	• S-Indicator	• CRT DF Indicator	• Keypad
• RF Gain	• Clarifier	• Dimmer	• Scanning
• 300 Memories	• Sweep	• Standby	• Up/Dn Tuning
• RS232 Port	• Spinner Knob	• AGC ON/OFF	

Specifications:

Coverage ...... 200 - 14000 kHz
Selectivity ...... 2 kHz -6dB.
Min. Field S. .. 3µV at 2 MHz.
Audio Out ...... 2 W 8 ohms
Modes .... AM/SSB/CW/MCW
D.F. Accuracy . ±1° at 1µV/m
Image Rej. ...... >60 dB

Accessories:

AC Power Supply
Gyro Compass Interface

Comments:

This receiver is designed for MF and HF direction finding. It is configured for commercial fishing, merchant and government ships. It includes a 31 inch (.8 meter) aluminum alloy cross loop antenna and 5 conductor cable. Note limited HF coverage.

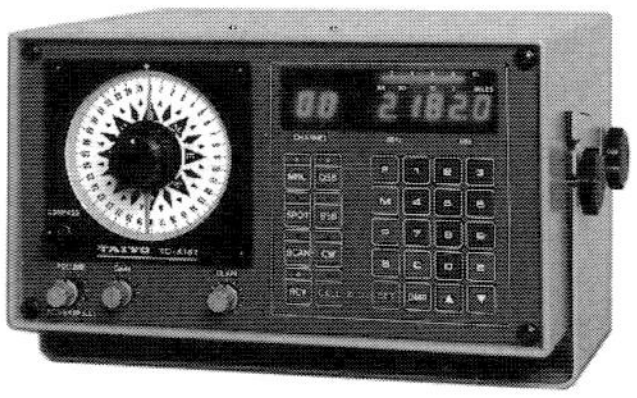

Variants:

The less expensive model **TD-A157** (shown right) covers only 200 to 4900 kHz, has 100 memories and lacks a rotary tuning knob.

Made In: Japan 1995-1998

Voltages: 24 VDC

Readout: 00000.1 Digital LED

Physical: 11.8x9.65x9.5" 29 Lbs.
300x245x242mm 13 kg

Status: Active Manufacturer
Active Model

Rarity: Typically Unavailable

	New	Used
Price:	①	⑤
Rating:	⑥	⑥

TAIYO MUSEN CO., LTD.

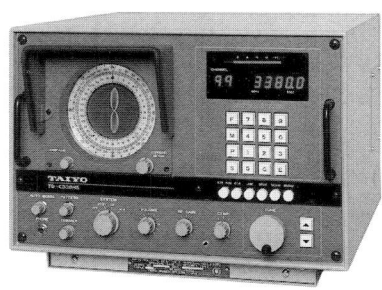

TD-C338HS

General Coverage Direction Finding Communications Receiver
Triple Conversion Superheterodyne. Solid State.

Features:
- ¼" Head. Jack
- S-Indicator
- CRT DF Indicator
- Keypad
- RF Gain
- Clarifier
- Dimmer
- Scanning
- 100 Memories
- Sweep
- Standby
- Up/Dn Tuning
- Rack Handles
- Spinner Knob

Specifications:

Coverage ......200 - 18000 kHz	Modes .... AM/SSB/CW/MCW
Selectivity ......2 kHz -6dB.	D.F. Accuracy . ±1° at 1µV/m
Min. Field S. ..2µV at 2 MHz.	Image Rej. ...... >60 dB
Audio Out ......2 W 4 ohms	IF .... 65.7, .455, 104.5 MHz

Accessories:
Gyro Compass Interface

Comments:
This receiver is designed for MF and HF direction finding. It is configured for commercial fishing, merchant and government ships. It includes a 31 inch (.8 meter) aluminum alloy cross loop antenna and 7 conductor cable. This receiver is very tolerant of long cable runs to the antenna.

Made In:	Japan	1984

Voltages:	100/110/115/220 VAC 50/60 Hz 60 W

Readout:	`00000.1` Digital LED

Physical:	16.1x11.6x17.7" 65 Lbs. 410x295x450mm 29.5 kg

Status:	Active Manufacturer Discontinued Model

Rarity:	Typically Unavailable

	New	Used
Price:	①	⑤
Rating:	⑥	⑥

 TAIYO MUSEN CO., LTD.

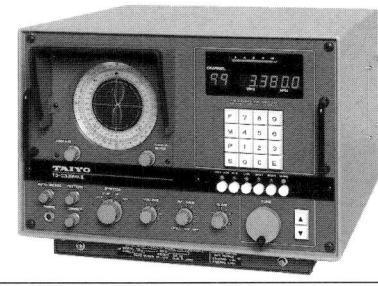

TD-C338MKII

General Coverage Direction Finding Communications Receiver
Triple Conversion Superheterodyne. Solid State.

Features:
- ¼" Head. Jack
- S-Indicator
- CRT DF Indicator
- Keypad
- RF Gain
- Clarifier
- Dimmer
- Scanning
- 400 Memories
- Sweep
- Standby
- Up/Dn Tuning
- Rack Handles
- Spinner Knob
- AGC ON/OFF

Specifications:

Coverage ......200 - 30000 kHz	Modes .... AM/SSB/CW/MCW
Selectivity ......2 kHz -6dB.	D.F. Accuracy . ±1° at 1µV/m
Min. Field S. ..2µV at 2 MHz.	Image Rej. ...... >60 dB
Audio Out ......2 W 8 ohms	IF .... 65.7, .455, 104.5 MHz

Accessories:
Gyro Compass Interface

Comments:
This is an updated version of the TD-C338HS. Note that the receiver coverage is up to 30 MHz. The direction finding capability, however, is accurate up to only 16 MHz. The 400 memories are grouped into four banks of one hundred each.

Variants:
The newer model **TD-C338MKIII** has replaced the TD-C338MKII.

Made In:	Japan	1994-1997

Voltages:	100/110/115/220 VAC 50/60 Hz 50 W

Readout:	`00000.1` Digital LED

Physical:	16.1x11.6x17.7" 65 Lbs. 410x295x450mm 29.5 kg

Status:	Active Manufacturer Discontinued Model

Rarity:	Typically Unavailable

	New	Used
Price:	①	⑤
Rating:	⑥	⑥

The **Technical Material Corporation**, or TMC for short, is recognized in over 128 countries worldwide for the quality of its design in RF transmissions systems. For fifty years TMC has created innovative products designed to make long range communications more efficient.

The TMC engineering and manufacturing sites are located just north of New York City. Field offices are located worldwide.

The Company was founded by Ray H. de Pasquale. The current President is Neil H. de Pasquale. T.M.C.'s more recent products include: receiver multicouplers, balun couplers, antenna tuners, antenna dummy loads, pre/postselectors and other communications products.

Although TMC no longer produces receivers, they remain in business to service their large international base of installed equipment.

Technical Material Corporation
700 Fenimore Rd.
Mamaroneck, NY 10543   1957-1998

TECHNICAL MATERIAL

GPR-90
(R-825/URR)

General Coverage Communications Receiver
Double Conversion Superheterodyne. 15 Tubes.

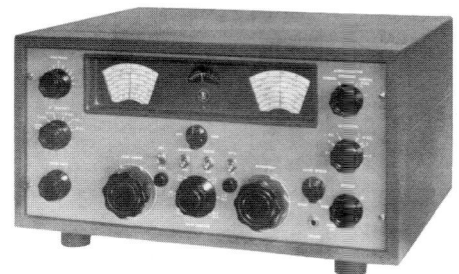

Features:
- ¼" Head. Jack
- S-Meter
- Antenna Trimmer
- Bandspread
- RF Gain
- BFO
- Mute
- Calibrator 100 kHz
- IF Out
- Dial Locks (2)
- NL
- Audio Filter

Specifications:

Coverage ......540 - 31000 kHz	Modes .... AM/CW/MCW
Selectivity ......6/2/1.5/1/.5/.2 kHz	Stability .... .002 - .003 %
Sensitivity ......<1µV (1.5-31 MHz).	IF Rejection 85dB 455, 100dB 3.955
Image Ratio ...>60 dB	Audio Out .... 2 W 4/8/16/600 ohms
IF ....3995, 455 kHz	

Circuit Complement:
6AB4 RF Amp, 6CB6 RF Amp, 6AU6 Mixer, 6AG5 Osc., 6AG5 BFO, 6BE6 Converter, (3) 6BA6 IF Amp, 6BA6 Buffer IF, 6AL5 Detector/ANL, 6V6 Audio, 0A2 Regulator and 5U4G Rectifier.

Accessories:
Speaker   GPR-D Diversity   GSB-1 SSB Adapter
Comments: Very early production units did not include the crystal calibrator. Single conversion below 3.2 MHz. Ranges: .54-1.4, 1.4-3.3, 3.2-5.6, 5.4-9.7, 9.4-17.8 and 17.3-31.5 MHz. Amateur bandspread: 1.8-2, 3.5-4, 7-7.3, 14-14.35, 21-21.45, 26.95-27.54 and 28-29.7 MHz. The GSB-1 SSB Adapter (shown right) was reviewed in QST March 1957.

Made In:	United States 1955-1960
Voltages:	105-125 AC 50/60 Hz 90W
Readout:	Analog
Physical:	20x10x15" 52 Lbs. 508x254x381mm 23.6 kg
Status:	Active Manufacturer Discontinued Model
Rarity:	Very Scarce
Review:	QST October 1955

	New	Used
Price:	$395-765	$380-650
Rating:	★★★★★	⑦

GPR-90RX
(R-840/URR)

General Coverage Communications Receiver
Double Conversion Superheterodyne. 17 Tubes.

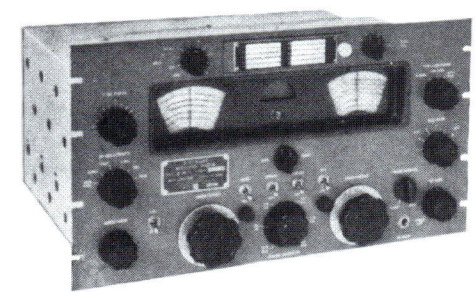

Features:
- ¼" Head. Jack
- S-Meter
- Antenna Trimmer
- 10 Crystal Positions
- RF Gain
- BFO
- Phono Input
- Calibrator 100 kHz
- Dial Locks
- Bandspread
- Audio Filter
- Mute
- NL

Specifications:

Coverage ......540 - 31500 kHz
Selectivity ......6/2/1.5/1/.5/.2 kHz
Sensitivity ......<1µV (1.5-31 MHz).
Image Ratio ...>60 dB
Audio Out ......2 W

Modes .... AM/CW/MCW
Stability .... .002 - .003 %
IF Rejection 85dB 455, 100dB 3.955
Environment ... 0° to +50° C. 90% Humid.
Antenna Input . 50/300 ohm Terminals

Circuit Complement:

6AB4 1st RF Amp, 6CB6 2nd RF Amp, 6AU6 1st Converter, 6BE6 2nd Converter & Osc., 6BA6 Buffer IF, 6BA6 1st IF Amp, 6BA6 2nd IF Amp, 6BA6 3rd IF Amp, 6AL5 Detector/ANL, 12AX7 AVC/1st Audio Amp, 6V6 2nd Audio Amp, 6AG5 Osc., 6AG5 BFO, 5U4G Rectifier, 0A2 Volt. Reg., 6CB6 Cal. Osc., and 6AG5 Crystal Osc. Amp.

Accessories: Speaker   GPR-D Diversity

Comments: The same as the GPR-90 but with ten adjustable fixed crystal positions available from the front panel of the receiver plus a rear deck input for an external high stability control oscillator or synthesizer. Ranges: .54-1.4, 1.4-3.3, 3.2-5.6, 5.4-9.6, 9.4-17.8 and 17.3-31.5 MHz. Amateur bandspread: 1.8-2, 3.5-4, 7-7.3, 14-14.35, 21-21.45, 26.95-27.54 and 28-29.7 MHz. Promoted as suitable for SSB when crystal (fixed) positions used.

Made In:	United States 1957-1960
Voltages:	105-125 AC 50/60 Hz 90W
Readout:	Analog
Physical:	19x10.5x15" 63 Lbs. 483x266x381mm 28.5kg
Status:	Active Manufacturer Discontinued Model
Rarity:	Very Scarce

	New	**Used**
Price:	①	$660-950
Rating:	⑥	⑦

GPR-90RXD

General Coverage Communications Receiver
Double Conversion Superheterodyne. 17 Tubes.

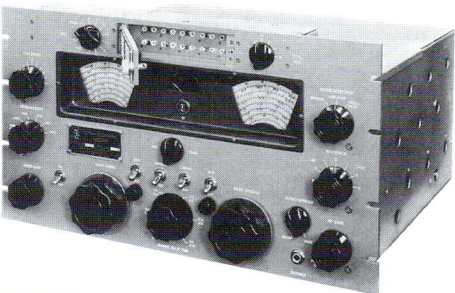

Features:
- ¼" Head. Jack
- S-Meter
- Antenna Trimmer
- Bandspread
- RF Gain
- BFO
- Phono Input
- Calibrator 100 kHz
- Mute
- Standby
- IF Out Jack
- 10 Crystal Sockets
- NL
- Ext. IFO In Jack
- Ext. BFO In Jack
- Ext. HFO In Jack
- Audio Filter

Specifications:

Coverage ......540 - 31500 kHz
Selectivity ......6/2/1.5/1/.5/.25 kHz
Sensitivity ......<1µV (1.7-31.5 MHz).
Image Ratio ...85 dB
Audio Out ......2 W 4/8/16 ohm

Modes .... AM/CW/MCW
Stability .... .002 - .003%
IF Rejection 85dB 455, 100dB 3.955
Noise Factor ... > 6 dB
Antenna Input . 75/300 ohm Terminals

Circuit Complement:

6DC6 1st RF Amp, 6BA6 2nd RF Amp, 6AU6 1st Converter, 6BE6 2nd Converter & Osc, 6BA6 Buffer, 6BA6 1st IF Amp, 6BA6 2nd IF Amp, 6BA6 3rd IF Amp, 6AL5 NL, 12AX7 Audio Amp, 6V6 Audio Amp, 6AG5 Osc, 6AG5 BFO, 5U4G Rectifier, 0A2 Voltage Reg, 6CB6 Calibrator and 6AG5 Crystal Osc. Amp.

Accessories: VOX-5 Diversity
MSR-4 SSB Adptr.   MSR-5 SSB Adptr.   CFA FSK Converter

Comments: Designed for 19" rack mounting. Bandspread: 1.8-2, 3.5-4, 7-7.3, 14-14.35, 21-21.45, 26.95-27.54 and 28-29.7 MHz. Similar to the GPR-90 but includes the ten crystal positions and connection for external oscillator.

Made In:	United States 1962-1971
Voltages:	115/230 AC 50/60 Hz 90 W
Readout:	Analog
Physical:	19x10.5x14" 63 Lbs. 483x266x355mm 28.5kg
Status:	Active Manufacturer Discontinued Model
Rarity:	Very Scarce

	New	**Used**
Price:	①	$700-1050
Rating:	⑥	⑦

GPR-91RXD

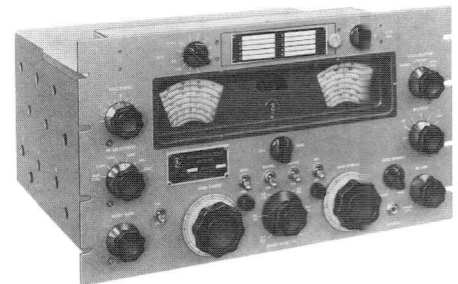

General Coverage Communications Receiver
Double Conversion Superheterodyne. 17 Tubes.

Features:
- ¼" Head. Jack
- S-Meter
- Antenna Trimmer
- Bandspread
- RF Gain
- BFO
- IF Out Jack
- Calibrator 100 kHz
- Mute Line
- Standby
- Phono In Jack
- 10 Crystal Sockets
- Audio Filter
- Ext. IFO In Jack
- Ext. BFO In Jack
- Ext. HFO In Jack
- NL

Specifications:

Coverage ......540 - 31500 kHz	Modes .... AM/CW
Selectivity ...... 15/2/1.5/1/.5/.25 kHz	Stability .... .002 - .003%
Sensitivity ......<1µV (1.4-31.5 MHz).	IF Rejection .... 85 dB 455, 100 dB 3.955
Image Ratio ...85 dB	Noise Factor ... > 6 dB
Audio Out ......2 W	Environment ... 0° to +50° C. 90% Humid.

Circuit Complement:
Same tube complement as the GPR-90RXD.

Accessories:
MSR-4 SSB Adptr.   MSR-5 SSB Adptr.   CFA FSK Converter
VOX-5 Diversity

Comments:
Similar to the GPR-90RXD, but with a 15 kHz bandpass position for ISB reception of 4 voice channels or up to 64 teletype channels. Designed for 19" rack mounting. Bands: .54-1.4, 1.4-3.3, 3.2-5.6, 5.4-9.6, 9.4-17.8 and 17.3-31.5 MHz. Amateur bandspread: 1.8-2, 3.5-4, 7-7.3, 14-14.35, 21-21.45, 26.95-27.54 and 28-29.7 MHz.

Made In: United States 1962-1971

Voltages: 115/230 AC 50/60 Hz
90 W

Readout: Analog

Physical: 19x10.5x14" 63 Lbs.
483x267x356mm 28.6 kg

Status: Active Manufacturer
Discontinued Model

Rarity: Very Scarce

	New	Used
Price:	①	$750-1100
Rating:	⑥	⑦

GPR-92

General Coverage Communications Receiver
Double Conversion Superheterodyne. 18 Tubes plus Semiconductors.

Features:
- ¼" Head. Jack
- S/AF Meter
- Bandspread
- Antenna Trimmer
- RF Gain
- BFO ±3 kHz
- NL
- Calibrator 100 kHz
- Mute
- AVC On/Off
- Squelch
- Rack Handles
- AVC Out Jack
- BFO Out Jack
- IF Out Jack
- Standby   • Tone

Specifications:

Coverage ......540 - 32000 kHz	Modes .... AM/CW/SSB/ISB/FSK
Selectivity ...... 15/7.5/3/2/1/.5 kHz	Sensitivity .... <1µV 15 dB S/N
Image Rej. ......>80 dB	

Circuit Complement: 6DC6 First RF, 6BA6 2nd RF, 6EW6 Isolation Amp, 6BA7 1st Mixer, 6AH6 High Freq. Osc., 6U8A 2nd Mixer, 6AU6 Crystal Osc., 6BA6 1st IF, 6BA6 2nd IF, 6AL5 NL, 6CB6 Calibrator, 6AZ8 IF/Squelch, 6U8A AVC/Detector, 6AB4 VFO Amp, 12AT7 BFO/Product Detector, 12AX7 AF Input/AF Driver, 6AQ5 AF Amp and 0B2 Regulator.

Accessories: TRX-1 Stabilized Crystal Osc. Diversity Adapter

Comments: Designed for 19" rack mounting. The GPR-92 was often sold in pairs for diversity applications. Ranges: .54-1.4, 1.4-3.3, 3.3-5.6, 5.6-9.5, 9.5-17.5, 17.5-32.3 MHz. Three bandspread dials were reportedly available: Ham, International Broadcast and Commercial. Reliable sources report that only 115 of these units were manufactured. This is a very highly collectable model.

Variants: Model **GPR-92C** (shown) features a cabinet. Model **GPR-92S** features a slide-out rack mount.

Made In: United States 1963-1966

Voltages: 115/230 AC 50-600 Hz
90 W

Readout: Analog

Physical: 19x8.75x17" 65 Lbs.
483x222x432mm 29.5 kg

Status: Active Manufacturer
Discontinued Model

Rarity: Extremely Scarce

Reviews: *Electric Radio* Nov. 1995

	New	Used
Price:	$920-1500	$1800-4400
Rating:	★★★★★	⑦

TELEFUNKEN

Daimler-Benz Aerospace

E2000LH1

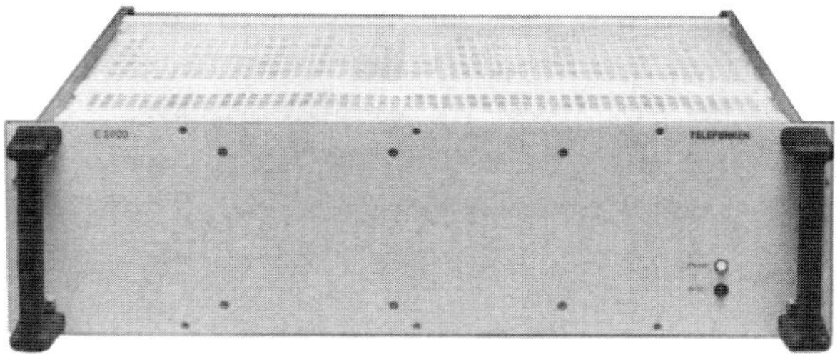

General Coverage Communications Receiver
Solid State.

Features:

• ¼" Head. Jack	• PBS ±5 kHz	• B.I.T.E.	• 1000 Memories
• RF Gain	• AGC	• Notch	• RS232/RS485
• Line Out Jack	• Ext. Spkr Jack	• BFO ±5 kHz	• Rack Handles
• Scan & Sweep	• IF Out Jack	• Mute	• DSP

Specifications:
Coverage ...... 0.3 - 30000 kHz   Modes .... AM/SSB/CW/FAX/FM/RTTY
Sensitivity ...... <.25 V CW 1-30 MHz
Selectivity ...... 43 bandwidths .1-10 kHz

Accessories:

UD2000 DSP Demod	VE2000 Recognition	TH2000 LAN Interface
AM2000 Audio Mixer	TCXO Option (2x10^{-8})	Windows™ Control Pgm.

Comments:
Note the very low frequency coverage of this receiver. The E2000LH/1 tunes down to 300 Hz (.3 kHz). Designed exclusively for computer control. 1 Hz tuning resolution is provided.

Variants:
Model **E2000LH3** is a three receiver version of the E2000LH1, still contained in the same 19 inch rack space. Model **E2000LH5** is a five receiver version.

Made In: Germany 1995-1998

Voltages: 115/230 VAC ±15%
50/60 Hz

Readout: 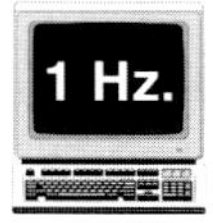

Digital on PC

Physical: 19" wide 33 Lbs
483mm wide 15 kg

Status: Active Manufacturer
Active Model

Rarity: Typically Unavailable

	New	Used
Price:	$18,000	⑤
Rating:	⑥	⑥

The **TEN-TEC** name was coined from Tennessee Technology. The company was founded in 1968 with the goal of designing and manufacturing amateur radio equipment. The first products were simple, low cost low power transceivers. The more advanced *Argonaut* QRP series was introduced in the early 1970's, staying in the product line with successive models until the early 1980's.

TEN-TEC pioneered solid state, no-tune 100 watt H.F. transceivers with the Triton II and IV in the mid 1970's. These early designs spawned development of 16 different H.F. amateur rigs over the next twenty years. TEN-TEC has produced linear amps for over 15 years. Most receiver hobbyists know TEN-TEC through their model RX-10 or RX-325 models. However, their most successful receiver is the RX-330A commercial, DSP model enjoying wide use in commercial and government markets. TEN-TEC currently offers two affordable "regen" shortwave radio kits. TEN-TEC will offer the RX-320 "black box" HF receiver in 1998.

Expanding several times, TEN-TEC currently has a 40,000 square foot plant with 100 employees. They are active in three businesses today. The largest division is the combined amateur, commercial and government electronics division. TEN-TEC has a tool and die division making molds for the die casting and plastic injection field. The third division makes standard and custom sheet metal for other manufacturers.

TEN-TEC, Inc.
Highway 411 East
Sevierville, TN 37862   1972-1996

TEN-TEC, Inc.
1185 Dolly Parton Pkwy.
Sevierville, TN 37862   1996-1998

315

Amateur Band Communications Receiver
Double Conversion Superheterodyne. Solid State.

Features:
- ¼" Head. Jack
- S-Meter
- RF Gain
- Calibrator
- Mute Line
- Speaker
- Preselector
- Sidetone Input Jack
- Dial Lamp

Specifications:
Coverage ...... Ham 80-10 Meters
Selectivity ...... 2.5 kHz -6dB.
Sensitivity ...... <.25 V
Modes .... CW/USB/LSB
Stability .... <100 Hz

Accessories:
Model 235 Audio CW Filter 300 Hz

Comments:
No AM mode. The tuning rate is approximately 25 kHz per revolution. The slide rule dial denotes band segments (0-5). The knob skirt reads kHz on a 0-100 scale. Dial accuracy is ±2.5 kHz.

Made In: United States 1972-1974

Voltages: 115 VAC 50/60 Hz

Readout: Analog Lin.

Physical: 13x4.5x7"
330x114x178mm

Status: Active Manufacture
Discontinued Mode

Rarity: Scarce

Reviews: *CQ* February 1974

	New	Used
Price:	$229	$70 90
Rating:	★★★★	

TEN-TEC

1253

Shortwave Communications Receiver Kit
Regenerative. Solid State.

Features:
- Head. Jack
- RF Gain
- Regen Tune
- Fine Tuning
- Speaker 3"
- Ext. Speaker Jack

Specifications:
Coverage ...... See Comments.   Modes .... AM/CW/SSB
Antenna Input Terminal

Circuit Complement:
A regenerative design with 3 integrated circuits and 5 transistors. Tuning and fine tuning is accomplished by varactor diodes.

Comments:
Ranges: Seven bands in the range of 1.8-2, 3.3-4.15 and 5.5-22 MHz. The rotary tuning knob is calibrated from 1 to 21 with additional calibration marks between each number. This 42 point scale is used in conjunction with charts in the Owner's Manual to interpolate the actual frequency. This kit takes three evenings to assemble. This is a very good looking set that is surprisingly stable.

Variants:
TEN-TEC also makes a smaller regenerative kit called the **1054** (shown to the right) which has 4 bands and sells for only $24. The 1054, however, does not include a case or speaker. The 1054 was reviewed in *Monitoring Times* February 1997.

Made In:	United States 1996-1998	
Voltages:	12 VDC or 8 C cells.	
Readout:	Indexed Scale 0-21	
Physical:	6x3.75x6" 152x95x152mm	
Status:	Active Manufacturer Active Model	
Rarity:	Scarce	
Reviews:	*NASWA* November 1996	

	New	Used
Price:	$59	$35
Rating:	⑨	⑨

TEN-TEC

RX-10

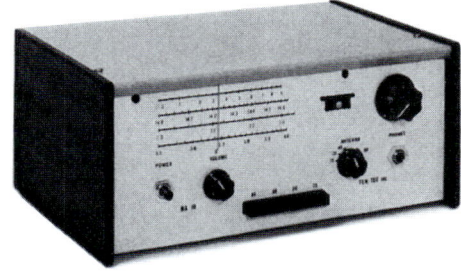

Amateur Band Communications Receiver
Single Conversion. Solid State.

Features:
- ¼" Head. Jack
- Built-in code oscillator & sidetone monitor.

Specifications:
Coverage ...... 80/40/20/15M   Modes .... CW(USB/LSB)
Selectivity ...... 2 kHz -6dB.   Stability .... <100 Hz
Sensitivity ...... <1μV   Antenna Input . 50-70 ohm

Accessories:
None

Comments:
Ranges: 3.5-4, 7-7.3, 14-14.6 and 21-21.9 MHz. The RX-25 was designed primarily as a beginner's CW receiver. A single conversion synchrodyne circuit eliminates images and birdies usually found in low-priced receivers. SSB and AM reception are possible, but not ideal.

Made In:	United States 1970-1973	
Voltages:	115 VAC 50/60 Hz 1/8 Amp or 12 VDC 35 ma	
Readout:	Analog	
Physical:	10.5x4.5x6.62" 2.25 Lbs. 267x114x168mm 1.1 kg	
Status:	Active Manufacturer Discontinued Model	
Rarity:	Scarce	
Reviews:	*Popular Electr.* May 1971 *QST* August 1971 *CQ* March 1971 *Ham Radio* June 1971	

	New	Used
Price:	$60	$40-50
Rating:	★★★	⑦

TEN-TEC

RX-325

General Coverage Communications Receiver
Double Conversion Superheterodyne. Solid State.

Features:
- Mini Head. Jack
- S-Meter
- Dimmer
- Attenuator -15 dB
- Dial Lock
- 25 Memories
- Memory Scan
- Memory Lockout
- Sweep
- Keypad Entry
- Noise Blanker
- 12/24 Clock-Timer
- AGC FST/SLO
- Record Jack
- Record Activation
- Ext. Speaker Jack
- Display Test
- Two Tuning Speeds

Specifications:
Coverage ...... 300 - 30000 kHz.	Modes .... AM/LSB/USB
Selectivity ...... 4/2.7/_ kHz -6dB.	Stability .... ±200 Hz after 1 min.
Sensitivity ...... <.4µV 2-30 MHz CW	IF Rejection .... >60 dB
Audio Out ...... 2W 10% distortion	Image Reject .. >60 dB
Accuracy .... ±50 Hz	Antenna Input . SO-239 & Terminals.

Accessories:
265 Speaker   266 SSB Filter   925 Battery Pack
267 Timer Relay

Comments:
Supplied with external AC wall supply. Memories store frequency only. Memories may be selectively locked-out. Three tuning steps: 50, 500 and 1000 Hz. The RX-325 was not a popular receiver because of its non-ergonomic design, marginal audio, RF-section problems and poor marketing. Not manufactured in large numbers (perhaps 200?), it now falls into the collector category.

Made In:	United States 1987-1988
Voltages:	115 VAC or 13.8 VDC .5A 9V battery for memory
Readout:	`00000.1` Digital Flors.
Physical:	9.5x3.25x7" 6 Lbs. 241x82x178mm 2.7 kg
Status:	Active Manufacturer Discontinued Model
Rarity:	Scarce
Reviews:	📖 Passport 1988 Monitoring Times Feb. 1986 Monitoring Times Jan. 1987

	New	**Used**
Price:	$549-629	$275-350
Rating:	★★	★

TEN-TEC

RX-330A

General Coverage Communications Receiver
Double Conversion Superheterodyne. Solid State.

Features:
- ¼" Head. Jack
- ¼" Jack Stereo
- PS Status LEDs
- Attenuator -15 dB
- B.I.T.E.
- 100 Memories
- Preamp 10 dB
- AGC FST/MED/SLO/OFF
- RS-232 Port
- BFO ±8 kHz
- Rack Handles
- Ext. Ref. In Jack
- IF Out Jack
- DSP IF
- Notch Filter
- Line Out 600 ohm (2)
- PBT
- Synchronous AM

Specifications:
Coverage ...... 500 - 30000 kHz.	Modes .... AM/LSB/USB/CW/FM/ISB
Selectivity ...... See Comments.	Stability .... ± 1 PPM per C°
IF .... 45.105 MHz, .455 kHz	IF Rejection .... >80 dB
Audio Out ...... 600 Ohm Line Out	Image Reject .. >70 dB
Antenna Input BNC 50 ohm	Environment ... 0 to +50°C 95% Humid

Accessories:
FM Squelch   TCVCXO High Stability   PBT CW/USB/LSB

Comments: The ¼" stereo headphone jack is for ISB output. Will tune 0 - 500 kHz with reduced performance. Memories store: frequency, mode, bandwidth, etc. S-Meter values are communicated to the RS232 Port. Designed strictly for computer control and rack mounting. The sophisticated RX-330A has been well received by the military, commercial and maritime market. Bandwidths: 16000, 15200, 14400, 13600, 12800, 12000, 11200, 10400, 9600, 8800, 8000, 7600, 7200, 6800, 6400, 6000, 5600, 5200, 4800, 4400, 4000, 3800, 3600, 3400, 3200, 3000, 2800, 2600, 2400, 2200, 2000, 1900, 1800, 1700, 1600, 1500, 1400, 1300, 1200, 1100, 1000, 900, 800, 700, 600, 500, 450, 400, 350, 300, 250, 225, 200, 175, 150, 125, 100 Hz -3dB.

Made In:	United States 1995-1998
Voltages:	115/230 AC ±15% 48-440 Hz 30W
Readout:	Digital on PC
Physical:	19x1.75x20" 12 Lbs. 483x45x508 5.43 kg
Status:	Active Manufacturer Active Model
Rarity:	Typically Unavailable
Reviews:	QST

	New	**Used**
Price:	①	③
Rating:	⑥	○

TEN-TEC

SP-325

General Coverage Communications Receiver
Double Conversion Superheterodyne. Solid State.

Features:
- ¼" Head. Jack
- S-Meter
- Display Test
- Attenuator -15 dB
- Dial Lock
- 64 Memories
- Memory Scan
- Memory Lockout
- Sweep
- Keypad Entry
- AGC FST/SLO
- Line Level Output
- Rack Handles
- Two Tuning Speeds

Specifications:

Coverage ......500 - 30000 kHz.
Selectivity ......6/2.8/.5 kHz -6dB.
Sensitivity ......<.5 µV 2-30 MHz CW
Audio Out ......2W 10% distortion
IF ....45 MHz, 455 kHz
Environment ..0 to +50°C 95% Humid

Modes .... AM/USB-CW/LSB-FSK
Stability .... ±50 Hz after 5 min.
IF Rejection .... >60 dB
Image Reject .. >60 dB
Antenna Input . 50 ohm N

Comments:
Memories store frequency, mode and bandwidth. Memories may be selectively locked-out. Three tuning steps: 50, 500 and 1000 Hz. The SP-325 was produced for the U.S. Navy under two contracts. It was designed as a training receiver for radio operators. It was supplied with racking mounting hardware for a standard 19" rack or could be simply used as a tabletop. Black metal cabinet with white lettering. The similar RX-325 consumer version followed. The SP-325 is beginning to show up at hamfests.

Made In: United States 1986-1987

Voltages: 105-115 VAC 60 Hz 6 W.

Readout: `00000.1` Digital Flors.

Physical: Cabinet Version:
12x3.5x10.5" 8.25 Lbs
305x89x267mm 3.7 kg
Rack Version:
19x3.5x10.5"
480x89x267mm

Status: Active Manufacturer
Discontinued Model

Rarity: Scarce

	New	Used
Price:	$980A	$380-490
Rating:	★★★	★★★

The **Thomson-CSF** company is one of Europe's largest suppliers of commercial, broadcast and military electronics and communications systems.

Thomson-CSF's current line of TRC620 receivers (not shown) are designed to meet electronic counter measure threats, frequency hopping and burst transmissions. The Model TRC621 covers 300 kHz to 30 MHz.

Thomson-CSF
Telecommunications Division
66 rue du Fosse Blanc, Gennevilliers
France   1979-1998

TRC241

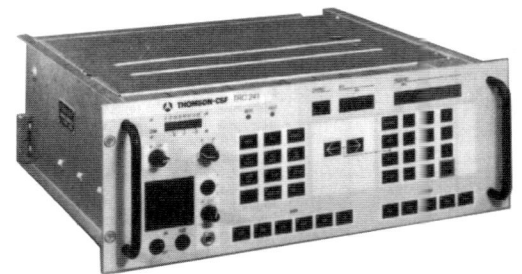

General Coverage Communications Receiver
Double Conversion Superheterodyne. Solid State.

Features:
- ¼" Head. Jack
- S-Indicator
- Keypad
- Modular Construction
- RF Gain
- Mute
- Rack Handles
- 40 Memories
- Speaker
- Scan
- Sweep

Specifications:

Coverage ...... 100 - 30000 kHz	Modes .... AM/LSB/USB/CW/ISB
Sensitivity ...... <.5µV CW	Stability .... 1×10^{-7}
Environment .. 0° to +55° C.	

Accessories:
VLF Option 10-100 kHz

Comments:
Designed for military applications. Memories store frequency, mode, BFO, bandwidth, squelch, AGC and attenuator settings. May be operated remotely. Later production included 100 instead of 40 memories.

Variants:
Model **TRC243** is later production with a similar appearance except for the addition of a large manual tuning knob in the center of the radio. The model **TRC251** is a VLF version covering 10 to 200 kHz in 1 Hz steps in CW, FSK and MSK modes (1987). They were produced for the French Navy in an effort to improve communications with submarines.

Made In: France   1936-1998

Voltages: 110/127/220/240 AC ±10%
                50/60 Hz

Readout: `00000.01` Digital LED

Physical: 19x4.6x18.3" 40 Lbs.
                483x117x465mm 18 kg

Status: Active Manufacturer
                Active Model

Rarity: Typically Unavailable

	New	Used
Price:	①	⑤
Rating:	⑥	③

THOMSON-CSF

TRC394/A

General Coverage Communications Receiver
Double Conversion Superheterodyne. Solid State.

Features:
- ¼" Head. Jack
- S-Meter
- Keypad
- Modular Construction
- RF Gain
- Mute
- Rack Handles
- Two Tuning Rates

Specifications:
Coverage ...... 400 - 30000 kHz
Selectivity ...... 6/3-.3/.25 kHz -6dB.
Sensitivity ...... <.25µV CW
Image Rej ...... >80 dB
Environment .. -20° to +55° C.
Modes .... AM/LSB/USB/CW
Stability .... $1x10^{-7}$
IF Rejection .... >80 dB
Antenna Input . 50 ohm

Comments:
Designed for military applications.

Variants:
Model **TRC394/B** uses decadic rotary switches for frequency selection (100 Hz steps) and has 12 memories. The TRC394/B model also supports remote operation.

Made In: France 1986-1988

Voltages: 127-220 VAC 50/60 Hz or 24 VDC 2A

Readout: 00000.01 Digital LED

Physical: 19x5.2x15.75" 33 Lbs. 480x132x400mm 15 kg

Status: Active Manufacturer Discontinued Model

Rarity: Typically Unavailable

	New	Used
Price:	①	⑤
Rating:	⑥	⑥

Trans World Communications produces an entire line of base transceivers, man-pack transceivers, amplifiers and other specialized communications equipment for the military sector.

The Trans World product line is popular with military and paramilitary concerns worldwide.

Trans World Communications
A Datron Company
304 Enterprise St.
Escondido, CA 92029

TRANSWORLD

TW100RX

General Coverage Communications Receiver
Double Conversion Superheterodyne. Solid State.

Features:
- ¼" Head. Jack
- Keypad
- Mute
- Speaker Switch
- S-Meter
- Speaker
- 100 Memories
- Memory Scan
- NB
- Squelch
- Atten. -12 dB
- Modular Construction
- Fine Tuning ±125 Hz
- Line Out 600 ohms

Specifications:

Coverage	200 - 30000 kHz
Selectivity	2.7/.3 kHz -6dB.
Sensitivity	<.3 µV 10 dB S+N
Image Rej	>80 dB
Audio Out	4 W 3.2 ohms
Fine Tune	±125 Hz.
Modes	LSB/USB/CW
Stability	±.0001%
IF Rejection	>80 dB
IF	75 MHz, 1650 kHz
Environment	-30° to +55° C.
Antenna Input	50 or 800 ohms

Accessories: AM mode is optional

Comments:
Tuning is by UP / DOWN buttons, keypad entry or memory recall. The TW100RX has excellent input protection (up to 100 W!).

Variants:
Rack mounted version (shown right).

Mfg. In: United States 1984-1998

Voltages: 115/230 VAC 50/60 Hz or 13.6 VDC 600 ma

Readout: 00000.1 Digital

Physical: 13.5x.4.4x14" 15 Lbs. 345x107x355mm 6.8 kg

Status: Active Manufacturer
Active Model

Rarity: Typically Unavailable

	New	Used
Price:	①	⑤
Rating:	⑥	⑥

101 Videoton-Mechlabor

The company **Videoton-Mechlabor Development and Manufacturing Ltd.** of Budapest Hungary produces state-of-the-art, world class communications products for the international military market.

The REV-400 family of high frequency receivers offers tremendous sophistication in a very compact and robustly built package. The different members of this family can be mounted singly or double; in a rack, portable box or desktop cabinet configuration.

Videoton-Mechlabor Development and Manufacturing Ltd.
Városligeti fasor 25-27
H-1071 Budapest, Hungary

VIDEOTON-MECHLABOR
Manufacturing & Development Ltd.

REV-401

General Coverage Communications Receiver
Triple Conversion Superheterodyne. Solid State.

Features:
- ¼" Head. Jack
- RF Gain
- Mute
- BITE
- Preamp 9 dB
- ISB Outputs
- S-Indicator
- Keypad
- 100 Memories
- Scan
- Dial Lock
- FM/AM Video
- Speaker
- BFO ±2.5 kHz
- AGC/MGC
- Sweep
- AFC
- Speaker Switch
- Attenuator -20 dB
- RS232 or IEC625
- Line Out 600 ohm
- 5 MHz Ref. I/O Jack

Specifications:
Coverage ...... 200 - 30000 kHz
Selectivity ...... 6/3/2/1.4/1/.7/.4/.1 kHz[1]
IF .... 72.2 MHz, 200, 8 kHz
Image Rej ...... >100 dB
Audio Out ...... 600 ohms 775 mV
3IP .... min +25 dBm w/ Pre.
Modes .... AM/LSB/USB/CW/FM/ISB
Stability .... ±0.1 PPM
IF Rejection .... >100 dB
Environment ... -10° to +60° C.
Antenna Input . 50 ohms

Accessories:
PR351 Panoramic Display

Comments:
[1] Additionally, a 2.4 kHz filter for single sideband only is provided. The 100 memories store frequency, bandwidth, mode and threshold. Channel lockout is available. The BFO tunes in 10 Hz steps. Two REV-401 receivers may be rack mounted side by side in a standard 19 inch rack.

Variants: Model **REV-400** previous model 1991-1995.

Made In: Hungary 1996-1998

Voltages: 220 VAC +10...-15% 40W or 24 VDC ±15%

Readout: 00000.001 Digital LED

Physical: 6.7x9.45x17" 33 Lbs. 170x240x430mm 15 kg

Status: Active Manufacturer
Active Model

Rarity: Typically Unavailable

	New	Used
Price:	①	⑤
Rating:	⑥	⑥

102 Vigilant Communications

The current status of Vigilant Communications is unknown by the author. Recent letters to the company have been returned marked "Gone Away". It is unclear whether they went out of business or were acquired by another company or moved.

Vigilant Communications Ltd.
Unit 5, Pontiac Works
Fernbank Road
Ascot, Berks, SL5 8JH
England

Vigilant

SR 500

General Coverage Communications Receiver
Double Conversion Superheterodyne. Solid State.

Features:
- ¼" Head. Jack
- RF Gain
- Mute
- AGC
- S-Indicator
- Dimmer
- Line Out Jack
- Speaker Switch
- BFO ±8 kHz
- Rack Handles
- Speaker
- 1 MHz Ref. In/Out
- Atten. -10/20 dB
- 6 Fixed Frequencies
- Speaker Out Jack

Specifications:
Coverage ...... 50 - 30000 kHz
Selectivity ...... 5.4/2 kHz -6dB.
Sensitivity ...... <1µV CW/SSB
Image Rej ...... >60 dB
Audio Out ...... 1 W 8 ohms
BFO Range ... ± 8 kHz.
Modes .... AM/USB/CW
Stability .... 1 part in 10^7 per ° C.
IF Rejection .... >70 dB
Ant. Jack .... BNC 50 ohm
Environment ... -15° to +55°C 95% Humid.

Accessories:
Cabinet (shown) FSK (SR521 only) Scan Fixed Channels Option

Comments: Frequency selection is by push-button decade switches. Modular construction. Designed for 19" rack mounting. Shown in optional cabinet. The 6 fixed channels are programmed by diode matrices. Automatic power switching from AC to DC if power failure.

Variants: Model **SR 501** $2795 new, is the same but has both USB and LSB for fixed communications. Model **SR 502** is similar but with 6 memories instead of 6 fixed crystal positions. Model **SR 521** is similar to the SR 501, but with 10 Hz synthesizer tuning. $4112 new. Model **SR 520** has 10 Hz synthesizer with USB only.

Made In: England 1981-1983

Voltages: 105-125/210-250 VAC
40/60 Hz and 15-35 VDC

Readout: 00000.1 Digital Mech.

Physical: 16.5x5.25x12" 22 Lbs.
420x133x330mm 10 Kg

Status: Inactive Manufacturer
Discontinued Model

Rarity: Extremely Scarce

	New	Used
Price:	$2768-2968	⑤
Rating:	⑥	⑥

Vigilant

SR 511

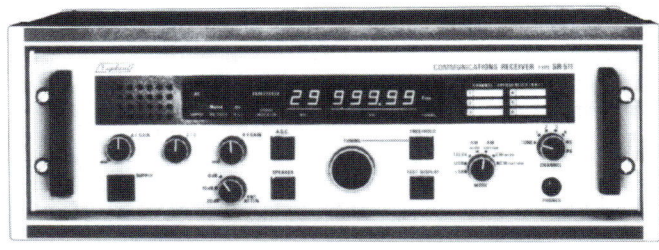

General Coverage Communications Receiver
Double Conversion Superheterodyne. Solid State.

Features:
- ¼" Head. Jack
- RF Gain
- Mute
- Dial Lock
- S-Indicator
- Dimmer
- Speaker
- APTR
- BFO ±8 kHz
- Rack Handles
- Line Out Jack
- Speaker Switch
- Atten. -10/20 dB
- 6 Fixed Frequencies
- AGC
- Modular Construction

Specifications:

Coverage ......50 - 30000 kHz	Modes .... AM/LSB/USB/CW/RTTY
Selectivity ......5.4/2 kHz -6dB.	Stability .... 1 part in 10^7 per ° C.
Sensitivity ......<1µV CW/SSB	IF Rejection .... >70 dB
Image Rej ......>60 dB	IF .... 35.4 MHz nom., 455 kHz
Audio Out ......1 W 8 ohms	Environment ... -15° to +55° C.
BFO Range ... ± 8 kHz.	Ant. Jack .... BNC 50 ohm

Accessories:
Cabinet

Comments:
Modular construction. Designed for 19" rack mounting. Shown in optional cabinet. The 6 fixed channels are programmed by diode matrices.

Variants:
Model **SR 510** $4530 new, is the same as the SR 511 but has only upper side band rather than upper and lower sideband.

Made In:	England	1981-1982
Voltages:	105-125/210-250 VAC 40/60 Hz and 15-35 VDC	
Readout:	`00000.01` Digital LED	
Physical:	16.5x5.25x12.5" 23 Lbs. 420x133x318mm 10.5 Kg	
Status:	Inactive Manufacturer Discontinued Model	
Rarity:	Extremely Scarce	

	New	Used
Price:	$4795-4995	⑤
Rating:	⑥	⑥

Vigilant

micon

SR 532

General Coverage Communications Receiver
Double Conversion Superheterodyne. Solid State.

Features:
- ¼" Head. Jack
- RF Gain
- Mute
- BITE
- S-Indicator
- AGC
- Speaker
- Keypad
- BFO ±8 kHz
- Rack Handles
- Speaker Switch
- Modular Construction
- 200 Memories
- Line Jack 600 ohms

Specifications:

Coverage ......50 - 30000 kHz	Modes .... AM/LSB/USB/CW/RTTY
Selectivity ......5.4/2 kHz -6dB.	Stability .... 1 part in 10^7 per °C.
Sensitivity ......<.5µV 10 dB S/N CW	IF Rejection .... >70 dB
Image Rej ......>70 dB	Environment ... -15° to +55°C 95% humid.
Audio Out ......1 W 8 ohms	Ant. Jack .... BNC 50 ohm
BFO Range ... ± 8 kHz.	

Accessories:
Cabinet   Dual Diversity Option   FSK Option

Comments:
Modular construction. Designed for 19" rack mounting.

Variants:
Model **SR 530** is the same as the SR 532 but has only upper side band rather than upper and lower sideband. Model **SR 531** is the same as the SR 532, but with 100 Hz steps rather than 10 Hz steps.

Made In:	England	1983-1988
Voltages:	110-125/220-240 VAC 47/60 Hz and 21-32 VDC	
Readout:	`00000.01` Digital LED	
Physical:	Cabinet Version: 20.25x7.3x16.5" 38 Lbs. 512x185x420mm 17.7 kg Rack Version: 16.5x5.25x15.5" 27 Lbs 420x133x395mm 12.2 kg	
Status:	Inactive Manufacturer Discontinued Model	
Rarity:	Typically Unavailable	

	New	Used
Price:	①	⑤
Rating:	⑥	⑥

The **Watkins-Johnson Company** has been a leader in high-technology products since 1957. This diversified corporation (NYSE: WJ) specializes in electronic products for the wireless telecommunications and defense industries. W-J is also a leading producer of semiconductors. Revenue in 1995 exceeded $380 million.

The company, headquartered in Palo Alto, California, employes over 2200 people. Engineering and manufacturing facilities may also be found in Scotts Valley and San Jose, California as well as Gaithersburg, Maryland.

The Telecommunications Group in Gaithersburg employes 450 people. They specialize in highly capable designs that mesh both DSP and RF technologies. This unique ability is the cornerstone of the Group's well earned reputation as a leading worldwide supplier of high-quality intercept receivers.

Most military and espionage agencies use Watkins-Johnson equipment. Although the main focus of W-J remains the military, government and commercial market, the recent introduction of the HF-1000 into the radio hobbyist arena has met with excellent success.

Watkins-Johnson Company
Corporate Headquarters
3333 Hillview Ave.
Palo Alto, CA 94304   1957-1998

Watkins-Johnson Company
Telecommunications Group
700 Quince Orchard Rd.
Gaithersburg, MD 20878   1966-1998

373A-2

General Coverage Communications Receiver
Triple Conversion Superheterodyne. Solid State.

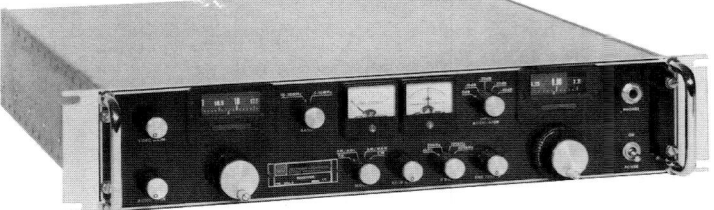

Features:
- ¼" Head. Jack
- S-Meter
- L.O. Output
- Center Tune Meter
- Rack Handles
- Fine Tuning
- Rack Handles
- Atten.-10/20/30/40dB
- Monitor Output
- Det Level Out
- Predet. Out 455 k
- Predet. Out 21.4 MHz

Specifications:
Coverage ...... 500 - 30000 kHz
Selectivity ...... 0.6/20/100/400 kHz
Sensitivity ...... <.5µV AM 6kHz 10 dB
Image Rej ...... >80 dB

Modes .... AM/FM/CW
IF Rejection .... >60 dB
IF .... 60 MHz, 21.4 MHz, 455 kHz
Antenna Input . 50 ohm (Two inputs)

Comments:
Frequency display is via two illuminated 26-inch steel tapes. Ranges: .5-10 and 10-30 MHz. Separate antenna inputs are provided for each of the two ranges. This specialty receiver is suitable for RFI detection and predetection recording. A bandwidth of 2 MHz is maintained through the RF tuners and is available for viewing at a signal monitor output jack. 60 MHz IF only above 10 MHz. The front-end of this receiver was intentionally wide for spectrum analysis and specialized monitoring. It is not suitable for conventional hobby DXing of SSB, RTTY, etc. Designed for 19 inch rack mounting.

Variants:
Model **373A-10** has bandwidths of 2/20/100/400 kHz.

Made In:	United States	1966-1984
Voltages:	115/230VAC 48-420 Hz 12W	
Readout:	Analog Lin.	
Physical:	19x3.5x18" 25 Lbs 483x89x457mm 11.3 kg	
Status:	Active Manufacturer Discontinued Model	
Rarity:	Extremely Scarce	

	New	Used
Price:	①	⑥
Rating:	⑥	⑥

HF-1000

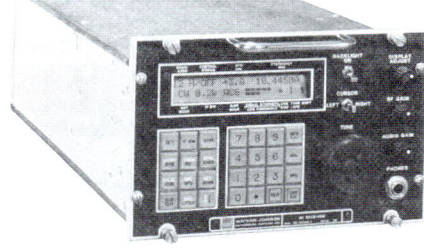

General Coverage Communications Receiver
Double Conversion Superheterodyne. Solid State.

Features:
- ¼" Head. Jack
- RF Gain
- Tune lock
- Keypad
- Scan/Sweep
- Rack Handles
- Notch

- S-Meter
- Squelch
- Speaker
- 100 Memories
- IF Out Jack
- Ext. Ref. Jack
- Local/Remote

- DSP IF
- DSP NB
- BFO ±8 kHz.
- Preamplifier
- Sync. Detection
- Speaker Out
- Channel Lockout

- AGC (Variable)
- RS232 port
- B.I.T.E.
- Atten.
- Passband Tuning
- Mute Line

Specifications:
Coverage ......5 - 30000 kHz
Selectivity ...... See Comments.
Sensitivity ...... <.35µV (.5-30 MHz)
3rd Ord Int. ....+30 dBm
Environment .. 0 to +50°C. 95% Humid.

Modes .... AM/CW/LSB/USB/FM/ISB
Stability .... 1 PPM
IF Rejection .... >80 dB
Antenna Input . BNC

Accessories: Suboctave Preselector (internal)

Comments: Independent speaker and headphone controls. Scan memories, scan with lock-outs, sweep frequencies, or sweep with lock-outs. Dwell time setable from .5-20 seconds. The ¼" headphone jack operates in stereo only during ISB reception. Sweep step setable from 1 Hz to 25 kHz. "Low cost" version of WJ-8711A. Supplied bandwidths: 8000, 7200, 6400, 6000, 5600, 5200, 4800, 4400, 4000, 3600, 3200, 3000, 2800, 2600, 2400, 2200, 2000, 1800, 1600, 1500, 1400, 1300, 1200, 1100, 1000, 900, 800, 750, 700, 650, 600, 550, 450, 400, 375, 350, 325, 300, 275, 250, 225, 200, 188, 175, 163, 150, 138, 125, 113, 100, 94, 88, 81, 75, 69, 63 and 56 Hz -3dB.

Made In: United States 1993-1998

Voltages: 97-253 VAC 47-440 Hz 35W

Readout: `00000.001` Digital LED

Physical: 19x5.25x20" 15 Lbs. 482x134x508mm 7 kg

Status: Active Manufacturer Active Model

Rarity: Scarce

Reviews: *Passport* 1995-1998
WRTH 1995
R.D.I. Whitepaper
Mon. Times November 1995
QST December 1994
Proceedings 1994-95

	New	Used
Price:	$3800-4000	$2700-2800
Rating:	★★★★★	★★★★★

WJ-8626A-4

General Coverage Communications Receiver
Double Conversion Superheterodyne. Solid State.

Features:
- ¼" Head. Jack
- RF Gain
- Dimmer
- Line Out Jack
- IF Out Jack

- Scan/Sweep
- Backlit LCD
- Keypad
- Ext. Ref. Jack
- Backlit Switch

- Scan/Sweep
- Atten.
- BFO
- Speaker Out
- AFC

- 100 Memories
- AGC
- RS232 port
- Mute Line

Specifications:
Coverage ......5 - 30000 kHz
Selectivity ......16/4/2/1/.2 kHz -6dB

Modes .... AM/CW/LSB/USB/FM
Antenna Input . BNC 50 ohms

Accessories:
Option   Sub-Octave Preselector   FSK Demod.

Comments: The 48 character, two line alphanumeric LCD that shows memory channel, control status, BFO offset, frequency, mode, bandwidth, AGC and signal strength bar graph. Can function with up to 34 WJ-9040 VLF HF/VHF/UHF hand-off receivers. Two of these receivers may be mounted side by side in a 19 inch rack. The WJ-8628A-4 model has an identical appearance, but covers 20 to 512 MHz.

Variants:
Model **WJ-8626A-1** (shown right) is a quarter-rack, hand-off receiver. A hand-off receiver must be operated by a control receiver or a receiver controller.

Made In: United States 1986-1990

Voltages: 115/230 VAC 48-420 Hz 25W

Readout: `00000.01` Digital LCD

Physical: 8x5.25x14.38" 21 Lbs. 203x133x366mm 9.7 kg

Status: Active Manufacturer Discontinued Model

Rarity: Typically Unavailable

	New	Used
Price:	①	②
Rating:	⑥	⑥

 **WATKINS-JOHNSON**

WJ-8700

Dual General Coverage Communications Receiver
Double Conversion Superheterodyne. Solid State.

Features:

• ¼" Head. Jack	• Scan/Sweep	• Scan/Sweep	• 100 Memories
• RF Gain	• Squelch	• Atten.	• AGC ON/OFF
• Tune lock	• Keypad	• BFO ±10 kHz.	• Self Test
• Line Out Jack	• Ext. Ref. Jack	• Speaker Out	• Mute Line
• IF Out Jack	• RS232 port	• Backlit LCD	• Spinner Knob

Specifications:

Coverage ......5 - 32000 kHz
Selectivity ...... 16/8/4/2/1/.5 kHz -6dB
Image Reject .>90 dB
3rd Ord Int. ....+30 dBm
Environment ..-20 to +60°C.

Modes .... AM/CW/LSB/USB/FM
Stability .... .7 PPM (0-50°C)
IF Rejection .... >90 dB
Antenna Input . BNC 50 ohms

Accessories:

Filter Sets   ISB Option   21.4 MHz Monitor Out   FSK Demod.

Comments:

All receiver functions are menu driven with softkey access to different menu levels via an eight line display. The BFO will tune in 1 Hz steps. Alternate filter sets are available. This is a dual receiver.

Variants: Model **WJ-8700S** is a single channel version of the WJ-8700. Model **WJ-8700-1** covers from 1 - 32000 kHz and has input low pass filter. Model **WJ-8700S-1** is a single channel version of the WJ-8700-1. Model **WJ-8700/NFP** has no front panel and must be controlled exclusively by computer.

Made In: United States 1989-1998

Voltages: 115/230 VAC 48-420 Hz
85W

Readout: `00000.01` Digital LCD
(8 Line)

Physical: 8.5x3.5x20" 18 Lbs.
210x89x508mm 8.1 kg

Status: Active Manufacturer
Active Model

Rarity: Typically Unavailable

	New	Used
Price:	①	②
Rating:	⑥	⑥

 **WATKINS-JOHNSON**

WJ-8709

General Coverage Communications Receiver
Double Conversion Superheterodyne. Solid State.

Features:

• ¼" Head. Jack	• AF/S-Meter	• RF Gain	• AGC FST/SLO/OFF
• BFO ±8 kHz	• Rack Handles	• IF Out 455 kHz	• 5 Tuning Speeds
• Modular			

Specifications:

Coverage ......5 - 30000 kHz
Selectivity ...... 16/5/3.2/1/.3 kHz -3dB
Sensitivity ......<.4µV 16 dB (S+N)/N CW
3rd Ord Int. ....+20 dBm
Audio Output .2W 600 ohm

Modes .... AM/FM/CW/MCW/USB/LSB
Stability .... 6x10⁻⁸ per day
IF Rejection .... >80 dB
Environment ... 0 to 50°C. 95% Humid.
Antenna Input . BNC 50 ohm

Accessories:

10 Hz BFO   ISB Option   LSB/USB Option Sub-Octave Preselector
Remote Module   Manual Control Module

Comments:

This is a half-rack version of the WJ-8718. BFO offset controllable by decadic BCD thumb switches to the left of the digital display.

Made In: United States 1982-1998

Voltages: 115/220 VAC 48-410 Hz

Readout: `00000.01` Digital LED

Physical: 9.5x5.2x21.7" 34.7 Lbs.
241x133x552mm 15.8 kg

Status: Active Manufacturer
Active Model

Rarity: Typically Unavailable

	New	Used
Price:	$5700	②
Rating:	⑥	⑥

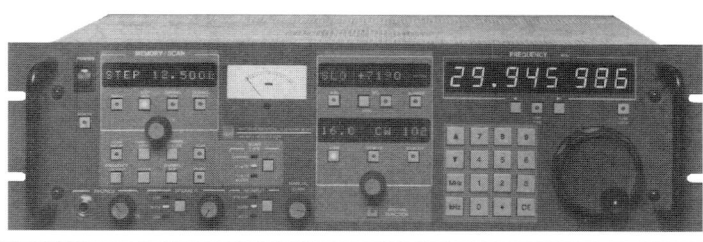

WJ-8711

General Coverage Communications Receiver
Double Conversion Superheterodyne. Solid State.

Features:

• ¼" Head. Jack	• S-Meter	• DSP IF	• AGC (2 pos.)
• RF Gain	• Squelch	• DSP NB	• RS232 port
• Tune lock	• Mute Line	• BFO ±8 kHz.	• Self test
• Keypad	• 100 Memories	• Preamplifier	• Atten.
• Scan/Sweep	• IF Out Jack	• Sync. Detection	• Passband Tuning
• Rack Handles	• Ext. Ref. Jack	• Speaker Out	

Specifications:

Coverage ......5 - 30000 kHz
Selectivity ......16/6/3.2/1/.3 kHz -3dB
Sensitivity ......<.35µV (.5-30 MHz)
3rd Ord Int. ....+30 dBm
Antenna Input BNC

Modes .... AM/CW/LSB/USB/FM/ISB
Stability .... .7 PPM (0-50°C)
IF Rejection .... >80 dB
Environment ... 0 to +50°C. 95% Humid.

Accessories:
Suboctave Preselector (internal)

Comments:
Replaces WJ-8718. Please also see the HF-1000.

Variants: Model **WJ-8711A** is later production supplied with the following bandwidths: 16000, 14400, 12800, 12000, 11200, 10400,9600, 8800, 8000, 7200, 6400, 6000, 5600, 5200, 4800, 4400, 4000, 3600, 3200, 3000, 2800, 2600, 2400, 2200, 2000, 1800, 1600, 1500, 1400, 1300, 1200, 1100, 1000, 900, 800, 750, 700, 650, 600, 550, 450, 400, 375, 350, 325, 300, 275, 250, 225, 200, 188, 175, 163, 150, 138, 125, 113, 100, 94, 88, 81, 75, 69. 63 and 56 Hz 3-dB.

Made In:	United States 1990-1998	
Voltages:	97-253 VAC 47-440 Hz 35W	
Readout:	`00000.001` Digital LED	
Physical:	19x5.25x20" 12 Lbs. 482x133x508mm 5.4 kg	
Status:	Active Manufacturer Active Model	
Rarity:	Very Scarce	
Reviews:	*NASWA* July 1993	

	New	Used
Price:	$4495-4795	$2800-3000
Rating:	★★★★★	★★★★★

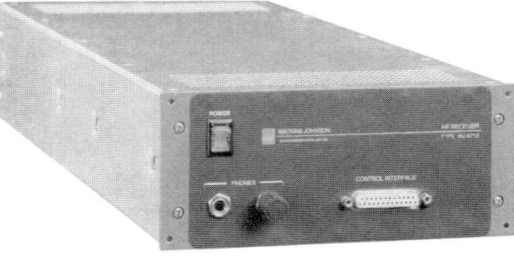

WJ-8712A

General Coverage Communications Computer Receiver
Double Conversion Superheterodyne. Solid State.

Features:

• ¼" Head. Jack	• DSP IF	• AGC (2 Pos.)	• Attenuator
• RF Gain	• Squelch	• DSP NB	• RS232 port
• BFO ±8 kHz.	• Self test	• 100 Memories	• Preamplifier
• Scan/Sweep	• IF Out Jack	• Sync. Detection	• Passband Tuning
• Ext. Ref. Jack	• Speaker Out	• Mute Line	

Specifications:

Coverage ......5 - 30000 kHz
Selectivity ......16/6/3.2/1/.3 kHz -3dB
Sensitivity ......<.35µV (.5-30 MHz)
3rd Ord Int. ....+30 dBm
Antenna Input BNC

Modes .... AM/CW/LSB/USB/FM/ISB
Stability .... .7 PPM (0-50°C)
IF Rejection .... >80 dB
Environment ... 0 to 70°C. 95% Humid.

Accessories:
Front Panel   Suboctave Preselector (internal)

Comments:
The WJ-8712A is the functional equivalent of the WJ-8711. This half-rack sized receiver is designed exclusively for computer control. Available bandwidths: 16000, 14400, 12800, 12000, 11200, 10400,9600, 8800, 8000, 7200, 6400, 6000, 5600, 5200, 4800, 4400, 4000, 3600, 3200, 3000, 2800, 2600, 2400, 2200, 2000, 1800, 1600, 1500, 1400, 1300, 1200, 1100, 1000, 900, 800, 750, 700, 650, 600, 550, 450, 400, 375, 350, 325, 300, 275, 250, 225, 200, 188, 175, 163, 150, 138, 125, 113, 100, 94, 88, 81, 75, 69, 63 and 56 Hz -3dB.

Made In:	United States 1992-1998	
Voltages:	97-253 VAC 47-440 Hz 35W	
Readout:	 `1 Hz.` Digital via computer.	
Physical:	8.25x3.5x20" 15 Lbs. 209x89x509mm 6.78 kg	
Status:	Active Manufacturer Active Model	
Rarity:	Typically Unavailable	

	New	Used
Price:	①	②
Rating:	⑥	⑥

WJ-8712P

General Coverage Communications Receiver
Double Conversion Superheterodyne. Solid State.

Features:

- Head. Jack
- RF Gain
- BFO ±8 kHz.
- Scan/Sweep
- Ext. Ref. Jack
- Keypad

- S Bar Graph
- Squelch
- Self Test
- IF Out Jack
- DSP IF

- AGC (2 Pos.)
- DSP NB
- 100 Memories
- Sync. Detection
- Speaker Out

- Attenuator
- RS232 port
- Preamplifier
- Passband Tuning
- Mute Line

Specifications:

Coverage ......5 - 30000 kHz
Selectivity ......16/6/3.2/1/.3 kHz -3dB
Sensitivity ......<.35µV (.5-30 MHz)
3rd Ord Int. ....+30 dBm
Antenna Input BNC

Modes .... AM/CW/LSB/USB/FM/ISB
Stability .... .7 PPM (0-50°C)
IF Rejection .... >80 dB
Environment ... 0 to 70°C. 95% Humid.

Accessories:

Suboctave Preselector (internal)

Comments:

Once again this receiver is the functional equivalent of the WJ-8711 in a half rack configuration. Unlike the WJ-8712A, this receiver has a full function front panel for traditional operation. Available bandwidths: 16000, 14400, 12800, 12000, 11200, 10400, 9600, 8800, 8000, 7200, 6400, 6000, 5600, 5200, 4800, 4400, 4000, 3600, 3200, 3000, 2800, 2600, 2400, 2200, 2000, 1800, 1600. 1500, 1400, 1300, 1200, 1100, 1000, 900, 800, 750, 700, 650, 600, 550, 450, 400, 375, 350, 325, 300, 275, 250, 225, 200, 188, 175, 163, 150, 138, 125, 113, 100, 94, 88, 81, 75, 69, 63 and 56 Hz -3dB.

Made In:	United States 1992-1998
Voltages:	97-253 VAC 47-440 Hz 35W
Readout:	`00000.001` Digital LED
Physical:	8.25x3.5x20" 15 lbs. 209x89x509mm 6.78 kg
Status:	Active Manufacturer Active Model
Rarity:	Typically Unavailable

	New	**Used**
Price:	①	②
Rating:	⑥	⑤

WJ-8718A

General Coverage Communications Receiver
Double Conversion Superheterodyne. Solid State.

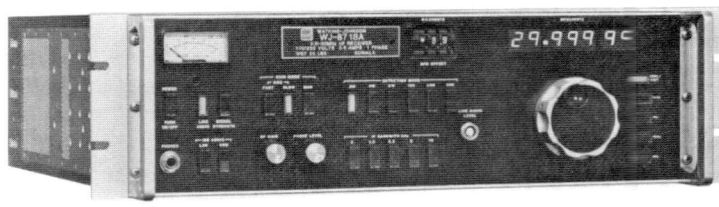

Features:

- ¼" Head. Jack
- BFO ±8 kHz
- B.I.T.E.

- AF/S-Meter
- Rack Handles
- Line Out Jack

- RF Gain
- 4 Tuning Steps
- IF Out Jack

- AGC FST/SLO/OFF
- 1 MHz Ref Jack

Specifications:

Coverage ......5 - 30000 kHz
Selectivity ......16/5/3.2/1/.3 kHz -3dB
Sensitivity ......<.4µV 16 dB (S+N)/N CW
3rd Ord Int. ....+20 dBm
Audio Output .2W 600 ohm

Modes .... AM/FM/CW/MCW
Stability .... 6×10^{-8} per day
IF Rejection .... >80 dB
Environment ... 0 to +50°C. 95% Humid.
Antenna Input . BNC 50 ohm

Accessories:

10 Hz BFO   ISB Option   LSB/USB Option Sub-Octave Preselector
Remote Module   1 Hz VFO Option Manual Control Module

Comments:

LEDs are red or green. BFO offset is adjusted by decadic thumbwheels. Also see model AN/URR-74(V)2 (shown later) which is a military version of the WJ-8718A with the Manual Control Module, ISB and Navy Environment options.

Variants:

Model **WJ-8718**, earlier production, was introduced in 1976.

Made In:	United States 1982-1990
Voltages:	115/220 VAC 48-440 Hz 70W
Readout:	`00000.01` Digital LED
Physical:	19x5.25x20" 35 Lbs. 482x134x493mm 15.8 kg
Status:	Active Manufacturer Discontinued Model
Rarity:	Extremely Scarce

	New	**Used**
Price:	$8500	②
Rating:	⑥	⑥

AN/URR-74(V)2

General Coverage Communications Receiver
Double Conversion Superheterodyne. Solid State.

Features:
- ¼" Head. Jack
- AF/S-Meter
- RF Gain
- AGC FST/SLO/OFF
- BFO ±8 kHz
- Rack Handles
- 4 Tuning Steps

Specifications:
Coverage ......5 - 30000 kHz
Selectivity ...... 16/5/3.2/1/.3 kHz -3dB
Sensitivity ......<.4μV 16 dB (S+N)/N CW
3rd Ord Int. ....+20 dBm
Audio Output .2W 600 ohm

Modes .... AM/FM/CW/MCW/ISB
Stability .... 6×10^{-8} per day
IF Rejection .... >80 dB
Environment ... 0 to +50°C. 95% Humid.
Antenna Input . N 50 ohm

Accessories:
10 Hz BFO   LSB/USB Option Sub-Octave Preselector
Remote Module   1 Hz VFO Option

Comments:
The LEDs may be red or green. The BFO offset is adjusted by decadic thumbwheels. The model AN/URR-74(V)2 is a military version of the WJ-8718A with the Manual Control Module, ISB and Navy Environment options provided as standard. The Navy Environment option consists of: mil.std. power connector, mil.std.. 13 pin audio connector, coated PCB, type N antenna jack, double fused AC circuit, DPST power switch, nickel-plated side panels, and mil. std. elapsed time and RF/AF audio meters.

Made In:	United States 1976-1990
Voltages:	115/220 VAC 48-410 Hz 70W
Readout:	`00000.01` Digital LED
Physical:	19x5.25x20" 35 Lbs. 482x134x493mm 15.8 kg
Status:	Active Manufacturer Discontinued Model
Rarity:	Extremely Scarce

	New	Used
Price:	①	②
Rating:	⑥	⑥

WJ-8718A/MFP

General Coverage Communications Receiver
Double Conversion Superheterodyne. Solid State.

Features:
- ¼" Head. Jack
- AF/S-Meter
- RF Gain
- AGC FST/SLO/OFF
- BFO ±8 kHz
- Rack Handles
- 4 Tuning Steps
- Keypad
- IF Out Jack
- Scanning
- 99 Memories
- BITE
- Sweep
- Dial Lock
- 3 Tuning Speeds

Specifications:
Coverage ......5 - 30000 kHz
Selectivity ...... 16/6/3.2/1/.3 kHz -3dB
Sensitivity ......<.4μV 16 dB (S+N)/N CW
3rd Ord Int. ....+20 dBm
Environment ..0 to +50°C. 95% Humid.
Audio Output .2W 600 ohm

Modes .... AM/FM/CW/LSB/USB/ISB
Stability .... 6×10^{-8} per day
IF Rejection .... >90 dB
Image Reject .. >90 dB
Antenna Input . BNC 50 ohm

Accessories:
10 Hz BFO   Sub-Octave Preselector   Microprocessor Front Panel
Remote Module   Manual Control Module

Comments:
LEDs are yellow in standard version. BFO offset is shown to the left of the frequency display. This is an enhanced version of the WJ-8718A. MFP stands for Microprocessor Front Panel. Additional features include: keypad entry, 99 memories, scanning and LED indicating push-button switches.

Made In:	United States 1986-1990
Voltages:	115/220 VAC 48-410 Hz 70W
Readout:	`00000.01` Digital LED
Physical:	19x5.25x20" 35 Lbs. 482x133x493mm 15.75 kg
Status:	Active Manufacturer Discontinued Model
Rarity:	Extremely Scarce

	New	Used
Price:	①	②
Rating:	⑥	⑥

WATKINS-JOHNSON

WJ-8721

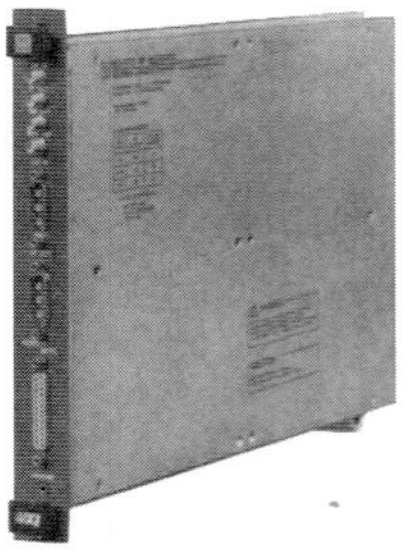

General Coverage Communications Computer Receiver
Double Conversion Superheterodyne. Solid State.

Features:
- Mini Head. Jack
- DSP IF
- Squelch
- BFO ±8 kHz
- Mute Line
- BITE
- Notch Filter
- Monitor Out
- PBT
- IF Out Jack
- Scan
- S Meter Function
- Line Out 600 ohm
- 100 Memories
- Sweep
- AGC FST/SLO/MAN
- Preamp & Attenuator
- Suboctave Preselect

Specifications:
Coverage ......5 - 30000 kHz
Selectivity ......16/6/3.2/1/.3 kHz -3dB
3rd Ord Int. ....+25 dBm (+30 typical)
Image Rej. .....>90 dB
Antenna Input 50 ohm
Modes .... AM/CW/LSB/USB/FM/ISB
Stability .... 0.7 PPM (0-50°C)
IF Rejection .... >85 dB (>90 dB Typical)
Environment ... 0 to +50°C.

Accessories:
UNIX Control Pgm. DOS Control Pgm.

Comments:
This receiver is designed to conform to the VXI computer bus. Up to 12 receivers may be fit in a single 19" rack. All receiver parameters, except power on/off are controlled by the computer bus. Mini headphone jack is stereo. Performance under specs. from 0 to 500 kHz. Available bandwidths: 16000, 14400, 12800, 12000, 11200, 10400,9600, 8800, 8000, 7200, 6400, 6000, 5600, 5200, 4800, 4400, 4000, 3600, 3200, 3000, 2800, 2600, 2400, 2200, 2000, 1800, 1600, 1500, 1400, 1300, 1200, 1100, 1000, 900, 800, 750, 700, 650, 600, 550, 450, 400, 375, 350, 325, 300, 275, 250, 225, 200, 188, 175, 163, 150, 138, 125, 113, 100, 94, 88, 81, 75, 69, 63 and 56 Hz -3dB.

Made In: United States 1992-1998

Voltages: N/A (off bus) 21W

Readout: 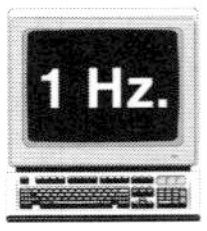

Digital via computer.

Physical: 1.2x9.2x13.4" 5 Lbs.
31x237x335mm 2.26 kg

Status: Active Manufacturer
Active Model

Rarity: Typically Unavailable

	New	Used
Price:	①	②
Rating:	⑥	⑥

WATKINS-JOHNSON

WJ-8888

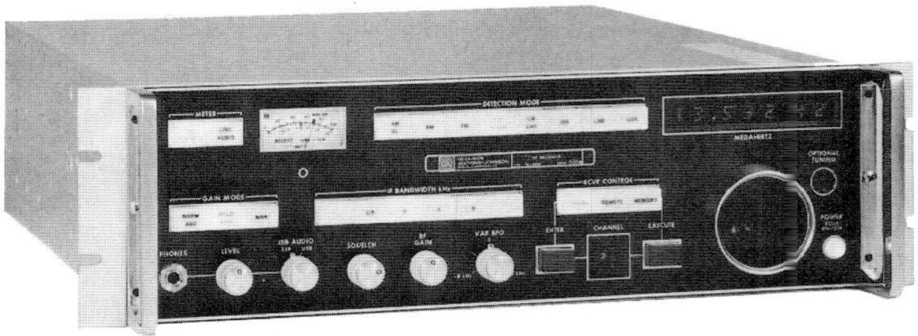

General Coverage Communications Receiver
Double Conversion Superheterodyne. Solid State.

Features:
- ¼" Head. Jack
- RF Gain
- Sync. Rem. Port
- BFO ±8 kHz
- S/AF Meter
- VRIT
- 1 MHz Output
- Spinner Knob
- Squelch
- Rack Handles
- IF Out 455 kHz
- Line Out 600 ohm
- ISB Mode
- AGC MAN/NORM/HOLD
- 4 Memories

Specifications:
Coverage ......500 - 30000 kHz
Selectivity ......8/4/2/.5/_/_ kHz -3dB
Sensitivity ......<.4μV 16 dB (S+N)/N CW
Image Rej ......>100 dB
IF Rejection ...>100 dB
Modes .... AM/FM/CW/LSB/USB/ISB
Stability .... $6x10^{-8}$ per day
Environment ... 0 to +50°C.
Antenna Input . BNC 50 ohm

Accessories:
.2 kHz Filter 1 kHz Filter 3 kHz Filter 6 kHz Filter
12 kHz Filter 16 kHz Filter 16 Chan Mem.

Comments: The WJ-8888 has three operating modes. In Local mode the receiver is tuned manually. In the Remote mode the receiver accepts and stores a digital word. This 64 bit word stores: frequency, mode, gain level, bandwidth, RF gain level, and BFO frequency. In Memory mode these same parameters may be stored and retrieved. 4 Memories are supplied. Audio is only available at line out levels or via headphone jack. Sub-octave filters are automatically switched for preselection. Referred to as the W-J "Quad 8". A sophisticated receiver and very collectable.
Variants: Model **WJ-8888A** 1975.

Made In: United States 1972-1974

Voltages: 115/220 VAC 48-62 Hz
80W

Readout: 00000.01 Digital LED

Physical: 19x5.25x19.5" 40 Lbs
483x133x495mm 18.1 kg

Status: Active Manufacturer
Discontinued Model

Rarity: Extremely Scarce

	New	Used
Price:	①	$950
Rating:	⑥	⑤

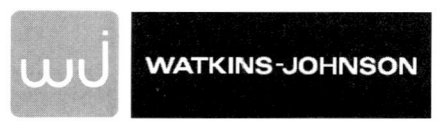

WATKINS-JOHNSON

WJ-8888B

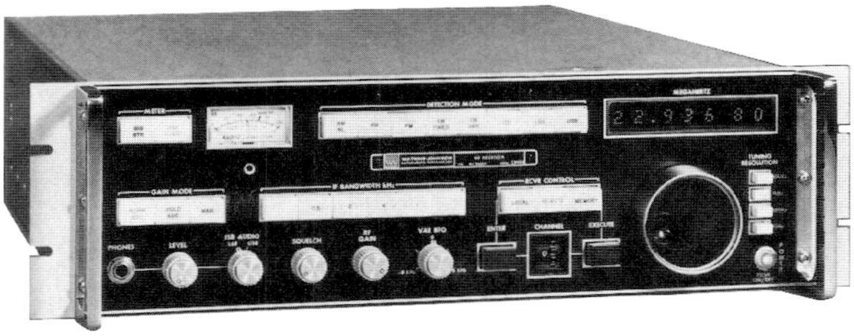

General Coverage Communications Receiver
Double Conversion Superheterodyne. Solid State.

Features:

• ¼" Head. Jack	• S/AF Meter	• Squelch	• ISB Mode
• RF Gain	• 4 Tuning Speeds	• Rack Handles	• AGC MAN/NORM/HOLD
• Sync. Rem. Port	• 1 MHz Output	• IF Out 455 kHz	• 4 Memories
• BFO ±8 kHz	• Line Out 600 ohm		

Specifications:

Coverage	500 - 30000 kHz	Modes	AM/FM/CW/LSB/USB/ISB
Selectivity	8/4/2/.5/_/_ kHz -3dB	Stability	6x10^{-8} per day
Sensitivity	<.4μV 16 dB (S+N)/N CW	Environment	0 to +50°C.
Image Rej	>100 dB	Antenna Input	50 ohm
IF Rejection	>100 dB		

Accessories:

.2 kHz Filter	1 kHz Filter	3 kHz Filter	6 kHz Filter
12 kHz Filter	16 kHz Filter	16 Chan Mem.	

Comments:
The later "B" model is very similar to the original version. Note the four buttons to the right of the main tuning knob. These select Tuning Resolution at 10 Hz, 100 Hz, 1 kHz or 10 kHz. Audio is only available at line out levels or via the headphone jack.

Made In: United States 1976-1980

Voltages: 115/220 VAC 48-62 Hz 80W

Readout: 00000.01 Digital LED

Physical: 19x5.25x19.5" 40 Lbs
483x133x495mm 18.1 kg

Status: Active Manufacturer
Discontinued Model

Rarity: Extremely Scarce

	New	Used
Price:	①	⑤
Rating:	⑥	⑥

In 1956 a young Japanese college graduate and amateur experimenter, Sako Hasegawa, read with enthusiasm about the development of Single Sideband by Arthur Collins. After constructing several SSB generators for himself and friends, he was soon receiving requests from radio amateurs all over Japan for complete transmitters. Yaesu Musen Company, Ltd. was thus founded and incorporated in 1959 to produce radio communications equipment for radio amateurs.

Beginning with the FT-20 SSB transmitter, the Yaesu brand rapidly became known over the air worldwide and in 1961 Yaesu Musen began exporting to Australia, followed soon afterwards by Switzerland, Germany and the United States. In 1966 Yaesu developed the first Japanese amateur transceiver using transistors, the FT-100 which was very popular. All profits were reinvested to expand the Company and the product line grew.

Yaesu became a household word among shortwave DXer's with the introduction in 1977 of the famous Yaesu FRG-7. Justifiably rated by the *1978 World Radio TV Handbook* as *"The best all around set"*. Yaesu is currently offering the FRG-100, which again, presents a good value for the money.

Yaesu Musen Company Ltd.
C.P.O. Box 1500
Tokyo, Japan   1970-1996

Yaesu Electronics Corporation
7625 E. Rosecrans Ave. #29
Paramont, CA 90723

Yaesu Electronics Corporation
15954 Downey Ave.
Paramont, CA 90723   1970-1980

Yaesu Electronics Corporation
6851 Walthall Way
Paramont, CA 90723   1982

Yaesu Electronics Corporation
17210 Edwards Rd.
Cerritos, CA 90701   1990-1996

FRdx400

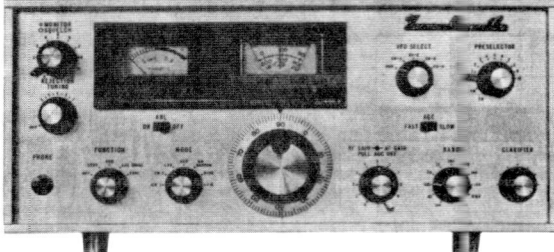

Amateur Band Communications Receiver
Double Conversion Superheterodyne. 10 Tubes plus Semiconductors.

Features:
- ¼" Head. Jack
- RF Gain
- Squelch
- ANL
- S-Meter
- Notch Filter
- Speaker Out
- BFO
- Standby
- Preselector
- Line Out
- Calibrator 100/25 kHz
- AGC SLO/FST/OFF
- VFO Out

Specifications:
Coverage ...... Ham. See Comments.
Selectivity ...... 4/2.4/1/_ kHz -6dB.
Sensitivity ...... <.5µV10 dB S/N SSB/CW
Dial Accur. ...... 1 kHz (after calib.)
Audio Out ...... 2W 4/600 ohms 5% dist.

Modes .... AM/LSB/USB/CW
Stability .... <100 Hz after warmup
IF Rejection .... >60 dB
Antenna Input . SO239 50-75 ohm

Accessories:
FC-6TR 6M Conv.   FC-2TR 2M Conv.   MF455-03AZ 600 Hz CW Filter
FM Mode Option   CFM455H 5 kHz AM Filter   CFM455B 24 kHz FM Filter
4 Chan. Fixed Crystal

Comments: Supplied coverage: 1.7-2.3, 3.5-4.1, 6.9-7.5, 13.9-14.5, 20.9-21.5, 27.9-28.5 and 28.5-29 MHz. Optional positions: 26.9-27.5, 29.5-30.1 and 9.9-10.5 MHz. The FRdx400 provides better than 1 kHz dial accuracy. Matches the FLdx400 Transmitter. Requires a speaker.

Variants: Model **FRdx400SD** includes the 6 and 2 meter converters built-in $399 new. The **Sommerkamp FRdx500**, sold in Europe, appears to be the same as the Yaesu FRdx400SD.

Made In:	Japan 1971-1974
Voltages:	100/110/117/200/220/234 VAC 50/60 Hz 50 W
Readout:	Analog Lin.
Physical:	14.5x6.5x11.5" 25 Lbs. 368x165x292mm 11.3 kg
Status:	Active Manufacturer Discontinued Model
Rarity:	Scarce

	New	Used
Price:	$299-399	$150-190
Rating:	★★★★	⑥

FR-101S

Amateur Band Communications Receiver
Double Conversion Superheterodyne. Solid State.

Features:
• ¼" Head. Jack	• S-Meter	• Dial Lock	• Calibrator 100/25 kHz
• RF Gain	• Squelch	• Preselector	• AGC OFF/FST/SLO
• Mute	• Standby	• Speaker Jack	• Side Tone Jack
• VFO Jack	• Record Jack	• Clarifier ±5 kHz.	• Fixed Crystal Jack
• NB	• Spinner Knob	• Atten. -10/20 dB	

Specifications:
Coverage ......Ham. See Comments.
Selectivity ......2.4/_/_ kHz -6dB.
Sensitivity ......<.3µV (@ 14MHz).
Dial Accur......1 kHz (after calib.)
Audio Out ......2 W 10% distortion

Modes .... AM/LSB/USB/CW/RTTY
Stability .... 100 Hz 30 min period
IF Rejection .... >60 dB
Antenna Input . SO239 50 ohms

Circuit Complement:
4 Integrated circuits, 20 Transistors, 12 FET and 33 diodes.

Accessories:
SP-101B Speaker   SP-101PB Patch/Speaker   FM-1 FM Detector
XF-30C .6 kHz Filter XF-30B 6 kHz Filter   FC-2 2 Meter Converter
XF-30D FM Filter   FC-6 6 Meter Converter

Comments:
Coverage: 1.8-2, 3.5-4, 7-7.5, 14-14.5, 21-21.5, 28-28.5, 28.5-29 plus optional shortwave bands. Dial accuracy is 1 kHz. Matches FL-101 Transmitter. A very good receiver, but analog readout and ham band only coverage keeps the value down.

Made In: Japan   1974-1978

Voltages: 100/110/117/200/220/234 VAC 50/60 Hz or 12 VDC

Readout: |␣␣␣␣␣␣| Analog Lin.

Physical: 6x13.5x12" 20 Lbs.
153x340x285mm 9 kg

Status: Active Manufacturer
Discontinued Model

Rarity: Scarce

Reviews: *Ham Radio* July 1974

	New	Used
Price:	$499-599	$150-160
Rating:	★★★★	★★★

FR-101SD

Amateur Band Communications Receiver
Double Conversion Superheterodyne. Solid State.

Features:
• ¼" Head. Jack	• S-Meter	• Side Tone Jack	• Calibrator 100/25 kHz
• RF Gain	• Squelch	• Preselector	• AGC OFF/FST/SLO
• Mute	• Standby	• Speaker Jack	• Clarifier ±5 kHz.
• VFO Jack	• Record Jack	• Dial Lock	• Fixed Crystal Jack
• NB	• Spinner Knob	• Atten. -10/20 dB	

Specifications:
Coverage ......Ham. See Comments.
Selectivity ......2.4/_/_ kHz -6dB.
Sensitivity ......<.3µV (@ 14 MHz).
Audio Out ......2 W 10% distortion

Modes .... AM/LSB/USB/CW/RTTY
Stability .... 100 Hz 30 min period
IF Rejection .... >60 dB
Antenna Input . SO-239 50 ohms

Accessories:
SP-101B Speaker   SP-101PB Patch/Speaker   FM-1 FM Detector
XF-30C .6 kHz Filter XF-30B 6 kHz Filter   FC-2 2 Meter Converter
XF-30D FM Filter   FC-6 6 Meter Converter

Comments:
Coverage: 1.8-2, 3.5-4, 7-7.5, 14-14.5, 21-21.5, 28-29 plus optional shortwave bands. Matches the FL-101 Transmitter. The digital frequency display only operates when the receiver is run from AC power. This receiver is also referred to as the FR-101D or FR-101 Digital. A good receiver. The used value is depressed because of the light demand for non-general coverage receivers.

Made In: Japan   1974-1978

Voltages: 100/110/117/200/220/234 VAC 50/60 Hz or 12 VDC

Readout: 00000.1 Digital LED

Physical: 6x13.5x12" 31 Lbs.
153x340x285mm 14 kg

Status: Active Manufacturer
Discontinued Model

Rarity: Scarce

	New	Used
Price:	$629-749	$250-300
Rating:	★★★★	★★★

FRG-7

General Coverage Communications Receiver
Triple Conversion Superheterodyne. Solid State.

Features:
- ¼" Head. Jack
- Tone (3 Pos.)
- Mute Line
- Spinner Knob
- S-Meter
- Preselector
- ANL
- Fiduciary Adj.
- Speaker
- Dial Lamp
- Record Jack
- Dial Lamp Switch
- Atten. 3 Pos.
- Carrying Handle
- Fine Tuning

Specifications:
Coverage ......500 - 29900 kHz	Modes .... AM/USB/LSB-CW
Selectivity ......±3 kHz -6dB.	Stability .... ±500 Hz 30 min period
Sensitivity ......<.25µV S/N 10 dB CW	IF .... 55, 3-2 MHz, .455 kHz
Audio Out ......2 W 10% dist.	Antenna Input . SO239 & Terminals

Accessories:
DC Kit   Intl. Battery Holder (8xD)

Circuit Complement:
2 Integrated circuits, 13 transistors, 9 FETs and 16 diodes. Outstanding stability is achieved through the use of the Wadley loop drift cancellation circuit developed by Dr. T. L. Wadley of South Africa.

Comments:
Ranges: .5-1.6, 1.6-4, 4-1 and 11-29.9 MHz. The Fine Tuning knob was not featured on early production. The analog linear dial yields ±5 kHz dial accuracy. An excellent, durable receiver. Perhaps the best used receiver value under $350. Also see Sears model 412.36380700.

Made In:	Japan	1976-1980

Voltages:	100/110/117/200/220/234 VAC 50/60 Hz or 12 VDC or 8 D cells optional holder

Readout:	Analog Lin.

Physical: 13.5x6.5x11.5" 15 Lbs. 340x153x285mm 7 kg

Status: Active Manufacturer Discontinued Model

Rarity: Common

Reviews: *WRTH* 1978, 1980
Popular Elec. June 1977
73 May 1977
73 June 1977

	New	Used
Price:	$290-370	$160-250
Rating:	★★★★★	★★★★★

YAESU

FRG-100

General Coverage Communications Receiver
Double Conversion Superheterodyne. Solid State.

Features:
- ¼" Head. Jack
- RF Gain
- Mute Line
- NB
- 50 Memories
- Carry Handle
- S-Meter
- Squelch
- Remote Jack
- Speaker
- Memory Scan
- Digital Clock Timer 12/24H
- Speaker
- Dimmer
- Record Jack
- CAT Jack
- Sweep
- Atten. -6/12/18 dB
- 3 Tuning Steps
- Dial Lock
- AGC FAST/SLOW
- Backlit LCD

Specifications:
Coverage ......50 - 30000 kHz	Modes .... AM/USB/LSB-CW
Selectivity ......6/4/2.4/_ kHz -6dB.	IF Rejection .... >70 dB 1.8-30 MHz
Sensitivity ......<.25µV SSB/CW 1.8-30 MHz	Antenna Input . SO239 & Terminals
Audio Out ......4- 8 ohm	

Accessories:
YF-100C 500 Hz Filt.   FIF Cat Interface   TCXO-4 High Stability
YF-100CN 300 Hz Filt.   FM Unit-100 FM Mode
BEEI Wired Keypad (Not made by Yaesu)

Comments:
Display resolution is selectable at 10 or 100 Hz. Many programmable features including: 10 Hz frequency display, memory sort, CW offset, continuous or noncontinuous back lighting, custom tuning steps, etc. Extremely stable, highly accurate and very sensitive. An excellent RTTY/FAX receiver. A third-party wired keypad, made in France by BEEI plugs directly into the CAT jack on the rear panel.

Made In:	Japan	1994-1998

Voltages:	11-12 VDC Supplied with AC PS

Readout:	00000.01 Digital LCD

Physical: 9.37x3.6x9.5" 6.6 Lbs. 238x93x243mm 3 Kg

Status: Active Manufacturer Active Model

Rarity: Common

Reviews: *Passport* 1994 1998
WRTH 1993
NASWA June 1993
ODXA May 1993
SW Magazine April 1993
QST January 1994

	New	Used
Price:	$590-650	$320-400
Rating:	★★★★★	★★★★★

FRG-7000

General Coverage Communications Receiver
Triple Conversion Superheterodyne. Solid State.

Features:
- ¼" Head. Jack
- RF Gain
- Mute Line
- ANL
- S-Meter
- Tone
- Remote Jacks
- Display ON/OFF
- Speaker
- Preselector
- Record Jack
- Carry Handle
- Attenuator
- Digital Clock Timer 24H
- Fine Tuning

Specifications:
Coverage ......250 - 29900 kHz
Selectivity ......6/3 kHz -6dB.
Sensitivity ......<.7µV S/N 10 dB
Audio Out ......2 W 4 ohm 10% dist

Modes .... AM/USB/LSB-CW
Stability .... ±500 Hz 30 min period
Antenna Input . SO239 & Terminals

Circuit Complement:
27 Integrated circuits, 15 FETs, 17 transistors and 44 diodes.

Comments:
The digital display and indicator lamps may be turned off to conserve power. Two remote jacks provide NO (normally open) or NC (normally closed) connections. The timer is one event with on and off. This first Yaesu digital general coverage receiver did not live up to the performance level of its predecessor, the venerable FRG-7.

Made In:	Japan	1977-1980
Voltages:	100/110/117/200/220/234 VAC 50/60 Hz 25W	
Readout:	00001.	Digital LED
Physical:	14.3x5x11.5" 16 Lbs. 360x125x295mm 7 Kg	
Status:	Active Manufacturer Discontinued Model	
Rarity:	Common	
Reviews:	*WRTH* 1980 *Australian DX News* July '78	

	New	Used
Price:	$599-655	$180-275
Rating:	★★★★	★★★

FRG-7700

General Coverage Communications Receiver
Triple Conversion Superheterodyne. Solid State.

Features:
- ¼" Head. Jack
- RF Gain
- Mute Line
- NB
- Carrying Handle
- S-Meter
- FM Squelch
- Remote Jacks
- Mem Fine Tune
- DX-Local Switch
- Speaker
- Dimmer
- Record Jack
- Tone Control
- Atten. Variable
- Digital Clock Timer 12H
- Acc Jack 5 pin DIN
- AGC FAST/SLOW

Specifications:
Coverage ......150 - 29999 kHz
Selectivity ......12/6/2.7 kHz -6dB [15FM]
Sensitivity ......<.5µV SSB/CW 2-30 MHz
Audio Out ......1.5 W 8 ohm 10% dist
Circuit Complement: 37 Integrated circuits, 107 transistors and 97 diodes.

Modes .... AM/USB/LSB-CW/FM
Stability .... ±300 Hz 30 min period
Antenna Input . SO239 & Terminals

Accessories:
FRV-7700 VHF Conv FRA-7700 Active Antenna   FF-5 Low Pass Filter
FRT-7700 Tuner   MU-7700 12 Ch Memory   DC7-700 DC Kit
The external FRV-7700 came in six versions: **A:** 118-130, 130-140, 140-150. **B:** 50-59, 118-130, 140-150. **C:** 140-150, 150-160, 160-170. **D:** 70-80, 118-130, 140-150. **E:** 118-130, 140-150, 150-160. **F:** 118-130, 150-160, 160-170 MHz.
Comments: The digital display utilizes yellow LEDs. Please note that the squelch works in FM mode only. Note that the Fine Tune knob only works on the (optional) MU-7700 memory channels. The MU-7700 Memory option is mounted internally without special tools and requires three AA cells.

Made In:	Japan	1981-1984
Voltages:	100/120/220/240 AC 50/60 Hz 33-39W	
Readout:	00001.	Digital Flor.
Physical:	13x5x10" 15 Lbs. 334x129x225mm 6 Kg	
Status:	Active Manufacturer Discontinued Model	
Rarity:	Common	
Reviews:	*WRTH* 1982 *Monit. Times* Sept. 1982 *73* September 1981 *QST* August 1981 *Buyer's Guide to Amateur R.*	

	New	Used
Price:	$400-550	$260-340
Rating:	★★★★	★★★

YAESU

FRG-8800

General Coverage Communications Receiver
Double Conversion Superheterodyne. Solid State.

Features:
- ¼" Head. Jack
- RF Gain
- Mute Line
- NB (NAR/WIDE)
- Keypad Entry
- CAT Jack

- S-Indicator
- Squelch
- Remote Jacks
- Speaker 4"
- Line Out 600 ohm
- 12 Memories

- Speaker
- Dimmer
- Rec Jack 50K ohm
- Fine Tuning
- Scan/Sweep
- Dual Digital Clock Timer 24H

- Atten. Variable
- Fast/Slow Tuning
- Fine Tuning
- AGC FAST/SLOW
- Carrying Handle

Specifications:
Coverage ...... 150 - 29999 kHz
Selectivity ...... 6/2.7 kHz -6dB. [12.5]
Sensitivity ...... <.4μV SSB/CW 1.6-30 MHz
Audio Out ...... 1.4 W 8 ohm 10% dist

Modes .... AM/USB/LSB-CW/FM
Stability .... <50 Hz after 30 mins.
Image Rej. ...... >60 dB
Antenna Input . SO239 & Terminals

Accessories:
FRV-8800 VHF Conv   FRA-7700 Active Antenna   FF-5 Low Pass Filter
FRV-7700 VHF Conv   FRT-7700 Tuner   DC-8800 DC Kit
FIF CAT Interface
Note: The FRV-8800 is an internal VHF converter covering 118-174 MHz.

Comments:
Three AA cells are required for memory backup. A small internal slide switch selects AUTO or MANUAL scanning. In MANUAL mode the Squelch is ignored. Tuning steps may be set for 25 or 500 Hz. The CAT Jack along with the optional FIF Interface Unit, permits control via computer, including interrogation of the S-Meter.

Made In:	Japan	¯985-1993
Voltages:	100/120/220/24ᴐ VAC	
	50/60 Hz and 3 ᴀA cells	
Readout:	`00000.1` Digital LCD	
Physical:	13x5x8.75" 13.ᴢ Lbs.	
	334x118x225mᴍ 6.1 Kg	
Status:	Active Manufacᴛurer	
	Discontinued Mᴐdel	
Rarity:	Common	
Reviews:	▤IBS-RDI Whitᴇpaper	
	▣*Passport* 19ᴇ5/86-1997	
	WRTH 1986	
	WRTH Buyer's ᴣuide 1993	
	Pop Comm Sepᴛ. 1985	
	Radio Elec.Worᴎd Aug. 1987	
	Proceedings 1ᴝ9	

	New	Used
Price:	$429-699	$ᴢ00-430
Rating:	★★★★	★★★

This Chapter includes receivers that fall into one or more of the following categories:
◆ An extremely rare model.
◆ Not significant enough to merit a full entry.
◆ Information lacking for a full entry.
◆ Limited frequency coverage.
◆ Late arriving information.
◆ Older models.

AKD — HF3 Target

General Coverage Communications Receiver
Digital   Solid State
Made in England   1996-1998 $260 new.
Features: Coverage is from 30 - 30000 kHz in 1 kHz steps. AM/USB/LSB, 6/3.8 kHz, 1 uV, IF: 45 MHz, 455 kHz, 4 tuning rates, one memory, attenuator, clarifier ±800 Hz, 2W audio, 3.5" speaker. Antenna input is a 70 ohm RCA phono jack. 7.25x2.5x7".
Comment: Operates from 12 VDC. Reviewed in *SW Magazine* Nov. 1996 and *NASWA* April 1997.
Variants: Lowe SRX-100 is the same. The **Target HF3M** marine version includes a FAX PC interface.

B&G — HOMER 5

General Coverage Communications Receiver
Digital   Solid State
Made in England   1980-1985 $2325 new.
Features:
Covers: 50-25000 kHz. 100 Hz LCD, 24-hour clock-timer, 18 memories, keypad, 4W audio, recorder activation, VRIT and dial lamp. Double conversion. AM/CW/LSB/USB. Water and corrosion resistant.
Comment: A receiver designed for sailing yachts, power boats, fishing vessels and commercial craft. Sold through Brookes & Gatehouse Ltd. 9x5x4.4" 4.6 Lbs. 12/24 VDC. Reviewed in *WRTH* 1983

Boulevard Electronics

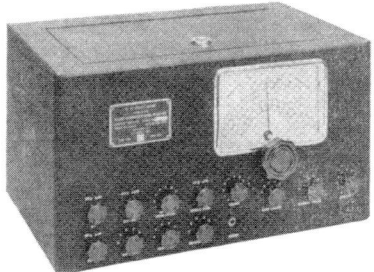

General Coverage Communications Receiver Kit
Analog   11 Tubes & Rectifier
Made in United States   1952-1953 $59 new.
Features:
Covers: .2-18 MHz. Has BFO, Hinged Top Cover, AVC ON/OFF, NL, Antenna Trimmer, Crystal Filter, RF Gain, ¼" Headphone Jack and Standby.
Comments: Uses (3) 6SG7, 6SB7Y, 6H6, 6SRF7, (2) 6SL7GT, (2) 25L6 and VR75. This radio designed was originally produced for the Coast Guard. Model number is unknown, if any. Also sold fully assembled for $89.50. 19.5x10.5x12.5"

cai communication associates, inc. — CR-70

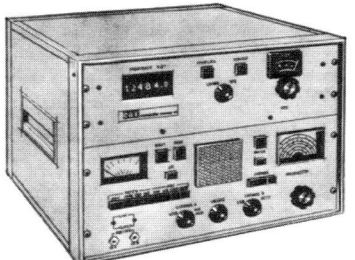

General Coverage Communications Receiver
Digital   Solid State
Made in U.S.A.   1970-1971

Features:
Covers: 2-30 MHz. Has: Preselector, S-Meter, Speaker, ANL, BFO and ¼" Headphone Jack. Tunes in 100 Hz steps via decadic switches or VFO. AM/LSB/USB/CW/FSK. 117 VAC.

Comments:
A commercial receiver. Scarce.

CODAR Multiband-6

General Coverage Communications Receiver Kit
Analog   Solid State
Made in England   1973   £13.2 new.

Features:
Bandspread.

Comments:
Coverage is 550 - 30000 kHz. Other details are unknown.

CODAR CR45

Shortwave Broadcast Receiver Kit
Analog   3 Tubes
Made in England   1964   £7.5 new.
Features:
¼" Headphone Jack, Bandspread and Regeneration Control.
Comments: Ranges: 1.7-5, 4-11.5 and 10.5-30. Provides 3 watts audio for 2-3 ohm external speaker. Tubes: 12AT7, EL84 and EZ80. 12x5.25x7" 200-250 VAC 50 Hz. Two optional coils (.15-.6 and .6-1.8 MHz) were also available.
Variants: This model was also sold fully assembled.

CODAR CR66

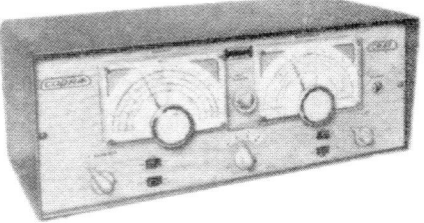

General Coverage Communications Receiver Kit
Analog   6 Tubes
Made in England   1964   £19.3 new.
Features: BFO, Bandspread, ANL, Standby, Speaker Switch, Regenerative IF stage.
Comments: Coverage is 540 - 30000 kHz in four bands. Operates from 220/240 VAC. This receiver requires a speaker. Tubes: ECH81, EF89, ECC81, EL84, EX80 and EM84. Provides 3 watts for 2-3 ohm external speaker. 16x6.5x8.75" Silver gray cabinet.
Variants:
Also sold fully assembled for £21.4.

CODAR CR70A

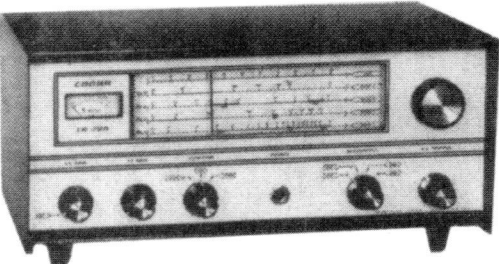

Shortwave Broadcast Receiver
Analog   4 Tubes
Made in England   1967-1974 £19.1 new.
Features:
¼" Headphone Jack, S-Meter, Dial Lamp and Record Output Jack.
Comments: Coverage is 550 - 30000 kHz in four bands. Operates from 220/240 VAC. This receiver requires a speaker. Tubes: ECH81, EF183, ECC81 and BY100. Made by Codar Radio Co., Bank House, Southwick Square, Southwick, Sussex, England.

COMMUNITRONICS LTD. 6010

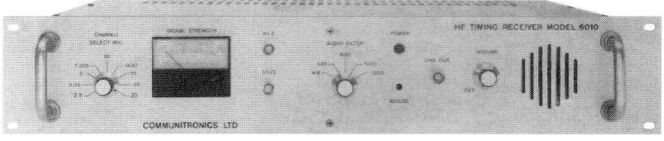

HF Time Station Fixed Receiver
Channel Display.
Made in U.S.A.
Features:
The model 6010 is an updated version of the 6006 WWV Receiver. It receives WWV time signals at 2.5, 5, 10, 15, 20 and 25 MHz (CHU frequencies with optional local oscillator unit). AM mode, 4 kHz -3 dB, 1 µV sensitivity, 2 W audio out, S-Meter, and audio filter. BNC type antenna input (50 or 300 ohms). Designed for 19" rack mounting. 19"x3.5x14" 115 VAC ±10.

500

Amateur Band Communications Receiver Kit
Analog   4 Tubes Scarce
Made in United States   1966 $30-40 used.
Features: Ranges: 3.5-4, 7-7.3 and 20.8-21.5 with the 80, 40 and 15 meter novice amateur bands highlighted. Has BFO, speaker, RF gain, standby, antenna trimmer and ¼" headphone jack. Circuit: 6BE6 Converter, 6BZ6 1st IF, 6U8 2nd IF/FO, 6U8 1st Audio/Audio Output and 2 diodes.
Comments: This receiver kit was included as part of a National Radio Institute novice training course. 9.875x7.5x6.375" 7 Lbs. 115 VAC 60 Hz 18W.

1-504

Shortwave Broadcast Receiver
Analog   5 Tubes
Made in United States   1947 $35 new.

Features:
Ranges: .535-1.72, 2.2-7.2 and 6.9-23.5 MHz, AVC, 5 inch speaker.
Comment:
Circuit complement: 6BE6 Converter Osc., 6BA6 IF Am, 6AT6 Detector/AVC/Audio Amp, 6AQ5 Power Output and 6X4 Rectifier. Plastic case. 110-230 VAC. 20x15.5x9.75" 10 Lbs.

1-604

Shortwave Broadcast Receiver
Analog   6 Tubes
Made in United States   1947 $20 new.
Features:
Ranges: .540-1.75, and 6-21 MHz, 5 inch speaker.
Comment:
Circuit complement: 12SQ7GT Detector/1st IF, 12SK7GT RF Amp, 12SK7GT IF Amp, 12SA7GT Osc/Mixer, 35L6GT Audio Amp and 35Z5 Rectifier. Plastic case. 105-125 VAC or 105-125 VDC. 12x7x75x6.5" 8 Lbs.
Variants: Model **1-605** is the same but ivory case.

6F26W

Shortwave Broadcast Receiver
Analog   6 Tubes
Made in United States   1947 $30 new.
Features:
Ranges: .540-1.73 and 5.4-18.3 MHz, 5 inch speaker, AVC, dial lamp, tone and MW ferrite antenna.
Comment:
Circuit complement: 12SQ7GT Detector/1st IF, 12SK7GT RF Amp, 12SK7GT IF Amp, 12SA7GT Osc/Mixer, 35L6GT Audio Amp and 35Z5 Rectifier. Walnut veneer case. 105-125 VAC or 105-125 VDC. 14.5x9.375x7.75" 13 Lbs.

CONTRACTOR SWITCHGEAR ELECTRONICS

2AR

Amateur 160M Communications Receiver
Analog   Solid State
Made in England   1966
Features:
Covers 1.8-2 MHz (160M ham band) only! S-Meter, RF Gain, wide/narrow selectivity, AM/SSB/CW and BFO. Provides 1 uV 10 dB S/N and >50 dB image rejection. Operates from 12 VDC
Comment:
This British receiver was designed for the dedicated 160 meter amateur enthusiast and matches the 2A10 transmitter.

850/2

Longwave Communications Receive
Analog   11 Tubes
Made in England   1961-1969
Features:
Coverage: 10-600 kHz in 6 ranges. RF Gain, BFO, AF Filter, AGC, Antenna Trimmer, S-Meter, Logging Scale, Rack Handles, Dial Lamp, NL and 140:1 tune ratio. 110/125 200/250 VAC 40/60 Hz. Single conversion, 720 kHz IF. 16.87x8.75x15" 50 Lbs.
Comment: Designed for the marine market.
Variants: Models **850/3** (1963) and **850/4** (1965) feature fixed operation. **850/5** is an STC version.

909A

Marine Communications Receiver
Analog   7 Tubes
Made in England   1960
Features:
Coverage: 1.6-4.7 MHz. RF Gain. Operates from AC or DC. Single conversion. SO-239 75 ohm ant. input, 2182 preset, 140:1 Tune ratio.
Comment: Designed for the marine market.
Variants: Model **909A/1** Operation from external supply 24V 100-150V. Model **909A/2** and **909A/3** had an internal supply to operate from 24 VDC. Model **909** is Swedish version.

EICO

711 Space Ranger

General Coverage Communications Receiver
Analog
Made in United States   1968 $70 new wired.

Features:
Covers: .55-30 MHz. Ranges: .55-16, 1.5-4, 4-10 and 9.5-30 MHz. Has: Bandspread 0-100, Standby, S-Meter, Speaker, BFO and ¼" Headphone Jack. 117VAC. Also sold as a kit for $50.

Funkwerk stätten Bernberg

AQST

General Coverage Communications Receiver
Analog   Tubes
Made in East Germany   1945-1952
Features: Single conv. with two RF and two IF stages. The external P.S. features a speaker. Has S-Meter, NB, AGC, RF Gain, BFO. Uses plug in coils.
Comment: This receiver is an East German knock-off of the famous National HRO design. They were made primarily for the East German state postal and telecommunication authority. A run was also produced for China. Gray green cabinet. Reviewed *Radio Bygones* #39.

FFR-230
(R-1735/URR)

Single Channel Strip Receiver
Analog Display   Solid State
Made in U.S.A.   1969-1975 $750 new.

Features:
Single channel, crystal controlled 2 to 30 MHz receiver. Preselector, S/AF-Meter, BFO, Standby ¼" Headphone Jack, Rack Handles, .5μV 10 dB S/N sensitivity, 110/220 VAC. 8.375x5.93x14.875" 13 Lbs. 115/230 VAC 50-60 Hz. Two receivers may be mounted side-by-side in a 19" space as shown.

FFR-230/6

Six Channel Strip Receiver
No Display   Solid State
Made in U.S.A.   1971-75 $950 new.

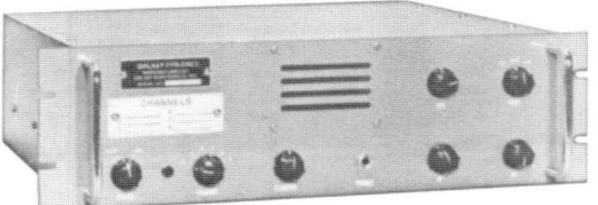

Features:
Six channel, crystal controlled 2 to 18 MHz receiver.
¼" Headphone Jack. BFO, Speaker, LSB/USB/CW,
RF Gain. .5μV 10 dB S/N sensitivity, 2.1/1.5 kHz, 4/
600 ohm. 110/220 VAC.
Comment:
Modular construction.   Reviewed in *Ham Radio*
October 1971.

G 512

Shortwave Broadcast Receiver
Analog   5 Tubes
Made in Italy   1953-1955
Features:
Phono input, dial lamp.
Comments: Ranges: .55-16, 2.3-7 and 7-21.4
MHz. This model was usually custom built into a
console. The large multicolored scale features Ital-
ian radio stations printed on the dial. 6BE6, 6BA6,
6AT6, 6AQ5 and 6X4.
Variants: Model **G 512-L** includes longwave.
Ranges: .15-.3, .55-1.6 and 5.8-21.4 MHz.

X-371

Shortwave Broadcast Receiver
Analog   11 Tubes
Made in United States   1954
Features:
Ranges: "9 shortwave bands, 2 general coverage
shortwave bands and one broadcast band"
Comment:
The X-371 has a 10 inch speaker. Other features
and specifications are unknown.

Variants: Model **X-372** is the same but with a Blond
Korina case.

2028

Shortwave Broadcast Receiver
Analog   5 Tubes & 1 Diode
Made in West Germany

Features:
Ranges: .51-1.62, 5.9-16 MHz plus FM 88-108 MHz.
Tone control, 2 Speakers, Tuning Eye, Black Forest
Walnut case, Dial Lamp, Aux. Input Jack, MW Ferrite
Antenna and Antenna Input Terminals.

Comment:
18.75x13x9" Operates from 110-220 VAC 60 Hz.

2120

Shortwave Broadcast Receiver
Analog   5 Tubes & 1 Diode
Made in West Germany

Features:
Ranges: .51-1.6, 5.9-16 MHz plus FM 87-108 MHz.
Tone control, 2 Speakers, Tuning Eye, Dial Lamp,
Aux. Input Jack, Plastic Case, MW Ferrite Antenna
and Antenna Input Terminals.

Comment:
20.5x12.25x8" Operates 115/220 VAC 60 Hz 50W.

GRUNDIG 3165

Shortwave Broadcast Receiver
Analog   6 Tubes & 3 Diodes
Made in West Germany

Features:
Ranges: .51-1.6, 5.9-16 MHz plus FM 88-108 MHz. Tone control, 3 Speakers, Tuning Eye, Black Forest Walnut case, Dial Lamp, Aux. Input Jack, MW Ferrite Antenna and Antenna Input Terminals.

Comment:
22.25x13x8.75"

GRUNDIG Classic 960

Shortwave Broadcast Receiver
Analog   Solid State
Made in China   1996-1998
Features:
Ranges: .53-1.1, 2.27-7.4 and 9.25-22.2 MHz plus FM 87.5-108 MHz. Tone control, 3 Speakers, Polished Wood Case, Dial Lamp, Aux. Input Jack, MW Ferrite Antenna and Antenna Input Terminals.
Comment: This is a modern recreation of the famous Grundig 960 produced in the 1950's. The Classic 960 provides more ambiance than performance. 15.25x11.25x6.5" 120 VAC 60 Hz.

hallicrafters CR-3000

Shortwave Communications Receiver
Analog   Solid State
Made in Japan   1968-70 $230 new.
Features:
Covers: .19-.4, .55-1.6, 2-4, 5.85-10.3 and 11-18 MHz and stereo FM 88-108 MHz. Has: 10 watts per chan. out., AGC, BFO and ¼" stereo head. jack. Inputs for ceramic and magnetic phono, tape deck & aux. Vinyl covered steel cabinet 15.125x5x12.25".
Comment:
A difficult radio to classify. Perhaps best described as a home stereo receiver with shortwave capability.

the hallicrafters co. R-96A/SR

General Coverage Communications Receiver
Analog   11 Tubes
Made in United States   1944
Features:
Ranges: 135-260, 255-510, 1485-3030, 2970-6060 and 5940-12120 kHz. Four fixed crystal positions are available for the range 1700-8700 kHz. Operates from 115 VAC 50-60 Hz or 115 VDC. AVC ON/OFF, RF Gain, Dial Lock, ¼" Headphone Jack, Speaker, Speaker Switch, Rack Handles, Standby.
Variants:
R-96/SR made by **Radiation Products**, is similar

HAMMARLUND RDF-10

Radio Direction Finding Receiver
Analog   Solid State
Made in United States   1962 $180 new.
Features:
Ranges: .19-.4, .55-1.6 and 1.7-3.4 MHz. RF Gain, Null Meter, Tone, Antenna Trimmer, Carry Handle.
Comment: The RDF-10 is designed to provide bearings for location of the marine craft on which it is used. A calibrated compass scale on the top of the receiver may be set by the shielded ferrite loop antenna. Note limited HF coverage. Operates from six "C" cells.

  **DF-3**

Radio Direction Finding Receiver Kit
Analog   Solid State
Made in United States   1961 $100 new.
Features: Ranges: .2-.4, .55-1.6 and 1.7-3.4 MHz.
RF Gain, Null Meter, Tone, 4x6" speaker, Antenna
Trimmer and Dial Lamp.
Comment: Designed for direction finding on a
marine craft. A calibrated compass scale on the top
of the receiver may be set by the shielded ferrite loop
antenna. Note limited HF coverage. Operates from
six "D" cells. 9x9.75x6.7"
Variants: Model **DFW-3** was assembled.

 **EKB-2B**

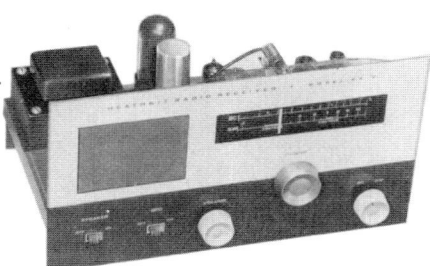

Shortwave Receiver Kit
Analog   Tubes
Made in United States   1961-1967 $20 new.
Features: Ranges: .55-1.6 and 3-10 MHz., BFO,
Speaker, Speaker Switch, Earphone Output.
Comment:
This basic receiver kit was integral to a two part basic
radio course. Upon completion of Part 2 you had a
regenerative EK-2A AM broadcast receiver. After
completing Part 2, the radio became the EKB-2B
superheterodyne AM broadcast receiver with short-
wave from 3 to 10 MHz. The AK-8 case was optional.

 **MR-11**

Radio Direction Finding Receiver Kit
Analog   Solid State
Made in United States   1962 $110 new.
Features: Ranges: .188-.41, .535-1.62 and 1.65-
3.45 MHz. RF Gain, BFO, Null Meter, Tone, 4x6"
speaker, Antenna Trimmer and Dial Lamp.
Comment: Designed for direction finding on a
marine craft. A calibrated compass scale on the top
of the receiver may be set by the shielded ferrite loop
antenna. Note limited HF coverage. Operates from
six "D" cells. 9x9.75x6.7"
Variants: Model **MR-21** similar successor model.

 **RA-1**

Amateur Band Communications Receiver Kit
Analog   8 Tubes
Made in England   1962   £39 new kit.
Features: Ranges: 1.7-2, 3.5-4, 7-7.3, 14-14.45,
21-21.5 and 28-30 MHz. RF Gain, BFO, S-Meter,
Antenna Trimmer, Calibrator (CL-1 option), Calibra-
tor Adj., BFO and Dial Lamp. 110-240 VAC 50-60 Hz.
Comment: The RA-1 and RG-1 were Anglicized
versions of similar American sets. Tubes: EF183,
ECH81, EF183, ECF82, EB91, ELC86, EX81 and
OA2. 13.75x6.5x11.5" Sold assembled for £52.5.
Variants: The **RG-1** was a general coverage version.

 **INTERAD LTD.**   **7802-3**

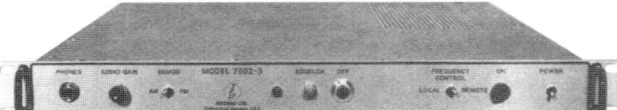

HF-VHF Fixed Receiver
No Display.   Solid State
Made in U.S.A.

Features:
In local mode, any single frequency from 10 to 500
MHz may be selected via internal switches to a 1 kHz
resolution. AM/FM detection, Squelch, ¼" headphone jack. Rack mounted. Optional remote via
RS232, RS422 or IEEE488. BNC antenna jack. 115/
230 VAC 50-400 Hz. 19"x1.72"x24" 25 Lbs.

AR-2000

Amateur Band Communications Receiver
Digital   Solid State
Made in United States   1974-1975 $ 250 new.
Features: Covers the 160 to 10M amateur bands. Has S-Meter, dual VFOs, Preselector, Spinner Knobs, RF Gain, PBT, NB, AGC OFF/FST/SLO, Notch and Peak Filters, Crystal Calibrator and RIT. Has .15µV sensitivity, 2.1 kHz BW and is modular.
Comments: The AR-2000 matches the AT-2000 transmitter. Produced by International Telecommunications Corp. of Torrance, Cal. The AR-2000 appears similar to the Signal One CR-1200.

Kantronics
8040-B

Amateur Band Morse Code Receiver
Analog   Solid State
Made in United States   1978 $30-40 used.
Features:
Covers only the 80 and 40 meter CW amateur bands (3.65-3.75 and 7.05-7.15 MHz). 6:1 vernier tuning, RF gain and headphone jack. Has 1µV sensitivity, 1 kHz bandwidth.
Comments: Only 3x5x7" 24 ounces. Operates from two 9V batteries. Reviewed in *QST* June 1978. May be found with the Kantronics "Rockhound" QRP transmitter. $90 new.

KARADIO
80-C

Communications Receiver
Analog   6 Tubes
Made in United States   1952 $65- 10 new.

Features:
Ranges: .54-1.65, 2.8-7.5 and 4.8-17.5 MHz. RF Gain.
Comment:
Designed for mobile amateur market. Circuit complement: 6BA6 RF, 6BE6 Mixer, 6BA6 IF, 6AT6 Demod/1st Audio, 6AQ5 Output and 6X5GT Rectifier. 6.675x4.5x6".

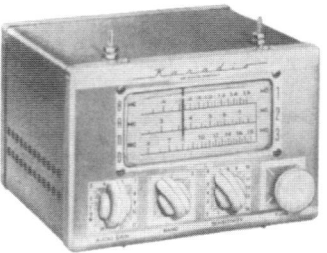

K/CR/12

Amateur Band Communications Receiver
Analog   Tubes
Made in Australia   1946
Features:
Covers the 40, 20, 10 and 6 meter amateur bands. An alternate version covering 80, 40, 20 and 10 meters was also produced. Has S-Meter RF Gain, NL, dial lamp and BFO.
Comments:
This single conversion receiver uses a 1.9 MHz IF. The tube complement is: 6J6, ECH35, 6SK7, 6SQ7, 6H6, 6V6, 6SHGT and VR150.

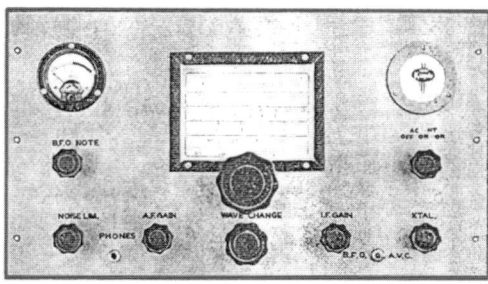

Kneisner+Doering Elektronik GmbH
KWZ-30

General Coverage Communications Receiver
Digital   Solid State
Made in Germany   1997-1998 $1950 new.

Features:
Covers: 50-30000 kHz. 1 Hz LCD, Clock-Timer, 250 Memories, Keypad, NL, PBT, Notch, Squelch, .5 µV Sensitivity, RS-232 port. Bandwidths: 9/6-4.8/3.6/3/ 2.6/2.3/2/1.8/1/.5/.3/.2/.05 kHz.
Comment:
DSP Technology. Operates from 12 VDC. 12x4.13x8.25" 8.8 Lbs. Referenced in *WRTH* 1997.

KNIGHT KIT

740 Ocean Hopper

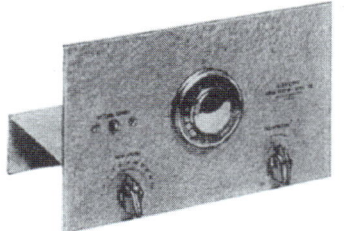

Shortwave[1] Broadcast Receiver Kit
Analog   3 Tube   Scarce
Made in United States   1953 $11 new.
Features: Antenna trimmer, regeneration control, speaker output, bandspread (0-100) and headphone tip jack. Circuit: 12AT6, 50C5 and 35W4.
Comments: [1]The 740 included only one coil covering the MW band (530 to 1900 kHz). Optional coils: S-741 .155-.47, S-742 1.65-4.1, S-743 2.9-7.3, S-745 7-17.5 and S-744 15.5-35 MHz. Allied #83S740.
Variants: Similar **749 Ocean Hopper** (1963) sold for $17 new and included cabinet. Allied #83Y749.

 **knight**

DX'er

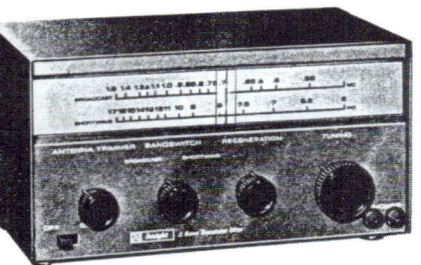

Shortwave Broadcast Receiver Kit
Analog   Solid State   Scarce
Made in United States   1963 $15 new.

Features:
Coverage: .54-1.5 and 6-17 MHz. Has headphone tip jack and antenna trimmer. Gray metal cabinet.

Comments:
Regenerative, 3 transistor circuit. 9x5x5.75" 5 Lbs. Allied #83Y943J.

KOBAYASHI

DH-66S

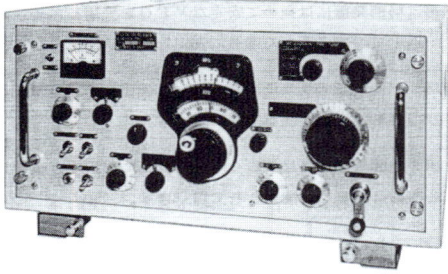

General Coverage Communications Receiver
Analog Linear   Solid State
Made in Japan
Features:
Covers: .09-32 MHz. Has: Motorized Tuning, Antenna Trimmer, RF Gain, Standby, S-Meter, Speaker, BFO, Calibrator, Rack Handles, AGC ON/OFF, ¼" Headphone Jack, IF Output and built-in Pretuner for .09-2 MHz. Bandwidths: 4.4/2.4/1/.3 kHz -6 dB. Audio output at 8 and 600 ohms 2W. 110/110/200 VAC 50/60 Hz 50 VA.
Comments: Large right knob selects 0-31 MHz.

LAFAYETTE

17F7805WX

Shortwave Broadcast Receiver
Analog   Solid State
Made in Japan   1966 $32-33 new.

Features:
Covers AM, FM broadcast band and shortwave from 3.9-12 MHz. Has Tone Control, FM AFC, dual 6x4" Speakers, Speaker Output Jack, Band-in-use Pilot Lights and Dial Lamp. Black plastic case. 24.25x6.25x5" 7 Lbs.
Comment:
Touted as the "Longest radio we have ever offered".

LORCH

HR-240

Communications Receiver
Partial Digital Display.

Features:
Digital display of MHz and 100 Hz position via Nixie tubes. Remaining digits via mechanical knob. S-meter.

Comment:
The author would welcome further information from readers on this interesting model.

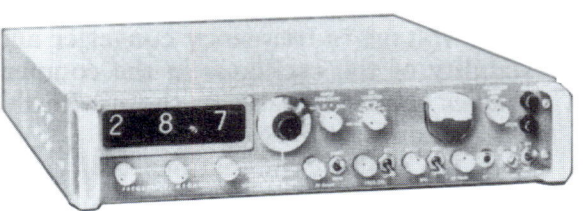

M3000
Oceanic

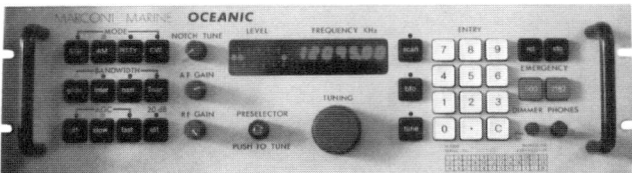

General Coverage Communications Receiver

Digital | Solid State
Made in England | 1982

Features:
S-Indicator, RF Gain, Notch, BFO, Memories, Attenuator, AGC, 4 bandwidths, Scanning ¼" Headphone Jack, Atten. -20 dB, Dimmer, Keypad and Rack Handles. Coverage is from 15 - 30000 kHz to 10 Hz in: AM/CW/SSB/RTTY/FAX. Presets for 500 & 2182 kHz. Has Line and IF out. Modular construction.

Comment: For maritime use. See Dansk M3000.

MAX FUNKE   RX-57

Amateur Band Communications Receiver

Analog | Tubes
Made in Germany | 1958   795DM new.

Features:
Coverage: 3.5-3.8, 7-7.15, 14-14.35, 21-21.45 and 28-30 MHz. 80:1 Tune Ratio, S-Meter, Rack Handles.

Comment:
110-150, 205-245 VAC, 50-60 Hz.

MAX FUNKE   RX-60

Amateur Band Communications Receiver

Analog | Tubes
Made in Germany | 1958   970DM new.

Features:
Coverage: 3.5-3.8, 7-7.15, 14-14.35, 21-21.45 and 28-30 MHz. 80:1 Tune Ratio, S-Meter, Rack Handles. IF Rejection >80 dB.

Comment:
110-150, 205-245 VAC, 50-60 Hz.

14-15

General Coverage Communications Receiver

Analog | 15 Tubes
Made in U.S.A. | 1938

Features:
Coverage is 540 to 32000 kHz. Tuning Eye, Hinged Top Cover and BFO.

Comments:
472 kHz IF.

802

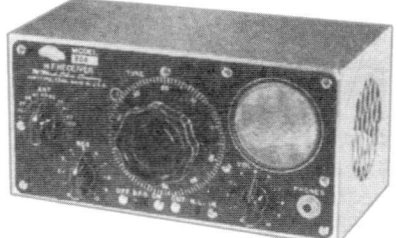

Amateur Band Communications Receiver

Analog | 8 Tubes
Made in U.S.A. | 1947 $39 new.

Features:
Single conversion superhet. Coverage includes: 80, 40, 20, 16, 11-10 and 6 meters. Antenna trimmer, ¼" Headphone Jack, Speaker and BFO.

Comment:
Tubes, power supply and coils were extra. The 802 was last McMurdo Silver receiver produced.

Variants:
Model **801 Atom-X** was a 4 tube regen .4-60 MHz.

Traffic Master

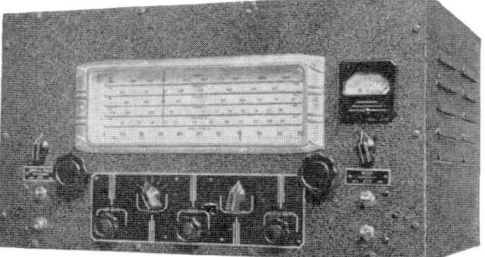

General Coverage Communications Receiver Kit
Analog   14 Tubes
Made in United States   1939-1943 $82-89 new.
Features:
Covers 530-32400 kHz. Bandspread 0-100, AVC, RF Gain, Tone Control, ¼" Headphone Jack, Crystal Filter, NL and BFO.
Comments: Ranges: .53-1.57, 1.51-4.6, 4.2-12.5, 7.3-18.5 and 11.5-32.4 MHz. 456 IF.
Variants:
The **Traffic Scout** was a less expensive kit (new $65), utilizing only 9 tubes and without the meter.

MFJ-8100

Shortwave Receiver Kit
Analog   Solid State
Made in United States   1992-1998 $60 new.

Features:
Ranges: 3.51-4.31, 5.95-7.4, 9.55-12.05, 13.2-16.4, 17.5-22 MHz. Regeneration Control, 6:1 Vernier Tuning, RF Gain Trimmer, Dual Mini Headphone jacks.
Comments: Operates from a 9 volt battery. Reviewed *QST* June 1994.
Variants: Model **MFJ-8100W** is prewired.

MS1000

Wideband Receiver
Digital LCD to 5 kHz   Solid State
Made in Japan   1991 $400A new.

Features:
Coverage: .5-600 and 800-1300 MHz in AM, FM-N and FM-W. 1000 Memories, squelch, attenuator, keypad or knob tuning, scan, sweep. Operates from 12 VDC. 4.7x2x8" 1.6 Lbs

Comment:
Sold in England.

W637 Norette

Shortwave Broadcast Receiver
Analog   6 Tubes
Made in West Germany   1954

Features:
Ranges: "3 Shortwave Bands". Speaker, Tone and AVC. This model operates from 90-110/125/150/220/240 VAC

Comment:
Other features and specifications are unknown.

W648Z Universal

Shortwave Broadcast Receiver
Analog   6 Tubes
Made in West Germany   1954

Features:
Ranges: "Medium Wave, Longwave and 2 Shortwave Bands". Speaker, Tone and AVC. This model operates from 90-110/125/150/220/240 VAC or a 6 volt car battery. Brown plastic case.

Comment:
Other features and specifications are unknown.

ONIIP — M3142 Brigantina

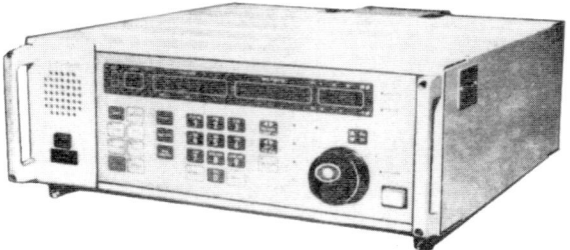

General Coverage Communications Receiver
Digital   Solid State
Made in Siberian Russia   1990-1998 $9000 new.
Features: Covers 10-30000 kHz. Has keypad entry,
100 memories, spinner knob, membrane front panel
and rack handles.   220 VAC or 24 VDC.
22.2x7x17.5" 51 Lbs. (-10° to +65° C.).
Comment: Manufactured by the Omsk Research
Institute of Communications & Electronics (formerly
the PRIBOR company) in Omsk, Russia.
Variants: Model **M18006 Brigantina-M** is similar
($7000 new).

ONIIP — M18005 Rumb

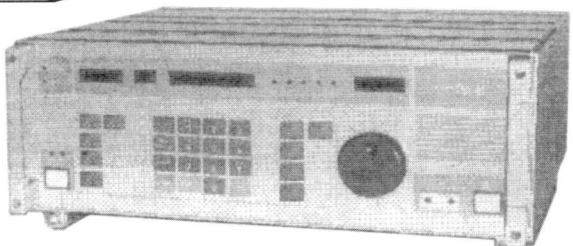

Wideband Communications Receiver
Digital   Solid State
Made in Siberian Russia   1990-1998 $7000 new.
Features: Covers 10-32000 kHz. Has keypad entry,
100 memories, scan, spinner knob, membrane front
panel, 10 Hz display, BITE, and rack handles. 220
VAC or 24 VDC. 16x6.1x17.6" 22 Lbs. (-10° to +55°
C.).
Comment:
Manufactured by the Omsk Research Institute of
Communications & Electronics. Designed for mari-
time use.

ONIIP — Olkhon-Gelios-215

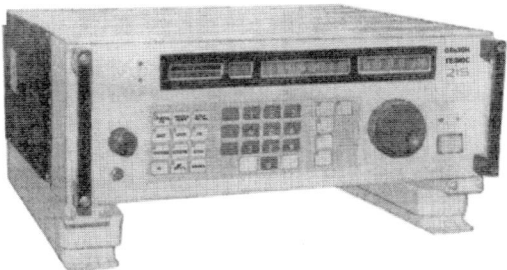

General Coverage Communications Receiver
Digital   Solid State
Made in Siberian Russia   1990-1998 $4000 new.
Features: Covers 10-30000 kHz plus 65.8-74 and
87-108 MHz.. Has keypad entry, 100 memories,
scan, spinner knob, membrane front panel, 1 Hz
display, click, BITE, and rack handles. 220 VAC or
24 VDC. 17.6x7.1x21.5" 44 Lbs. (-10° to +50° C.).
Comment:
Manufactured by the Omsk Research Institute of
Communications & Electronics. Designed for land
use.

Petersen — RM2/4

Four Channel Fixed Receiver HF/VHF
No Display   Solid State
Made in United States   1969-71 $70-75 new.

Features:
Up to four channels of fixed reception from 2 - 200
MHz in AM or FM mode. Uses frequency plug-in
cards. Speaker, Squelch. 3.375x5.75x7.5" 120
VAC or 12 VDC.

Variants:
Model **RM2/1** was a one channel version.

Philmore — 7001CR

Shortwave Broadcast Receiver
Analog   3 Tubes
Made in United States   1969-1970

Features:
3.5" Speaker.
Comments:
This two band regenerative set covers .55-1.65 and
3-13 MHz. It operates from 105-125 VAC or 105-125
VDC. The kit included the hardware only. The user
had to supply the wire, solder and tubes. The
required tubes were: 50C5, 12AU6 and 35W4.

RA87

Four Channel Communications Receiver
No Display   Tubes
Made in England   1967
Features: Four channel crystal coverage in the range of 3 to 15 MHz. AM/CW/LSB/USB, 1µV sensitivity, may be rack mounted. The BFO is ±3 kHz. The supplied bandwidth is 3.1 kHz -6dB. With rack handles, speaker, speaker out (3 ohms 1W) and ¼" Headphone Jack. 20.5x12x15" 50 Lbs.
Comments:
Single conversion superhet with 1400 kHz IF. Operates from 100-125 or 200-250 VAC 45-60 Hz 95W.

RA915

Four Channel Communications Receiver
No Display   Solid State
Made in Australia
Features: Four channel crystal coverage in the range of 2 to 30 MHz. Speaker, and line out 600 ohms. LSB *or* USB at customer's selection, 1µV sensitivity, may also be rack mounted. The supplied bandwidth is .3-3.1 kHz -6dB. 16x3.5x17" 21 Lbs.
Comments:
Operates from 21-32 VDC. Options include: PU7908 AC Power Supply, PU7909A 24/48 VDC Power Supply and 1370A Diversity Module.

RA1205

Single Channel Communications Receiver
No Display   Solid State
Made in England   1968
Features: Single channel crystal coverage in the range of 1.6 to 24 MHz. Possible modes of reception include: LSB/CW, USB/CW, AM/CW or FSK/CW. Audio output is to ¼" Headphone Jack on front and rear and 600 ohm line out. BFO range is ±8 kHz.
Comments:
Up to eight receivers and their associated MA606 case and PU1150 P.S. may be mounted "book case" style in a standard 19" rack. (Six shown in photo).

RA3721

Wideband VHF Communications Receiver
Digital   Solid State
Made in England   1994-1998
Features:
Covers 20-1000 MHz (see Comments). 500 memories, signal display, computer control, keypad
Comments:
This is primarily a VHF-UHF receiver. However, it is often configured with the optional H.F. module which extends coverage down to 10 kHz.
Variants: RA3722 is dual version, **RA3724** is dual, no front panel and **RA3723** is single, no front panel.

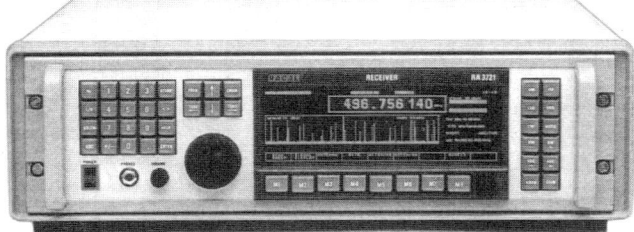

RA7915

Six Channel Communications Receiver
No Display   Solid State
Made in Australia   1972-1976
Features: Six channel crystal coverage in the range of .04 to 30 MHz. AM/CW/LSB/USB, 1 µV sensitivity, may be rack mounted. The BFO is ±500 Hz. The supplied bandwidth is 6/2.75 kHz -6dB. With rack handles, speaker, line out, speaker out (1W) and ¼" Headphone Jack. 12.5x3.5x12.5" 50 Lbs.
Comments: Operates from 12 VDC.
Variants: Model RA7943 has a carrier and a modulation meter.

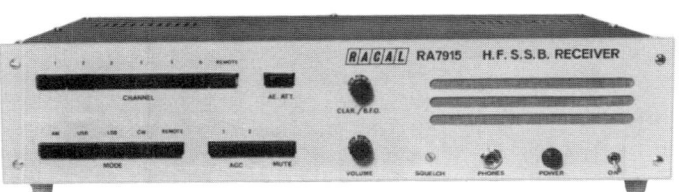

ACR-111

General Coverage Communications Receiver
Analog   16 Tubes
Made in United States   1937-1938 $190 new.

Features:
Coverage is from 540-32000 kHz. Tuning Eye, NL, ¼" Headphone Jack and Spinner Knobs. 2 ohm audio output. Two tuned RF stages and two tuned IF stages.

Comment:
460 kHz IF.

ACR-136

General Coverage Communications Receiver
Analog   7 Tubes
Made in United States   1934-1938 $69 new.

Features:
Coverage is from 540 to 18000 kHz. Hinged Top Cover, Tone Control, AVC ON/OFF, BFO, Speaker 5", Logging Scale, Headphone Jack, built-in Power Supply and 2W audio output.
Comment:
460 kHz IF. 110 VAC. Superheterodyne. Included semi-mounted external speaker. 22x10.5x11.5"

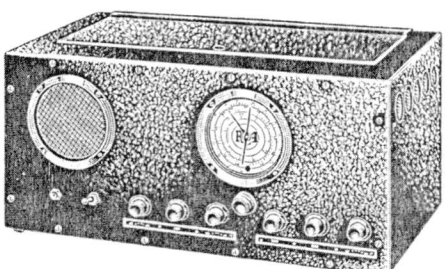

ACR-155

General Coverage Communications Receiver
Analog   9 Tubes
Made in United States   1936-1938 $75 new.
Features: Coverage is from 520 to 22000 kHz. Hinged Top Cover, Tone Control, Spinner Knob, AVC ON/OFF, BFO, Speaker 6", Logging Scale, Headphone Jack, built-in Power Supply.
Comment: Superheterodyne. Two-tone gray wrinkle finish. 460 kHz IF. 3.2 ohm audio. 6K7 RF, 6L7 Det., 6K7 IF, 6H6 2nd Det/AVC, 6F5 Audio, 6F6 Output, 6J7 Osc., 6J7 BFO and 5W4 Rectifier.
Reviews: *Radio* February 1937

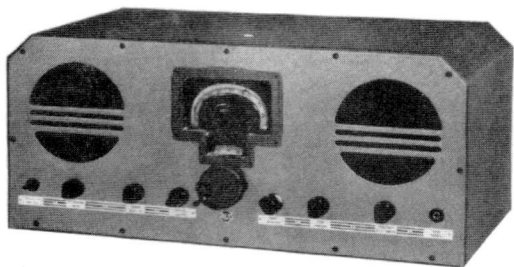

ACR-175

General Coverage Communications Receiver
Analog   11 Tubes
Made in United States   1936-1938 $119 new.
Features: Coverage is 500 to 60000 kHz. Tuning Eyes, Hinged Top Cover, Tone Control, AVC ON/OFF, BFO, Logging Scale, Headphone Jack, built-in Power Supply and 2W audio. 6K7 RF, 6L7 Det., 6J7 Osc., 6K7 IF, 6K7 IF, 6H6 2nd Det., 6E5 Tuning, 6J7 BFO, 6F5 1st AF, 6F6 Output and 5Z4 Rect.
Comment: 460 kHz IF. 110 VAC. Superhet. Included semi-mounted ext. speaker. 22x10.5x11.5"
Reviews: *Radio* July 1936

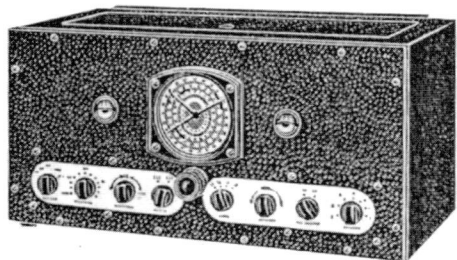

AR-60T

General Coverage Communications Receiver
Analog   11 Tubes
Made in United States   1936-1938 $475 new.
Features: Coverage is 1500 to 25000 kHz. Hinged Top Cover, Tone Control, AVC ON/OFF, BFO, ¼" Headphone Jacks, Audio Filter, RF Gain, Bandspread, built-in Power Supply and Crystal Filter.
Comment: 110/220 VAC 40-60 Hz. Superheterodyne. 22.25x12.5x17.875" Black wrinkle finish.
Variants: Model **AR-60S** is the same but with two-tone gray finish. **AR-60R** is rack mounted in black.
Reviews: *R-9* January 1936

AR-67

LF/MF Coverage Communications Receiver
Analog   9 Tubes
Made in United States   1938 $275 new.

Features:
Coverage is from 75 to 1500 kHz only.

Comment:
Please note this is an LF/MF receiver. It does not receive shortwave.

AR-8510

LF/MF Coverage Communications Receiver
Analog   5 Tubes
Made in United States   1945-1948
Features: Coverage is from 15 to 650 kHz only. Ranges: 15-38, 38-100, 100-250 and 250-650 kHz. Speaker, Speaker Switch, Trimmer, RF Gain, ¼" Headphone Jack, Line Output, 4 and 16 ohm output.
Comment:
Please note this is an LF/MF regenerative receiver. This receiver requires: 6.3 VAC/VDC 1.8 amps and 90 VDC 15 milliampere. Optional power units include: RM-2, RM-4, RM-23 and RM-37A.

Radio Marine Corporation
A Subsidiary of Electronics Assistance Corporation

CRM-R5A

LF/MF Coverage Communications Receiver
Analog   Solid State
Made in United States   1980's (?)
Features: Coverage is from 14 to 650 kHz only. Ranges: 14-33, 32-72, 70-160, 150-330 and 300-650 kHz. Speaker, Speaker Switch, Trimmer, RF Gain, ¼" Headphone Jack, Line Output, 4 and 16 ohm output. Requires 12 VDC at .25 amps.
Comment: Note this is an LF/MF receiver. This radio is interesting in that it is a solid state, commercial grade receiver with a regenerative detector stage. Clearly a descendent of the R.C.A. AR-8510.

RADIOVISION Ltd.   Commander

General Coverage Communications Receiver
Analog   Tube Type
Made in England   1948-1950 $261 new.
Features: Double conversion superhet covering 1.7-31 MHz in 5 bands. S-Meter, 3 position selectivity, RF Gain, BFO, NL, Standby, Antenna Trimmer, Osc. Trimmer, Bandspread 80-10M, 50:1 Tune Ratio and Hinged Top Cover. IFs: 1600, 100 kHz. 100-250 VAC, 50/60 Hz. 21x11.25x14" 46 Lbs.
Comment: Was imported and distributed in the United States by Radio Shack. Reviewed in *Radio Bygones* #25

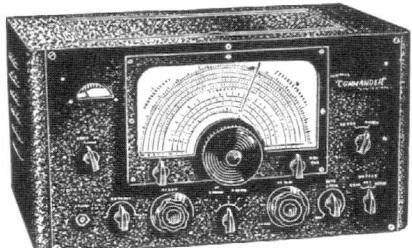

RME   RME 70

General Coverage Communications Receiver
Analog   10 Tubes
Made in United States   1938-1940 $139 new.
Features:
Coverage: .55-32 MHz in 6 bands. S-Meter, RF Gain, BFO, NL, Standby, Mute Line, Antenna Trimmer, Bandspread and Hinged Top Cover.
Comment:
Crystal filter. 465 kHz IF. Gray and black crinkle metal cabinet. An optional DB-20-70 Preselector and 510X-70 Frequency Expander (28-70 MHz) were offered.

ROHDE & SCHWARZ

ESMC

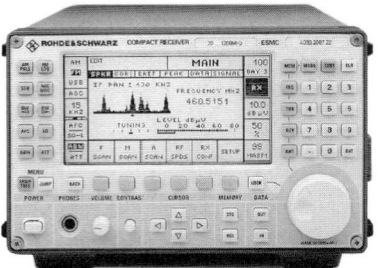

Wideband VHF Communications Receiver
Digital   Solid State
Made in England   1996-1998
Features:
Covers 20-1300 MHz (see Comments).   1000
memories, signal display, computer control, keypad,
1 Hz display, search, RF/IF spectral display, etc.
Comments:
This is primarily a VHF-UHF receiver. However, it is
often configured with optional H.F. module which
extends coverage down to 500 kHz. A highly ad-
vanced, commercial, multipurpose receiver.

Samyung

SYR-3000P

General Coverage Communications Receiver
Digital   Solid State
Made in Korea
Features:
S-Meter, RF Gain, BFO, Memories, Attenuator,
AGC, 3 bandwidths, Scanning, ¼" Headphone Jack,
Rack Handles. Coverage is from 90 - 30000 kHz in
the following modes: AM/CW/USB/LSB/FSK/FAX.
Has Line and IF out. Operation is from 110VAC or 24
VDC. Modular construction.
Comment:
$900-920 used.

SR-570

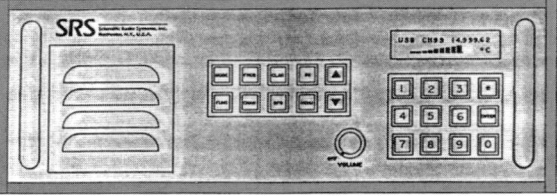

General Coverage Communications Receiver
Digital   Solid State
Made in United States   1987-1990
Features:
Covers 100-30000 kHz AM/USB/LSB CW. Band-
width: 2.4/_/_ kHz., Sensitivity 0.5 µV 10 dB. 10 Hz
display, stability 1x10^{-6}, audio is 2.5W 3.2 ohms 5%,
S-Indicator, RS-232, Rack Handles, Squelch, 100
Memories. 13.6 VDC. 13.5x4.7x15.5"
Comment:
The BFO clarifier is ±150 Hz in 10 Hz increments.
Optional filters and ten channel preselector.

SHARPER IMAGE

Tunemaster

PHOTO: Kurt Krozner/IBS, Ltd.

Shortwave Broadcast Receiver
Analog   Solid State
Made in Hong Kong   1989-1994
Features:
Coverage: .52-1.6, 2.4-7.5, 7.5-20.5 MHz and FM
broadcast 88-108.5 MHz plus VHF Air 10-137 MHz
and VHF public service band 144-164 MHz.
Comments:
Don't trade your R-8 in for this one. Rated the worst
tabletop ever tested in *Passport 1994.*
Reviews:
📖*Passport* 1990 & 1994

Sommerkamp

FR-100B

Amateur Band Communications Receiver
Analog Linear Display.   12 Tubes
Made in Japan   1967-1968 $250 new.

Features:
Dual Conversion. Covers 80 to 10 meters. Features:
RF Gain, Preselector, ANL, S-Meter, Standby, ¼"
Headphone Jack.
Comment:
Was imported for the American market by Barry
Electronics of New York. This radio was made by
Yaesu. Sells for $200 used.

SONAR   MR-3

Amateur Band Communications Receiver
Analog   8 Tubes
Made in United States   1953 $90 new.
Features:
This five band receiver covers: 80, 75, 20, 11 and 10 meters. Has dial lamp, BFO and NL.
Comment:
Designed for mobile amateur market. Requires 200-300 VDC 60-100ma and 6.3 VDC at 2.4 amps.
Variants:
Model **MR-4** covers: 80, 75, 40 and 20 meters.
Model **MR-5** covers: 80, 75, 20 and 10 meters.

SONAR   SR-9

Amateur Band Receiver
Analog   9 Tubes
Made in United States   1948-53 $73 new.

Features:
This radio was available in one of the following configurations: 2M, 6M, 10-11M or commercial band. AM only. Has NL.

Comment:
5.55x3.375x5.8". Designed for mobile ham market. An airport version was also produced.

S.P. RADIO A/S   R501

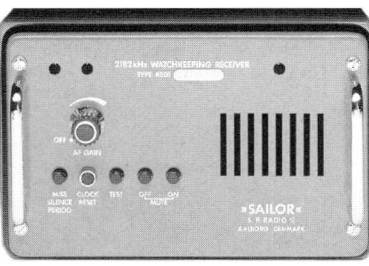

Marine Distress Receiver
No Display   Solid State
Made in Denmark   -1998
Features:
Automatic and manual volume control, speaker, rack handles, test function. 1μV sensitivity, 1 watt 8 ohms audio. Operates from 24 VDC, 110/127/220/237 VAC. 230x150x163mm 3 kg. (-15° to +55° C).
Comment:
The Sailor R501 is a single channel telephony receiver intended for reception of signals on the international emergency frequency of 2182 kHz.

 star-line   SR-200

Amateur Band Communications Receiver
Analog Display   8 Tubes
Made in Japan   1968 $120 new.

Features:
Single conversion, 160 to 10 meter coverage, S-Meter, ANL, ¼" Headphone Jack, Calibrator.

Comment:
Star was bought by Yaesu in 1968.

 star-line   SR-550

Amateur Band Communications Receiver
Analog Linear Display.   10 Tubes
Made in Japan.   1967-68 $150 new.

Features:
Double conversion, 160 to 6 meter coverage, Selectivity: 4/2.5/1.5/.5 kHz., S-Meter, ANL, ¼" Headphone Jack and Band Pass Filter.

 **star-line**   # SR-700E

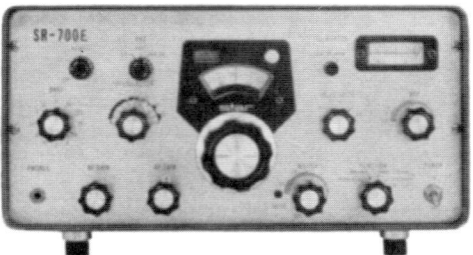

Amateur Band Receiver

Analog Linear Display	Tube Type
Made in Japan	1966-68 $325 new.

Features:
Triple conversion, 75 to 10 m. coverage. Selectivity: 4/2.5/1.5/.5 kHz., S-Meter, AVC OFF/FST/SLO, ANL, ¼" Headphone Jack, Preselector, RF Gain, Band Pass Filter, Notch, SO-239. 1 kHz dial accuracy.

Comment:
Matches ST-700E Transmitter. Reviewed in *QST* August 1967. 15.25x7.25x14.25" 30 Lbs. 115/230 VAC 50/60 Hz. Variants: **SR-700A** is 110/115 VAC.

TAPETONE   # 345 Sky Sweep

Amateur Band Receiver (VHF)

Analog	Tube Type
Made in United States	1959 $250 new.

Features:
A ¼" Headphone Jack, S-Meter, Antenna Trimmer, BFO, AGC, Standby. This radio receives 6 meters only! The dial is pre-calibrate for 2 meters, 220 MHz and 440 MHz external converters. Please note that this receiver has *no* HF coverage, even though it looks like an H.F. amateur band receiver.

TECHNICAL MATERIAL

FFR-49

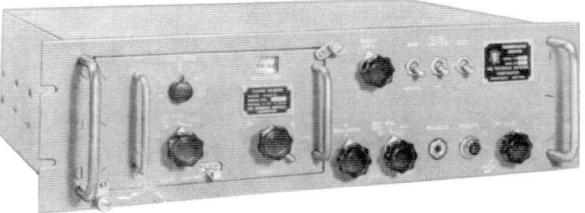

Semi-Fixed Communications Receiver

No Display	14 Tubes with Drawer
Made in United States	

Features: This receiver can cover a single band segment between 50 kHz to 32 MHz by means of a plug-in tuning drawer: FFRD-1 50-100 kHz, FFRD-2 100-200 kHz, FFRD-3 200-400 kHz, FFRD-3M 500 kHz ±10 kHz, FFRD-5 2-4 MHz, FFRD-6 4-8 MHz, FFRD-7 8-16 MHz and FFRD-8 16-32 MHz. Modes: AM/CW/MCW/SSB. Output is 8 and 600 ohms. Has RF Gain, BFO, AVC, NL, Rack Handles and ¼" Headphone Jack. 19x5.24x19" 21 Lbs.

TECHNICAL MATERIAL

SMR-1

Eight Channel Strip Communications Receiver

No Display	Solid State
Made in United States	1964-1966

Features:
Eight channel coverage in the range of 2 to 32 MHz. AM/CW/LSB/USB, 1µV 10 dB S/N sensitivity, 1 part per 10⁴ per day stability, for rack mounting. With S-Meter and ¼" Headphone Jack.

TOSHIBA   ## ZS-1447A

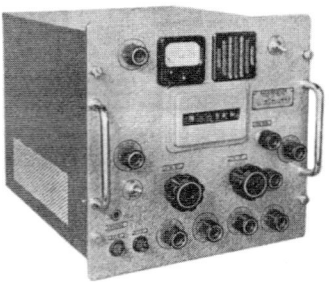

General Coverage Communications Receiver

Digital Mechanical	16 Tubes
Made in Japan	1958

Features: Coverage is from 400 to 32000 kHz. S-Meter, ¼" Headphone Jack, Rack Handles and Speaker. Bandwidths: 6/3/1/.3 kHz. Operates from 26 VDC 3 Amps. Odometer type frequency display. 13"x12.2"x15" 66 Lbs. Triple conversion superhet.

Comment:
Inspired by the Collins R-392? Few units made.

Variants:
Models **ZS-1446C** and **ZS-1446G** cover 2-25 MHz.

TRAVELLER TA-161

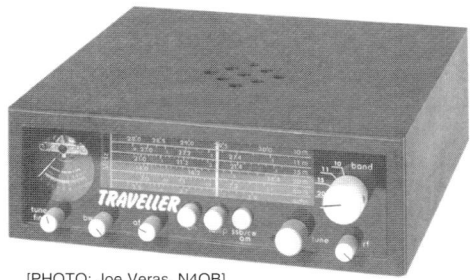

[PHOTO: Joe Veras, N4QB]

Amateur Band Communications Receiver
Analog   Solid State
Made in West Germany   1975-1976
Features: Covers: 3.5-3.9, 7-7.26, 14-14.4, 21-21.5, 26.9-27.5 & 28-30 MHz. Single conversion Superhet., Fine Tuning, S-Meter, Dial Lamp, RF Gain, Audio Filter, Crystal Filter, Speaker. Operates from 10 AA cells or 12 VDC.
Comments: Black metal cabinet with plastic front and back panels. 7.2x2.33x7" 2.5 MHz IF. This interesting set was designed for the European market. Note the incomplete 80/40 meter coverage,

---

⊠ T R U E T O N E DC 1270

General Coverage Communications Receiver
Analog Display   Solid State
Made in Japan
Features:
Covers 540 - 30000 kHz. Ranges: .54-1.6, 1.6-4.5, 4.8-11.5 and 11-30 MHz. Has BFO, 4" Speaker, ¼" headphone jack (on rear), ferrite MW antenna, bandspread 0-100, logging scale, dial lamp and standby. 105-125 VAC 50-60 Hz.
Comment:
Uses 9 transistors, 4 diodes and 2 thermistors. Has 455 IF. 12x5.375x4.75".

T. Withers Electronics MOBILE

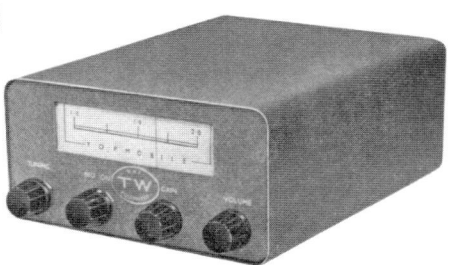

Amateur Band Communications Receiver
Analog Display   Solid State
Made England
Features:
Coverage: 1.81-2 MHz or 80 meters (exact 80 meter coverage is unknown).
Comments:
Only 6x2.75x6". Requires 12 VDC and speaker. Designed for mobile use.
Variants:
The model **TWO MOBILE** has the same specifications but has a 2 meter converter.

Unica UNR-30

General Coverage Communications Receiver
Analog Display
Made in England ?   1968

Features:
Covers 550 - 30000 kHz. Has BFO, Speaker and ¼" headphone jack. Operates from 220/240. Sold in England.

Utica 92-05

General Coverage Communications Receiver
Analog Display.   Solid State (?)
Made in Japan ?   1965-1966 $45 new.

Features:
Covers 550 - 30000 kHz. Ranges: .55-1.6, 1.6-4.4, 4.5-12 and 12-30 MHz. Has BFO, 3x5" Speaker, ¼" headphone jack, ferrite MW antenna, bandspread 0-100, dial lamp, telescopic antenna and standby.
Comment:
Similar to Midland model 11-500. Utica Communications Corp., 2917 W. Irving Park, Chicago, Ill.

Signal One CR-1200

Signal One CR-1500

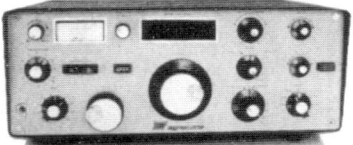

R.C.A.-Harvey Wells

[Photo by Joe Veras N4QB]

Many receivers were never manufactured in quantity for various reasons, including technical and financial. In some case only a concept was developed. In other case a non-operational advertising prototype was produced. This would be a totally nonfunctional unit, with no operating electronics, and usually with knobs just glued on. In other cases an actual working, or engineering prototype was produced. Finally some models were actually produced, but in very small numbers (typically less than ten).

Central Electronics 100-R
Collins 51J-5
Collins 51J-6 (evolved into the 51S-1).
Eska ESCOM 500 (working prototype).
Grove SW-100 (advertising prototype).
Grove SR-1000 (concept).
Hallicrafters SX-46 (advertising prototype).
Hallicrafters SX-112 (3 working prototypes).
Hallicrafters SX-1000A (advertising prototype).
Hammarlund HQ-66 (concept).
Hammarlund HQ-225 (3 working prototypes).
Jefferson-Travis CR-1
R.C.A. (G.C. version of Harvey Wells R9 - working prototype).
R.F.T. EKD 700 (working prototype).
Signal One CR-1200 (advertising prototype).
Signal One CR-1500 (advertising prototype).
Terra Nova FRR-1

Hammarlund HQ-66

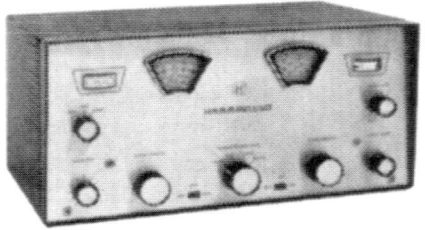

Jefferson-Travis CR-1

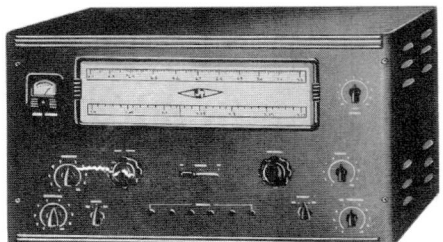

Collins 51J-6

Hallicrafters SX-112

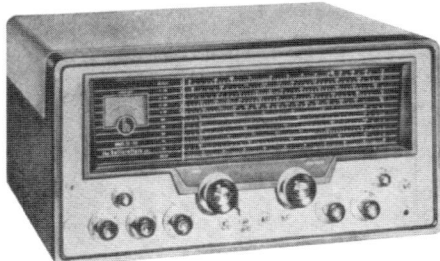

Grove SW-100

Grove SR-1000

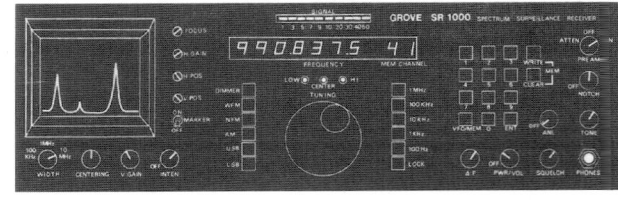

This chapter features important additional information that would not fit in the normal page format of this book.

Davco DR-30S

The Davco DR-30 amateur receiver is a scarce and desirable model. The matching **DR-30S** console is even more difficult to find. The DR-30S is a combination power supply, speaker and battery holder. This option allowed the DR-30 to be run from an AC source or it could hold eight D cells to permit portable operation. The front panel features a speaker, a ¼" headphone jack and a meter to indicate the voltage of the D cells. The DR-30S is the same size as the DR-30.

[Photo by Joe Veras N4QB]

Drake 1-A

The very earliest Drake 1-A receivers did not feature a crystal calibrator. This was added in later production. The calibrator was switched on and off by a switch ganged to the Antenna Trimmer. The Drake Company did offer a crystal calibrator upgrade kit to early 1-A owners. The kit sold for $20 and consisted of the calibrator with tube and crystal plus a new front panel control.

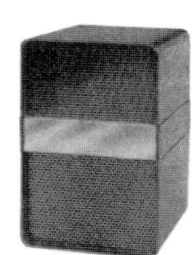

Another accessory for the 1-A was an **external speaker**. The matching metal enclosure housed a 5" x 7" oval speaker and sold for $15. It is shown to the right.

Hammarlund HC-10

The Hammarlund **HC-10** Converter is not a frequency converter. It is actually a "rear end" that connects to any receiver having an IF output from 450 to 500 kHz. The HC-10 has its own power supply and audio output system. It provides SSB reception and improved AM and CW reception for older receivers. It sold for $149 from 1958 to 1959. The Slot Filter function is quite effective.

Hammarlund SP-100 Series

Model	Type	Coverage	Filter	Speaker	Price New
SP-110	Cabinet	.54-20 MHz	Standard	8" Dynamic	$405
SP-110-X	Cabinet	.54-20 MHz	Crystal	8" Dynamic	$435
SP-120-X	Cabinet	.54-20 MHz	Crystal	12" Hi. Fidelity	$460
SP-110-S	Cabinet	1.16-40 MHz	Standard	8" Dynamic	$405
SP-110-SX	Cabinet	1.16-40 MHz	Crystal	8" Dynamic	$435
SP-120-SX	Cabinet	1.16-40 MHz	Crystal	12" Hi. Fidelity	$460
SP-150	Console	.54-20 MHz	Standard	15" Hi. Fidelity	$550
SP-150-S	Console	1.16-40 MHz	Standard	15" Hi. Fidelity	$550

Hammarlund SP-150 console is shown to the right.
It has simplified operation and fewer controls.

Hammarlund SP-200 Series

Model	Type	Coverage	Supplied Speaker	Price New
SP-210-X	Cabinet	.54-20 MHz	10" Jensen Dynamic	$465
SPR-210-X	Rack	.54-20 MHz	10" Jensen Dynamic	$482
SP-220-X	Cabinet	.54-20 MHz	12" Jensen Hi-Fidelity	$490
SPR-220-X	Rack	.54-20 MHz	12" Jensen Hi-Fidelity	$507
SP-210-SX	Cabinet	1.25-40 MHz	10" Jensen Dynamic	$465
SPR-210-SX	Rack	1.25-40 MHz	10" Jensen Dynamic	$482
SP-220-SX	Cabinet	1.25-40 MHz	12" Jensen Hi-Fidelity	$490
SPR-220-SX	Rack	1.25-40 MHz	12" Jensen Hi-Fidelity	$507
SP-210-LX	Cabinet	.15-20[1] MHz	10" Jensen Dynamic	$465
SPR-210-LX	Rack	.15-20[1] MHz	10" Jensen Dynamic	$482
SP-220-LX	Cabinet	.15-20[1] MHz	12" Jensen Hi-Fidelity	$490
SPR-220-LX	Rack	.15-20[1] MHz	12" Jensen Hi-Fidelity	$507

[1]In the LX model the .15-.3 MHz is *substituted* for the 2.5-5 MHz band.

In addition to the above variants, special export models for 50-60 Hz with a universal power supply tapped for 115, 125, 140, 230 and 250 ⌐AC were produced. A special 25 Hz power supply was also manufactured.

Hammarlund SP-600 Series

The famous Hammarlund Super-Pro 600 series had many variants. The following table was provided by Leslie A. Locklear.

Model	Date	Comment
SP-600-JX-1	Sept. 1951	Standard frequency range of 540 kHz - 54 MHz. Also designated R-274A/FRR. Signal Corps. order no. 1689-Phila-51-01. Part of AN/MRR-4, GRC-38A and B sets.
SP-600-JLX-2	Sept. 1951	Frequency range, 100-400 kHz and 1.35-29.7 MHz.
SP-600-J-3	Sept. 1951	Standard frequency range of 540 kHz - 54 MHz. No crystal frequency control.
SP-600-J-4	Sept. 1951	Standard frequency range of 540 kHz - 54 MHz. No crystal frequency control. Signal Corps. R-320A/FRC, order no. 19474-Pila-50-06. Also has a separate IF gain control. Part of OA-58B/FRC Set.
SP-600-J-5	Nov. 1951	Standard frequency range of 540 kHz - 54 MHz. No crystal frequency control. Equipped with 25-60 Hz. power supply. Signal Corps. R-483/FRR. Order number 21478-Phila-50. Also part of SCR-244D Set, had antenna and accessory items included and it was cabinet mounted.
SP-600-JX-6	Sept. 1951	Standard frequency range, 540 kHz - 54 MHz. BFO range ±10 kHz. U.S. Navy model R-274B/FRR, order number NObsr-52039 19 October, 1954. Navships manual 91661.
SP-600-JX-7	Sept. 1951	Standard frequency range, 540 kHz - 54 MHz.
SP-600-JX-8	Sept. 1951	Standard frequency range 540 kHz-54 MHz. Manufactured for Welch contract no. XG-479.
SP-600-JL-9	Sept. 1951	Frequency range, 100-400 kHz, 1.35-29.7 MHz. No crystal frequency control.
SP-600-JX-10	Nov. 1951	Standard frequency range, 540 kHz - 54 MHz. Replaces JX-7.
SP-600-J-11	Nov. 1951	Standard frequency range, 540 kHz - 54 MHz. No crystal frequency control. Replaces J-3.

SP-600-JX-12	Nov. 1951	Standard frequency range, 540 kHz - 54 MHz. Signal Corps. R-274A/FRR, order number 3376-Phila-52. Replaces JX-1.
SP-600-J-13	Nov. 1951	Standard frequency range, 540 kHz - 54 MHz. No crystal frequency control. Signal Corps. Order number 16838-Phila-51. Serial no's 52 to 67. Equipped with 25-60 cycle power supply. Replaces J-5.
SP-600-JX-14	April 1952	Standard frequency range, 540 kHz - 54 MHz. Signal Corps. R-274C/FRR, order number 3376-Phila-52. Also designated previously as R-542/FRR. Replaces JX-10.
SP-600-JLX-15	June 1952	Frequency range, 100-400 kHz, 1.35 - 29.7 MHz. Replaces JLX-2.
SP-600-JL-16	June 1952	Frequency range, 100-400 kHz. 1.35 - 29.7 MHz. No crystal frequency control. Replaces JL-9.
SP-600-JX-17	June 1952	Standard frequency rang e, 540 kHz - 54 MHz. Diversity receiver, manufactured for Air Material Command. The most common of all SP-600 receivers. Easily identified by "red metal knobs" on front panel.
SP-600-JX-18	June 1952	Standard frequency range, 540 kHz - 54 MHz. Manufactured for "GAUVREAU" contract. Replaces JX-10.
SP-600-J-19	Aug. 1952	Standard frequency range, 540 kHz-54 MHz. No crystal frequency control. Equipped with 25-60 cycle power supply. Replaces J-13.
SP-600-J-20	Aug. 1952	Standard frequency range, 540 kHz - 54 MHz. No crystal frequency control. Signal Corps. R-483-A/FRR, order number 3479-Phila-52-05. Equipped with 25-60 cycle power supply. Replaces J-19.
SP-600-JX-21	Feb. 1953	Standard frequency range, 540 kHz - 54 MHz. Replaces JX-10 and JX-14
SP-600-J-22	Feb. 1953	Standard frequency range, 540 kHz - 54 MHz. No crystal frequency control. Replaces J-11.
SP-600-JLX-23	Feb. 1953	Frequency range, 100-400 kHz, 1 35 - 29.7 MHz. Replaces JLX-15.

SP-600-JL-24	Feb. 1953	Frequency range, 100-400 kHz, 1.35 - 29.7 MHz. No crystal frequency control. Replaces JL-16.
SP-600-J-25	Feb. 1953	Standard frequency range, 540 kHz - 54 MHz. No crystal frequency control. Equipped with 25-60 cycle power supply. Replaces J-19.
SP-600-JX-26	Feb. 1953	Standard frequency range, 540 kHz - 54 MHz. Signal Corps. R-274C/FRR, order number 3376-Phila-52. Replaces JX-14 and JX-21.
SP-600-JLX-27	March 1953	Special frequency range, 200 - 400 kHz, 540 kHz - 29.7 MHz.
SP-600-JX-28	October 1953	Standard frequency range, 540 kHz - 54 MHz. Signal Corps. R-620/FRR, order no. 25693-Phila-53-61. Contract no. DA-36-039-SC-49453.
SP-600-JX-29	March 1954	Standard frequency range, 540 kHz - 54 MHz. Manufactured for CIA contract no. XG-1178.
SP-600-JX-30	Dec. 1954	Standard frequency range, 540 kHz - 54 MHz. Diversity receiver. Replaces JX-17.
SP-600-VLF-31	Dec. 1954	Special frequency range, 10 - 540 kHz. Crystal frequency control (4 position). Very low frequency receiver.
SP-600-JX-32	Dec. 1954	Standard frequency range, 540 kHz - 54 MHz. Black wrinkle finish front panel with white engraved lettering. Manufactured for Mackay Radio, their order no. M-41666, Hammarlund production order no. 2467-300. Internally the same as JX-21.
SP-600-JLX-33	Dec. 1954	Frequency range, 100-400 kHz, 1.35 - 29.7 MHz.
SP-600-JL-34	Aug. 1956	Special frequency range, 100 - 200 kHz, 540 kHz - 14.8 MHz. Manufactured for CIA, their contract no. XG-1765.
SP-600-JX-35	Aug. 1956	Standard frequency range, 540 kHz - 54 MHz. Crystal frequency control. BFO range ±10 kHz. U.S. Navy R-274B/FRR, order no. NObsr-71369. Navships manual 91661.

SP-600-JX-36	Oct. 1957	Standard frequency range, 540 kHz - 54 MHz. Crystal frequency control. Manufactured for the FBI, their order no. FBI-16876, Their contract no. J-FBI-3873. The same as JX-21 except for addition of audio input jack on rear of chassis.
SP-600-JL-24 "Special"	Oct. 1957	Frequency range, 100-400 kHz, 1.35 - 29.7 MHz. No crystal frequency control. U.S. Navy R-274B/FRR. The reason for the "Special" designation is not known.
SP-600-JX-37	March 1961	Standard frequency range, 540 kHz - 54 MHz. Crystal frequency control. 25 - 60 cycle power supply. Same as JX-21.
SP-600-VLF-38	March 1961	Special frequency range, 10 - 540 kHz. Crystal frequency control (4 position). Very low frequency receiver. Equipped with 25-60 cycle power supply. Otherwise, the same as VLF-31.
SP-600-JX-39	July 1961	Standard frequency range, 540 kHz - 54 MHz. Crystal frequency control. Manufactured for FAA contract no. FA-2338.
SP-600-JX-21A	1971-1972	Standard frequency range, 540 kHz - 54 MHz. Crystal frequency control. This was the last series of SP-600 receivers manufactured. It had 22 tubes, a product detector, LBS, USB, CW and MOD switch. The appearance was different from other SP-600's in that the knobs had no metal skirts. The front panel was engraved with markings for Xtal Phasing, Selectivity, BFO, Audio Gain, and RF Gain. Also, is engraved JX-21A on front panel below "Hammarlund Model SP-600". This is probably the rarest of the SP-600 series.

Japan Radio Co. NCM-515

The Japan Radio Company **NCM-515** wired keypad was designed exclusively for use with the NRD-515 receiver. This remote has its own liquid crystal display and it displays frequency information in tandem with the receiver's LED digital display. Direct frequency entry is supported. There is also an add and subtract function. If you are tuned to 15300.0 kHz and press +3050.0= you will move to 18350.0 kHz. The NCM-515 also provides 4 memories (storing frequency only and requiring a type #544 battery). You may also tune (slew) up or down from the remote at FAST or SLOW speed and in 1 kHz or 100 Hz steps. The NCM-515 was supplied with a CQE-515 Junction box that permits the remote to be used simultaneously with either the NDH-515 or NDH-518 memory units. The NCM-515, introduced in 1982, is now a scarce and collectable option.

Japan Radio Co. NDH-515 NDH-518 NVA-515

Japan Radio also produced two different external memory units for the NRD-515. The first was the **NDH-515**. This device could sit on top of the receiver and simply plugged into the buss on the back panel of the receiver. It stored 24 presets (frequency only). Later the similar **NDH-518** was produced. The NDH-518 stored four banks of 24 channels for a total of 96 presets. The four banks were selected by four push switches on the memory. Both memories required four AA cells for frequency retention during power loss. The NDH-518 is shown below with the optional **NVA-515** speaker to the left.

National HRO-5C

The National **HRO-5C** is a deluxe configuration of the HRO-5 with accessories. The HRO-5C consists of an HRO-5A1 receiver with SPC unit (power unit, coil container and loudspeaker) all mounted in a MRR Table Rack. The chromium-plated appearance strips and side strips yield a very continuous and custom look. This is an extremely collectable configuration.

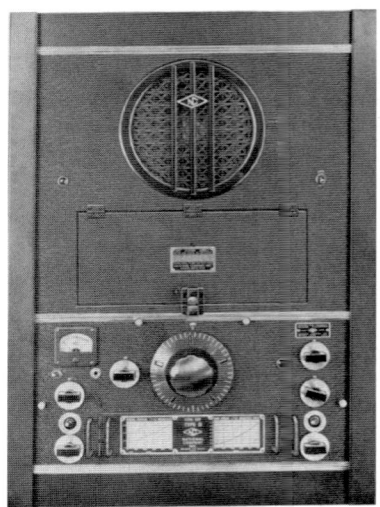

National SW-3 Series

The associated coils for the National SW-3 are as follows:

Range in Meters	SW-3U	AC SW-3	6DC SW-3	2DC SW-3
9 to 15	#30	#10	#60	#60
13.5 to 25	#31	#11	#61	#61
23 to 41	#32	#12	#62	#62
40 to 70	#33	#13	#63	#63
65 to 115	#34	#14	#64	#64
115 to 200	#35	#15	#65	#65
200 to 360	#36	#16	#66	#66
350 to 550	#37	#17	#67	#67
500 to 850	#38	#18	#68	#68
850 to 1200	#39	#19	#69	#69
1200 to 1500	#40	#20	#70	#70
1500 to 2000	#41	#21	#71	#71
2000 to 3000	#42	#22	#72	#72
Bandspread 10	#30A	#10A	#60A	#60A
Bandspread 20	#31A	#11A	#61A	#61A
Bandspread 40	#33A	#13A	#63A	#63A
Bandspread 80	#34A	#14A	#64A	#64A
Bandspread 160	#35A	#15A	#65A	#65A

To convert meters to MHz use the following formula. MHz = 300 / meters

This index will permit the reader to quickly find information on a particular model where the manufacturer is unknown. Radios with a model number and a name (S-20R Sky Champion) will be indexed by both. To cross reference by manufacturer and model number, please refer to the Table of Contents.
